# GROUNDWATER IN FRACTURED ROCKS

*Few tasks in hydrogeology are more difficult than locating drilling sites for water wells in igneous and metamorphic rocks.*

Davis SN and DeWiest RJM (1966)
Hydrogeology, p. 318.
Wiley and Sons, Inc., New York-London-Sydney.

*The hydrologist cannot blindly select a model, turn a crank, and accept the answers. He must devote considerable time and thought to judging how closely his real aquifer resembles the ideal.*

Ferris JG, Knowles B, Brown RH and Stallman RW (1962)
Theory of aquifer tests, p. 102.
*Geol. Surv. Water-Supply paper* 1536-E, Washington.

SELECTED PAPERS ON HYDROGEOLOGY

# 9

*Series Editor*: Dr. Nick S. Robins
*Editor-in-Chief IAH Book Series*
*British Geological Survey*
*Wallingford, UK*

INTERNATIONAL ASSOCIATION OF HYDROGEOLOGISTS

# Groundwater in fractured rocks

## Selected papers from the Groundwater in Fractured Rocks International Conference, Prague, 2003

Edited by

Dr. Jiří Krásný
*Associate Professor, Charles University Prague, Czech Republic*
*Scientific Programme Member and Chair of the Commission on Hardrock Hydrogeology, International Association of Hydrogeologists*

Dr. John M. Sharp, Jr.
*Carlton Professor of Geology, The University of Texas, Austin, USA*
*Treasurer, International Association of Hydrogeologists*

Taylor & Francis
Taylor & Francis Group
LONDON / LEIDEN / NEW YORK / PHILADELPHIA / SINGAPORE

**Cover photograph: Outcrop of the weathered granite in the Melechov region, Central Bohemia, Czech Republic.** Fractured granite merges upwards into its regolith with granite blocks and a thin soil cover on the top (Photo: Jiří Krásný). Deep parts of the Melechov Granite were selected as one of the places for possible radioactive waste disposal.

*Taylor & Francis is an imprint of the Taylor & Francis Group, an informa business*



Typeset by Charon Tec Ltd (A Macmillan Company), Chennai, India
Printed and bound in Great Britain by Antony Rowe Ltd (CPI Group), Chippenham, Wiltshire



Published by: Taylor & Francis/Balkema
P.O. Box 447, 2300 AK Leiden, The Netherlands
e-mail: Pub.NL@tandf.co.uk
www.balkema.nl, www.taylorandfrancis.co.uk, www.crcpress.com

Library of Congress Cataloging-in-Publication Data

Groundwater in Fractured Rocks International Conference (2003 : Prague, Czech Republic)

Groundwater in fractured rocks : selected papers from the Groundwater in Fractured Rocks International Conference, Prague, 2003 / edited by Jirí Krásný, John M. Sharp, Jr.

p. cm. – (Selected papers on hydrogeology ; 9)

ISBN 978-0-415-41442-5 (hardcover : alk. paper) 1. Hydrogeology–Congresses. 2. Rocks–Fracture–Congresses. I. Krásný, Jirí. II. Sharp, John Malcolm, 1944–III. Title.

GB1001.2.G78 2007
551.49–dc22 2007009818

ISBN13: 978-0-415-41442-5 (Hbk)
ISBN13 e-Book: 978-0-203-94565-0

# Contents

# Preface

Crystalline (igneous and metamorphic) and consolidated sedimentary rocks – so called hard or fractured rocks – are found in many regions of the world. They occur mainly in large areas – shields, massifs, and in cores of major mountain ranges. Their outcrops cover more than 20 per cent (i.e., approximately 30 million square kilometres) of the present land surface. In addition, these mostly old rocks form the basement of younger sedimentary rocks that are often concentrated into large basins. Thus, hard rocks represent at depths a continuous environment enabling extended, deep regional or even global groundwater flow.

During past decades adequate attention had not been paid to groundwater in this specific hydrogeologic environment. The exception is in some arid and semi-arid regions where, under typical non-availability of surface water resources, groundwater represents the major water resource. In temperate climatic zones, groundwater users had paid attention mostly to sedimentary basins that can host aquifers of high yield. These are economically and technically more profitable than scattered, often small-yield wells common in hard rocks.

In the last decade of the 20th century, however, there has been increasing interest in hardrock hydrogeology because of many theoretical and applied issues in both tropical and temperate climatic zones. These include:

- Natural groundwater resources of hardrock terrains, mainly those in mountainous regions, have been shown to be very important in maintaining flow of surface water courses in adjacent piedmont zones during dry periods. They may also provide significant recharge to adjoining sedimentary systems. Therefore, understanding these resources is indispensable for sustainable integrated water management.
- As water demands increase in many regions, adequately sited water wells or other water intake systems in hardrock areas typically can meet requirements on water supply for small communities, industry, and irrigation and for domestic water consumption. In some areas, groundwater abstraction possibilities are sufficiently high to supply even small cities. Siting of wells in open fracture systems or thick permeable overburden and the evaluation of permeability with depth to assess well yields are some of the important issues of applied hardrock hydrogeology. Social and economic considerations of groundwater development help to decide and justify whether development of groundwater in hardrock areas is preferable to other water-supply options.
- Knowledge of groundwater flow and transport in fractured rocks is vital in addressing groundwater pollution and environmental protection. Impacts of industrialisation and urbanisation, landfills, deep hazardous waste repositories, and agricultural chemicals must be studied and monitored in hardrock environments. Knowledge of the complex groundwater flow and contaminant transport in weathered and fractured zones of hard rocks is decisive for siting both surface and deep waste repositories and the assessment of future environmental impacts of potential contamination sources.
- Clear hydrogeologic understanding and quantitative hydrogeologic assessments are important for many geotechnical and engineering-geological activities, such as construction of tunnels, building foundations, and utility systems as well as in mining. Recent studies of deep repository sites for radioactive, toxic and other hazardous

wastes, many involving crystalline rocks, have extended our hydrogeologic knowledge of these environments to depths of hundreds or even thousands of metres. These data present an important and useful concepts to compare with results provided by other hydrogeologic methodologies and techniques.
- Results of recently deep boreholes in crystalline rocks, mostly connected with geothermal studies, have re-opened the issue of deep-seated groundwater flow and brine occurrences and discussions on the origins of mineral and thermal waters often associated with hard rocks.
- Increasing available data has stimulated efforts to regionalise and generalise results from different hydrogeologic environments. Knowledge of hierarchy of inhomogeneity elements and of a scale effect influencing spatial distribution of hydraulic properties might enable us to simplify real natural conditions when defining conceptual and numerical models of groundwater flow and solute transport.
- Data on hydraulic properties, groundwater availability, and water quality, obtained by different methodological approaches in different regions of the Earth, offer possibilities for correlative hydrogeologic studies so that conclusions can be drawn to understand hydrogeologic properties of fractured rocks both in local and regional scales.

Many international meetings have reflected the increasing attention paid to groundwater in hard rocks. The 24th Congress of the International Association of Hydrogeologists (IAH) held in 1993 in Oslo, focused directly on hardrock hydrogeology. This has been followed by several other professional meetings covering similar topics. Recent IAH Congresses in Cape Town, South Africa (2000), in Munich, Germany (2001), in Mar del Plata, Argentina (2002), in Zacatecas, México (2004) and many other international conferences have contributed significantly to hardrock hydrogeology.

Acknowledging the importance of groundwater in the hardrock environment, IAH established the Commission on Hardrock Hydrogeology to stimulate international co-operation and facilitate exchange of information between hydrogeologists and other specialists on groundwater issues in hard rocks (Commission web site: www.natur.cuni.cz/iah or www.iah.org/). The Commission assembles specialists from about 50 countries all over the world. In Europe, four regional working groups regularly convene workshops on different topics of hardrock hydrogeology starting from 1994. Thus, so far eleven workshops have been organised and all have published proceedings. Recently, a new South Asian regional working group was established.

In 2003 the IAH International Conference on "GROUNDWATER IN FRACTURED ROCKS" was held in Prague. The Conference provided an effective professional forum for presentations and discussions of many issues of fractured rock hydrogeology. The conference proceedings (Krásný et al., 2003) published 206 extended abstracts selected from over 286 submissions from 52 countries. Primarily out of these materials, the editors of this volume requested 43 papers, taking into account their goals, originality, relevance, and technical quality. The selected papers cover a wide field of important issues of modern hydrogeology of fractured environments. These include sustainable groundwater development, protection and management, evolving methodological approaches, new concepts of hydrogeologic properties both on local and regional scales and both quantitative and qualitative aspects of groundwater flow. We also desired to consider different hardrock regions on all the continents. There could be discussion on whether or not all the published papers fulfil these criteria, but our intention was to select representative contributions on the issues presented at the 2003 Prague conference. Three years have passed since the conference, and the authors were asked to update and/or extend their original findings. Some of these revisions were very extensive. As in the conference, the selected papers are subdivided into 6 sections as follows:

1. Hydrogeologic environment of fractured rocks
2. Conceptual models, groundwater flow and resources in fractured rocks
3. Groundwater quality in fractured rocks
4. Investigation and interpretation methods in fractured environment
5. Anthropogenic impacts on fractured environment
6. Numerical modelling of fractured environment.

We would like to thank to the reviewers that supported us in enhancing the submitted manuscripts. These include G. Barrocu, A. Bath, A. Chambel, J. Chilton, J. Conrad, E. Custodio, A.S. Engel, T.T. Garner, A. Grmela, K. Howard, Z. Hrkal, P. Kralj, R. Marrett, H. Marszalek, A. Mayo, A. Moench, H.S. Nance, P. Neill, L. Ribeiro, W.R. Robertson, P. Rouhiainen, T.K. Rubbert, K. Rudolph-Lund, K.-P. Seiler, R. Senger, C. Simmons, J. Slezák, D.T. Slottke, E. Steinhauer, G. Stournaras, U. Tröger, F. Villarroya, A. Voronov, J. Vrba, D. Watkins, and H. Wu, as well as many reviewers who chose to remain anonymous.

We think that the volume is the representative monograph discussing key issues, methodologies, and techniques in the field of hydrogeology of fractured (hard) rocks as well as summarising the results achieved in recent years. It is our hope that this will stimulate new ideas and considerations and lead to advances in the hydrogeology of fractured hard rocks. It should be a valuable reference for studies in fractured rock hydrogeology worldwide.

Prague-Austin, August 2006

Jiří Krásný and Jack Sharp
Editors of the volume

# About the editors

**Jiří Krásný** has worked most of his professional career with the Czech Geological Survey and from 1991 with the Charles University Prague. He was at long-term hydrogeological missions in Iraq and Nicaragua. His main interests are regionalisation of hydrogeological data, hydrogeological mapping, mineral water investigations, and studies in different hydrogeological environments. He is the chairman of the IAH Commission on Hardrock Hydrogeology and the Scientific Programme Member of the IAH Council.

**John M. (Jack) Sharp, Jr.** is the Carlton Professor of Geology at The University of Texas. He and his students research a variety of hydrogeological issues, including flow in fractured rocks, free convection, karstic hydrogeology, decision support systems, urban hydrogeology, and application of hydrogeology to geological problems. He is the Treasurer of the IAH and the President of the Geological Society of America.

# Hydrogeology of fractured rocks from particular fractures to regional approaches: State-of-the-art and future challenges

Jiří Krásný[1] and John M. Sharp, Jr.[2]

[1] *Faculty of Science, Charles University Prague, Czech Republic*

[2] *Department of Geological Sciences, Jackson School of Geosciences, The University of Texas at Austin, Austin, Texas, U.S.A*

ABSTRACT: Groundwater flow and transport in fractured rocks are important for development of groundwater, petroleum, and geothermal fluids and also for understanding the movement of anthropogenic and natural solutes in geological settings. Relative permeabilities and storage properties differentiate fractured media into purely fractured media, fractured formations, double-porosity media, and heterogeneous media. *A priori* understanding of flow in these systems requires knowledge of geological controls on fracture orientations, connectivities, apertures, roughnesses, spacings/densities and skin effects. More geological input, geophysical techniques and the use of proxy data are needed to address challenges in fractured rock hydrogeology. We require greater insights into geological controls of flow and transport in fractured porous media; better predictions of solute transport and the influence of fracture flow on other fundamental geological problems; and how or if we can scale these properties. This compendium addresses these issues, presents new techniques, and evaluates groundwater in fractured rocks in a variety of hydrogeological settings.

## 1 INTRODUCTION

Fractured rocks present complex, heterogeneous, and anisotropic hydrogeological environments with irregular distribution of groundwater flow pathways. Fractured rocks and hard rocks are often used synonymously even though there is no general agreement among hydrogeologists on the precise definition of hard rock. It is commonly understood by geologists that hard rocks are crystalline (i.e., igneous and metamorphic rocks), and some hydrogeologists (e.g., Larsson et al., 1987) define "hard rocks" as igneous and metamorphic, non-volcanic, and non-carbonate rocks. However, these definitions are restrictive and, in many hydrogeological studies, "hard rock" is used in a wider, but not precisely defined sense. This is because other rock types, such as well-cemented sedimentary rocks often occurring with crystalline rocks in shields and massifs have hydrogeological properties similar to crystalline rocks. In these terrains, it is often impossible to define exact boundaries between metamorphic ("crystalline") and sedimentary rocks where metamorphic rocks grade into slightly metamorphosed sedimentary rocks. Similar situations occur where nonindurated deposits lose their primary (intergranular) porosity and transition into intensively cemented sedimentary rocks. Consequently, because of lithification and complex geologic processes, such as faulting and folding, intergranular primary porosity of clastic deposits can be altered to an intergranular-fracture (double) or to fracture-dominated porosity: Well-cemented claystones, shales, carbonates, and sandstones, occurring both in basins and in folded mountainous zones, can display hydrogeological properties comparable to crystalline rocks.

Accordingly, we suggest the acceptance of the definition of "hard rocks" proposed by Gustafsson (1993): *"Hard rocks include all rocks without sufficient primary porosity and conductivity for feasible groundwater extraction."* This implicitly highlights the most important hydrogeological property of hard rocks – fractures that are common to all hard rocks irrespective of their great variety of mineralogy, petrology, and stratigraphy. Hard rocks include magmatic rocks from granites to basic rocks, metamorphic rocks, and many sedimentary rocks, including fractured shales, graywackes, sandstones, conglomerates, and some carbonate rocks. Carbonate rocks, especially if karstified, and neovolcanic rocks (e.g., unwelded tuffs) may differ hydrogeologically from hard rocks, but there occur many common features and the methodologies used in carbonate and volcanic rocks can often be applied in hardrock environments.

Hard rocks occur in all the continents. They often extend in geologically continuous areas – shields, massifs, and in cores of many mountain ranges. The largest areas belong to Canadian, Guyana, and Brazilian shields in the Americas and to extended areas in Africa, Asia and Australia. Some of these areas belong, at least partly, to arid or semi-arid zones with acute scarcity of water resources. In Europe, hard rocks occur in the Baltic Shield and in numerous smaller areas. Hardrock outcrops cover more than 20 per cent (i.e., approximately 30 million square kilometres) of the land surface. The total extension and volume of all fractured rocks (partly of double porosity), however, is much larger as old crystalline rocks commonly form the basement beneath sedimentary rocks including those in large sedimentary basins.

Historically, hydrogeologists underestimated the occurrence and importance of preferential pathways for groundwater flow, represented by fractures and other heterogeneities. Simplified solutions were based on assumptions of homogeneity and isotropy. However, the importance of fractures and fracture networks on flow and solute transport is now recognised and has led in the last decades of the 20th century to a greater focus on the hydrogeology of fractured systems.

This increasing interest in the hydrogeology of fractured rocks, especially in hard rocks, has led to a number of review papers and monographs. These are too many to mention all of them but the first may have been Davis and DeWiest (1966, Ch. 9) and IAHS/UNESCO (1967). These were followed by two UNESCO monographs (Larsson 1987; Lloyd 1999). Marinov (1974, 1978), Geothermal Resources Council (1982), LaPointe and Hudson (1985), Black et al. (1986), Evans and Nicholson (1987), Barton and Hsieh (1989), Ehlen (1990), Chernyshev and Dearman (1991), Wright and Burgess (1992), National Research Council (1996), Pointet (1997), Krásný (1999), Singhal and Gupta (1999), Faybishenko et al. (2000), Robins and Misstear (2000), Stober and Bucher (2000), Olofsson et al. (2001), Cook (2003), and Neuman (2005) are some of the more recent compilations.

The 24th IAH Congress focused on the "Hydrogeology of Hard Rocks" (Banks and Banks, 1993); it initiated a world-wide scientific co-operation and led to eleven workshops convened by four European Regional Working Groups of the IAH Commission on Hardrock Hydrogeology (Krásný and Mls 1996; Bochenska and Stasko, 1997; Yélamos and Villarroya, 1997; Annau et al., 1998; Knutsson, 1998; Rohr-Torp and Roberts, 2002; Stournaras, 2003; Rönkä et al., 2005; Stournaras et al., 2005). Ten years after the 24th IAH Congress, the IAH Conference on "Groundwater in Fractured Rocks" in Prague (Krásný et al., 2003) demonstrated the variety of excellent research on fractured rock hydrogeology and led to this compilation of research papers that document the state of the art and recognition of future challenges in this field and that expand upon the concepts of hardrock hydrogeology to fractured rocks in general.

## 2 TOPICS

This compendium is grouped of papers that reflect the current status of the hydrogeological knowledge of hard or fractured rocks:

- Hydrogeological environment of fractured rocks
- Conceptual models, groundwater flow and resources
- Groundwater quality
- Investigative and interpretative methods
- Anthropogenic impacts, and
- Numerical modelling

In the first section, the *hydrogeological environment of fractured rocks,* the hydrogeology of different hard rock regions is summarized. Barrocu and Watkins discuss granitic rocks in Sardinia, Italy, and Cornwall, UK, respectively. Faillace presents a historical view and generalization of his many years developing groundwater supplies in Africa; Darko and Krásný evaluate hydrogeological properties and groundwater potential in Ghana; Bocanegra and Da Silva consider fractured rocks aquifers of South America and Brito Neves and Albuquerque discuss the relationships of tectonics and groundwater in northeast Brazil. European systems discussed include the Alentejo region of Portugal (Chambel et al.), the Aegean or Hellenic Arc encompassing Greece and the Aegean islands (Stournaras et al.), and the Koralm massif of Austria (Winkler et al.) where fracture network modelling is used in conjunction with discrete fracture models. Finally, Emilio Custodio characterizes groundwater in volcanic rocks in which flow is also fracture dominated.

In the section on *conceptual models of groundwater flow and water resources,* Issar and Kotze use environmental isotopes to establish a conceptual model for flow in fractured metamorphosed sedimentary rocks. Allen and Milenic evaluate the role of fractures in low-permeability sandstone confining units. Conrad and Adams use GIS to assess groundwater recharge in fractured rocks in South Africa in areas with a paucity of data, while Misstear and Fitzsimons estimate recharge in Irish fractured aquifers using soil moisture budgets and stream gauging. Marechal et al. characterize hard rock aquifers in India at the catchment scale. Finally, Brighenti and Macini conceptualize non-Darcy two-phase flow in fractures.

*Water quality* is a key water-resources issue in fractured media because of the speed and directions of solute transport. Bath interprets the evolution and stability of groundwaters in fractured rocks in England and Sweden as do Frengstad and Banks in Norway. Gaut et al. discuss factors controlling the microbial quality of such groundwaters. Two Portuguese studies discuss quality issues of mineral waters. Lourenço and Ribeiro evaluate trends in mineral water quality and Marques et al. review the evolution of $CO_2$-rich waters in granitic rocks. Morgan and Jankowski investigate the origin of salinity in fractured rocks that is affecting agricultural activities. Palcsu et al. use isotopic analyses to infer flow systems in a granite massif to evaluate its potential as a waste repository. Finally, Troeger reports on structural controls on several Brazilian thermal springs and their water chemistry.

*Methods of investigating* in fractured rock systems are also evolving. Bunker discusses the use of oriented core drilling of environmental site investigations. Moeck et al. relate fracture networks and aquifer characteristics as related to the current stress field. Carneiro uses Monte Carlo methods of modelling to delineate groundwater protection zones for fractured-rock aquifers. Brauchler et al. use an aquifer analogue approach to characterize fractured porous media with a travel-time based tomographic inversion. Le Borgne et al. compare discrete and continuous descriptions for characterizing flow in fracture networks. Love et al. present a new method of borehole dilution with tracer tests to estimate flow horizontal and vertical rates in fractured metasediments in Australia. Of course, pumping tests are still the prime technique for gathering hydrogeological data. Lods and Gouze present a new method for pump test analyses in fractured rocks. Koskinen and Rouhiainen characterize groundwater flow in fractured crystalline rocks with borehole flow meters.

*Human effects* on fractured rock systems are becoming increasingly important. Water quality issues are discussed in a previous section, but physical effects are also important. Betson and Robins use unit specific capacity to assess vulnerability of fractured aquifers to pollution. Loew et al. review unique case histories of the effects of tunnels on groundwater flow in crystalline rocks of the Swiss Alps. Rudolph-Lund et al. examine monitoring and remediation activities along tunnels through urban areas of Norway and Paul et al. discuss model prediction of the flooding of an underground uranium mine in Germany.

The last section of this volume has 7 papers on various aspects of *modelling*. Beyer and Mohrlok use a double continuum approach for contaminant transport in fractured porous media. Because waste repositories are often planned in rocks of low primary permeability, fractures are a vital assessment item. 3 papers deal with these sites. Blum et al. use fracture data from the Sellafield (UK) site to define 3 important hydrogeological issues for modelling flow in fractured rock. Fourno et al. use a smeared fracture approach to predict post closure conditions, and Fahrenholz et al. discuss modelling and performance assessment at a site in Siberia with few hydrogeological data. Mouri and Halihan compare averaging of hydraulic conductivities for heterogeneous layered and fractured aquifers. Garner et al. model the transport of solutes in fractures in granites with fracture skins using an analytical model. Finally, last but not least, Noriel et al. use X-ray microtomography to measure fracture properties in a rough fracture undergoing dissolution.

Even the wide range of topics and geographic areas included in this compendium do not exhaust the range of hydrogeological studies in fractured rock. For instance, we do not cover geophysical methods (surficial or borehole) or rock mechanics to any great extent. There is also no discussion of the role of flow in fractures in permafrost regions and its role in glacial processes. However, we think that this collection of articles is generally indicative of the wide and growing field of the hydrogeology of fractured and "hard" rocks. Many of the papers suggest future applied and theoretical research needs in this field of science.

## 3 GENERAL HYDROGEOLOGICAL FEATURES AND VERTICAL ZONING OF HARD ROCKS

Hardrock environment typically consists of three vertical zones, upper weathered, middle fractured and deep massive (e.g., Biscaldi 1968, Larsson et al. 1987, Chilton and Foster 1993, Rebouças 1993, Krásný 1996b, Faillace, this volume, and Maréchal et al., this volume). Hydrogeological knowledge and understanding of this intricate environment, based on sound interdisciplinary geological, hydrogeological, hydrochemical, and geophysical studies are indispensable to provide adequate basis for theoretical considerations, development of methodological and technical tools, implementation of conceptual and numerical models, and many practical applications.

*Vertical sequence of the three zones* is defined from the land surface downwards as follows:

- *Upper or weathered zone* formed by regolith, colluvium, talus, etc. often juxtaposed with alluvial, fluvial, glacial, and lacustrine (mostly Quaternary) deposits. Intergranular (interstitial) porosity prevails. The usual thickness is several metres but under special conditions, mainly along deeply weathered fractures or in residual tropical soils this zone may be much deeper.
- *Middle or fractured zone* usually represented by fractured bedrock to depths of some tens to hundreds of metres. Fracture aperture depends mostly on exogeneous geologic processes so permeability in this zone generally decreases with depth.
- *Deep or massive zone* in massive bedrock where fractures, faults, or fracture-fault zones are relatively scarce and fracture apertures are commonly less than in the middle zone. Deep fractures may act as isolated, more or less individual hydraulic bodies. In a regional scale, however, these inhomogeneities may form interconnected networks enabling extended and deep, regional to continental groundwater flow reaching depths of hundreds or even thousands of metres. Under suitable structural conditions, mineral and thermal waters ascend along deep faults.

This vertical sequence is common but, under specific conditions, the upper-weathered zone can be missing (e.g., in recently glaciated regions, the weathered zone and, in some cases, portions of the middle/fractured zone have been eroded).

Intergranular porosity is important in the upper zone; fracture porosity dominates in the deeper zones. However, varying petrographic and mechanical properties of hard rocks can create in different hydrogeological characteristics. Permeability variations may reflect lithologically distinct intercalations and form stratiform aquifers that are more characteristic of sedimentary basins. Folded, fractured, and sometimes karstified layers of crystalline limestones (and marbles) and intercalations of other rocks (e.g., quartzites) may be more transmissive than the surrounding hard rocks but form part of the same aquifer systems.

The upper and middle zones can form a regionally extended *"near-surface aquifer"* that is generally conformable to the land surface with a thickness of tens to more than 100 metres with permeability generally decreasing with depth. This aquifer usually offers the best groundwater abstraction possibilities. However, the thickness and character of this complex and heterogeneous aquifer changes spatially in relation to tectonic deformation (faulting and fracturing), lithologic facies, and weathering.

## 4 GEOMETRY AND PROPERTIES OF PARTICULAR FRACTURES

Fractures are the most important hydrogeological element in almost any setting and especially in hardrock environments. They control the hydraulic characteristics and solute transport. Without fractures, except for the upper-weathered zone, there is no significant groundwater flow in hard rocks. Therefore, understanding of the fluid-transmission properties in fractures is important for production of critical natural resources and environmental protection as well as understanding of natural processes, such as formation of mineral and petroleum deposits, sediment diagenesis, the cooling of plutons, mass wasting, and the movement of nutrients and chemicals in the soil zone. Finally, fluid flow and solute transport in fractures is important in many geotechnical and mining issues.

### 4.1 *Fractured media*

Fractured media can be classified into four gradational categories (Figure 1) depending upon the relative hydraulic properties of the fractures and the blocks or matrix between the fractures. In purely fractured media, the hydraulic conductivity and fluid storage are completely in the fractures. The rock matrix has virtually no porosity or permeability. Examples include unweathered plutonic and metamorphic rocks, such as granites, gabbros, schists, gneisses, and slates. Some volcanic rocks are also purely fractured. With respect to their transport properties, even purely fractured formations may have "diffusional" porosity (Norton and Knapp 1977; Neretnieks, 1980) created by microfractures and defects along

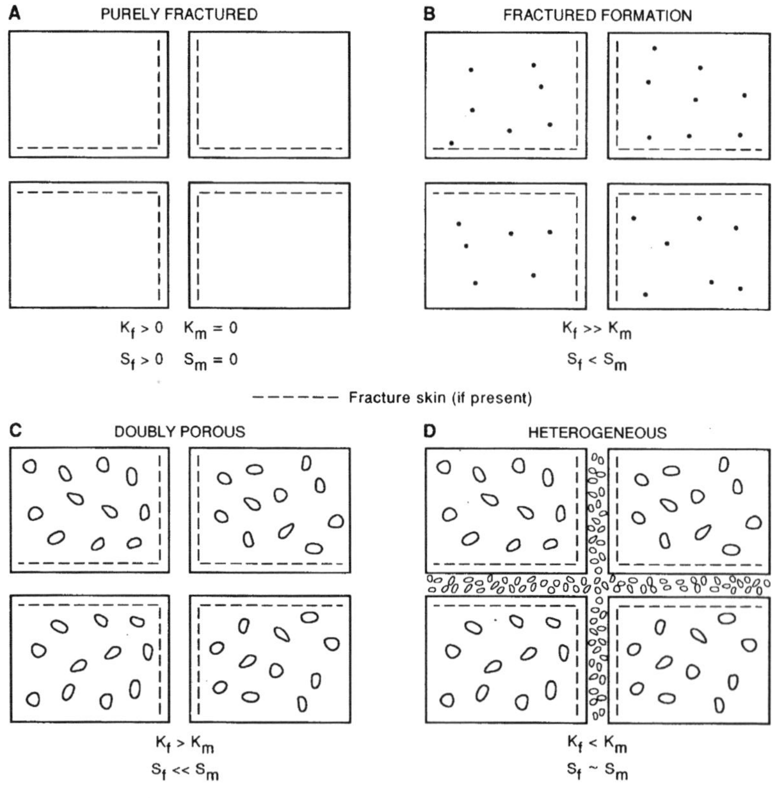


Figure 1. Hydrogeological classification of fractured media (after Streltsova, 1975; Sharp, 1993). $K_f$ and $K_m$ and $S_f$ and $S_m$ are the hydraulic conductivities and storativities of the fractures and matrix, respectively. A: purely fractured media. B: fractured formation. C: double porosity medium. D: heterogeneous formation. Especially, in cases A, B, and C, the fracture coating or "skin" may be in solute transport significant.

crystal boundaries that contribute significantly to transport on long time scales. In fractured formations, the flow is controlled by the fractures, but the fluid is stored primarily within the matrix. A fractured formation represents a common situation of great interest to the petroleum industry. Examples of fractured formations include "tight" gas sands, shales, and extrusive volcanic rocks, such as tuffs. In double porosity media, the relative permeability of the matrix can approach that of the fracture. Many aquifers, including sandstones, some basalts, and some carbonates can be considered double-porosity media. As in the fractured formation, most of the fluid is stored in the matrix. Double-porosity media constitute a particularly difficult challenge in modelling because flow in both the matrix and the fractures and interaction between the matrix and the fractures must be quantified. Finally, in some fractured rocks, the fractures are filled with material that is different in permeability than the matrix. This is termed a heterogeneous formation which may be modelled with standard equivalent porous medium techniques.

Also shown in Figure 1 is the fracture "skin" – a fracture surface or thin zone immediately beneath the surface which is altered by mineral deposition or coating with detrital or "infiltered" clays (Moench 1984, Fu et al. 1994). Fracture skins may possess significantly different hydraulic properties from the unaltered matrix (Fuller and Sharp 1992; Robinson et al., 1998; Phyu, 2002; Garner and Sharp, 2004; Garner et al., this volume). Skin permeabilities, porosities, sorptivities, and diffusion coefficients can be considerably different from the unaltered fracture surface or matrix and can be important in solute transport.

## 4.2 *Fracture characterization*

We conclude that hydrogeological characterization of fracture systems is vital to the understanding of flow and associated processes. Some characterization procedures are well documented, but for many fracture parameters, characterization remains uncertain because requires data on the orientation of the fractures, their frequency (fracture density), size, flow characteristics (aperture, fracture roughness, and channeling characteristics), and interconnectivity. We also need to evaluate how these properties vary with geological heterogeneity, scale, changing *in situ* stresses, and, in the case of hazardous waste disposal, how these properties evolve over centuries or millennia. This is so difficult that some (e.g., Voss, 2003) have suggested that hydrogeological characterization of fracture systems *a priori* might be impossible.

### 4.2.1 *Orientation*

In most cases, the assumption is made that a fracture is planar or curviplanar so that fracture orientation can be defined by strike and dip. Orientations can be illustrated by rose diagrams, stereonets, or directly on geological maps. A wealth of fracture data can be inferred from geological maps and aerial photographs (e.g., Mayer and Sharp, 1998). Uliana (2001) and Sharp et al. (2000) used length weighted rose diagrams to predict the directions and relative magnitude of permeability as a function of azimuth.

### 4.2.2 *Fracture density*

Fracture density quantifies the number fractures per unit length (along a scan line), in a unit area, or in a unit volume of rock. The number of fractures which cross a scan line or traverse is called the fracture spacing. The correlation of fracture spacing with fracture density has been discussed by LaPointe and Hudson (1985), among others. Recent studies by Marrett (1996), Marrett et al. (1999), and Ortega et al. (2006) extrapolate from thin sections or outcrop measurements to field scale.

#### 4.2.3 *Aperture*

Aperture is the distance between the fracture walls. The discharge and hydraulic conductivity of a fracture with smooth parallel walls are proportional, respectively, to the cube and square of aperture. Apertures may follow a power law scaling, but aperture variability within a fracture (Mouri, 2005) may overwhelm the scaling attempts. Natural fractures do not have smooth, parallel walls; the irregularities on the fracture surface are termed asperities. Asperities create roughness to reduce fluid velocity and create channels of preferential flow. This process is well known, but it is not yet commonly considered in studies of flow and transport.

#### 4.2.4 *Channeling*

Channeling in a fracture is where fluid flow takes a preferred path or channel so that flow velocities are highly irregular and the flow paths are difficult to predict. Channels may be anastomose or meandering, and channeled flow can occur without the saturation of the entire fracture. Channels are controlled by fracture geometry, roughness, the source and distribution of recharge to the fractures, and the hydraulic gradient. Tsang and Tsang (1987) suggest that, at depths of greater than 500–1000 metres, *in situ* effective stress causes all fracture flow to be channeled. The transport of solutes is also affected by channeling because effective porosity is reduced so that average linear velocity estimates for a given hydraulic gradient is greater (Mouri, 2005). Where diffusion into a porous matrix is an important attenuation mechanism for contaminants, channeling reduces the attenuation because the area for diffusive flux is reduced.

#### 4.2.5 *Fracture connectivity*

How fractures are interconnected is critical importance in fractured media. Longer fractures have a better of chance of intersecting another fracture. Barton et al. (1987) defined a ternary diagram to plot abutting, crossing, and blind fracture terminations. Another ternary classification was proposed by Laubach (1992) who noted that fracture terminations may splay and may interfinger with similar "diffuse" fracture ends; he grouped terminations into blind, diffuse, and crossing (which includes abutting). Abutting and crossing fractures provide connectivity; diffuse terminations provide limited connectivity; and blind fractures provide neither. Sets of blind fractures may, however, greatly influence the permeability tensor. Examples of sets of subparallel blind fractures commonly include neotectonic fractures (Hancock and Engelder 1989) and fracture sets in lenticular sand bodies (Lorenz and Finley 1989) that were subparallel to regional structure.

#### 4.2.6 *Fracture skins*

The movement of groundwater along fractures commonly alters the fracture surface. Fracture and vein fillings and altered fracture surfaces are observed in many rocks (Robinson et al., 1998); fracture skins are ubiquitous; they show that that fracture permeability and transport characteristics can change over time (Garner et al., this volume).

### 4.3 *Hydrogeological properties of fractures*

Key fractures parameters (including the hydraulic conductivity and the porosity both of the fractures and the fracture skins) are critical for estimating aquifer yields and transport of mass and energy. Methods for the estimation of fracture or effective porosity in the field are improving, but may not yet be adequate for the task of characterization because of variation in fractures properties. Halihan et al. (2005) discuss the effects on fracture connectivity on these types of analyses.

#### 4.3.1 *Hydraulic conductivity*

Conceptual models for hydraulic conductivity of fractured media follow well understood generalizations, but the quantitative estimation of the hydraulic conductivity tensor in such systems is not simple (Neuman, 2005). In hard crystalline rocks, valleys typically occur in areas of more intensive fracturing and, hence, higher permeability and fracture sets can impart a strong anisotropy. Greater numbers of fractures and more interconnected fractures will tend to reduce permeability anisotropy. Longer fractures, greater fracture densities, and greater apertures increase hydraulic conductivity, which also varies spatially (and temporally) because of geological constraints. The prediction of fracture domains, wherein fracture systems maintain some uniformity of hydrogeological properties and their hydraulic characteristics needs to be considered in simulation models, especially with regards to issues of upscaling.

#### 4.3.2 *Porosity*

Porosity can be subdivided in several ways. Fracture versus matrix porosity is one distinction, but this is simplistic even for double-porosity systems. In a classic study of cooling plutons, Norton and Knapp (1977) differentiated "flow (i.e., effective) porosity" (the porosity which controls fluid flow, mostly in fractures) from: 1) diffusion porosity that contributes to fluid and mass flux but fluid flow) and 2) residual porosity (isolated pores). This classification recognizes the nature of the fractured formation and extends it under geological scales to purely fractured formations. In a fractured medium, the effective porosity may assume tensor characteristics (Khaleel, 1992; Neuman, 2005). Finally, analyses may need to consider fracture porosities corresponding to different size fractures and fracture sets, as well as matrix and fracture porosities.

### 4.4 *Modelling of fracture systems*

Approaches for modelling of flow and transport in fractured media include: analytical solutions for flow between parallel plates (slot flow); equivalent porous medium models; discrete fracture models; theoretical or synthetic fracture models; double-porosity models; and an equivalent parallel plate method. Models hybridizing these various approaches also exist (Neuman, 2005).

#### 4.4.1 *Parallel plate models*

The discharge in fracture of a planar, uniform aperture per unit width (or height for vertical fractures) with no asperities is given by the cubic law (Lamb, 1932), expressed as:

$$Q = -\frac{b^3 \rho_w g}{12\mu} \nabla h \tag{1}$$

Where Q is discharge, b is aperture, $\rho_w$ is fluid density, g is gravitational acceleration, $\mu$ is dynamic viscosity, and $\nabla h$ is the hydraulic gradient in the plane of the fracture. The parallel plate can be applied to fractured media by integrating fracture densities and apertures with parallel fracture sets into the hydraulic conductivity tensor. This simple approximation can be useful, but it is strictly valid only for laminar flow. However, it is used in most of the approaches discussed below. Equation 1 must be modified for fractures with channeling and asperities by adding the appropriate empirical coefficients. For instance, Lomize's (1951) equation for hydraulic conductivity (K) is:

$$K = \frac{1}{f}\frac{b^2}{12\rho_w g} \tag{2}$$

where f is a friction or roughness factor, which is either determined by field tests or an empirical formula based upon roughness data. When dealing with upscaling issues, how does one upscale Equation 2?

#### 4.4.2 *Equivalent porous media*

The equivalent porous medium approach is commonly used in estimating flow and transport. It assumes a continuum and not ignores discrete fractures. Equivalent porous media models utilize estimates of hydraulic conductivity, storativity, and porosity that assume that a representative elemental volume (REV) can be defined at the appropriate scale. These models can yield adequate results when estimating discharge rates for a limited range of conditions, but they fail in calculating mass transport and, generally, in cases where the fractures are not efficiently interconnected. In this case, an REV may not exist (Shapiro and Bear, 1985). Scaling equivalent porous medium models is uncertain, but is attractive because it avoids data needed to characterize the actual hydraulic properties of the fractured media.

#### 4.4.3 *Discrete fracture models*

These models input as much detail as possible, based upon field data, about the geometry and properties of individual fractures, sets, and zones into 3D networks (Sahimi, 1995; Zhang et al. 2002; Neuman, 2005). Each transmissive fracture is usually assigned a uniform aperture and hydraulic conductivity. Connectivities must be assumed among other characteristics. Mapping of fractures for hydrogeological studies is rarely sufficient because mappable exposures and techniques are limited. Although quantitative field data on fracture spacings and hydraulic properties are generally lacking, semi-quantitative or qualitative estimates can be made of expected hydraulic properties, including upscaling based upon fractal or power-law relationships (e.g., Marrett et al., 1999). The only means to date for the characterizing of discrete fractures systems are by cross hole testing or tracer tests (e.g., Illman, 2003). These are expensive and time consuming, and extrapolation to adjacent areas or different scales is also difficult.

#### 4.4.4 *Theoretical statistical models*

Because of the limitations and disadvantages of the above approaches, theoretical models (e.g., Long et al. 1985; Blum et al., 1997; Hofrichter and Winkler, 2006) for flow in purely fractured and fractured formations have been developed to evaluate flow in fractures or fracture sets with synthetic distributions of apertures, orientations, spacings, and dimensions, centered on distributions of fracture nucleation points. However, application of these models to natural systems is limited because sufficient real data for input are generally lacking.

On the other hand, theoretical models increase our basic understanding of processes in fractured systems and we can use these models in an inverse-fashion to predict sets of more realistic conceptualizations.

#### 4.4.5 *Double-porosity models*

Double-porosity systems include many aquifers and petroleum reservoirs, and require calculation of the flow and transport in both fractures and matrix, as well as the interaction between them (e.g., Moench, 1984; Beyer and Mohrlok, this volume; Garner et al., this volume). This interaction is commonly treated as empirical transfer used to make model outputs reasonable, but the actual conceptualization of how this flow takes place is difficult, and experimental verification of this process uncertain (Moench, 1984). Both "pseudo-steady" flow between the matrix and the fractures and fully transient conditions, as well as fracture-skin effects, should be considered.

#### 4.4.6 *Equivalent parallel plate models*

Equivalent parallel plate or integrated fracture density of apertures, IFDA (Singhal and Gupta, 1999), applies geological data or inferences on fracture systems to models of flow and transport, particularly on a regional basis or over time spans where tracer tests or cross-hole testing are impractical. The precision of these models will depend on data availability, the hydrogeological conceptual model, and modeller insights. Aerial photography, remote sensing data, published and field geological maps, borehole and tunnel fracture studies, and data from tracer, cross hole, and geophysical tests, as well as other, standard hydrogeological analyses can provide inferences on fracture set properties which can then be input using the classical techniques of Snow (1968, 1969) into a variety of models. IFDA gives insight into the geological controls of fracture systems (e.g., Mayer and Sharp, 1998), although there are not always sharp distinctions between the above models. Again, scaling uncertainty and model evaluation are issues that need to be addressed.

## 5 UPSCALING, REV, HIERARCHY OF INHOMOGENEITY

Hydrogeological studies must often *scale data up* from laboratory tests and local site-studies to a regional scale. Rats (1967) first determined the relation between the magnitude of inhomogeneity elements of rocks and the extent of the study area. This was hydrogeologically interpreted by Rats and Chernyshev (1967) and, later, by Kiraly (1975) and Halihan et al. (2000) who focus on karstic terrains. Consequently, hydraulic parameters might differ considerably depending on methods of their determination. Conspicuous changes in permeability typically occur at a local scale. These variations are evidenced by different yields of near-by wells drilled in the same rocks that might reach several (usually up to three but sometimes even four) orders of magnitude (Figure 2a) because of variously permeable fractures. Distinct character and frequency of faults, fractures, joints and bedding planes of different orientation are observed at rock outcrops or inferred with borehole logging techniques. Identification of these inhomogeneities is important for local hydrogeological studies, particularly in analyzing preferential flow pathways and spreading of groundwater contamination or in designing remediation procedures.

By extending a tested area average permeability, based on either laboratory or aquifer tests, typically (Rovey, 1998; Schulze-Makuch and Cherkauer, 1998; Halihan et al., 2000)

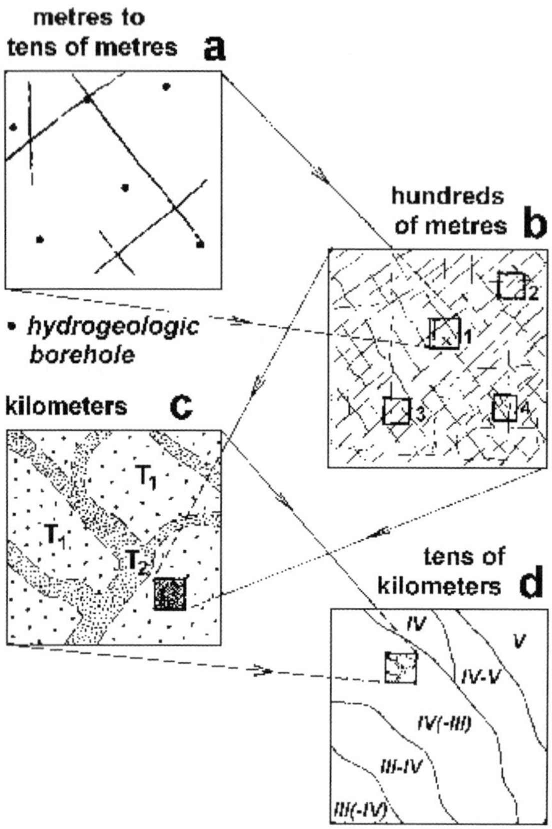


Figure 2. Relation of size of inhomogeneity elements to the extension of a study area (after Krásný 2000).
**a** – fractures and fracture zones in a local scale;
**b** – (sub-) regional more or less regular fissuring representing hydrogeological background; squares 1–4 represent different statistical samples characteristic of usually similar mean transmissivity magnitude and variation;
**c** – sub-regional inhomogeneities often following valleys with water courses: $T_1$ – lower prevailing transmissivity, $T_2$ – higher prevailing transmissivity;
**d** – regional changes in transmissivity caused by different neo-tectonic activity; roman numbers express the class of transmissivity magnitude after Krásný (1993a).

but not always (e.g., Robins, 1993) increases in spite of significant local variability. It has been suggested (Clauser 1991) that at some sufficiently large scale in crystalline rocks, the average permeability remains roughly constant, irrespective the position of the tested area within the whole environment (Figure 2b). If so, this represents a regional permeability/transmissivity background that corresponds to a *representative elemental volume* (REV), the smallest scale above which practically no change in mean values occurs. Then the average properties can be defined over scales larger than the REV. This is an important concept with consequences for groundwater protection and development as it suggests to what extent real, natural, or artificially influenced hydrogeological conditions can be schematized when implementing conceptual and numerical models. Whether or not this concept can be extended either to solute transport or to noncrystalline fractured rocks is unclear.

Occurrence of *larger inhomogeneities*, superposed upon a regional permeability background, however, may increase in permeability at greater than the REV scale. These are usually structural zones and/or belts of regional higher permeability along the river valleys. Important structural or tectonic features can form zones of higher permeability with transmissivities reaching 90–100 $m^2$/d or more, an order of magnitude higher than regional transmissivity background. These are of importance for groundwater abstraction in hardrock environment. Various authors have noted differences in the permeability of rocks depending on distinct topographical or morphological position (LeGrand, 1954; Krásný, 1974; Henriksen, 1995). Belts along river valleys may display 2 to 4.5 times higher permeabilities than surrounding areas represented by slopes and summits (Figure 2c). The position of valleys or depressions in tectonically-affected areas may be cause of this finding, and the position of recharge or discharge zones is another important factor (Krásný, 1998).

Consequently, hardrock terrains cannot be considered regionally homogeneous but rather a complex system where belts of regionally high permeability occur, usually following the valleys and depressions, in comparison with "intra-valley" blocks of lower regionally prevailing permeability. This is analogous to double-porosity medium, but at a large scale. Similar differences in regional permeabilities between valleys and slopes have been documented in other environments (Krásný 1998). Regional trends in transmissivity in some crystalline areas are reported in Norway (Rohr-Torp, 1994) and the Czech Republic (Krásný, 1996c) that reflect distinct intensity of neotectonic activities caused in Norway by isostatic uplift after Quaternary glacial retreat and in the Bohemian Massif by tectonic stress due to Alpine-Carpathian folding (Figure 2d).

Fractured environments are *intricate hierarchic systems* consisting of inhomogeneities on local, sub-regional, and regional scales and hydraulic parameters differing with different scale. Such a hierarchic system of permeability is expected in most, if not all, fractured media. Practical conclusions should be drawn for conceptual model implementation, groundwater flow and solute transport modelling, safe yield assessment, well siting, and studies on groundwater vulnerability and protection.

*Fracture properties also change with time*. Permeability of fractures and of fault zones may decrease because of processes as hydrothermal alteration, mineral precipitation and mechanical clogging (e.g., Mazurek 2000). Soluble rocks as carbonates, gypsum and salt deposits are the exceptions to this general rule; their permeability typically increases with time and in karst can be very high. In fractured rocks, geologically young fractures are typically the most permeable. Differences in hydrogeological position (clogging in recharge zones and outwash of fine particles in discharge zones) can also change fracture permeability over time (Krásný 1998).

## 6 EXACT AND COMPARATIVE HYDRAULIC PARAMETERS AND CLASSIFICATION OF TRANSMISSIVITY DATA

Most modern hydrogeological projects are of *local or site-specific character*. These include studies of water supply, contaminant and remediation, and groundwater problems in environmental planning, civil engineering, and mining. However, *regional studies* are indispensable for administrators, decision-makers, and hydrogeologists because regional studies influence land-use planning and determine conditions for integrated groundwater/surface water management, sustainable groundwater use, and groundwater protection. Based on results of regional hydrogeological studies, reasonable strategic decisions in regional, state, or even continental scales can be made. Regional studies enable us to compare hydrogeological conditions in different areas and to draw generalized conclusions. Small-scale maps offer excellent possibilities to exchange findings at regional, national and international levels. Data regionalisation is important in hard/fracture rock terrains, but such studies are not plentiful.

Hydraulic parameters, such as hydraulic conductivity, transmissivity, storage, different types of porosity, etc., can be designated as exact parameters and are determined by laboratory and field tests, but few data, usually on hydraulic conductivity and transmissivity, are commonly available in quantities useful for analyzing regional distribution (Chambel et al., this volume). Sometimes only well yields are available. These data may not be precise enough for use in regional studies or statistical treatment (Fahrenholz et al., this volume).

Although the data may not be sufficient to estimate the exact hydraulic parameters, they might be sufficient to provide general models of regional permeability. Therefore, *comparative or regional parameters*, expressing permeability and transmissivity, were introduced by Jetel (1964) and Jetel and Krásný (1968). Assuming that permeabilities and transmissivities are log-normal distributions indexes of permeability Z and of transmissivity Y were defined. Using these parameters simplifies data analysis. The two logarithmic parameters completed the system of comparative or regional parameters expressing permeability and transmissivity (Table 1). The specific capacity index or unit specific capacity is discussed by Betson and Robins (this volume).

Although *permeability*, especially in hardrock terrains, is variable at even close distances, different averaged, estimated, or prevailing values of hydraulic parameters are used in hydrogeological studies. Hydraulic conductivity, originally derived to characterise granular porous media, is used to interpret results of aquifer tests in any setting. In fractured rock, this approach can be used, but should consider scale differences in pathways of groundwater flow and objectives of hydrogeological studies (Krásný 2002). In a typical well with a depth of several tens of metres in fractured rock, only few fractures might have significant permeability so the hydraulic conductivity represents an ideal non-existing mean value between highly permeable fractures and a lower, sometimes negligible, permeability matrix.

Table 1. The system of hydraulic parameters expressing permeability/hydraulic conductivity and transmissivity (modified from Jetel and Krásný 1968).

| Property of aquifer or hydrogeological environment | Exact hydraulic parameters | Comparative/regional parameters | |
|---|---|---|---|
| | | Non-logarithmic | Logarithmic |
| Transmissivity | Coefficient of transmissivity $T = kM$ | Specific capacity $q = Q/s$ | Index of transmissivity Y $Y = \log(10^6 q)$ |
| Permeability | Coefficient of hydraulic conductivity k | Specific capacity index* $q' = Q/sM$ | Index of permeability Z $Z = \log(10^6 q')$ |

*The term used by Walton and Neill (1963); Q = well yield; s = drawdown; and M = thickness of aquifer/length of open section of a well.

In contrast, *transmissivity* expresses the property of the entire thickness of an aquifer. Because the main objective of many hydrogeological studies is the characterization of water yields or prevention of inflows during underground construction, transmissivity is a useful parameter. Where sufficient data exist for statistical analysis, sample populations delimited by different rock types, hydrogeological units, areas, structural, geomorphologic and hydrogeological position of water wells etc. can be treated to determine the mean and standard deviation. Transmissivity data can be graphically represented in a probability paper by cumulative relative frequencies of transmissivity values (Figure 3). Thus, the comparative parameter expression of transmissivity, the index of transmissivity Y, can be used with advantage.

To classify rock transmissivity, Krásný (1993a) presents a *classification system of magnitude and variation*. The range of transmissivity values is separated into six classes representing the orders of magnitude from very high (*I* class – transmissivity more than 1,000 $m^2$/day) to imperceptible transmissivity (VI class – less than 0.1 $m^2$/day) that indicate potential groundwater yields in different environments.

Another important property of these data is their variation. This reflects permeability heterogeneity and, consequently, indicates the character of the hydrogeological setting.

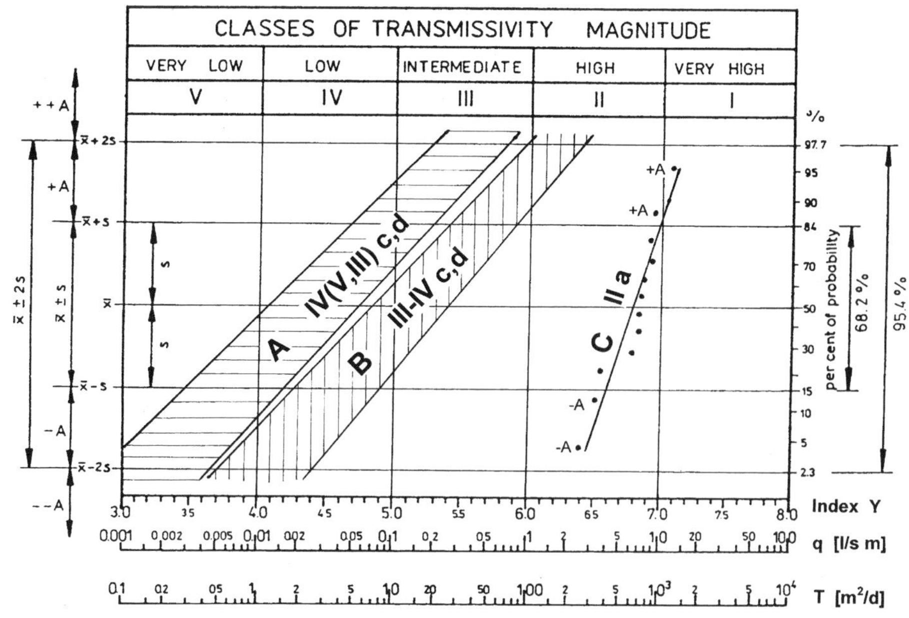


Figure 3. Prevailing transmissivity of hard rocks as fields of cumulative relative frequencies (after Krásný, 1993a, 1999). q = specific capacity in L/s m , T = coefficient of transmissivity in $m^2$/d, Index Y = index of transmissivity (Y = log $10^6$q), $\bar{x}$ = arithmetic mean, s = standard deviation, $\bar{x} \pm s$ = interval of prevailing transmissivity (transmissivity background) including approximately 68% of transmissivity values of a statistical sample, ++ A, +A, −A, −−A = fields of positive and negative anomalies (+A, −A) and extreme anomalies (++ A, −−A), resp. outside the interval $\bar{x} \pm s$ of prevailing transmissivity. A – field comprising transmissivity values of the majority of hardrocks, B – field of transmissivity values in crystalline limestones and/or in other hard rocks of higher prevailing transmissivity, C – cumulative relative frequency of transmissivity of fluvial deposits along the Labe River in the Czech Republic (for comparison).

Transmissivity variation is also classified by six classes, which are designated a to f, based on a standard deviation. This classification assesses aquifer capability to produce groundwater in different areas and the factors causing regional differences in transmissivity. A range of prevailing transmissivity values $\bar{x} \pm s$ ($\bar{x}$ = arithmetic mean, s = standard deviation of a statistical sample) represents the *transmissivity background* of a statistically-analyzed hydrogeological setting. Transmissivities outside the background are considered *anomalies* or extreme anomalies, both positive and negative.

## 7 PREVAILING TRANSMISSIVITY IN A REGIONAL SCALE, EXCEPTIONS AND DEPTH-RELATED CHANGES

Transmissivity distributions are similar in the *near-surface aquifer* of hard rocks in areas ranging from several to hundreds $km^2$, including Korea (Callahan and Choi, 1973), Sweden (Carlsson and Carlstedt, 1977), Poland (Staśko and Tarka, 1996), the Czech Republic (Krásný 1999), and Ghana (Darko and Krásný 1998, Darko 2001). Represented as cumulative relative frequencies, most transmissivity data form more or less parallel lines having fallen into a relatively narrow field A (Fig. 3). These samples belong to *classes IV(V,III) c,d* (i.e., very low to intermediate transmissivity with moderate to large transmissivity variation). Therefore, despite irregular local permeability/transmissivity changes usually scattered in a wide interval of several orders of magnitude, regionally prevailing values – transmissivity background encompasses the most frequent transmissivity values in *units $m^2/d$ up to slightly more than 10 $m^2/d$.* Comparison of regional prevailing transmissivity indicates only small differences in distinct hardrock areas. Therefore, except for certain rock types, influence of petrography on permeability and transmissivity spatial distribution is not significant because fracture distributions control the hydrogeology.

There are exceptions of this general rule; *crystalline limestones* (marbles) are one conspicuous exception. Because of their mechanical and chemical properties – suitability to fracturing and karstification – the prevailing transmissivity of limestones is usually half to one order of magnitude higher than in other crystalline rocks (field B in Figure 3). Higher permeability of limestone intercalations is often indicated by larger springs than in other hardrock terrains.

There are indications that relatively higher prevailing permeability/transmissivity may be expected in areas of *basic igneous rocks* (Havlík and Krásný 1998) that are subject to more fracturing. The gabbros complex of Beja in the south Portugal has been proved of water management importance (Chambel et al., this volume). On the other hand, *phyllites* may display lower regionally prevailing transmissivity (Krásný 1993b).

*Quartzite* hydrogeological properties differ depending on their extension. If reduced in thickness and forming part of a hardrock sequence, their prevailing transmissivity is usually comparable with their surroundings. On the other hand, where extended and exposed to intensive fracturing, quartzite bodies, due to their rigid geomechanical properties, can form highly transmissive aquifer systems with intensive and deep (as indicated by elevated water temperatures) groundwater flow (Issar and Kotze, this volume; Tröger, this volume).

Hydrogeological properties of *acidic igneous rocks* depend upon their position, geomechanical properties, weathering character. Depth-related decreasing permeability in granitic rock, orthogneisses, and migmatites is more significant than in metasedimentary rocks (Biscaldi and Derec, 1967: Krásný, 1975; and Havlík and Krásný, 1998). The regolith and weathered zone of these rocks, often sandy and coarse-grained, is permeable; metasediment regolith is typically more clayey and not as permeable. Fracturing in granites is usually not intense and reaches shallower depths than in other hard rocks. However, granite exfoliation fractures form a unique hydrogeological setting.

Depth-related changes in permeability can be used to predict productive depths of wells in hardrock near-surface aquifers (Davis and Turk, 1964; Read, 1982; and Darko and Krásný, 2000). Presence of generally more permeable products of weathering, regolith, debris and other juxtaposed young alluvial, fluvial, and eolian sediments commonly create higher transmissivity of wells where they overlay hard rocks.

Differences in geomechanical properties between granites and other hard rocks also exist in the *deep/massive zone* as indicated by thermal and mineral waters that typically occur in granitic or more generally, in acid igneous rocks, although uncertainty exists about the deep permeability distribution. The deep zone represents by far the largest volume of existing hard rocks, comprising the brittle part of the Earth's crust, i.e., extended hardrock regions below the near-surface aquifer, the basement of hydrogeological basins, and rocks underlying sedimentary nappes in intensively folded, mountainous areas. Traditionally, this zone has received little attention, but, in recent decades, applied projects as construction of deep waste repositories, construction of deep tunnels and underground cavities, deep mines, geothermal investigation of non-traditional energy sources, and few ultra-deep boreholes have been carried out (e.g., Boden and Eriksson, 1987; Stober and Bucher, 2000).

All these activities have extended our knowledge of hardrock environment up to the depths of many hundreds or even thousands meters. It has been found that hard rocks are permeable to some degree even in these depths. We do not know, however, whether any general regional scheme in permeability spatial distribution can be anticipated as depth-related changes or dependence of hydrogeological properties with respect to petrographic composition and structural position of rocks. If we admit that general decrease in permeability within the fractured zone of a near-surface hardrock aquifer is mostly connected with exogenous processes that enlarge aperture of originally existing tight fracture network, we can suggest that permeability in deep hardrock zone should not decrease considerably downwards if such a decrease occurs at all (Krásný 2003).

## 8 GROUNDWATER RESOURCES

In most hardrock terrains, the regionally extended near-surface aquifer is generally conformably with the land surface. This is where the most intensive groundwater flow and groundwater resources occur in hard rocks. However, local hydrogeological setting determines the geometry and anatomy of the hydrogeological units and their parameters. The availability of groundwater resources depends upon: 1) the hydraulic properties and their distribution, and 2) climatic conditions. For instance, because of limited precipitation and high evaporation in arid and semi-arid regions, groundwater resources are commonly limited. Intensive withdrawals, especially for irrigation, can cause regional water-table declines, deterioration of water quality, and serious environmental impacts even in hardrock terrains of low transmissivity (Marechal et al., this volume). On the other hand, in temperate climatic zones, there is abundant potential recharge, but limited transmissivity represents the controlling factor groundwater extraction. Global natural groundwater resources and recharge by Struckmeier et al. (2004) and by Zektser and Everett (2004) offer general information on ranges of natural groundwater resources in different parts of the world and in distinct climatic zones.

Depending on climatic and general hydrogeological conditions, a regional assessment of natural groundwater resources in hardrock terrains can be based upon:

- Analyses of available climatic and hydrological data resulting in water balance estimations.
- Analysis of records from river gauging stations – e.g., methods by Castany et al. (1970), Kille (1970) – partly modified by Köpf and Rothascher (1980) and field measurements of changes in total runoff along river courses in dry periods.
- Application of regionally prevailing transmissivity and morphometric characteristics of the area (Krásný and Kněžek 1977, Buchtele et al. 2003) can be used to check up results based on hydrological measurements.
- Implementation of deterministic precipitation-runoff models to simulate total runoff and its components (e.g., Sacramento SAC-SMA – Burnash 1995).
- Point/local measurement of direct or indirect groundwater recharge (e.g., Lerner et al. 1990).
- Interpretation of results of (long-term) monitoring of spring yield and groundwater level fluctuations.
- Use of environmental and radioactive tracers.

In the *temperate climates*, most streams are effluent or gaining and, under natural conditions, the separation of the baseflow component of streamflow is an acceptable method to assess groundwater resources. Investigations in Europe have revealed the importance of a hardrock environment in groundwater resources (e.g., Karrenberg and Weyer 1970, Krásný et al. 1982, Apel et al. 1996, Bocheńska et al. 1997, Kryza and Kryza 1997; and Watkins, this volume). Regional distribution of a long-term average groundwater runoff/baseflow within Central and East Europe was presented by Konopljancev et al. (1982).

In hardrock terrains, groundwater resources depend on present recharge which, under comparable hydrogeological conditions, is controlled mainly by climatic conditions. The highest natural groundwater resources in hard rocks occur in mountainous areas due to favourable climatic and geomorphologic conditions. High precipitation and low evapotranspiration enable high and relatively equable recharge. Large hydraulic gradients of groundwater level often result in intensive groundwater flow in spite of prevailing low transmissivity of rocks (Figure 4). In summits of Centro-European mountain ranges, specific long-term average groundwater runoff can exceeds 10–15 L/s km$^2$ that corresponds to 300–450 mm/a of recharge; maximum recharge may reach 20–30% of average annual precipitation. The highest values, however, should be viewed with caution, especially if snowmelt is important. On the other hand, snowmelt can positively influence the time distribution of groundwater runoff.

The *residence time* based on isotope studies was estimated to be between half to one year for shallow flow systems and several up to ten years for deeper groundwater flow systems in crystalline rocks in the Krkonoše Mts. in Czech Republic and in the Bavarian Forest in Germany (Martinec 1975, Seiler and Müller 1996). Thus, these aquifers have limited groundwater storage, but can maintain relatively intensive and continuous groundwater flow under favourable climatic conditions that helps maintain stream flows in the adjacent piedmont zones. Because of decreasing precipitation, higher evaporation rates, and generally lower hydraulic gradient with decreasing elevation, groundwater runoff diminishes (Figure 4). Relationships between climatic and hypsometric conditions and groundwater runoff in hardrock areas under temperate Central European climatic conditions are given in Table 2, which can be compared with other hardrock regions with similar climatic conditions.

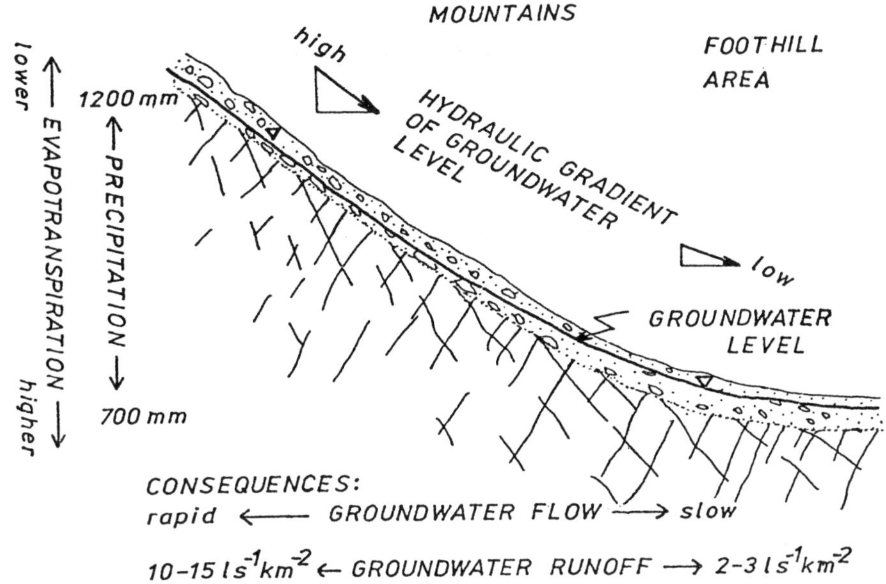


Figure 4. Groundwater flow and natural resources formation in the near-surface aquifer of mountainous hardrock terrains under temperate climatic conditions (after Krásný 1996a).

Table 2. Climatic and hypsometric conditions and groundwater runoff (baseflow) in hardrock areas of the Bohemian Massif (after Krásný 1996a).

| Morphological (hypsometric) unit | Approximate elevation (m a.s.l.) | Mean annual precipitation (mm) | Mean annual potential evaporation (estimation in mm) | Groundwater runoff (natural groundwater resources) (L/s km$^2$) |
|---|---|---|---|---|
| Mountains | 1200–1600 | 1000–1200 | 450 | 10–15 |
| Lower mountains | 800–1200 | 800–1000 | ↑ | 7–10 |
| Piedmont areas | 300–800 | 600–800 | ↓ | 3–7 |
| Flat areas, lowlands | less than 300 | 500–600 | 650 | 1–3 |

Importance of hardrock terrains for groundwater resources in temperate climatic zones is demonstrates by studies from the Bohemian Massif (Krásný 1996a) where hard rocks cover only 68% of the area, but more than 71% of total groundwater resources are recharged there.

In *semi-arid and arid regions* where few perennial streams exist, groundwater resources assessment is mostly based on water balance methods or on point/local measurements of direct or indirect groundwater recharge. Lerner et al. (1990), Struckmeier et al. (2004) and Zektser and Everett (2004) assess very low groundwater recharge that may be less than 20 mm/a in extended dry regions. It is even less than 5 mm/a in about 34% of the continent of Africa (Döll and Flörke 2005). In these arid climates, climatic zonality depending on surface elevation might be important as is shown in Table 2 for temperate conditions.

## 9 GROUNDWATER QUALITY IN HARD ROCKS

Groundwater quality in hardrock near-surface aquifers, under comparable hydraulic properties, depends mainly on climatic and geomorphologic conditions.

In the *temperate climates*, the dominant ions in groundwater are usually bicarbonate and/or sulphate and calcium. Total dissolved solids (TDS) are typically low, reaching up to 100–300 mg/L, somewhere even less. General decreases in TDS and pH (usually between 6.0 and 5.0) occur with increasing altitude, sometimes accompanied by higher relative content of sulphate that may be the result of acid rain contamination (Knutsson et al. 1995, Hrkal et al. 2006). Variations on this generalization can occur. Basic igneous rocks (gabbro, amphibolite, serpentine) are usually distinguished by higher magnesium and marbles (crystalline limestones) by pure bicarbonate-calcium types, higher hardness (and TDS), and pH about 7. This may be of practical importance for indication of limestone intercalations often representing zones of higher transmissivity. Higher sulphate and TDS are typical for some sedimentary rocks (especially shales) that may contain scattered sulphide minerals (e.g., pyrite). Hydrochemical or temperature anomalies appearing in some areas may indicate vertical upward flux of deep-seated mineral and thermal waters.

Elevated concentrations of some minority and trace constituents are typical of some hardrock areas (e.g., Frengstad and Banks, this volume). High contents of trace elements can indicate occurrences of mineral deposits and, in mining regions, water pumped from mines or outflows from mining galleries or tailings may represent a threat to the environment (Bocanegra and Cardoso, this volume). Near-surface hardrock aquifers are often exposed to bacteriological pollution from the surface as reported by Gaut et al. (this volume).

In *semi-arid and arid climatic zones*, groundwaters with low TDS contents occur only in areas with adequate recharge and relatively continuous natural groundwater discharge. Many hardrock regions in these climatic zones, however, have lost these processes because of human activities. Here (ground)water and soil are generally prone to salinisation.

In arid regions all over the world (Africa, the Americas, Asia and Australia) there are naturally closed basins without any (ground)water outflow. This is often the case of hardrock terrains where outcrops of these rocks along topographic divides represent low permeable zones. Then, negative water balance results in accumulation of salts. Similar "closed basins" can form where large amounts of groundwater are abstracted for irrigation or water-supply purposes so that lowering of groundwater levels impede groundwater outflow. Thus, under these conditions groundwater salinity can increase on long-term scales. This situation concentrates groundwater constituents and may change groundwater chemical composition from originally prevailing calcium-bicarbonate-(sulphate) type to different highly concentrated Na-Cl-rich saline waters or brines. Anthropogenic factors can also increase in contents of various contaminants from municipal, industrial, and agricultural wastes. The process can be accelerated in areas with intensive irrigation return flows, urbanization, or artificial recharge.

## 10 COUPLING GROUNDWATER FLOW AND QUALITY

Groundwater flow and water quality (its chemical composition, TDS content, temperature, etc.) in different hydrogeological environments are closely coupled. In accordance with general hydrogeological concepts, extension, depth, and rate of groundwater flow, *three vertical zones* reflecting main general features *of groundwater flow*, had been schematized in hydrogeological basins by many authors, starting from Ignatovitsh (1945) to Chebotarev (1955), Marinov et al. (1978), Tóth (1988, 1999) and many others. The zones were designated, from land surface downwards, as *local* (intensive, shallow), *intermediate* (retarded) and *regional* (slow or negligible, deep, stagnant) as shown on Table 3. Similar to the vertical zonation of groundwater flow (vertical hydrodynamic zonality), vertical changes in groundwater quality, expressed by its chemical composition, were defined as *vertical hydro(geo)chemical zonality* in sedimentary basins. In the shallow zone, which is characterized by the more intensive groundwater flow, calcium/magnesium and bicarbonate/sulphate groundwater composition typically prevails, except in arid zones where soil and shallow groundwater salinisation can occur. Groundwater composition commonly changes with depth from calcium-bicarbonate or sulphate types to sodium-bicarbonate and finally to sodium(-calcium)-chloride types in the deepest zones of basins. Depth limits between particular zones might differ considerably depending on hydrogeological conditions. With depth chemical changes are typically accompanied by increases in temperature and TDS and with possible occurrence of $CO_2$ and some hydrocarbon gases (Table 3).

Similar vertical zonations are documented in hardrock terrains – hydrogeological massifs, which were found permeable even to depths of more thousand metres and where also brines occur – mostly of sodium(-calcium)-chloride types (e.g., Gavrilenko and Derpgol'c, 1971 who designated this globally extended zone as a "hydrochlorosphere"; Collins 1975; Ingebritsen and Sanford, 1998; Stober and Bucher, 2000; and Krásný, 2003).

Regardless of near surface groundwater systems, varying petrography, and different climatic zones, we find common hydrogeological conditions globally that are manifested by vertical hydrodynamical and hydrochemical zonality at depths. Deep permeable zones enable flow groundwater over *geologic time-scales*. The origin of these deep-seated brines is a challenge for hydrogeologists. In general, there are two opinions regarding *brine origin: endogenous* and *exogenous*. Brines of an endogenous origin are considered to have ascended as different hydrothermal fluids limited to unstable zones of the Earth's crust. Water-rock interaction, often under specific pressure-temperature conditions and during

Table 3. Vertical scheme of combined hydrodynamic and hydrochemical zonality.

| Depth can reach up to | | Groundwater flow | Main groundwater constituents | TDS | General increase with the depth in |
|---|---|---|---|---|---|
| Basins | Massifs | | | | |
| hundreds of m | several tens of m | local (intensive, shallow) | $Ca(-Mg)$<br>$-HCO_3(-SO_4)$ | 0.0x–0.x g/L | temperature and gas content<br>↓ |
| few thousands of m | hundreds of m | intermediate (retarded) | $Na-HCO_3(-SO_4)$ | several g/L | ↓<br>↓<br>↓ |
| thousands of m | ↑<br>many thousands of m<br>↓<br>↓<br>↓ | regional (slow, deep, negligible – stagnant)<br>global (planetary) | $Na-Cl$<br>↓<br>↓<br>↓<br>$Na(-Ca)-Cl$<br>↓ | several hundreds g/L | ↓<br>↓<br>↓<br>↓<br>↓ |

long geologic periods, could result in brine origin in hardrock environments. In stable regions, often within the interior of tectonic plates, the exogenous concept can also explain origin of deep-seated brines.

During geologic history, because of past plate-tectonic movement, different regions have been exposed to variable climatic conditions (Schwarzbach 1974). Under arid climates, salts had been accumulated in soils and water during long past geologic periods, typically in extended closed continental basins as presently occur in arid regions. Gradual long-term downward percolation from saline soils and different surface water bodies containing brines had occurred. During very long intervals and under all the time changing natural conditions, when geologic time-scales are to be considered (i.e., up to hundreds of million years), a complex and dynamic groundwater system developed, mostly as a result of a gravity-driven and density-driven flow. After the original idea of Filatov (1956) the concept was supported and substantiated by Marinov et al. (1974, 1978) and Tóth (1999). Such a complex hydrogeological – hydrodynamic and hydrochemical system forms part of a *global (planetary) groundwater flow system* where deep-seated crystalline rocks play important, perhaps decisive role (Krásný 2003).

## 11 FUTURE CHALLENGES

Understanding flow and transport in fractured rock has proven challenging (Sharp, 1993; Voss, 2003; Neuman, 2005). This is largely because high transmissivity fractures control flow and transport and we are limited in our ability to predict where they occur and how they connect with other features, particularly other fractures. We are also limited in our understanding of the flow within a fracture and generally assign some "equivalent" permeability for an idealized uniform aperture fracture. Therefore our models are only predictive for a limited set of hydraulic conditions; different conditions can lead to vastly different results as is commonly observed in karstic aquifers. In addition, we need evaluate fracture effects in a variety of geologic scenarios, including free convection, petroleum migration, and multiphase flow. Sharp (1993) addressed future challenges. For some, progress has been made in the past decade, as is summarized in many of the papers in this volume. We have new compendia (e.g., Krásný et al., 2003; Faybishenko et al., 2000) and textbooks on fractured rock hydrogeology; new analytical and numerical models have been developed; there have been some field studies useful in testing numerical predictions; and new geophysical, GIS, and tracer methods have been developed that may yet prove fruitful. These recent studies have extended our knowledge up to the depths of hundreds or even thousands meters and it has been found that hard rocks are permeable to some degree even at great depths. However, many of the challenges posed over a decade ago remain largely unresolved.

Significant challenges include:

- Scaling – Can we scale up or down from measured or inferred properties at a given site?
- Field characterization of fracture domains – can we predict fracture properties *a priori* and the domain over which these properties are appropriate?
- Can we predict hydrogeological properties of fractures as a function of present and paleostress fields?
- Which are the depth-related changes in permeability of the deep/massive zone and what controls these changes?

- Can we predict geomechanical properties of rocks that determine their hydrogeological feature (e.g., fracture frequencies and apertures) as functions of depth?
- What techniques can determine hydraulically conductive fractures (distinguishing impermeable and permeable inhomogeneities)?
- Can we utilize geological, petrological, geophysical, well logging, or other proxy data, including data on paleoflow systems to provide insight on the extent and rates of flow and transport in fractures?
- Can we assess how fracture surfaces and skins control key processes?
- Can we assess the effects on fracture networks on density-driven flow and transport?
- How have paleohydrogeological conditions influenced the origin and evolution of deep-seated brines?
- What is role of endogenous processes in omnipresent and almost uniform vertical distribution of deep saline waters, occurring under extremely variable geological and geochemical conditions?
- Can we develop numerical codes that can simulate local-dominant effects in regional or aquifer studies?
- How can we apply fracture rock hydrogeology to:
  - Ore body mineralization
  - Petroleum migration
  - Diagenesis
- With respect specific lithologies:
  - Plutons and massifs:
    - Do fracture skins vary with fracture orientation?
    - How does fracture roughness control flow and channeling?
    - Do there exist important petrographic/rock-mechanical controls in forming fractured systems?
  - Tuffs:
    - Does pre-existing topography control fracture patterns?
    - Can we use petrographic and geological data, such as degree of welding, to make *a priori* estimates of hydrogeological parameters?
  - Carbonate systems – How and what rates do these systems change from a fractured to a karstic system?
  - Consolidated sandstones and other rocks of sedimentary origin:
    - Can microfractures observed in thin section be upscaled to aquifer or reservoir properties?
    - Can we predict attenuation properties in these double-porosity systems and estimate the effects of fracture skins?
    - What is the hydrogeological role of quartzites in hydrostratigraphic sequences?

Perhaps, we need to assess again whether or not some general regional schemes in permeability distribution can be anticipated as depth-related changes or dependence of hydrogeological properties with respect to petrographic composition and structural position.

## 12 CONCLUSIONS

Fractured aquifers often contain important water resources and the groundwater flow and solute transport in porous media are key processes in groundwater contamination, petroleum and geothermal energy production, as well as fundamental geological problems. Three major vertical zones are common: an upper or weathered zone, a middle or fractured zone, and a deep or massive zone. The upper zone and part of the middle zone may form a near-surface aquifer that is regionally extensive with important ramifications for stream flows and water resources. Despite high spatial variability in fractured rock properties, reasonable generalizations can be made about regional hydraulic and chemical properties. These aquifers are critical in many arid zones of the world and recent consideration of hard rocks for waste depositories has demonstrated how much more we need to know. Fractured geological materials are subdivided into four basic types depending principally upon the relative permeabilities of the fractures and the matrix (intrafracture materials). These are purely fractured rocks, fractured formations, double-porosity systems, and heterogeneous systems. Characterization of this flow and concomitant transport of solutes requires a knowledge of fracture orientations, apertures, asperities, flow channeling, fracture connectivity, and the nature of fracture skins, if present. The primary hydrogeological parameters, hydraulic conductivity and effective porosity, are tensor properties and the concept of a representative elemental volume for the characterization of the parameters is difficult. A variety of modeling approaches estimate flow and transport in fractured aquifers, reservoirs, and fractured geological materials of low permeability. Each has advantages and limitations. However, a major deterent to understanding of fracture systems has been reluctance or inability to obtain and apply realistic geological, geochemical, and hydrological data to these models.

Fractured rocks are vital water and environmental resources and they are being considered for new uses in waste disposal and in construction. There are many unanswered questions that require attention and, undoubtedly, new scientific questions will be posed in the next few decades. Administrative and legislative approaches must be based on general understanding and knowledge of hardrock hydrogeology and adequate professional terminology.

## ACKNOWLEDGMENTS

Acknowledgement is made to the National Science Foundation (Grant No. 0439806), the Petroleum Research Fund of the American Chemical Society (Grant No. 38949-AC9), the Geology Foundation of The University of Texas, the Grant Agency of the Czech Republic (GACR – Grant No. 205/05/177) and Grant Agency of the Charles University Prague (GAUK – Grant No. 364/2004/B-GEO/PPrF) for partial support of this research. Reviews by members of the Physical Hydrogeology graduate seminar at The University of Texas are appreciated.

## REFERENCES

Annau R, Bender S, Wohnlich S (eds) (1998) Hardrock hydrogeology of the Bohemian Massif. Proc. 3rd Internat. Workshop 1998, Windischeschenbach. Münchner Geol. Hefte B8: 79–86.

Apel R, Klemm A, Rüdiger F (1996) Grundlagen zum Wasserwirtschaftlichen Rahmenplan, Naab-region, Hydrogeologie. Bayerisches Geologisches Landesamt.

Banks D, Banks S (eds) (1993) Hydrogeology of hard rocks. Memoires, 24th Congress, Int Association Hydrogeologists, v. 24, 2 parts.

Barton CC, Hsieh PA (1989) Physical and hydrologic properties of fractures. Field Trip Guidebook T385, International Geological Congress, Washington, DC.

Barton CC, Larsen E, Page WR, Howard TM (1987) Characterizing fractured rock for fluid-flow, geomechanical, and paleostress modeling: Methods and preliminary results from Yucca Mountain, Nevada. US Geological Survey Open-File Report 87.

Biscaldi R, Derec G (1967) Un example d'application des méthodes statistiques en hydrogéologie. B.R.G.M. DS. 67 A.150.

Biscaldi R (1968) Problèmes hydrogéologiques des régions d'affleurement de roches éruptiv et métamorphiques sous climat tropical, Bulletin du BRGM III, 2: 7–22.

Black JJ, Alexander J, Jackson PD, Kinbell GS, Lake RD (1986) The role of faults in the hydrogeological environment. British Geological Survey, Natural Environmental Research Council FLPU 86–9.

Blum P, Mackay R, Riley MS, Knight JL (1997) Performance assessment of a nuclear waste repository, upscaling coupled hydro-mechanical properties for far-field analysis. Int. Jour. Rock Mechanics and Mining Sciences, 42: 781–792.

Bocheńska T, Marszalek H, Staśko S (1997) Hard rocks of the Sudety Mts. as a groundwater collector and its vulnerability. In: Bocheńska T, Staśko S (eds) (1997) Hydrogeology, 2nd Workshop on hardrock hydrogeology of the Bohemian Massif. Acta Univ. Wratislaviensis, 2052:11–19.

Bocheńska T, Staśko S (eds) (1997) Hydrogeology. 2nd Workshop on hardrock hydrogeology of the Bohemian Massif. Acta Univ. Wratislaviensis, 2052.

Boden, A., and Eriksson, K.G. (eds.), 1987, Deep drilling in crystalline bedrock. Springer-Verlag, Berlin.

Buchtele J, Čurda S, Hrkal Z, Krásný J (2003) New approach to using GIS in groundwater runoff assessment: the Krušné Mts, Czech Republic. In: Krásný J, Hrkal Z, Bruthans J (eds) Proceedings, Int Association Hydrogeologists, International Conference on Groundwater in fractured rocks, UNESCO IHP-VI, Series on Groundwater 7, Prague.

Burnash JCR (1995) The NSW River forecast system – catchment modelling. In: Singh VP (ed) Computer models of watershed hydrology. Water Resources Publications, Highlands Ranch, Colorado, 311–366.

Callahan JT, Choi SI (1973) Development of water from fractured crystalline rocks, Republic of Korea. Proceedings, 2nd International Symposium on Ground Water, Palermo 299–316.

Carlsson L, Carlstedt A (1977) Estimation of transmissivity and permeability in Swedish bedrock. Nordic Hydrology 8: 103–116.

Castany G, Margat J, Albinet M, Delarozière-Bouillin O (1970) Evaluation rapide de ressources en eaux d'une région. In: Atti Convegno Internaz. sulle acque sotterranee, Palermo.

Chebotarev II (1955) Metamorphism of natural waters in the crust of weathering. Geochimica et Cosmoschimica Acta 8: 198–212.

Chernyshev SN Dearman WR (1991) Rock fractures. Butterworth-Heinemann, London.

Chilton PJ, Foster SSD (1993) Hydrogeological characterisation and water-supply potential of basement aquifers in tropical Africa. In: Banks S, Banks D (eds) Hydrogeology of hard rocks. Memoires, 24th Congress, Int Association Hydrogeologists 1083–1100.

Clauser C (1991) Permeability of crystalline rocks. EOS, 73(21): 233–238.

Collins AG (1975) Geochemistry of oilfield waters. Elsevier.

Cook PG (2003) A guide to regional groundwater flow in fractured rock aquifers. CSIRO Land and Water. Australia.

Darko PK (2001) Quantitative aspects of hard rock aquifers: Regional evaluation of groundwater resources in Ghana,. PhD, Charles University Prague.

Darko PK, Krásný J (1998) Comparison of hardrock hydraulic parameters in distinct climatic zones: the Ghana and the Bohemian Massif areas. In: Annau R, Bender S, Wohnlich S (eds) . Proc. 3rd Internat. Workshop 1998, Windischeschenbach. Münchner Geol. Hefte B8: 3–10.

Darko PK, Krásný J (2000) Adequate depth of boreholes in hard rocks: A case study in Ghana. In: Sililo O et al (eds) Groundwater: Past achievements and future challenges, Proceedings, 30th Congress, Int Association Hydrogeologists 121–125.

Davis SN, Turk LJ (1964) Optimum depth of wells in crystalline rocks. Ground Water 22: 6–11.

Davis SN, DeWiest RJM (1966) Hydrogeology. John Wiley and Sons, New York.

Döll P, Flörke M (2005) Global scale estimation of diffuse groundwater recharge. Frankfurt Hydrol. Paper D3. Inst. of Phys. Geography.

Ehlen J (1990) Fractures in Rock: An Annotated Bibliography. US Army Corps of Engineers Report ETL-0555.

Evans DD, Nicholson TJ (1987) Flow and transport through unsaturated fractured rock. American Geophysical Union Monograph 42.

Faybishenko B, Withewrspoon PA and Benson SM (eds) (2000) Dynamics of fluids in fractured rock. American Geophysical Union Monograph 22.

Filatov KB (1956) Gravitational hypothesis of formation of groundwater chemical composition in platform depressions [in Russian]. Izdat. AN SSSR.

Fu L Milliken KL, Sharp JM Jr (1994) Porosity and permeability variations in the fractured and liesegang-banded Breathitt Sandstone (Middle Pennsylvanian), eastern Kentucky: Diagenetic controls and implications for modeling dual porosity systems. Journal of Hydrology 154: 351–381.

Fuller CM, Sharp JM Jr. (1992) Permeability and fracture patterns in welded tuff: Implications from the Santana Tuff, Trans-Pecos Texas. Geol. Soc. America Bulletin 104: 1485–1496.

Garner TT, Sharp JM Jr. (2004) Hydraulic properties of granitic fracture skins and their effects on solute transport. In: Proceedings, 2004 U.S. EPA/NGWA fractured rock conference: State of the science and measuring success in remediation, National Ground Water Association, Dublin, Ohio 664–678.

Gavrilenko ES, Derpgol'c VF (1971) Deep-seated Earth hydrosphere [in Russian]. Naukova Dumka. Kiev.

Geothermal Resources Council (1982) Fractures in geothermal reservoirs. Special Report 12.

Gustafsson P (1993) SPOT satellite data for exploration of fractured aquifers in southeastern Botswana. In: Banks S, Banks D (eds) Hydrogeology of hard rocks. Memoires, 24th Congress, Int Association Hydrogeologists 562–576.

Halihan T, Mace RE, Sharp JM Jr. (2000) Flow in the San Antonio segment of the Edwards aquifer: matrix, fractures, or conduits? In: Sasowsky ID and Wicks CM (eds) Groundwater flow and contaminant transport in carbonate aquifers 129–146.

Halihan T, Love A, Sharp JM Jr. (2005) Identifying connections in a fractured rock aquifer using ADFTs. Ground Water, 43(3): 327–335.

Hancock PL, Engelder T (1989) Neotectonic joints. Geol. Soc.America Bulletin 101: 1197–1208.

Havlík M, Krásný J (1998) Transmissivity distribution in southern part of the Bohemian Massif: Regional trends and local anomalies. In: Annau R., Bender S. and Wohnlich S. (eds.): Hardrock Hydrogeology of the Bohemian Massif. Proc. 3rd Internat. Workshop, Windischeschenbach. Münchner Geol. Hefte, B8: 11–18.

Henriksen H (1995) Relation between topography and well yield in boreholes in crystalline rocks, Sogn og Fjordane, Norway. Ground Water, 33(4): 635–643.

Hofrichter J, Winkler G (2006) Statistical analysis for the hydrogeological evaluation of fracture networks in hard rocks. Environmental Geology, 49: 821–827.

Hrkal Z, Prchalová H, Fottová D (2006) Trends in impact of acidification on groundwater bodies in the Czech Republic; an estimation of atmospheric deposition at the horizon 2015. J. Atmosph. Chemistry 53: 1–12.

IAHS/UNESCO (1967) Hydrologie de roches fisurées. Actes du colloque de Dubrovnik, October 1995

Ignatovitsh NK (1945) To the question of oil deposits origin and conservation [in Russian]. Dokl.Akad.Nauk SSSR 46(5: 215–218.

Illman WA, Neuman SP (2003) Steady-state analysis of cross-hole pneumatic injection tests in unsaturated fractured tuff. Journal of Hydrology, 281: 36–54.

Ingebritsen SE, Sanford WE (1998) Groundwater in geologic processes. Cambridge University Press.

Jetel J (1964) Application of specific capacity and new derived parameters in hydrogeology [in Czech]. Geol. Průzk 5: 144–145.

Jetel J, Krásný J (1968) Approximative aquifer characteristics in regional hydrogeological study, Věst. Ústř. Úst. geol. 43(5): 459–461, Praha.

Karrenberg H, Weyer KU (1970) Beziehungen zwischen geologischen Verhältnissen und Trockenwetterabfluss in kleinen Einzugsgebieten des Rheinischen Schiefergebirges. Z. Deutsch. Geol. Gessell. 27–41.

Khaleel R (1992) Equivalent porosity estimates for colonnade networks. Water Resources Research 28: 2783–2792.

Kille K (1970) Das Verfahren MoMNQ, ein Beitrag zur Berechnung der mittleren langjährigen Grundwasserneubildung mit Hilfe der monatlichen Niedrigwasserabflüsse. Z. Deutsch. Geol. Gessell, 89–95.

Király L (1975) Rapport sur l'état actuel des connaissances dans le domaine des caractères physiques des roches karstiques. In: Hydrogeology of karstic terrains. IAH Paris, 53–67.

Knutsson G, Bergström S, Danielsson L-G, Jacks G, Lundin L, Maxe L, Sandén P, Sverdrup H and Warfinge P (1995) Acidification of groundwater in forested till areas. Ecological Bulletin 44: 271–300.

Knutsson G (ed) (1998) Hardrock hydrogeology of the Fennoscandian shield. Proc Workshop on Hardrock Hydrogeol, Äspö, Sweden. May 1998, Nordic Hydrol. Programme Report 45:37–47.

Konopljancev AA et al. (1982) Podzemnyj stok territorii Centralnoj i Vostočnoj Evropy [in Russian]. VSEGINGEO Moskva.

Köpf E, Rothascher A (1980) Das natürliche Grundwasserdrgebot in Bayern im Vergleich zu den Hauptkomponenten des Wasserkreislaufes. Schriftenreihe Bayer Landesamt für Wasserwirtschaft, 13.

Krásný J (1974) Les différences de la transmissivité, statistiquement significatives, dans les zones de l'infiltration et du drainage. In: Memoires, 10th Congress, Int Association Hydrogeologists 204–211.

Krásný J (1975) Variation in transmissivity of crystalline rocks in southern Bohemia. Vest. Ústr. Úst. geol. 50(4): 207–216, Praha.

Krásný J (1993a) Classification of transmissivity magnitude and variation. Ground Water 31(2): 230–236.

Krásný J (1993b) Prevailing transmissivity of hard rocks in the Czech part of the Krkonoše and Jizerské hory Mts. In: Poprawski L, Bocheńska T (eds) Sb. Sympozia "Wspólczesne problemy hydrogeologii VI" Polanica Zdrój 79–86.

Krásný J (1996a) State-of-the-art of hydrogeological investigations in hard rocks: the Czech Republic. In: Krásný J, Mls J (eds) (1996) 1st Workshop on hardrock hydrogeology of the Bohemian Massif. Acta Universitatis Carolinae Geologica, 40(2): 89–101.

Krásný J (1996b) Hydrogeological environment in hard rocks: an attempt at its schematizing and terminological considerations. In: Krásný J, Mls J (eds) 1st Workshop on hardrock hydrogeology of the Bohemian Massif. Acta Universitatis Carolinae Geologica,40(2):115–122.

Krásný J (1996c) Scale effect in transmissivity data distribution. In: Krásný J, Mls J (eds) (1996) 1st Workshop on hardrock hydrogeology of the Bohemian Massif. Acta Universitatis Carolinae Geologica 40(2):123–133.

Krásný J (1998) Groundwater discharge zones: sensitive areas of surface-water – groundwater interaction. In: Van Brahama J, Eckstein Y, Ongley LK, Schneider R, Moore J (eds) Gambling with groundwater – physical, chemical, and biological aspects of aquifer-stream relation. Proceedings joint meeting 28th Congress, Int Association Hydrogeologists and annual meeting American Institute of Hydrology, Las Vegas, Nevada, 28 September–2 October 1998, 111–116.

Krásný J (1999) Hard-rock hydrogeology in the Czech Republic. Hydrogéologie, 2:25–38.

Krásný J (2000) Geologic factors influencing spatial distribution of hardrock transmissivity. In: Sililo, O., et al. (eds) Groundwater: past achievements and future challenges, Proceedings 30th Congress, Int Association Hydrogeologists 187–191.

Krásný J (2002) Understanding hydrogeological environments as a prerequisite for predicting technogenic changes in groundwater systems. In: Howard KWF, Israfilov RG (eds) Current problems of hydrogeology in urban areas, urban agglomerates and industrial centres. Proc. Advanced Research NATO Workshop, May 2001, 381–398.

Krásný J (2003) Important role of deep-seated hard rocks in a global groundwater flow: possible consequences. In: Krásný J, Hrkal Z, Bruthans J (eds) (2003) Proceedings, Int Association Hydrogeologists, International Conference on Groundwater in fractured rocks, UNESCO IHP-VI, Series on Groundwater 7: 147–148, Prague.

Krásný J, Kněžek M (1977) Regional estimate of ground-water run-off from fissured rocks on using transmissivity coefficient and geomorphologic characteristics. Jour Hydrological Sciences Warszawa, 4(2): 149–159.

Krásný J, Kněžek M, Šubová A, Daňková H, Matuška M, Hanzel V (1982) Groundwater runoff in the territory of Czechoslovakia [in Czech]. Český hydrometeor úst Praha.

Krásný J, Mls J (eds) (1996) 1st Workshop on hardrock hydrogeology of the Bohemian Massif. Acta Universitatis Carolinae Geologica 40(2).

Krásný J, Hrkal Z, Bruthans J (eds) (2003) Proceedings, Int Association Hydrogeologists, International Conference on Groundwater in fractured rocks, UNESCO IHP-VI, Series on Groundwater 7, Prague.

Kryza J, Kryza H (1997) Renewal rate of groundwater in the Sudety Mts. – its disposable and exploited resources. In: Bocheńska T, Staśko S (eds) (1997) Hydrogeology. 2nd Workshop on hardrock hydrogeology of the Bohemian Massif. Acta Univ. Wratislaviensis, 2052: 223–227.

Lamb H (1932) Hydrodynamics (6th edition). Dover Publications, New York.

LaPointe PR and Hudson JA (1985) Characterization and interpretation of rock mass joint patterns. Geol. Soc. America Special Paper 199.

Larsson I. et al. (1987) Les eaux souterraines des roches dures du socle. UNESCO Études et rapports d`hydrologie 33.

Laubach S (1992) Fracture networks in selected Cretaceous sandstones of the Green River and San Juan Basins, Wyoming, New Mexico, and Colorado. In: Schmoker JW, Coalsen, EB, Brown, CA (eds) Geological studies relevant to horizontal drilling: Examples from Western North America, Rocky Mt. Assoc. Geologists 61–73.

LeGrand HE (1954) Geology and ground water in the Statesville area, North Carolina: North Carolina Dept. of Conservation and Development, Div. of Mineral Resources Bull, cited in Davis SN, DeWiest RJM (1966) Hydrogeology. John Wiley, New York.

Lerner DN, Issar AS, Simmers I (1990) Groundwater recharge. A guide to understanding and estimating natural recharge. Int Associaton Hydrogeologists. International Contribution to Hydrogeology 8.

Lloyd JM (ed) (1999) Water resources of hard rock aquifers in arid and semi-arid zones. UNESCO Studies and Reports in Hydrology 58.

Lomize GM (1951) Flow in fractured rocks [in Russian]: Gosenergoizdat, Moscow.

Long JCS, Endo KH, Karasaki K (1985) Hydrologic behavior of fracture networks. In: Hydrogeology of Rocks of Low Permeability, Memoires, 17th Congress, Int Association Hydrogeologists 449–462.

Lorenz JC, Finley SJ (1989) Differences in fracture achracteristics and related production: Mesaverde Formation, northwestern Colorado. SPE Formation Evaluation, March, 11–16.

Marinov NA et al (1974) Hydrogeology of Asia [in Russian]. Nedra, Moscow.

Marinov NA et al (1978) Hydrogeology of Africa [in Russian]. Nedra, Moscow.

Marrett RA (1996) Aggregate properties of fracture populations. Journal Structural Geology 18: 169–178.

Marrett RA, Ortega OJ, Kelsey CM (1999) Extent of power-law scaking for natural fractures in rock. Geology, 27: 799–802.

Martinec J (1975) Subsurface flow from snowmelt traced by tritium. Water Resources Research 3: 496–498.

Mayer JM, Sharp JM Jr. (1998), Fracture control of regional ground-water flow in a carbonate aquifer in a semi-arid region. Geol. Soc. America Bulletin 110: 269–283.

Mazurek M (2000) Geological and hydraulic properties of water-conducting features in crystalline rocks. In: Stober I, Bucher K (eds) Hydrogeology of crystalline rocks. Kluwer Academic Publishers 3–26.

Moench AF (1984) Double-porosity models for a fissured groundwater reservoir with fracture skin. Water Resources Research 20: 831–846.

Mouri S (2005) Comparison of varying aperture calculations. Geol Soc. America, Abstracts with Programs 37(3): 8.

National Research Council (1996) Rock fractures and fluid flow. National Academy Press, Washington, DC.

Neretnieks I (1980) Diffusion in the rock matrix: An important factor in radionuclide retardation? Journal Geophysical Research, 85(B8): 4379–4397.

Neuman SP (2005) Trends, prospects and challenges in quantifying flow and transport through fractured rocks: Hydrogeology Journal 13: 124–147.

Norton D, Knapp R (1977) Tranport phenomena in hydrothermal systems: Cooling plutions. Americn Journal Science, 277: 913–936.

Olofsson B, Jacks G, Knutsson G, Thunvik R (2001) Groundwater in hard rock – a literature review. Kassam Report. Stockholm, Sweden.

Ortega OJ, Marrett RA, Laubach SE (2006) A scale independent approach to fracture intensity and average spacing measurement. American Assoc. Petroleum Geologists Bulletin 90: 193–208.
Phyu T (2002) Transient modeling of contaminant transport in dual porosity media with fracture skins. MS, University of Texas, Austin, Texas.
Pointet T (ed) (1997) Hard rock hydrosystems. Proceedings, Int. Symposium S2 1997. Rabat, Morocco. IAHS Publication 241.
Rats MV (1967) Inhomogeneity of rocks and of their physical properties [in Russian]. Nauka, Moskva.
Rats MV, Chernyshov SN (1967) Statistical aspect of the problem on the permeability of the jointed rocks. Proc. Dubrovnik symposium on hydrology of fractured rocks . IAHS Publication 73: 227–236.
Read RE (1982) Estimation of optimum drilling depth in fractured rock. In: Papers of the groundwater in fractured rocks AWRC conference, Series 5: 191–197.
Rebouças A da C (1993) Groundwater development in the Precambrian shield of South America and West Side Africa. In: Banks S, Banks D (eds) Hydrogeology of hard rocks. Memoires 24th Congress, Int Association Hydrogeologists 1100–1114.
Robins NS (1993) Reconnaissance survey to determine the optimum groundwater potential of the Island of Jersey. In: Banks D and Banks S (eds) Hydrogeology of hard rocks. Memoires 24th Congress, Int Association Hydrogeologists 327–337.
Robins NS and Misstear BDR (eds) (2000) Groundwater in the Celtic Regions: studies in hard rock and Quaternary hydrogeology. Geol. Soc. Special Public. 182.
Robinson NI, Sharp JM Jr, Kreisel I (1998) Contaminant transport in sets of parallel finite fractures with fracture skins. Jour. Contaminant Hydrology, 31: 83–109.
Rohr-Torp E (1994) Present uplift rates and groundwater potential in Norwegian hard rocks. Geological Survey of Norway, Bulletin 426: 47–52.
Rohr-Torp E, Roberts D (eds) (2002) Hardrock hydrogeology – proceedings of a Nordic Workshop. Geological Survey of Norway 439: 7–14.
Rönkä E, Niini H, Suokko T (eds.) (2005) Proceedings Fennoscandian 3rd regional workshop on hardrock hydrogeology. Finnish Environment Institute. Helsinki, 1–9 June 2004.
Rovey CW (1998) Digital simulation of the scale effect in hydraulic conductivity Hydrogeology Journal 6: 216–225.
Sahimi M (1995) Flow and transport in porous media and fractured rock. VCH, Weinheim, Germany
Schulze-Makuch D, Cherkauer DS (1998) Variations in hydraulic conductivity with scale of measurement during aquifer tests in heterogeneous, porous carbonate rocks. Hydrogeology Journal 6: 204–215.
Schwarzbach M (1974) Klima der Vorzeit. Ferdinand Enke Verlag, Stuttgart.
Seiler KP, Müller K (1996) Discharge and ground water recharge in crystalline rocks of the Bavarian Forest, Germany. In: Krásný J, Mls J (eds) (1996) 1st workshop on hardrock hydrogeology of the Bohemian Massif. Acta Univ. Carolinae Geologica 40(2): 205–219.
Shapiro AM, Bear J (1985) Evaluating the hydraulic conductivity of fractured rock from information on fracture geometry. In: Hydrogeology of rocks of low permeability, Memoires, 17th Congress, Int Association Hydrogeologists, 463–472.
Sharp JM Jr. (1993) Fractured aquifers/reservoirs: Approaches, problems, and opportunities. In: Banks D, Banks S (eds) Hydrogeology of hard rocks: Memoires 24th Congress, Int Association Hydrogeologists 23–38.
Sharp JM Jr., Halihan T, Uliana MM, Tsoflias GP, Landrum,, MT, Marrett R. (2000) Predicting fractured rock hydrogeological parameters from field and laboratory data. In: Sililo, O., et al. (eds) Groundwater: past achievements and future challenges, Proceedings 30th Congress, Int Association Hydrogeologists 319–324.
Singhal BBS, Gupta RP (1999) Applied hydrogeology of fractured rocks. Kluwer Academic Publishers, Dordrecht, The Netherlands.
Snow DT (1968) Fracture deformation and changes of permeability and storage upon changes of fluid pressure. Quarterly Jour Colorado School of Mines 63: 201–244.
Snow DT (1969) Anisotropic permeability of fractured media. Water Resources Research 5: 1273–1289.

Staśko S, Tarka R (1996) Hydraulic parameters of hard rocks based on long-term field experiment in Polish Sudetes. In: Krásný J, Mls J (eds) (1996) 1st Workshop on hardrock hydrogeology of the Bohemian Massif. Acta Universitatis Carolinae Geologica 40(2): 167–178.

Stober I, Bucher K (eds.) (2000) Hydrogeology of crystalline rocks. Kluwer Academic Publishers, Dordrecht, The Netherlands.

Stournaras G (ed) (2003) Proceedings,1st workshop on fissured rocks hydrogeology. Athens 12–14 June 2002.

Stournaras G, Pavlopoulos K, Bellos Th (eds) (2005) 7th Hellenic Hydrogeological Conference and 2nd MEM workshop on fissured rocks hydrogeology, Athens.

Streltsova TD (1975) Hydrodynamics of groundwater flow in a fractured formation. Water Resources Research, 12: 405–413.

Struckmeier W et al (2004) WHYMAP and the groundwater resources map of the world at the scale 1:50 000 000. Special edition for the 32nd Internat. Geol. Congress, Florence, Italy, August 2004.

Tóth J (1988) Ground water and hydrocarbon migration. In: Back, W, Rosenshein JS, Seaber PR (eds) Hydrogeology, the geology of North America, Geol. Soc. America ,O-2: 485–502.

Tóth J (1999) Groundwater as a geologic agent: An overview of the causes, processes, and manifestations. Hydrogeology Journal 7: 1–14.

Tsang YW and Tsang CF (1987) Channel model of flow through fractured media. Water Resources Research 23: 467–479.

Uliana MM (2001) Delineation of regional groundwater flow paths and their relation to structural features in the Salt and Toyah Basins, Trans-Pecos Texas. PhD, The University of Texas, Austin, Texas.

Voss CI (2003) Announcing a Hydrogeology Journal theme issue on "The Future of Hydrogeology" Hydrogeology Journal 11: 415–417.

Walton WC, Neill JC (1963) Statistical analysis of specific capacity data for a dolomite aquifer. Cited in Davis S.N., DeWiest R.J.M. (1966) Hydrogeology. John Wiley and Sons, New York.

Wright EP, Burgess WG (eds) (1992) The hydrogeology of crystalline basement aquifers in Africa. Geol. Soc. Special Publ. 66.

Yélamos JG, Villarroya F (eds) (1997) Hydrogeology of hard rocks: Some experiences from Iberian Peninsula and Bohemian Massif. Proceedings, Workshop of Iberian Subgroup on Hard Rock Hydrogeol.

Zektser IS, Everett LG (2004) Groundwater resources of the world and their use. UNESCO IHP-VI, Series on Groundwater 6.

Zhang X, Sanderson DJ, Barker AJ (2002) Numerical study of fluid flow in deforming fractured rocks using a dual permeability model. Geophys Jour. International 151: 452–468.

# Hydrogeological environment of fractured rocks

# CHAPTER 1

# Hydrogeology of granite rocks in Sardinia

Giovanni Barrocu
*Professor of Engineering Geology, University of Cagliari, Department of Land Engineering, Faculty of Engineering, Piazza d'Armi, Cagliari, Italy*

ABSTRACT: Hydrogeological investigations consisting of structural, petrographical and geomorphological studies, geophysical and drilling operations and pumping tests, have demonstrated that the groundwater in the granite rocks in Sardinia may occur in fracture zones, weathered parts of the rock, and microgranite dikes. It has been confirmed that structural analysis is a valid method for determining the local fracture pattern in the rock. Tensional, shear, and overthrust fractures were produced by lateral stresses related to the different tectonic phases that have affected Sardinia during the Caledonian, Hercynian and Alpine orogeneses. The state of openness is strictly dependent upon the type of fracture, and is maximum along the tensile cracks and overthrust zones, and is strongly influenced by weathering and hydrothermal alteration. The dimensions of the aquifers may be limited by the presence of lamprophyre dikes, which act as flow barriers between contiguous granite sectors. In cases where these dikes are frequent, the granite aquifers may have good hydraulic conductivity but their storage capacity is poor. The aquifers in the granite rocks of Sardinia may contribute decisively to solving local water supply problems.

## 1 INTRODUCTION

The problem of developing groundwater in hard rocks in Sardinia was systematically tackled in the late sixties by the Institute of Engineering Geology, at the University of Cagliari. The results of early investigations were discussed with scientists and technicians from different countries at the international symposium held in Cagliari in October 1971 (Montaldo et al., 1971). In 1973 the Regional Government of Sardinia financed a pilot investigation in different areas of Sardinia, which involved a group of researchers in Engineering Geology, Cagliari and the Department of Land Improvement and Drainage, Stockholm. The results were summed up in the report by Barrocu et al., (1974) and presented in two official meetings, in Stockholm, in April 1974, and in Cagliari, in October 1974. The cooperation between the Faculty of Engineering of the University of Cagliari and the Royal Institute of Technology, Stockholm, continued into a second project financed by the Regional Government of Sardinia and the Swedish Board for Technical Development. The results of these investigations were the subject of the applied part of the International Seminar on Groundwater in Hard Rocks (Stockholm Cagliari, September 22–October 7 1977). The seminar was convened by UNESCO and organized in cooperation with the Royal Institute of Technology, Stockholm and the University of Cagliari, with the financial support of the Swedish International Development Authority and the Technical Cooperation Department of the Italian Ministry for Foreign Affairs and in collaboration with FAO and the UNESCO-IHP Committees of Sweden and Italy, for the benefit of participants from developing countries.

The two essentially scientific research projects aimed to identify the best methodology for the development of groundwater in hard rocks under the climatic conditions of Sardinia. A research methodology was adopted which has since proven effective and has become routine in groundwater investigations carried out so far in granite rocks, mainly for the benefit of local communities.

The local interest for this type of investigation is evident if we consider that hard rocks, mainly consisting of granites, make up about one quarter of the total surface area of Sardinia (i.e., about 6,000 km$^2$ of its 23,833 km$^2$ or 24,089 km$^2$ including the small coastal islands). There is a pressing need for rationally developing all the island's water resources. Average annual rainfall for the region as a whole is not particularly low (about 660 mm/year in the period 1921–2002). Precipitation is concentrated in the cold periods and is highly variable from one year to another. In fact Sardinia has a typical Mediterranean climate. Because of the small rural centers and of the widely dispersed population in a rather rugged and scantily populated region (45 inhab/km$^2$, disregarding the provincial capitals), it is economically important to exploit all available groundwater resources, especially where the user points are scattered to meet the needs of shepherds, farmers and seasonal vacation communities.

## 2 GEOLOGICAL OUTLINE

The granite rocks in Sardinia are mainly muscovite-biotite- and biotite-granites, granodiorites, diorites, gabbrodiorites, aplites, and their frequent microgranular, porphyritic and pegmatitic varieties. The granite bodies are often intersected by swarms of lamprophyre and microgranite dikes. They generally belong to the crystalline basement of the island, which from the late pre-Cambrian up to the present time has been affected by all the geodynamic events that have taken place in the Mediterranean basin, namely the Caledonian, Hercynian, and Alpine orogenesis. As indicated in the geotectonic map of Figure 1, the granite rocks make up the northern and northern-central part as well as the south-eastern corner of Sardinia. Other major outcrops are present in the south-west and along the east coast. On the whole, they are associated with the Caledonian and especially the Hercynian orogenesis. Observations on the relationships between the different rock types within the granite bodies have suggested that the latter might have been reactivated in some way during the different orogenic phases.

The crystalline basement of Sardinia and Corsica, partially covered by Mesozoic and Cenozoic deposits, has also been strongly affected by the Alpine orogenesis. According to a number of authors (Zarudski et al., 1971; Montaldo et al., 1971; Van Bemmelen, 1972; Cocozza and Jacobacci, 1975; Pala et al., 1982 a, b; Ciminale et al., 1985; Egger et al., 1988), the Corsican-Sardinian block, which at the end of the Hercynian cycle formed part of Spain and France, in the late Cenozoic was separated along the present French-Spanish coast, and rotated 50° anticlockwise to its present position. This megatectonic process opened up two major tectonic trough structures, the western Mediterranean and the Campidano graben, in south-west-central Sardinia, interpreted as subparallel features of crustal extension. It has been suggested that the Campidano tectonic trough, as well as the Tyrrenian one, was produced by tensional actions (Barrocu and Larsson, 1977). Other secondary tectonic

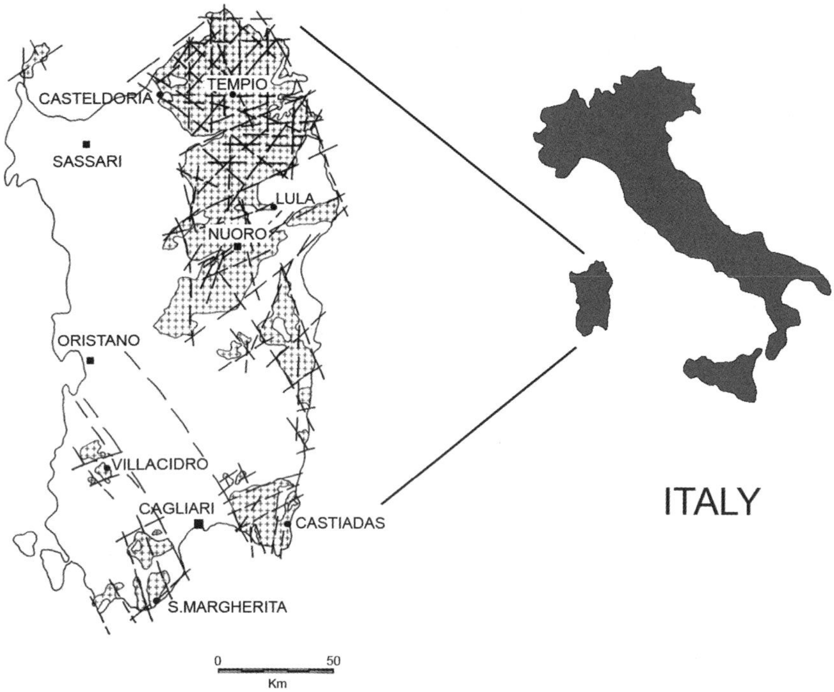


Figure 1. Granite rocks in Sardinia.

troughs formed on both sides of the Campidano, to the north-east (Chilivani – Ozieri), central east (Tirso, Marreri-Isalle, Cedrino) and to the south-west (Cixerri). These physiographic features are the most representative of the tectonic pattern of Sardinia.

Because of the up- and downfaulting of the different tectonic blocks during the Alpine orogenesis, the physiography of the crystalline basement has been rejuvenated and afterwards partly leveled by a strong denudation. At present, it is characterized by partly weathered, mostly rounded and almost flat-topped massifs, often separated by deep and intractable valleys. The scree covered slopes are bordered at the foot by piedmont deposits in the valleys and tectonic troughs.

## 3 METHODOLOGY OF GROUNDWATER INVESTIGATIONS

The hydrogeological investigations systematically carried out in different areas of Sardinia have consisted chiefly in determining the geological, petrographical, and geomorphological characteristics, structural analysis (field studies and photogeological interpretation), geophysical investigations, drilling operations, logging and test pumping, and hydrochemical analysis.

In order to determine the different parameters of the hydrological cycle (rainfall, surface run-off, infiltration and evapotranspiration) in relation to the structural pattern under different climate conditions, the elementary basins of Santa Margherita ($0.45\,km^2$), Castiadas ($3.37\,km^2$) and Nuoro-Monte Ortobene ($0.43\,km^2$) were initially taken into consideration

(Fig. 1). The first two basins are situated by the sea, respectively on the south-west and south-east coast, and the third near the top of a mountain in central Sardinia, at about 700 m a.s.l. All basins were equipped with recording rain gauges and V-notch and sharp crested weirs installed in concrete barrages. Water level recorders were placed in all weirs and recently also in two wells at Santa Margherita.

Geophysical investigations (refraction seismic, earth resistivity, magnetic and VLF electromagnetic methods) were carried out only where strictly necessary, because of the high costs involved. The results indicated that the combined VLF electromagnetic and magnetic surveys, can provide a useful and economical support in interpreting the anomalies identified by seismic and earth resistivity measurements, namely in determining the nature of low velocity zones (Barrocu and Ranieri, 1977).

Wells were drilled with a compressed air powered drilling rig operating with down-the-hole hammer. As the rig was mounted on a wagon drill, it was possible to drill wells even in rather rugged areas, down to a maximum depth of 18 m in the weathered granite and 85 m in the fractures. Internal diameters ranged from 75 to 100 mm in the exploration wells and from 100 to 165 mm in the productive wells. The latter were generally steel or plastic cased in order to prevent them from collapsing.

Structural analysis combined with geophysical surveying was generally adopted in subsequent investigations to identify the most favourable drill sites. Wells are generally drilled down to 50–80 m, but fresh groundwater was found down to a few hundred metres. In south-western Sardinia, in a tunnel of an old molybdenite mine, a constant fresh groundwater yield of 4.0 L/s was measured at −700 m in a shear zone crossing several tension fractures (Barrocu and Vernier, 1971).

Using an iterative procedure, the conceptual model defining hydrogeological and hydrochemical processes should be defined and progressively updated in relation to the local structural pattern. Hydrostructure model definition generally requires expensive field investigations, and for this reason problems need to be carefully examined so as to minimize costs and maximize benefits.

## 4 GROUNDWATER OCCURRENCE

Preliminary hydrogeological surveys carried out in the early seventies indicated that the granites in Sardinia first underwent plastic deformation during their intrusion, prior to being completely crystallized, and subsequently post-crystalline ruptural deformation (Barrocu and Larsson, 1977; Barrocu et al., 1974).

The plastic deformation may be indicated by the orientation, sliding and rotational movements of certain crystals, especially biotite and quartz, and generally by tension cracks "en echelon" parallel to the main compression stress and perpendicular to the deformation axis (Sander, 1948; Larsson, 1968, 1972; Barrocu and Larsson, 1977).

Tensional, shear and overthrust fractures were recognized as having been produced by lateral stresses associated with the different tectonic phases that affected Sardinia during the Caledonian, Hercynian, and Alpine orogenesis (Barrocu and Larsson, 1977).

According to the deformation model proposed by Larsson, 1972 (Fig. 2), the post-crystalline deformations due to tangential stresses acting on a rock body consist of: (1) tensional joints, generally vertical and perpendicular to the axis of deformation,;

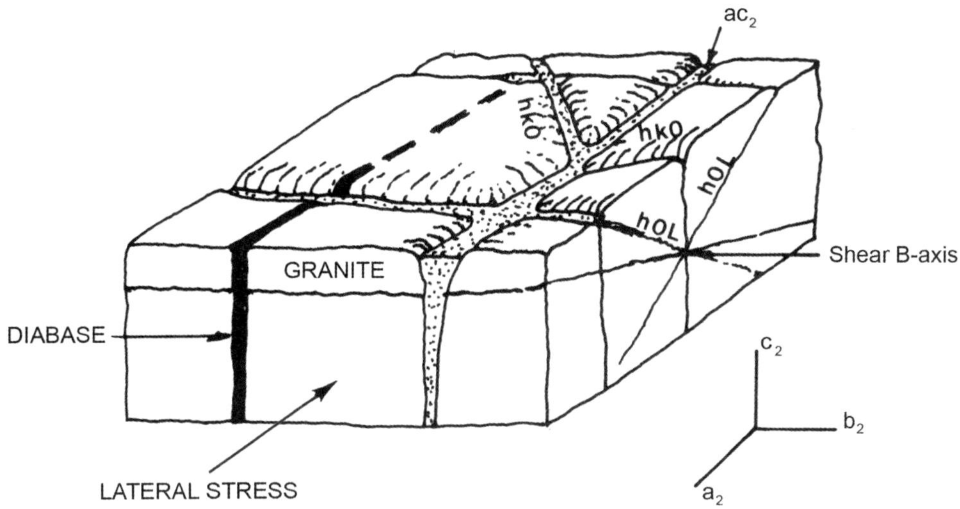


Figure 2. Granite deformation model representing tension fractures ($ac_2$), shears (hk0), and overthrusts (h0l) (Larsson 1972, slightly modified).

(2) shear fractures, usually vertical and oblique in relation to the tensional joints, often with slickensides, breccias, and mylonites; and (3) overthrust shear fractures from compression movements along sub-horizontal planes. Openness of fractures strongly depends on fracture type, and attains a maximum along the tensile cracks and overthrust zones.

Since tensional fractures are well developed both lengthwise and depthwise, they tend to create empty spaces enabling the intrusion of lamprophyre, aplite and microgranite dikes and thus generally indicate tensional forces in the earth crust.

Overthrust shear fractures, which indicate a shortening of the granite block in the direction of the main stress, may be distinguished by joints due to load release and cooling effects of the intrusive mass. They are often accompanied by slickensides and cataclastic phenomena, and their effective porosity consists of empty spaces clearly due to fracture gouge erosion produced by groundwater flow.

Aquifer dimensions may be limited by the presence of lamprophyre dikes, consisting of hornblende-plagioclase-spessartite with minor interstitial orthoclase and quartz. We have found in several areas under different conditions that they usually act as underground barriers between contiguous granite sectors. The damming effect seems to be caused by: the sealing of open fractures by the intrusion of lamprophyric magma, and especially, the intense weathering of the lamprophyres, which are generally much more argillified than the granites of the wall rock. In areas where these dikes occur, granite aquifers between them may have good hydraulic conductivity but their storage capacity is poor, depending on fractured granite zone dimensions. The closer together the dikes are the fewer the possibilities of finding a good aquifer (Barrocu and Vernier, 1975).

In conclusion, groundwater in the granite rocks of Sardinia may be found in fracture zones, weathered parts of the rock, and microgranite dikes. It may be difficult to determine the mutual relationship of the different type of aquifers because the groundwater reservoirs and the catchment areas do not often coincide (Barrocu, 2005).

## 4.1 *Aquifers in fracture zones*

Larsson (1968) demonstrated that the tension cracks formed in the plastic stage, on account of their shortness and "en echelon" development, represent poor aquifers unless their original pattern has been disturbed by later deformations. On the other hand, the post-crystalline tension fractures, which are much longer, are potentially good groundwater reservoirs and collectors of the water stored in the smaller fractures communicating with them.

Overthrust zones have high storativity as well, both because of their intense fracturing and of the fact that they are good conductors of surface infiltration, owing to their orientation in relation to the topography. The storativity of the shear fractures may vary considerably, depending on the state of weathering of the fractured rock. Fractured zone porosity may be greatly increased by relative rotational movements of the different blocks separated by faults as observed in Santa Margherita in the fracture zones between blocks A and B (Fig. 3). These phenomena may be surveyed using detailed structural analysis to determine the differential movements of the parts of a rock body. Similar block rotation effects, creating voids of hydrogeological interest, have also been recognized in other areas in northern and central Sardinia.

When exploring for groundwater in granite rocks, one should understand that the structural model may be much more complicated than expected, especially if the granite bodies have been affected by several phases of post-crystalline deformation. For instance, at Castiadas and Santa Margherita, respectively on the south-east and south-west coast of

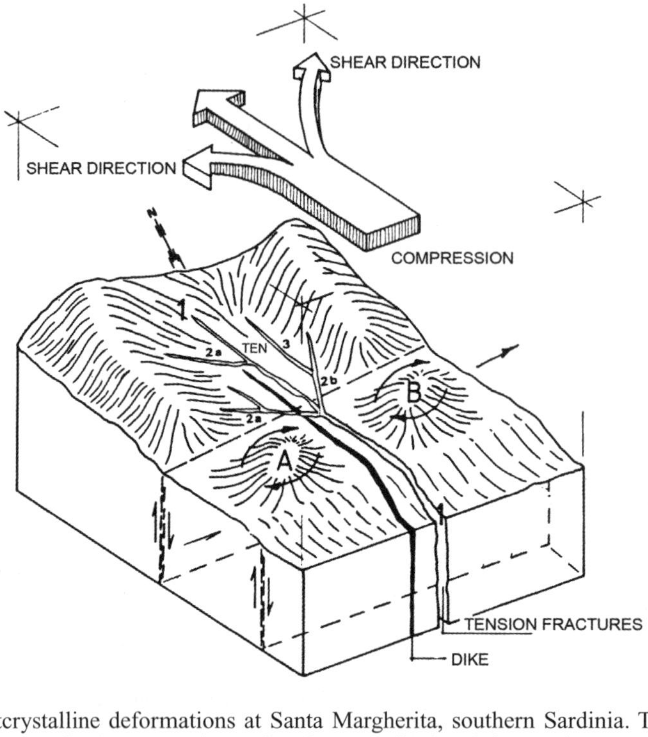


Figure 3. Postcrystalline deformations at Santa Margherita, southern Sardinia. Two ENE-WSW normal faults, perpendicular to a main tension fracture, displaced the southern part of the block in the two sub-blocks A and B, displaced en echelon, and sloping down into the coastal plain. Sub-blocks A and B, were rotated clockwise by nearly 10° (Barrocu and Larsson, 1977).

Sardinia, the granite bodies are crossed by swarms of N60°W–S60°E trending vertically dipping microgranite dikes, which, on account of their length (up to 1–2 km) and thickness (up to 40–50 m) seem to indicate an intense phase of post-crystalline deformation, which affected the whole crust at depth. It is possible that the intrusion of these dikes represents a reactivation phase of the granite batholiths. The lamprophyre dikes intersect them almost perpendicularly and may be associated with a successive deformation phase.

Yearly rainfall is in the order of 1000–1100 mm in the highlands of northern and central Sardinia, and ranges between 500 and 600 mm in southern Sardinia on the coast. Surface water is rapidly absorbed into fracture zones. A surface runoff coefficient of 0.02 was measured in 1976 at the catchment of Santa Margherita with a yearly runoff of nearly 18,000 $m^3$. In granite areas, potential evapotranspiration values ranging between 636 and 862 mm/year have been calculated. An average evaporation of 2,353 mm/year has been measured in the Lake Omodeo, in central Sardinia, and of 2,720 in the coastal pond of Santa Gilla, in southern Sardinia.

Several wells drilled in the Rio Pagghiolu valley demonstrated that the most productive aquifer is hosted in a large shear zone, which crosses several granite sectors and acts as their drain (Barrocu et al., 1974). An inclined well drilled through this fracture in the summer of 1973, gave a yield of about 2.1 L/s (7.6 $m^3$/h), and this value was confirmed in a pumping test in summer 1977. Further systematic *in situ* and laboratory investigations were carried out in the area, with a view to finding correlations between fracturing parameters, geomechanical characteristics, and permeability data measured down to −50 m below the land surface (Barrocu and Manca, 1979). A number of productive wells were drilled and groundwater has since been developed for a well known oligomineral water in Sardinia. The single granite sectors between the lamprophyre dikes, even if heavily fractured, form hydrological units quite independent of one another, with different water levels and a yield of only a few L/h. In an investigation carried out in the outskirts of Tempio to study the hydrogeology and hydrochemistry of the catchment area of the Rinaggiu oligomineral spring, the damming effect of the lamprophyre dikes was confirmed with several exploration and productive wells drilled spaced a few meters apart (Barrocu, Jacks and Vernier, 1976).

In the area of Nuoro, in central Sardinia, which has average annual rainfall of 800 mm, the tectonic pattern is dominated by rather tight shear fractures. A storativity of 0.05% was determined. This figure would correspond to a bed storativity of about 4,000–7,000 $m^3/km^2$. The yield of a number of drilled wells ranged from 0.10 to 0.86 $m^3$/h, the specific capacities being from $4.1 \times 10^{-2}$ to $2.9 \times 10^{-2}$ $m^2$/h (Rosén, 1974).

In the granite area of Santa Margherita di Pula, southern Sardinia, the main aquifer is represented by a tension fracture zone draining smaller cracks. Applying the Theis, Jacob and Cooper, and Chow methods, coarse fractures showed transmissivity T ranging from $8 \times 10^{-6}$ to $3 \times 10^{-5}$ $m^2$/s and storativity S from 0.3 to 2.2%. In the fine fractures transmissivity ranged between $1.5 \times 10^{-6}$ and $4.0 \times 10^{-6}$ $m^2$/s, and storativity between 0.15 and 0.8%. The fracture pattern had an average storativity S = 0.5%. Assuming a general coarse fracture depth of 10–15 m a storage capacity of 27,000–40,000 $m^3$ was obtained in the experimental catchment area of 0.45 $km^2$. Very rough calculations give storativity of about 50,000–80,000 $m^3/km^2$ for the fracture pattern of the granite. It was observed that the lag effect of precipitation on the drawdown during the recharge after the removal of water by pumping from the wells was very low, varying in relation to the soil moisture content according to previous rainfall and rain intensity. With a relatively dry soil the lag effect is longer. The discharge is strongly affected by rainfall, which is rapidly absorbed in the fractures. Three wells drilled in the most open fractures yielded 0.4, 0.14 and 1.50 $m^3$/h, the corresponding specific capacity being 0.02, 0.0067 and 0.063 $m^2$/h. Unfortunately, in the area of Santa Margherita the groundwater drains into the sea by the same fracture zone along which the rainfall and surface runoff are absorbed.

## 4.2 *Aquifers in weathered granites*

Most of the granite rocks in Sardinia are covered with a weathered mantle, especially in the flat or gently sloping plateaux and valley bottoms. The maximum thickness of this mantle does not generally exceed 15–20 m, but in heavily fractured zones weathering effects have been observed at a depth greater than 60 m, especially along overthrust fractures (Barrocu, 1971).

In the area of Milizzana, Tempio, northern Sardinia, several wells were drilled down to the bed-rock, at 15–17 m, in different semiconfined aquifers in weathered granite. Owing to the damming effect of the lamprophyre dikes, the mutual influence between the different wells, drilled very close to each other, is negligible. Yields ranging between 0.5 and 2.30 L/s (1.8–8.3 $m^3$/h) have been measured in different periods in 1975–1977. The water has been found in heavily weathered fractured zones 30–40 m thick, interpreted as overthrusts. Barbieri and Vernier (1977) calculated a transmissivity ranging between $1.37\times10^{-4}$ and $1.67\times10^{-4}$ $m^2$/s and a storativity S = 0.33–0.49%. A hydraulic conductivity value K ranging from $2.74\times10^{-4}$ to $3.34\times10^{-4}$ m/sec was calculated.

At Lula, central Sardinia, one productive and three observation wells were drilled in June 1973 in a mantle of weathered diorite, down to a maximum depth of about 12 m. The productive well yielded 1.4 L/s (5.04 $m^3$/h) at a drawdown of 3.2 m (Barrocu et al., 1974). A hydraulic conductivity of about $5\times10^{-4}$ m/s was calculated, and a storativity of 2–3% was estimated. The average rainfall at Lula amounts to around 500 mm/year.

## 4.3 *Aquifers in microgranite dikes*

Investigations carried out at Santa Margherita, south-west coast, and Fonni, central Sardinia, showed that microgranite dikes can be locally important hydrogeological reservoirs (Barrocu and Vernier, 1975). The capacity of such aquifers is determined by the dike dimensions, the fracture pattern, the wall rock lithological and structural characters, and the recharge conditions, which can be continuous or discontinuous, depending on rainfall. Because of their brittleness, microgranite dikes generally appear to be highly fractured. On the other hand, they are not affected to any great extent by weathering. It was observed in different areas that they may act as good drains for wall rock fractures, unless hydrothermal effects related to the intrusion of the dikes are present. In such cases abundant clay products naturally seal the fractures in the granite, which may become even tighter if weathering occurs as well. Because of the rugged morphology and their peculiar shape, the recharge areas are quite small and do not generally coincide with the catchment areas.

In a well drilled at Santa Margherita in a microgranite dike down to a depth of 36 m, groundwater issued from several 30–40 cm thick fracture zones, interpreted by the authors as overthrusts. After pumping for 72 hours, the yield remained stable for one week at 2 L/sec (7.2 $m^3$/h) with a drawdown at 16 m, after which it slowly decreased down to a stable value of 0.25 L/sec, after one month. Two observation wells drilled at the sides showed that slabs of argillified granite were included in the dike. Their yield was insignificant.

Wells drilled for water supply in a microgranite dike in the area of Fonni, central Sardinia, at altitudes of 900–1400 m, after 48 hours of pumping tests exhibited a stable yield in the order of nearly 1–2 L/sec (3.5–7.5 $m^3/h$) with a drawdown of 17–20 m, again fairly stable. In both cases the granite country rock is rather disintegrated because of intense fracturing and weathering.

Values of transmissivity T ranging from $1.26 \times 10^{-4}$ to $1.72 \times 10^{-4} m^2/s$, storativity S=0.15–0.45%, and hydraulic conductivity K ranging from $2.52 \times 10^{-4}$ to $3.34 \times 10^{-4}$ m/s were calculated.

## 4.4 *Thermal groundwater*

Very few deep boreholes have been drilled in Sardinia granite rocks, namely in the immediate vicinity of the major fracture zones where the thermal springs of Casteldoria issue groundwater at the temperature of 73.3 °C. Between 1956–63 systematic investigations for geothermal research were carried out in the area where two major N-S and E-W trending orthogonal faults cross in a graben which abruptly limits the north-western border of the Hercynian granite block of Gallura. The granite basement in the tectonic trough is overlaid by Tertiary and Quaternary sedimentary and volcanic formations of Anglona (Fig. 1). Groundwater with a head of 12 m a.s.l. issuing from exploration wells drilled down to −100 m, located at 5 m a.s.l., was an early confirmation that it flowed from a deep syphon circuit. Drilling operations and borehole logging ascertained that the granite bedrock is very fractured down to −1,000 m, with repeated higher permeability levels, which can be interpreted as overthrusts. In two boreholes thermal groundwater with a temperature of 98 °C was found at a depth of −735 and −260 m below sea level, with a head of 18 m a.s.l. (Trudu, 1971).

Subsequent geophysical investigations carried out in Sardinia identified deep thermal anomalies with temperatures of up to 150 °C at 2,000 m depth (Loddo et al., 1982). Major faults control thermal water circulation, with a downward movement schematically interpreted as a multi-cell system with relatively fast upward movement due to gas pressure (Panichi, 1982). Mixing of shallow groundwater with the deep saline water is marked by increasing TDS, in the range of 6,500–9,100 mg/L, decreasing Ca/Mg ratios, Na/Cl dominant character, and higher concentrations of B, and Sr. The Na/Cl ratios are close to values observed in seawater; and Cl, Na, K, B, and Sr show high correlation coefficients, indicating an origin from diluted, modified seawater (Caboi et al., 1986; 1993). The origin of thermal waters salinity can be chiefly attributed to sea level fluctuations during geologic history, especially during the Quaternary. Seawater saturated the deep fractures of the crystalline basement and is presently remobilised by circulating fluids. Isotope analysis results confirm that faults should be considered as always active circuits controlling mobility for water and gases in cratonic areas. Former active areas or quiescent areas are still producing hot water and deep gases (Angelone et al., 2005).

## 4.5 *Conceptual models*

Only the upper crust, where recent groundwater mostly circulates, is generally taken into consideration in groundwater investigations, but the conceptual model of fissured rocks may be rather complex and requires different methods of investigation, depending on the targets and scales of observation.

It is known that meteoric waters may partly circulate in fractures at shallow depths, and partly infiltrate at depth along major fractures, along which old groundwater stored therein flows up towards the surface. The groundwater flow in major hydraulically connected vertical and horizontal fracture zones having considerable dimensions (several kilometers) can create flow systems which would be very difficult to recognize in fractured media.

Fractured media may be so heterogeneous and complex with preferential pathways for groundwater and pollutants and with highly variable hydraulic conductivity and storage capacity so that modelling is a difficult task. Conceptual model uncertainty grade level is also high because of uncertainties in structure, boundary conditions, limits, and properties.

Precipitation and surface waters that infiltrate through the soil and formations overlaying the basement, flow partly along the discontinuities of the upper crust issuing from springs and wells as fresh groundwater, and down in depth, where they are heated, to rise again along major fractures of the crystalline bedrock, issuing as thermal waters in a number of springs located along the major faults which dissected Sardinia into uplift and downlift tectonic blocks.

The degree of interconnection between the flow systems of the fractured bedrock and the overlaying porous deposits defines, at the regional scale, the nature of the entire flow domain, and is a function of the hydraulic properties of the two components. These properties include fracture-network distribution, orientation, apertures, connectivity, sediment porosity and hydraulic conductivities of the entire bulk system.

## 5 CONCLUSIONS

The results of investigations carried out to date and those still in progress in different granite areas of Sardinia confirm the reliability and validity of a geomechanical approach on a local scale. Structural analysis has proven a useful method for determining the local rock fracture pattern. In the areas taken into consideration, the geological and structural studies, with the support of the photogeological interpretation and the use of geophysical methods (refraction seismic, magnetic, VLF electromagnetic and earth resistivity surveys) have made it possible to reconstruct the local tectonic pattern.

Aquifer characteristics are strongly conditioned by the presence of dikes and effects of weathering and hydrothermal alteration. Dikes can be strongly weathered in depth, so that they act as underground barriers between contiguous granite sectors.

The granite rocks of Sardinia may represent a valuable source of groundwater for farms and small rural villages as well as for tourist resorts along the coast. A well yield of 0.3–0.5 L/s (2.6–4.3 $m^3$/day) may be sufficient to satisfy human and animal drinking water demand and provide irrigation of some hectares on the basis of a yearly demand estimated at 5 $m^3$/ha. However, it is essential that the exploitation of these water resources be scientifically controlled in terms of safe yield considering local conditions.

Data are not sufficient yet to evaluate the amount of effective infiltration which recharges fractured granite aquifers in the upper and lower crust. Land use and morphology strongly affect water budget parameters, particularly in fracture zone aquifers. Thus, they should be calculated with reference to representative areas kept in observation, especially as concerns effective infiltration and runoff.

In certain cases the granite rocks can create fairly good reservoirs. In others, because rainfall is not very abundant and especially not evenly distributed throughout the year, yield tends to be fairly high after rainfall events, then decreasing as the groundwater finds its way to the sea. The investigations carried out so far have indicated that in certain cases the fracture zones and the microgranite dikes might be usefully and easily sealed off at their outlet, in order to create groundwater reservoirs.

## REFERENCES


Angelone M, Gasparini C, Guerra M, Lombardi S, Pizzino L, Quattrocchi F, Sacchi E, Zuppi GM (2005) Fluid geochemistry of the Sardinian Rift-Campidano Graben (Sardinia, Italy): fault segmentation, seismic quiescence of geochemically "active" faults, and new constraints for selection of $CO_2$ storage sites. Applied Geochemistry, 20: 317–340, Elsevier.

Barbieri G, Vernier A (1977) Subsurface hydrology – Tempio Pausania. "Geohydrological Investigations of Ground Water in Granite Rocks in Sardinia", II 3:21A, Dept Land Impr Drain, R Inst Technology, Stockholm.

Barrocu G (1971) Idrogeologia sotterranea dei graniti cataclastici della valle del Rio San Gerolamo fra Monte su Sinzurru e Monte Pauliara, Capoterra (Sardegna meridionale).

Barrocu G (2005) – Groundwater investigation planning and management in fissured rocks. Keynote, Proc 7th Hellenic Hydrogeol Conf on Fissured Rocks Hydrology. Athens 2005, Oct. 4–6, II, 55–68, ISBN 960-88816-2-5.

Barrocu G, Jacks G, Houtkamp H, Larsson I, Montaldo P, Vernier A, Wiberg L (1974) Geohydrological Investigations of Groundwater in Granite Rocks in Sardinia. Dept Land Impr Drain, R Inst Techn, Stockholm, I 3:15: 1–74.

Barrocu G, Jacks G, Vernier A (1976) Hydrogeological and hydrochemical Investigations on the Spring of Rinaggiu, Tempio. Techn Report, Cagliari.

Barrocu G, Larsson I (1977) Granite tectonics at groundwater prospection in southernmost Sardinia. In: Geohydrological Investigations of Groundwater in Granite Rocks in Sardinia. Dept Land Impr Drain, R. Inst Techn Stockholm, II 3:21A: 3–27.

Barrocu G, Manca PP (1979) Geomechanical characteristics of a granite body at a dam site. In: Proc Int Symp IAEG on Eng Geol Probl in Hydrotechn Constructions, Tbilisi, URSS – Bull IAEG, 20: 32–35.

Barrocu G, Vernier A (1971) Acque nella miniera di Perd'e Pibera (Sardegna sud-occidentale). In: Proc Int Sym on Groundwater in Crystalline Rocks. Cattedra di Geologia Applicata Fac Ing Univ Cagliari, 142–149.

Barrocu G, Ranieri G (1977) Preliminary considerations on the utilization of very low frequency electromagnetic methods in hydrogeological investigations in a granite area "Atti Fac Ing Univ Cagliari", 8.

Barrocu G, Vernier A (1975) Notizie preliminari sulla idrogeologia di alcuni filoni di porfido in rocce granitiche della Sardegna. III Conv Int Acque Sott Palermo Nov 1975. In: "Atti Cattedra di Geologia Applicata Fac Ing Univ Cagliari", 1–9.

Caboi R, Cidu R, Fanfani L, Zuddas P (1986) Geochemistry of thermal waters in Sardinia (Italy). In: Proc 5th Int Symp WRI Reykjavik, Iceland, August 8–1, 1986.

Caboi R, Cidu R, Fanfani L, Zuddas P, Zanzari AR (1993) – Geochemistry of the high-p$CO_2$ waters in Logudoro, Sardinia, Italy. Applied Geochemistry, 8: 153–160.

Ciminale M, Galdeano A, Gibert D, Loddo M, Pecorini G, Zito G (1985) – Magnetic Survey in the Campidano Graben (Sardinia): description and interpretation. Boll Geof Teor Appl, 27: 221–235.

Cocozza T, Jacobacci A (1975) Geological outline of Sardinia. In Geology of Italy. The Earth Science Soc of the Lybian Arab Rep, Tripoli, II: 49–81.

Egger A, Demartin M, Ansorge J, Banda E, Maistrello M (1988) The gross structure of the crust under Corsica and Sardinia. Tectonophysics, 150: 363–389.

Larsson I (1968) Groundwater in Precambrian rocks in Southern Sweden. Proceedings of the International Symposium on Groundwater Problems. October, 1966, Stockholm.

Larsson I (1972) Groundwater in granite rocks and tectonic models. Nordic Hydrology, 3.

Loddo M Mongelli F, Pecorini G, Tramacere A (1982) Prime misure di flusso di calore in Sardegna. In: Ricerche Geotermiche in Sardegna con particolare riferimento al Graben del Campidano. PFE-RF 10, CNR-Rome 181–209.

Montaldo P (ed), Proc Int Symp on Groundwater in Crystalline Rocks Cattedra di Geologia Applicata, Fac Ing Univ Cagliari, 115–126.

Pala A, Pecorini G, Porcu A, Serra S (1982 a) Schema geologico strutturale della Sardegna. In: Ricerche Geotermiche in Sardegna con particolare riferimento al Graben del Campidano. CNR-PFE-SPEG-RF10, CNR-Pisa 7–24.

Pala A, Pecorini G, Porcu A, Serra S (1982 b) Geologia e idrogeologia del Campidano. In: Ricerche Geotermiche in Sardegna. CNR-PFE-SPEG-RF10, CNR-Pisa 87–103.

Panichi C (1982) Carta delle temperature sotterranee. CNR-PFE-RF 10, CNR, Roma.

Rosén B (1974) Subsurface hydrology. In: Geohydrological Investigations of Groundwater in Granite Rocks in Sardinia, Dept Land Impr Drain, R Inst Techn, Stockholm, II, 3:21A: 69–82.

Sander B (1948) Einführung in die Gefügekunde der geologischen Körper, I Springler Verlag, Wien Insbruch.

Trudu R (1971) Esami fisici in fori sonda per ricerche di aree termali in Regione Casteldoria (Sardegna). Proc Int Sym on Groundwater in Crystalline Rocks. Cattedra di Geologia Applicata, Fac Ing Univ Cagliari, 29–40.

Van Bemmelen RW (1972) Geodynamic models, an evaluation and synthesis. "Developments in Geotectonics", 2, Elsevier, Amsterdam.

Zarudski EFK, Wong HK, Morelli C, Finetti IR (1971) Regional NW-SE seismic reflection profile across Tyrrenian. "Trans Am Geophys Union", 52, 4.

CHAPTER 2

# Groundwater exploitation of fractured rocks in South America

Emilia Bocanegra[1] and Gerson C. Silva JR.[2]

[1]*CIC. Centro de Geología de Costas y del Cuaternario Casilla de Correo Mar del Plata – Argentina*

[2]*Departamento de Geologia, Instituto de Geociências – Universidade Federal do Rio de Janeiro. Ed. CCMN Sala Rio de Janeiro – RJ BRASIL*

ABSTRACT: South America has a generous natural supply of water although it also has a notorious heterogeneous spatial distribution. Groundwater is an important resource that occurs mostly in sedimentary basins occupying 70% of the territory and in areas of igneous and metamorphic rocks that include mechanically fissured water bearing units of low permeability. Some study cases of intensive exploitation in fractured aquifers are presented in this paper, with examples of the semi-arid northeast of Brazil, in the large metropolis of São Paulo (Brazil), in the basin of the Bogotá river (Colombia), and in the city of Cuzco (Peru), among others. The knowledge of the hydrodynamic and hydrochemical aspects of the fractured aquifers in South America is insufficient, so it is necessary to improve the study of their potentiality inasmuch as, in all cases, groundwater extracted from these aquifers constitutes an essential support for the development of economic activities.

## 1 INTRODUCTION

South America has a population of over 345 million living in an area of 17.8 million km$^2$. The continent has significant water resources, with some of the more important and voluminous river basins of the world and a mean precipitation of 1600 mm/year, by far the best-ranked continent in terms of absolute water availability. Rainfall shows, however, a notorious heterogeneity regarding its distribution, ranging from a few mm/year in southern Peru and northern Chile (the driest area on the planet), to more than 8000 mm/year in southern Chile and northeast Colombia.

Large river basins with elevated rain modules such as the Amazonas and the Paraná River basins, as well as large wetlands, characterize South America. Sedimentary basins occupy approximately 70% of the territory, but some regions with acute water scarcity are located in crystalline terrains, where groundwater resources take place only in thin weathering mantles and rock fracture systems. Part of the hydrogeological properties and resources of the sedimentary aquifers is given by fracture systems, and sometimes close hydraulic interaction occurs between igneous rocks and sedimentary aquifers, as for example in the vast basalt lava flows of the Paraná and Amazon sedimentary basins. For this reason, sedimentary units where fracture systems play a significant role will be also considered in the present work, although most cases correspond to igneous and metamorphic fractured aquifers – crystalline rock aquifers. Crystalline basement rocks are commonly used as a source of groundwater because of their wide extent but yields are typically small and the low storage makes boreholes prone to drying up during drought (Morris et al. 2003).

Figure 1 shows the distribution of igneous and metamorphic rocks across South America. Those types of rocks generally coincide with the distribution of the so called "hard rock

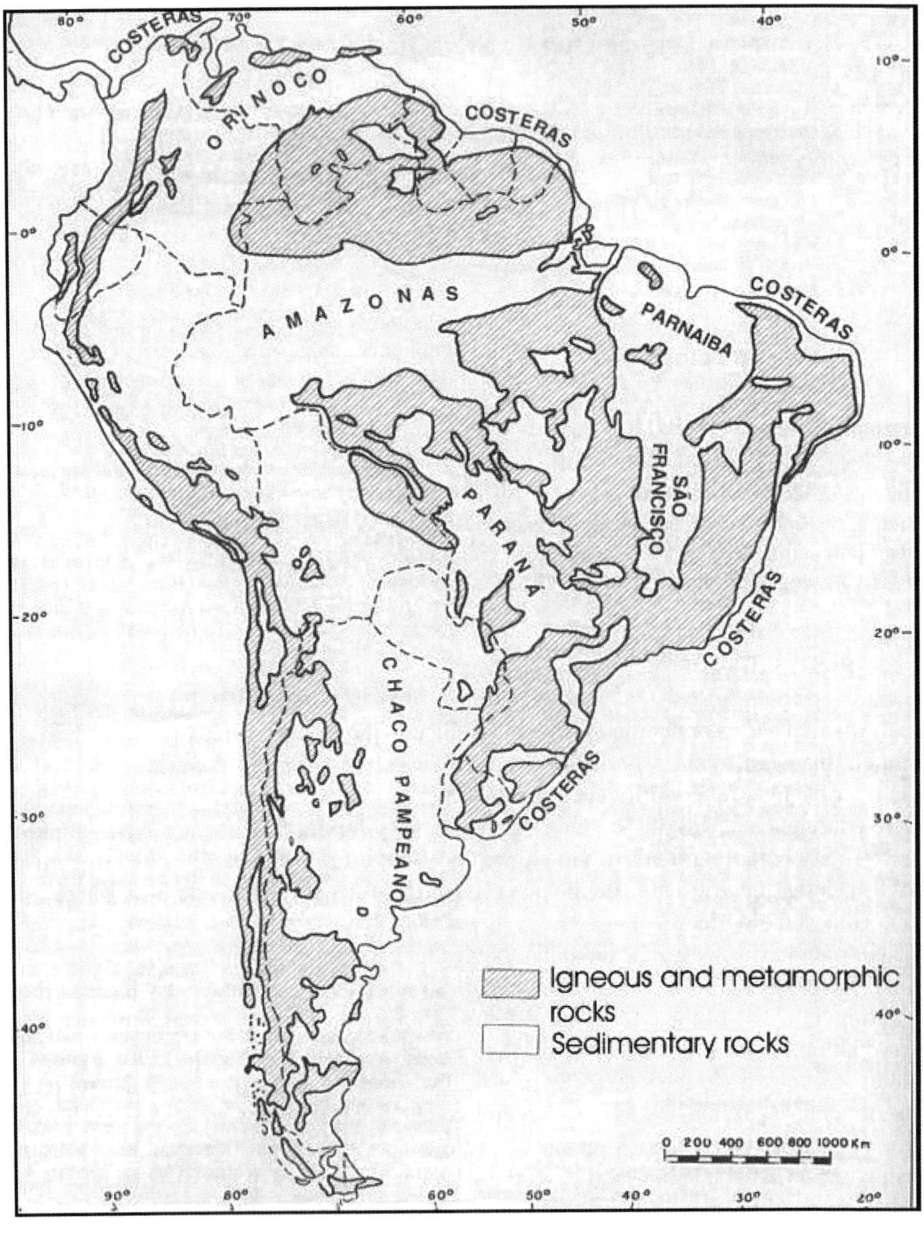


Figure 1. Distribution of igneous and metamorphic rocks across South America (adapted from UNESCO-IHP 1996).

aquifers" or "fractured aquifers" although there are important exceptions, like for example Andean volcanic ashes and tuffs, which are igneous rocks that may have an important primary porosity; and also some low grade metamorphic rocks which behave hydraulically more like sedimentary bodies than crystalline rocks. Important examples of this type of rock are the metasediments of Itabira, Andrelândia and Araxá Groups, covering large portions of the Brazilian Shield. Fractured rock aquifers are present in extensive areas of South America, and comprise a variety of igneous and metamorphic rocks generally with a relatively low permeability and small storage capacity, but of enormous social and economic importance for some regions, like Northeastern region of Brazil and Northern Chile. However, as pointed out by Anton (1993), when groundwater is considered as a potential source of water for large populations concentrated in comparatively small areas, the spectrum of possibilities to use this resource is considerably reduced, as frequently happens in crystalline areas in Latin America.

The proposal of this paper is to describe the location and main characteristics (extension, water quality and hydrodynamic aspects) of fractured aquifers in South America, with the presentation of some study cases of intensive exploitation of those resources.

## 2 FRACTURED AQUIFER OCCURRENCE AND USE IN SOUTH AMERICA

The hydrogeological map of South America prepared by UNESCO-IHP (1996) shows the identification of areas where fractured (also called "fissured") aquifers dominate (see Figures 2 and 3). A large portion of the continent corresponds to outcrops of crystalline rocks, consisting mainly of granites and gneissic or migmatitic suites, sometimes covered by relatively thin and narrow Cenozoic sediments (alluvial deposits, coastal sediments and residual or transported soils).

Those crystalline aquifers are present in the north portion of South America (Northern Shield), comprising eastern Colombia, southern Venezuela, northern Brazil, French Guyana, Suriname and Guyana; in the centre (Central Shield), encompassing the central portion of Brazil and eastern Bolivia, with small interesting reservoirs available only at a local scale; and also throughout the Andes Range, and along the Atlantic coast of Brazil (Oriental and Southern Shields).

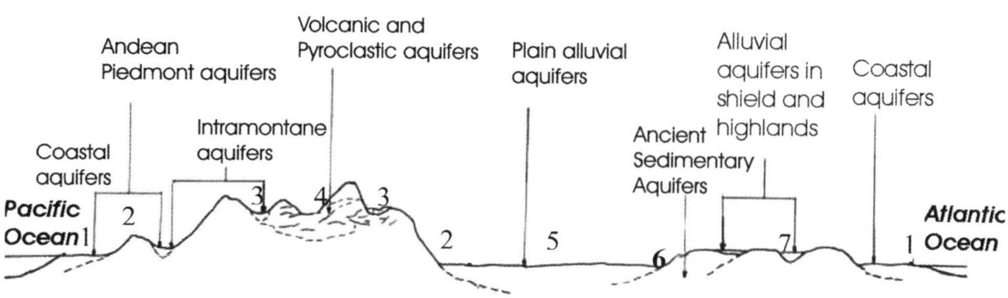


Figure 2. Generic cross-section of South America showing the location of various important cities and the main types of aquifers associated: 1. Lima, Iquique, Mar del Plata, Natal, Salvador, Maceió, Fortaleza and several other cities on the Brazilian coast; 2. Lima, Villavicencio, San Juan, Mendoza, Maracaibo, Santa Cruz; 3. Cochabamba, Valencia, Maracay, Querétaro, San Luis Potosí, Santiago; 4. Mexico City, Guatemala City, Managua, Quito; 5. Buenos Aires, San Nicolás; 6. Ribeirão Preto; 7. São Paulo, Santa Lucia (Montevideo) (adapted from Anton, 1993).

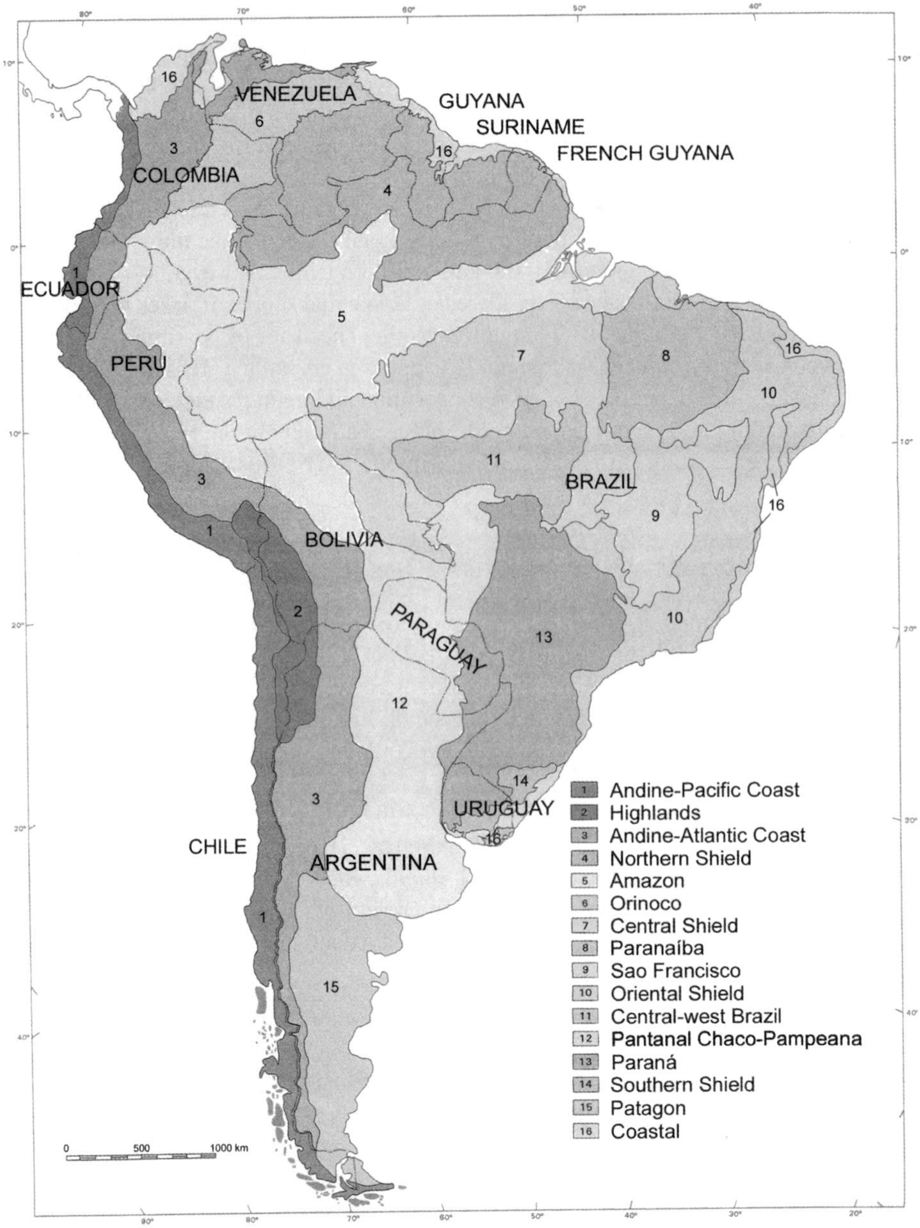


Figure 3. Hydrogeological units of South America (adapted from UNESCO-IHP 1996).

In the semi-arid northeast and humid southeast regions of Brazil (Oriental Shield), crystalline bedrock extends across approximately 3,000,000 $km^2$. Aquifer zones are associated to fractures and lineaments, exploitation levels ranging from 50 to 150 m depth, and specific capacity ranging from 0.001 to 7 $m^3h^{-1}m^{-1}$ (Celligoi and Duarte 1996).

The Guarani Aquifer System, in Paraná sedimentary basin, is overlain by basaltic lava flows (Serra Geral Formation) that stretches throughout approximately 1,000,000 $km^2$ in central southern South America. When it outcrops, forms a semi-confined aquifer with a low permeability and low salinity water, though sometimes with a high iron, fluoride and silica content. Exploitation levels range between 100 to 150 m and specific capacity varies between 0.01 to $10\,m^3h^{-1}m^{-1}$.

In the Argentine Patagonia, in the south of the country, aquifers of low permeability are exploited in some sectors.

Quaternary alluvial formations are found throughout the shield regions of South America, both in the valleys within the shields and at their periphery at the outlets of alluvial streams, with varying thickness (Figure 2). They are particularly well developed in the semi-arid Brazilian northeast following structural features like faults and important lineaments, and along the Brazilian coastal areas, as well as in the fluvial valleys of the Uruguayan-Río Grandense crystalline shield – the Southern Shield (Anton, 1993). Alluvial formation associated with fault zones are in fact the most important source of water in vast areas of the Brazilian shield, particularly in the dry Northeast, and the concept of the "fault-creek" has been useful for generations of groundwater prospectors: the association of fissured zones with the overlying alluvial deposits, formed by dissection of the shield and deposition of sediments in the more erodible fractured zones, enhances the possibility to obtain groundwater in both volume and quality acceptable in a region where none of those are easily achievable. It is important to consider that many important sedimentary units have their hydraulic behavior controlled by the fracture systems, since matrix permeability is low due to grain size or cementation of pores. In this regard, there are significant examples of those features in South America, even in some portions of the Guarani system, although unsatisfactorily mapped.

Some study cases of intensive exploitation in fractured aquifers are hereby presented, from drilling data in the semi-arid northeast of Brazil, in the large metropolis of Sao Paulo (Brazil), in the basin of the Bogota River (Colombia), and in the city of Cuzco (Peru), where the exploitation takes place by filtrating galleries.

### 2.1 *Brazil*

Brazilian territory contains about 12% of surface fresh water resources of the world (United Nations, 1997). Although groundwater data is much more uncertain, it is estimated that about the same percentage of global renewable groundwater resources is available in the Brazilian territory (permanent reserves are probably even bigger than this figure).

Nevertheless, huge contrasts in distribution prevail, with abundant and at times unexploited resources within the three big sedimentary basins of the country – units 5, 8 and 13 of Figure 3 – as compared to the densely populated areas of the oriental shield – unit 10 of Figure 3 – where groundwater resources are quite limited and insufficient to fulfil demand especially in the Northeast Region. For instance, the Northeast Region, with ca. 29% of the population, holds 3% of the country's water resources (Campanili, 2003).

Brazil is a nation of continental dimensions – more than 8.5 million $km^2$ – under humid equatorial/tropical conditions in its largest portion. Climate settings in tropical or equatorial areas include an annual rainfall between 1000 and 3000 mm/year and mild to hot temperatures. Bedrock of crystalline rocks covered by weathered mantle or detritic sediments, with a 30–50 m average thickness, occurs in approximately 4 million $km^2$.

Sedimentary deposits take place in 3.9 million $km^2$ of territory, with an average thickness close to 1000 meters. It results in perennial rivers in about 90% of the tropical territory. Nevertheless, in the central zone of Northeast Region, average rainfall varies between 500 and 800 mm/year and has a very irregular regime. Besides, combination of that climatic scenario with the domain of crystalline rocks reduces dramatically the permeability of the soil, resulting in a semi-arid climate, with periodic droughts.

Along the southern coast, there are few large rivers because the divides are not far enough from the ocean to allow development of extensive fluvial systems. Coastal rivers in Brazil tend to be short, with small basins, and average flows are rather limited in spite of high local levels of precipitation. This has promoted the use of groundwater in these areas, sometimes resulting in the intrusion or upwelling of saline water in aquifers.

### 2.1.1 *Northeastern Brazil*

Northeastern Brazil has an area of 1.6 million km$^2$, corresponding to 18.3% of the national territory and a population of 44.8 million (28.5% of the country population). In this region, the climate is semi-arid in 70% of the area. Regarding its groundwater resources, most of the area is dominated by a fractured aquifer in crystalline rocks with a very low permeability. Well yields are low, with a mean depth of 50 m, and discharges of about 1.5 m$^3$/h. Another serious problem in this aquifer is salinization, TDS values reaching 5000 mg/L due to the intense evaporation and low rainfall rates. Groundwater in fractured crystalline rocks has an important function in terms of water supply systems of the Northeastern Brazilian structural provinces (Borborema to the north, São Francisco to the south), the greater part of which under semi-arid climatic conditions (Neves and Albuquerque, 2004). The total number of deep wells drilled in this aquifer is believed to reach 150,000 during the last 150 years, although data is scarce and unreliable. Notwithstanding, 30% of them are deactivated due to low production and poor water quality.

Water consumptive use meets irrigation demands (59%), human consumption (22%), industrial and agricultural-industrial consumption (13%) and animal consumption (6%). Groundwater participation in this demand is complementary in some areas, although in others, mainly in small cities, is an almost exclusive source. In many areas of the Brazilian northeast region, intensive use of groundwater enhances risk of resources depletion, salinization and soil subsidence (Costa 2001).

The effects of scarcity of water are strongly felt in the Brazilian northeast semiarid areas. Nowadays, the surface water per capita is insufficient for the 15 million people that inhabit the rural area. In the central zone of the Northeast Region the combination of poor rainfall rates with the geological bedrock, formed by sub-outcropping crystalline rocks with a very low permeability, results in temporary rivers and conditions of semi-aridity over about 10% of the national territory (Rebouças, 2000). On the other hand, high temperatures and winds in the region provoke an intense evaporation, impairing long-term water reserves. Those annual average rainfall values can be concentrated in only one month in drought years or in the 3–5 months of the rainy season in the normal years. In practice, the drought in this context leads to the virtual impossibility of subsistence, unleashing migration to the southern and richer parts of the country, deteriorating social conditions of the country as a whole.

Northeast Region comprehends two different hydrogeological contexts: ca. 980,000 km$^2$ outcropping or sub-outcropping crystalline rocks, with extremely low permeability except in zones with fractured rocks; and 648,130 km$^2$ of sedimentary rocks, with large groundwater reserves (Rebouças, 1999). Sedimentary rocks are situated westwards from the "drought polygon" or in coastal zones of the region (units 8 and 16 in Figure 3). The crystalline area prevails in the inner portions in most the northeastern states, exactly where aridity is more pronounced (unit 10 of Figure 3).

Groundwater from crystalline rocks in Brazilian Northeast generally present poor quality due to a complex association of water circulation and continuous evaporation in discharge areas in fracture zones. They have frequently a high salt content, except when associated to alluvial deposits with significant depth and extent. Alluvial formations, which are frequently formed of quartzic or arkosic sandy or gravelly material, may contain significant volumes of groundwater and can deliver relatively high yields because of their porosity and hydraulic conductivity. Several cities of northeast Brazil and the Atlantic southern coastal plains use groundwater from this type of aquifer. In the Northeast Region, as stated, groundwater is widely used because of the lack of surface water.

Location of wells in the difficult conditions described above is a crucial component of the problem, especially when the high cost of perforations in hard rocks is considered. Many efforts have being carried out in the last few years to optimise well location and site selection, pursuing both quantity and quality. The best locations for water wells in crystalline terrains of semi-arid lands are narrow, densely fractured and altered belts, normally associated with fault zones and folding axes. Geophysical electrical methods have been tested and used with some success to prospect groundwater in this context (Medeiros and de Lima, 1990).

Coriolano et al. (2000) re-evaluated methods to locate water wells in crystalline terrains aiming to improve the success rate of the drilling procedures. The authors describe a new approach tested in Rio Grande do Norte (NE Brazil), with emphasis in modern concepts on rock deformation and the regional tectonic framework, finding correlations between productive fractured zones remotely interpreted lineaments following drainages. In other cases, they found that the best locations are related to opening of foliation surfaces by weathering and their intersection with fracture sets and favourable recharge conditions provided by locally thicker, alluvial covers, forming a through-like structure.

Silva and Jardim de Sá (2000) tried to set up relations between fracture chronology and neotectonic control of water wells in Brazilian northeast crystalline terrains (Equador region in Rio Grande do Norte State). The authors managed to establish apparently clear relations between fracture attitudes with different tectonic constraints to well productivity, in qualitative terms.

A significant recent example comes from Pernambuco State, where Tröger et al. (2003) studied the crystalline fractured groundwater bearing units from qualitative and quantitative standpoints. The authors studied more than 500 chemical analyses coming from 297 water wells and verified that deep wells mainly have a poor quality and low production. The water quality and the highest production are with some exceptions linked to the shallow wells with up to 50 m depth. Tectonic features and location apparently showed minor importance in conditioning productive zones, since wells in tectonic structures produced less water than expected. Groundwater from these wells mainly has a tendency to poor quality. When alluvial sediments accumulate to over 20 m thickness, shallow wells have good production and a good water quality. The origin of the salt content in groundwater is still a subject of controversy, according to the same authors, but they suggest it could be related to dissolution of the rock matrix and minerals. The most effective wells in the crystalline-rock areas of western Pernambuco are located in paleo-valleys in the weathered rock. The best conditions are found in places where the water circulation is high and the minerals are leached and transported away with the surface water (Tröger et al., 2003).

Although research efforts are important and enhance the comprehension of the fractured groundwater mechanisms and behaviour in the Northeast region, it is still a major task ahead to improve groundwater research methods and exploitation in semiarid crystalline rocks.

### 2.1.2 *São Paulo Metropolitan Area (Brazil)*

The São Paulo Metropolitan Area (SPMA) has 17.5 million inhabitants on a territory of 5680 km$^2$ (Figure 4) within the Atlantic Plateau, between two mountain ranges, the *Serra da Cantareira* (part of the Mantiqueira Range) in the north and the *Serra do Mar* in the south. Both Mantiqueira and Do Mar chains extend along thousands of km more or less parallel to Brazilian southeast shoreline. Groundwater resources play a central role in the complementary supply of the private sector in SPMA and constitute 11% of total consumption.

The SPMA lies in the geomorphologic compartment corresponding to the Atlantic Plateau province, in the following zones: the Paulistano Plateau (where most of the territory is situated, including the sub-zones, São Paulo Hills and Embú Mountains); the North Crystalline Zone; the Paraíba Valley, which includes the Crystalline Hills sub-zone and the Ibiúna Plateau. Most of the urban area in São Paulo Metropolitan Area comprises the São Paulo Sedimentary Basin, which is about 75 km long (NE direction) and 25 km wide (NW direction). The basement is composed of granite and gneiss. The overlying Tertiary sediments are comprised of alternating clay, silt and sand layers of Resende, Tremembé, São Paulo and Itaquaquecetuba Formations (Prado et al. 2001).

Two main aquifer systems occur in the SPMA: the Sedimentary Aquifer System and the Crystalline Aquifer System. The Sedimentary Aquifer System corresponds to the units of São Paulo Sedimentary Basin, behaving as free to semi-confined aquifers (specific capacity 0.5–0.9 m$^3$h$^{-1}$m$^{-1}$), and is the most exploited one, although covering only 25% of

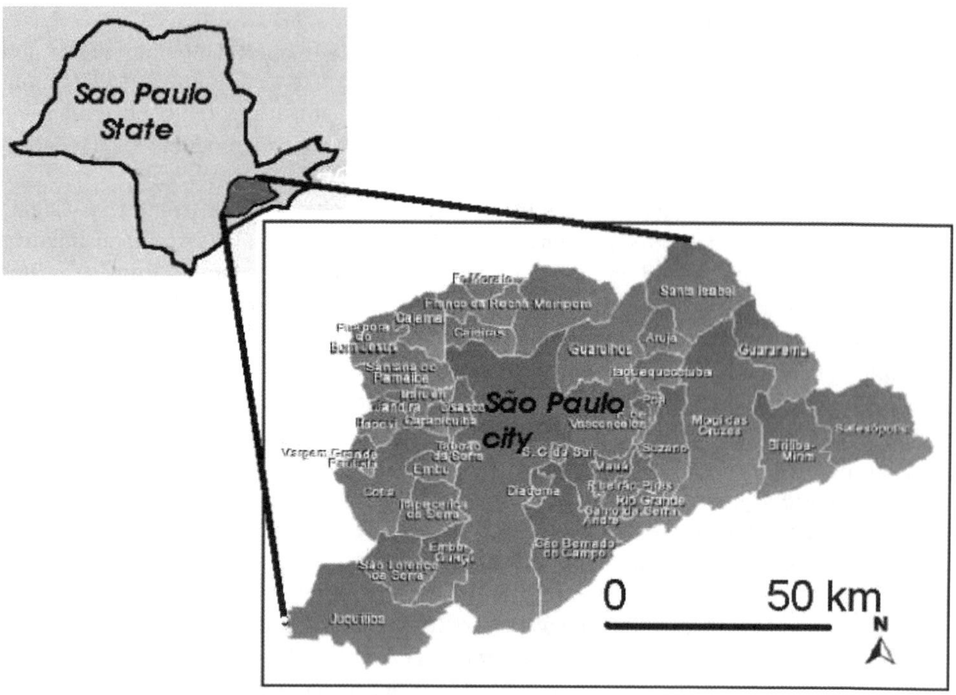


Figure 4. São Paulo Metropolitan Area (SPMA), Brazil, showing its location within São Paulo State and the various municipalities that comprise it.

the SPMA. Many wells, though, due to the relatively small depth of the sedimentary basin, are dug and screened in both aquifers.

The Crystalline Aquifer System is at or near surface along the borders of the São Paulo basin and in some portions of the inner territory and is made up of granitic as well as metamorphic rocks, with a variable permeability depending on the rock weathering (specific capacity 0.2–1.35 $m^3h^{-1}m^{-1}$). The Crystalline Aquifer System may be subdivided in the Upper Aquifer and the Lower Aquifer, the first one comprising the weathering mantle up to approximately 50 m depth and the second subsystem the fractured rocks properly, which is heterogeneous, anisotropic and free to semi-confined in character (Hirata and Ferreira, 2001).

It is estimated that a large number of industries, residential areas and others use aquifer resources through more than 9000 deep wells in operation, 2500 of which dug in the Crystalline Aquifer System. The number of hand dug wells is much higher, about 40,000 wells. There seems to be an increase at a rate of 1500 perforations per year due to the high fees on water service, which leads in some cases to amortize a 200 m perforation in only 6 months. Nowadays, there is interference between wells and a notorious decrease in the piezometric levels leading to a 50% reduction in the aquifer thickness. This has led to a rise in water costs, due to the greater electric energy consumption, perforation deepening and changes of pumping systems (Hirata et al. 2002). On the other hand, water from the crystalline aquifer in metropolitan area of São Paulo is considered as having a good or acceptable quality (the last one requiring some kind of treatment before consumption), except for some metals (Fe, Pb, Mn) and fluoride in a number of wells (Martins Neto et al., 2004).

The excessive groundwater withdrawal and contamination pressure over the crystalline rocks in São Paulo Metropolitan Area is immense. Still, its wells constitute more and more an invaluable source of good quality water to a huge population. Protection work must be accelerated in order to prevent its loss.

## 2.2 *Upper basin of the Bogota River, Colombia*

Upper basin of the Bogota River is an intra mountain range basin of 4,300 $km^2$, with a population of 848,000 inhabitants; aside from the 6,500,000 inhabitants living in the capital city of Colombia, Bogota.

The main aquifer is composed of Neogene-Quaternary fluvial-lacustrine sediments. It is underlain by the Guadalupe Aquifer, of secondary importance; and made up of consolidated sandstone and siltite, locally fractured and of Cretaceous age.

Guadalupe aquifer flow occurs mainly through fractures, and the transmissivity ranges from 5 to 536 $m^2$/day. Storage coefficients range from $1{,}10^{-2}$ to $9{,}10^{-7}$; though the most frequent values are in the range of $1{,}10^{-4}$–$1{,}10^{-5}$. In the south-western and central area of the basin, piezometric drawdown of as much as 20 m have been observed in these last 20 years; as well as an extension of the pumping cones towards the piedmont.

The waters of the Guadalupe Aquifer are calcium-bicarbonated at the piedmont and sodium-bicarbonated in the area overlain by Quaternary sediments.

The aquifer system exploitation counts on 5258 wells yielding 39.4 $hm^3$/year, of which 10.9 $hm^3$/year come from the Guadalupe Aquifer alone. Recharge in the Guadalupe aquifer is of 29.1 $hm^3$/year, defining thereby a positive difference between abstractions and recharge of 18.1 $hm^3$/year, the rest of it representing discharge from the aquifer. In the Quaternary Aquifer the difference between pumping and recharge is estimated at $-1$.

1 $hm^3$/year, indicating that it is going into intense exploitation (Castrillón Muñoz and Aravena 2002). This leads to significant changes in hydrochemical characteristics, like the dissolved organic carbon content rise in methane concentration and the estimated aquifer age. As commented before regarding the interaction between the Guarani aquifer and the overlying fractured basalt lava flow, it is evident the human impact in fractured aquifers due to actions in adjacent porous aquifers. They are particularly susceptible to changes in hydrodynamic and quality properties as a consequence of exploitation.

### 2.3 *Cuzco Valley. Peru*

The city of Cuzco is located in the inter-mountain valleys of the Peruvian Andes' Eastern Cordillera, at 3,400 m.a.s.l. The main sources of supply are 2 lagoons yielding 350 L/s, currently greatly exploited; they are reporting a water drawdown of as much as 5 m from their original level. The area is characterised by a rocky outcrop covering 90% of the total surface, where fissured aquifers of the Upper Cretaceous and Tertiary lie; and 10% shows fine recent deposits, of lacustrine- alluvial origin.

The main fissured aquifer is 6,000 m thick, with transmissivity ranging from 500 to several thousands $m^2$/day, the real velocity ranges from 100 to 800 m/year and the fracture density is in the order of 10 fr/$m^3$. Waters are of the calcium-bicarbonate type, great hardness, low pH and acceptable potability.

Fissured aquifers are the most important regarding urban and agricultural exploitation, using horizontal and slanted filtrating galleries whose length ranges from 150 to 300 m, and the flow rates varies from 20 to 65 L/s, allowing gravity supply, and minimising costs of maintenance and exploitation (Apaza et al., 1998). Once again the intrinsic characteristics of fractured aquifers make them prone to quality problems (bacteriological and chemical contamination) more easily than porous media.

### 2.4 *Other cases*

There are many other areas in South America, in addition to those cited in the present work, where fractured aquifers or fracture-conditioned groundwater circulation and recharge play a major role in supplying water to the population. It was not the aim of the work to make exhaustive descriptions of all areas, but it is important to mention some other relevant cases. In the extremely arid Northern Chile's Atacama, some works have shown the importance of the fracture system in supplying a fresh water source to porous aquifers, even at far distance from recharge areas (Margaritz et al., 1990; Houston, 2001). It has been suggested that recent fresh water, in the centre of the Pampa del Tamarugal may have originated through deep fractures and faults interconnected to the neighbouring highland aquifers. On the other hand, fault system is responsible for the deposition of alluvial fans that store a great amount of groundwater.

In Peru, as in Chile, Brazil and Bolivia, mining activities carried out in areas with occurrence of fractured aquifers has possibly imperilled groundwater resources due to potentially rapid propagation of contaminant plumes. Méndez (2005) has carried out extensive chemical analyses of water and sediment samples from the Rimac River Basin mining area in Peru. The results indicate low levels of contamination in river waters. However, significant levels of metal concentration in sediment samples were detected, putting eventually in risk groundwater resources.

## 3 FINAL COMMENTS

Fractured aquifers in South America have a considerable groundwater potential, sometimes representing the only source of potable water available. Nevertheless, exploitation undertaken without proper management and exploration techniques leads to a high number of unsuccessful wells. In some areas of South America, water supply is obtained from surface sources that are being gradually exhausted. Groundwater may become the main source for expansion of urban supply systems in, for example, Montevideo in Uruguay, Recife and Salvador in Brazil, among others (Anton, 1993). All those are cases of cities located total or partially upon crystalline rocks.

Groundwater from fractured rocks presents generally good quality, with low mineralization. Important exceptions are the crystalline rocks in the northeast arid region of Brazil, frequently with high TDS, and some waters from basalt lava flows in the western part of that country, which eventually have high fluoride contents, leading to fluorosis in some areas.

The knowledge of the hydrodynamic and hydrochemical aspects of the fractured aquifers in South America is insufficient, so it is necessary to improve the study of their potentials inasmuch as groundwater coming from these aquifers constitutes an essential support for the development of economic – productive activities.

## REFERENCES

Anton, D (1993). Thirsty cities: urban environments and water supply in Latin America. Ottawa, Ont., IDRC. iv + 197 p.: ill.

Apaza, D, Subias, C, Mathieu, H. (1998). Explotación de acuíferos fisurados del Valle Cuzco, mediante galerías filtrantes, para abastecimiento poblacional. In: Proc. 4° Congreso Latinoamericano de Hidrología Subterránea, Montevideo, Uruguay (2): 621–637.

Campanili, M (2003). No Brasil, há déficit no meio de abundância. Matéria publicada no jornal Estado de São Paulo, 16 de março 2003.

Castrillón Muñoz, F, Aravena, R (2002). Algunas evidencias de explotación intensiva de los acuíferos en la Cuenca Alta del Río Bogotá, Colombia. Boletín Geológico y Minero, 113 (3): 283–301.

Celligoi, A, Duarte, U (1996). Critérios hidrogeológicos na locação de poços tubulares em rochas basálticas da formação Serra Geral. Annals 2° Cong. Bras. Águas Subterrâneas – ABAS Salvador, Brazil. (*in Portuguese*)

Coriolano, ACF, Jardim de Sá, EF, Silva, CCN (2000). Structural and neotectonic criteria for location of water wells in semi-arid crystalline terrains: a preliminary approach in the eastern domain of Rio Grande do Norte State, Northeast Brazil. Revista Brasileira de Geociências 30(2):350–352.

Costa WD (2001). Água subterrânea no Brasil: um enfoque na regiao semi-arida. – Taller Aguas Subterráneas y Gestión Integrada de Recursos Hídricos. Global Water Partnership – INA. Mendoza. Argentina. 31 p.

Hirata RCA, Ferrari LC, Ferrari LMR, Pede M (2002). La explotación de las aguas subterráneas en la cuenca hidrográfica del Alto Tiete (Sao Paulo, Brasil): crónica de una crisis anunciada. Boletín Geológico y Minero. 113 (3): 273–282.

Hirata RCA, Ferreira, LMR (2001). Os aqüíferos da Bacia Hidrográfica do Alto Tietê: disponibilidade hídrica e vulnerabilidade à poluição. Revista Brasileira de Geociências. 31 (1):43–50.

Houston J. (2001). La precipitación torrencial del año 2000 en Quebrada Chacarilla y el cálculo de recarga al acuífero Pampa Tamarugal, norte de Chile. Revista Geológica de Chile Santiago. 28(2): 163–177.

Margaritz M, Aravena, R, Pena, H, Suzuki, O, Grilli, A (1990). Source of ground water in the deserts of northern Chile: evidence for deep circulation of ground water from the Andes. Ground Water, 28 (1): 513–517.

Martins Netto, JPG, Diniz, HN, Joroski, R, Okamoto, FS, França, VCÇ, Tanaka, SE, Silva, VHA (2004). A Ocorrência de Fluoreto na Água de Poços da Região Metropolitana de São Paulo e novas tecnologias para sua remoção. In: Congresso Brasileiro de Águas Subterrâneas, 13., Cuiabá, 2004. Cuiabá: ABAS. CD-ROM.

Medeiros, WE, de Lima, O (1990). Geoelectrical investigation for ground water in crystalline terrains of Central Bahia, Brazil. Ground Water. 28(4): 518–523.

Méndez, W (2005). Contamination of Rimac River Basin Peru, due to Mining Tailings. MSc Thesis. TRITA-LWR Master Thesis LRW-EX-05-23. Lima, Peru. 31 p + Ann.

Morris, BL, Lawrence, ARL, Chilton, PJC, Adams, B, Calow RC, Klinck, BA (2003). Groundwater and its Susceptibility to Degradation: A Global Assessment of the Problem and Options for Management. Early Warning and Assessment Report Series, RS. 03-3. United Nations Environment Programme, Nairobi, Kenya.

Neves, BBB, Albuquerque, JPT (2004). Tectônica e Água Subterrânea em Rochas Pré-Cambrianas do Nordeste do Brasil – A Diversidade do Sistema Aqüífero. Série Científica Revista do Instituto de Geociências – USP. São Paulo, 4(2): 71–90.

Prado, RL, Malagutti F°, W, Dourado, JC (2001). The use of shallow seismic reflection technique in near surface exploration of urban sites: an evaluation in the city of São Paulo, Brazil. Revista Brasileira de Geofísica. 19 (3): 112–120.

Rebouças (1999). Groundwater resources in South America. Episodes, 22(3): 232–237.

Rebouças (2000). O Potencial de água no semi-árido brasileiro: perspectives do uso eficiente. In: International Rainwater Catchment Systems Association Cong., 9, Recife, Brasil.

Silva, CCN, Jardim de Sá, EF (2000). Fracture chronology and neotectonic control of water wells location in crystalline terranes: an example from the Equador region, Northeasternmost Brazil. Revista Brasileira de Geociências. 30 (2): 346–349.

Tröger, U, Campos, JEG., Petersen, F, Vöckler, H (2003). Groundwater Occurrence in Fractured Metamorphic Rocks – Quality and Quantity, an example from North East Brazil. In: Krásný J, Hrkal Z, Bruthans J (eds) Proceedings, Int Association Hydrogeologists International Conference on ground water in fractured rocks, UNESCO IHP-VI, Series on Groundwater 7: 107–108.

UNESCO – IHP (1996). Mapa Hidrogeológico de América del Sur. CPRM.

United Nations (1997). Comprehensive Assessment of the World's Freshwater Resources. E/CN.17/1997/9.

CHAPTER 3

# Groundwater and tectonics – semiarid Northeast of Brazil

Benjamim B. de Brito Neves[1] and José do Patrocínio T. Albuquerque[2]

[1]*Instituto de Geociências da Universidade de São Paulo-SP, Brazil*

[2]*Centro de Ciências e Tecnologia – Universidade Federal de Campina Grande-PB, Brazil*

ABSTRACT: Groundwater in fractured crystalline rocks plays an important role for water supply of the Borborema and São Francisco geotectonic provinces of the northeastern Brazilian Shield, where semiarid climatic conditions prevail. In the last 150 years about 150,000 water wells have been drilled in this region that covers almost one million $km^2$. Only in the last 40 years have advanced processes for well location been used. This led to gradual improvement of results and decrease of the number of dry or abandoned water wells. The crystalline aquifer systems are very complex due to the varying geotectonical, physiographical and climatic conditions and different human interventions. Despite this diversity, however, certain combinations of structure and lithologic associations can be defined. This resulted in improved water well production in terms of their yield and groundwater quality. Thus, for this semiarid part of the Brazilian shield, seven different types of conditions (aquifer sub-systems) were distinguished. In spite of scarce hydraulic data available, prevailing specific capacity can be assessed ranging between 0.1–0.2 $m^3$/h/m (the worst conditions in the cratonic basements and interior fold belts) and 0.5–1 $m^3$/h/m (the best conditions in clastic cratonic covers).

## 1 INTRODUCTION

For several decades, wells have been drilled to provide water resources in different parts of eastern-northeastern South America in so called "Drought Polygon" (Fig. 1). This is a region of an area over 900,000 $km^2$ marked by the incidence of many droughts during the last two centuries. In these areas, specific nuances and substantial differences have been recognized to determine the ideal well locations. The hydrogeological experience in these areas and available literature (sparse until the 1960s) were valuable, but not always applicable in other areas. This quest to understand the diversity of aquifer systems, within the same geological region (which has often been treated in a generic and simplistic way) motivated us to cross the threshold of Geotectonic Regionalization field, combining the tectonic research with other variables.

Since the 1960s, importance of climate on groundwater quality and well yield has been recognised. As a result, several articles were written. These included a series of publications called "Basic Hydrological Inventory about Northeast" by SUDENE, from the end of the 1960s to the mid 1970s.

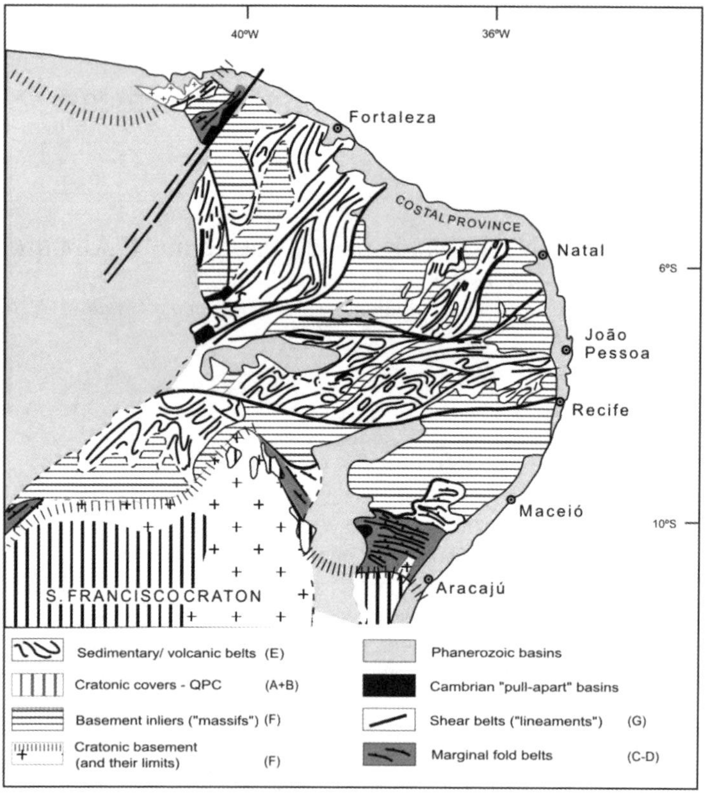


Figure 1. Geotectonic map of Northeast Brazil (for A, B, C, D , E, F, G see Table 2 and Figure 5).

For the past decades, the main concern was with lithological assemblages, i.e. with the quest of lithology considered "favourable" to groundwater. After those pioneer works (Siqueira 1963, Cruz and Melo 1968, DRN/Brasil/SUDENE 1974) where the lithological importance was consigned, the effects of climate in well yield and water quality started to be recognized. Equivalent rock assemblages could present completely different results in terms of water well quality and quantity under different climatic and tectonic domains. Several examples were documented, in the semi-humid coastal area, the semiarid interior, and inside the semiarid domain, which is characterized by an irregular rain distribution, in time and space (Figure 2), and high index of potential evaporation (Figure 3).

The notion of connections between morphology and groundwater was a knowledge that was gradually crystallizing. Thus, the correct handling of relief factor (the quest for more depressed areas, thalwegs, lowlands) became important. In this sense, the concern about thalweg lines commanded by fracturing (identification of "riacho-fenda" = fracture controlled creek) was initiated at the beginning of the 1960s by Siqueira (1963). From the point of view of groundwater quality, the paper and maps by Cruz and Mello (1968) are pioneer works. They identified and analysed the factors that influence water quality variation (Figure 4). Their maps are still valid.

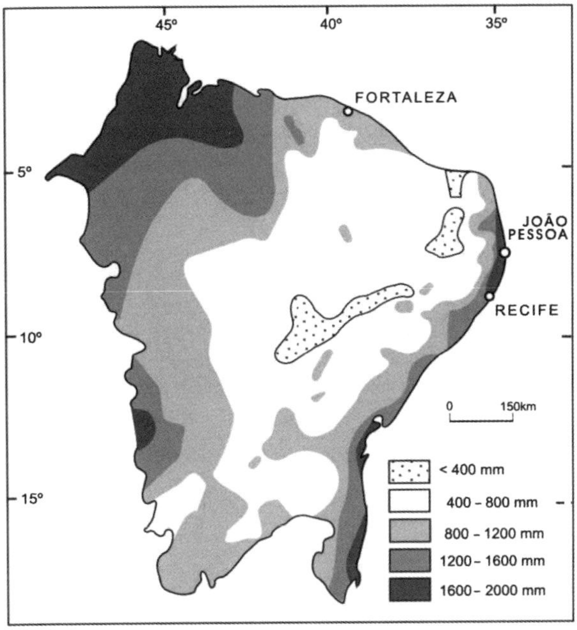


Figure 2. Annual rainfall rates (isohyets).

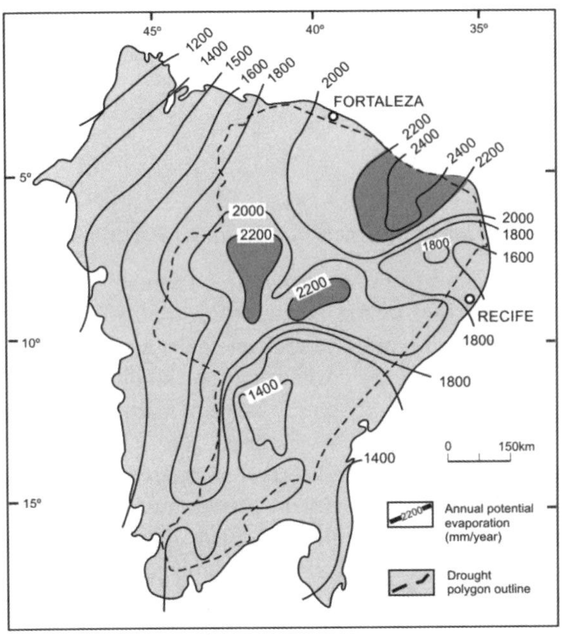


Figure 3. Annual potential evaporation rates (mm/year). Outline of the drought polygon.

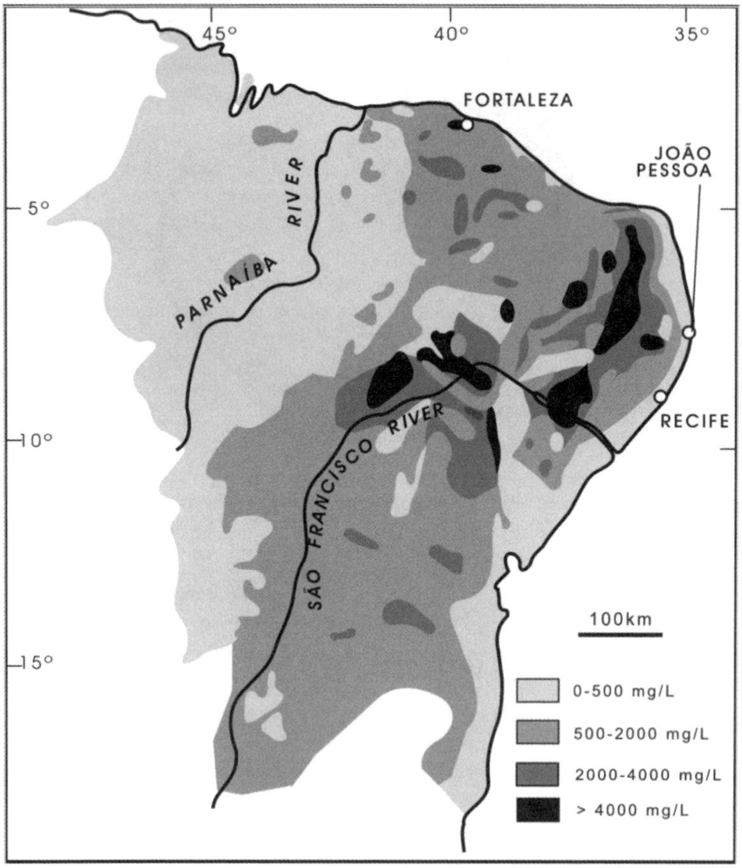


Figure 4. Groundwater quality, TDS in mg/L.

## 2 GEOTECTONICS

The Precambrian shield areas of Northeast of South America display two major elements:

(a) The São Francisco Craton (to the south), with basement of Archean to Paleoproterozoic age, composed by high-grade metamorphic rocks (gneisses, granulites, minor supracrustals) and low-grade rocks (greenstone belts and alike). There are two important groups of moderately folded cover rocks, of siliciclastic (Paleoproterozoic and Mesoproterozoic) and pelitic-carbonatic (Neoproterozoic QPC assemblages) natures (Figure 1).

(b) The Borborema Province (to the north) composed by a branched system of Neoproterozoic fold belts, separated by different kinds of basement inliers (high-grade gneissic-migmatitic terrains + granitic plutonism). There are two major types of fold belts: quartzitic- pelitic-carbonatic (QPC) assemblages, or "marginal belts", with low to medium grade of metamorphism, and distal (volcanic-terrigenous assemblages) belts, with many types of gneisses, meta-graywackes, mica-schists etc., pierced by Neoproterozoic granites.

Table 1. Main group of variables for crystalline aquifer systems.

| | |
|---|---|
| 1 – Geotectonics: Paleotectonics (internal dynamics) | Crustal type/ tectonic setting, Rock assemblages and their stratigraphic relationships, Structural Associations. |
| Neotectonics | Overprint of Cenozoic structural strain fields (including fracture controlled creeks). Modern sedimentary deposits |
| 2 – Climate and physiography (external dynamics) | Annual rainfall rates and distribution versus evapotranspiration rates.<br>Air humidity,<br>Runoff and recharge availability (infiltration),<br>Morphological features (e.g., valleys, lowlands). |
| 3 – General characteristics of water well (human interventions) | Selection of well site (better placement) + location (looking for the best hydraulic conductivity).<br>Well design, construction and development<br>Maintenance and usage |

An additional element to be considered is the net of shear belts or lineaments, some of them very wide (up to 6 km) and long (up to 700 km), poly-cyclical in character, with predominance of cataclastic and mylonitic rocks. These shear zones are products of "escape tectonics" of the final phase of consolidation (Ediacaran/Cambrian) of the basement of the Borborema Province. A long history of Phanerozoic sedimentation and superimposed tectonics (including Cenozoic stress fields) has to be considered in the history of this part of the South American shield. Further geologic and geotectonic information can be found in Brito Neves et al. (2000), among others. Based upon these geotectonic elements and groundwater analyses (field experiences and available data), it was possible to identify 7 (6+1) different aquifer (sub-)systems or hydrogeological environments for the crystalline aquifer as a whole. The defined aquifer systems are:

- The basement of the craton and fold belts – F;
- The cratonic covers, siliciclastic – A, and pelitic carbonatic – B;
- The thrust-and-fold belts and marginal belts – C/D and the distal belts – E.
- An experimental additional sub-system (shear belts G) is proposed, in terms of "expectation", up to now without critical mass of hydrogeological data.

The main litho-structural assemblages for these aquifer systems are described in Table 2 and schematically outlined in Figure 5. Additional information on tectonic setting, lithological assemblages, structural features etc., along with the discussion on the main hydrogeological features follows.

Concerning geotectonics (structural geology and neotectonics), there has been a considerable progress in the last three decades in Brazil and in the Northeast. Groundwater research in crystalline systems, however, have achieved few accomplishments but with undoubted gains in some local systems (Silva et al 2000, Albuquerque 2003).

A concern about the geotectonics diversity and the regional geologic conditions started in the 1960s (Brito Neves and Albuquerque 1977, 2004). Those preliminary and experimental classifications have been improved by continuous and systematic observations. The recognition of the importance of the geotectonics is a remarkable advancement (Table 1).

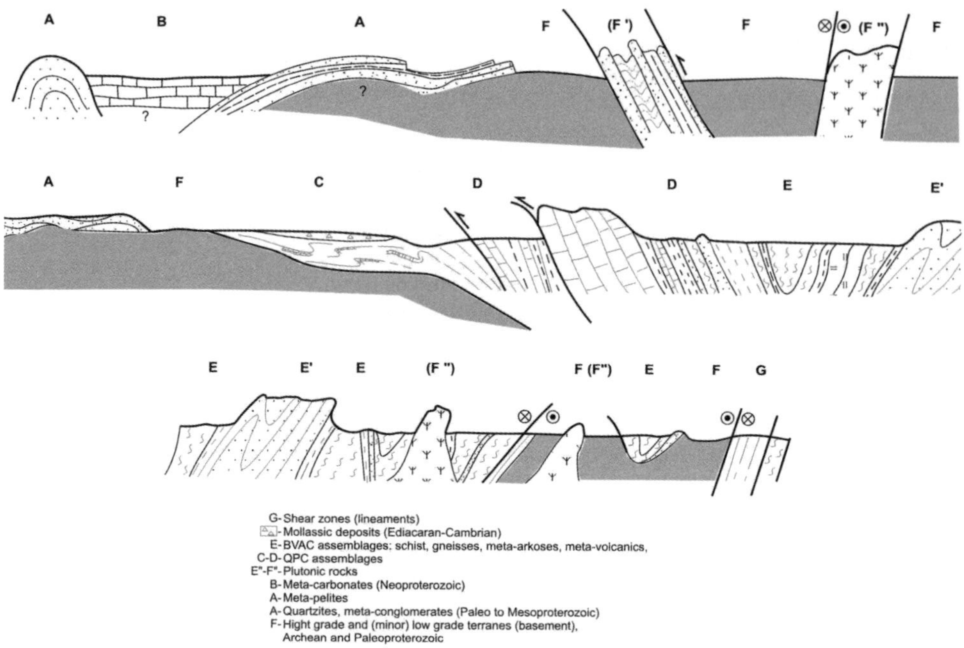


Figure 5. Idealized cross sections to show the relationships between:

- Cratonic basement (F, F′ and F″) and cratonic covers of Mesoproterozoic (A) and Neoproterozoic (B) ages;
- Marginal Belts and Thrust-and-Fold Belts (D, C) and the margin of the craton, as well as the gradational passage from Marginal belts (D) to interior belts (E / E′);
- An interior belt (E) with quartzitic (E′) and granitic (E″) ranges and high grade terranes as basement and a shear zone (G).

## 3 ANALYSIS OF FACTORS/VARIABLES

A single scheme of variables that affect water well productivity (Q/s, water quality) of the Crystalline Aquifer System (and sub-systems) is presented in Table 1.

Geotectonic regionalization (Figure 1, Table 1, Figure 5) is one of the conditioning factors of hydrogeological variability of the crystalline aquifer systems. The geotectonic factors (tectonic setting, lithological assemblages, structures) are the first group of variables that allow to define particular crystalline sub-systems. Thus, under similar climatic and physiographic conditions, it is possible to differentiate aquifer subsystems in terms of conditions of storage and groundwater circulation and to identify the ideal well sites and predict yields.

Climate and physiographic factors are also important. These include precipitation rates, evapotranspiration and morphological features. In many cases, even under ideal tectonic conditions, groundwater yields and quality are not sufficient if there are adverse climatic and physiographic conditions. On the other hand, even when the general tectonic conditions are not good, the results could be improved by favourable climatic and morphological (microclimates, altitude, proximity of the coastal areas) conditions.

These sub-systems are being proposed only for the Northeastern region. For similar geotectonic conditions, such as in the Southern and Southeastern Brazil where the climatic and morphological conditions are different, the general behaviour is different.

The third group of factors/variable is anthropogenic. Most selected well sites have not been supported by any kind of technical assistance. Many empirical and trial-and-error methods have been widely used. Well design is very rare. Maintenance services and rational usage count for only few local exceptions. Hence, it is necessary to take human intervention into consideration. This is part of a reality that we cannot rule out.

## 4 CRYSTALLINE AQUIFER SUB-SYSTEMS / HYDROGEOLOGICAL ENVIRONMENTS

Seven (6 + 1) aquifer condition types (Table 2, Figure 5) of the crystalline system are postulated for the study region, the north-eastern part of the Atlantic shield (Table 2, Figure 5). The geotectonic systematization of different kinds of crustal and litho-structural associations (Hobbs et al. 1976; Condie, 1982, and others) requires recognition/ identification of other aquifer conditions in the region. This evokes and paraphrases the classical assertion by Read (as regarding granites): there are aquifer conditions and aquifer conditions in crystalline system to be taken into consideration. This identification and proposition of seven distinct types of aquifers (sub-systems, from "A" to "G") is certainly a first approach to the problem.

The proposition about aquifer system diversity possesses both theoretical and practical implications. For instance, we desire to advance beyond the systematic generalizations of crystalline aquifers, and stimulate the usage of new kinds of analyses and exploitation.

### 4.1 *Siliciclastic Cratonic Covers – A*

Siliciclastic rocks (minor carbonatic intercalations) from Espinhaço super group in Bahia and Minas Gerais (Figure 1) represent the platform covers of Neopaleoproterozoic and Mesoproterozoic ages. They are lithologic assemblages of QP $\gg$ C type (quartzites, pelites >> carbonates, as per Condie, 1982) of varied palaeographic origins (fluvial, eolian, transitional, shallow marine).

These cover are found from moderately folded, forming extended landscapes of high plateaus ($\geqslant$ 800 m), to intensely deformed (forming quartzitic mountain ranges), as in the Northwestern Espinhaço portion. The latter is characterized by a degree of variable thermodynamic transformation, from anchimetamorphism up to green-schist facies (rare mesozonal facies).

Interstitial porosity, fractures, stratification joints, sedimentary contacts, and some local karstic features condition these aquifers. There are cases in which "real aquifers" (*stricto sensu*) are formed with other areas forming restricted aquifer zones. Regional water level is generally intercepted in valleys and depressions that dissect the orographic system; this facilitates the formation of many types of water sources (springs are very common). Unconfined aquifer conditions prevail, but there are locally confined aquifer levels (some wells with artesian discharge).

Along a longitudinal extension of about 1200 km (see Figure 1), this landscape of quartzitic plateaus crosses different climate circumstances, with variable annual pluviometrical rates

Table 2 Main crystalline aquifer (sub) systems and their hydrogeological characteristics – Northeast South America.

| Aquifer system | Main litho-structural types | Noteworthy features | Specific capacity (Q/s) $m^3/h/m$ | Water Quality TDS: mg/L | Additional Remarks |
|---|---|---|---|---|---|
| Siliciclastic cratonic covers (Paleo-Mesoproterozoic) (A) | QP > C assemblages: Quartzite-metapelite >>carbonate | Porosity of interstices, fractures, stratigraphic joints, contacts, unconformities | 0.5–1.0 to 25.0 | 0–600 | Shallow water tables are common. Locally, artesian conditions are possible |
| Neoproterozoic cratonic covers (B) | C > QP assemblages: Carbonates, meta-pelites, slates, quartzites | Porosity of fractures and karst dissolution. Local slaty cleavage | 0.5–0.8 to 4.0–6.0 | 500–2,000 | Q/s increases with the intensity of the folding and fracturing, as well as with karstification. |
| Thrust-and-fold belts and Marginal belts (C and D) | QPC assemblages + meta-diamictites, meta-pelitic-carbonatic rocks, metaturbidites | Same as above Meta-carbonates dissolution and slaty cleavage are very important. | 0.5–2.0 to 15.0–350.0 | 500–2,000 | Same as above. Local faulting very important. These could be the best aquifer sub-system. |
| Interior (distal) Fold Belts (E, E′, E″) | Schists, meta-arkoses, meta-volcanic (bimodal): BVAC assemblages >> QPC assemblages | Granitic and quartzitic hills and ranges are important (E′ and E″ ), as well as quartzites and carbonatic intercalations | 0.1–0.2 to 1.5 | 2,000–5,000 to 20,000 | Generally, the toughest conditions. Locally, orographic features and favourable lithology may attenuate the situation |
| Basement of cratons, "massifs" (inliers) and foldbelts." Tough" subsystems (F, F′, F″) | High grade terrains: gneisses, migmatites. Minor supracrustals | Granitic and quartzitic ranges may be locally important (F′, F″) | 0.1–0.2 to 1.5 | >2,000 to 10,000 | Same as above |
| Shear Zones/ Lineaments (Ediacaran-Cambrian Escape Tectonics) (G) | Cataclastic rocks s. l. up to 10 km wide. Polycyclical behaviour | "T" fractures are important. Overprint of modern tensional stress fields | – | – | Potential sub-systems. There are not sufficient data to characterize this sub-system. |

Obs: "to" = Maximum values.
QPC = Quartzites, pelites, conglomerates. BVAC = Bimodal volcanics, arkoses, conglomerates

(ranging from 600 mm to higher than 1200 mm). Their frequent occurrence in interfluve zones, in which catchment facilities are many, results in a low demand for tubular wells.

Groundwater is stored and circulates in this effective, diversified, and interlaced porous framework (interstitial porosity, joints etc.). Recharge occurs by infiltration of a small rate of rainfall and through losing steams that during some floods occasionally inundate the area. During drought periods discharge occurs in these same watercourses that become intermittent.

The unconfined aquifer well (Q/s) yields are always higher than $0.5\,m^3/h/m$ (0.5 to $1.0\,m^3/h/m$ values are usual) with excellent chemical quality (TDS between 50 and 600 mg/L). Therefore, with the improvement of criteria for location and those of well construction, it is possible to increase these Q/s values considerably.

Along the interaction zones with the cratonic cover and/or with the thrust-and-fold-belt (Figure 5, "A" subjacent to "B" or subjacent to "C"), there are other locals of interest for well locations, including some with the possibility of flowing artesian wells. There are the best characteristics/aquifer conditions (many types of empties) of "A", "B" and "C", complemented by the additional presence of the empties of an erosional and angular unconformity above "A".

### 4.2 *Neoproterozoic Covers – B*

São Francisco/Bambuí Supergroup mainly represents the Neoproterozoic covers, covering an area of $350{,}000\,km^2$ of San Francisco Craton and exhibits several gradations from unfolded and gently folded rock-units to intensely folded and faulted contexts (of thrust-and-fold-belts). The latter are mostly distributed along São Francisco craton peripheries that are related to centripetal vectors of vergence originated in the interior of the marginal fold belts.

Predominant lithology are carbonates and pelitic-carbonatic rock-units (marine sequences of Sete Lagoas, Santa Helena, Lagoa do Jacaré formations etc.), with subordinate occurrences of diamictites and quartzites in the base of the sequence (D-QPC = diamictite-quartzite, pelite, carbonate). These basal glacial deposits were products of the Sturtian glaciation. They were followed by shallow marine invasions during the late Cryogenian and Ediacaran periods.

The developed thicknesses vary because of basement erosional undulations and irregular faulting. The Neoproterozoic covers reflect that paleo-morphology, presenting values from some hundreds (more common) to some thousands of meters of thickness. Generally, these form unconfined aquifers where waters are stored by pluvial water infiltration that escape evapotranspiration and from surface runoff. The aquifer storage is function of an effective multi-interstitial porosity and fractured zones. Locally, there are confined conditions (lithological and structural injunctions), with occurrence of flowing wells.

Water wells drilled in moderately folded portions of these covers have great variation in Q/s values, because karstic dissolution factor is important, and very variable, causing the alternation of some local excellent discharges and others of mediocre production.

Values of Q/s between 0.5 to $0.8\,m^3/h/m$ can usually be considered representative averages for wells in this aquifer sub-system with TDS ranging from 500 to 1,000 mg/L. Climate is an important factor in these yields. Higher Q/s values are found in Minas Gerais (southern part of the region, where pluviometry ranges between 700–900 mm/year) in relation to those ones registered in central part of Bahia and northwards (annual pluviometry $\leqslant$700 mm/year).

Passages to moderately folded domains can also be defined in terms of Q/s values. Deformation degree of tectonic-stratigraphic context is an indicator that can be tracked from less deformed zones (smaller Q/s) to those in strongly-folded thrust-and-fold belt context (larger Q/s, ca. 6 $m^3/h/m$). Along the contacts between this sub-system ("B") with that subjacent ("A"), excellent storage conditions and groundwater production are usually obtained (see Figure 5, upper part).

### 4.3 *Thrust-and-Fold Belts – C and Marginal Fold Belts – D*

There are several types of transitions between unfolded and folded Neoproterozoic covers ("B"). The transitions can be observed with the thrust-and-fold-belts ("C", part of the marginal belt overthrusting the cratonic foreland) and marginal fold belts ("D") (Figure 5). The lithological assemblages of "B", "C" and "D" are from the same general origin and display similar rock types. In practical terms, "C" are allochthonous portions of "D" (marginal belts).

From the perspective of lithological assemblages, D-QPC (diamictites – quartzites, carbonates, and pelites) types prevail. Structural features of folds and centripetally vergent thrusts faults occur in low green-schist facies at shallow crustal levels. Deformation, folding, and fractures are seen at hand, outcrop, and regional geologic/mapping analysis scales.

In the cases of "C", there are drags with significant deformation of sedimentary sequences that are detached from the cratonic substratum. This deformation formed nappes and recumbent isoclinals folds that increased intergranular porosity (including porosity associated with slaty cleavage) and fracture porosity. Recharge is about 5% annual precipitation Discharge forms many perennial streams. Climatic conditions is also very important in well production in these domains. There are reliable data that demonstrate higher well yields under similar geologic-structural conditions with higher annual average pluviometric rates. From central Bahia area to north of Minas Gerais State (Rio Verde Grande Valley), there is an increase of average specific capacities, ranging from 0.5–1.0 $m^3/h/m$ to 2.0 $m^3/h/m$.

Besides presenting excellent crystalline aquifer conditions (wells in $500 < Q/s \leqslant 2{,}000$ L/h/m interval), there are local high water wells yielding up to 700.0 $m^3/h$ with small drawdowns (on the order of a few metres). In addition, there are the special cases of contact lines with the subjacent "A" – the siliciclastic covers. These usually provide the best general conditions for groundwater production. There are cases of deep wells exhibiting Q/s over 25.0 $m^3/h$ m. In general, groundwater quality is very good with TDS values reaching 500 and 10 00 mg/L.

From the point-of-view of lithological assemblages, sub-systems "C" and "D" are differentiated as special ("miogeoclinal assemblages"), among crystalline aquifer sub-systems. They are close to those of "B" (the almost autochthonous cratonic cover). They are superior aquifers that may be distinguished from the other environments/sub-systems ("E" and "F", with volcano-sedimentary assemblages).

Marginal belts display relatively simple folding systems (some are monophasic), verging to the craton, growing in intensity (together with metamorphism degree) with distance from the craton. Far away from the cratonic nuclei, the marginal belts gradually approach the litho-structural (polyphasic folding) and metamorphic conditions (amphibolite facies) of the interior belts ("E"). Fractures, karstic dissolution, slaty cleavage, and some interstitial porosity can lead to well yields in these fold belts of approximately 0.5 $m^3$/h/m and occasionally up to 2.0 $m^3$/h/m.

Some belts retain sparse Cambrian-Ordovician covers, and post-orogenic deposits; they are generally considered as "mollasses" (see Figure 5). These erosional remnants, clastic sedi-ments may function as zones of enhanced recharge and may constitute an important auxiliary of the sub-system context.

## 4.4 *Interior/Terrigenous/Volcano-Sedimentary Belts – E*

Interior belts, distal of cratons, are constituted by varied (meta)sedimentary associations (minor QPC types) with predominant "BVAC" contents = bimodal volcanic, arkoses, gneisses, schists, metagraywackes, local conglomerates etc. from shallow to deep waters, associated to diversified volcanism (acid, intermediate, and basic). They display variable metamorphic facies that range from green-schist up to high amphibolite facies. Migmatization is a common phenomenon. Poly-deformated metamorphic rocks crop out along areas of peneplains (some local pediplains) with little vegetation. They can be approximated as aquifuges (nearly impermeable rock units with minimum porosity).

The aquifer portions from these rocks are restricted to local fracture permeability and are independent of porosity of the weathered mantles or alluvial cover, which Albuquerque (1984, 2003) denominated the "aquifer zone", and Gustafson & Krásný(1994) called "hydraulic conductor". These interior belts (together with those of "F") are among the most difficult regimes (so-called "tough crystalline") in terms of groundwater occurrence and exploration. This occurs mainly in the zones where the climatic conditions are severe (low pluviometry, high evapotranspiration rates, etc.), which is the most common circumstances in the study area.

In this tectonic-stratigraphic context (metamorphic and igneous rocks), we commonly seek more favourable rocks (quartzites, limestones, meta-arkoses, meta-conglomerate, meta-felsites, etc.) in exploring for groundwater. However, outcrops of these rocks are restricted (QPC $<<<$ BVAC) in these interior belts. In these situations, we seek potential aquifer zones by employing geotectonics and structural geology insights, remote sensing, aerial photos, or geophysical surveys, in the search for favourable sites circulation. These are generally located underlying valleys and lowlands. The recharge to this subsystem is problematic and is conditioned by a presence of an alluvial or weathering cover.

These tectonic-stratigraphic domains/hydrogeological environments generally are of low hydrogeological productivity (Q/s varying from 0.1 to 0.2 $m^3$/h/m, in most of wells, see Table 2). There are also great number of practically dry and abandoned wells, primarily because of insignificant Q/s and high salinity. A high percentage of wells with TDS $>$1500 mg/L is observed, many wells have up to 60 00 mg/L, and a few have TDS of nearly 20,000 mg/L.

Despite of the generally unprolific nature of the Northeast experience, wells be successfully developed. Most existing wells were unfortunately located and drilled without considering technical criteria. Using intuition – even presumption–, and stubbornness (sometimes, two to three wells sometimes drilled at the same site with negative results. By and large, the lack of knowledge of all proposed basic fundaments of Table 1 prevailed. Had there been an experienced geologist adviser, more than half of these wells would not have been set or drilled.

In these "tough crystalline" domains ("E" + "F"), there are some special cases. There are areas in Northeast, where water wells with reasonable water quantity and quality are systematically obtained. This is particularly true for the Cabaceiras and Boa Vista zones (in Paraiba State), 150 km W of João Pessoa, under the realm of the more severe climatic conditions (see maps of Figure 2 and 3), and the case of Piancó Valley, in the far west of Paraiba State (>300 km W from João Pessoa). In both cases, litho-structural contexts (where mica-schists, metagraywackes and amphibolitic rocks prevail) are usually avoided for tubular wells. These are exceptional cases due to the maximization of general geotectonic factors (with emphasis of the neotectonic action (see Table 1) and the result of uncommon, persistent maintenance works ("good" human intervention), that for Northeast is important precondition as observed/highlighted in Table 1.

Other features of the "tough crystalline" are authentic oases (often forming local microclimates) associated with quartzitic (E′) and granitic (E″) outcrops. Recharge waters take a long time to reach the regional water table, and so, some important local water resources may includes depression springs, diffuse seepage, etc.

The difference between climate and physiography factors are also confirmed by the authors' experiences in other Brazilian provinces. Under the same general geologic-geotectonic circumstances, from the south of Bahia State to the southernmost states, Q/s values may double those found in Northeast (Costa, 1986) and possess with superior water quality. These substantial differences of quantity and quality are the result of climatic factors, especially a better distributed pluviometry that favours recharge, circulation, and discharge.

### 4.5 *Basement of Cratons and "Massifs" – F*

General geologic conditions of cratonic nucleus (São Francisco) and "massifs" (basement inliers, inter-distal fold belts, and even intra-fold belts (Figure 1 and Figure 5) are very similar. They are tectonic-stratigraphic terrains originated from the break-up of the same Mesoproterozoic supercontinent – Rodinia-. They are litho-structural realms/domains of mainly Archean and Paleoproterozoic age and, for the special cases of "massifs", there is frequent presence of granitic plutonism from the evolution of the Neoproterozoic Brasiliano Cycle. In general, high grade rocks (HGT) gneissic-migmatitic-granitic rocks prevail with mafic and ultramafic rocks subordinate, with some remnants of low-grade volcano-sedimentary (LGT) associations.

The general hydrogeologic conditions are generally poor like those ones of the interior mobile belts ("E"). These rock assemblages play a regional aquifuge role with significant permeabilities limited to fractured/faulted zones and weathered intervals – "aquifer zones" of Albuquerque (1984) –. This condition is exacerbated by adverse hydroclimatic factors. This results in low well yields with very low Q/s values in the range of 0.1 to 0.2 $m^3$/h/m and many dry and abandoned wells.

Groundwater quality is similar to that found in the interior mobile belts ("E"). TDS values are usually higher than 2000 mg/L. Due to the fact that great part of these high grade rocks domains is a more arid climate, it is common to find the highest values of TDS (Cruz and Mello 1986, Figure 4).

There are few exceptions to this generally adverse picture, such as oases in quartzitic ranges (F′), syenitic and granitic "highs" (F″), and local inselbergs. Particularly, it is worth mentioning the frequent presence of "granitic sierras" (Neoproterozoic in age) that pierce older (Paleoproterozoic) "massifs" and the interior orogenic belts of the Borborema Province. These are provide some good local good "lithological" and physiographic conditions for groundwater.

It is necessary to stress the different conditions from domains in semiarid Northeast to those similar (geotectonic) settings in the central and southern parts of Brazilian shield with, where temperate climates occur. For instance, in the "iron quadrangle", of the south of São Francisco craton, in Minas Gerais, groundwater is so abundant that it constitutes a serious problem in the mining industry.

### 4.6 *Lineaments / Shear Belts – G*

Important extrusion processes occurred in the end of the Brasiliano cycle (Ediacaran-Cambrian periods) that is responsible for the general geologic-geographical configuration of the province and for ordering the distribution of all interior belts and "basement massifs" (Figure 1).

First, tectonic analyses for the interior Paleo-Mesozoic sedimentary basins of the province show these basins were conditioned by extensional movements of considerable dimensions. The preferential sites for these vertical extensional displacements were older shear zones. This is true for coastal basins (of Atlantic margin) and for some Tertiary basins from interior of the region. Some newer Phanerozoic fault lines are associated to these older shear zones.

It is possible that some of these primary extensional fractures ("T"), associated to the main shear zones may create aquifer zones. This subject has never been object of systematic research, but there occur some wells located close to these zones with very high Q/s values. This regrettable lack of knowledge accounts for the separate treatment ("G") of these special domains.

## 5 HYDRODYNAMIC PARAMETERS

Determination of hydrodynamic parameters for crystalline aquifer systems is a major challenge due to the complex conditions in terms of heterogeneity and spatial discontinuity, especially for "E" and "F" environments. Unfortunately, in Northeast Brazil, Q and Q/s values are the most widely used "hydrodynamic parameters" and some of these values were obtained from highly questionable pumping tests.

In Brazil, Manoel Filho (1996) calculated these parameters based on Chaos Theory. He proposed a fractal model adapted to denominated "hydraulic conductor" (Gustafson and Krásný, 1994) that was composed by the set of well- rock blocks – associated fractures. Thus, the determined hydraulic characteristics are restricted to the fracture system (or hydraulic conductor) domain, not to the whole crystalline system. More than that, the values obtained from a pumping test are not representative of larger aquifer zones.

The values then obtained by Manoel Filho are higher than those many others had already determined for other regions of the world (see Table 3). We emphasize that only a very small number of water wells have adequate pumping tests (in terms of duration, systematic measurements of drawdown and recovery, etc.). The common routine is the drillers using empirical methods for the measurements of discharges and specific capacities. Manoel Filho could select only 63 confiding pumping tests, 53 in Borborema Province and 12 in the cratonic area. The calculated parameters were: transmissivity, permeability, effective porosity, and fracture opening.

Table 3. Average and range for transmissivity values ($m^2$/day) (Manoel Filhos, 1996).

| Sub-systems | Average | Range of values |
|---|---|---|
| Sililiciclastic cratonic cover (3 wells) | 3.65 | 1.33 to 5,961 |
| Neoproterozoic covers (9 wells) | 224.64 | 0.95 to 1,192.32 |
| Marginal belts (granites) (9 wells) | 2.74 | 0.54 to 6,670.08 |
| Interior/distal belts (7 wells) | 0.72 | 0.32 to 1.39 |
| Basement cratons/massifs (14 wells) | 1.46 | 0.07 to 6.10 |

These parameters were related to the rock units. However, significant relations are not developed in statistical terms because the small number of wells in each lithology and the sample sizes are different. This can affect "statistical measures". There was only a case (for crystalline limestones) where lithological influence was patent in well productivity (improving it). Manoel Filho did not investigate the correlation between well productivity and tectonic style or structural type at any scale.

Table 3 shows transmissivity values of crystalline aquifer sub-systems. Sampling values considered anomalous were culled statistics, however it does not mean such could not occur.

Although the number of selected wells was very small, analysis of transmissivities representative in productivity conditions of these tests, reveals that the best well yields are in Neoproterozoic covers, followed by siliciclastic platform covers, and marginal belts. The worst yields are associated with interior fold belts, the basement of cratons and "massifs" (the "tough crystalline" terrains). This corroborates with the conclusions obtained from the geotectonic analyses (Table 2). Unfortunately, parameters from wells drilled in thrust-and-fold-belts and shear zones sub-systems (lineaments) are not available.

## 6 ABOUT LOCATION CRITERIA

Criteria and technical care of well location in crystalline aquifer system are fundamental elements in locating wells with better yields and quality as well as better cost-benefit ratio. The diversity of hydrogeologic conditions in this and similar systems should be considered. Valid criteria for one sub-system may not be valid in others.

The closer the geologist's identification with variables from Table 1, the greater is the probability of water wells of high productivity. Lack of knowledge of or ignoring these criteria lead to many unsuccessful and abandoned drilled wells in this semiarid region. Lamentably, tentative methods or the lack of any scientific location methods is still practiced in many cases. This result is in frustrations of socio-economic demand supplying and enormous financial damages to public funds and private initiative.

In the cases of "A" and "B" sub-systems, there are some preserved characteristics of sedimentary aquifers (interstitial porosity, etc.), and the definition of ideal sites for location does not constitute a difficult problem. Hydrogeologic thematic maps (including inventory of water-wells, piezometric maps, maps of saturated thickness, etc.), and even geologic maps (lithostragraphic, structural, sand/clay ratio, etc.) are usual tools aiding successful well location.

For other crystalline aquifer sub-systems where it was difficult to access to hydrogeologic data, "riacho-fenda" ($\approx$ fracture-controlled creek) criterion proposed by Siqueira

(1963) was employed for well locations, regardless of other hydrogeologic conditions of the study area. It had long been assumed that this fracturing originated subsequent to the rocks ductile deformation event and did not considered the fact that other tectonic efforts (of Phanerozoic times) were superposed upon those events.

Cenozoic tectonism is present and occurring in shallow crustal levels. It is preponderantly extensional in character and is associated with post-Jurassic continental uplift and South American margin plate vectors. Albuquerque (1971) and Moraes Neto and Alkmim (2001) have related the origin of these fractures to a polyphasic tectonism, in which regional epirogenic uplifts play a fundamental role in the formation of aquifer conditions in crystalline basement domain.

"Riachos-fendas" are actually fractions of streams positioned/captured in rectilinear fracturing and faults. They can be identified in aerophotos in 1/70,000 and 1/25,000 scales, which should be considered essential tools for field geohydrological surveys (specially studies targeted to well site location). All available databases should be consulted, but a specific investigation of the characteristics of existing wells in the area is indispensable to arrive at the rational decision. In the study area, there are available maps of groundwater quality in the crystalline system (see Figure 4). These are good indicators of water salinity for the case of future wells.

In the field work, shape, nature, and intensity of fracturing (which used to be polyphasic) should be evaluated; the identification of the last phase of brittle deformation is the most important fact. In most cases, the best recharge conditions occur where alluvial sands cover the fractured rock unit ("riacho-fenda" = the aquifer zone).

Geophysics methods of research were employed in hydrogeological research with the objective of improving well site location in crystalline aquifer zones. Electrical resistivity, and in restricted cases, other methods (VLF) were employed in identifying fractures and formulating conceptual models. These studies have been complemented (in many phases) for important exercises of photo-interpretation and field mapping. In fact, the geophysical methods produced poor results, less than expected, but they should be improved and more explored for the future.

A great effort to develop groundwater from crystalline rock systems in Northeast Brazil was accomplished by SUDENE. It put into practice technical criteria about location, drilling and well construction, and well design in the 1960s. Since then, private enterprises and state agencies assumed this task, but they gradually abandoned technical criteria, replacing it for political (mostly) and economic problems.

The Research Centers and local universities are staffed with able people who can contribute to the issue of underground water in the region. The problem is that there are not enough funds/financial support from the state and federal agencies. Whenever those funds are obtained, they are little and provided in a discontinuous way.

## REFERENCES

Albuquerque, J. P. T. (1984) Os recursos de água subterrânea do trópico semi-árido do Estado da Paraíba. Campina Grande, Master dissert. 183p.-Departamento de Engenharia de Minas, Universidade Federal da Paraíba (in Portuguese).

Albuquerque, J. P. T. (1984) Contribuição ao conhecimento do sistema cristalino no Estado da Paraíba como meio aqüífero. In: Anais 3° Cong. Brasileiro de Águas Subterrâneas, Fortaleza-CE, pp. 489–506.

Albuquerque, J. P. T. (2003) Contribution to the understanding of crystalline systems as aquifers – the experience of the State of Paraíba, Northeast Brazil. In: Krásný J, Hrkal Z, Bruthans J (eds) Proceedings, Int Association Hydrogeologists International Conference on groundwater in fractured rocks, UNESCO IHP-VI, Series on Groundwater, 7: 21–22.

Brito Neves, B. B.; Albuquerque, J. P. T. (1997) Tectonics and Groundwater Research: Proterozoic Borborema Province, a semiarid region of Northeast Brazil. In: Proceedings Intern. Symposium on Engineering Geology and the Environment, 1997. Athens, Greece. Marinos, P. G.; Kouris, G. C. Tsiambaos, G. C.; Stournaras, G. C. (eds.), Bakelma/Rotterdam/ Brookfield: 1181–1186.

Brito Neves, B. B. and Albuquerque, J. P. T. (2004) Tectônica e Água Subterrânea em Rochas Pré-Cambrianas do Nordeste do Brasil – A diversidade do Sistema Aqüífero. Revista do Instituto de Geociências da USP, 4: 71–90.

Brito Neves, B. B.; Santos, E. J.; Van Schmus, W. R. (2000) The tectonic history of the Borborema province. In: Cordani, U. G; Milani, E. J.; Thomaz Filho, A.; Campos, D. A. (eds). Tectonic Evolution of South America. Rio de Janeiro, 31th International Geological Congress, 2000.: 151–182.

Condie, K. (1982) Early and Middle Proterozoic supracrustal sucession and their tectonic setting. American Journal of Science, 282: 341–357.

Costa, W. D. (1986) Análise dos fatores que atuam no aqüífero fissural – área piloto dos estados da Paraíba e do Rio Grande do Norte, Ph D, 225p. Instituto de Geociências da Universidade São Paulo (in portuguese).

Cruz, W. B. and Melo, F. A. C. F. (1968) Estudo geoquímico preliminar das águas subterrâneas do Nordeste do Brasil. Série Hidrogeologogia, 19: 147p (Recife, Brasil/SUDENE).

DRN/BrasilL/SUDENE (1974) Atlas de Recursos Naturais do Nordeste. Ministério do Interior, Superintendência de Desenvolvimento do Nordeste, Departamento de Recursos Naturais, Recife: 8 maps.

Guerra, A.M. (1968) Processos de carstificação e hidrogeologia do Grupo Bambuí na região de Irecê-Bahia. PhD, 132p. Instituto de Geociências da Universidade São Paulo (in Portugueses).

Gustafson, G. and Krásný, J. (1994) Crystalline rock aquifers: their occurrence, use and importance. Applied Hydrogeology, 2: 64–75.

Hobbs, B. E.; Means, W. D.; Williams, P. F. (1976) An outline of Structural Geology. J. Wiley & Sons, New York.

Mandelbrot, B. B. (1983) The fractal geometry of nature. Freeman, New York.

Manoel Filho, J. (1996) Modelo de dimensão fractal para avaliação de parâmetros hidráulicos em meio fissural. PhD 197 p. Instituto de Geociência da Universidade São Paulo (in Portuguese).

Moraes Neto, J. M. and Alkmim, F. F. (2001) A deformação das coberturas terciárias do Planalto da Borborema (PB-RN) e seu significado tectônico. Revista Brasileira de Geociências, 31: 95–106.

Passchier, C.W. and Trouw R. A. J. (1996) Microtectonics. Springer – Verlag, Berlin.

Pinto, A. S.; Santos, P. R. P.; Almeida, J. A. B.; Matos, I . V. (1982). Critérios de locação e aspectos práticos da perfuração em rochas calcárias da região central do Estado da Bahia. In: 2nd Cong. Brasileiro de Águas Subterrâneas, Salvador, pp. 447–461.

Santos, P. R. P. and Durães, A. M. M. (1986) Hidrogeologia do Grupo Chapada Diamantina e doGrupo Paraguassú. In: Anais 4th Cong. Brasileiro de Águas Subterrâneas, Brasília, pp. 106–119

Silva, A. B. (1984) Análise morfoestrutural, hidrogeológica e hidroquímica do aquífero cárstico de Jaíba, norte de Minas Gerais. PhD., 190p. Instituto de Geociências da Universidade São Paulo (in portuguese).

Silva, A. B. (1986) Contribuição da geologia estrutural na explotação de águas subterrâneas do Grupo Bambuí na região norte do Estado de Minas Gerais. In: 4th Cong. Brasileiro de Águas Subterrâneas, Brasília, pp. 251–262.

Siqueira, L. (1963) Contribuição da Geologia à pesquisa de água subterrânea no cristalino. Recife: Grupo de Trabalhos em Águas Subterrânea- GTAS, Departamento Recursos Naturais, DRN/ SUDENE (internal report), 53p.

CHAPTER 4

# Regional study of hard rock aquifers in Alentejo, South Portugal: methodology and results

António Chambel[1], Jorge Duque[2] and João Nascimento[3]

[1] *Geophysics Centre, Department of Geosciences, University of Évora, Apartado 94, Évora, Portugal*

[2] *Polytechnic Institute of Beja – High School of Agriculture of Beja, R. Pedro Soares, Apartado 6158, Beja, Portugal*

[3] *CVRM Geosystems Centre, Technical Superior Institute, Av. Rovisco Pais, 1096 Lisboa codex, Portugal*

ABSTRACT: Between 1997 and 2001 a Project called "Study of the Groundwater Resources of Alentejo Region" (ERHSA) was carried out in the region of Alentejo, southern part of Portugal. A great part of the area corresponds to igneous and metamorphic hard rocks. The areas of partly karstified crystalline limestones and a belt of sedimentary rocks on the western side. During ERHSA more than 7,000 data points were inventoried in hard rock aquifers. This extended study confirmed some results of previous studies, namely regarding the Gabbros of Beja Aquifer, and detected new productive or potentially productive areas within this region, namely the aquifers of the Charnokites of Campo Maior and Elvas, Pavia-Mora and the Évora-Montemor-Cuba. These last two systems mainly occur in gneissic or gneissic-migmatitic rocks, metamorphic volcanic-sedimentary complexes, and some tonalitic rocks. The difference between the hard rock aquifers and the less productive sectors are clearly related with the intensity of the fracturing net and the depth and type of the weathering layer. This distinguishes the gabbros and other basic or intermediate rocks (more productive) from the less productive granites. The weathering layers of the first ones can go to 30 m depth, against the normal 2 m in granites. The groundwater facies are mainly bicarbonate, except on the south part of Alentejo, where it is chloride-bicarbonate. The main cations are the magnesium, calcium and sodium, by this order, having most part of these waters a mixed cation composition between three or two of them. The northern part of Alentejo has waters with low levels of mineralization (average less than 400 μS/cm of EC). On the contrary, the south part has clearly the most mineralized waters (average more than that 1,300 μS/cm of EC). The future main task will be to conduct aquifer tests in order to have more hydrogeological parameters (transmissivity, permeability, and storage values), and try to understand relation between instant yields, steady state flow, abstraction and aquifer parameters.

## 1 INTRODUCTION

Between 1997 and 2001 a Project called "Study of the Groundwater Resources of Alentejo Region" (ERHSA, in Portuguese) was carried out in the region of Alentejo, southern part of Portugal (Figure 1). From the total area of Alentejo (26,931 km$^2$), 21,245 km$^2$ correspond to hard rocks. During the project more than 7,000 data points were inventoried in all hard rock aquifers of Alentejo. Most of them were wells or hand dug wells that are used basically

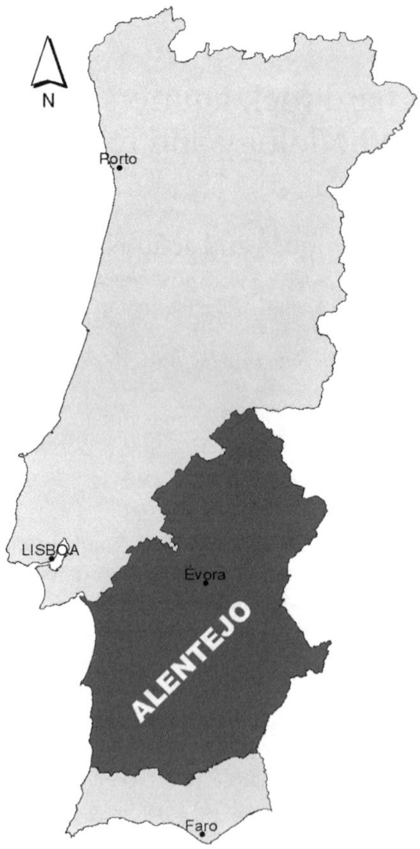


Figure 1. Location of Alentejo in Portugal.

for agriculture or domestic purposes. The most important abstractions in Portugal are related with agriculture. The actual number of public supply surface water points in Portugal is 459 (5% of the origins) and 8,900 groundwater points (95%). The total abstracted surface water volume is 627 $hm^3$ and the groundwater volume is 404 $hm^3$ (INAG 2006) for a total urban consumption of 529 $hm^3$. The difference is related with losses on the supply systems or other uses. The consumption per capita is 169 L/inhab/day. Figure 2 (adapted from Ribeiro 2004) shows the distribution of ground- and surface water consumption in Alentejo, by municipalities.

## 2 GEOMORPHOLOGY, CLIMATIC CONDITIONS AND INFILTRATION RATES

Alentejo is a flat area, at levels between 200 and 400 m a.s.l., with some little mountains, the highest being São Mamede, an elevation of about 1,200 m, on the northern part. All the other few mountains are less than 600 m high, and generally follow the Hercynian

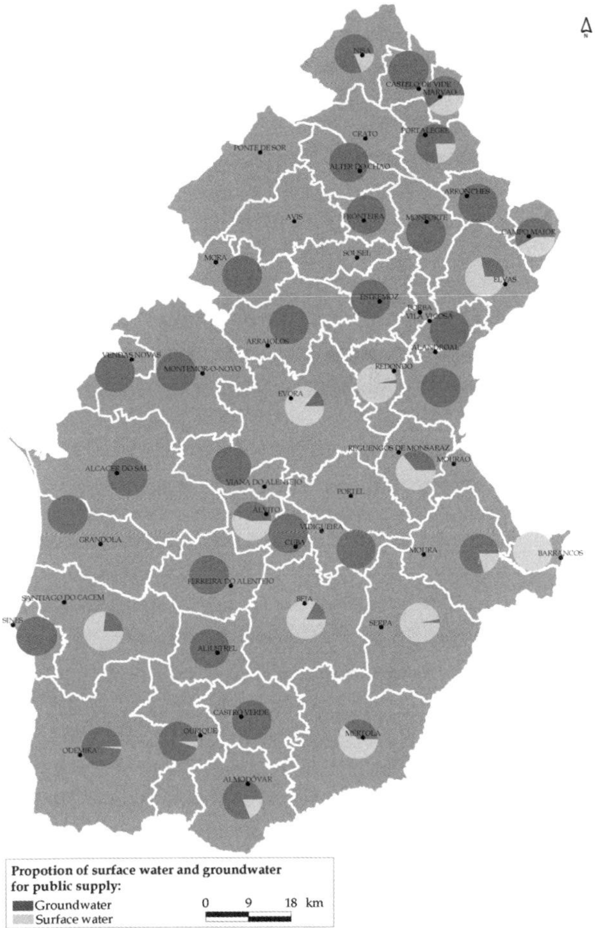


Figure 2. Percentage of groundwater uses for public supply in Alentejo (adapted from Ribeiro 2004).

Orogeny geologic main direction: NW-SE. This plain is the result of the erosion that follows the Hercynian Orogeny and flattened the Hercynian chain, letting some thin layers of an ancient erosion platform over the igneous and metamorphic Iberian craton.

The climatic conditions in Alentejo are Mediterranean type (except in the littoral west, which has the influence of the Atlantic climate), with the specific distribution of precipitation basically during winter, when the temperatures are low, and very low levels of precipitation in summer, when the temperatures are high (Figure 3).

Alentejo is a semi-arid region, with the precipitation ranging from about 450 mm per year until near or more than 800 mm on the highest mountains, as can be seen in Figure 4 (Chambel et al. 1998).

Mendes (1989) estimated the values of the Budyko Aridity Ratio (BAR), which represents the ratio between mean annual potential evapotranspiration and mean annual precipitation (Sankarasubramanian & Vogel 2003). Regions where BAR is higher than the unit

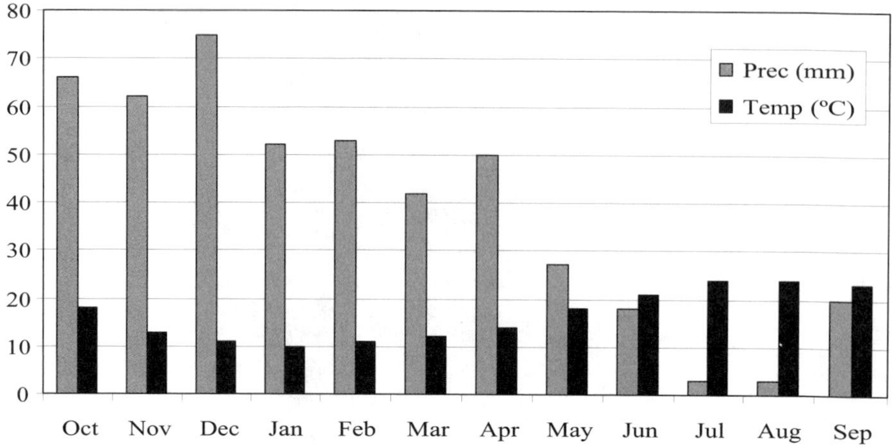


Figure 3. Average monthly precipitation and temperature distribution for the Alentejo climatic station of Vale Formoso, Mértola, South Alentejo (1971/72 to 1991/92 series) (Chambel 1999).

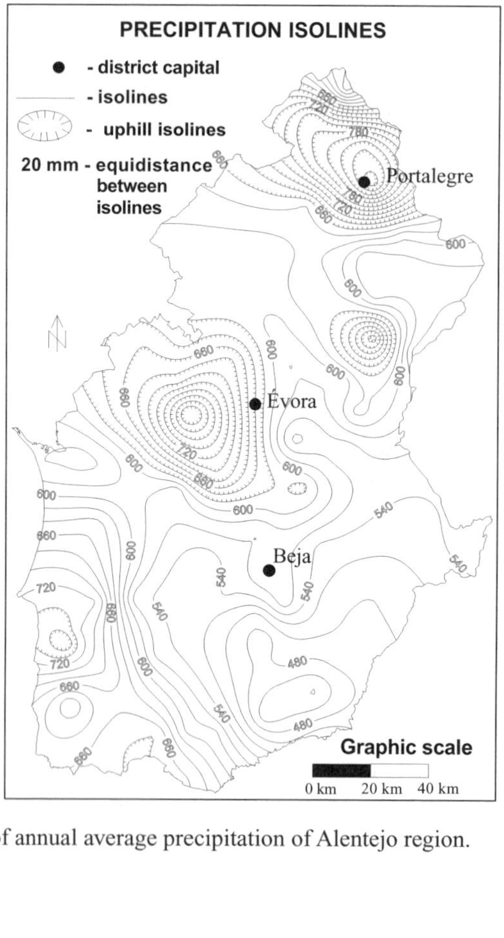


Figure 4. Projection of annual average precipitation of Alentejo region.

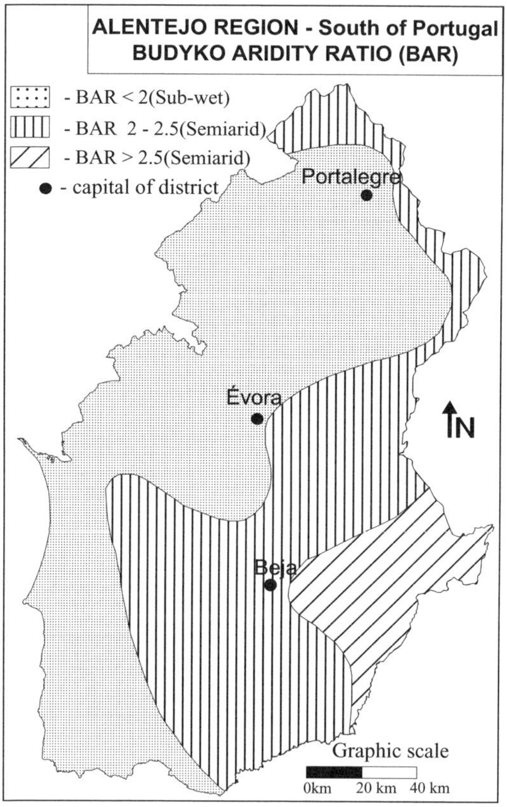


Figure 5. Budiko Aridity Ratio (BAR) distribution in Alentejo.

are boardly classified as dry, since the evaporative demand cannot be met by precipitation. Similarly regions where BAR is less than the unit are boardly classified as wet (Arora 2002). Figure 5 shows the BAR for Alentejo region. The littoral and the north are sub-wet. The continental area is semi-arid. This means that the desertification is taking place in the SE of Beja. The east region of Beja is one of the hottest areas of Europe continuous with the regions of Andaluzia and Extremadura in Spain.

Comparing the annual average precipitation isolines map with the BAR distribution map (Figure 4 and 5), it is possible to see the high correlation between the semi-arid climate and the precipitation deficit in the SE zone of Alentejo (Chambel et al. 1998).

The potential evapotranspiration is normally higher than 1,000 mm per year, causing a high water deficit in the soil.

It's a region of very warm and dry summer and the rainy season occurs on winter, with drought cycles sometimes during 2 or 3 consecutive years (Chambel et al. 1998). In summer (June, July and August) the rain corresponds to only about 4 to 5% of the annual precipitation. The agriculture requires intensive watering, due to high potential evapotranspiration during the productive cycle. The regional precipitation average in forty years, between 1958 and 1998 for all Alentejo, represented 604 mm (Chambel & Duque 1999). Henriques

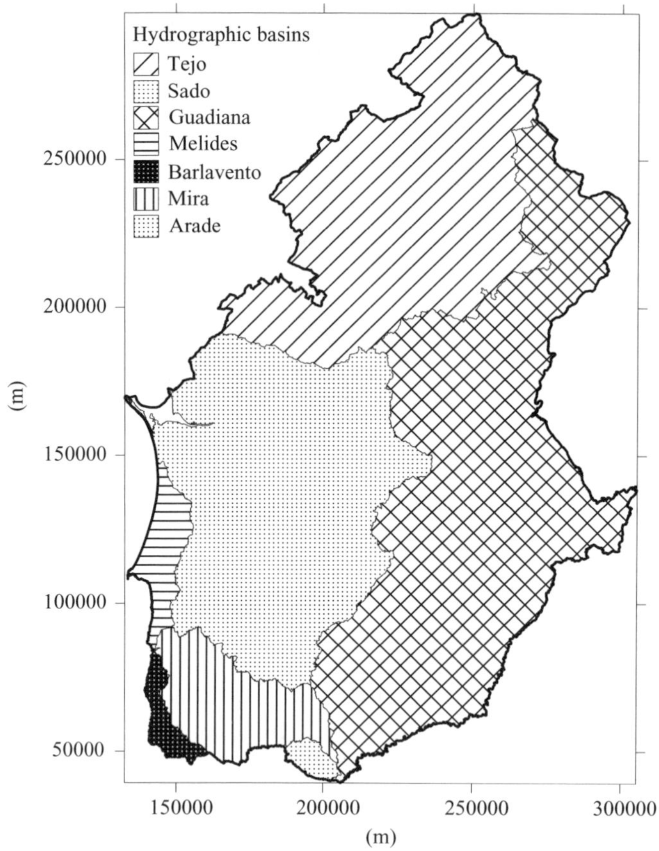


Figure 6. River basins in Alentejo region.

Table 1. Annual average of the precipitation, evapotranspiration and infiltration of the Alentejo main rivers basins (Henriques, 1985).

| River Basin | Area ($km^2$) | Precipitation (mm) | Evapotranspiration (mm) | Infiltration (mm) | Infiltration (% of precipitation) | Groundwater resources ($hm^3$/year) |
|---|---|---|---|---|---|---|
| Tejo | 9,612 | 686 | 438 | 50 | 7.3 | 481 |
| Sado | 8,217 | 678 | 480 | 51 | 7.5 | 411 |
| Mira | 1,689 | 682 | 481 | 19 | 2.8 | 84 |
| Guadiana | 11,855 | 581 | 426 | 28 | 4.8 | 593 |

(1985) had studied the first order river basins of Alentejo (Figure 6) and esteemed the average annual precipitation and real evapotranspiration for each one (Table 1). Also the infiltration values were calculated, showing that the average infiltration rates are between 2.8 and 4.8% of the total precipitation in the Mira and Guadiana river basins, and 7.3 to 7.5% in the Tejo and Sado River basins, involving in these last ones both sedimentary and hard rock aquifers.

Table 2. Annual base flow linear regressions over annual precipitation by watershed groups (adapted from Oliveira, in press).

| Predominant lithological groups (watershed code) | Equation | r |
|---|---|---|
| Igneous rocks (19N08; 17L/01) | Fb = 0.3987*(P − 428.1) | 0.952 |
| Metamorphic rocks (27I/01; 30G01; 22G/02; 06P/01) | Fb = 0.1866*(P − 405.9) | 0.736 |
| Igneous + metamorphic rocks (20I13; 18L01; 19N01) | Fb = 0.2394*(P − 458.0) | 0.894 |
| Igneous and/or metamorphic rocks + sedimentary (21F01; 18I01) | Fb = 0.2246*(P − 330.5) | 0.764 |
| All watersheds (including 03D/01; 05E/01) | Fb = 0.4703*(P − 521.0) | 0.955 |

r = correlation coefficient

These values seems to be underestimated by the last works of Oliveira (in press), who calculated the linear regression equation relating precipitation (P) with infiltration values in hard rocks using average annual base flow (Fb) values of 13 watersheds in Portugal:

$$Fb = 0.5210{*}P - 284 = 0.5210{*}(P - 546) \tag{1}$$

The correlation coefficient (r) for this equation is 0.989. Even so, the precedent equation considers two very high surface runoff/precipitation watersheds on the north of Portugal. More representative of the south conditions must be equation 2, that doesn't consider these two watersheds (Oliveira, in press):

$$Fb = 0.4107{*}P - 214 = 0.4107{*}(P - 520) \tag{2}$$

but on this one $r = 0.860$. This implies that results taken from these analysis are very much conditioned by the inexistence of a uniform distribution of watersheds' precipitations (Oliveira, in press).

In both cases (equations 1 and 2), the average infiltration values are approximately 8% of the average precipitation values in Alentejo. This seems to be a better approximation to real values, where the infiltration values on the less permeable hard rocks can be between 3 and 7% and in the more productive ones, around 10% of the annual precipitation values, as it was considered in ERHSA Project (ERHSA 2001), values that were also based on the preliminary works of Oliveira.

But Oliveira (in press) also calculated the linear regression equations relating precipitation with infiltration values in some groups of hard rocks, using average annual base flow values. The equations were calculated for basins with igneous rocks, metamorphic rocks, both igneous and metamorphic rocks in the same basin, and igneous or metamorphic and sedimentary rocks also on the same basin. The resulting equations are presented in Table 2, considering only the most representative values for Alentejo region, which can be the basis for the calculation of infiltration rates in this area. The original table has more equations, involving basins in the north of Portugal with high rates of precipitation, not representative of Alentejo conditions. Also weathering conditions of the rocks and the infiltration rates can be very different from north to south.

## 3 GEOLOGY

In terms of the geological features, the definition of the new aquifers was also based on the geo-structural divisions of Iberian Peninsula (Figure 7). The area is geologically complex, mainly the Ossa-Morena Zone.

The northern part of Alentejo (Centre-Iberian Zone) consists basically of schists and greywackes, with some quartzitic ridges, and highly compact granitic rocks. The schists and greywackes have a scarce soil, being many times skeletic, mainly on the slopes near the most important rivers. The fractures are open only when quartz fractured veins are present. The quartzites form extensive narrow ridges without soil covering and present a dense net of open and clean fractures. The igneous rocks cross the border between Centre Iberian and Ossa-Morena Zones and have an incipient fracture net, with granite products of spherical jointing caused by weathering covering the land, and only less than one to few meters of soil cover in some more weathered parts.

The Ossa-Morena Zone is formed by metamorphic and igneous rocks. The metamorphic ones are formed by schists, greywackes, gneisses, amphibolites, metamorphised volcanic rocks, and metamorphic limestones, in a great complexity, affected in some places by strong structural features that difficult the geological story interpretation of the area. The lithology of the igneous rocks is variable, going from granites to granodiorites, quartz-diorites, tonalitic rocks, diorites, gabbros, charnokites, andesites, etc. This knowledge is important on what concerns the analysis of the weathering layers deepness, directly associated to the aquifer volume, and to the water quality, that depends on the mineral constitution of the rocks: the most basic rocks have generally the most extended weathered layers.

The South Portuguese Zone has also its complexity. It is composed mainly by metamorphic rocks, including the so called "Pyrite Belt", a rich mineralised volcano-sedimentary complex. Most part of the area is formed by low metamorphic schists, greywackes, and conglomerates, with skeletic low productive soils.

## 4 INVESTIGATION METHODOLOGY

The lack of data on the hard rock aquifers in Portugal led to this study to establish differences of hydrogeological potential between different rock formations. This lack of data is more critical in hard rock aquifers because much drilling was done by companies characterized by lack of technical direction. In fact, this is due to the drilling techniques and to the economic lower values pay off that companies receive mainly when using the rotopercussion technique. On sandy aquifers, special techniques and more modern equipment are used and it is easier to find reliable reports.

Also much drilling was done without permission of legal authorities so that, in this case, the environment authorities did not receive the final reports and even when the reports were submitted, the information can be insufficient, erroneous, and even false.

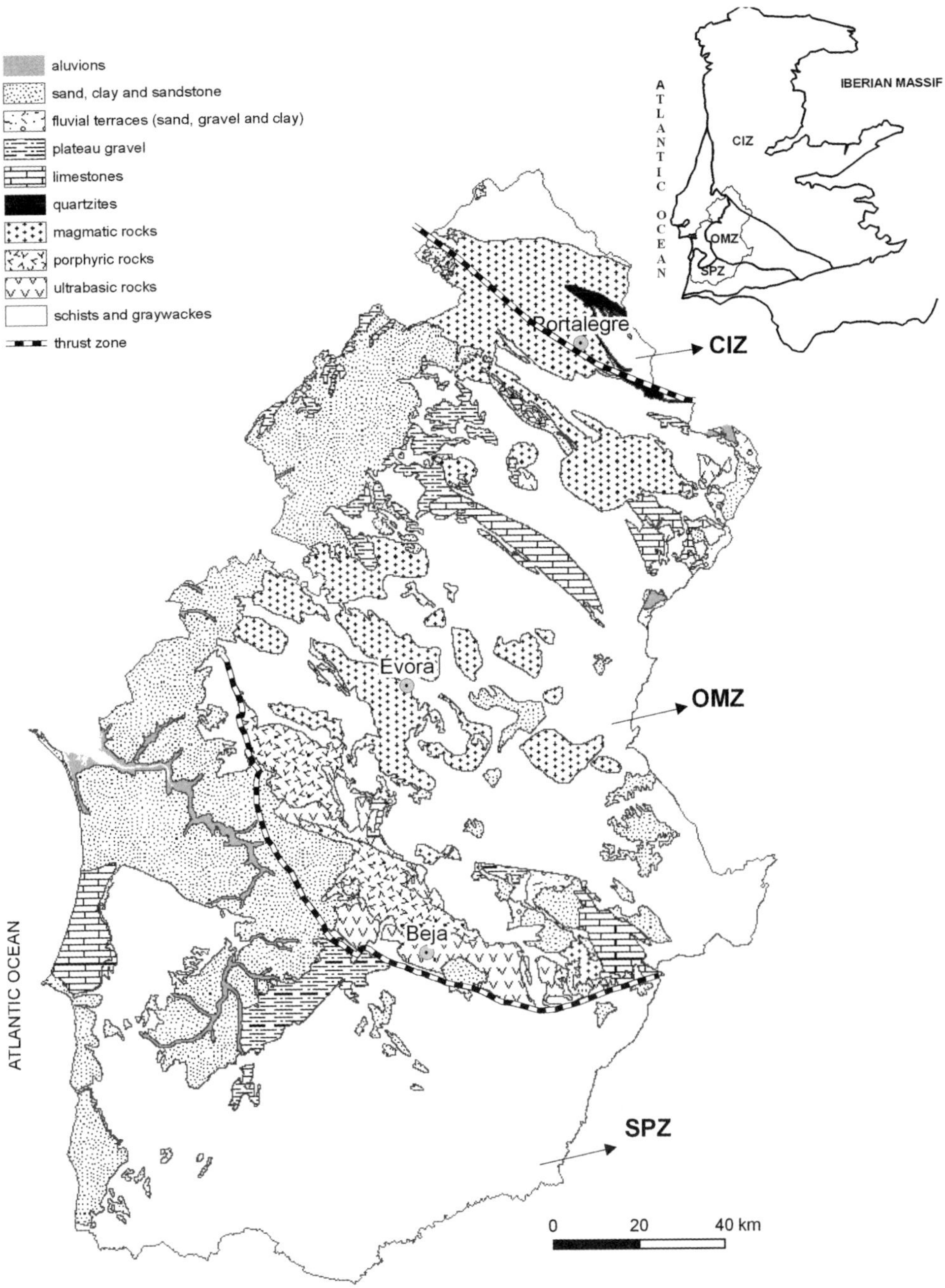


Figure 7. Simplified geology of Alentejo. CIZ – Centre-Iberian Zone; OMZ – Ossa-Morena Zone; SPZ – South Portuguese Zone.

Data should be collected and confirmed in the field in order to detect the misinformation or confirm the previous information. We collected data from governmental institutions and from drilling company archives, where it is sometimes possible to collect drilling data that were not recorded by the official authorities.

With the names of the land owners, locations, or phone numbers, it was possible to identify location in the field and, talking with the owner, sometimes to obtain information about the initial yields, other information received from the drilling companies, such as drilling depths. Depths of the wells and position of pumping systems were recorded, when possible. Information about water use and the yields produced were also registered. Well withdrawals could sometimes be measured and information about how many hours per day or per week this water was used were recorded. All this information permits correction or confirmation of suspicious original data, particularly related to yields, lithological logs, etc.. Some aquifer tests were performed, but this was not easy due to the quantity of ropes, electric wires, etc., that are normally inside the wells, the lack of piezometric tubes, and also the opposition of many owners.

In order to compare the different hard rock lithologies in terms of hydrogeological productivity, the yield values registered on the reports were considered very important. These yields were obtained using compressed air directed to the bottom of the well on the final of the drilling work and letting stabilize the abstracted yield. This stabilization can happen after some 5 to 15 minutes (in most of the cases) to some few hours in special cases. In some wells with low productivities, this method causes well dewatering so this is not a recommended exploration yield, and have been considered as instant yields by some Portuguese authors (Carvalho 2000; Mira and Chambel 2001), due to narrow time limitation of this tests, many times not supported by the longer conventional aquifer tests using submersible pumps. The practice show that normally the abstraction yields must not be higher than 30 to 50% of the measured instant yields. However, instant yield values can be used to compare with similar data on different aquifers. In this case, only the data from deep wells were considered. Unproductive wells were also inventoried, if data were available. This kind of information was obtained from company archives or directly from the land owners because these data are not usually registered by the public services.

Most of the ERHSA time and money was devoted to the data inventory, which was the basis for the final report and also for several papers and BSc, MSc and PhD theses. The database, which integrates hydrochemical data, maintains its importance for professionals studying the hydrogeology of the region.

The field work was organised with 5 to 7 hydrogeologists that coordinated the work on areas covering all Alentejo. More than 20 students participated on the data inventory and wrote their theses under instructions and supervision of the team. The task was to collect information on a minimum of one well in each 20 km$^2$ in homogeneous geologic formations and a maximum of wells in specific geologic complex areas and in new identified aquifers. In the first approach, all the possible construction data, yield, well deep and diameter, and the first field analysis of water temperature, EC and pH were recorded.

Based on this initial database, almost 2,000 data points were object of chemical and physical laboratory analyses. These represented deep wells, hand dug wells, and some springs.

EC and other hydrochemical parameters were studied to evaluate the spatial distribution of hydrochemical facies that could help to delineate new aquifers.

All the information was finally put on an easily accessible database to permit interface with hydrogeology software, data management, and use of geographic information systems

(GIS). The final project maps were organised on GIS, to allow the use of geographic bases in future development work.

## 5 RESULTS

Generally the hard rock aquifers of this area are considered of low productivity, with the recognized exception of the crystalline limestones, which are sometimes karstified. The metamorphic and igneous rocks have only a few meters of weathered layers (0 to 6 m deep) and a very incipient fracture system under these layers. More deeply, the bedrock has only a scarce net of fractures, where these fractures can be separated by tens or hundreds of meters under depths between 60 m (generally in igneous rocks) and more than 200 m (more likely in metamorphic rocks). In igneous rocks the wells have generally some water on the first 60 m deep, and it's rare to find new productive levels under this depth. In metasedimentary rocks the probability of occurrence of new productive levels with depth increases in relation with igneous rocks, being equally possible to get water on the first 100 m or on the following 100 ones.

Even so, an aquifer with special characteristics was already known in Alentejo (Figure 8): the Gabbros of Beja Aquifer (Paradela & Zbyszewski 1971) and it has been object of some thesis in the last years (Duque 1997; Duque 2005; Paralta 2001). Paralta is also preparing his PhD thesis on the same area. Here there is a median thickness of 26 m of weathered or fractured gabbro-dioritic rocks that can reach up to 80 m deep, with the groundwater flowing in fractures that are enlarged by chemical water attack. In 1971 in this aquifer, hand dug wells with a maximum of 20 m deep, much of them with 1.5 m high galleries in the bottom or at different levels under the piezometric level, and other wells with a maximum of 50 m deep, were used to abstract groundwater for human supply. The yields could reach 8 L/s, exceptionally 10 L/s, with the production clearly decreasing during the dry season. More recently, Duque (2005) has shown that the real productivity (including unproductive wells), for 1,067 well data, was about 3.34 L/s (average), 1.53 L/s (median) till a maximum of 36 L/s. The high productive levels vary between 1 and 78 m deep, being the major productive levels between the 12 and 20 m. Based on 1,030 well logs, the aquifer shows an average of 2 m of topsoil clays, a total weathered layer of 12 m and a fractured net till 27 m deep.

In fact, the difference between the less productive hard rock aquifers and some more productive ones are basically the fracture density and, more importantly, the total weathered thickness. More than 10–15 m of weathered layers create a more productive aquifer, based on a higher porosity and an easier interconnection between fractures. That is why the most ancient rocks, subject to more orogenic processes, and affected by more recent orogenies are the most productive hard rock aquifers in Alentejo.

During ERHSA studies, the information confirmed the existence of other aquifers with characteristics similar to those of the Gabbros of Beja (ERHSA 2001). One of the main conclusions was that gabbros, gneisses, migmatitic gneisses, some volcano-sedimentary, and tonalitic rocks were more productive than the other lithologies of the region. Three new aquifers were identified (Figure 8 and Table 3), now designated as:

- Charnokites of Campo Maior and Elvas
- Pavia-Mora
- Évora-Montemor-Cuba, subdivided in 5 different sectors: Évora, Montemor, Escoural, Cuba-S. Cristóvão, Vidigueira-Selmes

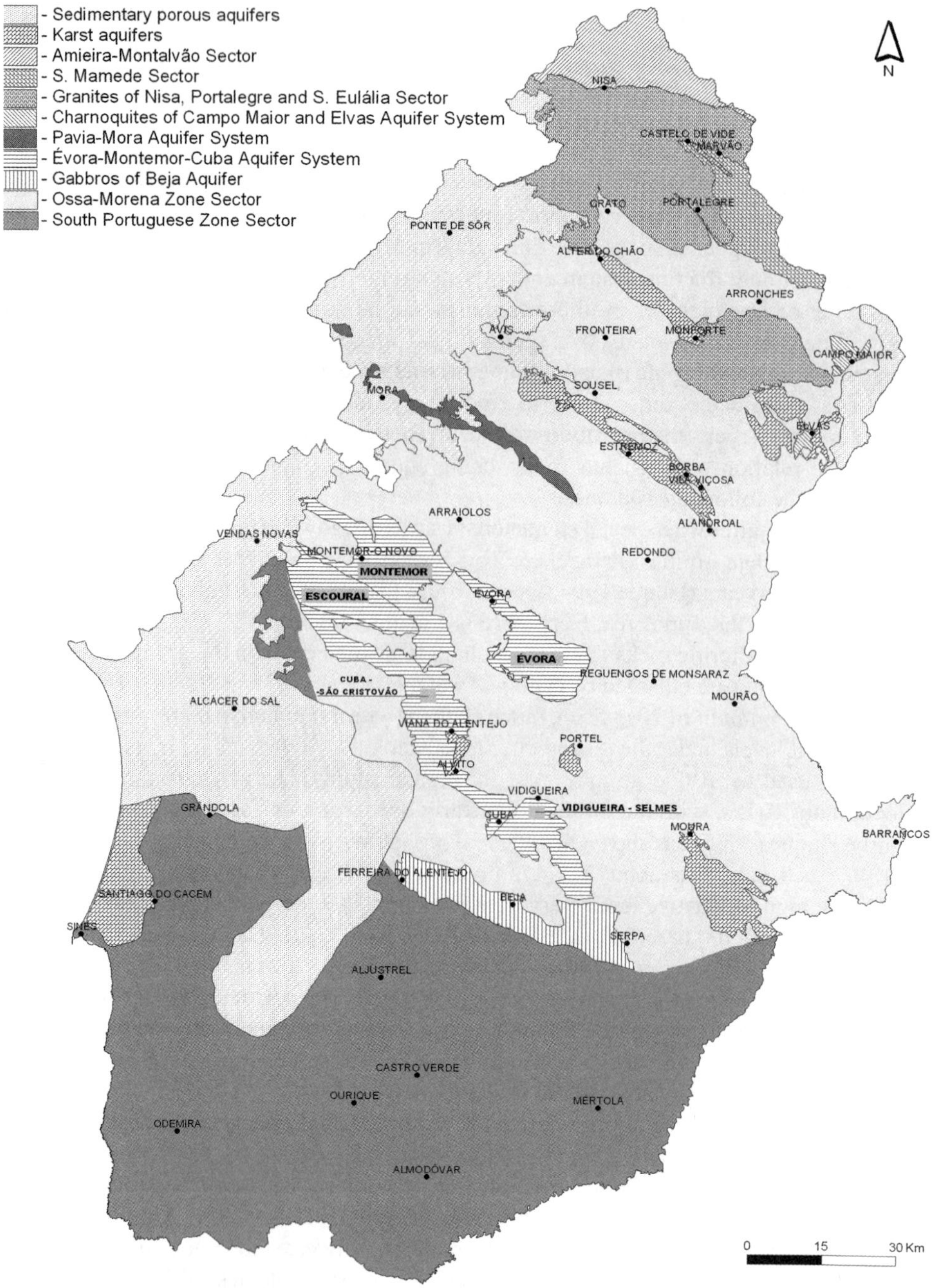


Figure 8. Hydrogeological cartography of Alentejo region, Portugal (ERHSA 2001).

Table 3. Main geologic and hydrogeologic characteristics of aquifers and low poductivity sectors of Alentejo region, South Portugal (ERHSA 2001).

| Aquifers and Hydrogeologic Sectors | Lithologic characteristics | Aquifer characteristics, weathering and fracturing |
|---|---|---|
| **Aquifers** | | |
| *Évora-Montemor-Cuba* | | |
| Évora Sector | Gneisses, migmatites, granodiorites and quartz-diorites | Maximum of 30 m, generally not more than 20 m of weathered and fractured rock, free to semi-confined aquifer |
| Montemor Sector | Migmatites, migmatitic gneisses, some granites, tonalites and hornfels | Similar to Évora sector |
| Escoural Sector | Leptinites, black quartzites, acid and basic meta-volcanites, meta-psamites, schists with high levels of quartz veins, calco-schists, crystalline limestones and amphibolites | Fractures occur until 80 m depth in some special rocks (quartzites, amphibolites), where the wells are more productive; low extension of the weathered zone; similar to the other sectors on the remaining lithologies |
| Cuba-S. Cristóvão Sector | Gabbrodiorites, granitic ortogneisses, a metapelitic-psamitic sequence, leptinites, black quartzites, metabasites and ortogneisses | Litologies are similar to other sectors of Évora-Montemor-Cuba Aquifer, but there are a big lack of information on this sector |
| Vidigueira-Selmes Sector | Basic volcanites, granodiorites, gabbrodiorites, hornfels | Fractures untill 20–30 m depth, free to semi-confined aquifer, with 4 to 8 m of weathered layer |
| Gabbros of Beja | Gabbros, diorites, serpentinites, meta-trondhejmites, meta-basalts, flasergabbros, amphibolites, piroxenites, dunites and peridotites | Average of 2 m of topsoil clays, an average of total weathered layer of 12 m and an average of the fractured net till 27 m deep; free to semi-confined aquifer |
| Charnokites of C. Maior and Elvas | Gabbro metamorphized rocks, piroxenites, anortosites, diorites, quartz-gabbros and hornblendites in two different spots | Extremely fractured and weathered rocks; no indications about the weathered zone extension |
| Pavia-Mora | Gneissic granitic and granitic gneissic rocks, granites, gabbros and diorites; the metamorphic rocks are less significant and represented by leptinitic gneisses, amphibolites, micaschists and crystalline limestones | Fractured media, with weathered layers generally less than 6 m deep |

(Continued)

Table 3. (Continued)

| Aquifers and Hydrogeologic Sectors | Lithologic characteristics | Aquifer characteristics, weathering and fracturing |
|---|---|---|
| **Low Productivity Sectors** | | |
| Amieira-Montalvão Sector | Mainly phylites, meta-greywackes, pelitic schists, meta-conglomerates and hornfels | Skeletic soils; great hydrogeologic importance of the presence of fractured quartz veins and faults filled with quartz |
| S. Mamede Sector | Mainly schists, lidites, quartzites, sandstone, clay schists, volcanoclasts, conglomerates and arkoses | Skeletic soils; great importance of the presence of fractured quartz veins and faults filled with quartz |
| Granites of Nisa, P. and S. E. Sector | Granitic rocks | At surface, the products of spherical jointing caused by weathering are visible, being the weathered zone in general terms less than 2 m deep and the fracture system very sparse |
| Ossa Morena Zone Sector | Mainly schists and greywackes, and different kinds of igneous rocks, granites, granodiorites, quartz-diorites, tonalites, diorites, andesites, some gabbros, but also volcanic or porphyritic rocks | Low extension of weathered layer and great importance of the presence of quartz fractured veins linked to groundwater production |
| South Portuguese Zone Sector | Mainly low metamorphic schists, greywackes and volcano-sedimentary complexes | In most of the area, skeletic soils, low extension of the weathered layer and a great importance of the presence of quartz fractured veins linked to groundwater production |

On the remainder areas, some low productive sectors were identified (Figure 8 and Table 3):

- Amieira-Montalvão
- S. Mamede
- Granites of Nisa, Portalegre and Santa Eulália
- Ossa-Morena Zone (OMZ)
- South Portuguese Zone (SPZ)

The characteristics of all the aquifers and low productive sectors are defined on Table 3.

Table 4 shows that the most productive aquifers have an average of instant yield values higher than 4.5 L/s and median values higher than 1.9 L/s (Évora and Montemor sectors of the Évora-Montemor-Cuba Aquifer).

The Escoural sector of the Évora-Montemor-Cuba Aquifer, the Gabbros of Beja Aquifer and also the new identified aquifer of the Charnokites of Campo Maior and Elvas and Pavia-Mora have average instant yield values of more than 3 L/s. Only the Cuba-S. Cristóvão and Vidigueira-Selmes sectors seem to have lower values with average values over 2.5 L/s.

The low productive sectors of Amieira-Montalvão, S. Mamede and of the Granites of Nisa, Portalegre and S. Eulália show average instant values less than 2 L/s, the last one with only 0.72 L/s.

The low productive sectors of Ossa-Morena Zone and South Portuguese Zone display average values respectively of 2.01 and 2.42 L/s, which means that there are, in these sectors, some productive areas linked to fractured zones, like the Pyrite Belt in South Portuguese Zone, that need more investigation.

Dividing the instant yield values by the depth of the wells, the same conclusion is possible (Table 5): the Évora, Montemor and Cuba-S. Cristóvão sectors of the Évora-Montemor-Cuba Aquifer and the Gabbros of Beja Aquifer have clearly higher productivities (more than 0.08 L/s/m drilled) when compared with Escoural and Vidigueira-Selmes sectors (more than 0.05 L/s/m drilled) or the Charnokites of Campo Maior and Elvas or Pavia-Mora sectors, under 0.03 L/s/m drilled. The other sectors, considered of low productivity, show average values under 0.03 L/s/m drilled.

The depth of the wells can be a measure of the depth of the weathering zone, but also an indication of the difficulties to find water during drilling. Because the drillings are normally not assisted by hydrogeologists, the decision to stop is responsibility of the land owner and commanded many times by the necessities on water, by economic reasons, or intuition.

As can be seen on Table 6, average values higher than 60 m deep are present of the sectors of Montemor, Escoural and Vidigueira-Selmes of the Évora-Montemor-Cuba Aquifer, on the Pavia-Mora Aquifer, on the less productivity sectors of Amieira-Montalvão, S. Mamede, OMZ and SPZ. The more productive sectors of Évora, Cuba-S. Cristóvão, the Gabbros of Beja Aquifer, the Charnokites of Campo Maior and Elvas and the low productivity sectors of the Granites of Nisa, Portalegre and Santa Eulália present average values lower than 50 m, the first ones probably due to high productivities on the first tens of meters and the last one because it is practically impossible to get water under some few tens of meters deep. The Gabbros of Beja present the lowest of all the medium values (38.1 m). The justification to these results is on the fact that the most productive systems have a weathered and fractured rock layer on the first 20–40 m. Thus, prospection below this depth is less productive. Also the low productive areas of granites have the water mainly on the first tens of metres, many times on the first 10 m, motif why the prospection

Table 4. Statistical results of instant value yields of the different sectors and aquifers in Alentejo region. The sectors of Évora, Montemor, Escoural, Cuba-S. Cristóvão and Vidigueira-Selmes are part of the Évora-Montemor-Cuba Aquifer. Values from ERHSA 2001, Ghira 2002, Espada 2003, Furtado 2004, Ramalho 2004, Duque 2005, Ferreira 2005, Monteiro 2005, Rodrigues, in press.

| Statistics of instant yield | Average (L/s) | Median (L/s) | Minimum (L/s) | Maximum (L/s) | Standard deviation | Number of deep wells with instant yield values | Number of deep wells | Total number of wells | Transmissivity values range ($m^2$/day) |
|---|---|---|---|---|---|---|---|---|---|
| **Aquifers** | | | | | | | | | |
| Évora Sector | 5.54 | 3.33 | 0 | 27.8 | 6.02 | 187 | 205 | 322 | 37–373 |
| Montemor Sector | 4.47 | 1.94 | 0 | 41.66 | 6.92 | 89 | 201 | 413 | |
| Escoural Sector | 3.97 | 1.53 | 0 | 26.39 | 6.07 | 30 | 63 | 144 | |
| Cuba-S. Cristóvão Sector | 2.75 | 1.11 | 0 | 26.39 | 4.19 | 280 | 313 | 417 | |
| Vidigueira-Selmes Sector | 2.61 | 1.11 | 0 | 28.75 | 4.76 | 88 | 112 | 229 | 40 |
| Gabbros of Beja | 3.33 | 1.52 | 0 | 36.1 | 4.82 | 1,067 | 1,059 | 1,798 | 1.7–432 |
| Charnokites of C. Maior and Elvas | 3.03 | 1.67 | 0.42 | 33.3 | 5.16 | 40 | 43 | 53 | |
| Pavia-Mora | 3.20 | 1.38 | 0 | 27.8 | 5.79 | 30 | 44 | 63 | |
| **Low Productivity Sectors** | | | | | | | | | |
| Amieira-Montalvão Sector | 1.64 | 1.39 | 0.28 | 6.94 | 1.62 | 15 | 19 | 53 | 0.8–60.6 |
| S. Mamede Sector | 1.46 | 1.25 | 0.06 | 4.00 | 0.88 | 31 | 56 | 74 | |
| Granites of Nisa, P. and S. E. Sector | 0.72 | 0.48 | 0 | 5.56 | 0.99 | 89 | 330 | 627 | 0.2–30.8 |
| Ossa Morena Zone Sector | 2.01 | 1.20 | 0 | 40 | 3.00 | 935 | 2,135 | 5,183 | 0.6–7 |
| South Portuguese Zone Sector | 2.42 | 1.49 | 0 | 19.4 | 2.68 | 538 | 927 | 1,402 | 0.07–18.3 |

Table 5. Comparative instant yield by metre drilled on the aquifer and other hydrogeologic sectors of Alentejo. The sectors of Évora, Montemor, Escoural, Cuba-S. Cristóvão and Vidigueira-Selmes are part of the Évora-Montemor-Cuba Aquifer. Values from ERHSA 2001, Ghira 2002, Espada 2003, Furtado 2004, Ramalho 2004, Duque 2005, Ferreira 2005, Monteiro 2005, Rodrigues, in press.

| | L/s/m drilled | Number of deep wells with both yield and depth values |
|---|---|---|
| **Aquifers** | | |
| Évora Sector | 0.1238 | 180 |
| Montemor Sector | 0.0864 | 80 |
| Escoural Sector | 0.0604 | 30 |
| Cuba-S. Cristóvão Sector | 0.0897 | 223 |
| Vidigueira-Selmes Sector | 0.0590 | 86 |
| Gabbros of Beja | 0.0874 | 1,059 |
| Charnokites of C. Maior and Elvas | 0.0275 | 11 |
| Pavia-Mora | 0.0197 | 18 |
| **Low Productivity Sectors** | | |
| Amieira-Montalvão Sector | 0.0210 | 8 |
| S. Mamede Sector | 0.0274 | 21 |
| Granites of Nisa, P. and S. E. Sector | 0.0110 | 66 |
| Ossa Morena Zone Sector | 0.0289 | 771 |
| South Portuguese Zone Sector | 0.0265 | 435 |

Table 6. Statistical results of the depth of the drillings and deep wells in the aquifer and sectors of Alentejo. The sectors of Évora, Montemor, Escoural, Cuba-S. Cristóvão and Vidigueira-Selmes are part of the Évora-Montemor-Cuba Aquifer. Values from ERHSA 2001, Ghira 2002, Espada 2003, Furtado 2004, Ramalho 2004, Duque 2005, Ferreira 2005, Monteiro 2005, Rodrigues, in press.

| Total depth | Average (m) | Median (m) | Minimum (m) | Maximum (m) | Standard deviation | Number of wells with data on depth | Number of deep wells | Total number of wells |
|---|---|---|---|---|---|---|---|---|
| **Aquifers** | | | | | | | | |
| Évora Sector | 44.87 | 40 | 8 | 130 | 21.08 | 180 | 205 | 322 |
| Montemor Sector | 61.47 | 60 | 15 | 151 | 28.63 | 138 | 201 | 413 |
| Escoural Sector | 77.81 | 75 | 35 | 150 | 28.05 | 36 | 63 | 144 |
| Cuba-S. Cristóvão Sector | 55.58 | 48.5 | 5 | 215 | 36.93 | 223 | 313 | 417 |
| Vidigueira-Selmes Sector | 72.12 | 70 | 20 | 177 | 28.70 | 107 | 112 | 229 |
| Gabbros of Beja | 38.1 | 34 | 2 | 140 | 18.30 | 1,059 | 1,059 | 1,798 |
| Charnokites of C. Maior and Elvas | 42.27 | 46 | 22 | 75 | 19.23 | 11 | 43 | 53 |
| Pavia-Mora | 68.63 | 67 | 31 | 108 | 19.70 | 26 | 44 | 63 |
| **Low Productivity Sectors** | | | | | | | | |
| Amieira-Montalvão Sector | 75.1 | 80 | 43 | 120 | 27.90 | 9 | 19 | 53 |
| S. Mamede Sector | 63.80 | 60 | 16 | 262.75 | 40.85 | 34 | 56 | 74 |
| Granites of Nisa, P. and S. E. Sector | 54.78 | 56.5 | 2 | 187 | 33.87 | 152 | 330 | 627 |
| Ossa Morena Zone Sector | 68.44 | 69 | 10 | 205 | 30.18 | 772 | 2,135 | 5,183 |
| South Portuguese Zone Sector | 65 | 65 | 0 | 200 | 25.16 | 662 | 927 | 1,402 |

Table 7. Hydrochemical characteristics of Alentejo groundwaters (ERHSA 2001), with the avarege values for the different parameters (n – number of samples).

| Aquifers and Hydrogeologic Sectors | Field EC | | Field pH | | Alcalinity | | Total Hardness | | $CO_2$ | | $SiO_2$ | | $NO_3$ | |
|---|---|---|---|---|---|---|---|---|---|---|---|---|---|---|
| | n | μS/cm | n | | n | mg/L | n | mg/L | n | mg/L | n | mg/L | n | mg/L |
| **Aquifers** | | | | | | | | | | | | | | |
| Évora Sector | 127 | 868 | 80 | 7.62 | 49 | 249 | 49 | 368 | 48 | 7.4 | 49 | 27.6 | 48 | 79 |
| Montemor Sector | 237 | 484 | 241 | 6.97 | 53 | 143 | 53 | 187 | 51 | 30.1 | 50 | 28 | 53 | 33.8 |
| Escoural Sector | 96 | 585 | 98 | 7.16 | 18 | 179 | 18 | 262 | 18 | 24.1 | 17 | 34.3 | 18 | 33.4 |
| Cuba-S. Cristóvão Sector | 118 | 849 | 122 | 7.45 | 36 | 231 | 35 | 323 | 35 | 18.5 | 32 | 27.1 | 36 | 42.4 |
| Vidigueira-Selmes Sector | 127 | 927 | 127 | 7.7 | 23 | 266 | 23 | 350 | 22 | 10.8 | 20 | 23.2 | 21 | 42.9 |
| Gabbros de Beja | 79 | 820 | 80 | 7.53 | 79 | 241 | 79 | 347 | 72 | 20 | 71 | 21.4 | 79 | 57.9 |
| Charnokites of C. Maior and Elvas | 17 | 816 | 17 | 7.78 | 6 | 265 | 6 | 343 | 2 | 8.6 | 6 | 36.6 | 5 | 64.9 |
| Pavia-Mora | 32 | 1,401 | 21 | 7.21 | 14 | 257 | 14 | 435 | 14 | 23.7 | 9 | 35.8 | 14 | 52.3 |
| **Low Productivity Sectors** | | | | | | | | | | | | | | |
| Amieira-Montalvão Sector | 35 | 372 | 35 | 6.6 | 13 | 56 | 13 | 58 | 12 | 5.8 | 13 | 29.5 | 13 | 2.5 |
| S. Mamede Sector | 28 | 168 | 28 | 6.47 | 19 | 87 | 19 | 87 | 18 | 20 | 19 | 13 | 16 | 12.3 |
| Granites of Nisa, P. and S. E. Sector | 319 | 253 | 319 | 6.5 | 102 | 226 | 101 | 263 | 98 | 15.4 | 102 | 31.5 | 104 | 14.8 |
| OMZ Sector | 2,754 | 745 | 2,438 | 7.37 | 718 | 214 | 714 | 295 | 661 | 17.6 | 704 | 25.9 | 722 | 39.2 |
| SPZ Sector | 870 | 1,311 | 775 | 7.04 | 577 | 190 | 577 | 367 | 547 | 23.8 | 577 | 16.2 | 530 | 25.7 |

| Aquifers and Hydrogeologic Sectors | Ca | | Mg | | Na | | K | | Cl | | $SO_4$ | | Corrected $HCO_3$ | | Facies |
|---|---|---|---|---|---|---|---|---|---|---|---|---|---|---|---|
| | n | epm | n | epm | n | epm | n | epm | n | epm | n | epm | n | epm | |
| **Aquifers** | | | | | | | | | | | | | | | |
| Évora Sector | 49 | 3.38 | 49 | 3.76 | 49 | 3.13 | 49 | 0.11 | 49 | 2.79 | 49 | 0.56 | 48 | 9.57 | C-Mg-Ca-Na |
| Montemor Sector | 53 | 1.84 | 53 | 1.84 | 50 | 1.68 | 50 | 0.04 | 53 | 1.52 | 53 | 0.32 | 52 | 5.46 | C-Mg-Ca-Na |
| Escoural Sector | 18 | 2.51 | 18 | 2.73 | 17 | 1.93 | 17 | 0.06 | 18 | 2.34 | 18 | 0.43 | 18 | 7.01 | C-Mg-Ca-Na |
| Cuba-S. Cristóvão Sector | 36 | 3.47 | 35 | 3 | 36 | 3.07 | 36 | 0.08 | 36 | 2.82 | 36 | 0.5 | 36 | 8.58 | C-Ca-Na-Mg |
| Vidigueira-Selmes Sector | 23 | 3.24 | 23 | 3.68 | 23 | 2.52 | 23 | 0.04 | 23 | 2.22 | 23 | 0.49 | 23 | 8.71 | C-Mg-Ca-Na |
| Gabbros de Beja | 79 | 3.45 | 79 | 3.5 | 71 | 1.81 | 71 | 0.02 | 79 | 1.43 | 79 | 0.64 | 79 | 9.62 | C-Mg-Ca |
| Charnokites of C. Maior and Elvas | 6 | 3.17 | 6 | 3.65 | 6 | 0.59 | 6 | 0.05 | 6 | 1.25 | 6 | 0.42 | 6 | 10.23 | C-Mg-Ca |
| Pavia-Mora | 14 | 3.27 | 14 | 5.1 | 14 | 5.19 | 14 | 0.09 | 14 | 4.81 | 14 | 0.35 | 14 | 9.87 | C-Na-Mg |
| **Low Productivity Sectors** | | | | | | | | | | | | | | | |
| Amieira-Montalvão Sector | 13 | 0.44 | 13 | 0.73 | 13 | 0.99 | 13 | 0.05 | 13 | 0.5 | 13 | 0.06 | 13 | 2.15 | C-Na-Mg |
| S. Mamede Sector | 19 | 0.86 | 19 | 0.63 | 19 | 0.48 | 19 | 0.03 | 19 | 0.43 | 19 | 0.09 | 19 | 3.41 | C-Ca-Mg-Na |
| Granites of Nisa, P. and S. E. Sector | 103 | 1.12 | 100 | 0.86 | 102 | 0.96 | 102 | 0.07 | 104 | 0.86 | 104 | 0.15 | 104 | 3.5 | C-Ca-Na-Mg |
| Ossa Morena Zone Sector | 718 | 2.75 | 712 | 3.08 | 706 | 2.3 | 706 | 0.06 | 718 | 2.91 | 718 | 0.35 | 717 | 8.05 | C-Mg-Ca-Na |
| South Portuguese Zone Sector | 577 | 2.66 | 576 | 4.58 | 577 | 6.25 | 577 | 0.05 | 577 | 8.05 | 577 | 0.87 | 577 | 6.92 | Cl-C-Na-Mg |

is not so deep in those areas. On the contrary, the schists can have productive fractures at any deep, at least on the first 200 m, so it's possible to expect these wells to be deeper.

Aquifer tests in some wells of these areas had show that the transmissivities in the gneissic-migmatitic complex of Évora Sector are between 37 and 373 $m^2$/day (Fialho et al. 1998; ERHSA 2001). In 227 aquifer tests in the Gabbros of Beja Aquifer the median value was 40.8 $m^2$/day and the maximum 432 $m^2$/day (Duque 1997; Duque & Almeida 1998; Duque 2005). Values between 0.6 e 7 $m^2$/day were registered in granodiorites and quartzodiorites of the low productive area of OMZ. Transmissivity on schists and greywackes of the low productivity area of SPZ are between 0.07 and 18.3 $m^2$/day in 9 aquifer tests on schist and greywacke aquifers on the south-eastern part of the area (Chambel 1999). More recent aquifer tests, on the low productive sector of Amieira-Montalvão, show transmissivities between 0.8 and 12 $m^2$/day on 5 aquifer tests on previous wells that were done without any scientific assistance, and between 15.2 and 60.6 $m^2$/day on aquifer tests of new three wells located with geologic and hydrogeologic investigation and geophysic methods. Three aquifer tests on the less productive sector of Granites of Nisa, Portalegre and Santa Eulália led to results between 0.2 and 5.3 $m^2$/day on the previous wells and 30.8 $m^2$/day on the only productive well in three new excavated boreholes (two were considered unproductive, in spite they have some water, between 1,000 and 2,500 $m^3$/h of instant yield values). Transmissivity values obtained in some common aquifer tests are also presented on Table 4, in order to compare with instant yield value statistics.

About 2,000 water-quality analyses were made on all the Alentejo hard rock area. The results show that the north area (Centre-Iberian Zone) has the less mineralised groundwaters (Amieira-Montalvão Sector, S. Mamede Sector and Granites of Nisa, Portalegre and S. Eulália Sector), and the South Portuguese Zone and Pavia-Mora Aquifer have the higher mineralization rates (Table 7). The groundwater facies change with the different geologic units (Table 7):

- In all the aquifers and less productive sectors, the facies are bicarbonate, except on the South Portuguese Zone, the Alentejo area with lower precipitation values and highest evapotranspiration rates, where the main anion is the chloride, immediately followed by the bicarbonate.
- Most part of the aquifers have magnesium as the main cation, but very similar contents of calcium and sodium occur. Magnesium is often related with basic igneous and metamhorphic rocks, as it can be seen by the crossing of information between Tables 3 and 7. In some cases, calcium is the main cation, but always closely followed by another one (sodium or magnesium); in two cases, sodium is the main cation, followed by magnesium (Pavia-Mora and South Portuguese Zone).
- Nitrate levels are high in practically all the aquifers because normally these correspond to the best soils and main irrigation areas in Alentejo. This is due to the fact that the better aquifers are those where weathering is more intense and the best quality soils exist. Together with the water availability, the nitrate use is more intensive there. The less productive sectors have generally lower nitrate average values.

## CONCLUDING REMARKS

This large study (ERHSA 2001) allowed to confirm some previous aquifer information regarding the Gabbros of Beja Aquifer and to detect new productive or potentially productive areas on this region, the aquifers of the Charnokites of Campo Maior and Elvas, Pavia-Mora and Évora-Montemor-Cuba. There are also some field indicators that some regions of porphyries, considered in the low productivity area of the Ossa-Morena Zone can also be very interesting aquifers, where fracturing is important.

The main aquifers occur in gabbros, gneissic or gneissic-migmatitic rocks, metamorphic volcanic-sedimentary complexes and some tonalitic rocks, particularly where these rocks were affected by the Hercynian Orogeny. Understanding the vertical extension of the weathering materials and the fracture networks is essential to define the best aquifers in the hard rocks of Alentejo and perhaps elsewhere. This paper also reflects all the new data collected after ERHSA, representing some hundreds of new well data.

The hydrochemical analysis shows that the groundwater facies are mainly bicarbonate, except on the South Portuguese Zone Sector where it is chloride-bicarbonate. This is probably due to the low level of metamorphism that has affected this area, permitting the presence of salts on the rock matrix of this previous marine environment and to the high concentrations of salts due to the high rates of evapotranspiration that affects this semi-arid environment (Chambel & Almeida 1998). The main cations are the magnesium, calcium and sodium, by this order, having most part of these waters a mixed cation composition between three or two of them.

For the future, the main task will be to conduct aquifer tests in order to have more hydrogeological parameters (transmissivity, permeability, and storage values), and try to understand the relation between instant yields, steady state flow, abstraction and aquifer parameters.

Quantitative and more precise structural studies are necessary to understand the relation between productivity, transmissivity, storage and groundwater pathways, which are mainly defined by the fracture network. Also the weathering layer thickness and type, clearly related with fracture network, are essential to understand and quantify recharge and storage coefficient. Storage and transmissivity which are essential to define groundwater exploration and use in such semi-arid areas as South Portugal.

## REFERENCES

Arora V (2003) The use of the aridity index to access change effect on annual runoff. Journal of Hydrology, Vol. 265: 164–177.

Carvalho J (2000) The impounding groundwater project in crystalline formations: some clues (in portuguese). University of Évora, Évora.

Chambel A (1999) Hydrogeology of Mértola Municipality [in Portuguese]. PhD. Thesis, Évora: University of Évora, Évora.

Chambel A, Almeida C (1998) Origin of highly mineralized waters in a semi-arid area of the South Portuguese Zone (Portugal). *In: Gambling with Groundwater – Physical Chemical and Biological Aspects of Aquifer-Stream Relations*, Brahana et al. (eds), Las Vegas, USA, 419–424.

Chambel A, Duque J (1999) Hard rock aquifers of Alentejo Region (South Portugal): contribution to the water and land use management. Proceedings of the XXIX IAH Congress, Hydrogeology and Land Use Management, Miriam Fendeková & Marián Fendek Eds., Bratislava, Slovakian Republic, 171–176.

Chambel A, Duque J, Matoso A, Orlando M (2006) Hydrogeology of continental Portugal [in portuguese]. Boletín Geológico y Minero, Madrid (Spain), 117 (1): 163–185.

Chambel A, Duque J, Fialho A (1998) Groundwater in a semi-arid area of South Portugal. In: Gambling with Groundwater – Physical Chemical and Biological Aspects of Aquifer-Stream Relations, Brahana et al. (eds), Las Vegas (USA) 75–80.

Chambel A, Duque J, Nascimento J (2002) Hydrogeology of the Crystalline Rocks of Alentejo: New Cartography based on the results of ERHSA Project [in Portuguese]. Proceedings of the 6th Water Congress – The Water is Gold, Porto, Portugal, CD-88.

Duque J (1997) Hydrogeological characterization and mathematical modelling of the Gabbros of Beja aquifer [in Portuguese]. Master Thesis, University of Lisbon, Lisbon, 220.

Duque J (2005). Hydrogeology of the Gabbros of Beja Aquifer System [in Portuguese]. PhD. Thesis, University of Lisbon, Lisbon, 420.

Duque J, Almeida C (1998) Hydrochemical characterization of the Gabbros of Beja Aquifer System (in portuguese). In: 4th Congress of Water, The Water as Structural Resource of Development, Reports and Abstracts, Abstract: 193–194, Paper: CD-164.

ERHSA (2001) Project "Study of groundwater resources of Alentejo region" (ERHSA, in portuguese). Coordination Commission of Alentejo Region, Évora.

Espada J (2003) Hydrogeological definition of the low productivity Sectores of Amieira-Montalvão and S. Mamede [in Portuguese]. Water Resources Engineering Final Thesis, University of Évora, 66.

Ferreira M (2005) Hydrogeological definition of the eastern half of the Cuba-S. Cristóvão Sector of the Évora-Montemor-Cuba Aquifer System [in Portuguese]. Water Resources Engineering Final Thesis, University of Évora, 76.

Fialho A, Chambel A, Almeida C (1998) Hydraulic characterization of fractured aquifers by double porosity models on Évora Municipality [in Portuguese]. In: Proceedings of the 4th Congress of Water, The Water as Structural Resource of Development, Reports and Abstracts, Abstract: 179, Paper: CD-128.

Furtado P (2004) Hydrogeological definition of the low productivity Sector of Granites of Nisa, Portalegre and Santa Eulália [in Portuguese]. Water Resources Engineering Final Thesis, University of Évora, 82.

Ghira L (2002) Hydrogeological definition of the western half of Cuba-S. Cristóvão Sector of the Évora-Montemor-Cuba Aquifer System [in Portuguese]. Water Resources Engineering Final Thesis, University of Évora, 82.

Henriques AG (1985) Evaluation of the water resources of continental Portugal. Contribution to territory management [in Portuguese]. Instituto de Estudos para o Desenvolvimento, Lisboa (Portugal).

INAG 2006. Instituto da Água (Water Institute). http://insaar.inag.pt/index2_noflash.htm

Mendes JC (1989) Desertification – Iberian problem? [in Portuguese]. Meteorologia y Climatologia Ibericas, Edition by C. Tomas & J. Labajo, Actas de las XVII Jornadas de la AME, Universidade de Salamanca, 21–25.

Mira F, Chambel A (2001) Productivity of hard rock aquifers in Alentejo region, South Portugal. In: Proceedings of the XXXI IAH Congress: New Approaches Characterizing Groundwater Flow, Seiler & Wohnlich (eds), Balkema, Lisse (Holand), 2:1035–1039.

Monteiro J (2005) Hydrogeological definition of Montemor Sector of Évora-Montemor-Cuba Aquifer System [in Portuguese]. Water Resources Engineering Final Thesis, University of Évora, 72.

Oliveira MM (in press) The estimation of groundwater recharge of fractured rocks in Portugal using surface flow measurements. Proceedings of the "2nd Workshop of the Iberian Regional Working Group on Hard Rock Hydrogeology", Portuguese Chapter of the International Association of Hydrogeologists (AIH-GP).

Paradela P, Zbyszewski G (1971) General hydrogeology of the centre and south of Portugal [in Portuguese]. Guide-Book of the excursion n. 9 of the I Hispano-Luso-American Congress of Economic Geology, General-Direction of Mines and Geological Services, Lisbon, 123.

Paralta E (2001) Hydrogeology and stochastic modelling of nitrate contamination in the Gabbro-dioritic aquifer in Beja region [in Portuguese]. Master Thesis, IST, Lisbon, 159.

Ramalho T (2004) Hydrogeological definition of the Vidigueira-Selmes Sector of Évora-Montemor-Cuba Aquifer System [in Portuguese]. Water Resources Engineering Final Thesis, University of Évora, 66.

Ribeiro L (2004) Groundwater in continental Portugal [in Portuguese]. In: Manual de Engenharia Águas Subterrâneas, GRUNDFOS, Lisboa (Portugal) 5–18.

Rodrigues R (2006) Hydrogeological definition of the Escoural Sector of Évora-Montemor-Cuba Aquifer System [in Portuguese]. Water Resources Engineering Final Thesis, University of Évora, in press.

Sankarasubramanian A, Vogel RM (2003) Hydroclimatology of the continental United States, Geophys. Res. Lett., 30(7), 1363, doi:10.1029/2002GL015937.

CHAPTER 5

# Groundwater in volcanic hard rocks

Emilio Custodio
*Dept. Geotechnical Eng., Technical Univ. Catalonia, Barcelona, Spain*

ABSTRACT: Volcanics include a large variety of rocks of magmatic origin with a wide range of chemical composition and emplaced on the Earth's surface as molten or partially molten material. Tephra may dominate near explosive volcanoes, while away from the effusion centres, lavas are the main constituents. Dykes play a variable role, from a barrier to groundwater to a water-collecting element. Fresh lava breccias may be one of the most permeable geological formations, while lava cores and welded tuff behave as some of the less permeable materials. Permeability is fracture dominated and tends to decrease with depth, yet there can be noticeable exceptions. Volcanics can be thermally weathered to a low permeability rock by water, especially if $CO_2$ is plentiful. Volcanic $CO_2$ diffusing from the deep cooling magma chambers is an important chemical reactant that may produce $Na\text{-}HCO_3$ waters.

## 1 INTRODUCTION

Volcanics represent a wide variety of rocks and formations that have taken their common origin from different types of magma poured on the Earth's surface. In addition to these effusive rocks, typically lavas and tephra, they include also intrusive formations, linked to volcanism, such as dykes and sills, and also cooled magma chambers, weathering of sediments and rocks derived from volcanic materials, and complex sedimentary bodies directly associated to volcanic activity. Such a variety results in extremely diverse hydrogeological conditions of these rocks.

Large volcanic formations have arisen at continents and islands over the Earth's long history. These range from flood basalts of low viscosity that cover extended areas (Walker, 1980, Macdougall, 1988, Lin and van Keken, 2005), large fissural effusions that can be traced along up to tens of km, to more acid rocks (andesites, rhyolites) that do not form widespread formations but may be large and thick along long strips, as in the western parts of the Americas and in Japan. Rates of eruption may be up to one $km^3$ of magma per year per volcano and several hundred $km^3$ per series of events (Swanson et al., 1975; Versey and Singh, 1982). Flood basalts correspond to intense events that can be able to produce enormous, thick plateaus of more than $10^6 km^3$ in a few thousand years at rates of up to 0.15 m/year (Sen et al., 2006).

The sea floor may be formed by thick volcanic formations, mostly basalts, resting on a densely injected dyke layer covering intrusive gabbro bodies (Wilson et al., 2006). In some cases, these formations now occur on continents or in islands after being tectonically uplifted (e.g., Cyprus).

There are numerous volcanic islands; most of them consist on submarine volcanics topped by subaerial effusions (e.g., archipelagos of Hawaii, the Canaries and Azores; Tahiti, Iceland, Reunion), where they form some of the highest Earth's mountains, up to some thousand metres from bottom to top. On the continents, volcanic materials pile up on older sediments and formations (Etna in Sicily, many of the Andean and Central America high volcanoes).

This paper comments on general geologic properties of volcanics, with a special focus on their hardrock character, to give a background on their main differential hydrogeological characteristics. It utilizes experience gained in diverse projects and his exchange of knowledge with many professionals and experts, not directly cited herein but to whom the author credits much information. This report addresses volcanic rock hydrogeology in a broad sense but emphasizes the Canary Islands, which represent a varied and extraordinary example of volcanic rock formations in which difficult water supply circumstances has lead to extensive drilling and tunnelling to develop groundwater.

Specific publications on volcanic rock hydrogeology are scarce and fragmental. What was presented in the UNESCO–PNUD sponsored meeting in Lanzarote Island in 1975 is summarized in Custodio (1978; 1987). The UNESCO publication on small islands (Falkland and Custodio, 1991) contains numerous references and comments on volcanic islands. Summary chapters and sections on the topic can be found in some classical textbooks (Davis and De Wiest, 1966; Freeze and Cherry, 1979; Walton, 1970; Custodio and Llamas, 1976; Kovalevsky et al., 2004). General information is not repeated here. Only specific aspects relevant to the understanding of the uniqueness of volcanic rock hydrogeology are analysed.

Regional reports are scarce and most of them include volcanics as one of the existing formations. Good references to the United States are in Back et al. (1988) and just to Central America can be found in Losilla et al. (2001), Krásný (1996) and Krásný and Hecht (1998). References to other parts of the world can be found in the different volumes on groundwater published by the United Nations in the 1980's under the guidance of R. Dijon (UN, 1979–1988). Very detailed studies have been carried out in the Yucca Mountains ignimbrites of Nevada, to evaluate the conditions for high-level radioactive waste disposal (e.g., Bodvarsson and Robinson, 2003).

## 2 VOLCANIC ROCKS: GENETIC ASPECTS OF HYDROGEOLOGICAL RELEVANCE

Magmas may arise from the deep Earth's mantle in convecting hot spots, the upper mantle along rift zones, or crustal material melting when pushed down at great depths due to the tectonic plate convergence along subduction areas. Magmas are complex mixtures formed by a liquid fraction laden with a crystalizing mass containing solids that may be partially separated by gravity to produce geochemical fractionation. The residual liquid may be progressively enriched in incompatible elements. The incorporation of sediments into the magma may increase the water and $CO_2$ content and may produce less warm, more viscous, explosive magmas near the land surface.

A wide variety of magmas and volcanic formations can be expected, even if coetaneous. Each volcanic formation is unique, although broad classes can be defined: hot-spot volcanism (Hawaii islands, the Canaries, Azores, Etna), plate ridge volcanism (Iceland, Kilimandjaro, submarine mid ocean ridges) and island–arc volcanism (Aegean islands, Japan, Java, the Andes, the Rocky Mountains, the two Sierra Madre in Mexico and California, Central America). Large flood basalts are probably due to hot-spot volcanism. Island arc volcanism is more prone to produce viscous lavas and large quantities of tephra.

There is a wide range of emplacement modes from the calmed pouring of molten rock (lavas) to highly explosive events that mostly produce fragments (pyroclasts or tephra) thrown into the air or into the water. Viscosity of the molten material and dissolved gas content (mostly water and carbon dioxide) play a key role in the eruption behaviour. Gases are released as pressure decreases. Their escape may be calm, producing scarce lava fragments, or may produce episodic violent explosions in high viscosity, almost solid effusions, in which most of the magma is fragmented and thrown away. Fragments (tephra) may vary from blocks and coarse pyroclasts, that remain and pile up close to the emission point, down to very fine dust particles that may be distributed over very large areas and even the whole Earth (Bindeman, 2006). The sudden and fast expansion of gas-rich magma may produce a fluidized, hot mass formed by gas, dust and particles that is able to flow on the land like lavas of very low viscosity and may cover large surface areas with tuff deposits (ash flows or ignimbrites). Ignimbrites may be intensively or partly welded due to the hot conditions during deposition, or remain more or less loose, with abundant vitreous particulate material. When ash-sized material cools before being deposited they form ash fall tuff, which is primarily nonwelded.

Volcanic formations can be thin layers on the Earth surface that may be later incorporated on into sediments, or up to hundreds and even thousands of metres thick. In volcanic islands there is a large piling of submarine volcanics, topped by subaerial formations that contain lava flows and water-quenched fragmental material (hyaloclastites or aquagene tuff). They can be observed in areas with large upward vertical crustal movements, as in Troodos Mountains, Cyprus (Boronina et al., 2003).

The extrusion of large magma volumes may be accompanied by subsidence if deep magma is not replaced; collapse calderas may be produced. The subsiding or collapsed areas and erosion gullies may be later on filled by reactivated effusions of magma and derived sediments. The ascending magma produces intrusive bodies and a dense swarm of dykes corresponding to the volcanoes' feeders. These often concentrate along the rings limiting collapse calderas or defined rift zones. In hot-spot volcanoes rifting may be radial with 3 or 4 branches.

Near the effusion centers, the structure of volcanic bodies is complex. Vertical dykes are abundant, and also large horizontal dykes (sills) may be found. Where formations are influenced by hot fluids derived from the magma, local rocks, including former volcanics, may be deeply altered and their properties intensively changed. Thermal convection of hot fluids may deeply alter the rock mass. With increasing distance from the effusion centers, complexity decreases in general and materials change into a less heterogeneous piling of lava flows or ignimbrites in which fall fragments proportion decreases, dykes are less frequent or absent, and the alteration effect of deep hot gases is small or nil.

Volcanic formations may be extended in the flow direction but laterally limited. In areas of slow flow or lava ponding a central core of dense but sometimes vesicular rock is typically found with cooling joints and top and bottom brecciated zones due to cooling and drag (at the bottom), with vesicles of gas, especially near the top. The upper part may be smooth or reduced to a thin layer of breccia (pahoehoe lavas), but other times it may be highly brecciated (aa lavas), especially near the lava front. In warm humid climates, weathering produces a lateritic cover. This is common in India (Uhl and Joshi, 1986) and Brazil, but in the Canaries and Madeira and also in Andean and Central America, chemical leaching is not enough and dark-brown, iron oxy-hydroxide and kaolinite-illite rich soils typically occur (Tejedor Salguero et al., 1985; Van der Weijden and Pacheco, 2003; Deutsch et al., 1982).

Occasionally a lava flow may keep a hot, partly degassed fluid core after the external part solidifies. A sudden rupture of the crust allows the fast emptying of the molten lava core to form a new lava flow, leaving a tunnel that may partly stand for some time. These are mostly ephemeral features since they collapse soon. These "lava tubes", as well as other features such as cavities created by overtopping slabs of solid lava crusts or burned tree trunks have being compared to caves in karst. However, they have no similarity to karstic features, and certainly they do not behave like them; these local heterogeneities lack a network of conduits. They are often dry, although they may concentrate groundwater flow and discharge under suitable, humid conditions.

## 3 GROUNDWATER FLOW IN VOLCANIC HARD ROCKS

From a hydrogeological point of view, in most cases volcanic rocks are a heterogeneous and anisotropic environment. They possess significant porosity except for some vitreous and densely welded formations. Volcanics can be modelled as more or less permeable blocks separated by a network of conductive features such as fractures, coarse layers, and even sedimentary interlayers. The result may vary from a formation dominated by fracture permeability, as in dense lavas or densely welded ignimbrites, to porous-like formations as in fresh, coarse ash-fall formations (Smyth and Sharp 2006). In large enough volumes, volcanics may approach the behaviour of a continuous, anisotropic, relatively high porosity medium. In some volcanic formations, stratification may be conspicuous with changing thickness and preferential orientation according to the flow direction. The result is a macroscopic permeability anisotropy that in detail depends on features such as open fractures, breccia layers, interflow sediments, dykes, and faulting effects.

The simple Jacob logarithmic method is the most commonly used interpretation tool for pumping tests with reasonable results for heterogeneous formations (Meier et al., 1998; Sánchez–Vila et al., 1999), combined with corrections for well capacity in case of large diameter wells (Cabrera et al., 2001; Sammel, 1974). Upscaling hydraulic properties to the volcanic formation cannot be done directly nor can they be downscaled to a given formation. Detailed scale permeability may vary by more than 7 orders of magnitude from one of the lowest known (densely welded ignimbrites) to one of the highest (recent lava flow breccias). Both porosity and permeability may be greatly changed by rock fracturing and especially by alteration. Alteration tend to smooth variations and to dramatically decrease hydraulic conductivity in zones that have been subject to hydrothermal fluids and magma injections.

The role of dykes ranges from low-permeability barriers to horizontal groundwater flow to a highly permeable features. This depends upon the characteristics of the wall rock, the importance of fracturing (van Everdingen, 1995), and the degree of mineral infilling of the associated fractures (Boonstra and Boehmer, 1986).

A compilation of data on hydraulic properties of volcanic hard rocks is not presented here, but some typical examples are instructive.

The Deccan traps of Western India are one of the extensively studied areas (Athavale et al., 1983; Deolankar, 1980). Regional transmissivity is about 3 $m^2$/day but may locally be to 15 $m^2$/day, and up to 500 $m^2$/day in exceptional cases for blocky and broken lava (Versey and Singh, 1982). Best conditions for flow are found in some of the vesicular and weathered layers separating the different lava flows. Boreholes often penetrate up to 10 lava flows and only 1 out of 6 interlayers yield exploitable water. Although productivity generally decreases with depth, good conditions are found deeper than 60 m (Uhl, 1979). Secondary fracturing seems to play a role in increasing transmissivity since dry wells are less frequent in valley bottoms then in flat uplands.

The younger (Pliocene–Pliocene) but similar to Deccan traps volcanics the north-western USA are an order of magnitude more productive. The widespread formations like the Paraná (Serra Geral) basalts in eastern part of South America behave as an aquitard with respect the underlying Botucatú sandstone formation of the Guaraní aquifer (Sarres Pessõa, 1982), but still is able to sustain flows averaging 2 L/s, sometimes decreasing to less than 0.6 L/s.

Large acidic volcanics are found along the Mexican transvolcanic belt with deep layers affected by brittle tectonics. Transmissivity may vary between 200 and 800 $m^2$/day for andesites with permeabilities of 0.03 to 1.0 m/day for unfractured and 0.8 to 40 m/day for fractured rocks (Carrcón–Frcyrc ct al., 2005).

Ignimbrites tend to be clearly less permeable volcanics. Data from boreholes in Yucca Mountain, Nevada (Hinds et al., 1999) show permeability values of $10^{-3}$ to $10^{-7}$ m/day that increase to $10^{-5}$ to $10^{-7}$ m/day for welded tuff. But fracturing greatly increases these values when considered at the massif scale.

In Central America extensive and varied volcanic formations are found, mostly from Tertiary to Recent (Losilla et al., 2001, Krásný 1996, Krásný and Hecht, 1998). They vary from basalts to andesites and even to rhyolites associated with large pyroclastic material deposits, both ash-fall and ash-flow, and frequent associated flows (lahars) and alluvium. They are up to several hundred metres thick. Permeability is primarily high in recent volcanics and decreases with depth. Transmissivity values can range up to 5000 $m^2$/day but vary areally and may be less than 50 $m^2$/day. The best conditions for water yield seem to be in thick, saturated, recent fall pyroclasts, but also recent lavas with brecciated tops.

Similar conditions are found in volcanic islands. Young volcanics may be highly permeable, as in the Hawaii islands: a flowing well yielded 180 L/s in confined young basalts in coastal Oahu, and water galleries 320 L/s in Maui and up to 200 L/s in Oahu. In Madeira Island some galleries (levadas) yield more than 100 L/s and some of them are used to produce hydroelectricity. In Tenerife (Canarian archipelago) one of the numerous water galleries produced 150 L/s after being extended. In Lanzarote island (Canarian archipelago) coastal recent volcanics allow pumping 500 L/s with a drawdown of some decimetres.

Pyroclastic material, except for very young to sub-recent deposits, is generally less permeable than young brecciated lavas.

Older, altered volcanics, unearthed by erosion, are generally of low to very low permeability, especially when they are thermally altered. Generally little is known about their hydraulic characteristics because commonly they are in high, steep, sparsely inhabited areas and in humid environments, except in the Canaries. The springs in them reflect the discharge of younger volcanics on the top of the fractured upper part of them, as in Hawaii and Reunion. In the densely populated Canary Islands (especially Gran Canaria, Tenerife and La Palma) deep exploration by means of galleries, deep shafts and boreholes, with thousands of km of penetration, allow a detailed observation of deep volcanic rocks. Permeability, although small, is still noticeable at 1000 m below land surface. Examples of

Table 1. Permeability values of volcanics in the Canary Islands.

| Island | Formation | m/day |
|---|---|---|
| **Tenerife (1)** | | |
| | Miocene (old) basalts | |
| | – close to the structural axis | 0.02 |
| | – far from the structural axis | |
| | * deep seated | 0.02 |
| | * shallow | 0.05 |
| | Plio–Quaternary (modern) basalt | |
| | – in general | 0.75 |
| | – shallow, in some areas | 2 |
| **Gran Canaria (2)** | | |
| | Extracaldera, Miocene (old) basalts | |
| | – "a–a" type | 0.01–0.04 |
| | – "pahoe–hoe" type | 0.1–1 |
| | Trachy–syenites and trachytes | 0.01–0.04 |
| | Phonolites and ignimbrites | 0.06–0.09 |
| | Phonolites of Telde area | 0.1–0.2 (up to 5) |
| | Phonolites of Amurga massif (selected wells) | 0.5–5 |
| | Caldera–rim materials | <0.002 |
| | "Roque Nublo" volcanic conglomerates | 0.15–0.45 |
| | Plio–Quaternary (recent) basalts, Telde area | 25 (up to 100) |
| | Quaternary (young) basalts | >1 |
| **Lanzarote (3)** | | |
| | Famara massif Miocene basalts | 0.05–0.02 |

(1) Modified from Custodio (1985). Adjusted by means of a numerical model
(2) Modified from Custodio (1985), Cabrera et al. (2001), Cabrera and Custodio (2003)
(3) Modified from Custodio (1985)

permeability values for volcanic materials in the Canarian archipelago are given in Table 1 to show a wide range of volcanic materials in this area.

Permeability of ophiolite complexes and crustal basalts is low, but it may be greatly changed by fracturing. About one third of the boreholes in old pillow lavas, sheeted dykes and gabbros are successful to yield water in Cyprus (Boronina et al., 2003), especially in the surroundings of fracture zones. Macroscopic porosity is about 5 ± 1% but varies from about 1.5 ± 0.5% for flows to 27 ± 3% for pillow lavas, or reaches 15 ± 2% if they belong to altered zone close to the sea bottom (Gillis and Sapp, 1977). Densely dyke-injected rocks properties are quite different from the original rock and may approach that of a densely fractured rock. Cold freshwater permeability of basaltic oceanic crust may be $10^{-9}$ to $10^{-4}$ m/day for the core of lava flows, and $10^{-5}$ to 1 m/day as determined by hydraulic tests in boreholes (Fisher, 1998; Manning and Ingebritsen, 1999) but can reach $10^{-3}$ to 100 m/day in highly fractured zones.

## 4 HYDROGEOLOGICAL IMPORTANCE OF VOLCANIC ROCK ARRANGEMENT

Properties of volcanic rocks vary with their position relative to the emission point or line. Distal formations tend to be free of pyroclasts and dykes, and are dominated by successions of lava flows with top and bottom breccias and intercalated sediments, fine ashes, and soils that may control groundwater flow. This occurs in large areas of flood basalts and at the periphery of large basaltic shield volcanoes and results in stratification of various hydrogeological bodies: aquifers and aquitards. Permeability typically decreases with depth due to alteration and compaction of scoriaceous layers.

Close to volcanic vents and outflow fissures structure becomes more complex due to the abundance of pyroclastic material that may dominate in zones of explosive volcanism. Viscous acidic lavas tend to form mounds and thick bodies, yet in many cases they became ignimbrites that may cover large areas and be very thick. They may show a regular stratigraphy, broken by faults, as in the Yucca Mountain, Nevada. In the eruption areas increased presence of dykes, intrusions, and thermally altered materials can be followed. This is often the area of highest original elevation and, consequently, is often subject to intense erosion. The permeability is variable, from relatively high in lavas and young pyroclastic deposits to low and even very low in thermally metamorphosed rocks.

The above also applies to volcanic islands. When these volcanic islands are the result of a hot spot a submarine mound is constructed, which often represents the largest part of the magmatic material, and forms a large shield volcano that finally may appear at the sea surface (Hildenbrand et al., 2005). Magma chemistry evolves from basalts to more acidic forms due to magma fractionation and contamination by crustal materials. Volcanic episodes become more violent, form abundant brecciated deposits and eventually large collapse calderas. Materials close to vents are highly altered by thermal and hydrothermal processes. The caldera collapse and large-scale landslides may trigger a reactivation of volcanic events with the formation of a second shield on the old one and the deposits may fill the caldera and even cover it. Around the caldera rim there is thermal metamorphism. This ends the eruptive period if new magma from the deep mantle is not added, and erosion progressively dismantles the island. Relatively minor volcanic events fed by remaining deep molten magma pockets may follow, although the results may still be spectacular by human standards, and of hydrogeological significance, as the six years of continuous activity in the "extinguished" Lanzarote Island (Canary Islands) in the 18th Century.

The hot spot may be more or less geographically stable with successive large events in the same place, or change its relative position when the Earth's tectonic plates move. In this case the successive supply of deep magma starts a new shield volcano at some distance. This is a well-known situation in Hawaii, Society, and Canarian archipelagos with old (drowned or uplifted) islands on one side and young islands on the other. Oahu is older than Hawaii, Reunion is older than Mauricius, and Lanzarote, Fuerteventura and Gran Canaria are older than La Palma and El Hierro. This migration can also be evident on a given island as in Reunion (Pic des Neiges relative to Piton de La Fourneuse) and La Palma (Cumbre Vieja relative to Cumbre Nueva). This has hydrogeological implications. The erosion of the older parts may expose the "core".

Vertical movements due to magmatism and local tectonics can also be important. In Hawaii Island subaerial volcanics were found down to 1800 m deep (García and Davis, 2001) or close to 1000 m in Lanzarote (Canary Islands), while in Fuerteventura and in Cyprus submarine lavas and oceanic crust crop out.

Often, in volcanic islands there is a core of low permeability volcanics consisting of the thermally-altered parts containing dykes and intrusions with possible shallow volcanic chambers, where the emission of volcanic gases from the cooling and consolidating deep magma chambers may continue. Surrounding the central core there is an apron of lava flows and ignimbrites that may include landslides and laharic deposits, which constitute the largest surface area and may contain quite permeable materials. This pattern is what appears in the Canary Islands but also in the Hawaiian Islands, even if the interpretation given by some authors (see Liu et al., 1983) differentiates between basal aquifers (in the apron) and dyke-impounded aquifers (in the core). Actually there is only one volcanic groundwater body with largely contrasting permeabilities. In the apron, the water table (piezometric surface) is typically close to the land surface and with a small hydraulic gradient. In coastal areas saline waters can underlie fresh-water bodies even if the latter are confined below recent coastal and offshore sediments, as in Pearl Harbor, Oahu. Low permeability in the core produces very high hydraulic gradients, especially if recharge is high. This creates springs in the valley bottoms and at the foot of cliffs. High elevation, perched springs may occur due to high recharge on local, low permeable heterogeneities. This also happens in continental basalts such as the Columbia Plateau. However, many of these springs are probably outflows from the high main groundwater body in the core through permeable layers or fissured upper part.

When volcanics overlay other pre-existing formations as in the case of the Etna in Sicily (D'Alessandro et al., 2004), the bedrock may play the role of the low permeability "core" or, on the other hand, behave as a drain, depending on relative hydrogeological properties. Thus, volcanics can behave as conductive and storage units or just as recharge collectors.

Along the coast freshwater–saltwater relationships follow common behaviour in other coastal areas (Custodio and Llamas, 1976; Custodio and Bruggeman, 1985). The coastal platform is often steep in volcanic islands, and the materials there are transmissive, recent ones. This means an inland penetrating seawater wedge and a floating layer of freshwater with a mixing zone from thin to thick (Thomas et al., 1996), depending on groundwater flow, heterogeneity, recharge fluctuations, and tidal effects. Vertical changes in volcanics and sedimentary formations that may develop there may produce conditions for confined flow, as is well known in Pearl Harbor, Oahu, Hawaii islands. When the core area is at the coast, the floating freshwater body on sea water is less developed as happens in the Canarian and Cape Verde islands and also in the older of the Hawaiian islands (Izuka and Gingerich, 2003). The extent, nature and elevation of the core relative to mean sea level and the apron play a key role in explaining the conspicuous differences among the different volcanic islands. Along the coastal areas there is often a thin layer of ground freshwater that becomes easily contaminated by marine water from below, as in Easter Island, Chile (Herrera et al., 2004).

Real situations in basalt-dominated islands are often more complex due to the existence of different volcanic centres and/or rift zones, of the same or different age (tree main volcanoes in Hawaii island, the north and south volcanoes of Reunion and La Palma islands) or recent volcanic stages covering older, deeply eroded previous ones (Gran Canaria and Tenerife islands). The core may include volcanic calderas, partly or totally filled with permeable lavas, as in Piton de la Fourneuse (Reunion), Las Cañadas (Tenerife), or the intra-caldera formations of Gran Canaria. The flow pattern may be changed by large landslides that are common in high islands (Carracedo 1999; Carracedo and Tilling, 2003). Groundwater in the core area is subject to high hydraulic gradients and important vertical head components. The deep submarine volcanics of many islands are of low but not nil permeability. They may still contain permeable features, as found in Hawaii (De Paolo et al., 2001). The general behaviour is little known.

## 5 GROUNDWATER RECHARGE IN VOLCANIC ROCKS

Groundwater recharge conditions in volcanic formations are not unique and can be compared with other formations (see Simmers, 1997) and depend on climate, soil, and landslope. In the bare or slightly vegetated land of recent volcanics, infiltration of rain and snow melt can be important, although when very permeable there is enhanced air convection that increase evaporative loses. In low permeability volcanics fracturing and faulting may play an important enhancing role, especially if they are covered by a cap rock able to store water.

Groundwater flow in the unsaturated zone is partly through fractures, and the space and time distribution may be quite complex (Faybishenko et al., 1991). How water penetrates is important, at least in the upper part of the unsaturated zone, where inflow may concentrate in fractures. Fractures become active for conducting water when inflow to them (e.g., from the surface or from the top soil) is greater than the inhibition capacity of the walls. This capacity is very low when the matrix is almost impermeable or is close to saturation. Otherwise most of the recharge flow is through the rock matrix. This is what seems to happen in volcanic rocks in many cases. Very detailed studies have been carried out in Nevada, in the arid Yucca Mountain (Fridrich et al., 1994; Wu et al., 2002; Finsterle et al., 2002; Flint et al., 2001; Sonnenthal and Bodvarsson, 1999; Bodvarsson et al., 2003) to study the feasibility of constructing a deep nuclear waste repository in the thick unsaturated zone of Miocene, densely fractured ignimbrites which are formed by a sequence of welded and non-welded ash-flow layers. Some faults with offsets from 10 metres to some hundred metres go across the whole formation and have a decisive influence in the unsaturated zone water movement. They may be barriers or conductive features, depending on the circumstances. The formations of limited extent, perched saturated levels on top of low permeability vitreous layers favour the down-flow through fault planes in wet periods. Faults are conductive when they cross welded tuff, but not in the non-welded parts (Hinds et al., 1999).

## 6 SPECIFIC HYDROGEOCHEMICAL ISSUES

General principles of hydrogeochemistry and environmental isotopes in volcanic hard rock are the same applicable to other formations (Appelo and Postma, 1993; Custodio and Llamas, 1976; Mazor, 1991; Mook 2001). The high relief of many volcanic areas favours intense erosion resulting in outcrops of fresh rock. Altitude-dependent isotope changes can often be followed there (Scholl et al., 1996; Custodio and Custodio, 2001).

Yet, volcanic rocks are often more chemically reactive than other hard rocks because of the presence of fine particles, large specific surfaces, and abundance of vitreous matter. Generally volcanic rocks change conspicuously with age for the most compact rocks when they have not been subject to high temperature environmental conditions. Groundwater flow enhances the weathering rate, which is faster the higher the temperature is. Rock weathering may be important in a few years in wet, tropical climate, especially for calcalkaline rocks.

Groundwater in volcanic rocks has typically a low content of chloride, sulphate, and other solutes, even if of marine origin. Chloride can be used to calculate recharge if rainfall contribution is known (Gasparini et al., 1990; Custodio, 1992; Alcalá, 2006). However, old marine water may be trapped in raised low permeability formations (Herrera and Custodio, 2002) or some chloride may be contributed in areas receiving deep warm gases.

High temperature alteration in zones subject to convective hot fluids produce new minerals and voids, and fractures may be totally or partially filled. Water and carbon dioxide, as well as other components, play a dominant role in these changes. Under low temperature conditions clay minerals are formed, especially if $CO_2$ from deep formations or from decaying vegetation help in lowering the otherwise high pH of water. Smectite is formed, or kaolinite if $CO_2$ availability is high. The rock frees $Na^+$ that mostly dissolves in the water, and part of the $K^+$ and $Mg^{++}$, depending on the type of clay that is formed. Na-$HCO_3$ waters are often found. $Ca^{++}$ may pass to the water or be precipitated as carbonate if pH is high, thus filling voids and fractures, and forming duricrusts (caliches) near the surface in dry climates, as in many dry areas of the Canarian and Cape Verde archipelagos.

Submarine volcanics may have trapped marine sulphate after being reduced and precipitated as pyrite. Also they may contain sulphate and sulphur from disproportionation of volcanic $SO_2$ (Embley, 2006). When subjected to subaerial weathering they release relatively high concentrations of $SO_4$.

Acid volcanism from very evolved residual magmas may concentrate incompatible heavy elements. When they are incorporated into the magma, they can be widely dispersed as part of the ash fall during the frequent highly explosive events of such type of volcanism. This may explain the association of rapidly weathered glassy ashes incorporated in sediments, producing poor groundwater quality, since ashes may be able to release relatively high concentrations of As and V, as it happens in large areas of the plains (pampas) of Argentina, or of F.

Except for the small quantities of trapped $CO_2$ in voids, volcanic rocks are carbon free, so they do contribute little to dissolved carbon. Thus, the recharge $^{13}C$ and $^{14}C$ contents are not affected. However, in areas with high flow of endogenic $CO_2$ or sluggish groundwater flow this volcanic carbon may influence C isotopic values, and $^{14}C$ dating and interpretation of $^{13}C$ values may be difficult, even more if calcite is present in fracture infilling. This may be a serious handicap for dating long turnover groundwater reservoirs. For short turnover cases, tritium has been a good tool, and probably will be $^{85}Kr$, $CFC_S$ and $SF_6$, but existing experience is limited.

## 7 CONCLUSIONS

Volcanics are partly hard rocks in which hydrogeological behaviour ranges from a typical fractured hard rock to a heterogenous anisotropic close-to-porous rock, depending on the genesis, degree of alteration, dyke density, effect of magmatic intrusions, presence of palaeosoils and intercalated sediments, and other features. They can form thick sequences of volcanic sediments that may preserve primary and secondary porosity and permeable features at great depth. This helps in producing potentially thick 3-D regional groundwater flow systems.

There can be large differences in mean (prevailing) regional permeability between volcanics far from and close to effusion centres. In the effusion centres low permeability material may dominate, especially if erosion has exhumed the more altered, deeper parts. This means high piezometric gradients and in wet areas the existence of high elevation springs from the unsaturated zone.

General hydrogeological, hydrogeochemical and isotopic principles apply with the specificity of the often high relief and intense erosive effects, and the high chemical reactivity of volcanics if leaching is intense and the supply of soil $CO_2$ is abundant. In some areas deep-originated $CO_2$ may play an important role. Na-$HCO_3$ waters are frequently found.

## ACKNOWLEDGEMENTS

Thanks to the comments received by Dr. Jiri Krásný and Dr. Peter Seiler and the improvements suggested by Dr. John Sharp.

## REFERENCES

Alcalá FJ (2006). Recarga a los acuíferos españoles mediante balance hidrogeoquímico. Doc. Thesis. Technical University of Catalonia. Barcelona.

Appelo CAJ, Postma D (1993). Geochemistry, groundwater and pollution. Balkema: 1–536.

Athawale RN, Chand R, Rangarajan, R (1983). Groundwater recharge estimates for two basins in the Deccan Trap basalt formation. Hydrol Sci J 28 (4); 525–538.

Back W, Rosenshein JS, Seaber PR (1988). The geology of North America; hydrogeology. Vol. 0–2. Geol. Soc. Am.; 1–524.

Bindeman IN (2006). The secrets of supervolcanoes. Sci. Am., June 2006: 26–33.

Bodvarsson GS, Ho CD, Robinson BA (eds) (2003). Yucca Mountain Project. J Contam Hydrol 62–63:1–750.

Boonstra J, Boehmer WK (1986). Analysis of data from aquifer and well tests in intrusive rocks. J. Hydrology 88: 301–317.

Boronina A, Renard P, Balderer W, Christodoulides A (2003). Groundwater resources in the Kouris catchment (Cyprus): data analysis and numerical modelling. Hydrol J 271: 130–149.

Cabrera MC, Custodio J, Custodio E (2001). Interpretación de ensayos de bombeo en el acuífero de Telde (Gran Canaria). En Las Caras del Agua Subterránea (Medina & Carrera eds). IGME Madrid. II: 609–614.

Cabrera MC, Custodio E (2003) Groundwater flow in a volcanic sedimentary coastal aquifer: Telde area, Gran Canaria, Canary Islands Spain. Hydrogeol Journal, 12: 305–320.

Carracedo JC (1999). Growth structure and collapse of Canarian volcanoes and comparisons with Hawaiian volcanoes. J Volc and Geotherm Research Special Issue 94 (1–4): 1–19.

Carracedo JC, Tilling RI (2003). Geología y volcanología de Islas volcánicas oceánicas. Canarias–Hawaii. Ser. Publ. Caja General de Ahorros de Canarias 293 (Varios 15): 1–73.

Carreón–Freyre J, Cerca M, Luna–González L, Gómez–González FJ (2005). Influencia de la estratigrafía y estructura geológica en el flujo de agua subterranean del Valle de Querétaro. Rev. Mex. Ciencias Geolog. 28(1): 1–18.

Custodio E (1978). Geohidrología de terrenos e islas volcánicas. CEDEX Madrid Publ 128: 1–303.

Custodio E (1985). Low permeability volcanics in the Canary Islands (Spain). In Hydrogeology of Rocks of Low Permeability. Mem. IAH, XVIII: 533–544.

Custodio E (1989). Groundwater characteristics and problems in volcanic rock terreins. In Isotope Techniques in the Study of the Hydrology of Fractured and Fissured Rocks. Intern Atomic Energy Agency, Vienna: 87–137.

Custodio E (1992). Coastal aquifer salinization as a consequence of aridity: the case of Amurga phonolitic massif, Gran Canaria Island. Study and Modelling of Salt Water Intrussion. CIMME–UPC. Barcelona: 81–98.

Custodio E, Bruggeman GA (1987). Groundwater problems is coastal areas. Studies and Reports in Hydrology no. 45, UNESCO, Paris: 1–576.

Custodio J, Custodio E (2001). Hidrogeoquímica del macizo fonolítico de Amurga (SE de la Isla de Gran Canaria). In Las Caras del Agua Subterránea (Medina & Carrera, eds). IGME, Madrid, II: 461–468.

Custodio E, Llamas MR (1976). Hidrología subterránea. Ed Omega Barcelona 2 Vols: 1–2350.

Davis SN, de Wiest JRM (1966). Hydrogeology. John Wiley, N.Y.: 1–520.

Deolankar SB (1980). The Deccan basalts of Maharashtra, India: their potential as aquifers. Ground Water: 434–437.

Deutsch WJ, Jenne EA, Krupka KU (1982). Solubility equilibria in basalt aquifers: the Columbia Plateau, Eastern Washington, USA. Chemical Geology, 36: 15–34.

D'Alessandro W, Federico C, Longo M, Parello F. (2004). Oxygen isotope composition of natural waters in the Etna area. J. Hydrol., 296: 282–289.

De Paolo DJ, Stolper E, Thomas, DM (2001). Deep drilling into a Hawaiian volcano. EOS 821 (13): 149–155.

Embley RW, Chadwich NW Jr, Becker ET, et al. (2006). Long–term eruptive activity at a submarine volcano. Nature, 441: 444–447.

Everdingen DA, van (1995). Fracture characteristics of the sheeted dike complex, Troodos Ophiolite, Cyprus: implications for permeability of oceanic crust. J Geophys Res 100 (B10): 19957–19972.

Falkland A, Custodio E (1991). Guide on the hydrology of small islands. Studies and Reports in Hydrology no. 49. UNESCO, París: 1–435.

Faybishenko B, Doughty C, Steiger M, Long JCS, Wood TR, Jacobsen JS, Lore J, Zawislanski PT (2000). Conceptual model of the geometry and physics of water flow in a fractured basalt vadose zone. Water Resour Res 36 (12): 3499–3520.

Finsterle S, Fabryka–Martin JT, Wang JS 4 (2002). Migration of a water pulse through fractured porous media. J Contaminant Hydrology, 54: 37–57.

Fisher AT (1998). Permeability within basaltic ocean crust. Reviews of Geophysics, 36 (2): 143–182.

Flint A, Flint LE, Bodvarsson GS, Kwicklis EM, Fabryka–Martin J (2001). Evolution of the conceptual model of unsaturated zone hydrology at Yucca Mountain, Nevada. J. Hydrology, 247: 1–30.

Freeze RA, Cherry JA (1979). Groundwater. Prentice Hall: 1–604.

Fridrich CJ, Dudley WW Jr, Stuckless JS (1994). Hydrogeological analysis of the saturated–zone ground–water system, under Yucca Mountain, Nevada J. Hydrology, 154: 133–168.

García MO, Davis MG (2001). Submarine growth and internal structure of ocean island volcanoes based on submarine observations of Mauna Lao volcano, Hawaii. Geology. Geological Soc. of America, 29(2): 163–166.

Gaspanini A, Custodio E, Fontes JCh, Jiménez J, Nuñez JA (1990). Example d'étude géochinique et isotopique de circulations aquifères en terrein volcanique sous climat semi–aride (Amurga, Gran Canaria, îles Canaries). J. Hydrology, 144: 61–91.

Gillis KM, Sapp K (1997). Distribution of porosity in a section of upper oceanic crust exposed in the Troodos ophiolite. J Geophys Res 102 (B5):10133–10149.

Herrera Ch, Custodio E (2002). Old marine water in Fuerteventura island deep formations. Proc 17th Salt Water Intrusion Meeting, Delft University of Technology, Fac Civil Eng. and Geosciences: 481–488.

Herrera Ch, Pincheira M, Custodio E, Araguás L, Velasco G (2004). El contenido en tritio de las aguas subterráneas de la Isla de Pascua, como una herramienta para calcular la recarga al acuífero volcánico. Bol. Geol. Min., Madrid, 115(esp): 299–310.

Hildenbrand A, Marlin Cln, Conroy A, Gillet PY, Filly A, Massault M (2005). Isotopic approach of rainfall and groundwater simulation in the volcanic structure of Tahití–Nui (French Polynesia). J. Hydrol., 302: 187–208.

Hinds JJ, Ge Sh, Fridrich Ch J (1999). Numerical modelling of perched water under Yucca Mountain, Nevada. Ground Water 37 (4): 498–504.

Izuka SK, Gingerich SB (2003). A thick lens of fresh groundwater in the southern Lihue Basin, Kauasi, Hawaii, USA. Hydrogeology Journal, 11:240–248.

Kovalevsky VS, Kruseman GP, Rushton KR (2004). Groundwater studies: an international guide for hydrogeological investigations. UNESCO–IHP–VI Series on Groundwater, 3, Paris. There is a chapter on Hydrogeology of volcanic rocks (E. Custodio): 395–425.

Krásný J (1996). El mapa hidrogeológico de la Zona Pacífica de Nicaragua. Asoc. Latino–Americana de Hidrología Subterránea para el Desarrollo, ALHSUD–Congreso México, 1996: 125–140.

Krásný J, Hecht G (1998). Estudios hidrogeológicos e hidrogeoquímicos de la Región del Pacífico de Nicaragua. INETER, Managua: 1–154 + An.
Lin SC, van Keken PE (2005). Multiple volcanic episodes of flood basalts caused by thermochemical plumes. Nature, 436: 250–252.
Liu CCK, Lau S, Mink JF (1983). Ground-water model for a thick fresh–water lens. Ground Water 21 (3): 293–300.
Losilla M, Rodríguez H, Schosinski G, Stimson J, Bethune D (2001). Los acuíferos volcánicos y el desarrollo sostenible en América Central. Editorial de la Universidad de Costa Rica. San José: 1–205.
Macdougall JD (ed) (1988). Continental flood basalts. Kluwer, Dordrecht.: 1–341.
Manning CE, Ingebitsen SE (1999). Permeability of the continental crust: implications of geothermal data and metamorphic systems. Reviews of Geophysiscs, 37 (1): 127–150.
Mazor E (1991). Applied chemical and isotopic groundwater hydrology. Helsted Press (J. Wiley): 1–274.
Meier PM, Carrera J, Sánchez–Vila X (1998). An evaluation of Jacob's method for the interpretation of pumping tests in heterogeneous formations. Water Resources Research, 34 (5): 1011–1025.
Mook WG (2001). Environmental isotopes in the hydrological cycle: principles and applications. UNESCO–IAEA, Paris (6 Vols.). Also Isotopes ambientales en el ciclo hidrológico: principios y aplicaciones. Instituto Geológico y Minero de España. Madrid: 1–596.
Sammel EA (1974). Aquifer tests in large diameter wells in India. Ground Water, 12: 265–272.
Sánchez–Vila X, Meier PM, Carrera J (1999). Pumping tests in heterogeneous aquifers: an analytical study of what can be obtained from their interpretation using Jacob's method. Water Resources Research, 35 (4): 943–952.
Sarres Pessõa M (1982). Banco de dados hidrogeológicos e análise estadística da vação dos poços do Estado do Rio Grande do Sul. Univ Fed Rio Grande do Sul, Inst Pesquisas Hidráulicas. M Thesis: 1–193.
Scholl MA, Ingebritsen SE, Janik CJ, Kauahikaua JF (1996). Use of precipitation and groundwater isotopes to interpret regional hydrology on a tropical volcanic island: Kilauea volcano area, Hawaii. Water Resources Research, 32(12): 3525–3537.
Sen G, Borges M, Marsh BD (2006). A case for short duration of Deccan trap eruption. EOS, 87(20): 197–200.
Simmers I (1997). Recharge of phreatic aquifers in (semi–) arid areas. Intern. Assoc. Hydrogeologists 19. Balkema, Rotterdam: 1–277.
Smyth RC, and Sharp JM, Jr. (2006) Hydrologic properties of tuff. In: Heiken G. (ed.) Tuffs – their properties, uses, hydrology, and resources. Geol. Soc. America, Special Paper 408 (in press).
Sonnenthal EL, Bodvarsson GS (1999). Constraints on the hydrology of the insaturated zone at Yucca Mountain, NV from three–dimensional models of chloride and strontium geochemistry. J Contam Hydrol 38: 107–156.
Swanson DA, Wright TL, Helz RT (1975). Linear vent systems (and estimated rates of magna production and eruption) for the Yakima Basalt of the Columbia Plateau Am J Sci, 275: 877–905.
Tejedor Salguero ML, Jiménez Mendoza C, Rodríguez Rodríguez A, Fernández Caldas E (1985). Polygenesis on deeply weathered Pliocene basalt, Gomera (Canary Islands) from ferrallitization to salinization. Catena Supplement 7. Volcanic Soils (E. Fernández Caldas & DH Yaalon, eds). Braunschweig: 131–151.
Thomas DM, Paillet FL, Conrad ME (1996). Hydrogeology of the Hawaii Scientific Drilling Project borehole KP–1, 2; groundwater geochemistry and regional flow patterns. J. Geophys. Res. 101(B5): 11683–11694.
Uhl VW Jr (1979). Occurrence of groundwater in the Satpura Hills region of Central India. J Hydrol 41: 123–141.
Uhl VW, Joshi VG (1986). Results of pumping tests in the Deccan trap basalts of Central India. J Hydrol, 86: 147–168.
UN (1979–1988). Natural Resources: Water Series (1 through 19). Dept. for Technical Cooperation Development. New York.
Van der Weijden CH, Pacheco FAL (2003). Hydrochemistry, weathering and weathering rate in Madeira island. J. Hydrol., 283: 122–145.
Versey HR, Singh BK (1982). Groundwater in Deccan basalts of the Betwa basin, India. J Hydrol 58: 279–306.

Walker GPL (1970). Compound and simple lava flows and flood basalts. Bull Volcanol. 35: 579–590.

Walton WC (1970). Groundwater resource evaluation. McGraw Hill, N.Y.: 1–664.

Wilson DS, Teagle DAH, et al. (2006). Drilling to gabbro in intact ocean crust. Science, 312: 1016–1020.

Wu YS, Pan L, Zhang W, Bodvarsson GS (2002). Characterization of flow and transport processes within the unsaturated zone of Yucca Mountain Nevada, under current and future climates. J Contaminant Hydrology 54: 215–247.

CHAPTER 6

# Regional transmissivity distribution and groundwater potential in hard rock of Ghana

Philip K. Darko[1] and Jiří Krásný[2]
[1]*CSIR Water Research Institute, Accra, Ghana*
[2]*Charles University Prague, Czech Republic*

ABSTRACT: Statistical assessment of specific capacity and transmissivity has made it possible to classify hydrogeologic units on a regional scale and to prepare regional transmissivity anomaly maps for the quantitative characterisation and prediction of borehole yields in hard rock aquifers in Ghana. In general, the rocks pertain to transmissivity classes IV and III, representing low to intermediate transmissivity magnitude. Mapping out areas of positive transmissivity anomalies delineates prospective zones for groundwater exploration. In such zones, the minimum expected yields range from 35 to 120 L/min in the various hydrogeological units. These yields are 6–15 times higher when compared with probable yields from regions of negative transmissivity anomalies (6 to 10 L/min). Detailed investigations in zones of positive transmissivity anomalies could reveal potential aquifers where large volumes of groundwater could be abstracted for local supplies. Natural groundwater resources, however, might represent an upper limit for regional groundwater development.

## 1 INTRODUCTION

In Ghana, many productive freshwater bearing aquifers are composed of weathered and fractured hard rocks, and increasing effort has been made to harness groundwater from these formations. The disparity in yield and transmissivity in boreholes drilled even in close proximity in similar hard rock types can be very considerable. This can be attributed to heterogeneity and anisotropy in related hydraulic characteristics that make predictions in hard rock areas very uncertain. Assessment based on statistical analysis of the transmissive capacity of aquifers could however, provide information on the general trends and conditions of groundwater resources. This information is essential for regional policy formulation and regulatory development of groundwater as a natural resource, and especially for targeting groundwater development and improving success rates in drilling programmes.

To address the need for comprehensive regional hydrogeological evaluation to support groundwater development in Ghana, statistical analysis of specific capacity and transmissivity data has been carried out. By comparing with similar evaluations from other areas, the prevailing regional trends in transmissivity have been determined, and local anomalies delineated to depict prospective zones for groundwater exploration. The effect of the transmissivity distribution on the probable yields of wells has been expressed, based on a generally available drawdown.

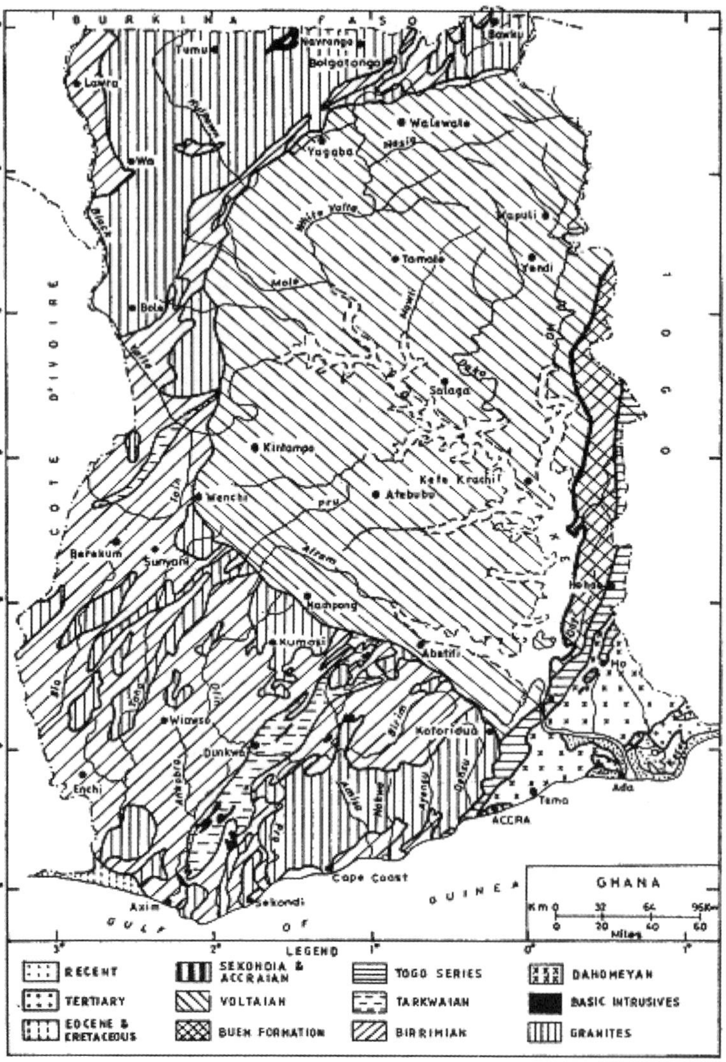


Figure 1. Geological map of Ghana (simplified after various sources).

## 2 HYDROGEOLOGICAL BACKGROUND

A complex of Precambrian igneous and metamorphic rocks underlies about 45% of Ghana. Late Proterozoic to early Paleozoic rocks of the Voltaian System cover approximately another 35% of the country. Together, these constitute the hard rocks in Ghana. The Precambrian rocks include the Dahomeyan (gneiss and migmatite with quartz and schist), the Birimian (metamorphosed sediments) with associated granitoid intrusives, and the Tarkwaian rocks (slightly metamorphosed sediments). The rest are the Togo Series (sandstone, quartzite, schist, shale and phyllite) and the Buem Formation (shale, sandstone and volcanic rocks). The Voltaian System consists of consolidated sedimentary rocks, mainly sandstone, shale and mudstone, conglomerate and limestone. The geological map of Ghana (Figure 1) shows the distribution of the various formations.

Table 1. Yields and depths of successful boreholes in Ghana.

| Geological formation | No. of boreholes | Yield (L/min) | | | | Depth (m) | |
|---|---|---|---|---|---|---|---|
| | | Mean | Median | Mode | Range | Mean | Range |
| Granite | 767 | 67 | 40 | 15 | 5–607 | 37 | 13–152 |
| Voltaian | 779 | 121 | 50 | 15 | 5–1200 | 42 | 22–355 |
| Togo & Buem | 212 | 94 | 60 | 20 | 7–525 | 44 | 21–146 |
| Birimian & Tarkwaian | 213 | 126 | 98 | 50 | 8–600 | 54 | 16–187 |
| Dahomeyan | 74 | 50 | 30 | 10 | 9–200 | 39 | 22–122 |

(Data source: WRI 1999)

The rocks are largely impervious but contain openings along joints and fractures, bedding and cleavage planes that enhance percolation of water to form limited groundwater reservoirs, which are structurally dependent and discontinuous in occurrence (Gill, 1969). Boreholes drilled in such hydrogeologic environments have a wide range of yields, with a significant failure rate (Darko 2001). The yield in the Voltaian rocks is highly variable and ranges from 5 to 1200 L/min, with an average yield of 121 L/min. The Dahomeyan gneiss and migmatite record the lowest mean yield of 50 L/min. Data on borehole yields and depths are provided in Table 1.

## 3 METHOD OF ASSESSMENT

Pumping test data from boreholes drilled across the country were compiled. On the basis of major lithology, age and location, the large data set was grouped into five main hydrogeologic units as given in Table 1. The data in each group was taken as a sample set for statistical analysis.

Specific capacity data, which were readily available, were used to assess the water bearing and yield potential of the aquifers. Since all the boreholes were pump-tested for 6 hours, the specific capacity data were used to make comparisons among the various geological formations. Defined as the volume of water pumped per unit time (yield) per unit drawdown in the pumping well, specific capacity is an important hydraulic parameter that indicates the transmitting properties of an aquifer. Among others, Brown (1963), Razack and Huntley (1991) and Huntley et al. (1992) have discussed the use of specific capacity to estimate transmissivity. It accounts for the loss in head that is associated with pumping, and is preferred to yield as a measure of well productivity (Knopman & Hollyday 1993). It also provides initial estimates of potential for water abstraction in the hydrogeologic environment (Krásný 1993).

Due to heterogeneity and anisotropy in hydraulic properties, the Theis equation and its numerous modifications often do not sufficiently represent the relationships between specific capacity and local and structural factors that are important in fractured media (e.g., local fault zones, degree of fracturing and fracture disposition, folding patterns, topography, etc.). A generalised deterministic model linking these features to hydraulic parameters does not exist. Nevertheless, flow in large enough volume of fractured medium may be reasonably represented by the equivalent continuum model, as demonstrated by Long et al. (1982), especially if the sample sizes tested are large enough and interest is mainly on volumetric flow, such as for groundwater supply. Still the differences in size and hydrogeologic character of inhomogeneities, such as fractures and weathered zones, mean that the representative elementary volume to be used for equivalent continuum models changes considerably in distinct hardrock environments (comp. e.g., Krásný 2000).

A statistical analysis of well records is the only practicable alternative for comparing and analysing the water-transmitting properties of fractured rocks and their causes on regional scales.

## 4 DATA ANALYSIS

The specific capacity was converted into an index of transmissivity Y, introduced by Jetel & Krásný (1968). The transmissivity index Y is given by the relation: $Y = \log(10^6 C)$, where C is specific capacity (l/s/m). The index has the advantage when used as a comparative regional parameter because most populations of specific capacity values are log-normally distributed (Knopman 1990). Hence by converting the specific capacity values into index Y, the data set becomes normally distributed to allow descriptive statistical analysis. From the index Y, the coefficient of transmissivity, T ($m^2$/day), was estimated by the conversion equation: $T = 86400\ (10^{Y-8.96})$.

To find the prevailing (or background) transmissivity, the mean ($\bar{x}$) and standard deviation (s) of the index Y values in each hydrogeologic unit were determined. The range ($\bar{x} \pm s$) of the index Y values (which accounts for about 68% of the data set) represents the prevailing transmissivity (transmissivity background) of the different units. The values outside this interval are considered as anomalies, which may be positive or negative. These anomalies have considerable practical importance in geological and hydrogeological investigations including environmental assessments. The interval between ($\bar{x} + s$) and ($\bar{x} + 2s$) is defined as the area of positive anomalies, and signifies more prospective zones for groundwater exploration when compared with the area of background transmissivity. On the other hand, the values between ($\bar{x} - s$) and ($\bar{x} - 2s$) indicate areas of negative anomalies and these represent less favourable zones. Such areas of very low transmissivity are needed when confronted with issues of waste disposal, and would require further investigation. Extreme anomalies can be found outside the intervals ($\bar{x} \pm 2s$). These areas may be important on one side when looking for sites for toxic waste disposal (extreme negative anomalies) and, on the other side as potential areas for groundwater exploitation (extreme positive anomalies).

The classification scheme of Krásný (1993), which is based on transmissivity magnitude and variation, was used to classify the various hydrogeological units (Table 2). The order of the transmissivity magnitude (or scale) depends on the percentage of the background transmissivity that belongs to a particular class, and the class variation is based on a standard deviation of interval 0.2. This classification scheme provides a practical quantitative method for evaluating the potential for groundwater abstraction in different areas. It also has the additional advantage of making it possible to express various regional hydrogeologic conditions and their comparison on hydrogeological maps.

## 5 TRANSMISSIVITY MAPS

The Kriging geo-statistical gridding method, which produces smooth surfaces and generates a good overall presentation, was employed to prepare the regional transmissivity maps

Table 2. Transmissivity classification of hard rocks in Ghana and their regional hydraulic characteristics (modified after Darko & Krásný 1998).

| Hydrogeologic unit (Number of wells) | Transmissivity index (Y) | | Transmissivity ($m^2$/day) | | | Transmissivity (class and magnitude) | Regional hydraulic characteristics |
|---|---|---|---|---|---|---|---|
| | ($\bar{x}$) | ($\bar{x} \pm s$) | ($\bar{x}$) | ($\bar{x} \pm s$) | Range | | |
| Granites (467) | 4.8 | 4.3–5.3 | 6.6 | 2.1–20.2 | 0.3–114.1 | IV-IIIc | Low to intermediate transmissivity of moderate variation in fairly heterogeneous environment. |
| Voltaian (332) | 5.1 | 4.4–5.6 | 11.9 | 2.5–39.6 | 0.3–266.8 | III-IVd | Transmissivity is intermediate to low in magnitude with a large variation. Hydrogeological environment is considerably heterogeneous. |
| Dahomeyan (69) | 4.7 | 4.3–5.1 | 4.5 | 1.7–11.9 | 0.3–42.2 | IV-(III)c | Moderate variation in a fairly heterogeneous medium with low transmissivity. |
| Birimian & Tarkwaian (152) | 4.9 | 4.3–5.5 | 7.4 | 2.0–27.3 | 0.2–118.7 | IV-IIIc | Low to intermediate transmissivity magnitude with moderate variation. Hydrogeological environment is fairly heterogeneous. |
| Togo & Buem (53) | 4.9 | 4.5–5.4 | 8.0 | 3.0–21.4 | 0.9–43.2 | IV-IIIc | Low to intermediate transmissivity with small to moderate variation in fairly heterogeneous environment. |

Explanation: ($\bar{x}$) = sample mean, (s) = sample standard deviation (Data sources: Gill 1969, WRI/CSIR, 1999).

by contouring the irregularly spaced transmissivity data. Figure 2 presents the spatial distribution of boreholes that were used. Due to the scale of the map, a single point on the map could represent up to four boreholes. To increase the likelihood that the original data points were applied directly to the grid file, a large number of grid lines was generated and the non-drift linear variogram system with unit anisotropy ratio was used in the exact interpolating format, while averaging duplicate values. The regional classification of the various units was then carried out based on the classification scheme outlined earlier. Figure 3 shows the regional picture of transmissivity distribution in Ghana.

The minimum curvature gridding method was used to prepare transmissivity anomaly maps for the respective hydrogeologic units. This method generates a weighted-average interpolated surface, which attempts to honour the data as closely as possible. The maximum residual parameter was set at 0.1 (i.e., 10% of the data precision) during iterations.

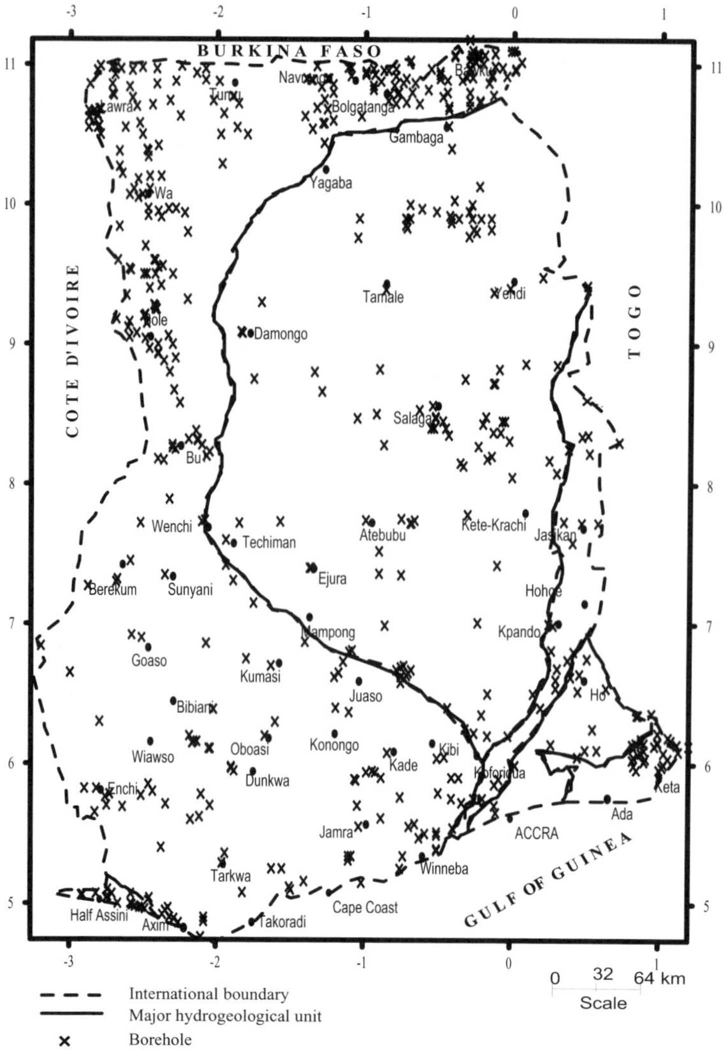


Figure 2. Map of Ghana showing location of selected boreholes.

By establishing the background transmissivity ($\bar{x} \pm s$), areas with positive and negative anomalies around a minimum of three boreholes were delineated. Figure 4 shows the areas of positive and negative transmissivity anomalies in the major hydrogeologic unit (Voltaian) in Ghana.

## 6 ESTIMATION OF YIELDS

The effect of the transmissivity differences (anomalies) on the probable yields of wells has been expressed based on a generally available drawdown of 5 m. The expected yield was computed by multiplying the specific capacity value of each well by the available drawdown.

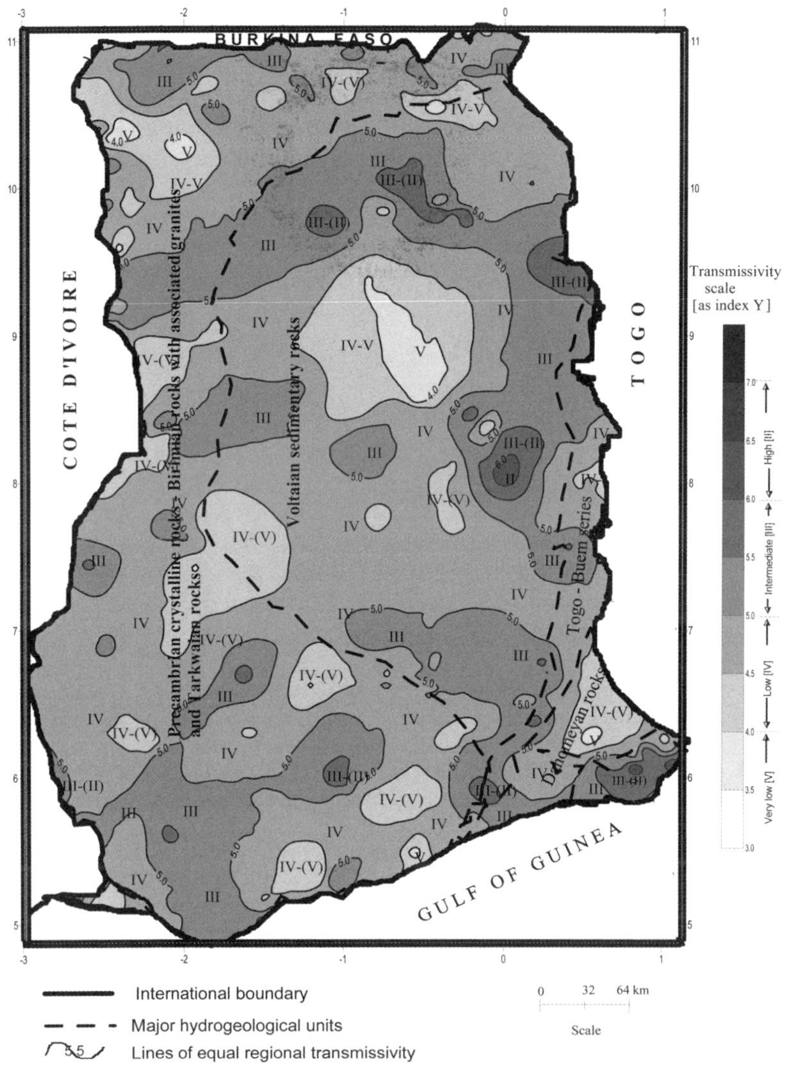


Figure 3. Regional transmissivity trends in hard rocks of Ghana. Roman numerals, e.g. V-IV: classes of transmissivity magnitude (based on Krásný [1993] classification scheme).

The 5 m drawdown has been considered as a practical value for rural, domestic hand-pump based water supply systems (Darko & Krásný, 2000). Table 3 gives the results of the expected discharges.

## 7 RESULTS

The transmissivity classification of the hydrogeologic units in Ghana is provided in Table 2. The Voltaian sedimentary rocks record the highest mean transmissivity coefficient of 12 $m^2$/day with values ranging between 0.3 and 270 $m^2$/day. These rocks are classified as

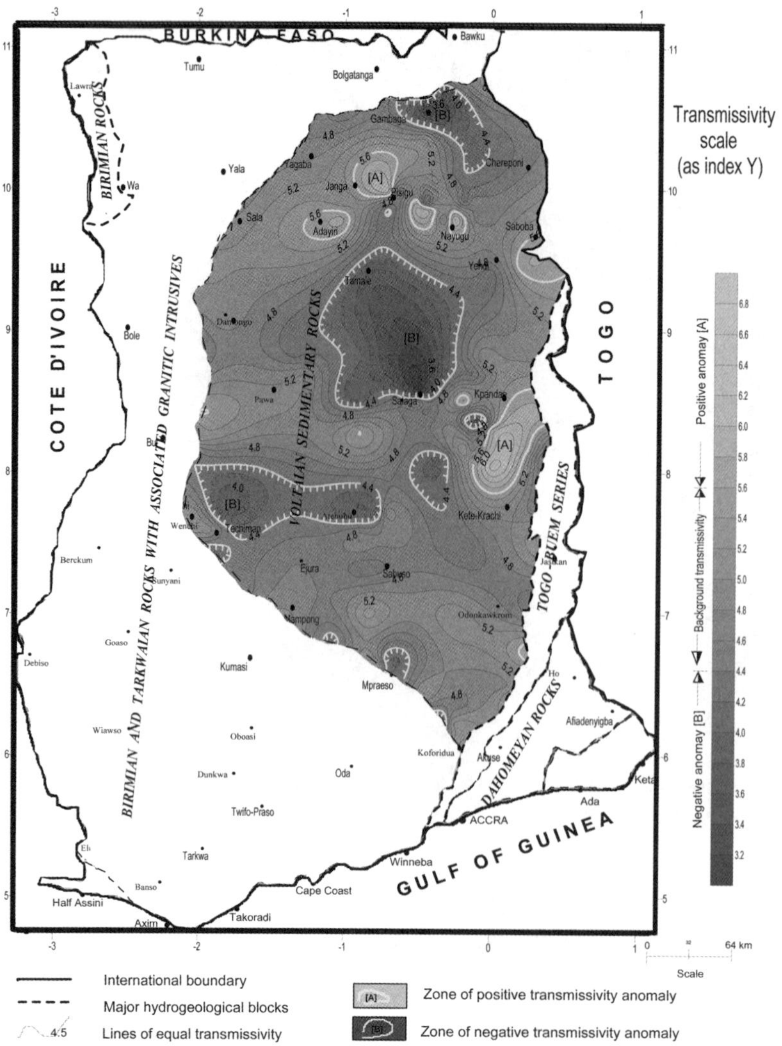


Figure 4. Map of Ghana showing positive and negative transmissivity anomalies in Voltaian consolidated sedimentary rocks (after Darko 2001).

Table 3. Minimum expected yields in zones of transmissivity anomalies in hard rocks of Ghana.

| Hydrogeologic unit | Expected minimum yield (L/min) | |
|---|---|---|
| | Negative anomalies | Positive anomalies |
| Voltaian | 8 | 120 |
| Granite | 6 | 60 |
| Birimian & Tarkwaian | 7 | 85 |
| Togo & Buem | 10 | 65 |
| Dahomeyan | 6 | 35 |

III-IVd. The Dahomeyan rocks have the lowest mean transmissivity coefficient of 4.5 $m^2$/day, and belong to transmissivity class IV-(III)c. In general, the rocks pertain to transmissivity classes IV and III, representing low to intermediate transmissivity in magnitude. Figure 3 shows the general regional trends in transmissivity. The transmissivity anomaly map for the Voltaian formation is presented in Figure 4.

The areas marked apparently by very low transmissivity of class IV-V are delineated as zones of negative anomaly. Generally, these areas are considered less favourable for groundwater exploration since probable yields would be less than 8 L/min (Table 3). The regions that belong to transmissivity class III-II represent areas of positive anomaly. These regions offer better prospects for groundwater abstraction with borehole yield expected to be more than 120 L/min.

The areas underlain by granites are associated mainly with low transmissivity, and belong to transmissivity class IV. The areas noted with very low transmissivity are marked as zones of negative anomalies, with expected yield of about 6 L/min. The yield at locations of positive anomalies is expected to exceed 60 L/min (Table 3).

The Birimian and Tarkwaian are characterised in many places by low to intermediate transmissivity of classes IV and III. The expected yield in zones of negative anomalies would be only 7 L/min or less. However, in those areas of positive anomalies the probable yields of wells would be more than 85 L/min.

The Togo and Buem rocks also give intermediate to low transmissivity classes III-IV in most places (Figure 2). The probable yields in areas of negative transmissivity anomalies would be less that 10 L/min, whereas in zones of positive transmissivity anomalies, the expected yield will be at least 65 L/min.

The groundwater potential of zones of negative anomaly in the Dahomeyan gneiss could be very low. The yield of wells would be less than 6 L/min, with little prospects for groundwater abstraction. At those areas marked as positive transmissivity anomalies yields are expected to be more than 35 L/min.

## 8 DISCUSSION

The analysis and results drawn from these specific capacity modifications are based mainly on statistical averaging on regional scales, without accounting for several other factors that could contribute to groundwater movement and storage in fractured and weathered rocks. Related variables such as topographic setting, structural factors, well construction and discharge, primary use of water and water quality should also be considered and investigated in detail before any conclusions on local scales could be made. The results presented are however, the sum total of the interplay of all these contributing factors on the accumulation and water-yielding properties of the fractured rocks on regional scale. The regional distribution of transmissivity and the transmissivity anomalies in the various geological formations generally follow the trend of borehole yields in Ghana, as obtained from yield maps reported by Dapaah and Gyau-Boakye (2000). The borehole yields are higher in zones of positive transmissivity anomalies and vice-versa.

The effect of the transmissivity differences on the probable yields of wells has been expressed based on a generally available drawdown of 5 m. Though this could result in an underestimation of maximum possible yield, it provides a useful and robust estimation of sustainable yield for the test to which the aquifers were subjected.

Using the same methodology, results obtained from transmissivity studies in hardrock areas in Ghana were compared with those from the Czech part of the Bohemian Massif (Darko and Krásný 1998). It was found out that the fields of prevailing transmissivity, expressed as cummulative relative frequences, of all the Ghanaian formations were shifted to the right, i.e. to higher values, compared with values from the Bohemain Massif. High transmissivity in some boreholes drilled in the Voltaian that are reported in hundreds of $m^2$/day might be attributed to occurrences of sandstones and limestones. Other differences determined in prevailing transmissivity of hard rocks in Ghana and in the Czech Republic, however, cannot be unambiguously explained. They could be caused by different duration of pumping tests (shorter in Ghana). Differences in the type of weathering could be an additional factor. Weathering in a tropical country like Ghana has been more intensive and therefore disintegration of rocks might be deeper and more extensive than within the Bohemian Massif. Due to enormous variety of natural conditions in such an extended country as Ghana, detailed analysis and additional studies would reveal significant differences within particular geological formations depending on their petrology, structure, way of weathering, topographic and hydrogeological position etc. In future, such analysis should be made as one of the important approaches towards a quantitative regionalisation of hydrogeological data. Therefore, the presented results can be considered only as the first attempt of regional treatment of transmissivity data.

## 9 CONCLUSIONS

The quantitative assessment of specific capacity and its modifications has enabled the delineation of local transmissivity anomalies to depict prospective zones for groundwater exploration, and to estimate the probable yield of wells in the various hydrogeological units in Ghana.

In zones of negative transmissivity anomalies, the probable yields range from 6 L/min in the Dahomeyan to 10 L/min in the Togo and Buem formation. These yields are not adequate to support a hand-pump based rural water supply since the minimum requirement is 13.5 L/min.

The expected yields in zones of positive anomalies range from 35 to 120 L/min in the Dahomeyan and Voltaian rocks respectively. The yields expected from areas of positive anomalies are about 6–15 times higher when compared with yields expected from areas of negative anomalies (6–10 L/min). Detailed investigations in zones of positive transmissivity anomalies could reveal potential aquifers where large volumes of groundwater could be abstracted.

Even though natural groundwater resources in the hard rock environment in Ghana usually do not represent a limit for groundwater development, in regions with high rock transmissivity recharge can be insufficient and might not support intense regional groundwater withdrawals. Under local climatic conditions in SW Ghana, only groundwater resources as high as 1.6 L/s/$km^2$ could be considered the threshold of groundwater exploitation as assessed in the Pra river catchment by conceptual hydrological modelling (Darko 2002, Darko & Krásný 2003).

The rainfall in Ghana, which grades continuously from the south to the north (annual totals changing from 2200 to 1000 mm), does not seem to strongly influence the regional transmissivity or distribution of yield. It can therefore be concluded that geology is much more important than climatic factors in rock transmissivity and groundwater potential.

## REFERENCES

Brown RH (1963) Estimating the transmissivity of an artesian aquifer from the specific capacity of a well. U.S. Geological Survey Water Supply Paper 1536-I, pp 336–338

Dapaah-Siakwan S, Gyau-Boakye P (2000) Hydrogeological framework and borehole yields in Ghana. Hydrogeology Journal. V 6, pp 405–416

Darko PK (2001) Quantitative aspects of hard rock aquifers; regional evaluation of groundwater resources in Ghana. PhD Thesis, Charles University, Prague

Darko PK (2002) Estimation of natural direct groundwater recharge in SW Ghana using water balance simulations. J.Hydrol.Hydromech, 50, 3:198–212

Darko PK, Krásný J (1998) Comparison of hardrock hydraulic parameters in distinct climatic zones: the Ghana and the Bohemian Massif areas. In: Annau R, Bender S, Wohnlich S (eds): Hardrock Hydrogeology of the Bohemian Massif. Proc. 3rd Internat. Workshop 1998, Windischeschenbach, Münchner Geol. Hefte, B8, 3–10, München

Darko PK, Krásný J (2000) Adequate depth of boreholes in hard rocks: a case study in Ghana. In: Sililo O. et al. (eds): Groundwater: Past achievements and future challenges. Proc. 30 IAH Congress, Nov.26–Dec.1,2000, 121–125. Cape Town. Balkema Rotterdam

Darko PK, Krásný J (2003) Quantitative limits for groundwater exploitation under natural conditions in southwest Ghana. In: Krásný J, Hrkal Z, Bruthans J (eds): Proceedings – IAH Internat. Conference on "Groundwater in fractured rocks", Sept. 15–19, 2003. Prague. 141–142, IHP-VI, Series on Groundwater 7. UNESCO

Gill HE (1969) A groundwater reconnaissance of the Republic of Ghana, with a description of geohydrologic provinces. Geological Survey Water-Supply Paper 1757-K, Washington

Huntley D, Nommensen R, Steffey D (1992) The use of specific capacity to assess transmissivity in fractured-rock aquifers. Ground Water, V 30(3), pp 396–402

Jetel J, Krásný J (1968) Approximative aquifer characteristics in regional hydrogeological study. Vest. Ust. Geol. V 43 (5), Prague, pp 459–461

Knopman DS (1990) Factors related to the water-yielding potential of rocks in the Piedmont and Valley and Ridge provinces of Pennsylvania. U.S. Geological Survey Water Resources Investigation Report 90-4174, pp 52

Knopman DS, Hollyday EF (1993) Variation in specific capacity in fractured rocks, Pennsylvania. Ground Water, V 31 (1), pp 135–145

Krásný J (1993) Classification of transmissivity magnitude and variation. Ground Water, V 31 (2), pp 230–236

Krásný J (2000) Geologic factors influencing spatial distribution of hardrock transmissivity. In: Sililo O et al (eds): Groundwater: Past achievements and future challenges. Proc. 30 IAH Congress, Nov.26–Dec.1 2000, 187–191, Cape Town, Balkema, Rotterdam

Long JCS, Remer JS, Wilson CR, Witherspoon PA (1982) Porous media equivalents for networks of discontinuous fractures. WRR, 18(3), pp 645–658

Razack M, Huntley D (1991) Assessing transmissivity from specific capacity in a large and heterogeneous alluvial aquifer. Ground Water, V 29 (6), pp 856–861

Water Research Institute (1999) Groundwater resources assessment of Ghana. WRI/CSIR, Accra, Ghana

CHAPTER 7

# Hydrogeology of hard rocks in some eastern and western African countries

Costantino Faillace
*Via Bari, Ciampino (Rome), Italy*

ABSTRACT: This paper outlines the drilling results and the hydrogeological conditions of some eastern and western African countries with special focus on Liberia, Uganda, Somalia, and Mauritania, with high, moderate, scarce and negligible rainfall, respectively. It also summarizes drilling results of a few other African countries. Rainfall regime and intensity, jointly with the geomorphologic conditions, have great importance in the occurrence of groundwater in hard rocks of the reported countries. Borehole yield from fractured rocks is generally low. It can, however, satisfy the water requirements of small villages and for watering livestock. The problems related to poor groundwater quality in hard rocks of arid and semiarid countries and the possible alternative solutions to secure the needy water are also indicated. Alternative solutions to solve the pressing water problems are to construct, wherever conditions are appropriate, underground dams across dry river courses, sand storage dams or rock catchments.

## 1 INTRODUCTION

Hard rock outcrops are widely extended in various parts of the world, and cover about 20% of the land surface. Until a few decades ago, not much importance was given to the development of groundwater from hard rocks due to the belief that their water potential was negligible.

The rapid increase in world population, especially in Africa and Asia, coupled with periodic droughts affecting extensive sub-desert areas of Africa, and the introduction of more efficient and cheaper drilling technology, boosted the well drilling programs in many countries. The United Nations have greatly contributed with numerous groundwater projects in more that 10 African countries, thus acquiring a better knowledge of the groundwater potential in many areas, especially those characterized by water scarcity.

The information contained in this paper derives mainly from the direct experience of the author in Somalia, Uganda, Kenya and Liberia during the 18 years he spent in those countries while working for international groundwater development projects.

## 2 HYDROGEOLOGICAL ASPECTS AND DRILLING RESULTS IN SOME AFRICAN COUNTRIES

Most of the results of groundwater exploration projects have shown that hard rocks neither receive nor transmit water unless they are fractured or weathered. The relation of groundwater to structure and topography is quite important because it may indicate promising hydrogeological conditions; the stream courses, in fact, may be active recharge and discharge zones, often marking fault and shear zones along which groundwater movement occurs. The concept of the aquifer is therefore related more to the structural aspects of the hard rocks than to their properties as groundwater conductors. The fractures, often highly developed along the fault lines, and the effect of climatic action, determine the hydraulic characteristics of these rocks, the primary porosity being nearly nil.

Therefore, the possibility that the igneous and metamorphic rocks act as aquifers can be correlated to the existence of fracture lines that intersect each other, and to highly weathered granite rock, which acts as a porous medium. A statistical evaluation of the type of fractures and their origin may help to identify possible groundwater occurrence. Any type of system of fractures and fissures often tends to decrease and ends up sealing itself off completely. There are, however, cases of deep fissures, especially those connected with regional faults and shear zones, along which the movement of groundwater can be turbulent. The flow is fed by a number of small fractures and fissures having a laminar flow, while the rock surrounding the fault zone may be dry or yielding little water.

Groundwater investigations in Africa, which included the inventory of the water sources, drilling, and geophysical methods, have shown that there are three main potential water-bearing zones in hard rocks:

(1) the upper sandy-clay zone with residual blocks of rock,
(2) the zone of weathered and fractured rock, and
(3) the zone of fresh, rarely fractured rock.

The electrical resistivity method has been extensively used in many African countries; it allows estimation of the approximate thickness of the weathered and fractured rock. The results, however, were not always of clear interpretation. The seismic refraction method, instead, gives more accurate information on the thickness and discontinuity between the fresh and fractured rock, and on the location of depressions in the compact, solid rock; in particular; it helps in identifying low-velocity zones, such as the shear zones, which are potentially water bearing.

Bernardi (1973) reported the relation between the thickness of the aquifer and the well yield in 32 boreholes drilled in Togo and Upper Volta in depressions of the basement detected by the seismic refraction method. The results were the followings:

- where the unfractured rock was deeper than 30 m, all boreholes were successful; in seven cases the yield was between 2 and 5 $m^3$/h, and in 10 boreholes the yield was between 5 and 15 $m^3$/h; and
- where the unfractured rock was less than 20 m deep, 4 out of 5 boreholes had a very low yield, generally less than 1 $m^3$/h.

The upper section is generally dry, but it may be water-bearing wherever the water table is close to the surface. The middle section may be divided into the upper deeply-weathered part and in the lower fissured one, which is, in most cases, the major water-bearing section of the crystalline rock. Groundwater in the underlying solid rock may occur only in deep, occasional fissures.

The main water-bearing areas are often characterized by the irregular distribution of the water level in wells and the rapid change in chemical composition in neighbouring wells.

Rainfall, with its three factors of regime, intensity, and values of evapo-transpiration, plays a very important role in the availability of groundwater.

This paper describes the hydrogeological conditions of hard rocks in four African countries: Liberia, Uganda, Somalia and Mauritania, with high, moderate, scarce and negligible rainfall, respectively, and reports the drilling results of some other African countries.

## 3 LIBERIA

Gneisses, shists and other metamorphic rocks, collectively referred as crystalline rocks, cover most of Liberia, West Africa. Rainfall ranges between 1,800 mm/year along the border with Guinea to 4,400 mm/year along the coast. The great availability of surface water obtained from various sources such as rivers and creeks, of springs and of shallow groundwater sources (e.g., in seepage zones and of various water holes and shallow hand-dug wells) has, to a certain extent, delayed the assessment of the groundwater conditions of the country. The bloody civil war that has been going on for more than 23 years has prevented the development of groundwater projects; consequently, the groundwater resources of the country have not yet been evaluated.

Groundwater in Liberia can be easily tapped by shallow hand-dug wells in high rainfall areas along the coast, even if located in watersheds, so far as they have penetrated thick laterite or weathered rock cover. The mantle covering the crystalline rocks ranges between 3 to 15 m.

### 3.1 *Hydrogeological conditions*

A few pump tests of the laterite aquifer have shown that the yield of dug wells varies from 2 to 20 L/min. It seems that the mantle above the coarse-grained rock, if thick enough, will always yield sufficient water for domestic water supply.

Crystalline rocks such as foliated coarse-grained granite rocks and quartz-feldspatic gneisses are generally water bearing. It is sometimes possible to increase the well yield by drilling to a depth of 100 m or more. Experience, however, has shown that in spite of the high rainfall, 50 out of 231 boreholes (21.6%) drilled in the late seventies for rural village water supply were reported dry. The successful boreholes, with depths ranging from 10 to 31 m, were located mainly in low-lying areas where water was met in laterite and in weathered and fractured rock. Boreholes with depths of less than 15 m found little water. Seventy-five percent of boreholes with an average depth of 30 m had a substantial yield, 15 of which had penetrated the bedrock; 25% of the boreholes yielded little water (Faillace, 1979). The four shallow test drillings for the Zwedru water supply, with depths ranging from 15 to 20 m, had a low yield, varying from 0.1 to 1.5 L/s (GKW, 1973).

Deeper boreholes, drilled for the water supply of six towns, gave better results. In fact, in 50–60 boreholes that penetrated deep into gneiss, shists and other metamorphic rocks that cover that country, groundwater conditions were rather good compared to other African countries. In particular, 16 boreholes with an average depth of 105 m, drilled in Bhuchanan for domestic and industrial purposes, had an average yield of 5,000 L/h.

In six additional exploratory boreholes drilled for the same town, the average yield was 7,400 L/h. Borehole No. 6, in particular, was pump-tested when its depth was 102 m, obtaining 9,500 L/h. Its yield increased to 18,200 L/h after it was deepened to 151 m, while its specific capacity rose from 190 and 260 L/m of drawdown. Groundwater conditions, however, vary from place to place, and even within short distances. The average specific capacity of wells in Bhuchanan was 0.423 $m^3$/h/m, with a range between 0.017 and 2.240 $m^3$/h/m (GKW, 1980). Not one of the deep boreholes was reported dry. The best results were obtained in coarse-grained, layered crystalline rocks where groundwater was found in more permeable layers acting as aquifers.

### 3.2 *Water quality*

Water quality is characterized by a low salt content. However, high iron content and low pH are the main water problems in Liberia. Groundwater in the laterite mantle contains iron in solution in the ferrous state and, when it comes in contact with the oxygen of the air is precipitated as ferric oxide. If not removed, people will not accept it; furthermore, it may partly plug pipes and clog the screen with consequent reduction of the yield and shortening the life of the well. Another major problem is the low pH of the water, in most cases below 6.5 and as low as 3.5. The low pH affects all the metallic parts of the pump, with severe corrosion and consequent malfunctioning and breakdown of some parts.

## 4 UGANDA

The groundwater location in fissured rocks of Karamoja District, Uganda, was investigated by a UNDP project, using methods including exploration drilling, correlation of data from previously drilled wells, geomorphology, photo-interpretation, and electrical resistivity and seismic refraction methods.

Karamoja covers an area of 30,000 $km^2$, with igneous and metamorphic rocks; its average yearly rainfall is 800 mm, ranging from 500 to 1,140 mm/year. The rocks underlying the District belong to the metamorphic suite, including biotite gneiss, granulite, quartzite, and marble. Much of the Karamoja featureless plain is covered by black heavy soil, which facilitates a fast run-off during torrential rains and limits infiltration into the underlying basement complex.

### 4.1 *Borehole location and drilling method*

The 92 exploration boreholes drilled in the District were sited using various methods such as the resistivity and seismic refraction, photo-interpretation, geological analysis of the results of previous drilling. Most of the boreholes were drilled to check such methods and to test geo-morphological features.

All boreholes were drilled by cable-tool rigs, which allowed the collection of very representative rock samples and enabled a precise detection of the sequence of water entries and bailer testing. All boreholes were wire-logged for temperature, resistivity, calliper, and gamma ray. Several boreholes were blasted to investigate the effect of blasting on the yield.

### 4.2 *Geophysical results*

Electrical resistivity profiling and sounding methods proved to be of little use as fissure systems could not be detected, and the true thickness of the weathered and fractured zone

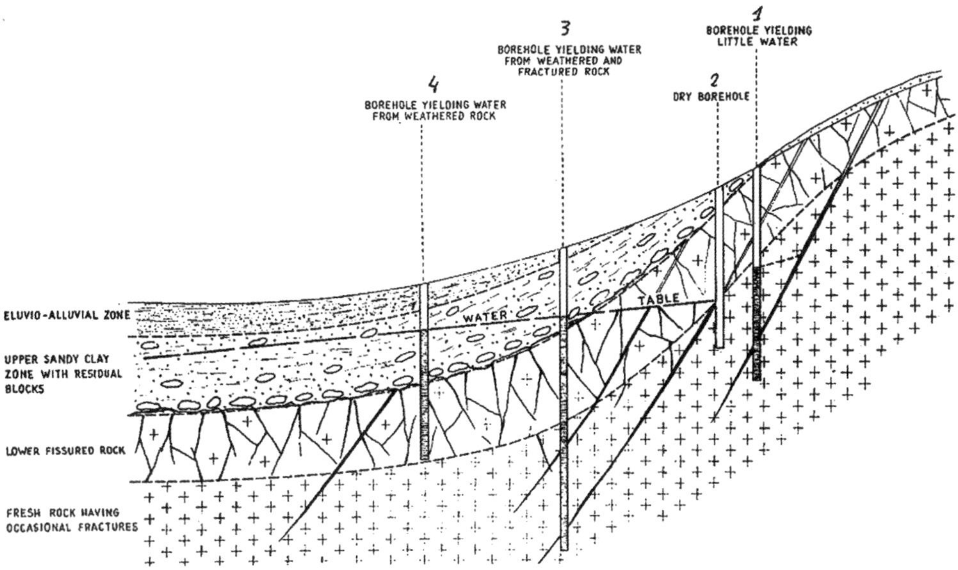


Figure 1. Schematic section representing vertical hydrogeological zonation of hard rocks in Karamoja, Uganda, also showing the hypothetical morphological location of boreholes and their yield results.
Borehole No. 1, located on the slope, struck a number of dry fissures and a deep, isolated low-yielding water-bearing fissure with water under sub-artesian conditions.
Borehole No. 2, located on a slope, met a number of dry fissures and solid rock without fractures.
Borehole No. 3, located in a lower position, struck a moderate amount of water in the weathered and fractured rock and deep fissures of the underlying solid rock.
Borehole No. 4, located at the base of the slope, struck a much higher amount of water from the highly weathered and fractured rock, which acts as a regular aquifer under sustained pumping.

could not be measured with any accuracy. Seismic refraction, instead, was quite useful to detect potentially water-bearing, "low-velocity zones" in metamorphic rocks.

## 4.3 *Hydrogeological conditions*

Three zones, potentially water-bearing under favourable geomorphological conditions, have been distinguished in hard rocks of Karamoja District (Figure 1):

- the upper sandy-clay zone with residual blocks, which is generally dry unless the water table is close to the surface due to local conditions;
- the middle weathered and fractured zone with its upper, deeply weathered part, and the lower fissured part. This is the main water-bearing section of the metamorphic rocks when located in bottom valleys, where it can be recharged directly by streams, rain water or lateral inflow. It is generally dry on divides, steep slopes and deeply eroded pediments; and
- the deep zone with fresh, rarely fractured rock, where water, under sub-artesian conditions, is occasionally found in fissures, often not interconnected. Groundwater exploration from fresh rock resulted extremely difficult in Karamoja as it is not possible to predict striking water-bearing fissures in any given point.

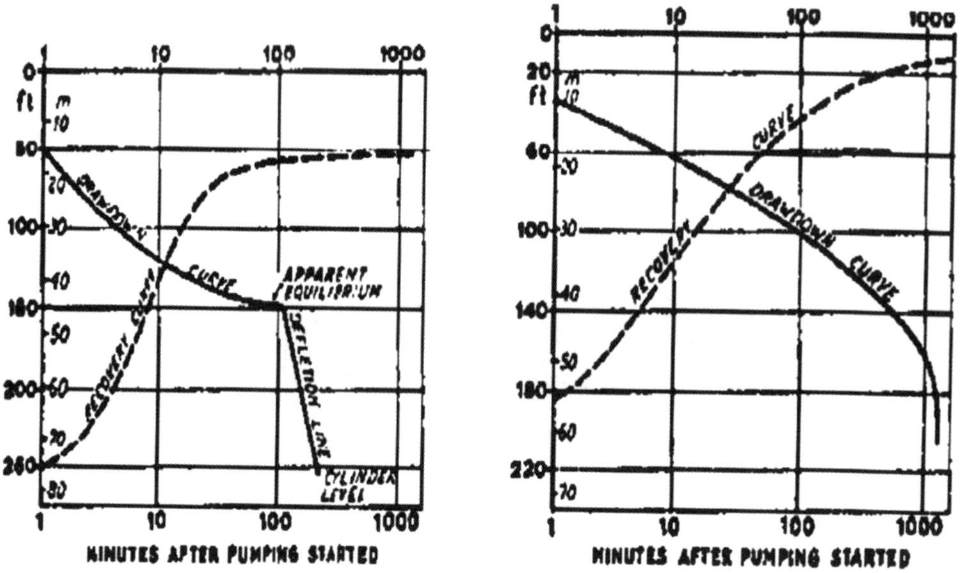


Figure 2. Examples of drawdown and recovery curves from pumped wells in metamorphic rocks in Karamoja, Uganda.
Well CD2340 (left): yield decreased from 109 to 89.5 L/min; the abrupt breakdown of the drawdown curve indicates that the initial equilibrium between the pumping level (drawdown) and discharge was only apparent.
Well No. 3645 (right): yield decreased from 156.6 to 147.5 L/min. The progressively steeping curve indicates that the aquifer is limited, bounded by barriers, and rapidly exhausted by the pumping.

It has been observed that when the dynamic water level is in the phreatic zone, the drawdown curve is smooth and gradual; but when this zone is dewatered, the drawdown in the underlying fresh rock drops rapidly to the point where the well cuts the first major deep fissure. This process will continue until the fissure system is de-watered, as in the case of borehole No. 3 in Figure 1.

Some examples of drawdown and recovery curves from pumped wells in metamorphic rocks are reported below and represented in Figures 2 and 3.

In most of the boreholes drilled in Karamoja, a decrease of specific capacity was observed when pumped at increased rates. Estimates of specific yield could be possible provided the water level was steady at points above the water entry. It has to be considered that the last water entry is the base of the active water-bearing section; it is therefore incorrect to estimate the specific capacity when the water level drops below this point.

In some boreholes the dynamic water level was steady for a given yield for many hours, but after increasing the discharge rate of only a few gallons, the level dropped abruptly to the pump cylinder with a consequent severe drop of the specific capacity (see graph of well CD2340 – Figure 2, left).

In borehole No. 4026 (Figure 3, left), the step drawdown pump test showed a drop of the specific capacity from 22.8 L/min during the first step to 18.8 L/min in the second step, with a further drop to 9.7 L/min in the third step.

In boreholes tapping alluvial deposits and weathered and fractured rock, the saturated section acts as a normal phreatic aquifer and, therefore, the specific capacity was nearly constant while pumping at increasing discharge rates.

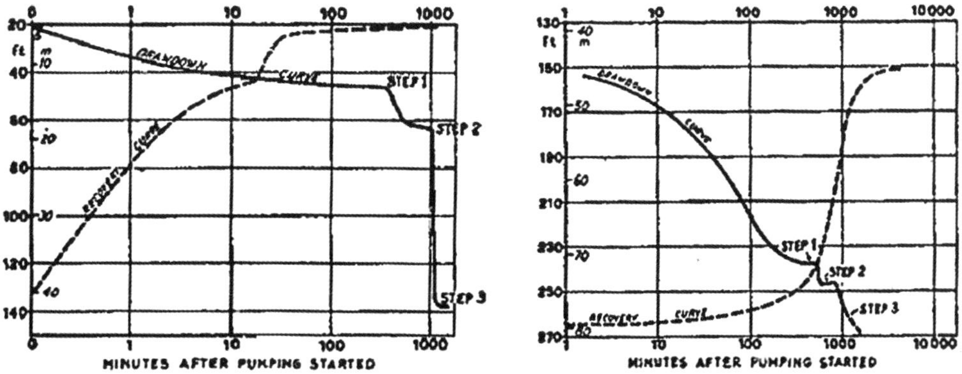


Figure 3. Examples of drawdown and recovery curves from pumped wells in Karamoja, Uganda. Well No. 4026 (left): step drawdown curve – the specific capacity decreased with increased yield. Well No. 1006 (right): step drawdown pump test – the specific capacity increased with increased yield.

There have also been a few boreholes in which specific capacity increased with the increased discharge rate. In Well No.1006 (Figure 3, right) during the first step with a sustained flow of 53.3 L/min, the specific capacity was 2.58 L/min/m of drawdown; during the second step, with 98.89 L/min of flow, it rose to 3.33 L/min/m and, finally, during the third step, with 133 L/min, the equilibrium was not reached. This anomalous behaviour is attributed to the dewatering of large, deep cavities in the hard rock. In this case the initial recovery of the water level is quite slow as can be observed in Figure 3 as well.

## 4.4 *Importance of geomorphology*

The UN groundwater project drilled 92 exploratory boreholes located under the following five different geomorphologic conditions:

(1) Flat Valley (F.V.) – well-defined bottom valley
(2) River Side (R.S.) – alluvial strip over basement along a river
(3) Flat Plain (F.P.) – open plain or flat pediment
(4) Pediment (P.) – junction between plain and hills
(5) Watershed (WS.) – in hilly country

The results are summarised in Table 1.

The best results of the 71 successful boreholes (78.2%) were obtained in the bottom of small valleys, where storage was combined with good permeability and good recharge. The water entries in almost all the successful boreholes were above 100 m, with the exception of one case with water entry below 100 m. The best average yield and specific capacities were obtained between 45 and 78 m; however, considering the overall results, the optimum depth of the wells in the metamorphic rocks of Karamoja lies between 30 and 90 m.

In most cases, the good results were due to deep weathering of the basement rock overlaid by alluvial cover. However, even in these conditions, well yields were very variable: large differences in yield occurred in boreholes located only a few meters apart, despite being drilled in similar conditions. This shows that the type of weathering and structural features, rather than rock lithology, determines water-bearing properties of metamorphic rock and crystalline rocks, in general.

Table 1. Comparison of results of 92 exploratory boreholes located under different geomorphologic conditions in Karamoja, Uganda.

| Number of boreholes and their location | Average yield per borehole (L/min) | Average depth per borehole (m) | Number of failures | Percentage of failures |
|---|---|---|---|---|
| 14 – F.V. | 228 | 85.96 | 1 | 7.1 |
| 45 – R.S. | 126 | 91.75 | 6 | 13.3 |
| 11 – F.P. | 65 | 96.32 | 3 | 27.2 |
| 14 – P. | 40 | 111.26 | 6 | 42.6 |
| 8 – WS. | 28 | 121.92 | 6 | 75 |

### 4.5 *Groundwater movement*

The movement of groundwater in Karamoja is generally restricted to small and irregular areas. This has been demonstrated by:

- the absence of any real groundwater chemical composition evolution along the hypothetical groundwater gradient;
- the impossibility to produce a regional water table map, as adjacent wells often have quite different water levels; and
- the large difference in yield occurring in adjacent wells.

## 5 SOMALIA

The basement complex in Northern Somalia covers extensive areas with a variety of metamorphic and intrusive rocks, often covered by an alluvial mantle through which rainwater seeps and recharges its fissures. Groundwater has a large variety of salinity due to the interrelation with bordering sedimentary rocks and to other local conditions. No systematic groundwater investigation, however, has been carried out in the basement complex.

The basement complex in Southern Somalia is noteworthy for the outcrops in the "Bur Area", a large peneplain of about 30,000 km$^2$, with a number of impressive inselbergs of granite and granite-gneiss rocks ("burs") which dominate the featureless plain. Most of the plain is covered by lateritic red sand and alluvial black clay overlying the eroded metamorphic rocks, which include quartzite, micashists and marble. The average rainfall of the area is 600 mm/year. Numerous large temporary rivers, with NNW-SSE direction, drain the area. The low infiltration due to alluvial black clay cover of large areas and the consequent fast runoff cause the over-flooding of large areas where water is lost by evaporation.

### 5.1 *Groundwater investigation*

Groundwater investigation was carried out by several projects and included the inventory of water sources, electrical resistivity surveys and exploratory drilling. About 90 boreholes were drilled during the period 1950–1990 for groundwater exploration, township water supply, or for watering stock. Drilling in the Bur Region has proved to be difficult and expensive. In fact, most of the boreholes were either dry or yielded brackish/salty water; only a few struck water of acceptable quality for the Somali standard.

The deepest borehole, 123 m, was drilled for Madax Marodi where water of poor quality was struck at 80 m. The small freshwater-bearing lenses at the base of the "burs" are reliable sources of water for small communities during severe water shortages.

Groundwater quality varies from place to place and even in the same locality, such as is the case of the following places:

- Buur Akaba: the E.C. varies from 850 to 8,100 micromhos/cm in hand-dug wells, and from 1,140 to 49,000 micromhos/cm in drilled wells;
- Diinsoor: in adjacent wells the E.C. values ranged from 2,000 to 16,000 micromhos/cm;

The Total Dissolved Solids (TDS) of 70 water samples collected from both dug and drilled wells are summarised as follows:

- <2 g/L in 51%
- 2 to 4 g/L in 20%
- 4 to 8 g/L in 23%
- >5 g/L in 6%

Sodium chloride type of water was predominant in 44 samples, followed by sodium bicarbonate in 8 samples and calcium chloride in 6 samples, and various other types for the remaining samples.

The temporary river courses store considerable amounts of water in their sand river beds; water quality, however, is highly variable, even within very short distances. The hundreds of rudimentary open holes and pits dug into their riverbeds are the main water sources for the nomadic people and their herds.

## 5.2 *Main findings*

The results of the groundwater investigation of the Bur Area concur with the findings that:

- The best hydrogeological conditions in the area are along the temporary watercourses, the "Wadis", where numerous open wells tap water from alluvial deposits and weathered rocks. Water quality varies from marginal to brackish, but people and livestock accept it. Water salinity increases rapidly just at short distances from the "sand river" beds.
- The irregular distribution of the water levels in wells and the rapid change in chemical composition in neighbouring wells reflect the discontinuity of these small, locally-recharged, groundwater bodies. In some cases, open wells in the headwaters of some streams yield water with a higher salinity than those located in the intermediate sections.
- Prospects to construct shallow open wells and infiltration galleries along the temporary watercourses, as well as to construct small earth dams (with the dual objective of impounding water to be used locally for watering stocks or for mini irrigation), are good. These structures are also useful for recharging the downstream shallow open wells.
- Additional water sources can be created by constructing:
  - (a) underground dams to stop the underground flow of temporary streams;
  - (b) small dams across rocky stream beds to store sand upstream of the dam, thereby creating small artificial aquifers. This type of dam is constructed in stages to allow, over the years, storage of enough water upstream of the dam. Stored water can be tapped by infiltration galleries to be constructed across the "sand river", with the intake on the riverbank.
  - (c) rock catchments, to impound runoff water in masonry or concrete walls, to be constructed in appropriate sites in selected Burs.

Table 2. Results of drilled boreholes from hard rocks of several African.

| Country | Borehole depth, m | Rock type | Yield, $m^3/h$ |
|---|---|---|---|
| Botswana | 40–240 | Granite, quartzite | 4–55 |
| Cameroon | 10–45 | Gneiss, granite-schist | 0.5–20 |
| Mali | 40–45 | Dolerite, schist | 0.5–6 |
| Mauritania | 10–70 | Weathered granite-gneiss | 0.5–4 |
| Togo | 17–70 | Shist, granite, granite-gneiss | 0.2–10 |
| Upper Volta | 20–55 | Granite, granito-schist | 0.5–6 |

## 6 MAURITANIA AND OTHER AFRICAN COUNTRIES

In Western Mauritania, characterized by extremely arid conditions with a rainfall of only 100 mm/year, four main sitings have been identified as favourable to the occurrence of groundwater: foothills, dykes, tectonic disturbances, and depressions located in closed drainage (Dijon 1977). The depth of the boreholes drilled mainly in granite and gneiss rocks ranged from 15 to 70 m, with an average of 30–40 m. Drilling results were very poor; in Amsaga only 15% of the boreholes struck water. Borehole yield ranged between 0.05 to 2 $m^3$/hour. Groundwater, in most cases, was highly saline. In these conditions water-well drilling is a very expensive operation.

Dijon also reports the drilling results of several other African countries, some of which are summarised in Table 2.

In Cameroon, where rainfall ranges between 500 and 800 mm/year, shallow exploratory boreholes with depths of 15 to 30 m were drilled in fractured and weathered granito-gneiss rocks. Their yield was up to 2 $m^3$/hour, while those drilled in sands overlying schists and granite ranged between 20 and 50 $m^3$/hour. Out of 98 slightly deeper drilled boreholes (35–40 m), only 53 were productive (53%); with yields ranging between 0.5 and 4.0 $m^3$/hour; in exceptional cases yields reached 10–15 $m^3$/hour (Dijon 1977).

## 7 CONCLUSIONS

From the overall analyses of the groundwater conditions prevailing in hard rocks covering many African countries, the following general conclusions can be drawn:

- Rainfall regime and intensity, jointly with the geomorphologic conditions, have great importance in the occurrence of groundwater in hard rocks of the reported countries.
- Regional hardrock aquifers, in most cases, do not exist.
- The groundwater occurrence is irregular and is related mainly to the weathering and fracturing of the rock, but not to its composition.
- Water in solid, unfractured rocks may be found in deep, isolated fissures, occasionally interconnected.
- The probability of striking water-bearing fissures decreases with the depth of the well.
- Borehole yield from fractured rocks is generally low. It can, however, satisfy the water requirements of small villages and for watering livestock.
- Depths of successful wells in hard rocks may range from 10 to 100 meters.
- Promising hydrogeological conditions are found along river banks of well-defined, narrow alluvial plains where high-yield boreholes may strike water-bearing alluvial

deposits and thick weathered and fractured hard rocks. Such boreholes, as well as large-diameter open wells and infiltration galleries, are the main water sources for the irrigation of narrow alluvial riverine belts in hard-rock terrain.

- Water-quality problems affect most of the areas characterised by arid and semi-arid conditions; alternative solutions to solve the pressing water problems are to construct, wherever conditions are appropriate, underground dams across dry river courses, sand storage dams or rock catchments.

## REFERENCES

Bernardi A. (1973) Ricerca di acqua nelle formazioni cristalline e metamorfiche dello zoccolo Africano – Possibilita' e limiti dei metodi sismico a rifrazione, elettrico e magnetico. Proceedings of the 2nd Convegno internazionale delle acque sotterranee, Palermo, Italy.

Biemi J. & Sarrot-Reynauld J. (1993) Infiltration and groundwater flow conditions in the fissured Pre-Cambrian basement (Ivory Coast). Proceedings of the XXIVth Congress of AH, Hydrogeology of Hard Rocks, Oslo, Norway.

Dijon R. (1977) Ground-water exploration and development in hard rocks in Africa, International Seminar on Ground Water in Hard Rocks, Stockholm 1977.

Faillace C. (1973) Location of groundwater and the determination of the optimum depth of wells in metamorphic rocks of Karamoja, Uganda, proceedings of 2° convegno internazionale sulle acque sotterranee, Palermo, Italy.

Faillace C. & Faillace E. (1987) Water Quality Data Book of Somalia – Hydrogeology and Water Quality of Southern Somalia, GTZ , Eschborn, Germany.

Faillace C. (1979) Evaluation of the rural water programme in Liberia. Internal report, UN Lir/77/004 Water Resources Policy, Management and Legislation, Monrovia, Liberia.

GKW Consulting Engineers (1973) Liberia Water Supply. Feasibility study for six towns, Zwedru. Unpublished report.

GKW (1980) Water sector study – Data collection and analysis. Report on groundwater resources. Planning for water supply system in Liberia. Unpublished report.

CHAPTER 8

# The fractured rocks in Hellas

George Stournaras[1], George Migiros[2], George Stamatis[2], Niki-Nikoletta Evelpidou[3], Konstantinos Botsialas[1], Barbara Antoniou[1] and Emmanuel Vassilakis[1]

[1]*University of Athens, Faculty of Geology and Geoenvironment, Dep. of Dynamic, Tectonic and Applied Geology, Panepistimioupoli Athens Greece*
[2]*Agricultural University of Athens, Lab. of Mineralogy-Geology, Iera Odos Athens Greece*
[3]*University of Athens, Faculty of Geology and Geoenvironment, Dep. of Geography and Climatology, Panepistimioupoli Athens Greece*

ABSTRACT: The contribution deals with geotectonic position and hydrogeological behaviour of hard-fractured rocks (mainly igneous, metamorphic, but not karstified carbonate rocks, as defined by the I.A.H Commission on Hardrock Hydrogeology). In particular geotectonic zones within the Hellenic territory their lithological and structural conditions are characterised. The main aquifers are defined with general features of groundwater flow related to the formation of the rock mass fractures. Five typical hardrock areas were selected for a detailed hydrogeological analysis.

## 1 INTRODUCTION

In Hellas, hard-fractured rocks are extended in many areas (Figure 1). From a geotectonic point of view the territory corresponds to the Aegean or Hellenic Arc, in which the African plate converges to the European. The alpine orogenetic system of Hellas, the so-called Hellenides, is composed by a set of geotectonic zones, extended primarily in N-S direction, changing southwards, in an E-W direction. Hellenides are subdivided to the following zones (Figure 2): (a) *Internal (eastern) Hellenides*: Rhodope Massif, Servo-Macedonian Massif (or zone), Axios Zone, Pelagonian Zone, Sub-Pelagonian Zone or Eastern Hellas Zone; (b) *External (western) Hellenides*: Parnassos Zone, Pindos Zone (or Pindos-Olonos zone), Gavrovo – Tripolis Zone, Ionian Zone and Paxi Zone.

At this point it should be mentioned that, in the frame of the Alpine orogenetic phase, some tectonometamorphic – tectonic units, have been created subparallel and transverse to Hellenides structures. The most significant of them is the Atticocycladic unit and the Olympos – Ossa and Almyropotamos tectonic windows, which geotectonically correspond to the external Hellenides. We find that the fractured rocks occur mainly in the Internal Hellenides related to metamorphic rocks and ophiolites (Mountrakis 1985, Ferriere 1982, Katsikatsos et al. 1986, Krásný 2002, Meyer & Pilger 1963).

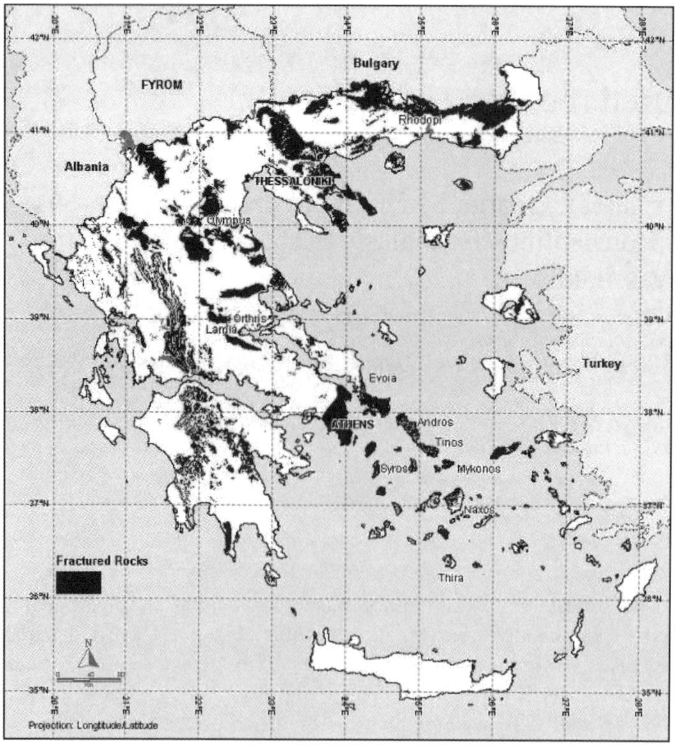


Figure 1. The fractured rock outcrops in Hellas.

## 2 THE CHARACTERISTICS OF THE GEOTECTONIC ZONES

(1) **Rhodope Massif** is characterized by the occurrence of gneiss (mainly orthogneiss), schists (mica-schist) and amphibolites, in alterations with marble horizons. The magmatic rocks are composed of Carboniferous to Cretaceous plutonic masses (granites, granodiorites, monzonites and diorites) and Eocene to Oligocene volcanic rocks (rhyolites, dacides, andesites and dolerites). The main aquifers of the Rhodope massif are related mainly with the metamorphic rocks, which are the typical fractured media included within the considered fractured rocks environment. Even in the case of the marble sequence, where karstification occurs, the horizons of schist and amphibolites into the sequence as intercalations could be included in the fissured rock sequence, and both plutonic and eruptive rocks are included within the considered fissured rocks.

(2) **Servo-Macedonian Massif (or zone)** is composed by a thick metamorphic formation of marbles, schists and gneisses, with intercalation of amphibolites, which is crosscut by Paleozoic to Oligocene granites. In the upper part of the sequence metagabbros, metadiabases, amphibolites and bodies of serpentinites are tectonically emplaced. The composition of the Servo–Macedonian corresponds to the typical hydrogeological environment of the fractured rock aquifers.

(3) **Axios Zone** is characterized by the intense tectonic events, where folding and thrusting occur and is subdivided (from E to W) in the sub-zones of: (a) Paeonia, (b) Paiko and (c) Almopia. The formations that occur are: (a) sedimentary and volcanosedimentary

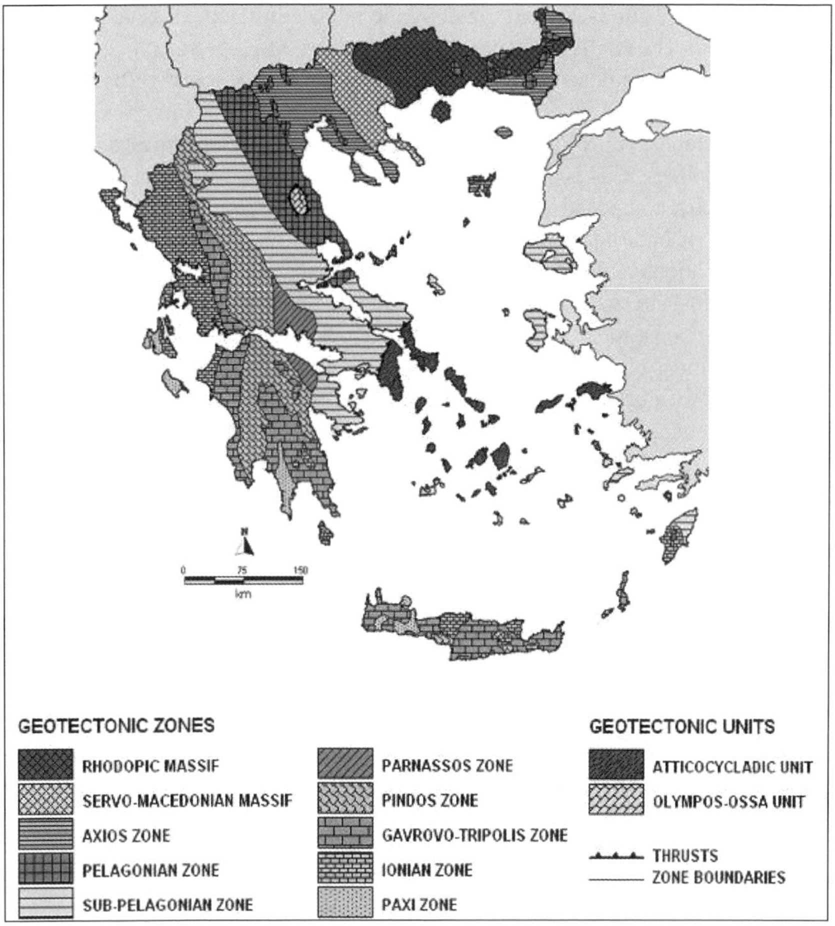


Figure 2. Geotectonic sketch map of Hellenides.

formations of Mesozoic age. These formations are composed of cherts, limestones, volcanic beds and intrusions, tuffs, and pyroclastics. (b) ophiolitic formations, which are mainly ultrabasic rocks, gabbros, pyroxenites, and basaltic lavas with intercalations of pelagic sediments and (c) formations of altered schists, sandstones, conglomerates, and limestones that are of Upper Jurassic-Cretaceous age. Due to the intense tectonics and to the nature of the formations, Axios zone is a classic example of the fractured rocks hydrogeological environment.

(4) **Pelagonian Zone** is composed by the following formations, from base to top: (a) crystalline basement consists of gneisses, granites/migmatites, schists and amphibolites; (b) Permian-Triassic meta-clastic sequence with volcanic rocks and tuffs; (c) Triassic-Jurassic carbonate sequences (marbles and dolomites) that correspond to the platform sedimentation of Alpine system. (d) ophiolitic formations that constitute an extent tectonic nappe; (e) Cretaceous formations composed of crystalline limestones and meta-flysch. Gneisses, schists, granites, and ophiolites are typical example of discontinuous media included within this zone.

(5) **Sub-Pelagonian Zone** is characterized by the intense lithological transformations over its whole extent. The stratigraphy of the zone, from base to top, is: (a) Neo-Paleozoic formations that are composed of sandstones, phylittes and schists; (b) Lower-Middle Triassic formations that are composed of sandstones, basic volcanic rocks and intercalations of limestones; (c) Middle Triassic-Upper Jurassic neritic limestones that include radiolarites, pellites, clay schists, etc; (d) Upper Jurassic-Lower Cretaceous clastic formations that are composed of radiolites, conglomerates, sandstones, and shales with olistostromes of ophiolites; (e) Pre-upper Cretaceous tectonic nappe, of volcano-sedimentary formations and ophiolites; (f) Upper Cretaceous limestone and flysch formations. The Neo-Paleozoic formations and the pre-upper Cretaceous formations of the tectonic nappe constitute the principal fracture rocks environment in this zone.

(6) **Parnassos Zone** is characterized by continuous Mesozoic neritic carbonate formations, typical karst configuration, and intense paleogeographic evolution, which fortmed three characteristic bauxite horizons. The carbonates pass upwards to flysch of Upper Cretaceous Lower Paleocene age. Pre-alpine basement is not recognized in this zone, and eruptive rocks are totally absent. The absence of crystalline or igneous rocks and the simultaneous presence of karst configuration in the neritic formations, excludes this zone from the typical hydrogeological environment of the fractured rock aquifers.

(7) **Pindos-Olonos Zone** is the "backbone" of the Hellenides Mountains and is characterized by deep sea sedimentary formations. Lithologically, the older deposits are dolomites and limestones of Middle-Triassic age, which in some cases also overlie clastic rocks (sandstones) of Middle-Triassic age. Upper-Triassic limestones of pelagic phase are present with intercalations of cherts and shales. During the Jurassic, alteration of cherts, clay schists, pelagic limestones, and sandstones occurred. Lower Cretaceous altered breccias, conglomerates, sandstones,and shales are the dominant rocks with characteristics similar to those of the flysch. A second sequence of thin-bedded pelagic limestones overlies the first flysch until the start of Paleocene, where the typical flysch (alterations of sandstones, shales, with intercalation of conglomerates) occurs. Despite the absence of crystalline and igneous rocks in Pindos zone, thin bedded limestones, radiolarites, and some horizons of flysch (sandstone), due to the intense folding and thrusting, cause quick drainage of the groundwater. The aquifers in those formations present very similar properties with the typical hard rock aquifer, especially because these limestones are not significantly karstified.

(8) **Gavrovo-Tripolis Zone** is composed by continuous neritic carbonates from the early stages to the end of Mesozoic. Limestones are thick-bedded to massive and are karstified. The carbonate rocks pass upward to the flysch formations of sand-stones, conglomerates, and shales. Groundwater flow in this zone is mainly controlled by large karstic conduits. Equivalent with this zone is the tectonic window of Olympos-Ossa.

(9) **Ionian Zone** is characterized evaporates above which there is a thick Triassic carbonate sequence. The overlying Jurassic consists of altered cherts with marly limestones and shales overlain by red pelagic limestones with intercalations of cherts. The flysch (Upper Eocene-Lower Miocene) of Ionian zone is composed of altered marls and sandstones in its lower parts. The upper parts are composed of altered marls, marly limestones, and conglomerates. Despite the fact that this zone is characterized by the absence of crystalline or igneous rocks, some sequences present typical characteristics of hard rock aquifers due to the intercalation of cherts and of marly limestones.

(10) **Attico-Cycladic Unit (Complex)** consists of concatenations of tectonic covers and windows, of ophiolites and metamorphic rocks (schists, gneiss-schists, amphibolites, greenschists, and glaucophane schists). Plutonic intrusions of granitic composition are Miocene age. The Attico-Cycladic Unit corresponds to the typical hydrogeological environment of the fractured rocks. This is very important for the Cyclades Islands, where the fractured rocks are the principal aquifer in an environment of limited precipitation, strong evaporative conditions, and limited island surface.

## 3 CASE STUDIES

**(a) Xanthi area.** The study area of the Xanthi region geotectonically is part of the Rhodope Massif and is characterized by marbles, gneisses, and schists (Figure 3) (Kronberg & Raith 1977, Meyer & Pilger 1963). The Rhodope Massif is subdivided into two tectonic units: (a) Sidironero with orthogneiss, micaschists, amphibolites, and intercalations of marbles and (b) Paggaeo, which consists of three sequences: (i) the lowest of orthogneiss, micaschists, schists and amphibolites (ii) the middle of marbles with intercalations of mica-schist and amphibolite, and (iii) the upper of schists and marbles. Plutonic rocks are granites, granodiorites, monzonites, and diorites. Their age ranges from Carboniferous to Cretaceous to Oligocene. The volcanic rocks of Rhodope are rhyolites, dacites, andesites, and dolerites,of Eocene-Oligocene age.

The orientation analysis of the tectonic and microtectonic regime Rhodope Massif is shown on Figure 3(a).

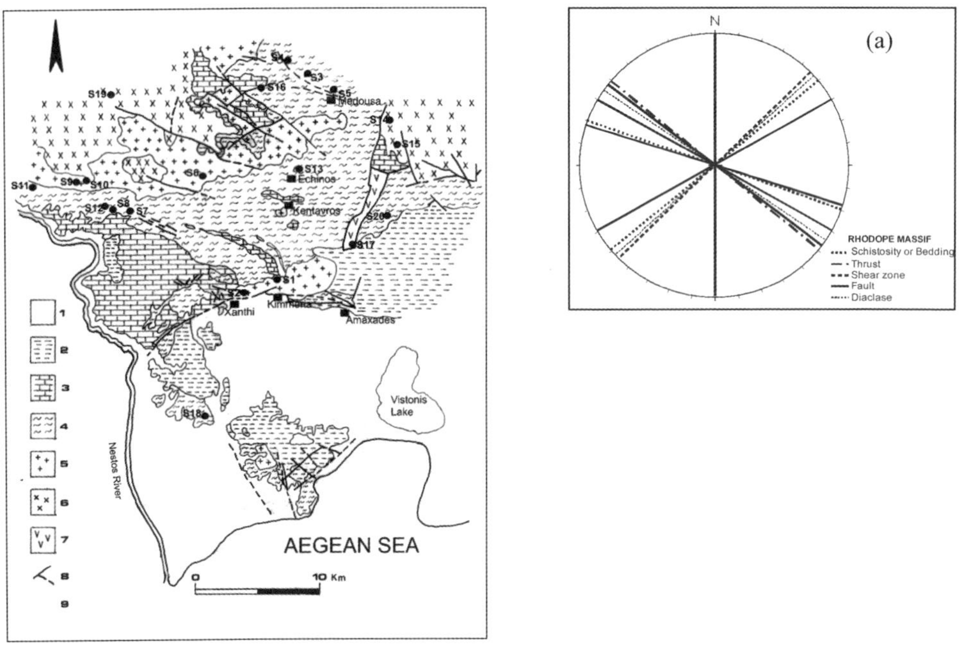


Figure 3. Geological map of Xanthi area (western Rhodope): 1. Quaternary formations; 2. Neogene formations; 3. marbles; 4. schists and gneisses; 5. granites; 6. volcanic rocks (andesites and cacites); 7. ophilites (mainly ultra basic masses); 8. faults; 9. springs (S). (a) Rose Diagram of tectonic elements.

The fractured rocks encountered in the massif of Rhodope are: (a) metamorphic rocks: gneiss, schist, amphibolites, migmatites, and marbles; (b) plutonic rocks: granites, granodiorite, monzonites, and diorites); and (c) volcanic rocks: rhyolites, andesites, dacites, and dolerites.

In the Xanthi study area springs emerge from marbles, volcanic rocks, granites, and schists. Thermal springs emerging from schists are also located (Table. 1, Figure 3).

*Springs emerging from marbles*. At the northeastern part of Xanthi, schists are characterized by lenticular intercalations of marbles from which springs issue. Their chemical characteristics are shown in Table 1. Geochemically these springs are classified as normal geo-alkaline, (mainly bicarbonate) and belong to the hydrochemical type Ca—Mg—$HCO_3$. The hardness ranges from 5.4 to 10.4°dH (German degrees), characterizing the water from soft to hard. TDS ranges between 179.6 and 350.0 mg/L with an average value of 250.0 mg/L. The dominant cations are the calcium and the magnesium the dominant anions are the bicarbonate and the sulfate. Generally, the waters are of good quality.

*Springs emerging from volcanic rocks*. Volcanic rocks springs are characterized by low values of dissolved solids, low hardness, and their alkaline character. These waters contain few metals and are classified as geo-alkaline, corresponding to the hydrochemical type Ca—Na—Cl—$HCO_3$.

*Springs emerging from granites*. The main characteristics of these springs are low dissolved solids, low hardness, and their alkaline character. They contain few metals and are classified as geo-alkaline, corresponding to the hydrochemical type Ca—Na—$HCO_3$.

*Springs emerging from schists.* The main characteristics of these springs are low dissolved solids, low hardness, and their alkaline character. The water is classified as normal geo-alkaline, corresponding to the hydrochemical type Ca—Na—Cl—$HCO_3$. The hydrogeological environment of these springs corresponds to the typical one of fractured rocks. Some of the springs emerge directly from fractures. In some other cases, groundwater percolates through the weathered mantle to the fractured zone and emerges through the intersection of the topographical surface and the fractured zone/bedrock contact.

*Thermal spring of Thermes*. The thermal spring of Thermes (Figure 3, spring S5) is located in Xeropotamos basin and issues from schists. Its occurrence is strongly related to fracture pattern of the surrounding area. The water temperature ranges between 33°C and 43°C. The "Thermes" spring has the hydrochemical type Na—Ca—HCO3 (Table 1). The amount of the dissolved solids reaches 1200 mg/l. The spring has alkaline characteristics and, according to the temperature, is classified in the category of mesothermal springs.

**(b) Olympos–Ossa area.** The geological composition of Olympos and Ossa mountains (Figure 4) is complicated. It is composed of several tectonic units, which are (Ferriere et al. 1998, Godfriaux 1962, Katsikatsos et al. 1982, Stamatis 1999, Stamatis & Migiros 2004):

*Olympos-Ossa Unit*, which is a series of crystalline limestones and dolomites equivalent to the Gavrovo Zone, from Triassic up to Eocene. This series ends in slightly metamorphic flysch.

*Ampelakia Unit*, which is mainly composed of schists, gneiss-schists, gneisses and prasinites, with some intercalations of marbles, meta-basalts, meta-greywacke and meta-pelites. The unit is characterized by blueschist phase metamorphism. From hydrogeological point of view, the formations of this unit present variable permeability connected to the presence of the gneisses or prasinites.

*Pelagonian Unit*, which is a complex, entirely metamorphosed unit of gneisses, granites/migmatites, amphibolites, schists, intercalations of marbles, and thrusts of metamorphosed

Table 1. Hydrochemical characters of the springs the Xanthi area.

| | Marble | | | Schists | | | Granites | | | Volcanic rocks | | | Thermal water |
|---|---|---|---|---|---|---|---|---|---|---|---|---|---|
| | Min | Max | Aver | Min | Max | Aver | Min | Max | Aver | Min | Max | Aver | |
| T °C | 14,7 | 17,3 | 15,6 | | | | | | | | | | 33–43 |
| pH | 7,8 | 8,7 | 8,3 | 8,0 | 8,1 | 8,0 | 8,8 | 9,1 | 9,0 | 7,5 | 8,1 | 8,2 | 7,3 |
| EC (μS/cm) | 235 | 484 | 332 | 126 | 185 | 156 | 11,5 | 19,5 | 15,5 | 61 | 130 | 96 | 1369 |
| $Ca^{2+}$(mg/l) | 24,1 | 54,5 | 43,1 | 22,8 | 27,6 | 25,2 | 8,8 | 13,6 | 11,2 | 6,0 | 15,1 | 10,7 | 136,4 |
| $Mg^{2+}$(mg/l) | 6,8 | 11,9 | 7,7 | 2,2 | 2,4 | 2,3 | 1,5 | 2,4 | 2,0 | 1,0 | 2,2 | 1,7 | 10,0 |
| $Na^{+}$(mg/l) | 4,7 | 12,0 | 10,5 | 3,1 | 4,0 | 3,6 | 9,2 | 18,0 | 13,6 | 5,2 | 6,7 | 5,8 | 171,8 |
| $K^{+}$(mg/l) | 1,0 | 2,2 | 1,4 | 0,7 | 1,3 | 1,0 | 4,1 | 4,3 | 4,2 | 0,9 | 1,4 | 1,1 | 19,6 |
| $HCO_3^-$(mg/l) | 79,3 | 201,3 | 140,5 | 67,1 | 88,4 | 77,8 | 32,9 | 34,2 | 33,5 | 12,2 | 54,9 | 35,1 | 652,7 |
| $Cl^-$(mg/l) | 7,8 | 26,5 | 13,6 | 10,6 | 12,4 | 11,5 | 5,3 | 7,1 | 6,2 | 5,7 | 10,6 | 7,5 | 64,5 |
| $SO_4^{2-}$(mg/l) | | | | 0 | 0 | 0 | 2,5 | 13,2 | 7,8 | 38,4 | 0,0 | 19,7 | 144,5 |
| $NO_3^-$(mg/l) | 0,5 | 32,0 | 4,7 | 0,2 | 17,0 | 8,6 | 0 | 0 | 0 | 0,0 | 17,0 | 4,3 | 1,0 |
| TDS (mg/l) | 179,6 | 350,0 | 250,0 | 125,0 | 135,0 | 130,0 | 81,5 | 93,6 | 87,6 | 37,9 | 103,8 | 75,4 | 1200 |
| Tot. Hard.* | 5,4 | 10,4 | 8,2 | 3,7 | 4,4 | 4,0 | 2,8 | 4,4 | 3,6 | 1,1 | 2,6 | 2,1 | 21,4 |
| Temp. Hard.* | 3,6 | 9,2 | 7,1 | 3,1 | 4,1 | 3,5 | 2,8 | 4,4 | 3,6 | 0,1 | 0,8 | 0,4 | 16,7 |
| Perm. Hard* | 1,8 | 1,2 | 1,1 | 0,6 | 0,3 | 0,5 | 0 | 0 | 0 | 1,0 | 1,8 | 1,7 | 4,1 |

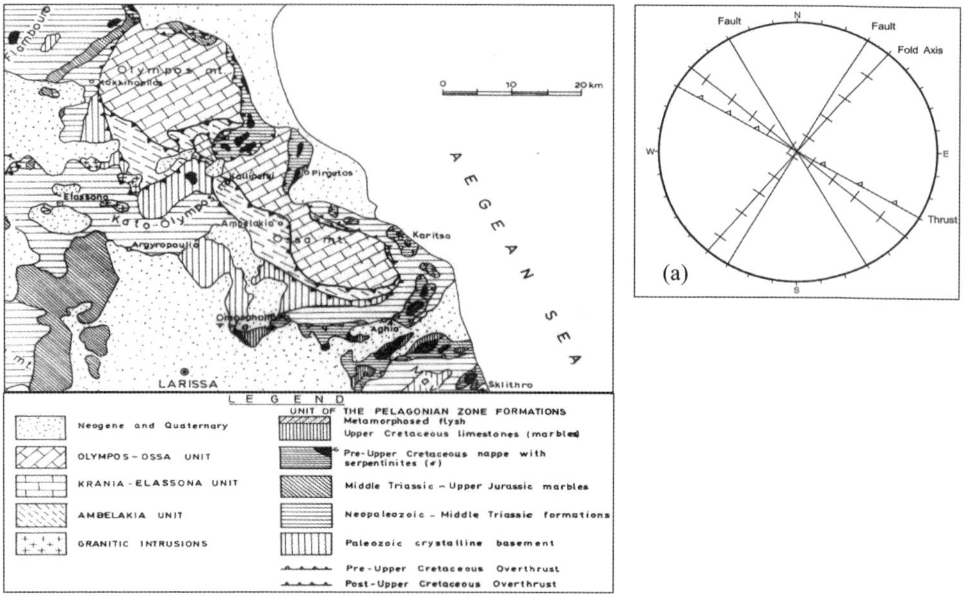


Figure 4. Olympos-Ossa geological map. (a) Rose diagram of major fault, thrust and diaclase orientations of Olympos-Ossa region.

ophiolites (basic and ultra-basic rock formations). The gneisss are connected with improvement of the hydrogeological behavior.

The main tectonic direction is N30°–60°W, which is combined with axes of structures, thrusts and faults. The direction N30°–70°E is a second major tectonic direction combined with axes of structures and faults. Related to the structural geology conditions that control groundwater flow within the fractured (hard) rocks of the given area, the rose diagram on Figure 4(a) shows that the orientation relation between folds and faults present a remarkable constancy.

Some hydrogeological and hydrochemical characters of the discussed groundwater manifestations are given in Table 2.

**(c) Othrys area.** Othrys is a mountain of Central Hellas consisting of Alpine formations that geotectonically belongs to the Sub Pelagonian zone. The fractured rocks are represented by the intercalations of basaltic lavas within a schist-chert sequence. Ophiolites and limestones are also present. The following sequences are encountered (Figure 5) (Ferriere 1982, Migiros 1990):

- *Chert-schist sequence.* This is a thin-bedded formation of chert, clayey schist, siltstone, that is intensively tectonised and folded.
- *Basaltic rock sequence.* These are volcanic rocks, fractured and alternated with chert and siltstone intercalations.
- *Ultrabasic rock sequence.* Peridotites with variable serpentinization that are intensively fractured. In the base of the system gabbros, basaltys, and amphibolites are encountered.

The area's faults and rupture surfaces are induced by the tensional post-orogenetic stresses.These are depicted in Table 4 and on Figure 5(a). Hydrogeologically, two groups of formations are distinguished at the base of their permeability. The first, mainly including

Table 2. Hydrochemical characters of the Olympos – Ossa area.

| | Marble | | | Gneisses | | | Ultrabasic | | | Metaflysch | | | Schists | | | Serpentinites | | | Thermal water |
|---|---|---|---|---|---|---|---|---|---|---|---|---|---|---|---|---|---|---|---|
| | Min | Max | Aver | Min | Max | Aver | Min | Max | Aver | Min | Max | Aver | Min | Max | Aver | Min | Max | Aver | |
| T°C | 11,7 | 14,9 | 13,63 | 11,5 | 17,6 | 13,83 | 6,5 | 12,3 | 9,8 | 9,6 | 13,6 | 11,98 | 11,4 | 14,6 | 13,2 | 7 | 7,2 | 7,1 | 14,5 |
| pH | 7,1 | 7,4 | 7,2 | 7,3 | 7,5 | 7,43 | 7,1 | 7,9 | 7,48 | 6,8 | 8,2 | 7,3 | 6,7 | 7,6 | 7,3 | 7,1 | 7,5 | 7,3 | 5,3 |
| EC (μS/cm) | 277 | 465 | 375 | 250 | 375 | 317 | 65 | 436 | 259 | 203 | 582 | 314 | 162 | 505 | 334 | 234 | 348 | 291 | 2447,0 |
| $Ca^{2+}$(mg/l) | 38,2 | 62,4 | 53,9 | 25,6 | 56 | 41,2 | 3,4 | 45 | 31,2 | 24 | 97,6 | 45,57 | 8,8 | 72,8 | 41,78 | 9,6 | 44,8 | 27,2 | 408,0 |
| $Mg^{2+}$ (mg/l) | 7,5 | 23 | 12,8 | 5,6 | 12,2 | 9,1 | 3,3 | 23,9 | 12,75 | 3,2 | 16,1 | 8,06 | 3,9 | 21,7 | 13,27 | 12,9 | 20,2 | 16,55 | 121,7 |
| $Na^+$ (mg/l) | 5,3 | 10,8 | 7,83 | 7,2 | 10,6 | 8,3 | 2,4 | 8,6 | 4,42 | 1,3 | 8,2 | 5,41 | 1,9 | 11,8 | 7,07 | 1,6 | 3,2 | 2,4 | 8,4 |
| $K^+$ (mg/l) | 0,4 | 1 | 0,6 | 0,1 | 0,7 | 0,35 | 0,3 | 0,5 | 0,37 | 0,3 | 0,8 | 0,48 | 0,3 | 1,4 | 0,88 | 0,3 | 0,5 | 0,4 | 1,5 |
| $HCO_3$ (mg/l) | 136,6 | 262,3 | 195,6 | 122 | 201,3 | 167,8 | 26,8 | 244 | 146,8 | 122 | 353,8 | 175,3 | 61 | 298,9 | 186,4 | 122,5 | 195,2 | 158,9 | 1750,7 |
| $Cl^-$(mg/l) | 7,1 | 24,8 | 15,2 | 14,2 | 24,8 | 18,6 | 3,5 | 17,7 | 12,4 | 5,3 | 17,7 | 12,3 | 5,3 | 28,4 | 14,9 | 5,3 | 7,1 | 6,2 | 14,2 |
| $SO_4^{2-}$ (mg/l) | 17,3 | 26,9 | 22,47 | 2 | 43 | 16,13 | 4 | 33,5 | 12,23 | 4,5 | 27,5 | 12,71 | 6,2 | 24,1 | 16,4 | 8,6 | 9,5 | 9,05 | 4,0 |
| $NO_3^-$ (mg/l) | 0,2 | 8,7 | 3,37 | 0,4 | 3,1 | 1,18 | 0,2 | 4 | 1,07 | 0,2 | 3,4 | 1,04 | 1 | 7,2 | 2,51 | 0,2 | 0,3 | 0,25 | 0,5 |
| TDS (mg/l) | 213 | 408 | 312 | 190 | 341 | 262 | 48 | 374 | 221 | 184 | 521 | 260 | 109 | 457 | 283 | 171,0 | 270,0 | 220 | 2309,0 |
| Tot. Hard. °dH | 7,1 | 14 | 10,5 | 6,2 | 9,5 | 7,8 | 1,3 | 11,8 | 7,2 | 5 | 17,4 | 8,2 | 2,8 | 14,6 | 9,1 | 5,6 | 9 | 7,3 | 85,2 |
| Temp. Hard. °dH | 6,3 | 12 | 8,9 | 5,6 | 9,2 | 7,7 | 1,3 | 11,2 | 6,7 | 5,6 | 16,2 | 8,1 | 2,8 | 13,7 | 8,5 | 5,6 | 9 | 7,3 | 80,4 |
| Perm. Hard. °dH | 0,8 | 2 | 1,5 | 0 | 0,6 | 0,3 | 0 | 1,1 | 0,4 | 0 | 1,2 | 0,4 | 0 | 2 | 0,6 | 0 | 0 | 0 | 4,8 |

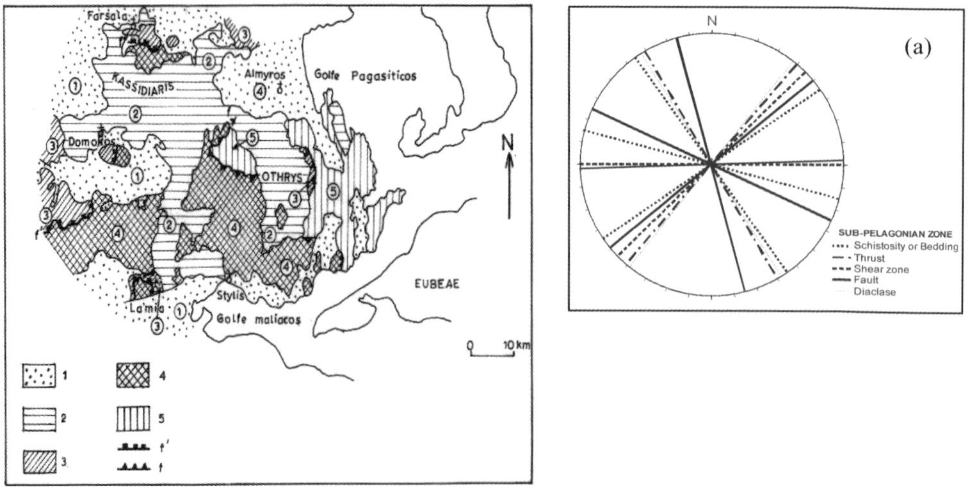


Figure 5. Geological map of Othrys Mt.: 1. Post Alpine formations, 2. Cretaceous formations, 3. Ultra basic masses, 4. Sedimentary-volcanosedimentary Triassic-Jurassic formations, 5. Neopaleozoic-Triassic formations, f-f′: Thrusts. (a) Rose diagram of tectonic elements.

Table 3. Hydrochemical characters of the springs in the Othrys area.

| | S1 A | S1 B | S1 C | S1 D | S1 E |
|---|---|---|---|---|---|
| Electrical conductivity (μS/cm) | 565 | 532 | 540 | 729 | 385 |
| pH | 11.1 | 11.1 | 11.0 | 10.9 | 10.8 |
| Water temperature (°O) | 22 | 20.8 | 19.2 | 24.2 | 23.5 |
| Air temperature (°O) | 28 | 28 | 28 | 28 | 29 |
| Discharge (l/h) | 226 | 226 | 1 | **960** | 50–100 |
| Permanent hardness °O dH | 6.18 | 6.12 | 5.5 | 2.13 | 4.55 |
| $Ca^{2+}$ mg/l/meq/l | 43.2/2.1 | 42.4/2.1 | 38.4/1.9 | 15.2/0.7 | 32.4/1.6 |
| $Mg^{2+}$ mg/l/meq/l | 0.5/0.04 | 0.7/0.06 | 0.5/0.04 | 0/0 | 0/0 |
| $Na^{+}$ mg/l/meq/l | 21.6/0.9 | 21.6/2.1 | 29.4/1.3 | 111.3/4.8 | 19.78/0.8 |
| $K^{+}$ mg/l/meq/l | 0.78/0.02 | 0.78/0.02 | 0.78/0.02 | 2.35/0.06 | 0.78/0.02 |
| $HCO_3^-$ mg/l/meq/l | **0/0** | **0/0** | **0/0** | **0/0** | **0/0** |
| $Cl^-$ mg/l/meq/l | 27.66/0.8 | 26.6/0.7 | 30.1/0.8 | 90.4/2.5 | 23.05/0.6 |
| $SO_4^{2-}$ mg/l/meq/l | 2.4/0.05 | 3.36/0.07 | 4.80/0.10 | 21.61/0.45 | 3.84/0.08 |
| $NO_3^-$ mg/l/meq/l | 0/0 | 0/0 | 0/0 | 6.20/0.10 | 0/0 |
| $NO_2^-$ mg/l/meq/l | 0/0 | 0/0 | 0/0 | 0/0 | 0/0 |
| $OH^-$ mg/l/meq/l | 25.16/1.5 | 23.80/1.4 | 23.46/1.4 | 22.44/1.32 | 13.7/0.8 |
| $SiO_2$ | 4 | 6 | 19 | 36 | 19 |
| TDS | 153 | 152.97 | 171.79 | 345.17 | 143.28 |

the serpentinized peridotites, represent the most permeable sequence of the fractured rocks. The second group, including the chert-schist sequence and the volcanic rocks, represent the relatively less permeable sequence of the fractured rocks. Data from the wells and the springs of the Othrys area given in Table 3.

**(d) Euboea Island.** Euboea Island is a complicated geological structure with the Pelagonian formations to the north, Sub-Pelagonian to the center and the tectonic window of Almyropotamos, equivalent to Olympos-Ossa, to the south. Therefore, we present the

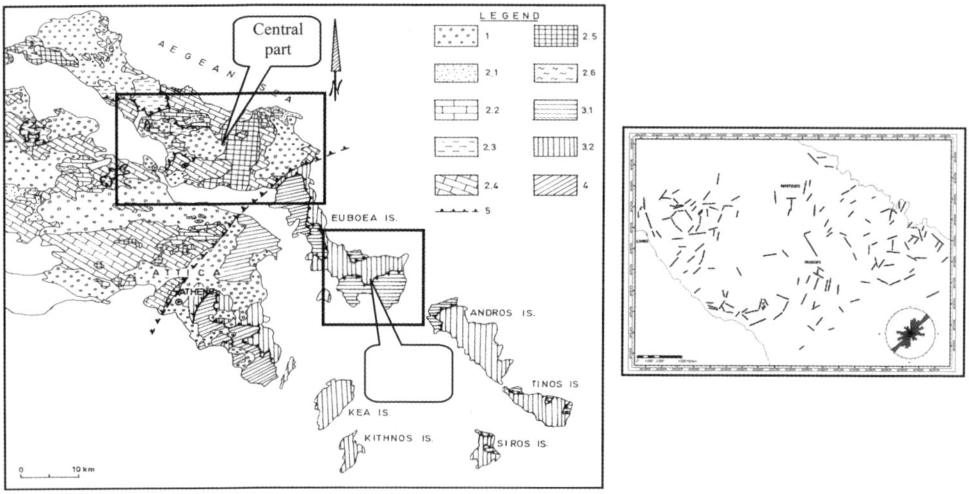


Figure 6. Geotectonic map of Euboea-Attica and N. Cyclades: 1. Neogene and Quaternary formations; 2. Pelagonian zone; 2.1. Flysch; 2.2. Upper Cretaceous limestones; 2.3. Eohellenic nappe formations; 2.4. Middle Upper Triassic-Upper Jurassic limestones and dolomites; 2.5. Neopaleozoic-Middle Triassic formations; 2.6. Crystalline basement; 3. Neohellenic nappe; 3.1 Ochi unit; 3.2. Styra unit; 4. Autochthonous system; 5. Overthrust. (a) Lineament map of Central part.

hydrogeological conditions of the fractured rocks in two different areas, central and southern parts (Katsikatsos 1977, Stamatis & Gartzos 1999, Stamatis et al. 2005).

**($d_1$) Central part of Euboea Island.** The central part of Euboea island corresponds to the typical structure of Sub-Pelagonian zone and consists of Palaeozoic basement covered by non metamorphic Mesozoic formations that present tectonic intercalations of ophiolites (Figure 6). The fractured rocks aquifers on the central (and north) Euboea mainly belong to the following sequences:

- The crystalline basement with a thickness over 800 m of gneiss and gneissic schist. The upper part is micaceous and amphibolitic schist; carbonate rocks are entirely absent.
- The Neopaleozoic sequence consists of sandstones, sandstones-schists, arkoses, graywackes, and clay schists.
- The Lower-Middle Triassic sequence consists of clayey-sandstones, basic volcanic rocks, and tuffs.
- The ophiolitic tectonic nappe is composed of volcano-sedimentary formations, ultra-basic bodies (serpentinized peridotites), gabbros, amphibolites, and basalts.

The main fractured aquifers are found in theophiolitic cover. Figure 6(a) depicts the lineaments in the ophiolitic cover, derived from a Landsat – 7 ETM+image, showing a NE-SW preferential orientation.

**($d_2$) Southern part of Euboea Island.** The southern part of the island constitutes the Almyropotamos tectonic window, equivalent to the the Olympos-Ossa tectonic window (Figure 6). The Ochi and Styra units, corresponding to the Ampelakia unit, are composed of marbles, cipolines, schists and quartzites. The hydrogeological conditions of the southern part (Figure 7, 8), show that spring locations are related to fracture patterns and to the drainage network. Some hydrochemical data are given in Table 4.

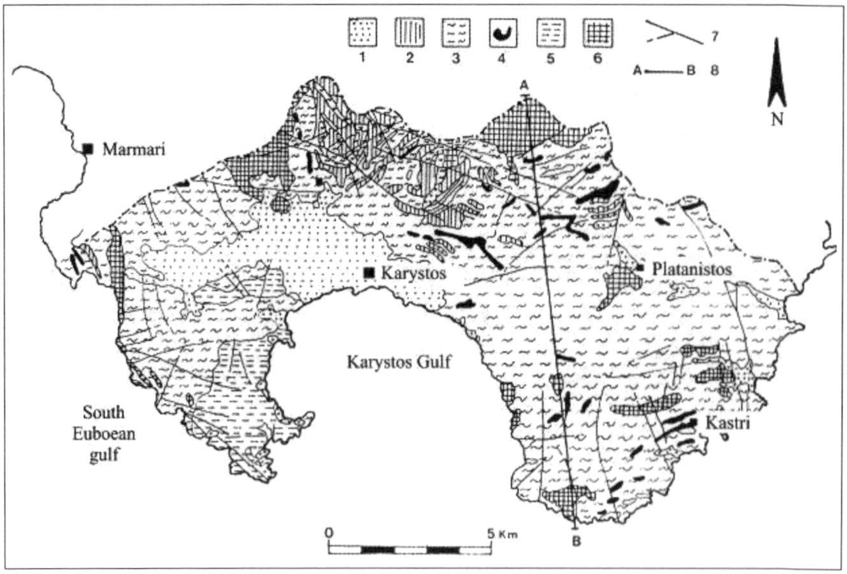


Figure 7. Geology of the study area 2: 1. alluvium, 2. marbles and sipolines, 3. schists 4. quartz schists 5. orthogneisses and 6. metamorphic basic rocks, 7. faults.

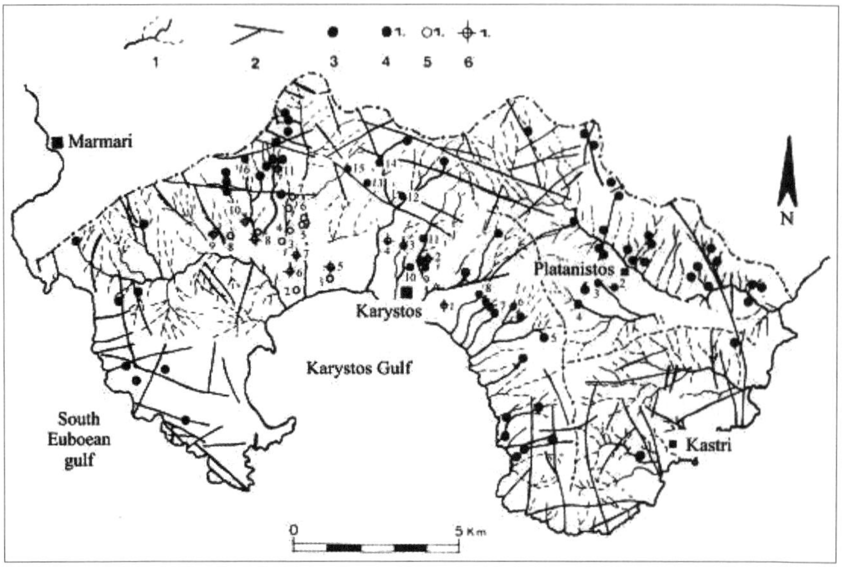


Figure 8. Drainage network and sampling points of the study area (1. Drainage network, 2. Faults, 3. Spring locations, 4. Sampling springs, 5. Sampling wells and 7. Sampling boreholes).

**(e) Cyclades Island Complex – Tinos Island.** The islands of Cyclades arein the central Aegean and constitute the Cycladic plateau. The largest islands of Cyclades (Naxos, Andros and Tinos) are located at the eastern part of the plateau and have the highest elevation. There are four islands, Kea, Kythnos, Serifos, and Sifnos, located at the western edge that

Table 4. Hydrochemical data of Southern Evoia.

| Parameters | Springs | | | | | | | | | Wells | | |
|---|---|---|---|---|---|---|---|---|---|---|---|---|
| | Marbles and Cipolines | | | Schists | | | Metabasites | | | | | |
| | min | max | aver | min | max | aver | min | max | aver | min | max | aver |
| T°C | 14.8 | 16.8 | 15.4 | 12.8 | 18.5 | 16.3 | 18.2 | 18.5 | 18.4 | 17.5 | 20.9 | 18.9 |
| $Q(m^3/h)$ | 4.8 | 35.0 | 17.9 | 1.7 | 60.0 | 28.1 | 0.1 | 2.0 | 1.1 | | | |
| pH | 7.7 | 8.1 | 7.9 | 7.6 | 8.2 | 7.9 | 7.5 | 7.6 | 7.6 | 7.2 | 8.4 | 7.9 |
| EC (μS/cm) | 413 | 489 | 451 | 289 | 617 | 386 | 830 | 848 | 839 | 587 | 1794 | 969 |
| $Ca^{2+}$ (mg/l) | 48.1 | 64.1 | 54.1 | 27.3 | 40.1 | 36.4 | 56.1 | 72.1 | 64.1 | 32.1 | 120.0 | 80.8 |
| $Mg^{2+}$ (mg/l) | 8.8 | 16.5 | 13.5 | 2.9 | 19.9 | 9.4 | 30.2 | 39.4 | 34.8 | 11.7 | 68.1 | 30.5 |
| $Na^+$ (mg/l) | 20.7 | 34.5 | 25.3 | 12.0 | 66.7 | 29.2 | 48.3 | 57.5 | 52.9 | 34.5 | 179.0 | 76.9 |
| $K^+$ (mg/l) | 0.8 | 2.3 | 1.4 | 0.4 | 0.8 | 0.7 | 0.8 | 0.8 | 0.8 | 0.8 | 5.9 | 2.5 |
| $HCO_3^-$ (mg/l) | 207.0 | 234.0 | 217.0 | 73.0 | 167.0 | 125.0 | 266.0 | 293.0 | 279.5 | 95.0 | 397.0 | 266.7 |
| $Cl^-$ (mg/l) | 28.4 | 53.2 | 36.4 | 21.3 | 141.8 | 50.5 | 88.6 | 117.0 | 102.8 | 53.2 | 195.0 | 99.3 |
| $SO_4^{2-}$ (mg/l) | 13.4 | 21.6 | 17.6 | 8.2 | 36.5 | 18.1 | 38.4 | 48.0 | 43.2 | 21.6 | 205.0 | 78.2 |
| $NO_3^-$ (mg/l) | 0.0 | 6.2 | 4.7 | 0.0 | 6.2 | 2.6 | 0.0 | 9.3 | 4.7 | 0.0 | 74.4 | 18.3 |
| $SiO_2$ (mg/l) | 7.6 | 14.9 | 10.2 | 10.1 | 14.6 | 12.8 | 20.3 | 20.3 | 20.3 | 12.2 | 34.5 | 20.9 |
| TDS (mg/l) | 347.6 | 418.2 | 380.1 | 207.7 | 366.0 | 278.2 | 581.0 | 604.8 | 592.9 | 436.7 | 814.8 | 674.2 |
| Total Hard. | 178.0 | 208.0 | 190.5 | 112.0 | 150.0 | 129.2 | 302.0 | 304.0 | 303.0 | 194.0 | 520.0 | 327.1 |
| Temp. Hard. | 170.0 | 192.0 | 178.0 | 60.0 | 137.0 | 102.5 | 218.0 | 240.0 | 229.0 | 78.0 | 325.0 | 219.2 |
| Perm. Hard. | 8.0 | 16.0 | 12.5 | 8.0 | 90.0 | 26.7 | 64.0 | 84.0 | 74.0 | 16.0 | 316.0 | 107.9 |

much smaller in size. The islands of Syros, Paros, Ios and Mykonos are located between these two groups of islands. The islands of Milos, Santorini and Amorgos are situated around the southern edge of the plateau and are peripheral expansions of the plateau (Botsialas et al. 2005, Leonidopoulou et al. 2005, Louis et al. 2005, Melidonis 1980, Melidonis & Triantaphyllis 2003, Stournaras et al. 2002, Stournaras et al. 2003).

**Geology.** The islands of Cyclades are generally characterized by metamorphic rocks such as mica-schists, marbles, gneisses, amphibolites, glaucophane schists and plutonian rocks (Figure 9). The principal metamorphic events occurred during the Tertiary (Eocene, Lower Miocene). The structures of metamorphic rocks are dominated by isoclinic folding, thrusts and re-folding during the Eocene–Oligocene.

**Tectonics.** The islands of Cyclades belong to the internal metamorphic zone of central Aegean that is part of the internal crystallic zone of the Hellenides. The recent tectonic history of Cyclades starts with the Eocene Alpine orogeny, a period of dominantly tectonic compression. This compression gave way to a tensile phase during Oligocene or Miocene; the result of this phase was the development of shallow normal faults. During Miocene, the sea depths at central Aegean were shallow with extensive emerging regions and small elongated basins, resulting from the intense compression. The compression resulted in the generation of a graben, which gave rise to the blueschists. During that period, low pressures transformed a mass of rocks into the greenschists. After the end of the faulting processes, the phase of low pressures was followed by granitic intrusions through the older faults. Volcanic activity in Cyclades took place mainly on the islands of Milos and Thira.

In order to describe the relationship between faults and fractures and lithological units, rose plots have been created for each lithological unit in Figure 9(a) and 9(b). These show that the dominant orientation for the faults of all lithologies is NW-SE. In the carbonate rocks there is a second dominant orientation of NEE–WSW; the soft rocks appear to have an important number of faulting zones which are oriented normal to the dominant faulting

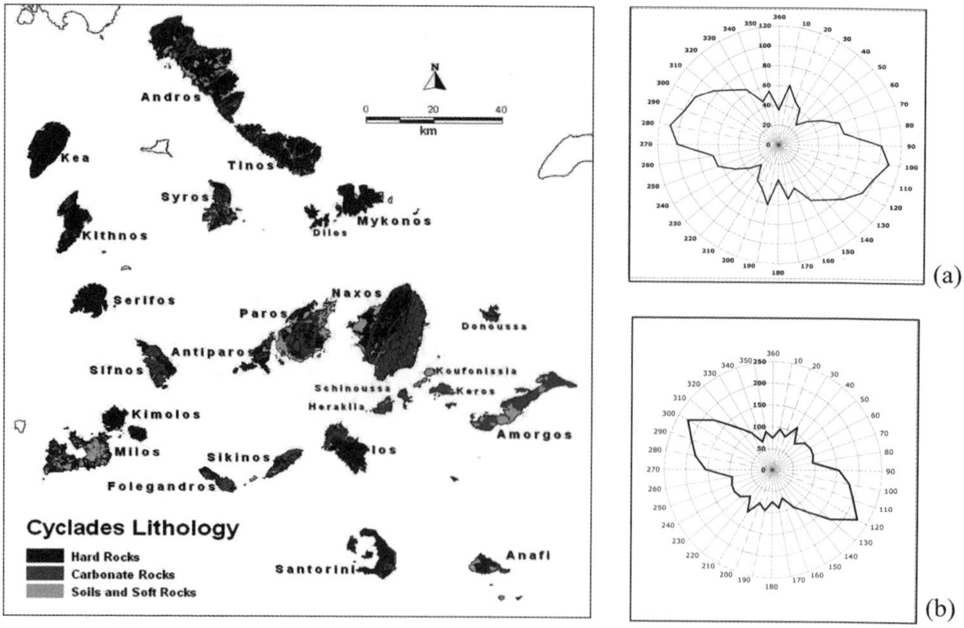


Figure 9. The main lithological units at Cyclades area. (a) Faults Rose Plot in soft rocks and (b) Faults Rose Plot in hard rocks in Cyclades area.

zones. The fissured rocks cover most of the Cyclades to explain the fact that the dominant faulting orientations are almost uniform for the whole area.

**Hydrogeology.** Cyclades islands are small in extent and are characterized by low annual rainfall, high medium annual temperatures, and high sunlight; the developed aquifers are of low capacity. Simultaneously most aquifers are unconfined and discharge to the sea. The main water-bearing rocks of the region, apart from carbonate rocks, are the hard, fissured rocks (granites, schists, gneisses, etc.).

**Tinos Island.** Tinos Island (Figure 10) belongs to the Attic-Cycladic Complex. Metamorphic rocks are classified into three tectonic units: (1) the Upper Unit composed of serpentinites, metagabbros, metabassalts, phyllites, and amphibolites, approximately 500 m thick; (2) Cycladic Blueschists more than 2000 meters thick and that cover the greatest part of the island. Meta-volcanic, clastic rocks and marbles also are present; (3) the Lower unit, derived from Mesozoic limestones, marls, shales, cherts, tuffs, basaltic vulcanite, and acidic rocks of probable volcanic origin. Magmatic rocks of the island are being classified into: (1) a complex of granite and granodiorite intrusion, which took place at early Miocene; (2) small outcrops of rocks of volcanic origin, with rhyolitic and andesitic composition.

A study area (Figure 11) was selected in Tinos to investigate, the hydrogeological environment of the hard rocks aquifers. From July to August of 2003, a spring inventory was conducted in the study area. Topographic maps at the 1/5.000 scale were used to plot 150 spring locations. Their spatial distribution shows that 95.3% of the springs, are located in hard rock formations (gneisses, gneisschists, glaucophane schists, greenschists, ophiolites, and prasinites).

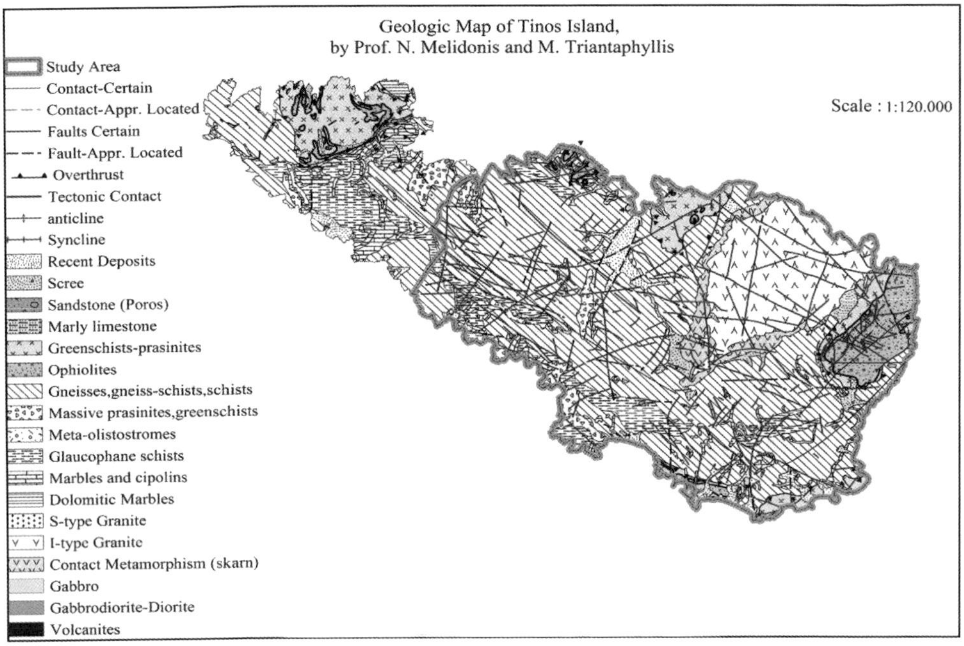


Figure 10. Geological map of Tinos Island.

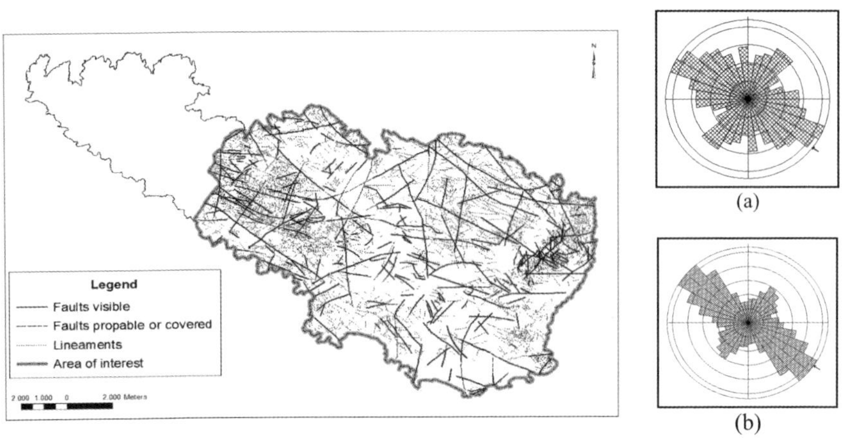


Figure 11. Faults and lineaments of the study area. (a) Faults Rose Plot and (b) Fractures Rose Plot.

Discharge, varies from 0.5 $m^3$/sec to 3.5 $m^3$/sec in the humid periods of the year. The majority of the springs (69.34%) are located on or near the streams of the drainage network, which is heavily influenced by the tectonic regime of the island. 82% of the springs are located very close to faults, fractures, or the intersection of two or more fractures. This is a strong evidence of the influence of the hard rock environment on the groundwater flow in the study area.

In order to obtain a reliable picture of the groundwater flow, it was necessary to depict the fractures systems of the study area. GIS and remote sensing techniques were integrated along with results from field study. A map of fractures (Figure 11) was found 3178 features that correspond to map-scale faults and lineaments from aerial photographs and satellite images. Most of these lineaments are easily identified in the field as steeply dipping to vertical large scale fractures and as meso-scale faults. The criteria of interpreting image lineaments and identify them as indicators of fractured zones of hydrogeological interest are: (i) length, (ii) directional distribution, (iii) the detection of anomalous directions, (iv) intersection, (v) the existence of a constant distance between lineaments of a directional group and (vi) relation between fracture density and the density of lineament intersections.

In most of the cases, the orientation of fractures is identical with the orientation of the preferential flow path. In Figure 11(a), the fractures rose plot shows that there are two sets of orientation. The main one has a NW-SE strike while the secondary one has NE-SW strike. The faults rose plot in Figure 11(b) reveals four orientation classes. The two main classes have NW-SE and NE-SW strike, and the secondary ones are of N-S and E-W strike.

The conclusions that were obtained from this study are:

(a) Four orientation classes of faults are located in the study area. The two main classes have NW-SE and NE-SW strike, while the secondary ones are of N-S and E-W strike.
(b) Lineaments are trending at the same strike with the main fault systems (NW-SE and NE-SW). Exceptions occur, where ductile tectonics affect the development of fractures.
(c) The occurrence of fractures is strongly linked with the proximity to the map scale faults. The majority of lineaments/fractures are located in the distance of 250 m to faults.
(d) The fractures density and degree of interconnection is depended on the combination of brittle and ductile tectonics, on the thickness of the weathered mantle, and on the lithology.
(e) Springs are strongly related or controlled by fracture location.

## ACKNOWLEDGEMENTS

We wish to thank the General Secretariat of Research and Technology (GSRT – Ministry of Development) for funding a part of the research leaded to this article (PENED Project).

## REFERENCES

Botsialas K, Vassilakis E, Stournaras G (2005) Fracture pattern description and analysis of the hard rock hydrogeological environment, in a selected study area in Tinos Island, Hellas, 7th Hellenic Hydrogeological Conference – 2nd Workshop on Fissured Rocks Hydrology Athens pp 91–100.

Ferriere J (1982) Paléogéographies et tectoniques superposées dans les Hellénides internes au niveau de l' Othrys et du Pélion (Grèce). Thèse sciences Univ. Lille 1982 et Soc. Géol. Nord. Publ. No 8 970 p. Lille.

Ferriere J, Reynaud J-Y, Migiros G, Proust J-N, Bonneau M, Pavlopoulos A, Houze, A (1998) Initiation d' un bassin piggy-back: Le sillon Mesohellenique. C.R.Acad. Sci. Paris Sciences de la terre et des planetes – Earth & Planetary Sciences 326:567–574.

Godfriaux I (1962) L'Olympe: une fenêtre tectonique dans le Hellenides internes. C. R. Acad. Sc. Paris 255 D 1761–1763.

Katsikatsos G (1977) La structure tectonique d'Attique et de l' île d' Eubée. VI Colloquium on the Geology of the Aegean Region I pp. 211–228.

Katsikatsos G, Migiros G, Vidakis M (1982) La structure géologique de la région de la Thessalie orientale (Grèce). Ann. Soc. Géol. Nord CI pp. 177–188.

Katsikatsos G, Migiros G, Triantaphylis M, Mettos A (1986) Geological structure of Internal Hellenides (E. Thessaly, SW Macedonia, Euboea-Attica-Northern Cyclades Islands and Lesvos). Geol. & Geoph. Res. Special Issue I.G.M.E. pp 191–212 Athens.

Krásný J (2002) Hard Rock Hydrogeology. 1st MEM Workshop on Fissured Rocks Hydrogeology Proceedings pp. 11–18 Athens.

Kronberg P, Raith M (1977) Tectonics and metamorphism of the Rhodope Crystalline complex in the Eastern Greek Macedonia and parts of Western Thrace. N. Jb. Geol. Palaont. Mh 11:697–704.

Leonidopoulou D, Stournaras G, Maroukian H (2005) Morphmetric analysis and geoundwater regime in Falatados-Livada drainage basin, Tinos Island, Hellas. 7th Hellenic Hydrogeological Conference – 2nd Workshop on Fissured Rocks Hydrology Athens. pp. 141–150.

Louis I, Botsialas K, Louis F, Stournaras G (2005) 2-D Resistivity imaging of weathering manles and fissured rock zones over metamorphic basement rocks of selected areas of Tinos Island, Greece. 7th Hellenic Hydrogeological Conference – 2nd Workshop on Fissured Rocks Hydrology Athens pp. 141–150.

Melidonis MG (1980) The geology of Greece: the geological structure and mineral deposits of Tinos island (Cyclades, Greece). Inst. Geol. Min. Explor. Athens 13:1–80.

Melidonis N, Triantaphyllis M (2003) Sheet Tinos–Yaros Islands. Institute of Geology and Mineral Exploration Geological Map of Greece 1:50.000 Athens.

Meyer, W, Pilger, A (1963): Zur Geologie de Gebietes zwischen Strymon und Nestos (Rhodopen massiv) in Griechisch Mazedonia. N. Jb. Geol. Palaont. Abh., 118, 272–280.

Migiros, G. 1990. The lithostratigraphical – tectonic structure of Orthris. (Central Hellas). Ann. Greek Geol. Soc., XXVI, p. 270.

Mountrakis, D. 1985 Geology of Greece. Univ. Studio Press, 207p., Thessaloniki.

Stamatis G. (1999): Hydrochemical groundwater investigations of Ossa/E-Thessaly (Central Greece), Hydrogeologie und Umwelt, Heft 18, 26s.

Stamatis G, E. Gartzos (1999): The silica supersaturated waters of Northern Evoia, and Eastern central Greece, Hydrological Processes, 13, pp. 2833–2845.

Stamatis, G, Migiros G (2004). Relation of brittle tectonics and aquifer existence of Ossa Mt. hard rocks (E. Thessaly, Greece). Bull. Of Geol. Soc. Vol. XXXVI. Proc. of 10th Int. Con., Thessalonica, (in press). (in Greek).

Stamatis, G Vitoriou A, Zaggana E (2005): Hydrogeological Conditions and quality of the groundwater, in the crystalline schist mass of Karystia area (S. Evoia). 7th Hellenic Hydrogeological Conference / 2nd Workshop on Fissured Rocks Hydrology, Athens, 2005. pp. 463.

Stournaras G, Alexiadou, M Ch., Koutsi, R. Athitaki Th. (2002). Main characters of the schist aquifers in Tinos Island (Aegean Sea, Hellas), 1st MEM Workshop on Fissured Rocks Hydrogeology Proceedings, pp.73–81.

Stournaras G, Alexiadou Ch., Leonidopoulou D (2003). Correlation of hydrogeologic and tectonic characteristics of the hardrock aquifers in Tinos Island (Aegean Sea, Hellas), International Conference on Groundwater in Fractured Rocks, Prague, pp. 103.

CHAPTER 9

# Determining a representative hydraulic conductivity of the Carnmenellis granite of Cornwall, UK, based on a range of sources of information

David C. Watkins
*Camborne School of Mines, University of Exeter, Tremough Campus, Penryn, Cornwall, UK*

ABSTRACT: This paper reviews the data available on the range, distribution and variation in bulk hydraulic conductivity (k) of the Carnmenellis granite outcrop in SW England based on information drawn from a number of studies and sources. It is suggested that the granite aquifer can be characterised by four vertically distributed zones: an uppermost extremely weathered intergranular flow zone of very high k, up to 2 to 3 m thick; a high permeability upper zone, 30 to 100 m deep with k of around $10^{-5}$ m/s and $10^{-6}$ m/s; a moderately permeable middle zone extending to a few hundred metres in depth and exhibiting k in the region of $10^{-7}$ to $10^{-9}$ m/s; beneath this the granite may be considered effectively impermeable, with k of around $10^{-9}$ and $10^{-10}$ m/s, though water flow at depth can still be considerable within major discontinuities, providing local zones with k of $10^{-5}$ m/s or more.

## 1 INTRODUCTION

The introduction in the UK of Integrated Pollution Prevention and Control (IPPC) licensing by the Environment Agency for operations such as landfill sites requires comment on, and sometimes modelling of, the local groundwater conditions and flow paths. Whether used for comprehensive numerical modelling simulations or for analytical calculations of residence times, some knowledge or estimate of the effective bulk hydraulic conductivity of the rocks beneath operations such as landfill sites is required.

The Carnmenellis granite outcrop in Cornwall, South West England, contains a number of abandoned and active quarries. Many closed quarries have been utilised as landfill sites, mainly for the disposal of inert waste materials. These still require IPPC licensing and associated hydrogeological studies. The introduction of IPPC requirements, coupled with generally increased environmental assessment of any new operations, means that accepted estimates of the hydraulic conductivity and of the nature of the groundwater flow conditions in the Carnmenellis granite are increasingly needed. At a recent planning inquiry into a landfill application, much of the hydrogeological argument was over whether the effective hydraulic conductivity of the granite was less than $10^{-9}$ m/s or whether it was many orders of magnitude higher. This argument was not resolved.

The Carnmenellis granite outcrop in SW England has been the subject of a number of hydrogeological investigations over many years, and for many different purposes. These include historic experience from the mining industry, geothermal energy exploration, deep borehole hydraulic testing, water well testing and recent environmental impact studies. Much of the information derived from these studies is recorded in relatively obscure publications and are not all readily available for reference.

The objective of this paper is to review available information related to the bulk hydraulic conductivity of the Carnmenellis granite, obtained from a number of different sources, and to provide a description of the groundwater flow system that may be of benefit to future hydrogeological and environmental studies or modelling. Although the granite is predominantly a hard rock fractured aquifer, it is a representative bulk hydraulic conductivity, as an equivalent porous medium, that is sought here.

## 2 GEOLOGY

The Carnmenellis granite forms a plateau, 150 to 250 m above sea level. The outcrop is almost circular in plan, 10 to 15 km diameter and covering an area of 125 km$^2$. It is located in the south west peninsular of Cornwall and its boundaries are within 1 km of the south coast, 6 km of the north coast and about 25 km from Land's End to the west, which marks the south-westernmost extremity of mainland Britain, Figure 1.

The Carnmenellis granite is one of a series of outcrops of the Cornubian batholith that underlies Cornwall and much of Devon in South West England, extending some 250 km

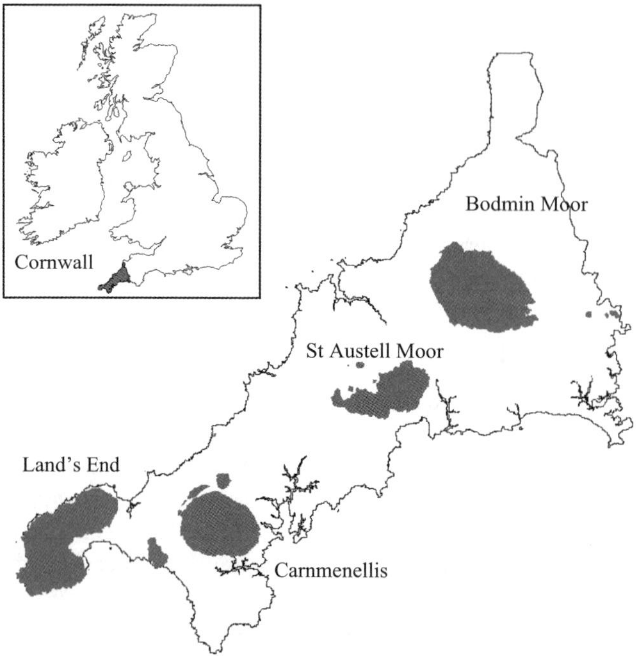


Figure 1. Location map showing granite outcrops in Cornwall.

from Dartmoor in the east to the Isles of Scilly in the west. The granite is Late Carboniferous in age and has been dated at around 290 Ma (Chen et al. 1993). A high degree of mineralization occurred during the Permian and this has provided the basis of extensive copper and tin mining industries, especially during the 18th and 19th Centuries.

The granite is generally coarse grained, though medium to coarse grained toward the centre of the outcrop and with some fine grained intrusions. It is of feldspar-rich porphorytic muscovite-biotite composition and in places has been altered by tourmalinisation, griesenisation and kaolinisation (Edmunds et al. 1990).

NW-SE trending features influenced deep weathering of the granite, and were probably initiated in late Tertiary times (Osman 1928). The most common set of steeply dipping joints trend NW-SE and ENE-WSW, parallel to major faults and mineral lodes, respectively (Ghosh 1934), and a subordinate pair trend WNW-ESE and NNE-SSW (Edmunds et al. 1990). In the uppermost 50 m, or so, horizontal joints are significant, though decreasing dilation with depth reduces their impact below this (Heath 1985).

During periglacial conditions in the Pleistocene the top of the granite was exposed to extreme weathering and Head deposits make up the top few metres in some places. The Head consists of frost-shattered angular granitic fragments in a sandy matrix.

## 3 HYDROLOGY AND HYDROGEOLOGY

Average annual rainfall varies from 1200 mm per year around the perimeter of the outcrop to 1500 mm per year at the centre. Potential evapotranspiration is around 530 mm per year (Smedley et al. 1989). The soils are generally thin and are mostly well-drained coarse sandy acid podzols. The high permeability of these soils results in high infiltration rates of 500–700 mm per year (Smedley et al. 1989), based on baseflow estimates derived from river and stream hydrograph separations.

Four main rivers (the Red, Kennal, Argal and Cober) drain the plateau, outward from the centre. The courses of each of these rivers follow major fault lines and lineaments in the granite, parallel to hydraulically transmissive fractures known in the mining industry as cross-courses, Figure 2. These appear to have formed as tension cracks, some of which appear to have reactivated as wrench faults and have typical fracture spacings between 0.5 km and 2.0 km (Randall et al. 1990). The cross-courses impact on groundwater flows to great depths, forming flow paths for thermal brines issuing into the South Crofty mine at 820 m depth. According to Smedley et al (1989) a dominance of flow in joints and cross-courses is aligned in a NW-SE direction due to the present-day horizontal stress pattern, the contemporary horizontal stress fields lying at an angle with the cross-courses and lodes such that those aligned NW-SE remain open.

Numerous elvan (rhyolite) dykes cross the area parallel to ENE-WSW striking mineral lodes. These contain microfractures that result in high permeabilities and preferential groundwater flow (Smedley et al. 1989).

Three public supply reservoirs are located on the outcrop, impounded by dams. The rivers tend to show very high baseflow indices, of around 70%. This is probably due to the high permeability of the soils and uppermost weathered zones, promoting recharge, and possibly the hydrological impact of historic mining and mine drainage and also the fact that the rivers connect with the aquifer through the highly permeable major lineaments.

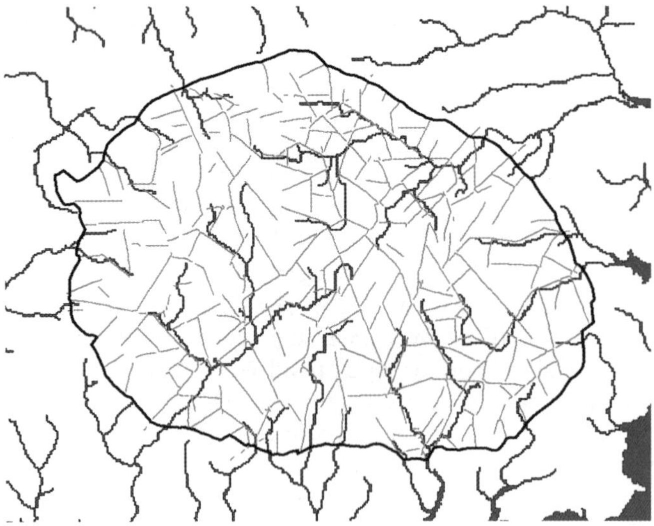

Figure 2. Major lineaments in the Carnmenellis granite outcrop (based on Edmunds et al. 1990), overlain by the courses of rivers and streams. The NNW-SSE trending features tend to be cross-courses whereas the SW-NE trending features tend to relate to mineral lodes.

The Carnmenellis granite is classified as being a minor aquifer, locally important for some private supplies (NRA 1992). The water table is rarely more than 10 m below the ground surface and is constrained by mine workings, springs and the elevations of streams and rivers. Borehole yields rarely exceed 100 $m^3$/day and average around 50 $m^3$/day with a 95% probability of striking water bearing fractures within 20 m of the ground surface (Edmunds et al. 1990).

## 4 HYDROGEOLOGICAL INFORMATION

Information on the hydrogeological characteristics of the granite were obtained from a number of sources, including:

### 4.1 *Mining*

The granite outcrop and surrounding area have been extensively mined, mainly for copper and tin. Over 3000 shafts are recorded in the area, mainly on the northern and southwestern parts of the outcrop. These reflect only a proportion of those actually present, as many of the older workings were not mapped. Alluvial and near-surface metal ores have been worked since the Bronze Age and deep mining is recorded from the 14th Century.

Traditionally, mines have been dewatered using adits (near-horizontal drainage tunnels bored from the lowest nearby locations in river valleys to intersect mineral-rich lodes and drain the groundwater by gravity). Surface water power was also used to drive pumps to dewater mines from below adit level. During the 18th and 19th Centuries, steam power replaced water power and enabled mines to be driven to much greater depths, the water being pumped up to adit level from where it discharged to watercourses.

Whilst water above adit level tended to provide a fairly diffuse fracture inflow, deeper seeps tended to be controlled by major fractures along discontinuities and major fault zones. The NNW-SSE trending cross-courses were notorious within the mining industry as sources of inflow and for creating hydraulic connections between the drainage systems of adjacent mines.

After the cessation of intensive mining activity at the end of the 19th Century, the majority of mines were abandoned and allowed to fill. Water now decants through the adit systems, into local watercourses. Historical mining activity still impacts strongly on the local groundwater flows through the lode structures, adits, shafts and cavities. Flooded mine workings affect the hydrogeology of the granite by increasing the overall storage and providing zones of preferential groundwater flow.

By the 1980s, only four deep mines were still working in the area: South Crofty, Pendarves, Wheal Jane and Mount Wellington. South Crofty and Pendarves are located on the northwest edge of the granite outcrop whereas Wheal Jane and Mount Wellington are situated on the southeast margin of the granite pluton. The combined groundwater pumping rate was 76,000 $m^3$/day (Burgess et al. 1982). The inflows to the working levels of Pendarves and Mount Wellington mines, at 200 m depth, were mainly through diffuse groundwater seepage, though discrete fissure flow is noted as well. The deeper working levels at Wheal Jane (300 m) and South Crofty (700–800 m) were drier with discrete inflows seeping through cross-courses and mineral lodes (Burgess et al. 1982), often described by miners as underground hot springs.

The last working deep mine in Cornwall was South Crofty, which penetrated deep into the Carnmenellis granite. Operations ceased in 1997 and the groundwater rebound took over three years until the mine water decanted at adit level (into the Red River) in 2000. Prior to closure, pumping rates from the different levels to various stages in the mine were as follows: 350 m depth = 70–120 litres/s; 600 m depth = 40–70 litres/s; 680 m depth = 30 litres/s; 750 m depth = 10 litres/s, averaging about 25,000 $m^3$/day in total. The ranges quoted above reflect seasonal variation, which reduces with depth. Whilst these inflows depend on the horizontal as well as the vertical extent of the mine workings, and cannot directly be related to permeability, they do show that significant inflows still occur at great depth and that lateral flow into the mine system occurs, as direct recharge is intercepted at the upper levels.

Adams and Younger (2002) numerically modelled the groundwater rebound of the South Crofty mine system. They applied a hydraulic conductivity of $10^{-10}$ m/s to intact rock regions and of $10^{-5}$ m/s to regions containing mine workings.

### 4.2 *Water wells*

Many small private water supply wells abstract from the granite for domestic and agricultural purposes. The wells are usually up to 100 m deep and there is rarely any benefit gained from drilling deeper.

Unfortunately, on the granite outcrop only two wells have time-drawdown data recorded and held by the Environment Agency. These two wells were drilled to 46 m and 73 m depth and exhibited specific capacities of 8.1 $m^2$/day and 1.3 $m^2$/day, respectively.

Analysis of the time-drawdown data using log-log and semi-log methods enable estimates of hydraulic conductivity to be made. The results show that for the two wells the hydraulic conductivities were around $10^{-6}$ m/s (0.08 m/day) and $10^{-7}$ m/s (0.01 m/day), respectively.

## 4.3 *Geothermal energy research*

A quarry close to the centre of the granite outcrop was the site of a deep drilling project to investigate the potential of hot dry rock (HDR) geothermal energy during the 1970s and 1980s. Two boreholes were drilled into the granite to depths of 2500 m. This has provided a wealth of information on the nature of the hydrogeology and fracturing of the granite at depth.

According to Batchelor (1978) the primary permeability of the granite blocks is in the order of $10^{-15}$ m/s to $10^{-13}$ m/s ($10^{-10}$ to $10^{-8}$ Darcy). Taking into account the secondary permeability, Durrance (1985) provides a bulk hydraulic conductivity of around $2 \times 10^{-7}$ m/s (0.02 Darcy).

Hydraulic testing of the deep boreholes showed an *in situ* hydraulic conductivity of around $1.5 \times 10^{-10}$ m/s at 2000 m depth (CSM 1984) though this was increased by deliberate hydrofracturing to $1 \times 10^{-8}$ to $5 \times 10^{-8}$ m/s between 2300 m and 2400 m depth (Parker 1989).

Randall et al (1990) mapped joint sets where the granite is exposed in nearby quarries. They report two steeply dipping sets striking NNW-SSE and ENE-WSW with a scatter in joint direction about these means of about $\pm 20°$, while the dip may be $\pm 30°$ either side of the vertical. They found that typical joint spacings were in the range 0.5 m to 2.0 m for the smaller joints, with larger joints every few tens of metres. Randall et al (1990) also mapped the joint sets in two deep boreholes using geophysical methods and confirmed the consistency of the joint orientation with depth. They used hydraulic tests to estimate the equivalent hydraulic aperture size and spacing. These indicated mean aperture sizes in the order of 50 μm and 100 μm, in comparison to 500 μm as estimated by Andrews et al (1986) from radon measurements. They also mapped mean joint spacings of 1.5 m and 1.9 m for the NNW-SSE joint set and 3.9 m and 6.0 m for the ENE-WSW striking set, in the two boreholes.

Burgess et al (1982) noted the significance of the deep penetrating cross-courses in the mechanisms for circulating and mixing of water between deep and shallow parts of the granite, resulting in thermal springs in deep mines. They base their observations on geochemical and isotopic data that indicates a mixture of young meteoric water with much older saline waters.

## 4.4 *Deep packer tests*

A series of deep boreholes were drilled up to 700 m depth at a quarry in the north west of the granite as part of a tracer test for research into fractured flow implications for underground nuclear waste repositories.

Six boreholes were drilled in the quarry within a floor area of just 20 m by 20 m. Four of these were packer tested using 1 m sections; three were tested to about 200 m depth whilst the fourth was tested to 650 m depth (Heath and Durrance 1985). The results are shown in Figure 3. They generally show hydraulic conductivities in the range $10^{-9}$ to $10^{-5}$ m/s, with $10^{-8}$ to $10^{-7}$ m/s being more typical. It is worth noting that some of the highest flows were encountered at the deepest levels.

## 4.5 *Shallow packer tests*

More recently, an environmental impact assessment was conducted in relation to the proposed conversion of a quarry in the southeast of the granite outcrop to a landfill site.

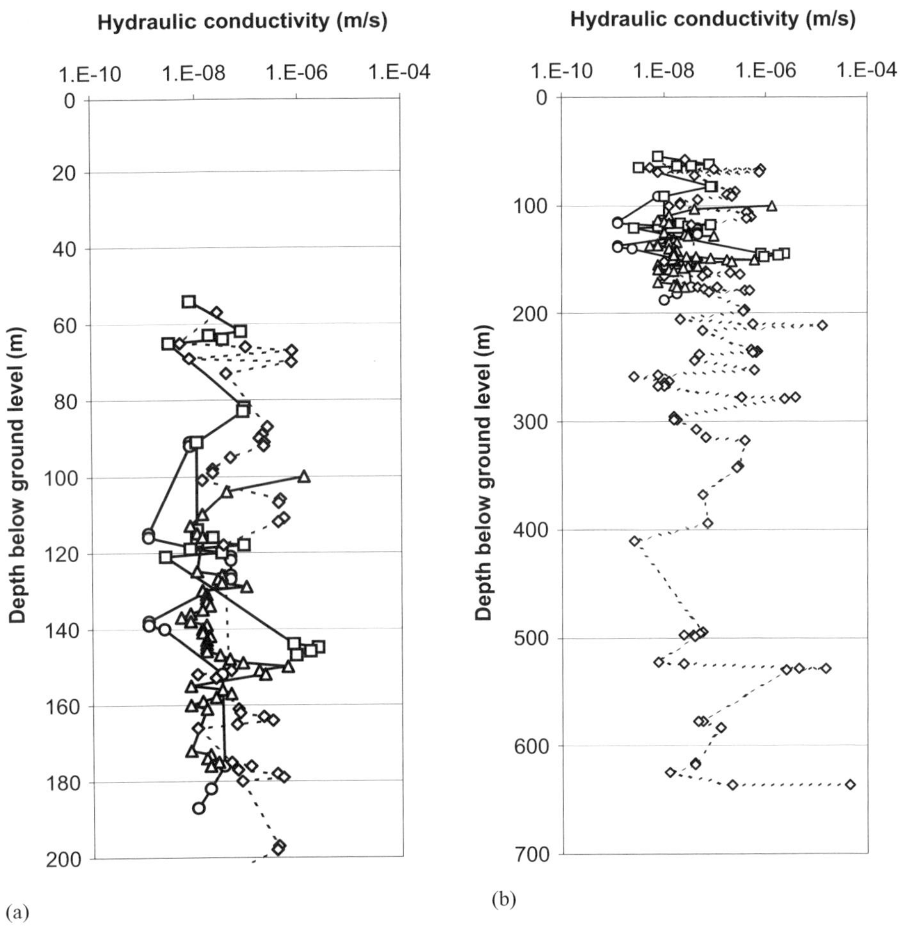


Figure 3. Deep packer tests, data from Heath and Durrance (1985). (a) to 200 m depth and (b) to 700 m depth.

As part of the hydrogeological investigations, boreholes were drilled and hydraulically tested and groundwater monitoring undertaken (Knight Piesold 2000).

As part of this, a series of comparatively shallow packer tests allow a profile in variation of hydraulic conductivity with depth and location to be made. The results are shown in Figure 4.

The analysis of time-drawdown data in terms of pumping, recovery and slug tests conducted in these boreholes provided bulk hydraulic conductivities in the range $10^{-8}$ to $10^{-5}$ m/s (Knight Piesold 1999).

## 4.6 *Numerical modelling*

In connection with the environmental assessment of the site described above, a numerical model of groundwater flow in the region around the quarry was constructed and calibrated against piezometer monitoring data (unpublished by the author 2001). The model simulated

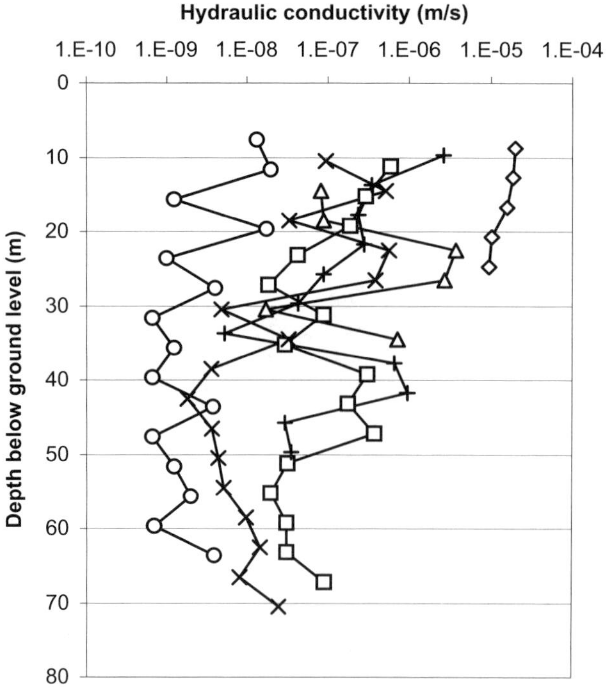


Figure 4. Shallow packer tests, data from Knight Piesold (2000).

the near surface movements of groundwater within the catchment area around the quarry, a region of 2 km × 2 km, between boundaries provided by river systems, a reservoir and a perceived catchment boundary, and in steady state only.

An areal infiltration recharge rate of 0.365 m/year was used (this value of 1 mm/day is considered to be conservatively low). By trial and error, a satisfactory fit with field data was obtained using a two-layer model with a 50 m thick upper layer (i.e., extending to 50 m below the surface topography) having a hydraulic conductivity of $10^{-5}$ m/s and a 350 m thick lower layer with a hydraulic conductivity of $10^{-7}$ m/s.

The values used for hydraulic conductivity are on the upper limits of those obtained from borehole tests, and yet the recharge value is considered to be conservatively low. A higher value of recharge could be accommodated and the model still calibrated by using even larger values of hydraulic conductivity or layer thickness.

The fact that the model requires values of hydraulic conductivity from the upper limits of those derived from small scale tests may reflect a scale dependent representative elemental volume due to the presence of sparse, but large, discontinuities – a common problem when applying an equivalent porous medium approach to fractured aquifers. The presence of major discontinuities in the form of cross-courses, lodes and dykes in addition to the joint sets may imply a multi-modal fracture distribution.

Further numerical modelling work was undertaken in order to assess the potential hydrogeological implications of "unabandoning" and dewatering the South Crofty mine (unpublished by the author 2004). A 3D finite element model was used to represent a 5 km × 4 km

Table 1. Properties of the vertically distributed zones used in a numerical model of the region around South Crofty mine.

| Zone | Description | Location | Hydraulic conductivity (m/s) |
|---|---|---|---|
| 1 | Soil/alluvium or head | Ground surface to 10 m depth | $1 \times 10^{-4}$ |
| 2 | High permeability upper zone | 10 m depth to 100 m depth | $1 \times 10^{-6}$ |
| 3 | Moderate permeability middle zone | 100 m depth to 500 m depth | $1 \times 10^{-8}$ |
| 4 | Unweathered lower zone | 500 m depth to 1000 m below sea level | $1 \times 10^{-10}$ |

region of the Red River catchment and the mine workings within it, from ground surface to 1000 m below sea level.

Calibration of the model was achieved using 4 zones for the unworked parts of the domain as listed in Table 1.

### 4.7 *Hydrochemistry*

Shallow groundwater has a low salinity with total dissolved solids (TDS) of around 150 mg/l, a pH of 4.3 to 6.9 and a temperature of 10–12° C (Smedley et al. 1989, Edmunds et al. 1990).

Deep groundwaters in the form of thermal springs encountered in mines have salinities up to 30,000 mg/l and temperatures up to 45° C (Smedley et al. 1989, Edmunds et al. 1990). The chemical composition of the deep groundwaters is dominated by mixing between shallow fresh waters and more saline deeper waters that have evolved without a marine influence (Burgess et al. 1982), the high salinities resulting from water-rock interactions.

The complex mixing makes dating of the thermal waters difficult. Based on isotope studies, Burgess et al (1982) conclude that the age of the saline component must be in excess of $5 \times 10^4$ years and could be up to $10^6$ years.

## 5 CONCLUSIONS

The granite aquifer is dominated by two main sets of sub-vertical joints and one horizontal set. The horizontal fracture plane tends to be non-transmissive beneath depths of 50 to 100 m whilst the sub-vertical joints may be quite permeable at depths of hundreds of metres. Small scale fractures tend to have joint spacings of 0.5 to 2 m and larger fractures occur at about 10 m spacings. Large scale discontinuities, due to mineral lodes, faults and dykes, may be highly permeable, especially those oriented in a NNW-SSE direction. These have spacings at intervals of about 0.5 to 2 km. They are responsible for discrete inflows into mines and for mixing between shallow meteoric water and deep saline waters.

Experience from mine dewatering indicates a generally regular inflow into the upper levels from uniform joint sets. With depth, however, the diffuse nature of the inflow diminishes and seeps become dominated by discrete inflows from faults and major fractures. At great depths, up to 800 m, the mines are relatively dry but with issues of thermal and saline water from cross-courses and other major discontinuities.

Water wells provide minor abstractions from the top 50 to 100 m of the granite. The limited information available suggests a hydraulic conductivity of $10^{-6}$ to $10^{-7}$ m/s for the granite into which they penetrate.

Research from HDR geothermal energy exploration shows that the fracture patterns encountered at surface extend to great depths. At 2000 m depth the hydraulic conductivity was found to be around $10^{-9}$ to $10^{-10}$ m/s.

Hydraulic testing of deep boreholes as part of a nuclear waste industry research study showed hydraulic conductivities between 50 m and 200 m depth varying between $10^{-9}$ m/s and $10^{-6}$ m/s, mainly in the range $10^{-8}$ to $10^{-7}$ m/s. A deep test also showed similar values to 650 m but with zones at 210 m, 275 m, 550 m and 650 m that reached hydraulic conductivities in the order of $10^{-5}$ to $10^{-4}$ m/s.

Shallower hydraulic tests on boreholes, conducted as part of an environmental impact assessment show hydraulic conductivities in the range $10^{-9}$ to $10^{-5}$ m/s in the top 70 m.

A numerical modelling exercise demonstrated that field values of phreatic levels and groundwater flow patterns may be simulated using a two layer model with the upper layer, 50 m thick, having a hydraulic conductivity of $10^{-5}$ m/s and the lower layer, 350 m thick, having a hydraulic conductivity of $10^{-7}$ m/s. The generally high conductivity values, in relation to those exhibited by borehole tests may indicate a scale influence, the distributed model encompassing a more representative range of fractures than those encountered in vertical boreholes.

Further numerical modelling of a 3 dimensional block incorporating a deep mine system resulted in the use of four vertical distributed zones: (i) 0–10 m below ground surface with hydraulic conductivity, k, $= 10^{-4}$ m/s; (ii) 10–100 m depth, $k = 10^{-6}$ m/s; (iii) 100–500 m depth, $k = 10^{-8}$ m/s and; (iv) 500 m depth to 1000 m below sea level, $k = 10^{-10}$ m/s.

## 5.1 *Summary of results*

In an attempt to provide a general description of the hydraulic conductivity of the granite aquifer, it may possibly be described in terms of four depth-dependent zones depending mainly on the degree of weathering:

1. Extremely weathered uppermost zone
   Flow here is intergranular due to the high degree of mechanical, periglacial, weathering of the granite resulting in frost shattered fragments *in situ* in a sandy matrix. This zone may be virtually absent or 2 to 3 m in depth. Its high permeability promotes groundwater recharge but has little other effect on the regional hydrogeology as it is generally above the water table.
2. High permeability upper zone
   Flow here is secondary through fractures but can generally be described as diffuse. This zone extends from surface (or the base of zone 1) to between 30 and 100 m depth. The hydraulic conductivity is generally around $10^{-5}$ m/s and $10^{-6}$ m/s. This is the dominant zone for shallow groundwater flow and for water abstractions from wells. Flow is dominated by closely spaced fracture sets including the horizontal plane.
3. Moderate permeability middle zone
   This zone extends from the base of Zone 2 to around a few hundred metres depth. The hydraulic conductivity is generally in the region of $10^{-7}$ m/s to $10^{-9}$ m/s, though regions of both higher and lower permeability exist within it. Horizontal fractures are generally closed or non-transmissive due to overburden pressures.

4. Low permeability lower zone
   This zone extends down from the base of zone 3. For most intents and purposes, this zone may be considered impermeable; the hydraulic conductivity is around $10^{-9}$ m/s and $10^{-10}$ m/s. However, deep discontinuities and fault zones exist within it allowing circulation and interaction with the upper zones. These zones may exhibit local hydraulic conductivities of $10^{-5}$ m/s or more, even at depth.

The zones described above may provide a useful start in attempting to generalise the hydrogeology of the granite, particularly when embarking on regional numerical modelling studies, but may not be the most pertinent approach for all situations. As a note of caution the distinction between zones is not always clear and the zones do not always exist at all locations. Indeed, the middle zone (3) appears to crop out in some quarry exposures. Also, the ranges in hydraulic conductivity found may overlap between zones, or one zone may grade into another.

For most regional modelling studies an equivalent porous medium approach can be taken, but (i) a higher value of hydraulic conductivity than those derived from hydraulic tests in boreholes may be required to account for scale-dependent fracture distributions and (ii) preferential flow paths caused by faults, mineral lodes and dykes may be present.

## REFERENCES


Adams R, Younger PL. 2002. *A physically based model of rebound in South Crofty tin mine, Cornwall.* In: Younger PL, Robins NS (Eds) Mine water hydrogeology and geochemistry, Geological Society of London Special Publication 198:89–97.

Andrews JN, Hussain N, Batchelor, AS, Kwakwa K (1986) Rn solution by circulating fluids in a hot dry rock geothermal reservoir. *Applied Geochemistry* 1:647–657.

Batchelor AS. 1978. *The engineering properties of the south western granites related to artificial geothermal exploitation.* Proc. Ussher Soc. 4:355–361.

Burgess WG, Edmunds WM, Andrews JN, Kay, RLF and Lee DJ. 1982. *Investigation of the geothermal potential of the UK. The origin and circulation of groundwater in the Carnmenellis granite: the hydrogeochemical evidence.* Institute of Geological Sciences, NERC, London, 69 pp.

Chen Y, Clark AH, Farrar E, Wastneys HAHP, Hodgson MJ, Bromley AV. 1993. Diachronous and independent histories of plutonism and mineralization in the Cornubian Batholith, southwest England. *Journal of the Geological Society* 150:1183–1191.

CSM. 1984. *CSM geothermal energy project. Low flow rate hydraulic tests Phase 2 report no. 2–52.* Camborne School of Mines, UK, 110 pp.

Durrance EM. 1985. Hydrothermal circulation and isostacy, with particular reference to the granites of southwest England. In *High heat production (HHP) granites, hydrothermal circulation and ore genesis.* Inst. Min. Metall., London: 71–85.

Edmunds WM, Davenport MF, Smedley PL, Kay RLF, Robins NS, Collier C, Burfitt KA. 1990. *The Carnmenellis Granite – hydrogeological, hydrogeochemical and geothermal characteristics.* British Geological Survey, Map.

Ghosh PK. 1934. The Carnmenellis granite: its petrology, metamorphism and tectonics. Quarterly *Journal of the Geological Society of London* 90:240–276.

Heath MJ. 1985. Geological control of fracture permeability in the Carnmenellis granite, Cornwall: implications for radionuclide migration. *Mineralogical Magazine* 49:233–244.

Heath MJ, Durrance EM. 1985. *Radionuclide migration in fractured rock: Hydrological investigations at an experimental site in the Carnmenellis granite, Cornwall.* Commission of the European Communities: Nuclear science and technology, EUR 9668EN, 68 pp.

Knight Piesold. 1999. *Carnsew Quarry proposed landfill development planning application, Volume 3: environmental statement.* Knight Piesold Limited, Ashford, UK. Held by Cornwall County Council.

Knight Piesold. 2000. *Carnsew Quarry proposed landfill development planning application: supplementary information.* Knight Piesold Limited, Ashford, UK. Held by Cornwall County Council.

NRA. 1992. *South West regional appendix to the National Rivers Authority groundwater protection policy.* Environment Agency, UK.

Osman CW. 1928. The granites of the Scilly Isles and their relation to the Dartmoor Granite. Quarterly *Journal of the Geological Society of London* 84:258–290.

Parker RH. 1989. *Hot dry rock geothermal energy, Phase 2B final report of the CSM Project, Vol 1.* Permagon Press, 677 pp.

Randall MM, Lanyon GW, Nicholls J, Willis-Richards J. 1990. *Evaluation of jointing in the Carnmenellis granite.* In: Baria R (ed) Hot Dry Rock, Camborne School of Mines International Conference. Robinson Scientific Publications, London:108–123.

Smedley PL, Bromley AV, Shepherd TJ, Edmunds WM, Kay RLF. 1989. *Geochemistry in relation to hot dry rock development in Cornwall, Volume 4. Fluid circulation in the Carnmenellis Granite: hydrogeological, hydrogeochemical and paleofluid evidence.* British Geological Survey Research Report SD/89/2, 117 pp.

CHAPTER 10

# Estimating the change of fracture volume with depth at the Koralm Massif, Austria

Gerfried Winkler[1], Hans Kupfersberger[2] and Elmar Strobl[3]

[1] *Institute of Earth Sciences, Karl-Franzens University, Graz, Austria*
[2] *Institute of Water Resources Management, Joanneum Research GmbH, Graz, Austria*
[3] *ZT Neubauer, Graz, Austria*

ABSTRACT: The fracture network and its fracture sets predominantly determine the permeability and storage capacity of fractured aquifers. The hydrogeological effectiveness of fracture sets can be statistically characterised by analysing the geological spatial attributes (trend, plunge, frequency) and the hydraulic relevant attributes (aperture, length, linear degree of separation, termination) of fractures. Fracture attributes are recorded at an exposure and along a borehole in the area of Koralm, Austria, both belonging geologically to the Plattengneis unit of the Koralm Complex having three fracture sets. It could be shown with discrete fracture modelling that for this geological unit the fracture set volumes correlate with depth by an exponential function.

## 1 INTRODUCTION

For numerical modelling of fractured aquifers it is necessary to determine and quantify the hydraulic attributes and the spatial distribution of the fracture network. At exposures the fracture network can be recorded with the scanline method and subsequently statistically analysed. As a result the volume and subsequently the hydraulic effectiveness of the individual fracture sets can be estimated. Exposures and their fracture network are predominantly influenced by weathering and slope tectonics. Boreholes can give important complementary information for the interpolation of the fracture network and its attributes with depth. In this work it is tried to combine fracture data observed at an exposure with fracture data obtained along a borehole to improve the estimation of a three dimensional fracture network. With discrete fracture modelling it is tried to extrapolate the modification of the fracture set volumes with depth.

## 2 LOCATION AND GEOLOGIC SETTING OF STUDY AREA

The investigation area is located in the east of the Speikkogel which is the peak of Koralm Massif. The Speikkogel is situated on the border of Carinthia and Styria both federal countries of Austria. The Koralm Massif lies about 50 kilometres to the south-west of Graz. Figure 1 shows the location of the study area.

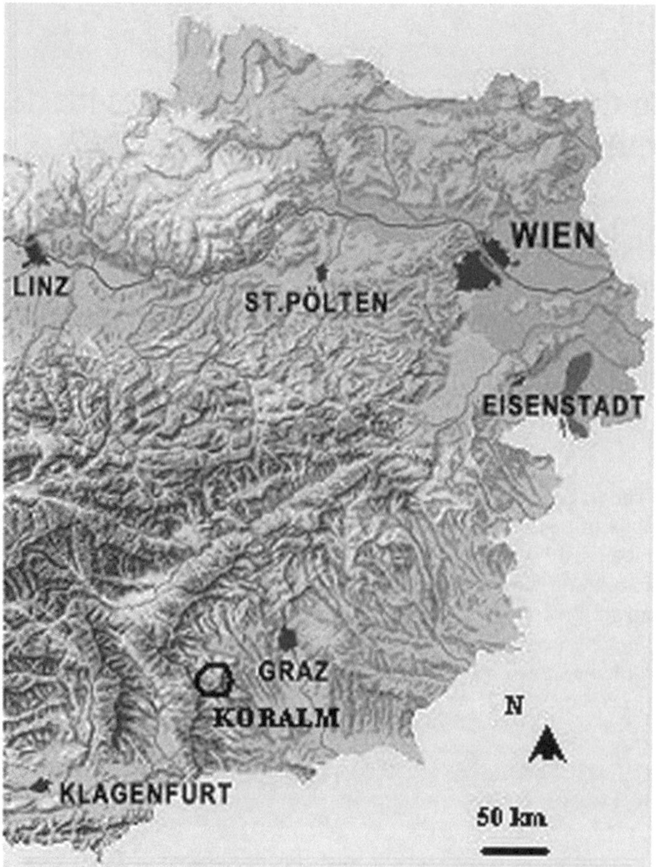


Figure 1. Geographical position of the investigation area Koralm.

The investigation area belongs geologically to the Plattengneis of the Koralm Complex (Eastern Alps). The Koralm Complex is part of the Koriden Unit within the Middle Austroalpine nappe complex of the Eastern Alps incorporated into the Austroalpine nappe stack during the Early Cretaceous (Kurz et al. 2002). This collisional event formed several regionally important shear zones. The Plattengneis represents one of the major shear zones within the Koralm Complex. Its main extension can be observed in the Koralm Complex and the Plankogel Complex. The thickness of the Plattengneis reaches up to 500 metres.

The Plattengneis is a schisted dark-grey to grey-brown gneiss. The weathering influenced rocks at the exposures tend to the grey-brown variation. Parallel to the foliation bright feldspar and quartz layers alternate with brown-grey biotite, garnet and muscovite layers. The thickness of the layers vary from a few millimetres to a few centimetres. The gneiss shows a distinct planar fabric (penetrative foliation) enhancing a perfect tabular fissility parallel to this foliation. Locally the Plattengneis tend to a variation of biotite gneiss. Its colour vary between dark-grey and grey-brown and the transitional zones are characterized by distinctly fewer feldspar and quartz layers.

The Plattengneis is observed at exposure KOR20, located at an altitude about 1500 metres above Adria extending below a gravel road down to the altitude of 1440 metres. TB02/00 is drilled down at an enlargement of the gravel road at the top of exposure KOR20. The structural-geological analysis show three main fracture sets forming the fracture network of the Plattengneis in the investigation area as it is shown in the structural plot in Figure 2e.

## 3 METHODOLOGY

In a first step the fracture sets and their volumes at exposure KOR20 were detected and specified. The results of the exposure were carried forward to TB02/00 and were combined with the data recorded along the borehole. Both data sets were used to generate a three dimensional DFN (discrete fracture network). Between the surface and a depth of 300 m no additional data were available.

### 3.1 *Data recording*

The hydraulic effectiveness of fracture sets can be inferred from two groups of fracture attributes. The first group contains the geological spatial attributes like trend, plunge and the frequency of fractures. The other group includes the hydraulic relevant attributes like aperture, trace length and linear degree of separation of the fractures.

At exposures the attributes of the fractures are recorded with the scanline sampling technique based on Priest (1993). In advance some recording guidelines were defined:

- the aperture should be measured on several points along one fracture and then averaged.
- fractures where the aperture cannot be measured are marked separately with "n.m." (not measurable) for later statistical analysis.
- the total length is the complete measurable stretch of a fracture.
- the linear degree of separation is the sum of the trace length sections of one fracture where an aperture can be observed.

### 3.2 *Statistical data analysis – orientation related fracture volumes*

#### 3.2.1 *Pre-processing*

In advance to the statistical analyses some biases have to be corrected. Biases can be caused by a) the sampling technique itself (Kulatilake et al. 1993) b) the immeasurable attributes (aperture, length) of fractures forming the exposure and c) the varying orientation of scan lines to the orientation of the intersecting fractures.

The maximum trace length is standardized by 2 metres and the maximum averaged aperture by 1 centimetre inferred from existing data. The fractures are weighted by their angle between the fracture and the scan line with

$$g_i = \frac{1}{l_s} \min\left( (1/\left|\langle x_s, x_i \rangle\right|, 5\right), \qquad (0)$$

where $l_s$ is the length of the scanline on which the fracture is detected, $x_s$ and $x_i$ are the vectors of the scanline and the fracture pole on the unit sphere surface respectively. The immeasurable fractures attributes on the exposed rock face are estimated using the attributes of similar orientated fractures (Harum et al. 2001).

Another goal of the fracture analysis is the determination of the fracture set volume. Therefore the individual fractures are weighted by the attributes aperture $o_i$ and linear degree of separation $d_i$. We assume as a first approximation that the attributes recorded on the surface continue through the whole rock mass. So the area defined by the aperture and linear degree of separation can be regarded as a representative weighting value. Considering these assumptions the weights $gv_i$ are given by

$$gv_i = o_i * d_i. \tag{1}$$

Combining the two weights $g_i$ (0) and $gv_i$ (1) the total weight $v_i$ for one fracture yields

$$v_i = g_i * gv_i. \tag{2}$$

This total weight describes the hydrogeological importance of a fracture based on the fracture volume.

### 3.2.2 *Cluster analysis*

The cluster analysis classifies objects into groups (clusters). The classification is based on the similarity and dissimilarity of the objects' attributes. So we have to define the similarity or dissimilarity (= distance) between the objects which can be defined in many ways. The cluster analysis describes the similarity of two objects by the distance between the two objects. This means: the stronger the similarity the smaller the distance. The definition of different statistical distances is well discussed by many authors including Hammah & Curran (1998 and 1999) and Steinhausen & Langer (1977). For the spherical data we consider the sine-square distance and the Mahanalobis distance of spherical data. The common distance of numeric data is the Euclidean distance. To combine the different kinds of distances, Steinhausen & Langer (1977) propose a **mixed distance** $d_M$. Let $d_r$ be the distance of the direction (sine-square or Mahalanobis of spherical data) and $d_e$ the Euclidean distance, then $d_M$ is defined as

$$d_M = \frac{m_r d_r}{m} + \frac{m_e d_e}{m}, \tag{3}$$

where $m_r$ is the number of the variables representing the direction (in the discrete demand $m_r = 3$), $m_e$ is the number of variables included in the Euclidian distance (in the discrete demand $m_e = 2$) and out of that $m = m_r + m_e$.

The **agglomerative hierarchical clustering** leads to an exact partition of objects. This means: that one object is sorted into exactly one group. At the beginning every cluster is defined by one object. The two clusters with the lowest distance are combined to a new cluster. So the total number of clusters is reduced by one. This step is repeated till the predefined number of clusters is reached. For the distance between two clusters we use the "mixed distance" as described above. After the first step one cluster can include one or more than one object. So we derive the distance between two cluster with the centroid-method calculating the "mixed distance" between the mean of the clusters.

The **fuzzy c-mean clustering** differs from the hierarchical clustering in two points. The number c of cluster have to be determined before starting the cluster analysis and each object is classified to each cluster with a certain degree of membership. That means that the objects are not classified to one cluster but can be assigned to the cluster with the smallest distance.

The results of both methods can now be compared.

The result of the clustering is the definition of groups containing fractures with similar spatial and hydraulic attributes which can be regarded as homogenous groups (clusters).

#### 3.2.3 *Cluster attributes*

The **center of gravity** describes the mean orientation of a fracture set (Wallbrecher 1986). The uncertainty of the center of gravity can be described by cones of confidence (95% and 99%). In the structural geology the cone of confidence is parametrically estimated. Therefore the data are assumed to follow a certain spatial distribution. One approach which is not bound on a certain spatial distribution is based on the nonparametric bootstrap method (Davison & Hinkley 1997). The advantage of that method is that the shape of the cone of confidence is a result of the empirical distribution of the data. So the shape of the cone of confidence is not bound on the shape of the assumed spatial distribution.

The orientation related **fracture volumes** define the hydrogeological effectiveness of fracture sets. Multiplying the aperture and the standardized linear degree of separation (1) leads to the standardized area (equal to the weight $gv_i$). Correcting the standardized area like the weights in (2) leads to a weighted area that is proportional to the fracture volume. So the sum of the weighted areas can be regarded as a first estimation of the fracture set volume.

## 4 SAMPLING AND PRE-PROCESSING

### 4.1 *Sampling*

KOR20 and TB02/00 were selected to analyse the modification of the fracture network with depth. Both were assumed to belong to the same geological unit (Stainzer Plattengneis) with the same tectonic stress regime. The scanline sampling technique was applied for the data recording.

Along 3 scanlines with a total length of 19.4 metres 79 fractures are recorded at exposure KOR20. In addition to the geological spatial and hydraulic relevant fracture attributes the terminations of the fractures are also recorded. Thus the termination index (ISRM 1978, Priest 1993) of the fracture sets can be calculated. This parameter gives important information about the interconnections between the fractures and is needed later to generate a discrete fracture model with the software FRACMAN.

The TB02/00 extends to a depth of 651 metres below surface. TB02/00 was analysed with acoustic well-logging along the total length. 1495 structures and their spatial orientation could be identified with a Borehole Televiewer (BHTV).

In four different depth sections (310–360 m, 427–446 m, 492–540 m and 570–647 m) the drill core is analysed cinematically in detail for tectonic features. Within these sections 154 fractures were recorded as probably open fractures and the aperture of 34 of these fractures could be measured. Therefore the apertures of all 154 fractures are calculated statistically based on the measurable data. With the result of the combination of these two techniques we were able to define the geological spatial orientation and the aperture of the fractures.

### 4.2 *Pre-processing*

The later analyses are based on the assumption that KOR20 and TB02/00 belong to the same tectonic environment. So the exposure and the five depth sections (310–360 m, 434–448 m, 500–540 m, 570–615 m and 618–648 m) are analysed structural-geologically.

For further analyses the fracture data of KOR20 and TB02/00 are plotted separately in Figure 2. Combining the fracture data of KOR20 and TB02/00 the analyses yield three clusters with a similar spatial orientation as the clusters of KOR20.

The fracture sets CL01 and CL03 can be observed over the total length of TB02/00 with a slight variation in their plunge. It can be observed that along the borehole the number of open fractures parallel to the schistosity (CL02) is obviously reduced (Figure 2 b) and c)). This is considered as a result of the increasing overburden pressure. Additional to the three clusters of TB02/00 CL03a can be observed. This cluster is considered as a sub-cluster of cluster CL03 caused by stronger spatial distribution.

Comparing the structural-geological plots of the total amount of the fractures along the drilling TB02/00 and the exposure KOR20 it can be derived that the fracture networks are similar (Figure 2). Both show three main clusters with the same spatial orientations.

Therefore it can be derived that both KOR20 and TB02/00 are deformed by the same tectonic stress regime considering that the increasing depth cause an increasing overburden

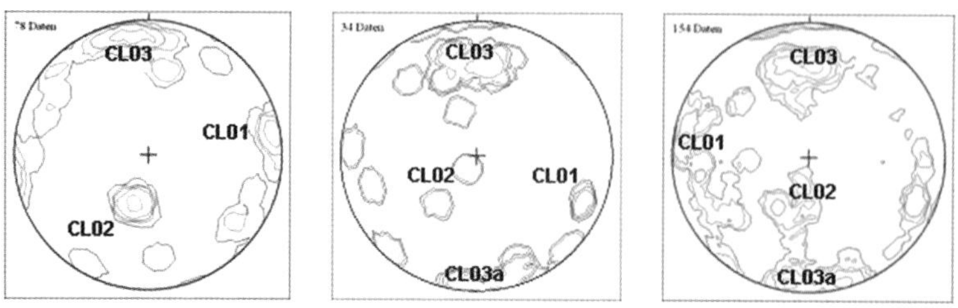


(a) KOR20: structural data (b) TB02/00: structural data of measured/ open fractures (c) TB02/00: structural data of potential open fractures

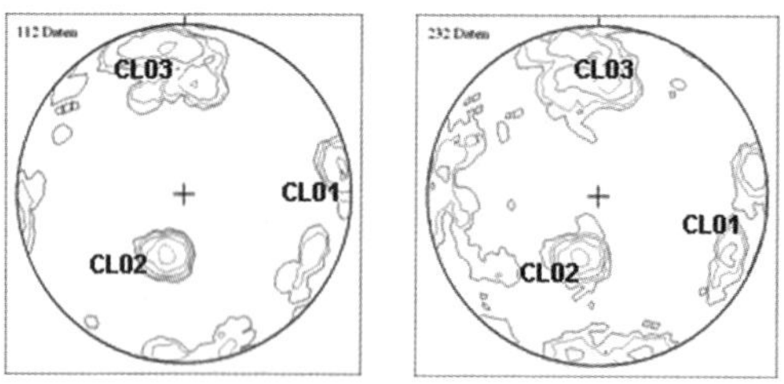


(d) KOR20 + TB02/00: structural data of measured/ open fractures (e) KOR20 + TB02/00: structural data of all fractures

Figure 2. Structural plots of KOR20 and TB02/00; distributions of the fracture sets, three main cluster CL01, CL02 and CL03 can be shown. TB02/00 shows an additional cluster CL03a as a sub-cluster of CL03.

pressure. Thus the number of the open fractures parallel to the schistosity decrease but it can be ascertained that there are still open fractures below a depth of 600 metres.

## 5 BOREHOLE ANALYSES

### 5.1 *Clustering analyses and estimation of the fracture volumes*

The data of KOR20 are analysed statistically using fuzzy c-mean clustering. The analyses yield three different fracture sets (clusters). Two fracture sets dip nearly perpendicular striking N-S and W-E, the third fracture set (foliation) dip into NNE with an angle about 30°. Subsequently the fracture set volumes are calculated/estimated. Additionally the termination index, the mean length (linear degree of separation), the shortest and the longest fracture of the set and the mean aperture is calculated (Table 1).

It can be seen that cluster CL01 has the largest weighted mean aperture and mean trace length and subsequently the biggest fracture volume (Table 1). Furthermore the termination index shows the best persistence and connectivity for CL01 compared to the others.

The data logging of the fractures along the borehole yield the apertures and the spatial orientation of each fracture but no information about the fracture length, their linear degree of separation and the frequency distribution of the fracture length within one fracture set.

The influence of weathering and slope tectonics cause in many cases the maximum loosening of the rock mass at the surface exposure. Therefore it is considered that trace lengths at exposures on the surface reach their maximum extension. So it is assumed that the exposure fractures are as large as the fractures along the borehole and that the trace length distribution within a fracture set is used. The frequency distribution of the aperture and the linear degree of separation of KOR20 is best fitted by an exponential function. So using the results of the data analyses of KOR20 it is possible to estimate statistically the lengths of the fractures of TB02/00.

This enables to determine the spatial orientation, the linear degree of separation (length) and the aperture of the fractures for further investigations and calculations. So all essential fracture attributes are available for the clustering method and the estimation of the fracture set volumes.

The cluster analyses of KOR20 and TB02/00 yield three clusters and their volumes for 5 depth sections between 300 m and 600 m and at the surface (Table 2). In all sections CL01 has the biggest fracture volume ranging between 0.06% and 0.02% and the other clusters

Table 1. Results of the statistical clustering of KOR20; three clusters with their spatial orientation, the types of trace terminations: $N_a$ = fracture terminates at another fracture, $N_o$ = termination is obscured and $N_i$ = terminates at intact rock material, the calculated termination index (ISRM 1978) and the weighted mean apertures and lengths.

| | Trend | Plunge | $N_a$ | $N_o$ | $N_i$ | Termination index | Mean length (cm) | $L_{min}$ (cm) | $L_{max}$ (cm) | Mean aperture (cm) | Fracture set volumes |
|---|---|---|---|---|---|---|---|---|---|---|---|
| CL01 | 269 | 84 | 34 | 6 | 2 | 4.8 | 158.3 | 30 | 400 | 0.303 | 0.32 |
| CL02 | 15 | 30 | 45 | 2 | 9 | 16.1 | 156.7 | 2 | 700 | 0.050 | 0.05 |
| CL03 | 171 | 76 | 32 | 5 | 21 | 36.2 | 80.5 | 3 | 600 | 0.081 | 0.05 |

Table 2. Results of the clustering and fracture volume analyses of KOR20 and the five sections of TB02/00.

| | Cluster 01 | | | Cluster 02 = foliation | | | Cluster 03 | | | Cluster 04 | | | Sum |
|---|---|---|---|---|---|---|---|---|---|---|---|---|---|
| | Trend | Plunge | Vol% | Trend | Plunge | Vol% | Trend | Plunge | Vol% | Trend | Plunge | Vol% | Vol% |
| Section 01 (310–360 m) | 280 | 83 | 0.06 | – | – | – | 167 | 63 | 0.01 | – | – | – | 0.07 |
| Section 02 (434–448 m) | – | – | – | 3 | 31 | 0 | 167 | 72 | 0.01 | – | – | – | 0.01 |
| Section 03 (500–540 m) | 90 | 83 | 0.02 | – | – | – | 177 | 55 | 0.01 | 7 | 77 | 0.01 | 0.04 |
| Section 04 (570–615 m) | 106 | 71 | 0.02 | 29 | 21 | 0 | 189 | 46 | 0 | 13 | 73 | 0.01 | 0.03 |
| Section 05 (618–648 m) | 94 | 83 | 0.02 | 46 | 51 | 0.01 | – | – | – | – | – | – | 0.03 |
| KOR20 | 269 | 84 | 0.32 | 15 | 30 | 0.05 | 171 | 79 | 0.05 | – | – | – | 0.42 |
| KOR20 + TB02/00 | 94 | 88 | 0.12 | 17 | 33 | 0.03 | 176 | 70 | 0.02 | – | – | – | 0.17 |

are obviously smaller (about 0,01%). The predominance of CL01 can also be observed at KOR20. CL02 which is orientated parallel to the foliation has the lowest volumes.

For a continuing quantification of the fracture network modification with depth it is necessary to get additional data within the first 300 metres below surface which is approached by discrete fracture modelling.

## 5.2 *Generating discrete fracture networks*

The idea is to generate a discrete fracture network (DFN) of the recorded data to interpolate a DFN and its fracture set volumes in sections that are not observed. The data of the recorded and analysed fracture sets of KOR20 and TB02/00 are the data basis for the generation of the DFN. Furthermore it is controlled if the DFN and especially its fracture set volumes are reproducible with the sampling technique applied at the exposure KOR20 and the borehole TB02/00. Combining the interpolated DFN and the recorded data improves the estimation of the change of the fracture set volumes correlated to depth.

### 5.2.1 *Methodology*

For this approach the model region should contain the borehole TB02/00 and the exposure KOR20. The modelled DFN can be generated with the data of TB02/00 and KOR20. Subsequently the DFN is analysed with the same sampling techniques as the techniques used for the data recording at exposure KOR20 and TB02/00. Scanlines can be aligned on theoretical exposed rock faces and boreholes along which the intersecting fractures and their attributes can be recorded. Subsequently the recorded data is analysed statistically with the clustering method and subsequently the fracture set volumes can be calculated.

So it is possible to calibrate the discrete fracture model iteratively at the observed sections of KOR20 and TB02/00 using the volumes of the fracture sets for calibration. Based on the calibrated DFN additional depth sections in the first 300 metres below surface can be imposed.

The calculated fracture set volumes which are listed in Table 2 show a modification correlated to depth. The reduction of the fracture set volumes are based on polynomial functions. The few data points (5 depth section) of each fracture set enhance the good fitting of the polynomial function ($R^2$ range between 0.93 and 0.99). The fracture set volumes below 400 metres have approximately constant values ranging between 0.02% and 0.01% for each fracture set regardless of their orientation. CL01 is considered as the dominant fracture set with the highest hydraulic effectiveness.

It can be observed that some fractures below a depth of 300 metres have apertures exceeding 5 millimetres. Therefore it can be deduced that the fracture apertures do not generally decrease with depth. So the reduction of the fracture set volumes seems to be predominantly a result of the reduced number of open fractures.

### 5.2.2 *Discrete Fracture Network (DFN) Model*

For the discrete fracture network generation the BART-Model ("Enhanced Baecher Model" in Dershowitz et al. 1999) is selected because of its parametrisation. The BART-Model uses a Poisson-point process to locate fracture centres in space randomly. No further input is required for the location model. The general shape of the fractures is defined by a sexangle. This sexangle is defined by the radius of a circular area which has the same dimension as the sexangle. The recorded trace lengths and their frequency distribution are the basis for the determination of the diameters of the circles and subsequently of the sexangles. The area of each fracture is dissolved into a finite element mesh where the finite elements are triangular shaped. With this refinement it is possible to determine and to quantify the interconnecting of the fractures.

The DFN is generated with the software FRACMAN (Dershowitz et al. 1998 and 1999). The software version used in this work is based on an evaluation licence and has some limitations. Thus it was not possible to generate a DFN model for a region dimensioned by a slab of $30 \times 30 \times 650$ metres surrounding the total length of TB02/00. So the model region is split into 6 DFN-cubes including 5 depth sections and one at the surface having $30 \times 30 \times 30$ metres respectively. The DFN-cube at the surface simulates the exposure KOR20 to calibrate the generation of the DFN iteratively by comparing it with the fracture network of KOR20. The other 5 DFN-cubes are generated at the depth sections 30 m, 60 m, 100 m, 250 m and 400 m below surface. The depth sections 30 m, 60 m and 100 m below surface are selected for a better quantification of the influence of the weathering and slope tectonics near the surface. The depth section 400 metres below surface is regarded as another calibration depth of the DFN.

The fracture network of each DFN-cube can be generated by data based on results of previous analyses (Tables 1 and 2). The previous analyses result in fracture sets with different geological-spatial and hydraulic relevant attributes. Each fracture set can be characterised by its mean aperture, mean linear degree of separation (trace length), the minimum and the maximum length, its spatial orientation, the termination index and its calculated volume related to a unit volume. The feature intensity is to be specified by the volume percentage $\{P_{33}\}$ regarded to an unit volume (Dershowitz et al. 1999). Thus the fracture set volumes control the fracture intensity.

In advance some assumptions have to be met to generate the individual fracture networks:

- As a result of the previous cluster analyses all the fracture networks consist of three fracture sets (Figure 2). The spatial orientations of the fracture sets are taken from the clusters of KOR20 and it is assumed that they persist for all the depth sections.

- Based on the structural-geological analyses all fracture sets (cluster) are characterized by a Fisher distribution generated in FRACMAN with the constant value "k = 50" (Dershowitz et al. 1999).
- The fracture sets of all DFN-cubes are generated with the same fracture attributes determined in Table 1.
- The trace lengths of a fracture set are defined by the lowest, largest and mean values and an exponential distribution.
- The termination index is calculated for each fracture set and is constant for all depth sections.
- The fracture apertures of each fracture set are considered to be constant. So the weighted mean aperture is calculated for each fracture set.
- The reductions of the fracture set volumes with depth are based on polynomial functions. Because of the default trace length and apertures (described above under points 3. and 4.) the reduction of each previous calculated fracture set volume is controlled only by the intensity of open fractures.

Considering these assumptions the fracture networks of 6 DFN-cubes are generated, bound on their fracture attributes and on the volumes of the fracture sets. The intensity of the fractures depends predominantly on the calculated volumes. So the highest intensity of fractures is generated in DFN-cube 00 simulating exposure KOR20 with a total calculated fracture network volume of 0.42%.

The lowest intensity is considered to be at DFN-cube 05 (the depth section 400 metres below surface) generated under respect of the total fracture network of 0.03% (Table 2).

### 5.2.3 *Data recording*

For the reconstruction of the DFN of the cubes their fractures have to be recorded with the same sampling techniques as the fractures at exposure KOR20 and TB02/00. The first logging is done by the cube 00 for the iterative calibration with KOR20. Then the fractures are recorded at the other cubes like the fractures of TB02/00 that means along simulated boreholes.

The data recording of the generated DFNs is based on 4 scanlines at theoretical exposed rock faces of DFN-cube 00 and boreholes carried down in all 6 DFN-cubes. The boreholes are simulated as vertical scanlines with a length of 30 metres through the midpoints of the DFN-cubes. Figure 3 shows a cross section as an example of a theoretical exposed rock face of DFN-cube 00. Scanlines are aligned for the data recording at the "exposed rock face" (tracemap). The orientations of the scanlines are defined by the same aspects as at a natural exposure.

The data recording of the top cube (cube 00 = representing KOR20) result in 102 fractures along 48 metres of scanlines. Additionally 202 fractures are observed along the borehole in the 5 simulated cubes. The data of cube 00 comprise the apertures, the trace lengths (linear degree of separation) and the spatial orientation of the fractures. The data of the simulated boreholes comprise the aperture and the spatial orientation of the fractures. Therefore the trace lengths of the fractures along simulated boreholes have to be estimated deduced from the data of the scanlines at the simulated rock faces of cube 00.

The clustering methods (fuzzy c-mean, hierarchical) are applied for the analyses of the data observed at cube 00 and subsequently their fracture set volumes are estimated. In further consequence the fracture set volumes of the 5 other cubes are calculated statistically in the same way.

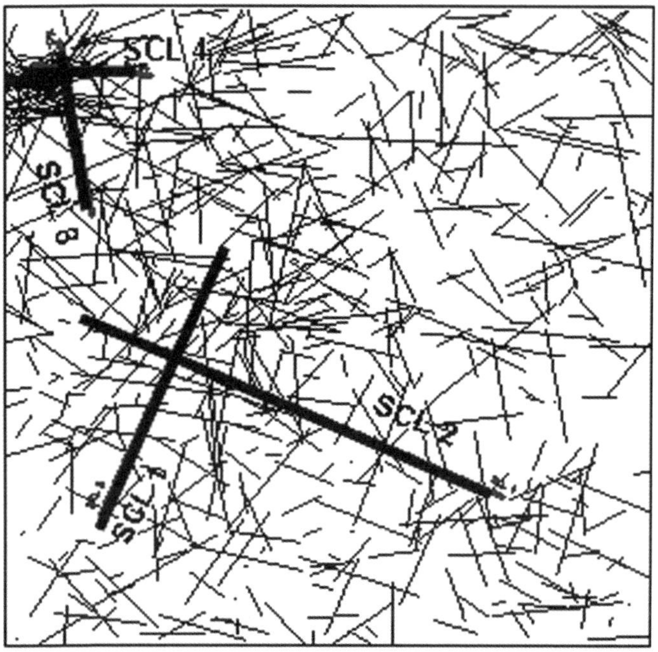


Figure 3. An example of a simulated exposed rock face of DFN-cube 00 with 4 scanlines for data logging.

## 6 ANALYSES AND INTERPRETATION

Within this approach it was tried to reproduce the fracture set volumes deduced from the polynomial approximation calculated from data of KOR20 and TB02/00. Therefore the fractures and their attributes of the model region are recorded and subsequently analysed.

The analyses of the required depth sections (30 m, 60 m, 100 m, 250 m and 400 m) result in three clusters respectively. The calculated fracture set volumes (C.V.) are listed in Table 3. Additionally the table includes the fracture set volumes (O.V.) based on the polynomial approximation deduced from recorded data of KOR20 and TB02/00.

The results of the clustering analyses and the volume estimations show that the new calculated fracture set volumes (C.V.) differ from the original fracture set volumes (O.V.) approximated by a polynomial function. The total volumes are characterized by approximately similar values and correlation factors (F) about "1" (Table 3).

Combining the new calculated fracture set volumes with the recorded data of KOR20 and TB02/00 the modification of the fracture set volumes correlated to depth can be specified by an exponential function. The data are displayed in (Figure 4).

Figure 4 shows that the highest reduction of the cluster volumes can be observed for cluster 2. Between the depth of 300 meters and the surface the variation of the individual cluster volumes comparing for each depth is much higher than below. This is strongly influenced by the persistent higher volumes of cluster 2 (black triangles in Figure 4). Below the depth of 300 metres all the cluster volumes range between 0.01 and 0.02%. So the cluster volumes have a different distribution below the depth of approximately 300 metres than above.

Table 3. Depth sections of the simulated DFN cubes with the volumes of the individual fracture sets based on the polynomial function (O.V.) and the calculated values (C.V.) by the data of the simulated scanlines and boreholes. Additionally the correlation factor (F) is listed.

| Depth | CL01 | | | CL02 | | | CL03 | | | Total volume | | |
|---|---|---|---|---|---|---|---|---|---|---|---|---|
| | O.V. | C.V. | F | O.V. | C.V. | F | O.V. | C.V. | F | O.V. | C.V. | F |
| −30 | 0.043 | 0.11 | 2.55 | 0.288 | 0.18 | 0.626 | 0.046 | 0.09 | 1.94 | 0.377 | 0.38 | 1.008 |
| −60 | 0.038 | 0.04 | 1.06 | 0.257 | 0.16 | 0.621 | 0.044 | 0.04 | 0.916 | 0.339 | 0.24 | 0.708 |
| −100 | 0.031 | 0.1 | 3.22 | 0.221 | 0.12 | 0.547 | 0.040 | 0.05 | 1.241 | 0.291 | 0.27 | 0.928 |
| −250 | 0.001 | 0.02 | 16.7 | 0.107 | 0.08 | 0.746 | 0.003 | 0.05 | 16.37 | 0.112 | 0.15 | 1.346 |
| −400 | 0.001 | 0 | 0 | 0.040 | 0 | 0 | 0.025 | 0.03 | 1.186 | 0.066 | 0.03 | 0.454 |

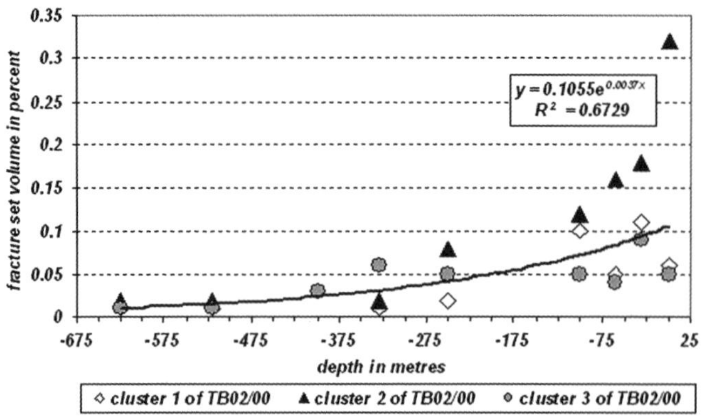


Figure 4. Correlation of the fracture set volumes to depth based on the simulated exposure and boreholes surrounding TB02/00.

Weathering and slope tectonics influence surface-exposures by loosening the fracture network. So the reduction of the fracture set volumes following the exponential function seems to be a consistent approach considering the geological environment.

## 7 DISCUSSION

The aim of this approach is to combine fracture data observed at an exposure with fracture data obtained along a borehole to improve the estimation of a three dimensional fracture network.

As a first result it was possible to detect and to specify fracture set (clusters) volumes along drilling sections deduced from the combined analyses of a surface-exposure and the drilling. So the clustering method developed in the first task could be applied successfully for interpreting borehole data.

In further consequence the various fracture set volumes of different depth section along the borehole enhance the estimation of the volumes correlated to depth. Because of the limited data set it was not possible to detect the fracture set volumes in the depth between 0 and 300 metres. The generation of a DFN with a discrete fracture model (FRACMAN) and its calibration at KOR20 enhanced to interpolate fracture set volumes for additional depth section.

Considering the recorded data at KOR20 and TB02/00 and the deduced information it can be approximated that the fracture set volumes are reduced exponentially with depth (Figure 4).

It can be summarized that because of their linear character boreholes can be integrated into the estimation of a three dimensional fracture network as additional information. It can be worked out that the modification of the fracture set volumes is based on an exponential function in the required investigation area.

It has to be considered that this is a first approach so the approximation is based on various assumptions which have to be investigated in more detail in the future considering various aspects including:

- persistent data logging along the borehole
- to detect the fracture aperture along the boreholes more precisely with e.g. optical televiewer
- to control the assumption that the trace length recorded at the surface-exposure can be carried forward to the borehole fractures
- to compare various generating algorithms for the development of the DFN

## 8 SUMMARY AND CONCLUSION

Fracture networks have to be considered in the definition of hydrogeological units. So the first step of the presented approach was developed at exposures to characterize the individual fracture sets and to estimate their volumes. This approach combines the geological spatial fracture attributes with the hydraulic relevant attributes. Two different clustering methods, hierarchical and fuzzy c-mean, were applied to determine the fracture sets. The scan line sampling allows to integrate borehole data in the calculation and interpretation. The reduction of the fracture set volumes correlated to depth was modelled using a stochastic discrete fracture network model and showed an exponential trend for the investigation area Koralm.

It could be shown that it is possible to determine and to quantify the modification of the fracture network with depth. This approach is based on several assumption that will be analysed in more detail in the future. Nevertheless this method can help in creating conditioned fracture networks including the spatial distribution of fracture attributes as an input for numerical modelling of flow in fractured aquifers.

## ACKNOWLEDGMENTS

This methodology was developed within a research project supported by the Federal Ministry of Transport, Innovation and Technology of Austria and JOANNEUM RESEARCH Forschungsgesellschaft mbH

## REFERENCES

Davison, A.C. & Hinkley, D.V. (1997) Bootstrap Methods and their Application. Cambridge Series in Statistical and Probabilistic Mathematics.

Dershowitz, W., Lee, G., Geier, J., Hitchcock, S., LaPointe, P., & Cladouhs, T. (1999) FracWorks95 Discrete Feature Simulator. User Documentation. Version 1.3. Golder Associates Inc., Seattle, Washington. 139 Seiten.

Dershowitz, W., Lee, G., Eiben, Th., & Ahlstrom, E. (1998) Meshmaster Discrete Fracture Mesh Generation Utility. User Documentation. Version 1.55. Golder Associates Inc., Seattle, Washington. 44 Seiten.

Hammah, R.E. & Curran, J.H. (1998) Fuzzy Cluster Algorithm for the Automatic Identification of Joint Sets. Int. J. Rock Mech. Min. Sci. Vol. 35, No. 7, pp. 889–905.

Hammah, R.E. & Curran, J.H. (1999) On Distance Measures for the Fuzzy K-means Algorithm for Joint Data. Rock Mech. Rock Eng. 32 (1), 1–27.

Harum, T., Hofrichter, J., Kleb, U., Kupfersberger, H., Poltnig, W., Reichl, P., Reinsdorff, S., Ruch, C., Schön, J., Strobl, E., & Winkler, G. (2001) Erfassung von Fließprozessen zur hydrogeologischen Bewertung von klüftigen Festgesteinen: unpublished report, Joanneum Research GmbH, Graz, S. 124.

ISRM International Society for Rock Mechanics (1978) Suggested methods for the quantitative description of discontinuities in rock mass. International Journal of Rock Mechanics and Mining Sciences Vol. 15, 319–368.

Kulatilake, P. H. S. W., Wathugala, D. N., & Stephansson O. (1993) Joint network modelling with a valiation exercise in Stripa Mine, Sweden: International Journal of Rock Mechanics and Mining Sciences, Vol. 30, pp. 503–526.

Kurz, W., Fritz, H., Tenczer, V., & Unzog, W. (2002) Tectonometamporhic evolution of the Koralm Complex (Eastern Alps): constraints from microstructures and textures of the Plattengneis shear zone: Journal of Structual Geology, v. 24, p. 1957–1970.

Priest, S.D., (1993) Discontinuity Analysis for Rock Engineering: Burg St. Edmunds, Suffolk, Edmundsbury Press Ltd.

Steinhausen, D. & Langer, K. (1977) Clusteranalyse Einführung in Methoden und Verfahren der automatischen Klassifikation. Walter de Gruyter, Berlin New York.

Wallbrecher, E. (1986) Tektonische und gefügeanalytische Arbeitsweisen: Enke-Verlag, Stuttgart.

# Conceptual models, groundwater flow and resources in fractured rocks

CHAPTER 11

# Fractured Old Red Sandstone aquifers of the Cork Harbour region: Groundwater barriers between adjacent karstic systems

Alistair Allen[1] and Dejan Milenic[2]

[1]*Dept. of Geology, University College Cork, Cork, Ireland*

[2]*Institute of Hydrogeology, Faculty of Mining and Geology, University of Belgrade, Belgrade, Serbia and Montenegro*

ABSTRACT: The Cork Harbour area of SW Ireland consists of E-W trending Variscan anticlines and synclines, the former cored by Upper Devonian Old Red Sandstone (ORS), and the latter by karstified Lower Carboniferous limestones. The area is cut by vertical, N-S trending Variscan compartmental faults. Overburden, mostly glacial till of variable thickness overlies bedrock, but sand and gravel deposits also occur. Aquifers exist in both the bedrock and overburden.

The ORS anticlinal ridges are fractured due to the Variscan deformation and are regarded on a regional scale as a fractured rock aquifer, forming a barrier between adjacent karstic systems. However, the N-S faults may provide interconnections between the two limestone aquifers. The storativity of the ORS sequence is relatively limited, characteristic well yields ranging from $<100$ to $500\,m^3$/day. The chemical composition of groundwater from the fractured aquifer is $HCO_3$-Ca type, with total dissolved solids ranging up to 300 mg/l.

## 1 INTRODUCTION

Ireland, situated on the NW fringe of Europe (Figure 1), consists of Precambrian to Lower Palaeozoic crystalline rocks around its generally mountainous rim, with a lowland interior largely underlain by Lower Carboniferous limestone. Late Palaeozoic, Mesozoic and Tertiary rocks are absent, apart from in the NE corner of the island, where they are preserved beneath the basalt plateau of the Tertiary North Atlantic Igneous Province associated with the opening of the North Atlantic. However, there is evidence that these were also deposited over much of the rest of the island, but were stripped away by the intense erosion and peneplanation that accompanied the opening of the North Atlantic.

All of the Lower Palaeozoic and older rock units in Ireland have been deformed and metamorphosed during the ~400 Ma Caledonian orogeny, so all of these rock units are highly recrystallised and have little primary porosity. The Lower Carboniferous limestones, underlying $>50\%$ of the country, are primarily massive bioclastic and biomicritic reef limestones formed in marine shallow shelf environments, and are mainly buried beneath thick Holocene overburden deposits. Because of their predominantly biomicritic character, they also tend to have a poor primary porosity. However, all of the limestones are heavily karstified and have good secondary fissure porosity, particularly where the karstification process was promoted by the presence of more intense fracturing of the massive limestones such as in the southwest due to the effects of the Variscan orogenic episode.

In addition, Upper Devonian sandstones and shales (ORS), derived mainly from erosional products of the Caledonian orogenic belt are widespread in the south of Ireland. In the southwest, Upper Palaeozoic units have been subjected to the Variscan Orogeny and have undergone extensive recrystallisation, so primary porosity has again been lost. Thus, most of the Irish bedrock has little primary porosity and groundwater generally resides in fractures and fissures.

Ireland has a relatively wet maritime climate and does not depend primarily on its groundwater resources. Groundwater supplies only about 25% of the water requirements of the country, although in some rural areas, this may rise to more than 80%. Due to the absence of Permian or younger rocks, none of Irelands bedrock aquifers possess significant primary porosity, so the best bedrock aquifers are the karstified Carboniferous limestones, particularly where karstification has been enhanced within deformation zones. Intergranular overburden aquifers, in the form of Pleistocene fluvioglacial sand and gravel deposits, are locally important, and in some areas may even be of regional significance.

In the southwest of Ireland (Figure 1), the dominant rock unit is ORS sandstone and shale, which forms higher ground, with locally intervening limestone underlying more lowland

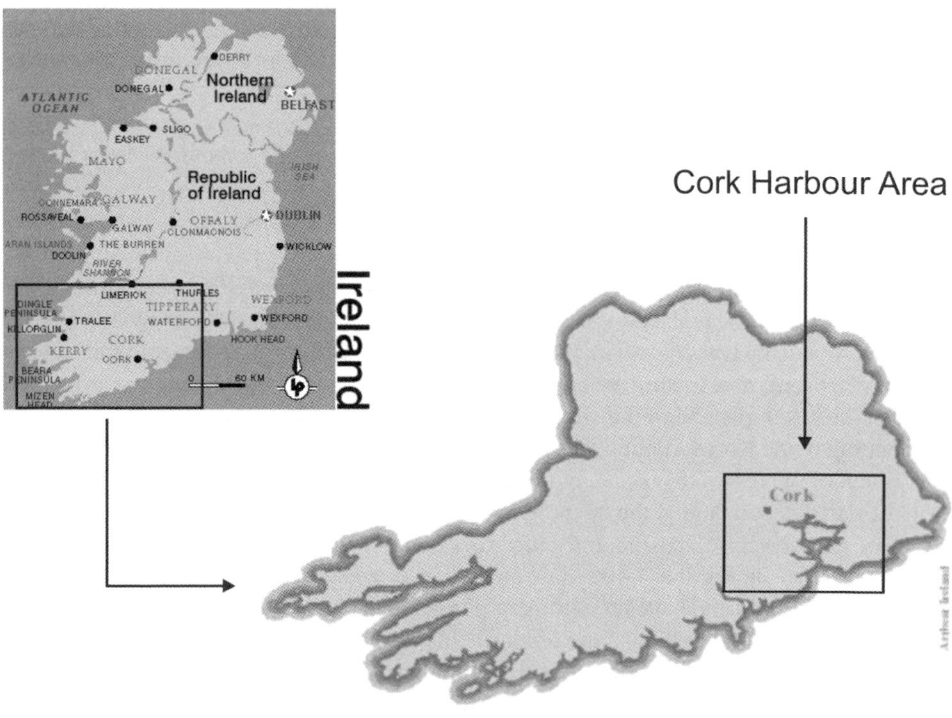


Figure 1. Location and geographical setting of the Cork Harbour area.

areas. Aquifers are relatively vulnerable to pollution as overlying confining layers are thin (0–3 m). As the EU Water Framework Directive requires national governments and local authorities to protect all surface and groundwater resources, it is incumbent to understand the relationships between the different water bodies. This paper, which arises from a research project funded by two local authorities to assess groundwater resources and quality within the Cork Harbour area (Allen & Milenic 2001, Milenic 2004), considers the hydrogeological characteristics of the ORS fractured aquifers, and their role as groundwater barriers between adjacent karst aquifers.

## 2 REGIONAL SETTING

Cork Harbour is one of the largest natural harbours in the world, a glacially eroded, almost completely enclosed body of water, connected by a narrow entrance to the Atlantic Ocean. Cork, situated at the mouth of the River Lee, which drains into Cork Harbour, is the second city in Ireland with a population of approximately 250,000 within its environs. The Cork area has undergone rapid industrial growth over the last two decades, posing a significant risk of groundwater pollution. Thus, the present investigation, covering an area of the order of 1,000 $km^2$ encompassing Cork City and the Harbour area (Figure 1), was instigated.

## 3 GEOLOGY

The Cork Harbour area lies within the low-grade Rheno-Hercynian fold-thrust terrane of the late Carboniferous Variscan Orogenic Belt (Gill, 1962). The area is characterised by a series of E-W upright horizontal anticlines and synclines (Gill 1962; Cooper et al, 1986), the former cored by U. Devonian sandstones and shales and the latter by massive Lower Carboniferous reef limestones (Figure 2). The folded sequence is cut by low angle E-W thrusts and steep N-S compartmental faults (Cooper et al, 1986). Two major limestone-cored synclines occur in the Cork Harbour area, the Cork-Midleton Syncline and the Cloyne Syncline, which are separated by the intervening sandstone/shale-cored Great Island Anticline. Other sandstone/shale-cored anticlines, the Rathpeacon Anticline and the Church Bay Anticline occur to the north and south of the Cork-Midleton Syncline and Cloyne Syncline respectively. Overburden of variable thickness mostly glacial till, overlies bedrock, but sand and gravel deposits, particularly infilling deep buried valleys, also occur.

The topography of southwest Ireland is controlled by the geological structure. The anticlines form uplands, ranging up to 1000 m in height and the synclines occupy broad lowland areas 0–50 m high (Figure 3). This geomorphological pattern developed during the Pleistocene glaciation (1.8 Ma–10 ka). Prior to this, the Devonian–Carboniferous succession was buried beneath Mesozoic cover, and the Tertiary land surface sloped gradually southwards, giving rise to southwards-flowing drainage systems (Nevill, 1963). Extreme Tertiary erosion completely stripped away the Mesozoic cover and exposed the U. Palaeozoic rocks, initiating a process of intense karstification of the Carboniferous limestones, and leading to the establishment of an entrenched N-S drainage pattern.

This Tertiary drainage was truncated at the onset of Pleistocene glaciation by glaciers advancing outwards from the mountainous regions of western Ireland, preferentially exploiting the weaker karstified limestones forming the cores of the synclines. This resulted in the

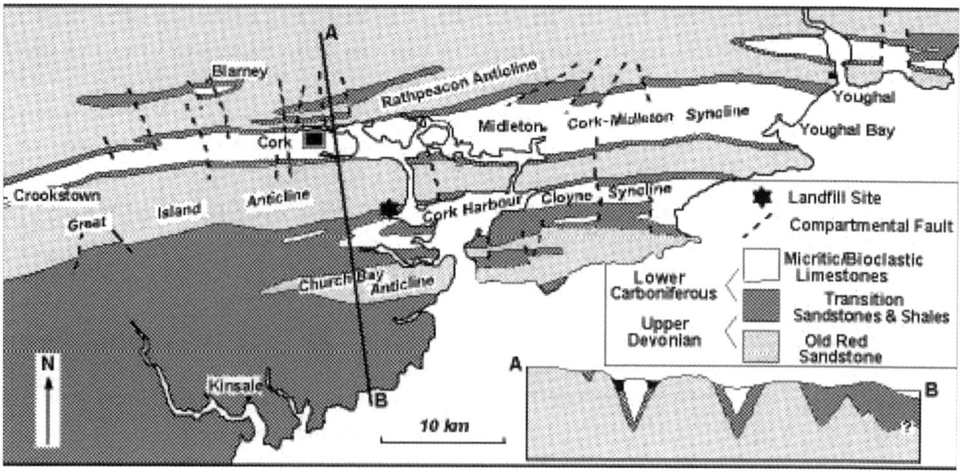


Figure 2. Simplified geological map of the Cork Harbour area.

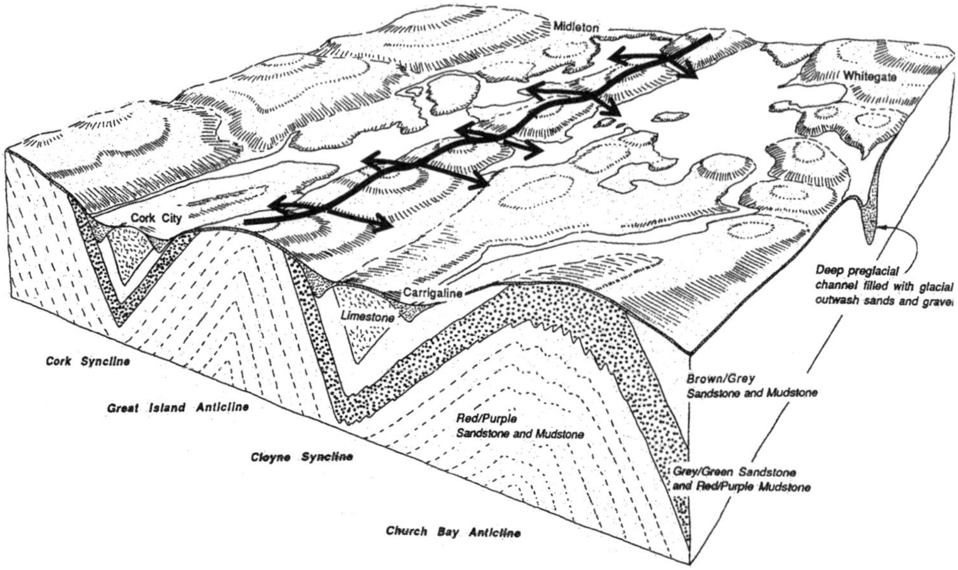


Figure 3. Block diagram of the Cork Harbour area viewed from the SW showing the Great Island Anticline with its watershed and consequent N-S streams rising and draining into the bounding limestone-cored synclines (modified after MacCarthy, 2001).

development of a number of broad u-shaped lowland areas 3–4 km wide that truncated the pre-existing N-S drainage patterns. Superimposed on these u-shaped lowlands, are a number of deep buried valleys infilled with sand and gravel, that formed at the peak of the last glaciation 15–18,000 years ago, when sea level fell to an estimated 130 m lower than at present (Mitchell, 1976; Pirazzoli, 1996). In the Cork Harbour area, these valleys represent important local ribbon aquifers. Cork Harbour was also created, probably also during the late

Pleistocene, by glacial scouring. In post-Pleistocene times, river capture of many of the E-W rivers created during the Pleistocene, by the N-S pre-Pleistocene river systems, has given rise to a trellised drainage pattern in the Cork Harbour area.

The Old Red Sandstone is a thick heterogeneous sequence of terrestrial sandstone, ranging lithologically from grit to fine sandstone, with subordinate siltstone and shale. The unit was deposited in an alluvial plain environment by braided river systems, with intermittent finer-grained overbank deposits reflecting flooding episodes (MacCarthy, 1990). The ORS grades upwards through a transition sequence of shallow marine sandstones and calcareous shales into the L Carboniferous limestones of the Cork-Midleton and Cloyne Synclines. The ORS displays cross- and graded bedding and other sedimentary structures (MacCarthy, 1990). Deformation structures include a spaced fracture cleavage in the coarser beds and a penetrative slaty cleavage in the argillaceous layers.

Primary porosity in both the ORS and the limestone was obliterated by recrystallisation associated with the Variscan deformation, but a secondary porosity developed due to intense fracturing, which in the limestones controlled later karstification. Three approximately orthogonal sets of fractures were developed, steep E-W fractures parallel to the axial planes of Variscan folds, steep N-S fractures parallel to the compartmental faults, and subhorizontal fractures, possibly reflecting stress release associated with thrust-related uplift and unloading. In the ORS, intensification of the E-W fracturing could be expected in the cores of the anticlines, and of the N-S fractures adjacent to the compartmental faults.

## 4 HYDROGEOLOGY

In southwest Ireland, the ORS is commonly considered a relatively low permeability fractured aquifer on a regional scale, so the anticlinal structures cored by the ORS represent hydrogeological barriers between the intervening karstic systems. However, the late Variscan N-S compartmental faults cut through both the ORS-cored anticlines and the limestone-cored synclines, so the faults may provide local leakage between the limestone aquifers. In the Cork Harbour area, an investigation has been made of the ORS-cored Great Island Anticline and its relationship with the two limestone synclines, the Cork-Midleton and Cloyne Synclines, which border it to the north and south.

Major groundwater aquifers occur in both the bedrock and in the overburden of the Cork Harbour area. The Cork-Midleton and Cloyne Synclines are cored by intensely karstified limestones, which have significant storage capacity and transmissivity properties. Overlying these bedrock aquifers, are important ribbon aquifers represented by the gravel-infilled buried valleys. These aquifers are a major resource.

## 5 THE FRACTURED AQUIFERS

### 5.1 *Aquifer delineation*

Within the study area, the Great Island Anticline is 60 km in length and 2–5 km in width. It forms a steep-sided relatively flat-topped ridge rising to a maximum height of about 200 m. Soil cover is relatively thin, and overlies an approximately 10–15 m thick weathering zone.

ORS cores the anticline, and transitional marine sandstones and shales occupy the steep margins of the ridge adjacent to the synclines.

A fracture analysis of the ORS within the Great Island Anticline indicates that the dominant fracture set parallels the compartmental faults and is approximately perpendicular to the orogenic trend with a mean strike of 345°–165°. Less important folding-related fractures parallel to the regional strike have a mean orientation of 075°–255°. These two sets of fractures, which formed during the late Carboniferous Variscan Orogeny, provide the main storage for groundwater in the ORS and also control rates and directions of groundwater movement. Other groundwater reserves are located in the relatively thick weathered zone of this fractured aquifer. Because of the negligible primary porosity of the ORS sandstones and shales, storativity of the ORS is relatively low. The fractured nature of the aquifer suggests, however, probable deep circulation through the fracture network.

Based on the fracture analysis, linear and areal coefficients of fracture density for the ORS have been determined. Surface values of up to 1% are indicated. Taking into account the effects on surface fractures of physical and chemical weathering processes, and the closure of fractures with depth due to confining pressure, an overall effective porosity for the fractured aquifer of around 0.1% is estimated. However, more intensely fractured and deeply weathered zones are associated with the compartmental faults, which have a spacing of about 1 km and are usually exploited by streams; effective porosity in these zones is probably considerably enhanced.

## 5.2 *Recharge*

The dominant recharge process in the fractured aquifers is through direct recharge by precipitation. The climate of the Cork area is dominated by the maritime influence of the Atlantic Ocean and the prevailing south-westerly wind flow. Rainfall is greatest in the mountainous areas adjacent to the Atlantic coast and decreases dramatically eastwards. At Cork airport, situated within the Great Island Anticline, the mean annual rainfall and evapotranspiration values measured over an almost 40 year period up to 2000, were 1222 mm and 486 mm respectively (Milenic, 2004).

The trellised drainage pattern in the Cork area, defined by N-S consequent and E-W subsequent streams, reflects the dominant control of geological structure and the history of drainage development on present day drainage patterns. Surface run-off is strongly dependent on topography. In the Cork area the ORS-cored anticlines form relatively steep-sided ridges, with incised valleys. Although precise measurement of the proportion of run-off is lacking, the large number of streams, which rise in the Great Island Anticline and feed the river systems of the synclines to the north and south suggest that the proportion of run-off is high, and it is estimated that 75% of the available rainfall may run off at the surface. Thus, for the Great Island Anticline fractured aquifer, the effective infiltration ($i_{ef}$) is estimated as 184 mm/yr.

However, the fractured aquifer is overlain by a variable thickness of relatively impermeable clay-rich glacial till. Thus, percolation through the glacial till is slow and a significant proportion of the infiltrated precipitation may be held in the till as soil moisture so the proportion of infiltrated rainwater, which ultimately recharges the water table, may be quite small. The ORS forms the highest topography in the Cork Harbour area, so it is unlikely that there is any recharge from the adjacent limestone synclines and in fact it is probable that the fractured aquifer feeds the karst aquifers.

### 5.3 *Groundwater discharge sources*

Discharge from the Great Island Anticline fractured aquifer (Figure 3) occurs by a variety of processes:

- Minor springs – representing the sources of a number of streams rising in the Great Island Anticline (Figure 3), with discharges of the order of 0.1–1.0 l/s.
- Private water wells – over 150 registered wells, ranging in depth up to 130 m, with discharges of 0.1–6.7 l/s.
- Baseflow – most of the streams draining from the Great Island Anticline into the Cork-Midleton and Cloyne Synclines are fed under normal low flow conditions by groundwater – up to 50% of the streamflow is probably derived from groundwater sources.
- Recharge of the karst aquifers – discharge into the adjacent karst aquifers of the Cork-Midleton and Cloyne Synclines is probably significant, particularly via the N-S compartmental faults.

### 5.4 *Estimation of groundwater resources*

Groundwater reserves were delineated applying the classical water balance formula. The fractured aquifer of the Great Island Anticline is subdivided into two zones, an upper weathered zone 10 m thick with an estimated porosity of 1% and a lower zone with a thickness of 100 m with a fracture porosity of 0.1%. Taking the length of the Great Island Anticline as 60 km and the average width of the aquifer as 3.5 km, the storage of the weathered zone of the fractured aquifer is $2.1 \times 10^9 \, m^3$, and the storage of the fractures in the lower zone is also $2.1 \times 10^9 \, m^3$, giving a total storage for the fractured aquifer of the order of $4.2 \times 10^9 \, m^3$. Increased porosity within the compartmental fault zones, on average about 1 km apart within the Cork Harbour area, has not been included in this calculation so groundwater reserves may be greater than the above estimate.

### 5.5 *Hydrochemistry*

Hydrochemical data, representing a compilation of over 50 analysed samples from the Great Island Anticline, are presented in Table 1. Groundwater composition is $HCO_3$-Ca type, with total dissolved solids ranging up to 300 mg/l (Figure 4). Most chemical components are present at low concentrations well within the maximum admissible concentrations (MAC) as specified by the Irish EPA. On the basis of oxygen and deuterium isotopic analyses, a meteoric origin is indicated for the groundwater of the fractured aquifers.

### 5.6 *Groundwater vulnerability and protection*

Rapid urban and industrial growth in the Cork area has led to an increased risk of groundwater pollution. In particular, urban spread of Cork City to the south into the northern flank of the Great Island Anticline, has created a huge potential for groundwater pollution both from housing estates and from commercial and industrial complexes. In line with other European countries subject to the requirements of the EU Water Framework Directive, the Geological Survey of Ireland in cooperation with the Irish EPA and the Dept. of the Environment has developed a general groundwater protection scheme methodology (DELG/EPA/GSI, 1999).

Table 1. Chemical composition of groundwater from the fractured aquifers

| Parameter | Unit | MAC | Range of values |
|---|---|---|---|
| Temperature | °C | – | 12.0–14.0 |
| pH | – | 6–9 | 6.2–7.9 |
| $Na^+$ | mg/l | 150 | 8.50–24.77 |
| $K^+$ | mg/l | 12 | 1.23–8.33 |
| $Ca^{2+}$ | mg/l | 200 | 25.57–48.72 |
| $Mg^{2+}$ | mg/l | 50 | 3.86–13.51 |
| $HCO_3^-$ | mg/l | – | 62.83–250.10 |
| $SO_4^{2-}$ | mg/l | 250 | 6.17–27.74 |
| $Cl^-$ | mg/l | 250 | 17.25–34.59 |
| $NO_3^-$ | mg/l | 50 | 3.38–32.27 |
| Hardness | $CaCO_3$ | – | 82–224 |
| EC | μS/cm | 1500 | 246–492 |
| Total Diss. Solids | mg/l | 1000 | 162–325 |

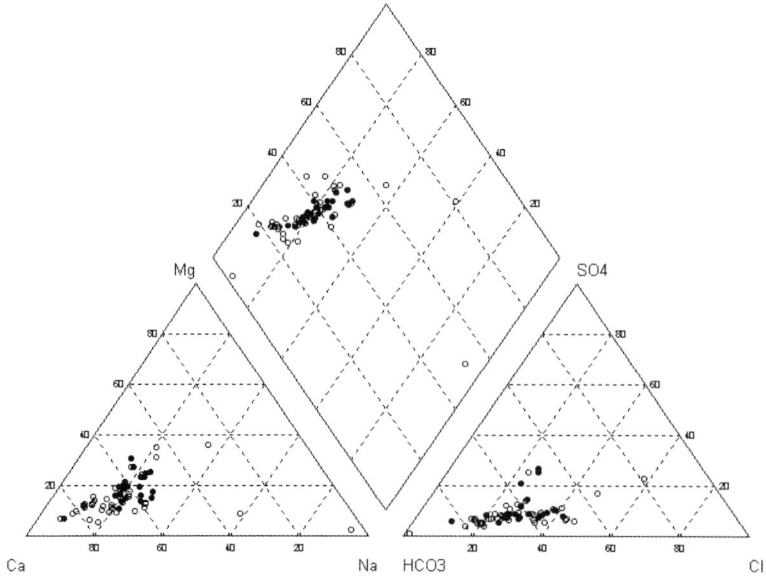


Figure 4. Chemical composition of groundwater in fractured aquifer – Piper Diagram (black points – fractured aquifer, transparent points – other aquifers).

Currently, the GSI is in the process of producing a detailed groundwater vulnerability map on a 1: 50,000 scale for the South Cork region.

Fractured aquifers are particularly vulnerable to pollution. Although they are blanketed by a layer of glacial drift, in the upland areas underlain by these aquifers, the drift is generally relatively thin. The upland areas are largely given over to farming; poor animal husbandry practices and land fertilisation are the major potential sources of groundwater contamination. However, the shallow aquifers are also vulnerable to pollution from a large number of private septic tanks, many of which are inappropriately sited too close to bedrock. The presence of total and faecal coliforms has been detected in some water samples from the fractured aquifers, demonstrating human sewage pollution. In addition a landfill has operated in the southern part of the Great Island Anticline for the past 20 years, although it is now in the process of closure. The landfill is sited in a valley exploiting a compartmental fault, and occupies a quarry used to extract fill for a deepwater port facility in Cork Harbour, so highly fractured bedrock is exposed, and the potential for pollution of groundwater is high.

### 5.7 *Connection zone or barrier?*

Because of negligible primary permeability, the fractured ORS aquifers of the Cork Harbour area behave on a regional scale as flow barriers between the karst aquifers they separate. Although, the local fracture network of the fractured aquifers could facilitate groundwater exchange between the karst aquifers, the anticlinal ridges are topographically higher than the intervening lowland areas underlain by the karstified limestones, so groundwater watersheds exist within the upland areas between the various limestone-cored synclinal structures. This is particularly so within the Great Island Anticline separating the Cork-Midleton and Cloyne Syncline karst aquifers, and coincides with the surface water watershed shown in Figure 3. Thus, local shallow groundwater flow between the two synclines almost certainly does not take place.

The numerous N-S compartmental faults cutting the E-W Variscan folds of the Cork Harbour area may also provide possible pathways for groundwater flow between the karstic aquifers. Although again the groundwater watershed in the Great Island Anticline probably inhibits shallow groundwater movement between the limestone synclines via the compartmental faults, they may facilitate deep regional circulation between the two karst aquifers.

## 6 CONCLUSIONS

Devonian Old Red Sandstone bedrock in SW Ireland was recrystallised, folded and fractured during the Carboniferous Variscan Orogeny, and lost its primary porosity, but simultaneously developed a relatively strong secondary fracture porosity. In addition numerous late compartmental faults cutting the E-W Variscan folds imparted a further degree of secondary porosity. Although the least productive of the aquifers in the Cork Harbour area, the ORS fractured aquifers nevertheless provide some groundwater storage – in excess of $4.2 \times 10^9 m^3$ within the Great Island Anticline alone, and within the Cork Harbour area, supports a large number of private wells with yields of up to 6.7 l/s.

These fractured aquifers represent barriers to groundwater movement on a regional scale, particularly since they form upland areas, so both surface water and groundwater divides occur within them. Thus, no shallow groundwater movement between intervening karst aquifers occupying synclinal structures can take place. However, late Variscan N-S compartmental faults cutting through the Variscan E-W folds, may possibly allow deep regional circulation between karst aquifers separated by the ORS-cored anticlines.

## REFERENCES

Allen AR, Milenic D (2001) Preliminary Assessment of Groundwater Resources and Groundwater Quality in the Cork City/Harbour Area, Ireland In: Seiler K-P, Wohnlich S *New Approaches to Characterising Groundwater Flow*. Balkema, Rotterdam, Vol 2, pp 1119–1123.

Cooper MA, Collins DA, Ford M, Murphy FX, Trayner PM, O'Sullivan M (1986). Structural evolution of the Irish Variscides. J Geol Soc Lond, 143, pp 53–61.

DoELG., EPA., GSI. 1999. Groundwater protection schemes. Government Printing Office, Dublin.

Gill WD (1962) The Variscan Fold Belt in Ireland. In: Coe K (ed) Some Aspects of the Variscan Fold Belt, Manchester Univ Press, pp 49–64.

MacCarthy IAJ (1990) Alluvial sedimentation patterns in the Munster Basin, Ireland. Sedimentology, 37, pp 685–712.

MacCarthy IAJ (2001) The geological history of Cork City and Harbour region. Dept. of Geology, National University of Ireland-Cork Report Series, 01/1, 23 pp.

Milenic D (2004) Evaluation of groundwater resources of the Cork Harbour area. Unpublished Ph.D thesis, National University of Ireland-Cork, Ireland, 486 pp.

Mitchell GF (1976) The Irish landscape. Collins, London, 240 pp.

Nevill WE (1963) Geology and Ireland: with physical geography and its geological background. Allen Figgis & Co., Dublin, 263 pp.

Pirazzoli, PA (1996) Sea level changes-the last 20,000 years. John Wiley & Sons.

CHAPTER 12

# Non-Darcy two-phase flow in fractured rocks

Giovanni Brighenti and Paolo Macini
*University of Bologna, Department of Chemical, Mining and Environmental Engineering, Viale del Risorgimento, Bologna, Italy*

ABSTRACT: The paper recalls the main cases in which the two-phase flow in fractured and porous rocks does not follow the classical extension of Darcy's law to multiphase flow. In particular, here are highlighted some applications in the field of hydrogeology and of hydrocarbon reservoir engineering. Apart from the case (typical also of single-phase flow) of "high-velocity" flow, where inertial forces cannot be neglected with respect to viscous ones, the paper analyzes the typical cases of two-phase flow, where the interactions between the two phases can influence the flow behavior. It is also recalled that Darcy's law does not hold in case of two-phase dispersed flow. Generally, it is pointed out that both natural porous media and fractured rocks may have very different features of the porous network. Natural porous media range from cemented sandstones to unconsolidated sand or loose gravel, with large variation of the geometry and interconnection of the porous network. Fractured rocks are characterized by a matrix (occasionally porous and permeable, depending on lithological types) intersected by a network of fractures whose opening can range between a few microns to several centimeters or more. In both cases, the dimensions of the flow channels may be very different, and hence the ratios among the forces driving the fluid flow (i.e., viscous, gravitational, inertial and capillary forces). It would therefore be of practical interest to study the laws of fluid flow in the above cases not based on the geological distinction between porous media and fractured rocks, but on the values of the ratio between viscous, gravitational, inertial and capillary forces driving fluid flow inside the porous or fractured network. Finally, it is pointed out that, notwithstanding the apparent differences between porous and fractured media, the study of fluid flow behavior inside both media can be performed with the same physical approach.

## 1 INTRODUCTION

Non-Darcian multi-phase flow has been investigated since long ago, with many studies, both theoretical and experimental. However, a complete and unitary theory on the subject still does not exist. In particular, the extension of Darcy's law to multi-phase flow has not been yet completely probed either in the case of Darcian flow. For this reason, the present study is limited to the examination of the two-phase flow in Darcian and non-Darcian conditions.

Until a few years ago, the study of underground multiphase flow was of practical interest only for reservoir engineers, both in the geothermal field and in that of oil and gas production, and for agricultural engineering applications, to study water flow in the unsaturated zone. However, these investigations have recently acquired practical importance also in the field of hydrogeology, to study aquifers in presence of gas, or in case of aquifer pollution by

Non-Aqueous Phase Liquids (NAPL), for the subsequent modeling, design and planning of the possible remedial operations.

Depending on the circumstances, the fluids contained in the void spaces of natural rocks may have very different characteristics, and different may be the characteristics of the rocks where the fluids flow. When the void spaces are hydraulically connected, these are generally described as "porous" or "fractured" network. Here it is possible to distinguish between natural porous media (unconsolidated or cemented granular materials of sedimentary nature) and fractured rocks. In these latter, the fracture opening can range from micro-fractures of a few micrometers in size, to mega-fractures which can cover thousands of meters and can have openings sometimes several centimeters wide. Consequently, the factors affecting fluid flow can impact differently in different geometrical situations, following different flow laws, and giving rise to different fluid flow behaviors.

At first sight, it would seem obvious that there can be different models for fluid flow in porous media and in fractured rocks. It can be seen, however, that the flow laws are strongly affected not only by fluids properties and their field of velocity, but also by the size of the flow paths, a size that may be more similar between a porous medium and a fractured rock than between two very different porous media or two fractured rocks. For example, the fluid flow within a fractured rock characterized by openings of a few millimeters is more similar to the flow inside a porous medium made up by coarse gravel than to that occurring inside a rock with micro-fractures.

Also in this simple case, the validity of the extension of the Darcy's law to the two-phase flow poses very complex issues, especially in presence of a porous fractured rock, or a fractured rock with two different families of fractures neatly distinguished by the size of the openings (and hence characterized by different capillary pressure) and by their spatial direction; here, the fluid behavior can be furthermore complicated by non negligible inertial and/or coupling effects.

In the light of the above problems, it seems more correct and appropriate to examine the flow in fractured rocks and in porous media in a unitary manner, based on the size of the flow paths or, rather, on the relative importance (or the weight) of the forces governing fluid flow. In particular, the forces most important to drive fluid flow behavior, apart from wetting phenomena, can be classified into four main types: a) Viscous forces, proportional to $\mu\cdot v$ [$N\cdot m^{-1}$]; b) Gravitational forces, proportional to $\rho \cdot g \cdot D^2$ [$N\cdot m^{-1}$]; c) Inertial forces, proportional to $\rho\cdot v^2\cdot D$ [$N\cdot m^{-1}$]; d) Capillary forces, proportional to $\sigma$ [$N\cdot m^{-1}$], where $\rho$ and $\mu$ are the fluid density [$kg\cdot m^{-3}$] and dynamic viscosity [$Pa\cdot s$], respectively, v is the mean fluid velocity [$m\cdot s^{-1}$], D is a distinctive size of the flow channels [m], $\sigma$ is the surface tension [$N\cdot m^{-1}$] and g is the acceleration of gravity [$m\cdot s^{-2}$].

The effect of each force on the fluids flow can be more precisely quantified in terms of three well-known dimensionless numbers, and namely: 1) Reynolds number (Re), 2) capillarity number (Nca) and 3) Bond or Eötvös number (Bo). These numbers express the ratio (and thus the relative importance) of inertial to viscous forces (Re), viscous to capillary forces (Nca), and gravitational to capillary forces (Bo), respectively. Many different expressions have been used in the literature to define these dimensionless numbers. The most common ones are the following:

$$Re_i = \frac{\rho_i v_i D}{\mu_i} \qquad Nca_i = \frac{\mu_i v_i}{\sigma_{ij}} \qquad Bo = \frac{(\rho_i - \rho_j) g D^2}{\sigma_{ij}}$$

Table 1. Typical values of Re, Nca and Bo in gas-liquid two-phase flow in coarse porous media, compared with values calculated for consolidated porous media (typical of oil recovery processes) and in piping. D represents pipe or particle diameter (from De Santos et al., 1991, modified).

| Solid support | Re | Nca | 1/Bo |
|---|---|---|---|
| Coarse porous media ($D = 10^{-3}$–$10^{-2}$ m) | $10^{-2}$–$10^{3}$ | $10^{-1}$–10 | $10^{-1}$–10 |
| Consolidated porous media ($D = 10^{-7}$–$10^{-4}$ m) | $10^{-9}$–$10^{-2}$ | $10^{-7}$–$10^{-3}$ | $10^{2}$–$10^{9}$ |
| Piping ($D = 10^{-2}$ m) | 10–$10^{5}$ | 10–$10^{2}$ | $10^{-3}$–$10^{-1}$ |

Here the subscripts i and j refer to the scalar and vectorial properties of the i-th and of the j-th phase, respectively. Values of these three parameters bring out the contrasts and can help to discriminate between two-phase flow through pipes, consolidated and coarse porous media. Typical values of the above dimensionless numbers in these cases are reported in Table 1 (De Santos et al., 1991), comparing Re, Nca and Bo values for both coarse porous media, consolidated porous media and piping.

The effect of each driving force is particularly important when the fluid flow does not follow Darcy's law extended to multiphase flow, i.e. when the inertial forces are not negligible compared to the viscous ones, or the capillary forces are not prevalent compared to some of the other forces and when the flow of the different phases is not co-current.

Several studies have been carried out on this research field until recent times. However, due to the complexity of the phenomenon, so far it has not been possible to identify generally valid relationships, but only empirical ones, having limited validity. In many cases, the very dynamics of the phenomenon are still not completely clear. The present study is aimed to briefly recall the main cases in which the two-phase flow does not follow the usual extension of Darcy's law to multiphase flow. The investigation takes into account some practical cases typical of the hydrocarbon reservoir engineering which, with proper modifications, can be applied to hydrogeological studies, by means of a systematic investigation and comparison between theoretical and experimental relationships. This study analyzes the simple effects due to the presence of not negligible inertial forces, the extension of Darcy's law to the two-phase flow and the discussion of the Forchheimer relationship, one of the oldest and most utilized for this application. Finally the study analyzes the cases in which it is necessary to take into account also the direction of flow and the presence of a fluid in non-continuous phase.

## 2 EXTENSION OF DARCY'S LAW TO TWO-PHASE FLOW

It is well known that for the case of laminar isothermal two-phase flow of Newtonian fluids in a uncompressible, homogeneous, isotropic porous medium, when it is possible to consider as null or negligible the coupled phenomena due to the presence of variations of electric potential, temperature and chemical concentration, when fluid phases have no chemical interaction with the solid matrix, are continuous and in condition of no-slippage (i.e., the Klinkenberg effect can be considered as negligible), then the following relationship is held to be valid:

$$u_i = -\frac{k_{ri}k}{\mu_i}\rho_i \mathrm{grad}\Phi_i, \qquad i = 1, 2 \tag{1}$$

where $u_i$ is the volumetric fluid flow per surface unit of the porous medium of the i-th phase (also known as Darcy's velocity) $[m \cdot s^{-1}]$, k is the (intrinsic) permeability $[m^2]$, $k_{ri}$ is the relative permeability of the i-th phase (i.e., the ratio between effective permeability $k_i$ and intrinsic permeability, dimensionless), $\rho_i$ is the density of the i-th phase $[kg \cdot m^{-3}]$, and $\Phi_i$ is the potential per unit mass $[m^2 \cdot s^{-2}]$, also known as Hubbert's potential, referred to the i-th phase. In this simple two-phase case, in a first and basic approximation, the relative permeability $k_{ri}$ can be considered as a function of wettability, saturation $S_i$ of each phase and local saturation history, i.e. the sequence of drainage or imbibition processes the porous medium has undergone (Honarpour et al., 1987). The linear relationship between the Darcy's velocity $u_i$ and the potential gradient expressed by eq. (1) is adequate in ordinary underground fluid flow, where the inertial forces are negligible compared to the viscous ones, and the capillary forces are prevalent compared to all the others. In this case, a fluid cannot move until it has formed a continuous phase, and the ratio of the volumetric flow of the phases depends on the ratio $M_{ij}$ of the mobility $\lambda_i$ of each phase, where $\lambda_i = k_i/\mu_i$, and mobility ratio $M_{ij} = \lambda_i/\lambda_j$.

In underground fluid flow, it is common practice to extend the notion of intrinsic and relative permeability to fractures and fractured rocks as well (Bear and Berkowitz, 1987; Bear, 1993). Nevertheless, while in the case of porous media the relative permeability in two-phase flow can be measured by routine laboratory measurements or by on-site tests, still very little is known about relative permeability in fractures and fractured rocks. The simple hypothesis proposed by Romm (1972), considering the two-phase flow inside a fracture as a segregated flow with $k_{ri} = S_i$, seems today acceptable only in few particular cases of fractures with smooth walls and constant openings. In general, Kazemi and Gilman (1993) do not consider the validity of these hypotheses, and proposes a dispersed flow model, where the relative permeability of each phase is expressed as:

$$k_{ri} = \frac{S_i}{\sum_i S_i \mu_i} \qquad i = 1, 2 \tag{2}$$

It is worthy reminding that in all probability the shape of relative permeability curves in fractured rocks is generally similar to the one measured in porous media, due to the presence of possible filling materials and of distinctive wall roughness (Sahimi, 1995).

## 3 INERTIA FORCES NOT NEGLIGIBLE COMPARED TO VISCOUS FORCES

For many years Darcy's law has been considered the fundamental equation that governs liquid flow in porous media. However, it has been known for over a century that, in the case of single phase fluid flow (both in porous media and in fractures), in laminar flow condition, but at high velocity (or high Re), Darcy's law no longer holds. It is a common practice to extend the criteria to discriminate between laminar and turbulent flow valid inside constant-section rectilinear pipes (i.e., based on the values of the Reynolds number Re), to porous and fractured media. However, it is interesting to recall that inside curved or variable-section pipes, the Reynolds number criterion does not hold any more. In fact, in this case, at high velocity the effect of inertial forces become significant, and the relationship between flow rate and head loss is no more constant. The same phenomenon is likely to happen also inside porous media and fractured rocks, where true "turbulent" flow initiate for Re number at least one order of magnitude larger than the one marking the deviation from Darcian flow.

Many researchers have used indifferently the term "turbulent" and "non-Darcy" to describe visco-inertial flow at high velocities in the vicinity of production and injection wells, especially gas wells in fractured reservoirs and in condensate reservoir. Although turbulence might be conceivable in gas saturated media, it can develop in liquid saturated media only under very special conditions. Liquids under similar conditions of "turbulence" most likely will flow in a laminar non-linear mode. Early researchers supposed that high velocity flow effects were due to turbulence. This opinion has been recently rejected by investigators who related these effects to inertial effects. The main causes are attributed: 1) to the convective acceleration and deceleration of the fluids traveling through pores and/or fractures, and 2) to movement through the tortuous path of the porous or fractured network (Firoozabadi and Katz, 1979; Ma and Ruth, 1997; Garrouch and Ali, 2001). This cannot be considered as true turbulence, and is defined as "non-Darcy" flow (Scheidegger, 1974; Belhaj et al., 2003), and can be justified by recalling that Darcy's law does not consider the inertial forces (in this case no longer negligible), due to the continuous variations of pore and throat (and fractures) cross section and to the tortuous, winding and interconnected nature of the flow paths. More precisely, in this case the potential gradient is no longer a linear function of the flow rate, as expressed in eq. (1). To take into account these phenomena, many empiric and semi-empiric relationships have been studied. One of the simplest and most utilized is the well-known relationship proposed by Forchheimer (1901). In the case of single-phase flow, it is possible to write the Forchheimer's equation as:

$$-\mathrm{grad}\Phi = \frac{\mu}{\rho k}u + \beta u^2 \tag{3}$$

where $\beta$ is the Forchheimer's coefficient [$m^{-1}$]. Different names have been used for the coefficient $\beta$, and it has been called the turbulence factor, the coefficient of inertial resistance, the velocity coefficient, the non-Darcy flow coefficient, the Forchheimer flow coefficient, the inertial coefficient, the Beta factor, *etc.* The coefficient $\beta$ has the unit of inverse length, and so its dependency was attributed on media characteristics.

It is interesting to recall that the Forchheimer's coefficient of eq. (3) can be expressed as:

$$F_{nd} = \left(1 + \frac{\beta k \rho}{\mu}u\right)^{-1} \tag{4}$$

In this case, it is possible to re-write eq. (3) into the following form:

$$u = -\frac{k}{\mu}\rho F_{nd}\,\mathrm{grad}\Phi \tag{5}$$

The above equation is formally equivalent to eq. (1) with the addition of the non-Darcy flow coefficient $F_{nd}$.

Many attempts have been made to derive eq. (3) from a theoretical point of view, in order to evaluate the coefficient $\beta$. These studies showed that $\beta$ can be related in various ways to petrophysical properties rather than fluid properties, such as permeability, porosity, tortuosity, specific surface area, grains and pore size distribution, surface roughness, *etc.* (Belhaj et al., 2003). The coefficient $\beta$ may be determined on-site by the analysis of multi-rate pressure test results, by calculating the skin factor at different flow rates, or by using correlations and determination discussed in the literature (Li and Engler, 2001). Various empirical and theoretical approaches have been presented, and many empirical correlations are available for single and two phase flow, although there is no theoretical evaluation for multiphase flow. Examples of some correlations proposed for the single-phase non Darcian flow are:

$$\beta = \frac{c}{k^{0.5} n^{1.5}} \qquad \text{(Irmay, 1958)} \tag{6}$$

where: c = constant, k = permeability, n = porosity. This relationship is valid for a parallel set up, i.e. a medium composed of a bundle of straight parallel capillaries of uniform diameter. For natural or artificial consolidated porous media, also tortuosity must be kept in account. In this case, other relationships are:

$$\beta = \frac{c'\tau}{kn} \qquad \text{(Geertsma, 1974)} \tag{7}$$

where: $c'$ = constant, $\tau$ = tortuosity

$$\beta = \frac{3.1 \times 10^{-15} \tau^{1.943}}{k^{1.023}} \qquad \text{(Cooper et al., 1999)} \tag{8}$$

In the above correlation the permeability is expressed in millidarcy [mD] ($1\,\text{mD} = 0.987 \cdot 10^{-15}\,\text{m}^2$), and the $\beta$ coefficient is expressed in [$\text{m}^{-1}$].

In the case of two-phase flow, many empirical relationships have been proposed for both porous media and fractured rocks, with special reference to propped hydraulic fractures (Ma and Ruth, 1997; Ucan and Civan, 1996; Wang and Mohanty, 1999; Jin and Penny, 2000, *etc.*). For example, for a gas-liquid system where the liquid phase is immobile (e.g., in presence of residual water saturation $S_{wr}$), Geertsma (1974) assumed that the $\beta$ coefficient depends only on the voids occupied by the gas, and hence on porosity *n*, on gas saturation $S_g = 1 - S_{wr}$, and on gas effective permeability $k_g = k_{rg}k$, proposing the relationship:

$$\beta_g = \frac{0.5}{k^{0.5} n^{5.5}} \frac{1}{(1 - S_{wr})^{5.5} k_{rg}^{0.5}} \tag{9}$$

where again the units of permeability k is in [millidarcy], and the $\beta$ coefficient is in [$\text{m}^{-1}$]. Further researches established similar empirical relationships, measured on-site or in laboratory, valid also in the case of flow of both phases. Some of these expressions have been specifically validated for porous media or for fractured rocks, while other expressions are considered of general validity (Ma and Ruth, 1997; Ucan and Civan, 1996; Wang and Mohanty, 1999; Jin and Penny, 2000; Li and Engler, 2001, and references herein contained). In all the above relationships, the coefficient $\beta_i$ is always a function of porosity n, saturation $S_i$, effective permeability $k_i$ or of some of these terms. However, the expressions are different both in form, exponents and coefficients.

Accordingly to Li and Engler (2001), these differences might depend on the different geometry of the flow paths (mainly on dimensions and tortuosity), on the lithology, and possibly also on the fact that the parameters under consideration vary from case to case. Narayanaswamy et al. (1999) found that differences between laboratory and on-site measurements are probably due to heterogeneities of the considered media, bearing in mind also that heterogeneities increase with the increasing reference volume influencing the measurements. In the case of two-phase flow through beds of coarse particles many relationships have been proposed. The following is the one by Schulenberg and Muller (1987):

$$-\text{grad}\Phi_i = \frac{\mu_i}{k_{ri} k \rho_i} u_i + \frac{1}{\eta_{ri} \eta \rho_i} u_i^2 \qquad i = 1, 2 \qquad (10)$$

where $\eta = 1/\beta$ is the inverse of Forchheimer's coefficient, and $\eta_{ri}$ are coefficients depending on particle shape, on effective permeability $k_i$ and on saturation $S_i$, to be determined experimentally.

Many numerical models and theoretical studies have been developed for modeling single-phase and multi-phase non Darcian flow through heterogeneous porous and fractured rocks. In particular, Wu (2002) developed a formulation incorporating the Forchheimer's equation, based on an integral finite-difference or a control volume numerical discretization scheme. The proposed model was then implemented into a three-dimensional, three-phase flow simulator and is applicable to both single-porosity porous media and fractured rocks. For flow in fractured porous media, fracture-matrix interactions are handled using an extended dual-continuum approach, such as double- or multiple-porosity, or dual-permeability methods. The model, though formally valid, need to be verified for the assumptions and simplifications adopted in the physics of fluid flow.

## 4 COUPLING EFFECT

In the usual extension of Darcy's law to two-phase flow (where relative permeability depends on fluids saturation and their distribution inside the medium), it is considered that each fluid flows through continuous flow paths and the flow of one phase interferes with the flow of the other only insofar as it takes up part of the flow paths, reducing the flow area of each phase. This does not take into account the fact that the two phases are flowing side by side in the porous medium and offer complex fluid contact surfaces. Thus, one phase can influence the flow of the others, and vice-versa (coupling effect). This phenomenon was first put in light by Yuster (1953), though still today many researchers consider it as negligible in almost all cases, or simply ignore it. A comprehensive discussion of this matter with an accurate bibliography can be found in Rose (1991, 1999, 2000), Rose and Rose (2004), Ayub and Bentsen (1999), Bentsen (2001).

Today the importance of the coupling effect seems proved by several theoretical and experimental studies. Accordingly to the majority of researchers, this phenomenon is consequent to the reciprocal viscous drag (in term of shear stresses at the fluid contact interface) that may affect significantly the flow of the contiguous fluid. Especially in the case of high velocity laminar flow, if the fluids have different velocities (in intensity, direction, or both), at the contact interfaces they deliver to each other tangential forces whose effect is all the more sensitive, the greater is the ratio of viscous to capillary forces, expressed by the capillary number. In this situation of coupled flow, the following relationships are supposed to be valid:

$$u_i = -(\lambda_{ii}\,\mathrm{grad}\Phi_i + \lambda_{ij}\,\mathrm{grad}\Phi_j) \tag{11}$$

$$u_j = -(\lambda_{ji}\,\mathrm{grad}\Phi_i + \lambda_{jj}\,\mathrm{grad}\Phi_j) \tag{12}$$

where $\lambda_{ij}$ is the coupling coefficient, and:

$$\lambda_{ii} = \frac{\rho_{ii} k_{ii}}{\mu_{ii}}, \qquad \lambda_{jj} = \frac{\rho_{jj} k_{jj}}{\mu_{jj}} \tag{13}$$

The coefficients $\lambda_{ii}$, $\lambda_{jj}$, and $\lambda_{ij}$ are to be determined experimentally. Generally it is here assumed to hold the reciprocity relationship $\lambda_{ij} = \lambda_{ji}$ (Onsager, 1930).

In granular porous media this effect is important for capillary numbers larger than $10^{-6}$ or $10^{-5}$ (Hughes and Blunt, 1999; Mott et al., 1999, 2000); in fractured rocks this effect has not been completely studied, but it might be significant because of the high flow velocity that can be attained in fractures, increasing the capillary number Nca. Hughes and Blunt (1999) report that in porous reservoirs usual values of capillary number are $Nca < 10^{-7}$ (with $Nca \sim 10^{-3}$ in proximity of the borehole); on the other hand, in fractured reservoir usual values are $Nca < 10^{-5}$. So, it is clear that high flow velocity makes viscous and inertial forces dominant with respect to capillary forces.

Experimental studies have clearly proved this phenomenon in the case of flow in both porous and fractured media. Here it has been found that the volumetric flows increase considerably when both fluids move in the same direction (co-current flow) compared to the case of countercurrent flows (Bourbiaux and Kalaydjian, 1990; De Santos et al., 1991; Fourar et al., 1993; Persoff and Preuss, 1995; Mott et al., 1999).

Figure 1 reports an example of the different relative permeability curves measured through capillary displacement of oil by water (imbibition, in the case of co-current and countercurrent flow) in a porous medium, considering valid the conventional Darcy's law. For the case of gas and gas-condensate co-current flow, Figure 2 reports the values of gas relative permeability $k_{rg}$ vs. capillary number in a particular case of high rate flow and constant mobility ratio. It is worthy reminding that Houpeurt (1974) noticed that the curves of relative permeability measured on cores in routine laboratory tests, by imposing the same direction of the flow of the phases, are not applicable to the case of non co-current flow. This is typical for example, of oil reservoir producing for solution gas drive, where during the flow towards the well, the oil tends to flow downward and the gas tends to slip upward.

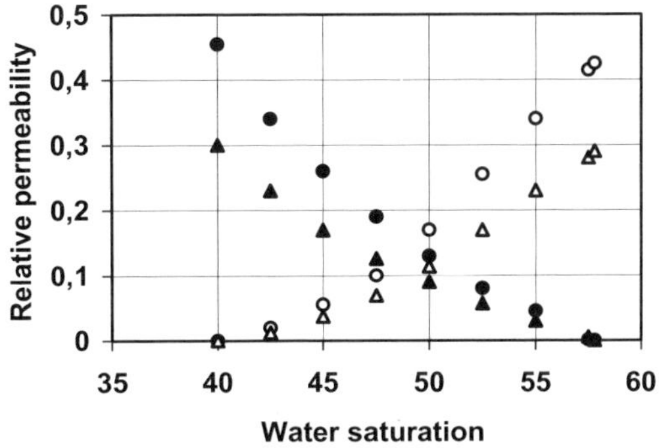


Figure 1. Co-current and countercurrent relative permeability curves in imbibition tests. Solid dots represent oil relative permeability $k_{ro}$ ($k_{ro}$, ● co-current, ▲ countercurrent), while empty dots represent water relative permeability $k_{rw}$ ($k_{rw}$, ○ co-current, △ countercurrent) (from Bourbiaux and Kalaydjian, 1990, modified).

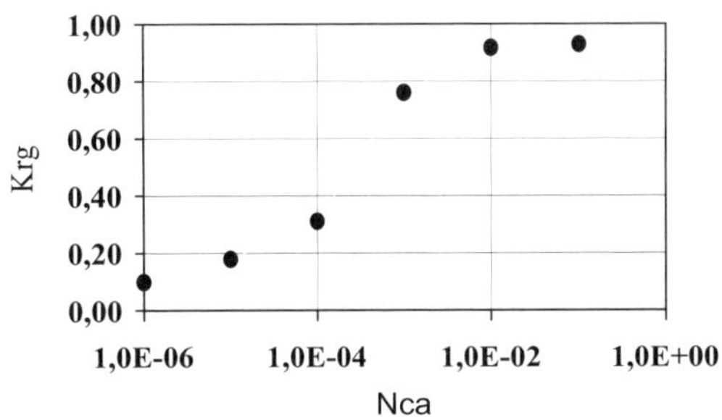


Figure 2. Gas relative permeability $K_{rg}$ measured vs. different capillary number Nca at constant mobility ratio $M_{go}$, in condition of co-current flow of gas and gas-condensate (from Mott et al., 1999, modified).

## 5 INERTIAL AND COUPLING EFFECTS

While studying the flow of gas in the immediate vicinity of a gas-condensate well, where there is a strong saturation of liquid and where both the gas and the condensate move in the same direction, Mott et al. (2000) found that at a high velocity there are two competing phenomena which make the gas effective permeability rate-dependent. In this case, on one hand the gas effective permeability increases due to positive coupling (or stripping velocity) and on the other decreases due to the inertial effect. In this case, these authors have suggested introducing into eq. (3) an effective permeability of the gas:

$$k_{g.eff} = kF_{nd}\,k_{rg}(S_g, \mathrm{Nca}) \tag{14}$$

where, recalling eq. (4):

$$F_{nd} = \left(1 + \frac{\beta k \rho_g}{\mu_g} u_g\right)^{-1} \tag{15}$$

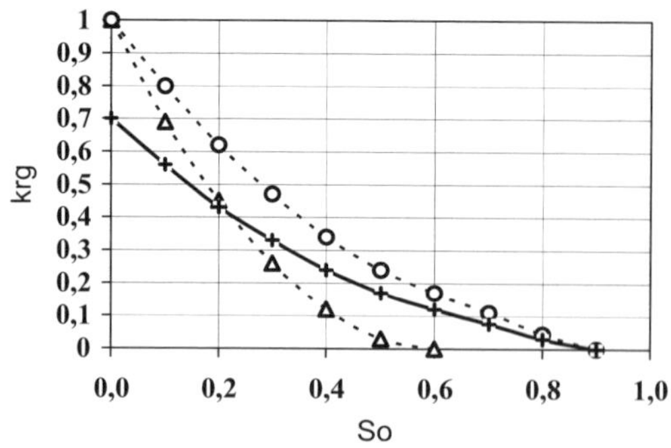


Figure 3. Gas relative permeability at different velocity: Δ = curve "a", low velocity; ○ = curve "b", high velocity excluding inertial effects; + = curve "c", high velocity including inertial and capillary number effects (from Mott et al., 2000 modified).

Here, $F_{nd}$ takes the inertial flow effects into account, while $k_{rg}$ takes the coupling effect into account and is determined as a function of gas saturation $S_g$ and of capillary number Nca. An example of these conflicting effects is illustrated in Figure 3, where the curve "a" is the usual gas relative permeability (at low velocity and low capillary number), while the curve "b" is the high-velocity gas relative permeability excluding inertial effect and finally the curve "c" is high-velocity gas relative permeability including inertial and capillary number effect. At high velocity, while at low liquid saturation, the inertial effect dominates, at high saturations the capillary number effect is the most important, so that gas relative permeability becomes larger than the one at low velocity and low capillary numbers.

Similar results have been obtained by Henderson et al. (2000), always in the case of gas and gas-condensate flow in the vicinity of a wellbore. In this particular study, it has been performed more than 90 tests in limestone and sandstone samples, with permeability ranging from 10 to 550 millidarcy and gas-liquid interfacial tension ranging from $0.015 \cdot 10^{-3}$ to $0.7 \cdot 10^{-3} N \cdot m^{-1}$. In particular, here it has been highlighted a transition from inertia-dominated relative permeability curves with increasing velocity at low condensate saturation (low mobility ratio), to conditions where the positive coupling effect was dominant as the condensate saturation and mobility ratio increase. Relationships are also given to calculate relative permeability accounting for both inertial and positive coupling effects.

Again, in the case of gas and gas-condensate flow, some researchers (Ali et al., 1997-a, 1997-b; Wang and Mohanty, 1999; Amili and Yortsos, 2002) noticed that by increasing Nca, the condensate turns from a continuous phase attached to the pore wall (with condensate ganglia) to a non Darcy flow condition. In this case, the liquid can move both as a continuous phase adherent to the medium wall surfaces, and as in the form of insulate droplets drawn by the gas flow (mist flow), whose dimensions are smaller than the pore throats, as shown in Figure 4. Droplets can be formed for both blow-up of the ganglia (Figure 4-a), and, as gas velocity increases, for viscous drag inside the liquid accumulated around the pore or by the fracture wall (Figure 4-b). By increasing furthermore the velocity, the flow attains the limit case of two-phase flow inside pipes, both in coarse porous media, and in fractures. In most of these cases it is likely that true turbulent flow conditions exist, though they are not explicitly indicated (Fourar et al., 1993; De Santos et al., 1991).

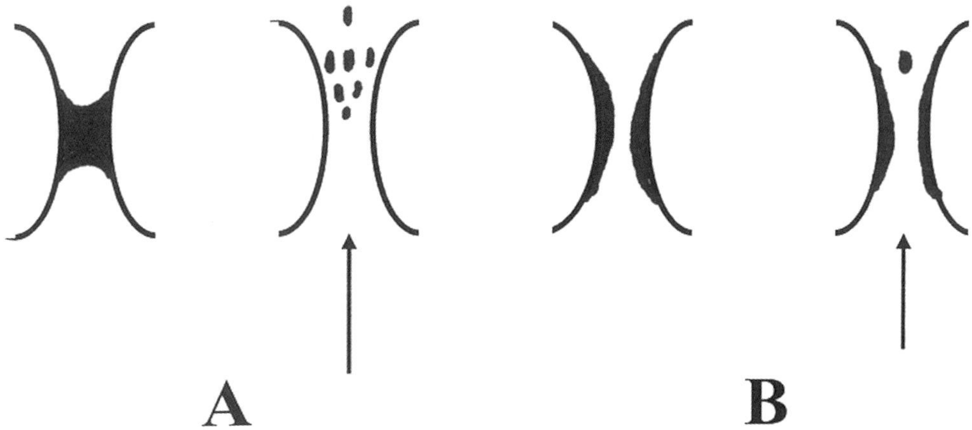


Figure 4. Representation of the formation of small oil-condensate droplets in proximity of pore throats, occurring when liquid ganglia blows-up. This mechanism is possible when there is a high velocity gas flow. Obviously, droplets dimension is smaller than the pore throats (modified from Wang and Mohanty, 1999).

## 6 FOAMY-OIL FLOW

It has been already mentioned that, according to Darcy's law extended to two-phase flow, each fluid cannot move until it has formed a continuous phase. In this case the rate of the volumetric flow of the phases depends on the ratio $M_{ij}$ of the mobility $\lambda_i$ of each phase. However, it has been noticed that this does not happen in many reservoirs of viscous oil which produce by dissolved gas drive. Here the wells produce a stable foam emulsion of gas in oil for a long time, and the gas to oil ratio (GOR) is much lower than it should be according to Darcy's law, with a positive effect on the primary oil production of the reservoir (Maini, 2001; Kamp et al., 2001-a, 2001-b, *etc.*).

According to Maini (2001, and references herein enclosed), the different behavior is associated not so much to super-saturation of dissolved gas in the oil, as to the distribution of the fluids in the medium. In fact, in the usual light oil reservoirs, the ratio of viscous to capillary forces is low and the distribution of the phases, which depends on the capillary forces, is the same under static and flowing conditions and is not influenced by the local pressure gradient. On the contrary, in this kind of reservoir, due to the high viscosity of the oil and to the high drawdown pressure used in cold production (i.e., during the production of heavy oil with no thermal enhanced processes), the ratio of viscous to capillary forces is high and such as to mobilize the micro-bubbles (i.e., bubbles much smaller than average fracture or pore-throat size). An alternative hypothesis is that the larger bubbles can break during the migration with the oil. The phenomenon becomes easier to detect when the gravitational forces do not induce rapid segregation of the phases (i.e., when there is a low Bond number) and the pore-body/pore-throat ratio is small, as, for example, in the case of well-sorted unconsolidated sands.

Evidence indicates that, for all these cases, the classical Darcy's law does not adequately describe the oil and gas flow. Several conceptual models have been proposed to study physical mechanism driving the production of these reservoirs. The simplest model considers that the local oil and gas velocities are the same and the emulsion is homogeneous (homogeneous emulsion flow), and is characterized by an average density. In this case, the single-phase Darcy's law still holds, considering the intrinsic permeability of the medium and the viscosity of the emulsion. Another model considers that part of the gas flows as a continuous phase in some of the flow paths, and the rest as a dispersed phase in the oil. Finally, a third model considers the flow of two continuous phases. Probably the first model is valid for low gas saturation values, and the second one for gas saturation values sufficient to allow the formation of a continuous gas phase in some flow paths. At last, the third model should be valid for high gas saturation values, where eq. (1) probably still holds. Many researchers have studied the issue and have suggested several approaches to it, but many questions still remain unsolved.

## 7 CONCLUSIONS

Non-Darcian multi-phase flow has been investigated since long ago, with many studies, both theoretical and experimental. However, a complete and unitary theory on the subject still does not exist. In particular, the extension of Darcy's law to multi-phase flow has not been yet completely probed either in the case of Darcian flow. For this reason, the present study is limited to the examination of the two-phase flow in Darcian and non-Darcian conditions.

The considerations discussed in this paper highlight the fact that, for two-phase laminar flow in porous media or in fractures, the generalized Darcy's law does not apply in cases where:

a) forces of inertia are no longer negligible compared to viscous forces;
b) the flow of one phase is influenced by the forces transmitted by the other phase across the contact surface, when both phases are in motion (coupling effect);
c) the distribution of the phases inside the medium does not depend so much on the capillary forces as on the viscous forces and gravity ones, a condition that takes place in foamy-oil flow when there is a high pressure gradient (and, hence, high velocity flow).

The same conditions generally take place in the case of flow inside unconsolidated porous media or fractured media with small pore-body/pore-throat ratio, where there are different flow models, depending on the velocity of both phases and on the surface tension. It is believed that all the above hold for all porous and/or fractured media, since the flow model does not depend on the particular network of flow paths (pores and/or fractures), but mainly on the characteristics of the flow paths themselves (i.e., dimensions, tortuosity, cross-section variations, *etc.*). Flow paths inside a coarse-grained porous rock (gravel, coarse clean sand, *etc.*) are probably more similar to those of a fractured rock with aperture of a few millimetres than the flow paths inside a fine-grained cemented porous rock.

In conclusion, it clearly appears that the flow model does not depend so much on whether the medium consists of porous or fractured rocks, but on the ratio between the forces driving the fluid flow (i.e., Reynolds number, capillary number and Bond number). When such ratios vary, the flow models move from those taking place in a capillary network to models that take place in a network of pipes.

## REFERENCES


Ali JK, McGauley PJ, Wilson CJ (1997-a) Experimental studies and modeling of gas condensate flow near the wellbore. SPE 39053, 5th Latin American and Caribbean Petroleum Engineering Conference and Exhibition, Rio de Janeiro.

Ali JK, McGauley PJ, Wilson CJ (1997-b) The effects of High-velocity flow and PVT changes near the wellbore on condensate well performance. SPE 38923, SPE Annual Technical Conf and Exhibition, San Antonio.

Amili P, Yortsos Y (2002) Darcian Dynamics: a new approach for the mobilization of ganglia in porous media. SPE 75191, SPE/DOE Improved Oil Recovery Symp, Tulsa.

Ayub M, Bentsen RG (1999) Interfacial viscous coupling: a myth or reality? J Petroleum Science & Engineering 23:13–26.

Bear J, Berkowitz B (1987) Groundwater flow and pollution in fractured rocks aquifers. In Novak (ed), Development in hydraulic engineering, 4, Elsevier, London.

Bear J (1993) Modeling flow and contaminant transport in fractured rock. In Bear, Tsang, Chin-Fu, De Marsily (eds) Flow and contaminant transport in fractured rock, Academic press, San Diego.

Belhaj HA, Agha KR, Nouri AM, Butt SD, Vaziri HF (2003) Numerical simulation of non-Darcy flow utilizing the new Forchheimer's diffusivity equation. SPE 81499, SPE 13th Middle East Oil Show & Conference, Bahrain.

Bentsen RG (2001) The physical origin of interfacial coupling in two-phase flow through porous media. Transport in Porous media 44:109–122.

Bourbiaux BJ, Kalaydjian FJ (1990) Experimental study of cocurrent and countercurrent flow in natural porous media. SPE Reservoir Engineering 3:361–368.

Cooper JW, Wang X, Mohanty KK (1999) Non Darcy flow studies in anisotropic porous media. SPE Journal 4:334–341.

De Santos JM, Melli TR, Scriven LE (1991) Mechanics of gas-liquid flow in packed-bed contactors. Ann Fluid Mech 23:233–260.

Firoozabadi A, Katz DL (1979) An analysis of high velocity gas flow through porous media. J Petrol Techn 31:211–218.

Forchheimer P (1901) Wasserbewewung durch Boden. ZWDI 45:1781–1901.

Fourar M, Bories S, Lenormand R, Persoff P (1993) Two-phase flow in smooth and rough fractures: measurement and correlations by porous-medium and pipe flow models. Water Resources Research 29:3699–3707.

Garrouch AA, Ali L (2001) Predicting the onset of inertial effects in sandstones. Transport in Porous Media 44:487–505.

Geertsma J (1974) Estimating the coefficient of inertial resistance in fluid flow through porous media. SPE Journal 10:445–450.

Henderson GH, Danesh A, Tehrani DH, Al-Karusi B (2000) The relative significance of positive coupling and inertial effects on gas condensate relative permeability at high velocity. SPE 62933, SPE Annual Technical Conf and Exhibition, Dallas.

Honarpour M, Koederitz L, Harvey AH (1987) Relative permeability of petroleum reservoirs, 2nd edn., CRC Press, Boca Raton.

Houpeurt A (1974) Mécanique des fluides dans les milieux poreux. Technip, Paris.

Hughes RG, Blunt MJ (1999) Pore-scale modeling of multiphase flow in fractures and matrix/fracture transfer. SPE 56411, SPE Annual Technical Conf and Exhibition, Houston.

Irmay S (1958) On the theoretical derivation of Darcy and Forchheimer formulas. Trans Amer Geoph Union 39:702–707.

Jin L, Penny GS (2000) A study of two-phase non-Darcy gas flow through proppant pacs. SPE Production & Facilities 15:247–254.

Kamp AM, Henry C, Andarcia L, Lago M, Rodriguez A (2001-a) Experimental investigation of foam oil solution gas drive. SPE 69725, SPE Thermal Operations and Heavy Oil Symp, Poriar, Venezuela.

Kamp AM, Joseph DD, Bai R (2001-b) A new modeling approach for heavy oil flow in porous media. SPE 69720, SPE Thermal Operations and Heavy Oil Symp, Poriar, Venezuela.

Kazemi H, Gilman JR (1993) Multiphase flow in fractured petroleum reservoirs. In Bear, Tsang, Chin-Fu, De Marsily (eds) Flow and contaminant transport in fractured rock, Academic press, San Diego, 267–323.

Li D, Engler TW (2001) Literature review on correlations of the non-Darcy coefficient. SPE 70015, SPE Permian Basin Oil and Gas Recovery Conf, Midland.
Ma H, Ruth DW (1997) Physical explanations of non-Darcy effects for fluid flow in porous media. SPE Form Eval 12:13–18.
Maini BB (2001) Foamy-oil flow. J Petroleum Technology 53:54–64.
Mott R, Cable A, Spearing M (1999) A new method of measuring relative permeabilities for calculating gas-condensate well deliverability. SPE 56484, SPE Annual Technical Conf and Exhibition, Dallas.
Mott R, Cable A, Spearing M (2000) Measurements and simulation of inertial and high capillary number flow phenomena in gas-condensate relative permeability. SPE 62932, SPE Annual Technical Conf and Exhibition, Dallas.
Narayanaswamy G, Sharma MS, Pope GA (1999) Effect of heterogeneity on the non-Darcy flow coefficient. SPE Reservoir Evaluation & Engineering 2:296–302.
Onsager L (1930) Reciprocal relations in irreversible processes. Phys Rev 37:405–426, 38:2265–2279.
Persoff P, Pruess K (1995) Two-phase flow visualization and relative permeability measurement in natural rough-walled rock fractures. Water Resources Research 31:1175–1186.
Romm ES (1972) Fluid flow in fractured rocks, Blake, Bartlesville.
Rose W (1991) Critical questions about the coupling hypothesis. J Petroleum Science & Engineering 5:299–307.
Rose W (1999) Relative permeability ideas then and now (Richards to Leverett to Yuster, and beyond). SPE 57442, SPE Eastern Regional Meeting, Charleston.
Rose W (2000) Myths about later-day extension of Darcy's law. J Petroleum Science & Engineering 26:187–198.
Rose W, Rose DM (2004) "Revisiting" the enduring Buckley-Leverett ideas. J Petroleum Science & Engineering 45:263–290.
Sahimi M (1995) Flow and Transport in Porous Media and Fractured Rock, VCH, Weinheim.
Scheidegger AE (1974) The physics of flow through porous media, 3rd edn., University of Toronto Press, Toronto.
Schulenberg T, Muller U (1987) An improved model for two-phase flow through beds of coarse particles. Int J of Multiphase Flow 13:87–97.
Ucan S, Civan F (1996) Simultaneous estimation of relative permeability and capillary pressure for non-Darcy flow steady-state. SPE 35271, Mid-Continent Gas Symposium, Amarillo, 155–163.
Wang X, Mohanty KK (1999) Multiphase non-Darcy flow in gas-condensate reservoirs. SPE 56486 SPE Annual Technical Conf and Exhibition, Houston.
Wu YS (2002) Numerical simulation of single-phase and multiphase non Darcy flow in porous and fractured reservoirs. Transport in Porous Media 49:209–240.
Yuster ST (1953) Theoretical considerations of multiphase flow in idealized capillary systems. 3rd World Petroleum Congress II:437–445.

CHAPTER 13

# GIS based assessment of groundwater recharge in the fractured rocks of Namaqualand, South Africa

Julian Conrad[1] and Shafick Adams[2]

*[1]GEOSS – Geohydrological and Spatial Solutions International (Pty) Ltd, TechnoPark, Stellenbosch, South Africa*

*[2]Earth Science Department, University of the Western Cape, Private Bag X17, Bellville, South Africa*

ABSTRACT: For protection and management of groundwater resources it is essential that groundwater recharge is calculated. In semi-arid Namaqualand (South Africa), a recharge assessment was conducted using a geographical information system (GIS). The approach taken was based on assigning recharge probabilities, based on field experience, to a number of data sets and then combining these layers. The data sets included: geological lineaments; lithology; land cover; soil type; soil thickness; soil texture; slope and depth to groundwater. A groundwater recharge percentage map was produced and the results were applied to the mean annual precipitation to obtain the spatial distribution of recharge. As the data processing is GIS based, new data sets can easily be included and the processing swiftly completed. The GIS based recharge results correlated closely with point recharge calculations. Although the approach simplifies groundwater recharge mechanisms, knowing the spatial distribution of recharge enables improved resource protection and management.

## 1 INTRODUCTION

The quantification of the rate of groundwater recharge is vital for efficient groundwater resource management or sustainable management of groundwater resources (Simmers, 1996, Bredenkamp et al., 1995). The prediction of sustainable yields of aquifers is dependent on the amount of water recharging the aquifers. Recharge is vital in basement aquifers that commonly have small storativities, and it becomes even more important if the basement aquifers occur in arid regions. Basement aquifers here include both the weathered zone and fractured rock aquifers. As a result, the need to assess recharge in these regions, using suitable methods, arises. The definition of recharge used in this report is: the portion of rainfall that reaches the saturated zone, either by direct contact in the riparian zone or by downward percolation through the unsaturated zone (Rushton and Ward, 1979).

There are numerous recharge estimation techniques, and no two methods, even if applied to the same area, yield identical recharge rates. These methods are often limited to particular environments (i.e., humid or arid), and the availability of data (especially long-term monitoring data). Techniques are not often transferable from one environment to another and recharge estimation is one of the most difficult hydrogeological variables to

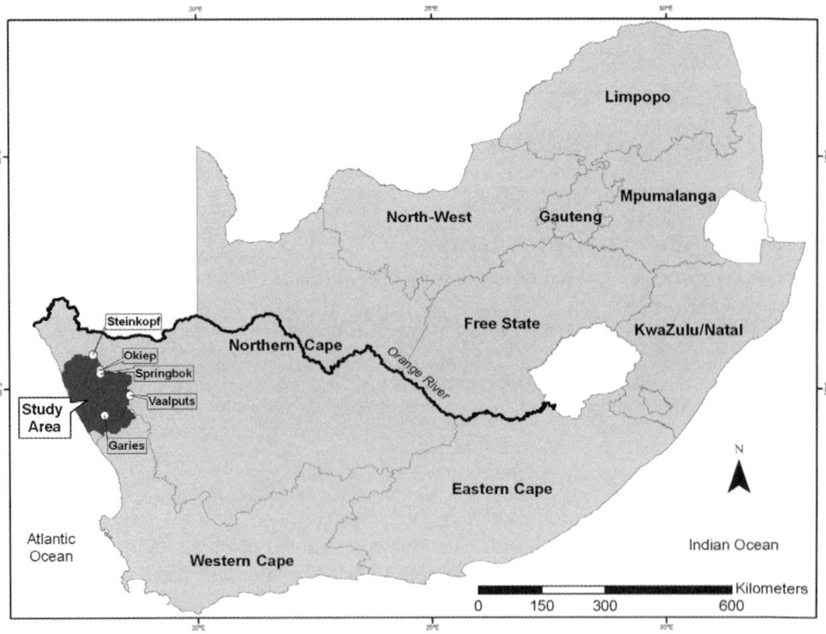


Figure 1. The study area located in Namaqualand, western South Africa.

determine. Recharge is estimated by using known variables that are directly measurable to some degree of accuracy. Recharge can be estimated by determining the fluxes in the unsaturated zone or by estimating the net contribution to the saturated zone. Methods that are applicable in the different zones can be classified as chemical and isotopic methods, physical methods, and combination methods that can be integrated into mathematical and GIS models. Crucial to the estimation of groundwater recharge is a proper understanding of the aquifer system being investigated.

Estimating recharge in the Namaqualand area is difficult due to the paucity of data. The study area is shown in Figure 1. This paper provides a base for future groundwater resource assessments. The approach of this paper is to use existing and easily obtainable data to estimate recharge in the area.

## 2 BACKGROUND

The South African climate is generally divided into a wet east coast and a drier west coast. The study area (Figure 1), in Central Namaqualand, is situated on the west coast of the subcontinent and experiences an arid to semi-arid climatic regime, brought about mainly by the topography of the area, as well as the climatic systems operating on the subcontinent. No perennial river systems occur in the study area. The Orange River, to the north of the study area, is the only perennial river in the primary catchment area. Most rural communities rely on groundwater for their existence. The Orange River is a major source of piped water supply for the larger towns (Springbok, O'kiep and Steinkopf) and for the large-scale diamond and base-metal mining activities.

Groundwater development has been dominated by the need for rural community water supply subsidised by the government, the private water supply for domestic and agricultural activities and the exploitation of groundwater for mining activities. The development of the groundwater resource is a complex task given the complexity of the aquifer systems. Crystalline rocks are inherently poor aquifers because of low storage capacity and water-quality problems. In this study area we can divide the basement aquifers into weathered zone aquifers and fractured rock aquifers.

Groundwater hydrology in South Africa has been chiefly concerned with supply of water to areas where surface water was not sufficient or too expensive to meet demand. Vegter (2001) gives a detailed account of the various stages through which South African groundwater hydrology passed. The current trend in groundwater studies is to focus on the field of integrated management of water resources which has arisen as a result of the National Water Act promulgated in 1998.

Poor understanding of the aquifer systems and rates of aquifer recharge in Namaqualand led to poor management practices. The assessment of the groundwater systems and the socio-economic dynamics of the area have been the focus of research by the Department of Earth Sciences at the University of the Western Cape. One of the main constraints identified was that the processes and rates of aquifer replenishment/recharge were not fully understood. This paper reviews a study initiated to investigate the recharge characteristics of meteoric water to the aquifers of Central Namaqualand.

## 3 STUDY AREA

The study area (Figure 2) is located in the Northern Cape Province of South Africa, and is situated in an area called Namaqualand. Copper and diamond mining occurs in the area. The area is also well known for its extensive display of flowers in the spring season. The area is classified as arid to semi-arid and the climate of the area is controlled by altitude, topography and distance from the sea. The mean annual precipitation (MAP) varies from 44 mm in the coastal zone area to 480 mm in the Kamies Mountains, for the period June 1999 to August 2001. Higher rainfall in the higher lying areas is a result of orographic rainfall. Rainfall occurs predominantly during the winter months. Snow in the Kamies Mountains is not uncommon during the winter. Potential annual evapotranspiration is between 12–15 times the MAP. The vegetation of the area is characterised by the Nama Karoo biome, which is dominated by a mixture of grasses and low shrubs. The area is characterised by three geomorphologic regions, (1) a coastal lowland, (2) an escarpment zone, which comprises of highland and lowlands within this zone and (3) the Bushmanland Plateau. Due to low rainfall over the semi-arid to arid region, most of the rivers are ephemeral. The most important ephemeral rivers in the area are the Buffels River, Groen River and Swartlintjies River. The area includes tertiary catchments F30, F40 and F50.

Namaqualand can be subdivided into three major geological provinces (Tankard et al., 1982). These are: the basement rocks of the Namaqua Province; the volcano-sedimentary rocks of the Gariep Complex (Visser, 1989) in the northwest; and a Phanerozoic cratonic cover (Table 1). The geology of the area has been discussed in detail by Titus et al. (2002) and Titus (2003).

The study area is predominantly underlain by Proterozoic crystalline basement rocks of the Namaqua Province. Cover rocks of the Nama group overlay the basement rocks to the north

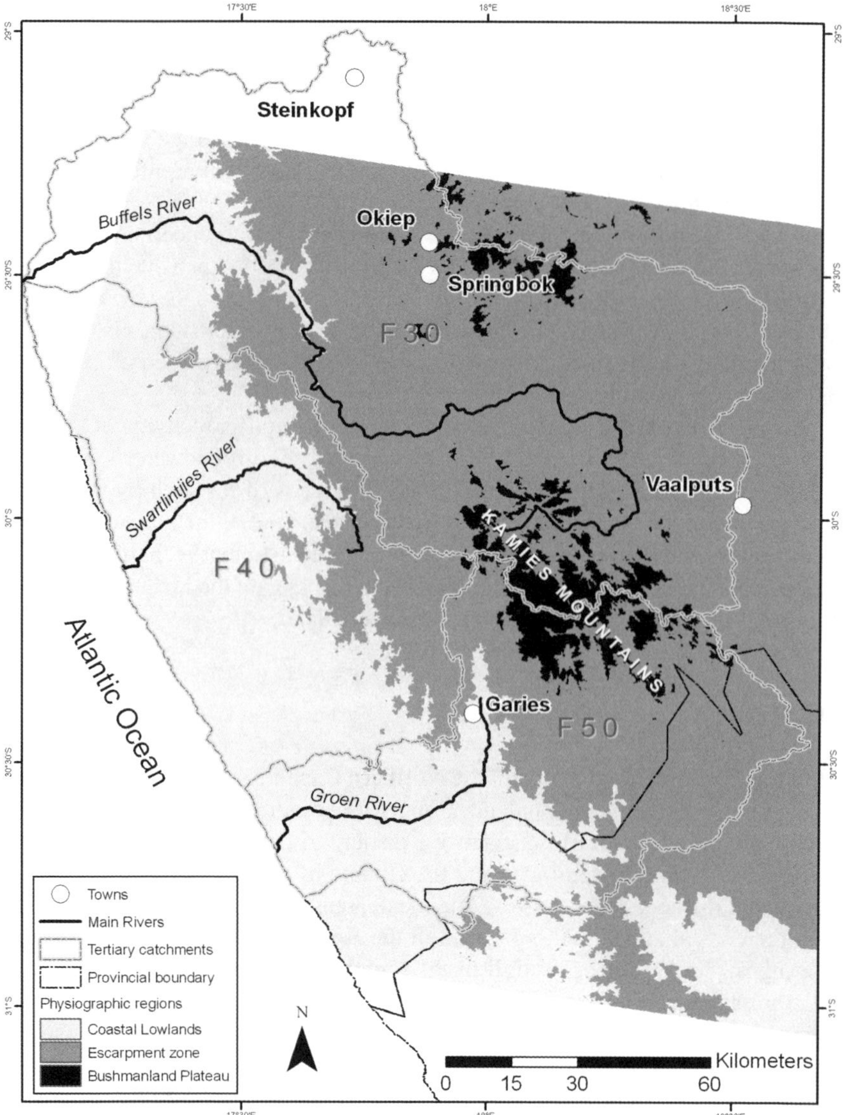


Figure 2. Study area and relevant features.

of the Buffels River catchment. The margins of the Namaqua Province are also obscured by younger cover rocks of the Gariep Complex to the north as well as by Cenozoic surficial sediments to the east and west. In the south, rocks of the Nama Group and the lowermost units of the Karoo Supergroup cover the rocks of the Namaqua Province (Albat, 1984).

Groundwater occurs in three different aquifer systems in the Namaqualand area, they are: (1) fractured bedrock; (2) the weathered zone or regolith; and (3) the sandy/alluvial aquifers. The fractured bedrock and weathered zone aquifers are hydraulically linked; the weathered zone acts as a reservoir that is able to recharge the bedrock aquifers. The weathered zone

Table 1. Classification of major geological provinces (Tankard et al., 1982 and Visser, 1989).

| Geological province | Group | Age | Hydrogeological characteristics |
|---|---|---|---|
| *Cover Rocks* | Sand, Alluvium and Calcrete | Late Phanerozoic (Cenozoic) | Shallow to deep sand deposits – variable yield and quality |
| | Nama Group | Late Proterozoic (Late Namibian) | Shale, limestone and quartzite, low yielding, variable quality |
| *Gariep Complex* | Gariep Complex | Late Proterozoic (Early/Middle Namibian) | Quartzites mainly – variable yield and quality |
| *Namaqua Province* | | | |
| Central Zone | Namaqua Metamorphic Complex | Middle Proterozoic (Middle Mokolian) | Various metamorphic and igneous rocks – variable yields and quality |
| Western Zone | Vioolsdrif Intrusive Suite | Middle Proterozoic (Early Mokolian) | Granitic rocks – low yields and poor quality |

aquifers are generally considered to be the most productive groundwater zones (e.g., Tindimugaya, 1995; Acworth, 1997).

Superimposed on the basement aquifers are the alluvial aquifers associated with the ephemeral rivers, paleochannels and the coastal plain. Alluvial aquifers associated with the river systems are usually very shallow (1–15 m) and in the coastal areas, tens of metres. Boreholes and large diameter wells drilled in Namaqualand are usually found in river courses and alluvium filled valleys. The alluvial aquifers are the main source of water supply for most villages. The alluvial aquifers are efficiently recharged during the rainy season. The alluvial aquifers, in turn, recharge other hydraulically connected aquifers.

## 4 PREVIOUS WORK

The Department of Water Affairs and Forestry (DWAF) and Toens and Partners did most of the community water supply projects for the area, which led to numerous consultancy reports and monitoring data being produced. The Atomic Energy Corporation (AEC), now the National Energy Corporation of South Africa (NECSA), did hydrogeological work in and around the Vaalputs area (a radioactive waste repository) to the east of the study area. A detailed description of the groundwater resources and hydrogeology of the study area can be found in Titus et al. (2002). The recharge manual by Bredenkamp et al. (1995) gives an overview of recharge in South Africa, except for the crystalline basement aquifers. No detailed recharge studies, either published or unpublished, exist for the specific area. Toens and Partners (2001) estimated recharge in the Bitterfontein and Rietfontein areas to be between 0.9% (1.03 mm/yr) and 2.2% (3.62 mm/yr) of precipitation. Vegter (1995) cited recharge rates for the towns of Springbok and Garies of 7.3 mm/yr and 2.9 mm/yr respectively. Recharge in the Komaggas area was estimated at 9.6 mm/yr (DWAF, 1990). Verhagen and Levin (1986), using environmental isotopes in the Vaalputs area, noted that recharge is "minimal", occurring only periodically during periods of above-normal rainfall.

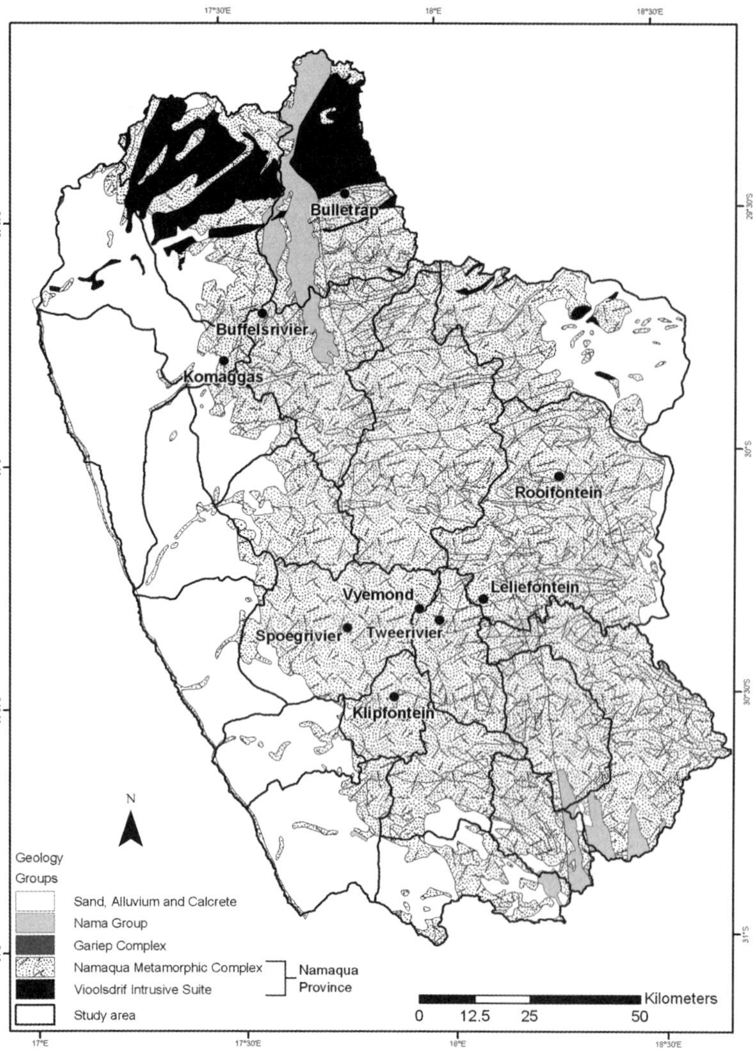


Figure 3. Geology of the study area.

## 5 METHODOLOGY

A number of recharge measurement methods were used in the study area (Adams et al., 2004). The chloride mass balance (CMB) method was used to estimate recharge on a local and regional scale using the saturated zone approach (Adams et al., 2004). The stable isotopes oxygen-18 and deuterium and the radiogenic isotope $^{14}C$ were coupled with factor analysis and the GIS assessment method to delineate and constrain recharge. The saturated volume fluctuation (SVF) method (Bredenkamp et al., 1995) was used to estimate storage

Table 2. A listing of the data sets used in the GIS approach.

| Data |
|---|
| Geology map (1: 1,000,000) |
| 1. Lithology |
| Land Type (1 : 250,000) |
| 1. Soil type and clay content |
| 2. Soil thickness |
| 3. Land cover |
| Landsat TM image (Analysis scale 1: 250,000) |
| 1. Lineaments |
| 2. Lineament intersections |
| 3. Digital Elevation Model |
| 4. Slope |
| Topographical maps (1: 50,000) |
| 1. Drainage density |
| Climatological Data (1: 250,000) |
| 1. Rainfall distribution and amount |
| Hydrogeological data (Point data) |
| 1. Water levels |

coefficients and recharge (Adams et al., 2004). The cumulative rainfall departures (CRD) method (Bredenkamp et al., 1995) was used to estimate the recharge rates to the aquifers (Adams et al., 2004). This paper focuses on the GIS method followed. In validation of the GIS approach, however, comparison was made to the recharge values obtained using the other approaches mentioned above.

The GIS approach involves the evaluation of different surface and subsurface features and their influence on natural groundwater recharge. The approach is mainly of a subjective nature where probability weights are assigned to features that may have a positive or negative effect on groundwater recharge. The approach is similar to the methodology of creating groundwater vulnerability maps. Different data sets that may be used in defining a recharge potential map were collated. The data sets are shown in Table 2 with overlays that can be derived.

All data sets were imported into ArcView GIS and standardised to the UTM 34 (WGS84 datum) projection. UTM was chosen as all directions, distances, shapes, and areas are reasonably accurate to within 15° of the central meridian. As this projection is conformal, shape and angles within any small area are essentially true. The different components of the recharge potential map are subjectively weighted according to the probability of recharge occurring in the particular area.

### 5.1 *Lithology*

The geology of the area is considered the most important factor in that all the surface and subsurface features of the area are controlled by the different lithologies. The major drawback is the scale of the available geological map of the area (1: 1,000,000) compiled by the Council for Geoscience. The lithological ratings and recharge probability are shown in

Table 3. Lithology and recharge potential found within the study area.

| Geology – Lithology | Recharge probability |
|---|---|
| Gneiss | 65 |
| Granite | 65 |
| Limestone | 85 |
| Quartzite | 65 |
| Sediments | 90 |
| Schist | 65 |
| Sedimentary | 75 |
| Shale | 55 |

Table 3. Recharge probability is the likelihood of direct groundwater recharge occurring for a particular feature. A value of "100" indicates that all water reaching the upper surface will pass through that feature, whilst a value of "0" indicates complete impermeability.

### 5.2 *Soil characteristics*

The characteristics of the soils were derived from a 1: 250,000 land type data set obtained from the Institute of Soil, Climate and Water. The land types incorporate terrain form, topography, and microclimate. Although the data set does not include a soil map it does have information pertaining to the soil thickness, percentage clay and soil texture within each land type. The depth of the soil zone and its clay content are important factors due to the high evapotranspiration rates of the area. The soil layers over most of the coastal area are not considered significant aquifers due to the poor quality and quantity of the groundwater. An exception are the alluvial aquifers associated with the main drainage systems. Thick soil covers will inhibit the movement of water to the water table. A linear relationship is assumed, i.e. if no soil cover is present it is rated as having a 100% probability of recharge. The maximum soil thickness recorded in the land type data is 1.2 m and this is rated at 50%.

Due to the arid nature of the area, soil moisture deficits are high and considerable rainfall is needed to initiate vertical groundwater recharge in thick soils. The soil characteristics over most of the area are a function of *insitu* weathering and transported material along rivers and paleochannels. The soil thickness map indicates that the soil depths vary across the mountainous areas.

### 5.3 *Lineaments*

Geological lineaments provide an indication of tectonic activity and may indicate fault or shear zones or geological weaknesses along which volcanic rock types have intruded. They may be zones of preferred groundwater flow or may be so weathered and fine grained that very little flow occurs. In the context of the Namaqua Province, they are considered significant and are thus taken into account. The process followed included a visual assessment of geological lineaments, mapped from the Landsat TM image (p176r081), acquired in December 1999. The Landsat TM scene did not cover the entire area of interest, but the bulk of the relevant lineaments did occur within the core region of the satellite

Table 4. The classification system used for lineaments.

| Type of lineament | Code | Description |
|---|---|---|
| Lineaments | 1 | Any linear feature that could not be classified |
| Shear zones | 2 | A wide zone where lithological displacement is evident |
| Dykes | 3 | Zones of igneous intrusion |
| Faults | 4 | A narrow zone where lithological displacement is evident |
| Geological contacts | 5 | Where differing lithologies come into contact, yet no displacement is apparent |

Table 5. Lineament intersection values.

| Intersection | Recharge probability % | Buffer distance (m) |
|---|---|---|
| Low (50) + low (50) | 50 | 50 |
| Low (50) + medium (70) | 60 | 75 |
| Low (50) + high (90) | 70 | 100 |
| Medium (70) + medium (70) | 70 | 100 |
| Medium (70) + High (90) | 80 | 150 |
| High (90) + high (90) | 90 | 200 |

image. The lineaments were mapped at a scale of 1:250,000. The classification used is listed in Table 4.

Once the lineament mapping had been completed, the entire image was reviewed at a scale of 1:125 000 and lineaments missed, inaccurate, or misclassified were updated. Only visible lineaments were mapped and no inferences were made if the lineament could not be seen. Many of the valley floors are filled with cover alluvial material and the linear nature of such valleys suggests that they are fault controlled. However, if the lineament could not be seen it was not mapped.

In an attempt to assess the groundwater flow potential of each lineament, a normalized difference vegetation index (NDVI) and a soil adjusted vegetation index (SAVI) was generated from the Landsat TM image. The SAVI was suggested by Kellgren et al. (2000) as a better index than the NDVI to use for this purpose, however for this particular area the difference between the NDVI and the SAVI was minimal. Lineaments associated with higher NDVI values were assumed to have more groundwater than is available for plants. The lineaments were also visually assessed in terms of groundwater flow potential and in conjunction with the NDVI, a qualitative value of high, medium and low were assigned to the lineaments. For the high flow lineaments a buffer distance of 200 m was assigned and this zone given a 90% probability for recharge potential. The medium flow lineaments were buffered by 100 m and assigned a 70% recharge probability and the low flow potential lineaments buffered by 50 m and given a 50% recharge probability. The buffering approach, which is the delineation of a zone, around a particular feature was used to give more "importance" and width to a feature that is considered to be a more favourable recharge zone.

Based on field observations the occurrence of lineament intersections is considered important, as the probability of recharge occurrence is considered higher at these zones. The ratings used for the intersections are shown in Table 5.

Table 6. Land cover types occurring within the study area and assigned recharge potential.

| Land cover | Recharge probability (%) |
|---|---|
| Barren rock | 100 |
| Cultivated: temporary – commercial dryland | 90 |
| Cultivated: temporary – commercial irrigated | 93 |
| Mines and quarries | 100 |
| Shrubland and low Fynbos | 95 |
| Thicket and bushland (etc) | 70 |
| Unimproved grassland | 97 |
| Urban/built-up land: residential | 85 |
| Waterbodies | 100 |
| Wetlands | 100 |

Table 7. The relationship used between surface topography slope and percentage recharge potential.

| Slope (degrees) | Recharge probability |
|---|---|
| 0–5 | 100 |
| 5–10 | 95 |
| 10–20 | 75 |
| 20–50 | 50 |
| 50–90 | 25 |

### 5.4 *Land cover*

For assessing the amount of precipitation interception and evapotranspiration, the national land cover data set was used. The land cover types are based on mapping done by the Agricultural Research Council and the CSIR from 1:250,000 Landsat TM images. The land classes occurring within the study area and associated subjective recharge probability ratings are given in Table 6.

### 5.5 *Topography and slope*

The slope was calculated from the digital elevation model. It is assumed that the steeper the slope is, the greater the rainfall runoff and the smaller the groundwater infiltration. Table 7 shows the values used in association with slope.

### 5.6 *Depth to water level*

The thickness of the unsaturated zone is considered a relevant factor in calculating recharge potential. The deeper the groundwater the more likely the adsorption of infiltrating surface moisture within the unsaturated zone and the more effective groundwater recharge is reduced. The values used for the depth to groundwater parameter are given in Table 8.

The input data for the depth to groundwater level parameter needs to be carefully assessed before being used to generate groundwater level maps, so as to ensure that pumped water levels were not included in the same data set as rest water levels. The rest

Table 8. The relationship used between depth to water level and recharge probability.

| Depth to water level (m) | Recharge probability |
|---|---|
| 0–5 | 90 |
| 5–10 | 80 |
| 10–15 | 70 |
| 15–20 | 60 |
| 20–26 | 50 |

Table 9. A listing of the data sets used.

| Data set |
|---|
| 1. Lineaments + flow potential (confirmed by NDVI/SAVI) |
| 2. Lineament density |
| 3. Lineament intersections |
| 4. Drainage density |
| 5. Land cover |
| 6. Soil type (clay content) |
| 7. Soil thickness |
| 8. Soil texture (% rock vs % soil) |
| 9. Slope (degrees) |
| 10. Depth to groundwater |
| 11. Lithology (geology) |
| 12. Rainfall distribution |

water levels were used as the basis for generating the depth to groundwater surface. An Inverse Distance Weighting (IDW) algorithm was used to generate the depth to water level map (Shepard, 1968).

## 6 RESULTS

For the generation of the final recharge map all the input coverages were converted to 50 m grids with the same area of extent. Each grid was then reclassified according to the classification table (Table 9) and then subsequently weighted according to how significant the individual factor was considered to be in contributing to groundwater recharge potential, based on the conceptual understanding of recharge for the study area.

The recharge potential map gives an idea of direct vertical infiltration. Overlaying a map with the rainfall distribution and amounts, produces a map of recharge from direct rainfall (Figure 4).

## 7 COMPARISON OF A NUMBER OF INDEPENDENT APPROACHES FOR RECHARGE CHARACTERIZATION AND ESTIMATION

Assessing recharge to any aquifer depends on the type of area under investigation, the availability of data, the distribution of available data, and the ability to obtain meaningful

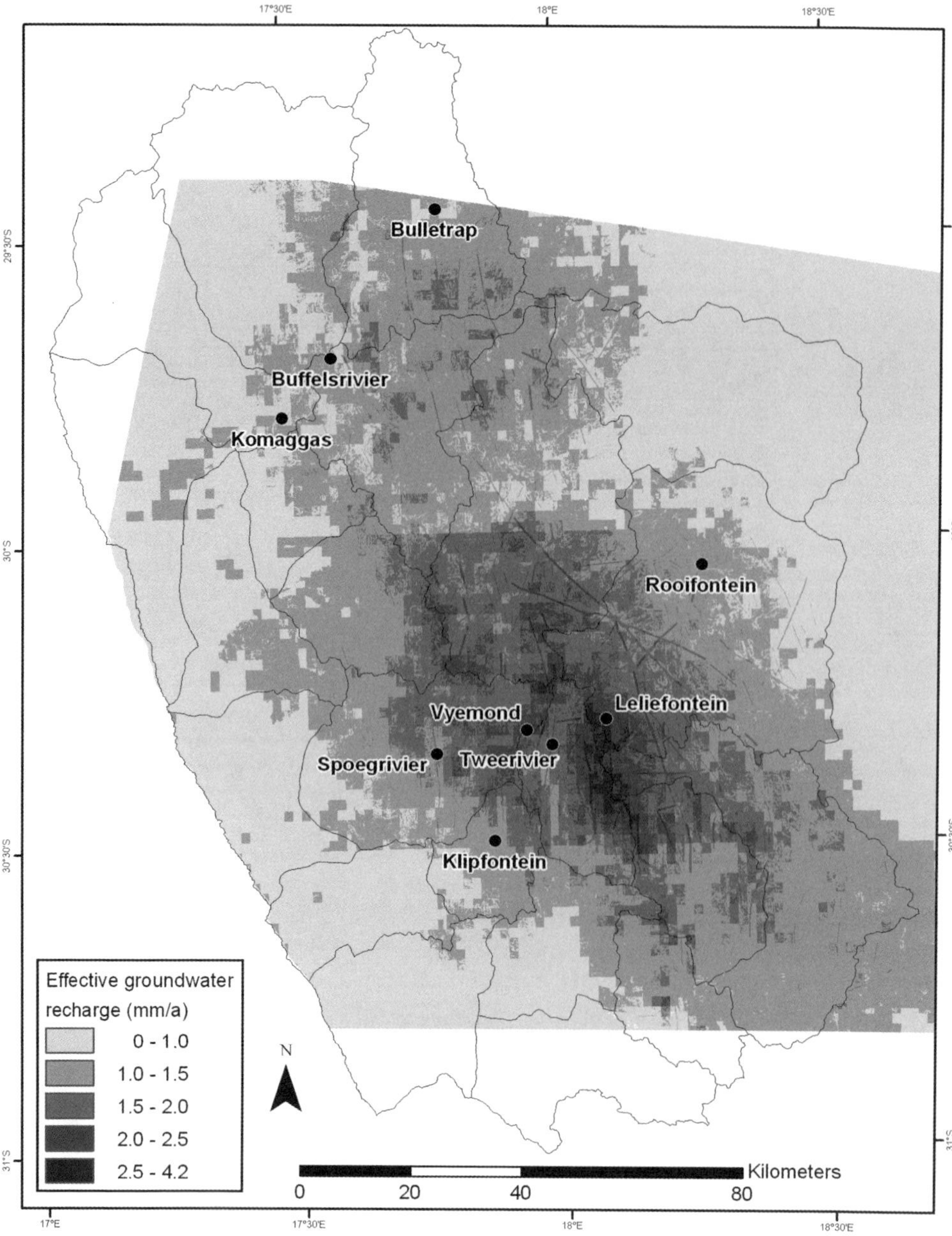


Figure 4. Recharge potential map for the Central Namaqualand region.

data. Groundwater recharge rates over large areas are difficult to estimate due to problems associated with upscaling and data distribution. Two approaches were followed in this study whereby recharge was qualitatively and quantitatively assessed. The qualitative assessment involved using existing data from the area and applying statistical and spatial techniques to assess recharge processes and patterns. Applying the CMB method and water level to rainfall relationships gave quantitative estimates of recharge.

Table 10. Comparison of results between the CMB, SVF, CRD and GIS methods.

| Site | Ave. rainfall (mm/yr) | CMB (mm/yr) | SVF (mm/yr) | CRD (mm/yr) | MRT (yrs) | GIS (mm/yr) | Aquifer |
|---|---|---|---|---|---|---|---|
| Buffelsrivier | 188 | 0.7 | 5.3 | 6.4 | Modern | 1.2 | Alluvial/ Basement |
| Bulletrap | 172 | 1.1 | 0.3 | 0.1 | | 1.0 | Basement |
| Klipfontein | 196 | 0.2 | 0.3 | 0.1 | >30,000 | 1.2 | Basement |
| Komaggas | 229 | 1.0 | 4.4 | 0.7 | Modern | 1.1 | Alluvial/ Basement |
| Leliefontein | 395 | 2.2 | 1.9 | 2.3 | > 5000 | 2.8 | Basement |
| Rooifontein | 138 | 0.4 | 2.6 | 3.06 | | 1.3 | Alluvial/ Basement |
| Spoegrivier | 200 | 0.2 | 0.2 | 0.1 | >2000 | 1.3 | Basement |
| Tweerivier | 296 | 3.5 | 0.6 | 1.8 | | 1.6 | Basement |
| Vyemond | 239 | – | 0.05 | 0.09 | | 2.3 | Basement |

The qualitative assessments of recharge, using statistical analysis and the GIS assessment techniques, identify areas that are receiving recharge and are favourable for potential artificial recharge. The statistical analysis indicates areas of localised artificial recharge through agricultural and domestic activities. The results of the quantitative analysis are shown in Table 10. Recharge was calculated as the average annual recharge. However, it is well recognized that only certain rainfall events during the wet winter season contribute significantly to recharge. Table 10 shows that higher rates of recharge occur in areas where groundwater is exploited from a combination of the alluvial and basement aquifers. Basement aquifers include both the weathered zone and fractured rock aquifers. The mean residence time (MRT) of groundwater is also indicated. The CMB method under-estimates recharge to alluvial aquifers associated with the ephemeral rivers in the vicinity of Buffelsrivier and Rooifontein). This may be explained by additional chloride input from run-on during periods of recharge. The current calculation of the CMB method lacks chloride data from run-on. The recharge rates also generally correlate with the $^{14}C$ ages from individual boreholes; the Leliefontein sample does not fit the expected age and may be due to a sampling or analytical error. The scatter of estimates is probably due to the fact that the CMB, CRD and SVF methods measures recharge at different temporal scales. The CMB method estimates recharge over an extensive period, as opposed to the CRD and SVF methods that estimate recharge over shorter periods. The CMB method also depends on the degree of groundwater mixing within the borehole. It has been shown that the groundwater quality generally deteriorates with depth and $^{14}C$ evidence suggests stratification of groundwater of different ages. The uncertainties in the degree of mixing and rainfall chloride can have a considerable effect on the error associated with an estimate. The recharge values obtained using the GIS approach compare well with the other results obtained, except where the groundwater age is relatively high.

## 8 SUMMARY

The GIS approach is an excellent tool for visualisation and spatial analysis. However, the method is more descriptive, less quantitative and subjective than rigorous analytical approaches to recharge calculations. The main shortcoming of the GIS approach is that it only assumes direct vertical infiltration and a number of assumptions are needed to arrive at a groundwater recharge probability map. The fracture-flow dominated mechanisms complicate the determination of effective groundwater recharge. However, by adopting a best estimate approach, an acceptable prediction of groundwater recharge can be arrived at and the spatial distribution and variation of recharge presented. The GIS generated results compare well with point estimates, using several different methods for calculating groundwater recharge. The groundwater recharge map provides valuable information for groundwater resource protection and management.

## ACKNOWLEDGMENTS

The Water Research Commission is thanked for their support and funding to carry out this work.

## REFERENCES

Acworth RI (1987). The development of crystalline basement aquifers in a tropical environment. Quaterly Journal of Engineering Geology, 20, 265–272.

Adams S, Titus R and Xu Y (2004). Groundwater recharge assessment of the basement aquifers of Central Namaqualand. WRC Report No. 1093/1/04. Water Research Commission, Pretoria.

Albat HM (1984). The Proterozoic granulite facies terrane around Kliprand, Namaqualand Metamorphic Complex. Bulletin 33, Precambrian Research Unit. Dept. of Geology, University of Cape Town.

Bredenkamp DB, Botha LJ, Van Tonder GJ and Van Rensburg HJ (1995). Manual on quantitative estimation of groundwater recharge and aquifer storativity. WRC Report TT 73/95.

DWAF (1990). Grondwaterondersoek in die Komaggas-kleurling gebied, distrik Namakwaland. Tegniese Verslg GH 3654. Department of Water Affairs and Forestry.

Kellgren N, Sander P, Blomquist-Lilja N, Groenewald J, Smit P and Smith L (2000). Remote sensing data for effective borehole siting. Progress report No. 3. Department of Geology, Chalmers University of Technology, Göteberg, Sweden.

Rushton KR and Ward C (1979). The estimation of groundwater recharge. Journal of Hydrology, 41, 345–361.

Shepard, D. (1968) A two-dimensional interpolation function for irregularly-spaced data, Proc. 23rd National Conference ACM, ACM, 517–524.

Simmers I (1996). Challenges in estimating groundwater recharge. WRC: Groundwater – Surface Water Issues in Arid and Semi-arid Areas. Pretoria.

Tankard AJ, Jackson MP, Eriksson KA, Hobday DK, Hunter DR and Minter WEL (1982). Crustal evolution of South Africa: 3.8 Billion years of earth history. Springer-Verlag, New York.

Tindimugaya C (1995). Regolith importance in groundwater development. 21st WEDC conference. Kampala, Uganda.

Titus R (2003). Hydrogeochemical characteristics of the basement aquifers in Namaqualand. PhD Thesis, University of the Western Cape.

Titus RA, Pietersen KC, Williams ML, Adams S, Xu Y, Colvin C and Saayman IC (2002). Groundwater assessment and strategies for sustainable resource supply in arid zones – The Namaqualand case study. WRC Report 721/1/02. Water Research Commission, Pretoria.

Toens and Partners (2001). Regional geohydrological assessment of the Bitterfontein and Rietpoort rural areas. Report No. 2001251.

Vegter JR (1995). Groundwater resources of South Africa. An explanation of a set of National Groundwater maps. WRC Report No. TT 74/95. Water Research Commission, Pretoria.

Vegter JR (2001). Groundwater development in South Africa and an introduction to the hydrogeology of groundwater regions. WRC Report No. TT 134/00. Water Research Commission, Pretoria.

Verhagen B Th and Levin M (1986). Environmental isotopes assist in the site assessment of Vaalputs radioactive waste disposal facility. In Ainslie LC (ed) Proceedings volume – Conference on the treatment and containment of radioactive waste, and its disposal in arid environments. Atomic Energy Corporation of SA, Pretoria.

Visser DJL ed. (1989). Explanation of the 1:1,000,000 geological map, fourth edition. Department of Mineral and Energy Affairs. Government Printer. ISBN 0-621-12516.

CHAPTER 14

# The use of environmental isotopes to establish a hydrogeological conceptual model in a region with scarce data: the Table Mountain Group of South Africa as a case study

Arie S. Issar[1] and Johanita C. Kotze[2]

[1]*Ben Gurion University of the Negev, Beer Sheva Israel, Jerusalem, Israel*
[2]*Environmental Resources Management, Southern Africa Pty Ltd, Gallo Manor, Johannesburg, South Africa*

ABSTRACT: The methodology of testing and verification of a hydrogeological working hypothesis in regions with scarce hydrological data is demonstrated in the case of the Table Mountain Group (TMG) of the Cape Provinces of the Republic of South Africa (RSA). The rocks of the TMG group have no primary porosity and were considered to be water bearing only in the uppermost shallow part near the surface. The existence of thermal springs along regional faults suggested an alternative conceptual model, in which the fault zones are fractured permeable media, enabling flow into great depths over long distances. Testing and verifying this conceptual model were done by sampling and testing for the environmental isotopic composition (oxygen 18/16, hydrogen 2/1) of the perched water table springs along a section from the sea to the mountains. This proved that the mountains are the areas of recharge of the thermal springs. The high temperature of the water is a function of flow to great depths along the regional fracture zones.

## 1 INTRODUCTION

In recent decades groundwater development has spread over many regions. In most areas the shallow aquifers are over-exploited by the local landowners. This worldwide trend, described as the "silent revolution", was encouraged by the fact that irrigation by groundwater is more efficient, technically and economically, than irrigation by surface water (Llamas and Martinez-Santos, 2005). Moreover in certain regions, as the Western Cape Province of South Africa, inconsistency of climate, such as a strong El Ninõ effect, causes drought and the failure of the water supply relying on dams. To allow the appropriate management of the over-exploited aquifers and also to provide additional groundwater resources, it is essential to explore alternative deeper aquifers. Such exploration involves the investment of time and money, which calls for the development of methods, which will shorten the research period, and of a number of exploration wells as suggested by Issar (1973) and Issar and Gat (1981) while investigating the water resources in regions with scarce hydrological data. The authors and their colleagues from the Department of Water Affairs and Forestry of the Republic of South Africa (DWAF RSA) have applied and further developed these methods in the case of the Cape Province of the RSA.

## 2 GENERAL DESCRIPTION OF THE REGION

The Table Mountain Group (TMG) comprises most of the Cape Provinces of the Republic of South Africa (Figure 1). It crops out over an area of about 9,000 square kilometers, mostly mountainous, reaching elevations of 2500 meters. It has a Mediterranean type climate of winter rainfall, while summers are mostly dry. The precipitation falling on the outcrops varies extremely. On its northern borders the climate is arid, its annual precipitation does not exceed 200 mm, while in the southwestern region it may reach 2000 mm/y. The total thickness of the TMG is about 4000 meters (de Beer 2002) consisting of quartzitic sandstones of Ordovician to Silurian age.

## 3 THE DEVELOPMENT OF THE HYDROGEOLOGICAL CONCEPTUAL MODEL

### 3.1 *Conventional hydrogeological model*

Until the late 20th century, the conventional hydrogeological model adopted by the official RSA water authorities was that only the upper-most part of the TMG, not exceeding a few tens of meters, should be considered as water bearing and be worth drilling. It was reasoned that because of the tectonic processes of intense folding which this region underwent since the Paleozoic Era the rocks are highly consolidated and of zero primary hydraulic conductivity. It was maintained that the rocks were permeable only near the surface where relaxation and weathering processes brought about fracturing and fissuring.

### 3.2 *Preliminary observations suggesting a different conceptual model*

A preliminary survey, invited by the DWAF RSA, was carried out by the senior author during the mid seventies. The survey included visits to the main thermal springs, which emerge from the TMG, as well as a study of the satellite images of the region. This survey showed that the thermal springs, with temperatures ranging between 37 to 64 degrees centigrade, are located along regional faults breaking up the TMG province into structural blocks (Figure 1, Table 1).

No recent volcanic phenomena, which could explain these thermal anomalies was known in the TMG province. As these observations could not be explained by the conventional hydrogeological mode, mentioned above, an alternative model had to be considered.

The hypothetical new model suggested that rather wide zones of fractured rock exist along the regional faults, as well as along the secondary faults that branch off from the major regional fault systems. Water infiltrating into these zones in the highland regions flows down to great depths along these fractured systems, which extend in some regions to more than a thousand meters, as depicted by the high temperature of the thermal springs. They emerge along these faults due to a combination of hydraulic and thermo-dynamic

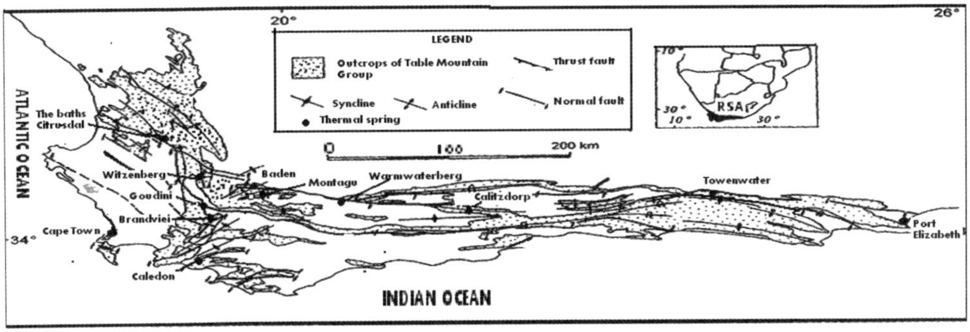


Figure 1. Extension of outcrops of the Table Mountain Group, major faults, and location of thermal springs (Based on de Beer 2002).

Table 1. Partial list of thermal springs in the TMG province of RSA. (Based on Harris and Diamond 2002).

| Spring | Temp. °C | Quantity $m^3/h$ | Altitude m. above MSL | Geology |
|---|---|---|---|---|
| Baden | 37 | 137 | 280 | TMG–Bokkveld Group contact near regional fault |
| Brandviei | 64 | 455 | 220 | Ditto on regional fault |
| Caledon | 50 | 33 | 360 | TMG–Bokkveld Group contact near regional fault |
| Calitzdorp | 52 | 97 | 200 | Ditto |
| Citrusdal | 43 | 104 | 250 | Fault in TMG |
| Goudini | 40 | 40 | 290 | Regional fault in TMG |
| Montagu | 43 | 130 | 280 | TMG–Bokkveld Group contact near regional fault |
| Towenwater | 44 | 40 | 800 | Regional fault in TMG |
| Warmwatenberg | 45 | 32 | 500 | Ditto |

forces. Thus the regional fault system should be considered from the hydraulic point of view as a permeable fractured media. As a result of these considerations the TMG fault zones should be regarded, as regional aquifers.

## 3.3 *Testing of hypothesis*

In order to test our hypothetical model samples of water were taken from fresh water springs emerging along sections assumed to be the recharge areas of the thermal springs from the sea to the mountainous regions. All the samples were analyzed for the environmental isotopes of oxygen 18 and deuterium. The results (Issar 1995) have shown that the water of the thermal springs is rather depleted in these isotopes and is similar to that of the local cold springs in the mountainous regions. These results were later confirmed by various other isotopic surveys, as for instance that by Harris and Diamond (2002) shown in Figure 2.

As can be seen from the data, the environmental composition of the water of the thermal springs is definitely depleted in deuterium and a little heavier in oxygen 18 in relation to the composition of weighted average of the rainwater, which is near the Global Meteoric Line. Yet, as the oxygen 18 compositions are not heavier than many rain samples. Thus no special

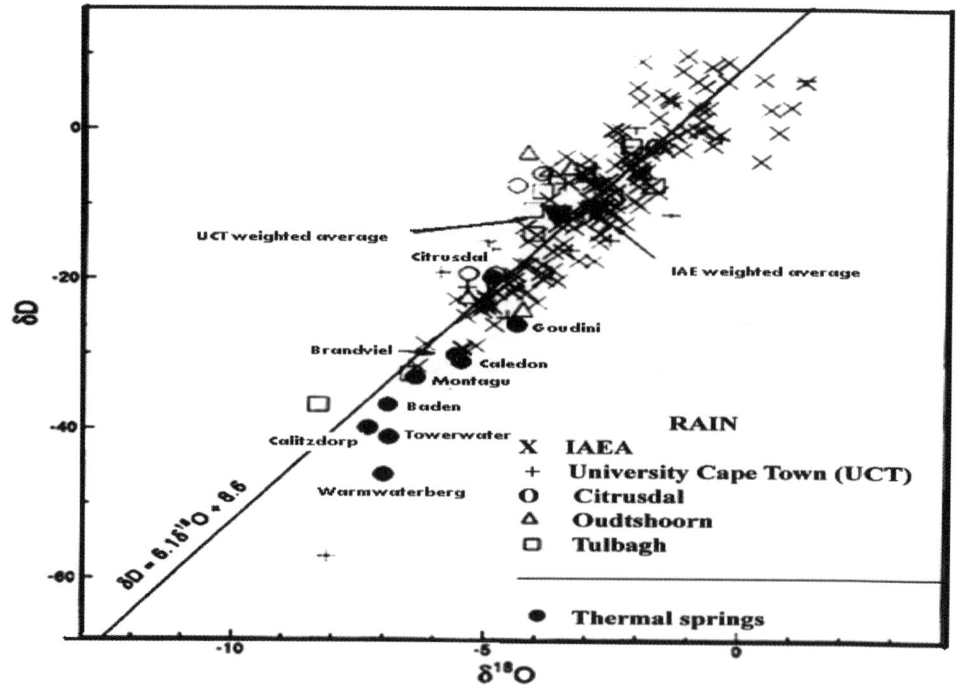


Figure 2. Isotopic composition of rainwater and thermal springs TMG province RSA (According to Harris and Diamond 2002).

explanation is needed for an oxygen shift due to temperatures exceeding 80 degrees centigrade, as, for example, in the Central Plateau of Mexico, where a similar trend was observed, (Issar et al. 1984). The depletion in deuterium is explained by the altitude effect, as the recharge regions extend over the mountainous areas. This is confirmed by the work of Harris and Talma (2002) who have shown that the oxygen 18 composition of the thermal springs correlates with the composition of borehole water located at altitudes exceeding 600 meters above MSL. (Figure 3)

These results, as well as results of the research on the chemical and isotopic composition of the thermal springs by Mazor and Verhagen (1983) enabled the formulation of a more detailed conceptual model. Accordingly the TMG aquifer was suggested to consist of three interconnected zones of fissured and fractured rocks, namely:

Zone 1. Local shallow systems of fissures related to the mechanical failure of the rocks, due to pressure release, and rock tension joints connected to local folds and faults. The recharge takes place mainly in the high altitude regions where precipitation is maximal and evaporation is low, as evidenced by the depleted composition of the environmental isotopes. The rainwater and surface flow, which infiltrates into these fractures form a shallow perched aquifer, which either emerges as small local springs or flows into Zone 2.

Zone 2. A regional system of fissures and fractures, along the faults and folds, which break the tectonic blocks into secondary blocks. From this system the water flows into Zone 3.

Zone 3. Inter-regional systems of fractures connected to the mega-faults along the major tectonic lines dividing the TMG province into tectonic blocks (Figure 1 & Figure 4). At

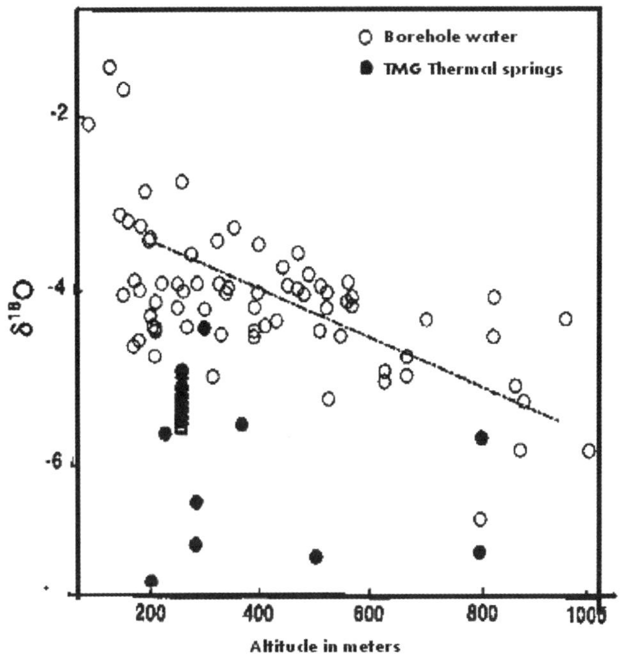


Figure 3. Oxygen 18 composition of water of boreholes and thermal springs versus altitude (Harris and Talma 2002).

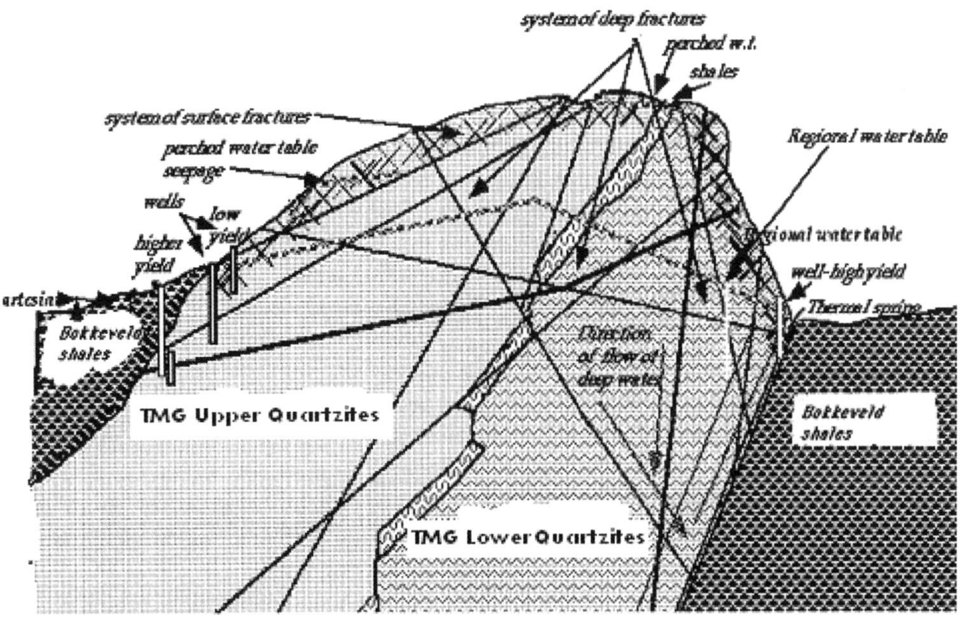


Figure 4. Schematic conceptual hydrogeological model of the TMG.

certain low altitudes, outflows take place either as thermal springs or directly to the sea along the mega-faults.

## 3.4 *Verification of working hypothesis*

In accordance with Issar's recommendations the Department of Water Affairs and Forestry of RSA adopted a long-term program of research, exploration and development of the TMG. One of the regions in which intensive hydrogeological and hydrochemical research was carried out was the Klein Karoo province in the central part of the TMG belt (Kotze 2001). Simultaneously, intensive investigation proceeded in other regions. The results verified the hypothetical conceptual model, as for example at Citrusdal in the Western Cape Province, where a series of wells was drilled as a precaution against drought in case of a strong El Ninõ effect. Wells drilled to depths of between 174 and 348 meters yielded quantities of 35 to 450 $m^3/h$. Oxygen 18 and $^2H$ (deuterium) analyses have shown that the composition becomes lighter with depth (Hartnady and Hay 2002), confirming the above-mentioned conceptual model.

One of the main problems faced during the quantitative stages of the hydrogeological research was the determination of the hydraulic parameters of the aquifers. This is because the classical methods of deriving these parameters from pumping tests were based on the assumption of laminar flow in a porous homogenous media. The TMG was neither porous, nor homogenous and the flow in the open fractures in the vicinity of the pumping wells could have become turbulent. Yet the various investigations have shown that on the average, conventional pumping tests, coupled with other method of evaluation such as analysis of spring flow, together with data available from engineering tests of hydraulic properties for dam foundations etc. give a range of hydraulic coefficients. The average of these values could be applied to calculate preliminary regional hydrological balances. This could provide the order of magnitude of the volume of water flowing through a certain section and stored. Thus TMG hydraulic conductivity coefficients (K) ranged from 1.99 m/d to 0.002–0.005 m/day, while transmissivity (T) values of a few hundred $m^2$/day characterized productive fractured zones. When it comes to storativity coefficients (S), 0.001 was found to be "a fair estimate for the bulk storativity of the Peninsula and Nardouw formations" (Weaver et al., 2002), which are the two most arenitic formations and thus with the highest potential of being fractured and becoming groundwater bearing. On the basis of these values Rosewane estimated that there are a few tens of billions of cubic meters of groundwater stored in the TMG aquifer. Yet, in the opinion of the present authors this may be an overestimation and further investigation, as detailed in the following paragraphs, is needed to reach a more precise value.

The impact of groundwater extraction on the surface water is another aspect which has to be taken into account when the reduction of available reserves is planned. In other words, the possibility has to be considered that massive pumping will dry up many springs in the Cape Mountains. On the other hand, field observations by the present authors have shown that there are sections along the western coastline of the Cape Province where TMG rocks continue into the sea. Along these sections groundwater which could be tapped inland.

These uncertainties and observations call for a massive investment in an intensive program of investigations aimed to quantify in a more dependable way, the TMG aquifer coefficients of storage, permeability and recharge. These investigations should include the additional drilling of deep exploration wells along the regional fault zones above regional spring outlets, as well as along the regional fractures extending into the sea. These wells will be tested for their hydraulic coefficients and will be part of a national hydrological observation network. In addition it is recommended to map the temperature, salinity and environmental isotope composition of the seawater along the shoreline where TMG rocks outcrop.

## 4 CONCLUSIONS AND RECOMMENDATIONS

Data concerning the environmental isotopes (especially oxygen 18 and deuterium) composition of the water at various stages of the hydrological cycle along its way from recharge to discharge zones enables a general understanding of the characteristics of this cycle. Thus in order to develop a basic conceptual hydrogeological model to optimize investments in research budgets and cost of exploration drilling, data about these isotopes have to be collected and processed. This together with the geological data, which may be obtained from analysis of satellite images, and preliminary hydrological data obtained from the survey of springs and wells, and detailed geological mapping and cross-sections will enable the development of a preliminary framework for long-term aquifer management policies.

The next stage of the hydrogeological investigation must aim at the construction of computerized flow models. This stage should be promoted despite the constraints already discussed concerning determination of the hydraulic parameters. These constraints can be solved through trial and error procedures while calibrating the models to fit field data.

## ACKNOWLEDGEMENT

The permission of Prof. Chris Harris, Department of Geological Sciences University of Cape Town, RSA, to present Figures 2 and 3 is thankfully appreciated. The editorial remarks of the reviewers and editors as well as the help of Ms. Marcia Ruth in editing the article are thankfully acknowledged.

## REFERENCES

De Beer Ch (2002) The Stratigraphy, Lithology and Structure of the Table Mountain Group. In Pietersen K, and Parsons R. (eds.) A Synthesis of the Hydrogeology of the Table Mountain Group – Formation of a Research Strategy. WRC Report No TT 158/01, Water Research Commission, Pretoria, R.S.A 9–18.

Hartnady CJH and Hay ER (2002) Boshkloof Groundwater Discovery. In: Pietersen and Parsons (eds.) ditto. 168–177.

Harris C and Diamond RE (2002) The Thermal Springs of the Table Mountain Group: A Stable Isotope Study. In: Pietersen and Parsons (eds.) ditto. 230–235.

Harris, C and Talma AS (2002). The application of isotopes and hydrochemistry as exploration and management tools. Water Research Commission (SA) TMG Special Publication 59–67.

Issar A 1973. Methodology of groundwater investigation in regions with scarce hydrological data. Proc. of the International Symposium on Water Resources Development, Madras, India, pp. 1–6.

Issar A, Gat J 1981. Environmental Isotopes as a tool in hydrogeological research in an arid basin. Ground Water, 19(5):490–494.

Issar JL, Quijano JL, Gat J, Castro M (1984). The isotope hydrology of the groundwater of central Mexico. J. of Hydrology, 71:201–224.

Issar AS (1995) On the Regional Hydrogeology of South Africa. Unpublished Report no. 3. The Directotrate of Geohydrology DWAF SRA.

Kotze JC (2001) Hydrogeology of the Table Mountain Sandstone Aquifer in the Little Karoo. Ph D Thesis, University of Orange Free State, Bloemfontein.

Llamas MR and Martinez-Santos P (2005) Values and Rights in the Silent Revolution of Intensive Groundwater Use. Proceedings International Water Conference, Ramallah, Palestinian Authority, May 2–5 2005.

Mazor E, Verhagen BT and Rosewarne P (2002) Hydrogeological Characteristics of the Table Mountain Group Aquifers In: Pietersen K, and Parsons R. (eds.) etc. ditto. 33–44.

Weaver JMC, Rosewarne P, Hartnady CJH, Hay, ER (2002) Potential of Table Mountain Group Aquifers and Integration into Catchment Water Management. In: Pietersen and Parsons (eds.) etc. ditto. 239–240.

CHAPTER 15

# Hard rock aquifers characterization prior to modelling at catchment scale: an application to India

Jean-Christophe Maréchal[1], Benoît Dewandel[2], Shakeel Ahmed[3] and Patrick Lachassagne[1]

[1]*BRGM, Water Division, Resource Assessment, Discontinuous Aquifers Unit, Montpellier, France*

[2]*BRGM, Water Division, Resource Assessment, Discontinuous Aquifers Unit, Indo-French Centre for Groundwater Research, Hyderabad, India*

[3]*National Geophysical Research Institute, Indo-French Centre for Groundwater Research, Hyderabad, India*

ABSTRACT: The structure and the hydrodynamic properties of the weathering profile of a granitic area (53 $km^2$ Maheshwaram catchment, state of Andhra Pradesh, India) were characterized in detail and mapped from observations on outcrops, vertical electric soundings, and data from borewells. The structure of the weathering profile results from a multiphase process: an ancient weathering profile was partly eroded down to its fissured layer. It was later re-weathered more or less parallel to the current topographic surface. The hydrodynamic properties of the weathered-fractured layer of the profile are characterized using hydraulic tests at different scales and a comprehensive hydrodynamic model is proposed. A water budget approach is developed to estimate specific yield and natural recharge in the aquifer with significant seasonal water table fluctuations. Water table fluctuations are due to distinct seasonality in groundwater recharge. These techniques constitute a first step towards the modelling at catchment scale in hard-rock aquifers.

## 1 INTRODUCTION

In India hard rock constitutes more than two-thirds of the total surface and is a great example of the nexus between water scarcity and the occurrence of hard-rock aquifers. The groundwater boom during the Green revolution of the seventies lead to a complete inversion of the irrigation scenario with groundwater now sustaining almost 60% of irrigated land (Roy and Shah, 2002). As discussed by Bredehoeft (2002), sustainable groundwater development usually depends on the dynamic response of the system to development. The principal tool to evaluate the sustainability of management scenarios for hydrosystems is groundwater modelling. Major prerequisites for establishing the reliability of such models are an accurate knowledge of the geometry and the hydrodynamic properties of the aquifer, and a reliable estimate of boundary conditions. This constitutes a challenging task in hard-rock aquifers given the heterogeneity of these environments.

This summarizes a set of methodologies applied on a pilot watershed located in rural India to overcome these difficulties:

- geological and geophysical mapping for aquifer geometry determination;
- hydraulic tests at various scales for hydrodynamic properties characterization;
- groundwater budget and water table fluctuation for boundary conditions estimation.

## 2 GEOMETRY OF THE WEATHERING PROFILE

The Maheswaram pilot watershed (Figure 1), 53 km$^2$ in area, is located 35 km south of Hyderabad, State of Andhra Pradesh, India. The area is characterized by relatively flat topography with altitudes between 670 and 590 m above sea level and the absence of perennial streams. The region experiences a semi-arid climate controlled by the periodicity of the monsoon (rainy season: June to October). Mean annual precipitation is about 750 mm of which more than 90% falls during the Monsoon season. The mean annual temperature is 26°C, but in summer (March to May) the daily maximum temperature approach 45°C.

The geology of the area (Figure1) is composed of Archean granites. The weathering profile of these granites is observed in hundreds of dugwells (up to 20 m in depth) located throughout the watershed area. Dugwells provide mainly information on the upper part of the weathering profile (i.e., the laminated layer) and cuttings from 45 observation boreholes drilled during the research project (down-to-the-hole hammer), up to 64 m in depth, provide deeper information.

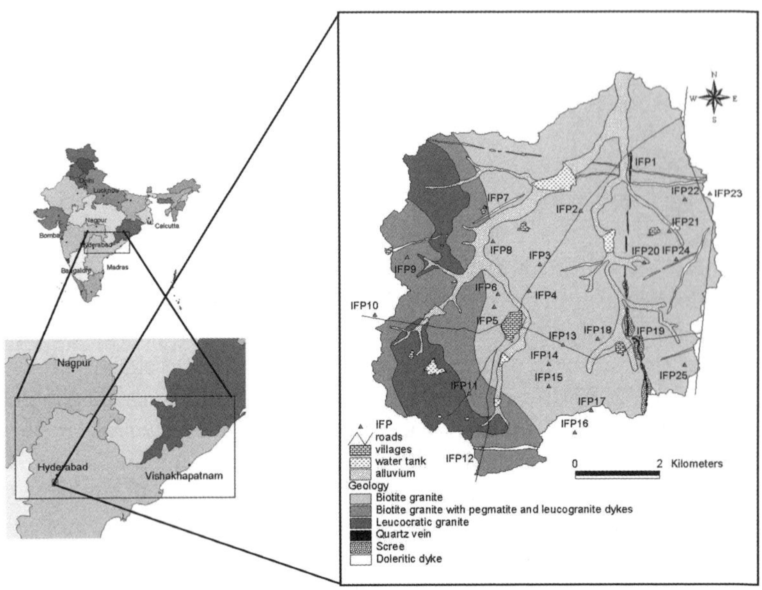


Figure 1. Simplified geological map of the study area with location of observation wells (IFP).

From the top to the bottom, the profile generally shows (Figure 2):

- a thin layer of red soil (10–40 cm),
- a 1–3 m thick layer of sandy regolith, which is locally capped by a lateritic crust (<50 cm in thickness),
- a 10–15 m thick layer of laminated saprolite. This layer is characterized by a penetrative millimetre-spaced horizontal laminated structure and an unusual network of preserved mainly subhorizontal and some subvertical fissures partially filled up by clayey mineral. Both sandy regolith and laminated saprolite layers contain a few granite corestones (preserved fresh rock). While such fresher blocks are only occasionally present in the biotite granite, they are very common in the leucocratic granite,
- the fissured fresh granite that occupies the next 15–20 m, where weathered granite and a few clayey minerals commonly partially fill up the fissures,
- the unfissured granite (bedrock).

Because of the very low ratio sandy regolith/saprolite, 0.1–0.2 instead of 0.5–0.7 in the "classical" profile, the small thickness of fissured layer compared to that of saprolite (ratio of 1 instead of 2) and the presence of fissures (numerous) within the laminated saprolite, it differs from the "classical" weathering model and questions the regional weathering processes in South India.

Geophysical investigations using eighty vertical electrical soundings and geological observations have been used to build cross sections and map the spatial distribution of the layers constituting the weathering profile at the watershed scale (Dewandel et al. 2006). They show that bottoms of laminated and fissured layers more or less follow the current topography and thus that their thicknesses are relatively constant at the watershed scale (Figures 2 and 3). This particular structure also differs from the classical stratiform weathering model.

It appears that the structure of the weathering profile results from a multiphase process: an ancient weathering profile was partly eroded, down to its fissured layer. It was later re-weathered more or less parallel to the current topographic surface. This peculiar structure

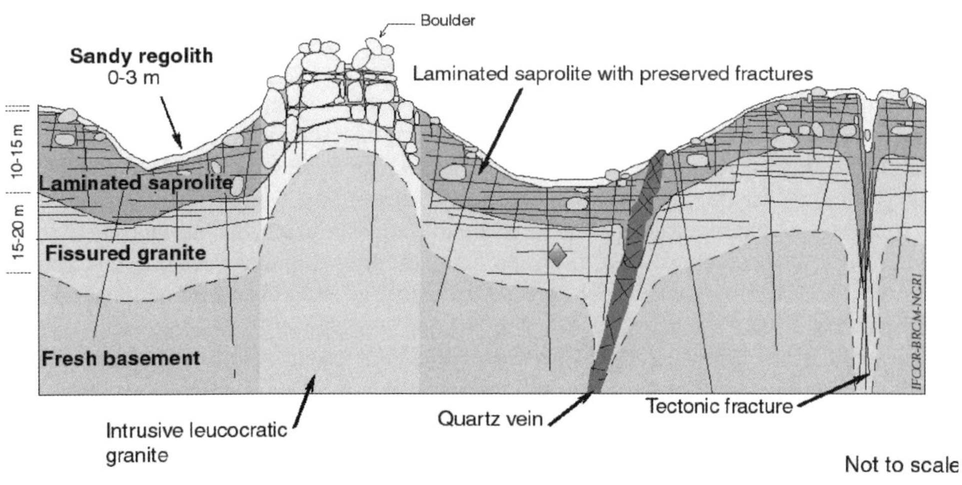


Figure 2. Weathering profile in Matheshram watershed (from Dewandel et al. 2006).

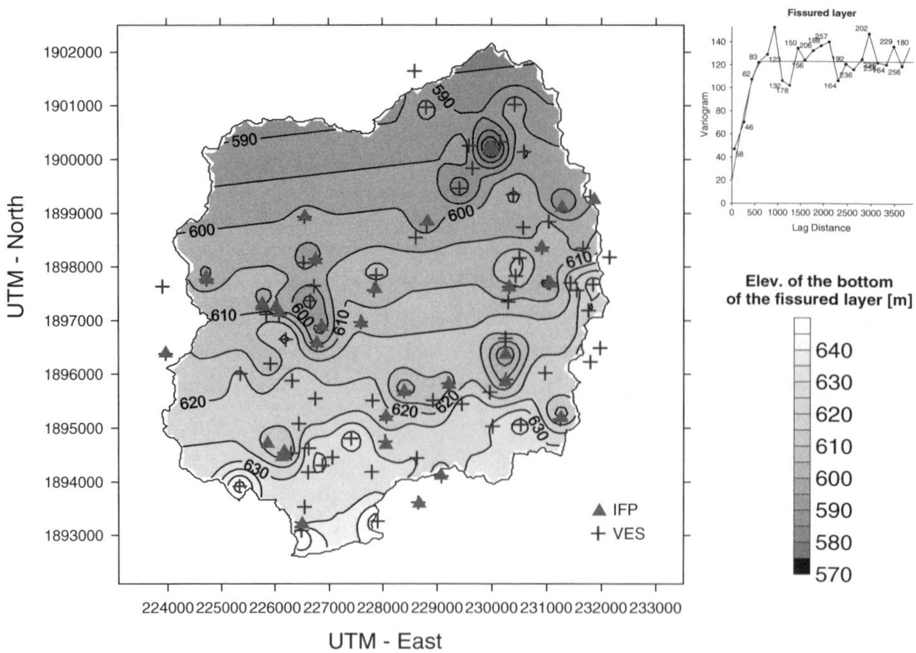


Figure 3. Kriged maps (45 borewell lithologs and 80 VES interpretations) of elevation (in m above sea level) of the bottom of the fissured layer (or the top of fresh and unfissured basement).

is linked to the geodynamic history of the Indian Peninsula that underwent alternative weathering and erosion-dominated phases.

## 3 HYDRODYNAMIC PROPERTIES OF THE WEATHERED-FRACTURED LAYER

Hydraulic tests at different investigation scales were used to determine the hydrodynamic properties of the weathered-fractured layer. Slug tests, injection tests, flowmeter tests, and pumping tests are interpreted using specific techniques for fractured media (Maréchal et al. 2004). Hydraulic conductivity, storage coefficient, anisotropy, connectivity, fracture density, conductivity and radius of the fractures are investigated using various analytical solutions (Table 1). Only flowmeter tests and long duration pumping tests using double porosity and anisotropy methods are here below described. For other applications, refer to Maréchal et al. (2004).

The application of flowmeter profiles during injection tests determines the vertical distribution of conductive fracture zones and their permeabilities. Seventeen flowmeter tests were conducted in the most permeable observation wells according to the sensitivity of the flowmeter, the total injection flow during the test being higher than 25 l/min. In the absence of core samples in such a fractured aquifer, it is not possible to determine if the identified conductive layers (which ones where the net radial flow is not nil) are constituted

Table 1. Characteristics of techniques used for the interpretation of hydraulic tests.

| Hydraulic test | Interpretation method | u/c[1] | t/s[2] | Parameters obtained (Scale effect) |
|---|---|---|---|---|
| Slug test | Solution for unconfined aquifers | u | t | Local permeability |
| Injection test | Classical Dupuit (1848, 1863) solution | u | s | Permeability |
| Flowmeter test | Classical Dupuit (1848, 1863) solution for confined aquifers corrected for unconfined aquifers | u | s | Permeability and density of CFZ[3] |
| Pumping test | Double porosity (Warren and Root, 1963) | c | t | Bulk permeability and storage of blocks and fractures |
| Pumping test | Anisotropy (Neuman, 1975) | u | t | Permeability and storage, degree of permeability anisotropy |
| Pumping test | Single fracture (Gringarten, 1974) | c | t | Radius of fractures |
| Pumping test | Fractional dimension flow (Barker, 1988) | c | t | Flow dimension, generalized transmissivity and storage |

[1] hypothesis on aquifer property in the analytical solution (u: unconfined, c: confined)
[2] analytical solution type (t: transient, s: steady-state)
[3] CFZ: conductive fracture zones (see later for more information)

by a single fracture, several fractures or a fractured zone. Therefore, in order to avoid any confusion, the results are referred to "conductive fractures zones (CFZ)" without any hypothesis concerning their exact nature (single fracture, multi-fractures, etc.). According to the geometry of the well and the observed drawdowns, the sensitivity of the flowmeter limits the identification of fractures zones to those with a hydraulic conductivity higher than $1 \times 10^{-5}$ m/s (hydraulic transmissivity $T > 5 \times 10^{-6}\,m^2/s$), corresponding to a net radial flow of about 5.5 l/m/min in a 0.5 meter-thick layer. This means that the technique gives information only on the most conductive fractures zones. The geometric mean of available data is $K_{CFZ} = 8.8 \times 10^{-5}$ m/s.

The vertical profile of fracture zones distribution (Figure 4a) shows the occurrence of conductive fractures between 9 and 39.5 meters, with higher concentrations of fractures between 15 and 30 meters. However, observations are limited at shallow depths by the presence of unperforated casings in the boreholes, and by the bottom of the well at deeper depths. Thus, it is necessary in the interpretation to consider a statistical bias of the data introduced by unequal representation of the depth ranges for the investigated boreholes. The curve on Figure 4b shows that observations between 15 and 42 meters can be considered as representative, the percentage of investigated wells being higher than 50% (for a given aquifer portion, ratio between the number of available observations and the total 17 possible measurements, Figure 4b). The low number of observations at shallow depths (down to 15 meters) suggests that the apparent decrease in fractures density above 15 meters is an artifact due to a lack of observations (ratio decreasing much, Figure 4b). Consequently, it can be assumed that the weathered-fractured layer extends above the upper limit of observations, which is consistent with field and well geological observations.

The bottom of the weathered-fractured layer is better constrained by the quality of the available observations. The number of fractures starts to decrease between 30 and 35 meters (Figure 4a) and is followed by an absence of fractures below 35 meters while the quality ratio of observations remains high down to 39 meters (Figure 4b). This depth roughly corresponds to the top of the fresh basement as identified by geological observations during the drilling of the wells, and also to a sudden decrease in drilling rates at this depth. Figure 4c shows a profile of the arithmetic average of the hydraulic conductivities (including those less than $1 \times 10^{-5}$ m/s, which are considered nil) obtained from flowmeters for each 0.5 meter-thick layer in all tested wells. It demonstrates the effect of the number of fractures on hydraulic conductivity, with the more highly transmissive zone located mainly between 9 and 35 meters. These observations mean that in the watershed area the weathered-fractured layer has a limited vertical extent (e.g., from the bottom of the weathered mantle down to maximum 35 meters depth).

For each well, the vertical density of the fracture zones is the ratio of the number of conductive fracture zones to the thickness of the investigated weathered-fractured layer (equal to the difference between the final water level during the injection test and the bottom of the weathered-fractured layer as defined above). The density varies from $0.15\,m^{-1}$ to $0.24\,m^{-1}$. This range of density of conductive fracture zones in such a weathered-fractured layer leads to a vertical persistence (e.g., the distance between two consecutive CFZ, equal to the inverse of the density) between 4.2 and 6.8 meters.

The "U" shape of the derivatives curve during long duration pumping tests suggests that the application of the double porosity method (Warren and Root 1963) is justified for this data set. The results of the application of this model to all the pumping tests are summarized in Table 2. The fracture network commonly used to match the observed data is a horizontal fracture set cross-cut by a vertical one ($N = 2$), suggesting the existence of a network complementary to the horizontal one as observed in the dugwells. The (geometric) mean of the hydraulic conductivity of the fracture network is $K_f = 2.1 \times 10^{-5}$ m/s, 400 times higher than that of the blocks (i.e., $K_b = 5.1 \times 10^{-8}$ m/s). This last value is similar to the lowest of the hydraulic conductivities estimated from slug-tests. These low values were measured in three wells probably drilled in an unfractured area where only the hydraulic conductivity of the blocks contributes to the flow. However, it is quite probable that a primary network of fractures (PFN) also affects the matrix, accounting for the high value of the hydraulic conductivity ($10^{-8}$ m/s) of the blocks compared to those measured on rock samples of the same lithology (matrix permeability $K_m = 10^{-9} - 10^{-14}$ m/s, de Marsily, 1986). At this stage, the use of the term "primary" refers only to the small scale (block-level) of the fracture network and does not imply any consideration on the specific origin of the PFN. That point will be considered in the final discussion. Together, the horizontal and the vertical sets of fractures observed in dugwells and designated in the double-porosity model constitute the secondary fracture network (SFN) at the borehole scale. The total storage coefficient (specific yield in this unconfined aquifer, obtained from the sum of fracture ($S_f$) and block ($S_b$) storage coefficients) is equal to $S_y = 6.3 \times 10^{-3}$, a value quite consistent with those evaluated from other interpretation methods: Neuman technique (see below) or the water-table fluctuation techniques (see below). Storage in the primary fractures network (PFN) affecting the matrix accounts for the bulk

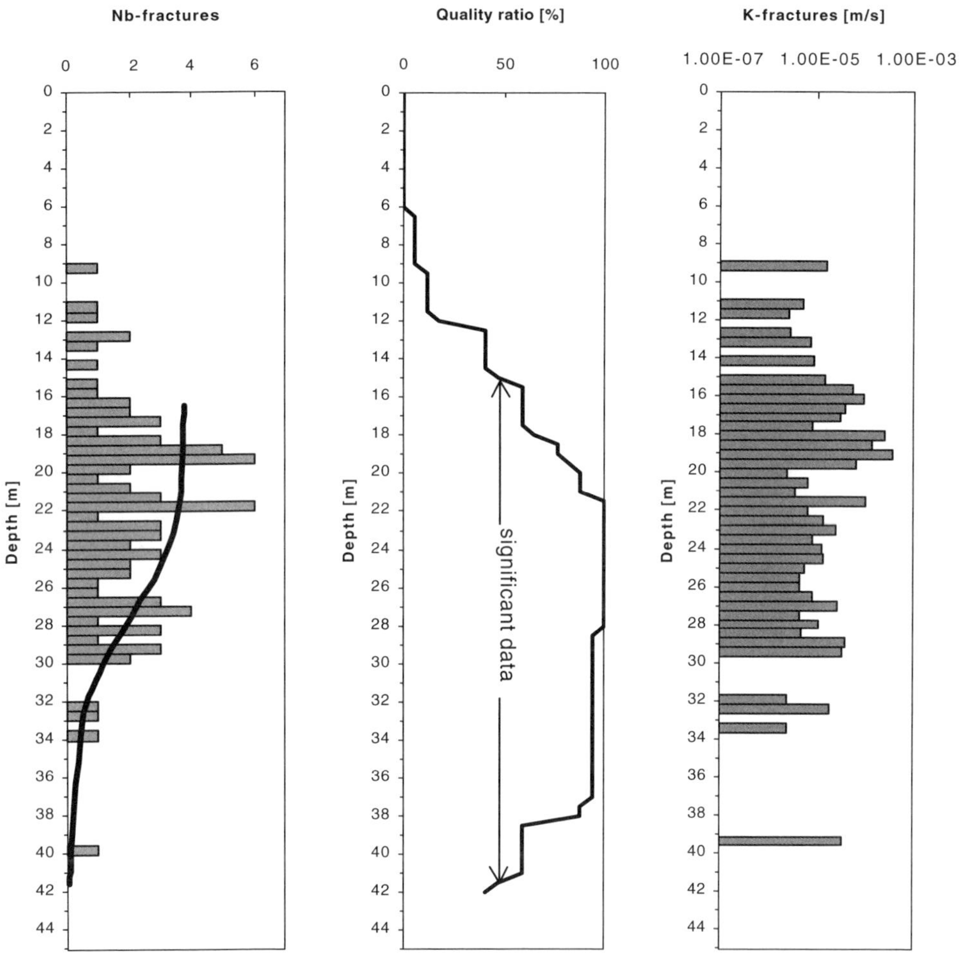


Figure 4. Interpretation of flowmeter-measurement profiles in 0.5-meter-thick layers in the aquifer from 17 wells. (a) The number of hydraulically conductive fractures identified. (b) Quality of observation expressed as the ratio between the number of available observations for each aquifer portion and the total 17 measurements (c) Averages of the fracture permeabilities.

(91%) of the total storage of the aquifer; the secondary fracture network (SFN) contributes the rest (9%).

In accordance with geological observations, using the method of Neuman (1975) for anisotropic unconfined aquifer, the interpretation of the data from the observation wells (Table 3) arises an anisotropy of the permeability tensor: the horizontal permeability $K_r$ is systematically higher (9 times on average) than the vertical $K_z$. This result is consistent with the observation of many horizontal fractures in dugwells (Maréchal et al., 2003). The average value of the vertical permeability ($K_z = 1.3 \times 10^{-6}$ m/s) is much higher than the permeability of the blocks obtained from the application of the double-porosity model. This confirms that the second set of sub-vertical fractures belonging to the secondary

Table 2. Values of matrix and fracture media hydrodynamic properties calculated using the double-porosity (DP) model (l: mean distance between fractures, determined from flowmeter tests; $T_f$: transmissivity of the fracture network; $T_b$: transmissivity of the blocks).

| Known parameters | | | Parameters calculated by adjustment on DP model | | | | | |
|---|---|---|---|---|---|---|---|---|
| Well | l (m) | N | $T_f$ ($m^2$/s) | $K_f$ (m/s) | $S_f$ | $K_b$ (m/s) | $S_b$ | $S_y$ |
| IFP-1 | 7.1 | 2 | 6.0E-5 | 2.8E-6 | – | 2.6E-8 | – | – |
| IFP-1/1 | 7.1 | 1 | 2.3E-5 | 1.0E-6 | 9.0E-4 | 1.3E-7 | 3.4E-3 | 4.3E-3 |
| IFP-1/2 | 7.1 | 2 | 2.6E-5 | 1.2E-6 | 1.0E-6 | 6.9E-10 | 2.6E-3 | 2.6E-3 |
| IFP-8 | 2.3 | 2 | 1.2E-3 | 7.4E-5 | – | 1.7E-8 | – | – |
| IFP-9 | 5.5 | 2 | 7.6E-4 | 1.0E-4 | – | 9.6E-9 | – | – |
| IFP-9/1 | 5.5 | 2 | 7.3E-4 | 1.0E-4 | 3.0E-4 | 8.2E-7 | 5.0E-3 | 5.3E-3 |
| IFP-16 | 0.9 | 2 | 1.4E-3 | 5.9E-5 | – | 2.6E-7 | – | – |
| IFP-20 | 1.9 | 2 | 9.2E-4 | 5.1E-5 | 1.0E-4 | 3.9E-8 | 4.4E-3 | 4.5E-3 |
| BD-8 | 2.4 | 2 | 1.8E-3 | 7.5E-5 | 2.0E-3 | 2.1E-7 | 1.8E-2 | 2.0E-2 |
| BD-10 | 2.4 | 2 | 5.6E-4 | 2.3E-5 | 2.0E-4 | 1.7E-7 | 1.2E-3 | 1.4E-3 |
| Mean[1] | 3.7 | – | 3.5E-4 | 2.1E-5 | 5.8E-4 | 5.1E-8 | 5.7E-3 | 6.3E-3 |

Table 3. Permeability and degree of anisotropy determined at the observation wells using anisotropic (AN) model (r: radial distance from pumping well, b: initial saturated thickness of aquifer.

| Observation well | Known parameters | | Parameters calculated by adjustment on SF model | | | | |
|---|---|---|---|---|---|---|---|
| | r (m) | b (m) | S | Sy | $K_r$ (m/s) | $K_z$ (m/s) | $1/K_D$ |
| IFP-1/1 | 28 | 21.8 | 7.0E-5 | 1.6E-3 | 8.5E-7 | 5.2E-7 | 1.7 |
| IFP-1/2 | 27.5 | 21.8 | 3.7E-5 | 1.5E-3 | 8.0E-7 | 1.0E-7 | 8.0 |
| IFP-9/1 | 30.7 | 7.3 | 7.1E-4 | 3.4E-3 | 8.9E-5 | 3.0E-6 | 29.5 |
| IFP-20 | 14.6 | 18 | 1.1E-3 | 7.8E-3 | 3.7E-5 | 5.6E-6 | 6.6 |
| BD-8 | 85 | 23.9 | 8.3E-4 | 9.7E-3 | 7.0E-5 | 4.4E-6 | 15.8 |
| BD-10 | 50 | 23.9 | 1.7E-4 | 1.9E-3 | 1.4E-5 | 1.3E-6 | 10.9 |
| Mean[1] | – | – | 4.8E-4 | 4.3E-3 | 1.1E-5 | 1.3E-6 | 8.7 |

fracture network (VSFN) is less permeable than the horizontal and that it also affects the entire aquifer by ensuring its vertical permeability and drainage.

The advantage of the Neuman technique is the ability to determine both specific yield ($S_y$) and elastic storage coefficient (S). The values obtained for the specific yield ($Sy = 5 \times 10^{-3}$ as an average) are consistent with an unconfined aquifer of low drainage porosity, as also suggested by water-table fluctuations analysis at the watershed scale (see below). The storage coefficient ($S = 5 \times 10^{-4}$) is one order of magnitude lower than the specific yield. Because the observed drawdowns are most of the times low (<25%) compared to the thickness of the aquifer, the application of analytical solutions for confined aquifer to the unconfined aquifer of the study area is possible without introducing any inaccuracy. This is confirmed by the comparison between the values obtained for specific yields and hydraulic conductivity using double porosity model (confined aquifer solution)

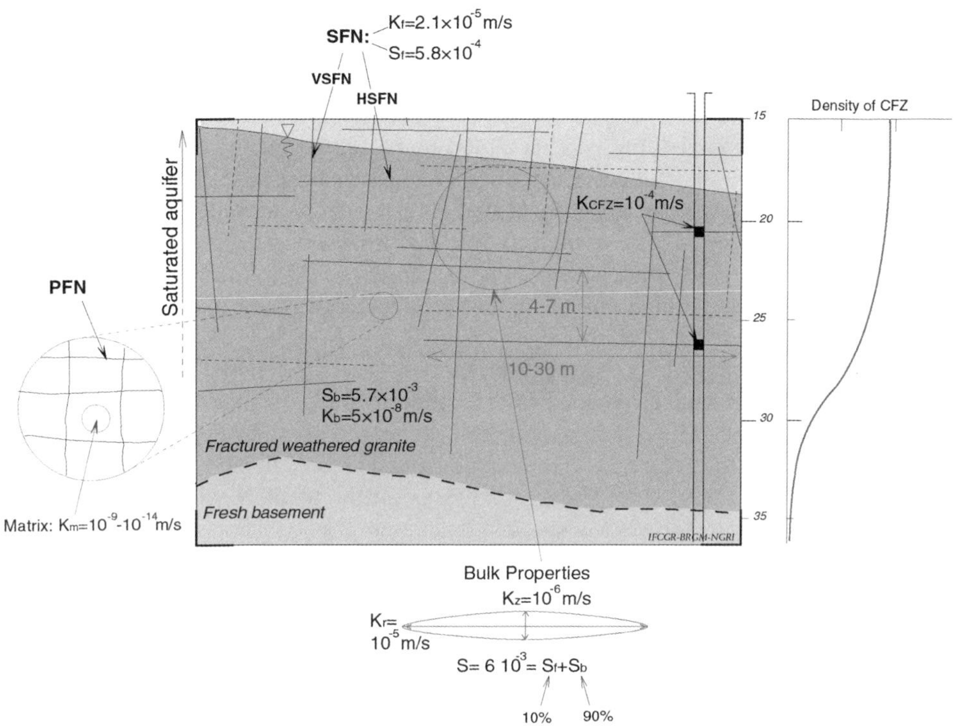


Figure 5. Hydrodynamic model of the weathered-fractured layer of hard rock aquifer. (from Maréchal et al. 2004).

and anisotropic model (unconfined aquifer solution). The difference between the average values is less than 30% for both parameters, which is very low compared to uncertainties usually existing in the determination of storage and permeability. Consequently, although the aquifer is clearly unconfined, a few techniques for confined aquifers are applied in this study.

The good connectivity of fractures networks is shown by fractional dimension flow solutions (Maréchal et al. 2004). The absence of scale effect in the study area suggests that the hydraulic conductivity at the borehole scale is laterally homogeneous. Many geological and hydrogeological indicators suggest that a continuous and laterally homogeneous weathering process is responsible for the origin of the fractures and permeability encountered in the aquifer. These results confirm the major role played by weathering in the origin of fractures and on resulting hydrodynamic parameters in the shallow part of hard-rock aquifers.

Gathering all the hydraulic tests results, a comprehensive hydrodynamic model of the weathered-fractured layer of the hard rock aquifer is presented at Figure 5. Two different scales of fractures networks are identified: the primary fracture network (PFN), which affects the matrix on a decimeter scale by contributing to an increase in the permeability and storage capacity of the blocks, and the secondary fracture network (SFN), which affects the blocks at the borehole scale. SFN is composed of two sets of fractures. The main set of horizontal fractures is responsible for the sub-horizontal permeability of the weathered-fractured layer. A second set of less permeable sub-vertical fractures insures the connectivity of the aquifer at the borehole scale.

## 4 GROUNDWATER BUDGET AT CATCHMENT SCALE

The Maheshwaram watershed is a representative Southern India catchment in terms of overexploitation of its hard-rock aquifer (more than 700 borewells in use), its cropping pattern (rice fields dominating), rural socio-economy (based mainly on traditional agriculture), and agricultural practices. Groundwater resources face a chronic depletion that is observable by the drying-up of springs and streams and a declining water table. Water table is now 15 to 25 m deep and is disconnected from surface water : no spring, no baseflow, no regular infiltration from surface streams beds is observed. Therefore, the knowledge of boundary conditions (pumping rate, natural recharge and irrigation return flow) is essential prior modeling.

The employed methodology is based on applying the water table fluctuation (WTF) method in conjunction with the groundwater basin water budget method. The water budget method focuses on the various components contributing to groundwater flow and groundwater storage changes (Figure 6). Changes in groundwater storage can be attributed to recharge, irrigation return flow and groundwater inflow to the basin minus baseflow (groundwater discharge to streams or springs), evapotranspiration from groundwater, pumping, and groundwater outflow from the basin according to the following equation adapted from Schicht and Walton (1961):

$$R + RF + Q_{on} = ET + PG + Q_{off} + Q_{bf} + \Delta S \quad (1)$$

where R is total groundwater recharge (sum of direct recharge $R_d$ through unsaturated zone and indirect and localized recharge $R_{il}$ respectively from surface bodies and through local pathways like fractures), RF is irrigation return flow, $Q_{on}$ and $Q_{off}$ are groundwater flow onto and off the basin, ET is evaporation from water table, PG is the abstraction of groundwater by pumping, $Q_{bf}$ is baseflow (groundwater discharge to streams or springs) and $\Delta S$ is change in groundwater storage.

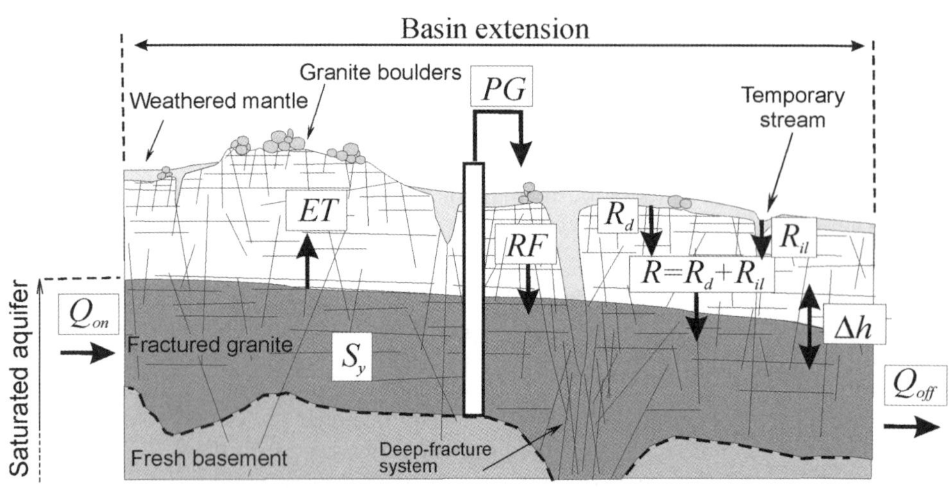


Figure 6. schematisation of flow components of the groundwater budget in a depleted unconfined aquifer (modified after Maréchal et al. 2004).

Due to the significant thickness of the unsaturated zone overlying the unconfined aquifer in the Maheshwaram basin – on average more than 17 m – the following simplifications can be made to the water budget:

- Groundwater discharge to surface water, $Q_{bf}$, via stream discharge or springs does not exist ($Q_{bf} = 0$). All groundwater discharge is via groundwater pumping.
- Transpiration from the water table is negligible due to large depth to groundwater higher than the depth of trees roots evaluated to maximum 10 m in this area from borewells and dugwells observation. Therefore, this flow can be neglected and the evaporation (E) from the water table has been estimated according to the water table depth using the relation proposed by Coudrain-Ribstein et al. (1998) for semi-arid areas.

Equation (1) can be rewritten:

$$R + RF + Q_{on} = PG + E + Q_{off} + \Delta S \tag{2}$$

The main advantage of the groundwater budget method compared to the classical hydrological budget is that evapotranspiration from the root zone of soils – already included in the natural recharge – which usually constitutes a major component with large associated uncertainties is not present in Eq (2).

The methodology used to determine the unknown groundwater storage is the Water Table Fluctuations method (WTF), which links the change in groundwater storage $\Delta S$ with resulting water table fluctuations $\Delta h$:

$$\Delta S = S_y.\Delta h \tag{3}$$

where $S_y$ is the specific yield (storage) or the fillable porosity of the unconfined aquifer.

The WTF method, applicable only to unconfined aquifers, is best applied to shallow water tables that display sharp water-level rises and declines. Deep aquifers may not display sharp rises because wetting fronts tend to disperse over long distances (Healy and Cook, 2002). In the study area, the monitoring of water table between 2000 and 2003 using ten automatic water level recorders shows that the aquifer displays well-identified large seasonal water-level fluctuations due to percolation of water during monsoon period through a rather thick unsaturated zone and small daily fluctuations due to pumping cycles. The Kharif season, during which the water table level rises several meters due to rainfall recharge, is followed by the Rabi season during which the water level drops due mainly to groundwater pumping. Therefore, the hydrological year can be divided into two distinct seasons each with a distinct water level rise or decline. To each of these seasons, the WTF method can be applied separately.

Combining the water budget equation (2) with the WTF method expressed in (3), we obtain:

$$R + RF + Q_{on} = PG + E + Q_{off} + S_y \Delta h \tag{4}$$

As is typical for semi-arid basins with irrigated agriculture, two terms that cannot be evaluated independently without extensive in situ instrumentation are the basin-average natural recharge rate, R and the basin-average, effective specific yield, $S_y$. By applying (4) separately to the dry season, during which R = 0, and to the wet season, we obtain two equations with two unknown parameters:

$$RF^{dry} + Q_{on}{}^{dry} = PG^{dry} + E^{dry} + Q_{off}{}^{dry} + S_y\Delta h^{dry} \tag{5}$$

$$R + RF^{wet} + Q_{on}{}^{wet} = PG^{wet} + E^{wet} + Q_{off}{}^{wet} + S_y\Delta h^{wet} \tag{6}$$

which can be solved sequentially, first by obtaining $S_y$ by solving (5), then by solving (6) for R, given the season-specific values for the known parameters:

$$S_y = \frac{RF^{dry} + Q_{on}^{dry} - E^{dry} - PG^{dry} - Q_{off}^{dry}}{\Delta h^{dry}} \tag{7}$$

$$R = \Delta h^{wet}.S_y - RF^{wet} - Q_{on}^{wet} + E^{wet} + PG^{wet} + Q_{off}^{wet} \tag{8}$$

Equation (7) known as the "water-budget method" for estimating $S_y$ (Healy and Cook, 2002), which is the most widely used technique for estimating specific yield in fractured-rock systems, probably because it does not require any assumptions concerning flow processes (Healy and Cook, 2002). Maréchal et al. (2006) described the methods used for obtaining the "known" parameters in equations (7) and (8), which are needed for the estimation of $S_y$ and R. The flow components are considered to be spatially distributed throughout the groundwater basin on a 200 × 200 meter cells-length grid with measurements taken from June 2002 until June 2004. A Geographical Information System is used to compute all parameters in (7) and (8) cell by cell which are then aggregated at the groundwater basin scale. $Q_{on}$ and $Q_{off}$ are reliably determined only at the larger basin scale through the basin boundaries; hence $S_y$ and R can only be computed at groundwater basin scale.

The WTF method requires a very good knowledge of the piezometric level throughout the entire basin. This could be achieved owing to a very dense observation network (99 to 155 wells, Figure 7a) provided mainly by defunct or abandoned agricultural borewells. Standard deviation of the error on the water table fluctuation measurement has been calculated by geostatistics. Admitting a Gaussian statistical distribution of errors, it defines the 66% confidence interval of the error. The relative error on water table fluctuation logically decreases with the increasing number of measurements. Water table elevations are computed by difference between ground elevation from a Digital Elevation Model obtained by a couple of satellite images stereoscopy treatment (grid resolution: 30 m; accuracy: 1 meter) and water depth obtained from piezometric measurements. The water table maps were then interpolated using the kriging technique. The map was then critically evaluated. The automatic interpolation technique gave satisfactory results owing to the very dense observation network and to the fact that there is no surface water capable of locally modifying the water table. The map for June 2002 (Figure 7a) shows that the water table roughly follows the topographic slope, as is usually observed in flat hard-rock areas. However, local water table depletion is observed in highly pumped areas where natural flow paths are modified by groundwater abstraction.

Water table levels fluctuate between 610 and 619 m, which indicates that the water table is always in the fissured aquifer layer.

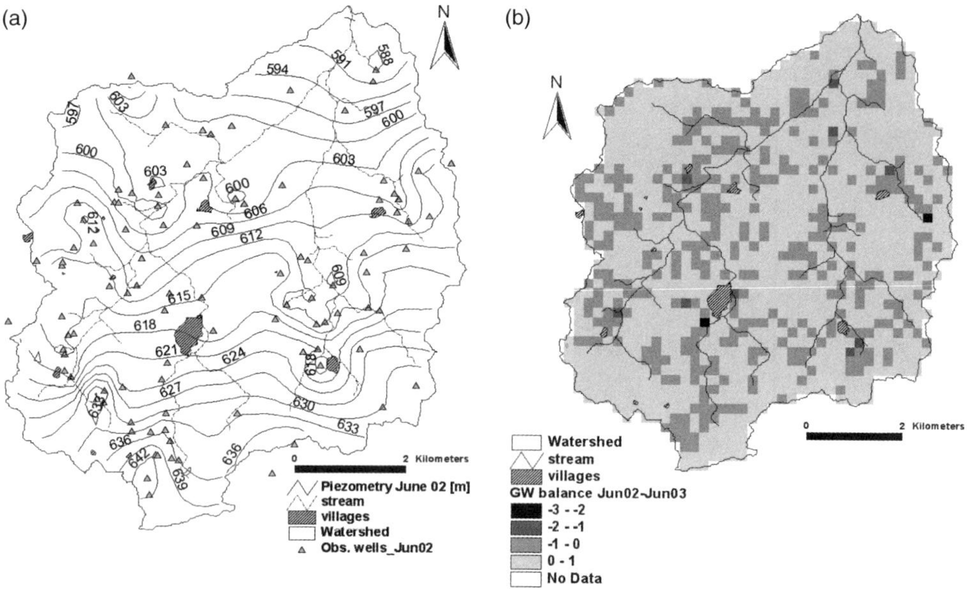


Figure 7. Water table map (a) and annual (June 02 to June 03) groundwater budget expressed as water table fluctuations.

Basin-wide effective specific yields obtained from (7) were 0.014 ± 0.003 for both dry seasons (Table 4). Because these values reflect an effective basin-wide process, they are insensitive to local heterogeneities in the fractured rock aquifer system, in comparison with locally obtained values using lab samples or local aquifer testing, which are highly variable and relatively unreliable (Bardenhagen, 2000). Therefore, for water resource assessment at the watershed scale, this methodology for specific yield estimation is much more sound than the aforementioned punctual techniques. Error on specific yield (~20%, Table 4) has been computed cumulating all the sources of errors described above.

The specific yield obtained is realistic for fissured granite and is of the same order of magnitude as values estimated at the well scale using pumping data in the fissured layer itself. Higher values obtained with the water budget method can be explained by the fact that the upper part of the weathering cover (saprolite with specific yield much higher than in the fissured zone) can be partially saturated in some areas after Monsoon, which increases the global storage at the watershed scale. Heterogeneity effects can also explain this apparent increase of $S_y$ with scale.

It is generally assumed that specific yield varies with depth–especially in hard-rock aquifers where fracture density and porosity change with depth, namely between the different layers constituting the aquifer (Maréchal et al., 2004). Water budget results in 2002 and 2003 indicate that the specific yield does not vary. In fact, the water table is located mainly in the fissured layer of the aquifer and water table fluctuations are small enough so that the water table remains in the same portion of the aquifer, characterized by a constant specific yield.

Equation (8) was used to estimate natural recharge (Table 5). Natural recharge is determined at the watershed scale, not cell by cell like other budget components, and is therefore

Table 4. Groundwater budget during the Rabi seasons, estimation of specific yield and absolute errors.

| Season | Date | $\underline{RF}^{dry}$ (mm) | $PG^{dry}$ (mm) | $E^{dry}$ (mm) | $Q_{on}^{dry} - Q_{off}^{dry}$ (mm) | $\Delta h^{dry}$ (m) | $S_y$ [–] |
|---|---|---|---|---|---|---|---|
| Rabi 2003 | Nov02–Jun03 | 37.9 ± 3.2 | 99.3 ± 5 | 0.6 ± 1 | −0.3 ± 1 | −4.4 ± 0.35 | 0.0140 ± 0.0029 |
| Rabi 2004 | Nov03–Jun04 | 53.7 ± 3 | 123.8 ± 6.2 | 1.3 ± 1 | 1.0 ± 1 | −5.1 ± 0.23 | 0.0138 ± 0.0027 |

Table 5. Groundwater balance during monsoon seasons, estimation of natural recharge and absolute errors.

| Season | Date | $RF^{wet}$ (mm) | $PG^{wet}$ (mm) | $E^{wet}$ (mm) | $Q_{on}^{wet} - Q_{off}^{wet}$ (mm) | $\Delta h^{wet}$ (m) | R (mm) | Seasonal rainfall (mm) | Rainy days | $R/P$ (–) |
|---|---|---|---|---|---|---|---|---|---|---|
| Kharif 2002 | Jun02–Nov02 | 31.0 ± 4.6 | 84.2 ± 4.2 | 0.5 ± 1 | 0.0 ± 1 | 1.2 ± 0.27 | 70.5 ± 15.8 | 543 | 43 | 0.13 ± 0.03 |
| Kharif 2003 | Jun03–Nov03 | 32.6 ± 4.6 | 70.8 ± 3.5 | 1.0 ± 1 | −1.2 ± 1 | 8.3 ± 0.32 | 156.5 ± 37.5 | 824 | 54 | 0.19 ± 0.05 |

Table 6. Groundwater balance during two hydrological cycles.

| Year | Annual rainfall | R (mm/yr) | $RF_{TOT}$ (mm/yr) | $PG_{TOT}$ (mm/yr) | $E_{TOT}$ (mm/yr) | $Q_{onTOT} - Q_{offTOT}$ (mm/yr) | BAL (mm/yr) | $\Delta h_{TOT}$ (m) |
|---|---|---|---|---|---|---|---|---|
| 2002–2003 | 613 | 70.5 ± 15.8 | 68.9 ± 7.8 | 183.5 ± 9.2 | 1.1 ± 2 | −0.3 ± 2 | −45.5 ± 9 | −3.2 ± 0.62 |
| 2003–2004 | 889 | 156.5 ± 37.5 | 86.3 ± 7.6 | 194.6 ± 9.7 | 2.30 ± 2 | −0.1 ± 2 | +45.8 ± 8 | +3.2 ± 0.55 |

not spatially distributed. Relative error on natural recharge (22–24%, Table 5) has been computed cumulating all the sources of errors described above.

At Table 5, the recharge is compared to precipitation (P) during the monsoon (seasonal rainfall) between June and November. During both hydrological years of monitoring, the recharge coefficient $R/P$ varies between 0.13 and 0.19. This is similar to recent results obtained in India under the same climate conditions. Its fluctuation, year to year, depends mainly on the intensity and temporal distribution of rainfall events during the monsoon. Notice that the recharge coefficient increases with the number of rainy days during the monsoon (Table 5).

The "double water table fluctuation method" consists in aggregating dry and rainy season water budgets. The annual groundwater balance was calculated from June 2002 to June 2004 (Table 6) and we see a respective deficit and excess of water due to discrepancies between annual rainfall and an average rainfall of about 740 mm/year (average in Maheshwaram since 1985). Considering the uncertainty on the components of the budget, this suggests that the balance should be lightly negative for an average rainfall. Historical water level data shows a global depletion of the aquifer at a rate of about one meter per year in pumped areas, confirming that the overexploitation threshold has been reached in such areas. Moreover, given the abstraction rate in the basin, any deficient monsoon (the 2002 monsoon, for example) causes a significantly negative balance followed by a drop in the water table, which can be fully or only partially replenished by the next heavy monsoon. In spite of the fact that the pumping areas represent only 25% of the 1324 cells of the basin (Figure 7b), the entire balance is negative.

## 5 CONCLUSION

Classical distributed models are often not considered adaptable to heterogeneous and anisotropic hard rock aquifers. A hard-rock system has been fully characterized in order to provide, at the watershed scale, the necessary elements for a classical modelling approach. Heterogeneity and geometry of the aquifer are regionalised using geological and geophysical investigations. The hydrodynamic properties of the weathered layer are characterised in details. The groundwater budget at the watershed scale allows determination of important flows and specific yield at the relevant scale.

Further developments have been done on models to simulate various groundwater exploitation scenarios and to develop a decision support system for groundwater management. In the specific context of high pressure of population on the resources, the careful use of this tool can be associated to socio-economic scenarios for impact assessment within integrated approaches.

## REFERENCES

Bardenhagen I (2000) Groundwater reservoir characterisation based on pumping test curve diagnosis in fractured formation. In: Sililo O. (ed) Groundwater: past achievements and future challenges, Cape Town, South Africa Balkema, Rotterdam: 81–86.

Barker JA (1988) A generalized radial flow model for hydraulic tests in fractured rock. Water Resources Research 24(10): 1796–1804.

Bredehoeft JD (2002) The water budget myth revisited: why hydrogeologists model. Groundwater 40(4): 340–345.

Coudrain-Ribstein A, Pratx B, Talbi A and Jusserand (1998) Is the evaporation from phreatic aquifers in arid zones independent of the soil characteristics ? C.R. Acad. Sci. Paris, Sciences de la Terre et des Planètes 326: 159–165.

Dewandel B, Lachassagne P, Wyns R, Maréchal JC, and Krishnamurthy NS (2006) A generalized 3-D geological and hydrogeological conceptual model of granite aquifer controlled by single or multiphase weathering. Journal of Hydrology 330 (1–2): 260–284.

Dupuit J (1848) Etudes théoriques et pratiques sur le mouvement des eaux dans les canaux découverts et à travers les terrains perméables [Theoritical and practical studies on water flow in open channels and through permeable terrains], 1st ed., Dunod, Paris.

Dupuit J (1863) Etudes théoriques et pratiques sur le mouvement des eaux dans les canaux découverts et à travers les terrains perméables [Theoritical and practical studies on water flow in open channels and through permeable terrains], 2nd ed., Dunod, Paris.

Gringarten AC and Ramey HJ (1974) Unsteady-state pressure distribution created by a well with a single horizontal fracture, partially penetrating or restricted entry. Trans. Am. Inst. Min. Eng. 257: 413–426.

Healy RW and Cook PG (2002) Using groundwater levels to estimate recharge. Hydrogeology Journal 10: 91–109.

Maréchal JC, Wyns R, Lachassagne P, Subrahmanyam K and Touchard F (2003) Anisotropie verticale de la perméabilité de l'horizon fissuré des aquifères de socle: concordance avec la structure géologique des profils d'altération [Vertical anisotropy of hydraulic conductivity in fissured layer of hard-ricj aquifers due to the geological structure of weathering profiles.], C.R. Geoscience, 335: 451–460.

Maréchal JC, Dewandel B and Subrahmanyam K (2004) Use of hydraulic tests at different scales to characterize fracture network properties in the weathered-fractured layer of a hard rock aquifer, Water Resources Research, 40 (11): Art. No. W11508.

Maréchal, JC, Dewandel B, Ahmed, S, Galeazzi, L and Zaïdi FK (2006) Combined estimation of specific yield and natural recharge in a semi-arid groundwater basin with irrigated agriculture, Journal of Hydrology, 329, 1–2, 281–293, DOI: 10.106/jjhydrol.2006.022.

Marsily G de (1986) Quantitative Hydrogeology, San Diego, California, Academic Press.

Neuman SP (1975) Analysis of pumping test data from anisotropic unconfined aquifers considering delayed gravity response, Water Resources Research 11(2): 329–342.

Roy AD and Shah T (2002) Socio-ecology of groundwater irrigation in India, Intensive Use of Groundwater: Challenges and Opportunities. R. Llamas and E. Custodio. Lisse, A.A. Balkema: 307–335.

Schicht RJ and Walton WC (1961) Hydrologic budgets for three small watersheds in Illinois, Illinois State Water Surv Rep Invest 40, 40 p.

Warren JE and Root PJ (1963) The behaviour of naturally fractured reservoirs. Soc. Petroleum Eng. J. 3: 245–255.

CHAPTER 16

# Estimating groundwater recharge in fractured bedrock aquifers in Ireland

Bruce D.R. Misstear[1] and Vincent P. Fitzsimons[2]

[1] *Department of Civil, Structural and Environmental Engineering, Trinity College Dublin, Ireland*
[2] *Scottish Environmental Protection Agency, Clearwater House, Heriot Watt Research Park, Edinburgh, EH14 4AP, UK (formerly of the Geological Survey of Ireland, Beggar's Bush, Hadington Road, Dublin 4, Ireland)*

ABSTRACT: Recharge assessment has become increasingly important in Ireland in response to recent European legislation. The reliability of groundwater recharge estimates is influenced by the complexity of the geology – fractured bedrock aquifers with a thick covering of superficial deposits, generally glacial tills, often of low permeability – and by the relative scarcity of hydrometric and hydrogeological data. The two main approaches applied to recharge estimation are the soil budgeting method and river baseflow analysis. A recent study of the soil moisture budgeting approach showed that the recharge estimates are more dependent on the characteristics of the till cover (especially hydraulic conductivity) than on the particular soil budgeting method applied or the assumptions concerning land use and crop root constant (readily available water). A sensitivity analysis of baseflow separation parameters highlighted the importance of having a good conceptual model of the various subsurface pathways contributing flow to a river in order to achieve realistic baseflow results. The baseflow separation can be improved by comparing the flow data against borehole water level hydrographs, and by also calibrating the output against both major summer recessions and short term winter recessions

## 1 INTRODUCTION

The main impetus for the current study of recharge in the fractured bedrock aquifers of Ireland was the introduction of the European Union Water Framework Directive (European Commission, 2000). Prior to the directive, recharge received little attention in Ireland, primarily because the total groundwater usage was small (in national terms), the resource was not regarded as being under pressure and because there was no permitting system for well abstractions. Due to the implementation of the directive, with its emphasis on an integrated approach to managing and protecting water resources and dependent ecosystems, there is now a greater emphasis on acquiring estimates of recharge. Reliable estimates of recharge are needed for many purposes, including: quantifying groundwater resources within river basins; issuing abstraction licences; determining groundwater vulnerability; assessing the groundwater contributions to rivers (baseflow) and to sensitive wetland habitats, and hence protecting these resources. The paper considers the two main approaches used for estimating groundwater recharge in Ireland: soil moisture budgets and river baseflow analysis. The results obtained are examined both in relation to the assumptions made in the analysis, and also with regard to the influence of the geology.

## 2 LANDSCAPE, CLIMATE AND GROUNDWATER OCCURRENCE

In general terms, Ireland's topography has a saucer-like shape, with a mountainous rim and a flat or gently undulating centre. The coastal mountains rise to around 1,000 m above sea level and consist of Precambrian and Lower Palaeozoic igneous, metamorphic and sedimentary rocks, often with a thin covering of peat or glacial deposits on the plateau areas and lower hill slopes. The lowland areas of central Ireland are mostly underlain by Carboniferous limestones and other sedimentary rocks. These are generally covered by a blanket of tills and other glacial deposits, which can reach thicknesses of several tens of metres (these superficial deposits are termed subsoils in Ireland). The country has a temperate maritime climate, with annual precipitation varying from less than 800 mm on the east coast to over 2,500 mm in the mountains along the western Atlantic coast. Potential evapotranspiration is typically between 450 and 500 mm, and the majority of the rural landscape is under grass pasture.

The bedrock aquifers in the Republic of Ireland are all characterised by fracture flow – there is negligible primary porosity. The main bedrock aquifers are found in the Carboniferous limestones and dolomites in the midlands, southern and western parts of the interior lowlands. The most important of these aquifers – termed "regionally important" in the Geological Survey of Ireland classification (Misstear and Daly, 2000) – show varying degrees of karstification, have transmissivities generally in excess of $500\,m^2 d^{-1}$ and specific yields of around 1 or 2%. Where karstification is well-developed, groundwater flow rates can be rapid – with measured velocities of up to $200\,m\,h^{-1}$ in some cases (Drew and Daly, 1993). Other regionally important bedrock aquifers occur in Ordovician volcanic and Devonian and Carboniferous sandstone formations. Though significant water strikes can occur to depths of 100 m, most flow occurs within the top 30 m to 40 m of bedrock. This is primarily due to the predominance of fracture flow and to the effects of karstification, which become decreasingly important with depth. With low specific yield allied to relatively high transmissivity, the regionally important fractured bedrock aquifers show much more pronounced water level fluctuations during recharge and recession periods than gravel aquifers, which have specific yield values in the region of 12 to 20% (Figure 1).

Not all of the bedrock aquifers are "regionally important". In over half of the country's area – comprising most of the upland areas as well as extensive parts of the lowlands – the bedrock aquifers are poorly productive, with transmissivities generally less than $100\,m^2 d^{-1}$. Nevertheless, these poorly productive aquifers are important in Ireland, as there are a great number of small private well sources, many of which tap such aquifers – it has been estimated that 138,000 Irish households have a private well source (Loughman, 2005), a large number of wells in a country with a population of only around 4 million. Again, groundwater flow in these poorly productive aquifers is via fractures, but the flow pathways are generally concentrated within the upper few metres of the aquifer, or in isolated fracture zones, and are more localised than those in the regionally important aquifers. The shallow nature of much of the groundwater flow in these bedrock aquifers poses difficulties when trying to estimate recharge by river baseflow separation methods, since it can be difficult to

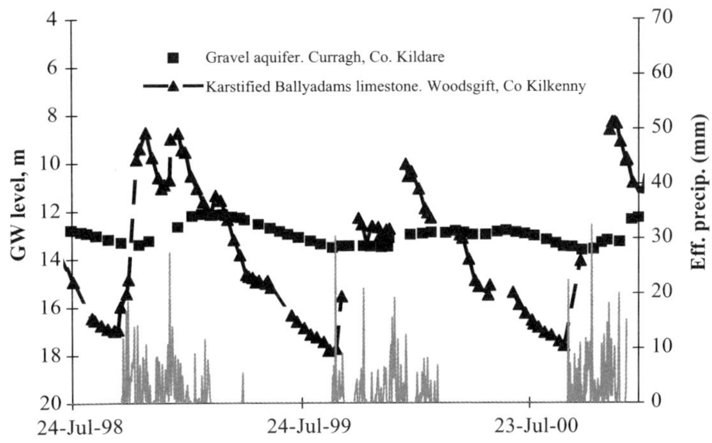


Figure 1. Water level hydrographs showing the rapid response to effective precipitation in a fractured limestone aquifer compared to a gravel aquifer.

distinguish between the contribution to total runoff from shallow groundwater and that from interflow or bank storage, as discussed later on in this paper. A further challenge is that there are usually few data available on the characteristics of the poorly productive fractured rock aquifers, including few pumping test or water level monitoring records.

## 3 GROUNDWATER RECHARGE ESTIMATION

Recharge in Ireland occurs both by direct infiltration of rainfall and by infiltration following surface runoff, the latter being especially important in karstic areas. Most recharge studies in Ireland have focused on direct recharge only, or have estimated the total recharge for a catchment without attempting to distinguish between the contributions from the direct or indirect sources. The approaches adopted can be categorised (Misstear, 2000) into:

1. Inflow methods, including soil moisture budget methods, Darcy flux analyses and tracer studies.
2. Aquifer response methods, including well hydrograph interpretation and throughflow calculations.
3. Outflow methods, specifically river baseflow analysis.
4. Catchment water balance and numerical modelling methods.

The majority of studies have been based on either soil moisture budgeting methods or river baseflow analyses (Daly, 1994; McCarthaigh, 1994; Aslibekian, 1999). The soil moisture budget methods provide an estimate of the effective precipitation. The effective precipitation is the excess of precipitation over actual evapotranspiration, and when the soil is at field capacity effective rainfall is available for either downward percolation or for surface runoff. Not all of the downward percolation will necessarily result in recharge to the water table; some of the infiltrating water may be diverted to lateral flow within subsoil layers, for example, and so be "lost" as interflow. Because of the important role of subsoils, the concept of a recharge coefficient has been introduced, which is defined as the proportion of effective precipitation that becomes recharge (Fitzsimons and Misstear, 2006). This is similar to the concept of recharge factors applied by Rushton (2005) to recharge through Drift (i.e., tills) for aquifers in the UK.

## 4 SOIL MOISTURE BUDGET APPROACH

In Ireland, effective precipitation predictions have been most commonly based on the estimation of soil moisture surplus through soil moisture budgeting techniques. The Penman-Grindley (Penman, 1949 and Grindley, 1970) and Aslyng (1965) models are two related budgeting techniques which have been used (Cawley, 1994; Daly, 1994). In both these models, the amount of soil moisture surplus or soil moisture deficit is estimated from precipitation and potential evapotranspiration (PET) data, the latter calculated from the Penman or Penman-Monteith formulae. A key challenge is the derivation of *actual* evapotranspiration (AET), which is assumed to equal PET unless the soil moisture deficit is below a critical threshold beyond which the vegetation has difficulty in transpiring, termed the *root constant*. For more information, the reader is referred to Lerner et al. (1990), Rushton and Ward (1979) and Fitzsimons and Misstear (2006).

Hulme et al. (2001) and Rushton (2003, 2005) have described changes to the Penman-Grindley model. These involve changes to the relationship of AET to PET during times of low soil moisture, the incorporation of evaporation from bare soil (where appropriate) and the replacement of the root constant concept with a similar concept of *readily available water*, using information about crop water requirements from the Food and Agriculture Organisation of the United Nations (FAO, 1998). Due to this closer linkage to agricultural science, this method has been widely adopted in the UK and is being increasingly applied in Ireland.

However, Fitzsimons and Misstear (2006) highlighted the relatively limited sensitivity of effective precipitation estimations to variations in the main soil budgeting parameters and methods. This was due to the combination of relatively wet climate and to the dominance of grassland in Ireland. This combination results in only limited development of soil moisture deficits in excess of the root constant in the summer months. It is during episodes when soil moisture deficits exceed the root constant (or the related parameter of readily available water) that variations in crop parameters and soil moisture budgeting methods are most evident. The authors derived a baseline estimate of average annual recharge for a set of parameters and meteorological data that were representative of grassland in the south-east of Ireland, one of the driest parts of the country. Using a sensitivity analysis based on a range of methods, and reasonable ranges in the key crop parameters, the authors went on to conclude that effective precipitation in conditions typical of Irish fractured aquifers is likely to be insensitive to the key soil moisture budgeting parameters, with simulated effective precipitation ranging from 101% to 95% of the baseline estimate. The authors indicated that recharge estimates in Ireland are much more sensitive to the permeability of the strata overlying the aquifer, particularly in the extensive areas of Ireland where the aquifers are overlain by glacial till. Reasonable ranges in till permeability and vertical hydraulic gradient produced a substantial range in simulated recharge coefficients from 80% to 2% of the baseline estimate of effective precipitation. This range reflected the range identified from a literature review of recharge estimates in till areas from the UK and Ireland. The greatest sensitivity to permeability lies in the range from $0.001\ \mathrm{m\,d^{-1}}$ to $0.01\ \mathrm{m\,d^{-1}}$ (Fitzsimons and Misstear, 2006).

## 5 RIVER BASEFLOW ANALYSIS

The assessment of the groundwater component of river hydrographs has developed in recent years from relatively simple baseflow separation techniques to the modelling of the "fate" of all waters in a catchment, including elements such as evaporation, rapid runoff, interflow and deep groundwater flow. Early examples of the latter can be found in Serenath and Rushton (1984), where relatively simple routing models were used. More recently, workers such as Bradbury and Rushton (1998) have incorporated these concepts into more complex numerical models. These types of analyses, when undertaken appropriately, are a clear improvement on simple baseflow separation exercises as they can replace the assumptions and generic constants inherent in these exercises with quantifications which are based more closely on the features of specific catchments.

However, there is still a role for baseflow separation, particularly in the early stages of conceptual modelling or in situations where complex numerical modelling is not feasible or not justified. These situations are common in Irish fractured rock aquifers, where there are often limited data available for modelling and where abstraction is rarely significant when compared to the overall recharge resource. In these scenarios, simple baseflow separation exercises can provide useful supporting evidence in the assessment of recharge. In suitable aquifers, and for appropriately located gauges, it may be reasonable to assume that the proportion of baseflow to total runoff (often termed the baseflow index) provides an approximation to the recharge coefficient described in previous sections of this paper.

There are several assumptions inherent in using runoff flow records to estimate recharge (Halford and Mayer, 2000):

1. There is a direct relationship between the timing of groundwater and surface runoff peaks.
2. The evapotranspiration from lakes, rivers and wetlands is insignificant compared to the total volume of groundwater baseflow.
3. Bank storage and river recharge effects are also insignificant.
4. There is minimal interaction between the shallow and deeper aquifers.

Most of these assumptions are generally valid in Irish fractured rock aquifers. For example, an examination of Figure 1 shows the typical rapid response of the hydrograph from a fractured aquifer in comparison with a gravel aquifer example. Figure 2 provides an example of how the responses of river flow and groundwater levels to effective precipitation can be similar (outside the main recession periods).

The assumption relating to bank storage and river recharge effects may require more detailed consideration, along with the related effects of groundwater releases from shallow subsurface pathways, such as peat, glacial tills, and weathered rock. In some cases, these shallow effects may be significant and may require careful consideration if we are to conceptualise adequately groundwater-surface water interactions on a catchment scale. Figure 3 provides some hypothetical examples of different subsurface pathways and their contribution to the overall hydrograph. The examples were derived from effective precipitation data, by calculating moving averages of 3 day, 10 day and 30 day duration. In this hypothetical example, overland flow is assumed to be insignificant and all contributions to the runoff hydrograph are assumed to come from the three subsurface pathways described above. The shorter duration moving averages were assumed to represent the more shallow subsurface pathways such as release from peat or interflow within glacial till. The longer duration averages were taken to represent a higher component of flow from the deeper aquifer pathways.

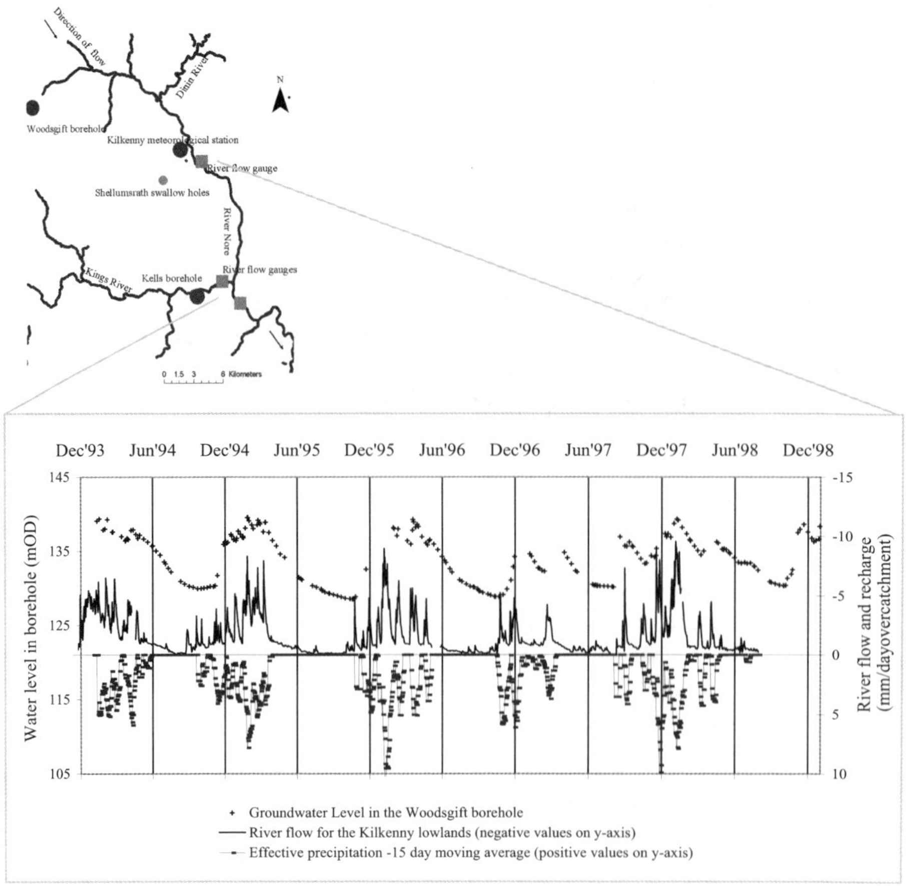


Figure 2. Nore River catchment between upstream and downstream gauges. A comparison of the response of surface water and groundwater to effective precipitation.

Note that Figure 3 provides an example of how subsurface pathways of varying duration can interact to produce a typical storm hydrograph. It does not represent an attempt to model the runoff hydrograph.

The meteorological and runoff data used in this example are from the Kilkenny lowlands in southeast Ireland (the area shown in Figure 2). The meteorological dataset was calculated using a soil moisture budgeting approach, with daily rainfall and potential evaporation data being taken from a weather station at Kilkenny City. More information is provided in Fitzsimons and Misstear (2006). The runoff data were taken from a hybrid dataset, calculated from the three gauges shown in Figure 2. Daily data were calculated from the difference in daily flows between the two upstream and one downstream gauge on the river. This hybrid dataset was generated in order to focus assessments on runoff generated in the catchment area between these three gauges. This area is relatively small (1140 $km^2$) compared with the catchment of the downstream gauge (2778 $km^2$). As a consequence, the data provide more hydrogeological control than is usually possible in Ireland, where the hydrogeology is highly variable. The hydrogeological system comprises a fractured limestone/dolomite aquifer

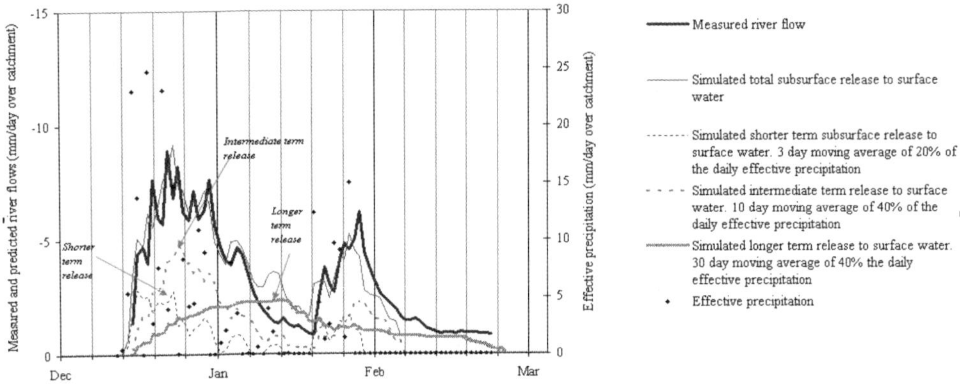


Figure 3. A hypothetical case depicting contributions of differing subsurface release pathways to the runoff hydrograph of a typical winter storm.

overlain by thin, moderate permeability tills. The river floodplain is occupied by a productive sand and gravel aquifer.

The simulations provided in Figure 3 follow a relatively obvious pattern. The early part of the overall storm hydrograph is dominated by short term subsurface releases (e.g., from bank storage or interflow). The main portion of the hydrograph is dominated by contributions from intermediate term releases (e.g., from weathered rock) while the latter part is dominated by longer term release (e.g., from deeper portions of a bedrock aquifer).

All three simulations presented in Figure 3 could be termed baseflow, but all are reflecting very different elements of the hydrogeological regime. Of the variety of automated baseflow separation methods available, most employ empirical parameters. Depending on the parameter selected, the baseflow result may actually incorporate a number of different subsurface pathway elements. This can be an advantage where the subsurface pathways have been adequately conceptualized and where it is clear which types of baseflow is under investigation. However, it can lead to significant errors where automated techniques are applied without due consideration of the subsurface pathways.

### 5.1 *Sensitivity analysis of the link between subsurface pathways and baseflow separation parameters*

A sensitivity analysis was carried out to examine the potential effects on baseflow estimates where there has been insufficient conceptualization of the groundwater system, and the various subsurface pathways present. A commonly used baseflow separation technique was employed across three catchments representing a range of hydrogeological regimes.

Workers such as Nathan and MacMahon (1990) point out that automated baseflow separation techniques are useful only for deriving broad indices and, as such, should not be used to derive recharge estimates from particular storm events. Nevertheless, a simple sensitivity analysis such as this is considered useful in helping to demonstrate the effects of mixing deep and shallow effects into the same baseflow estimate. The actual baseflow estimates derived using the sample methodology are of less importance – it is recognised that the use of different techniques would lead to a different set of estimates.

## 5.2 *Methodology*

Three catchments were selected for this study: the Scart, the Dinin and the hybrid catchment for the Kilkenny lowland described earlier in this paper (and shown in Figure 2). These represent a range in hydrogeological regimes, from "locally important" fractured sandstones overlain by thin glacial deposits, to locally important fractured sandstones overlain by valley gravels and low permeability glacial deposits, to "regionally important" karst limestone and dolomite aquifers overlain by valley gravels and moderately permeable glacial deposits.

The baseflow separation method used is the automated baseflow separation method developed by the Institute if Hydrology (now Centre for Ecology and Hydrology). Estimates were made over the period from 1990 to 1998. The method has been used previously in Ireland, and it is probably the most commonly used method in the UK, forming a key part, for example, of the national hydrological classification of soil types (Boorman et al., 1995). This method involves the identification of runoff minima at fixed time intervals and the selection of key "turning points" from these minima (IOH, 1989). The length of this time interval is termed the 'time base' for the purposes of this paper and can be considered to represent the number of days after a peak in the hydrograph when runoff is assumed to cease. The value of the time base is normally set to 5 days, but for the purposes of this simple sensitivity analysis, the time base was varied to reflect the relevant hydrogeological regime. The range in time base selected was constrained using three controls:

1. The time base was not allowed to vary by more than 2 days from that predicted from catchment size alone, using the equation $N = A^{0.2}$, as cited by Linsley et al. (1975). In this equation, N is the time base of surface runoff, and A is the drainage area.
2. For the Kilkenny catchment, a suitable borehole hydrograph was used to help judge the suitability of the various timebase responses. Borehole hydrographs can be used to identify recharge events which, in rapid flow systems such as those in Ireland, could be expected to correspond in time with discharge peaks in the baseflow hydrograph. Therefore, appropriate borehole hydrographs can potentially be used to help calibrate baseflow simulations, through a comparison of the timing of recharge events.
3. For the remaining catchments, borehole hydrograph data were not available and hence hydrogeological judgment was applied to eliminate those simulations that appeared unrealistic.

## 5.3 *Results of baseflow sensitivity analysis*

The results of the sensitivity analysis are presented in Table 1.

Even when using only one baseflow separation technique, there is a significant variation in predicted baseflow index across all three catchments. The largest sensitivity to time base parameter occurs in the Dinin Catchment, which is a consequence of its upland location, the poorly productive nature of the aquifers and the thick covering of low permeability subsoils. The runoff hydrograph is particularly "flashy", with significant variation in daily mean discharge values. In view of these factors, the estimated BFI range for the Dinin catchment is significantly higher than would seem reasonable. The lowest sensitivity to the time base parameter occurs in the hybrid Kilkenny catchment where the cover is generally thinner, and of moderate permeability, resulting in a smaller fluctuation in mean daily runoff values and hence in the range of derived BFI values.

Table 1. Examples of the sensitivity of groundwater baseflow to the time base parameter.

| Gauge | Hydrogeological setting | Catchment of gauge | Estimate of time base (N) from N = catchment Area$^{0.2}$ | Range in time base values (N) | Range in baseflow indices[1] |
|---|---|---|---|---|---|
| Hybrid flow record: derived from upstream and downstream gauges | Lowland. Typically moderate permeability till (~5 m thickness) and some high permeability gravel subsoils. | Nore (1140 $km^2$) | 4 | 3, 5 | 58% to 68% |
| Scart | Upland. Thin subsoils. Locally important fractured sandstone aquifer. | Kilkenny Blackwater (108 $km^2$) | 3 | 1, 3 | 60% to 76% |
| Castlecomer | Poor to locally important aquifers. Generally low permeability subsoils of varying thickness. Small gravel body occurs close to central river. | Dinin (153 $km^2$) | 3 | 1, 3 | 46% to 65% |

[1]Note: Baseflow index (BFI) is the ratio of average annual baseflow to average annual runoff. Baseflow was estimated by the IOH method (IOH, 1989).

Figure 4 provides an example of the baseflow separation outputs for the Kilkenny lowlands hybrid catchment. Using a time base of 5, the predicted baseflow in Figure 4 does not appear to represent the appropriate frequency of recharge events within the aquifer during the winter storms of early 1995 (as shown by the borehole water level data). The frequency of these events is better predicted using smaller time bases of 3 or even 1. In terms of the volume of baseflow, results from winter estimates appear to be more sensitive to time base than results from summer estimates. It appears that most sensitivity arises from the fact that each simulation is, in effect, modelling a different combination of the various subsurface pathways. Shallower pathways would be expected to release water more rapidly into the river system. It is therefore considered that the time base values of 1 and 3 provide a baseflow estimate that includes significant shallow as well as deep groundwater contributions. In this study area, the shallower pathway contributions might include bank storage release from the alluvial gravels, interflows in the glacial till and groundwater within the upper and highly fractured portion of the bedrock aquifer. The time base value of 5 includes a larger contribution from deeper groundwater sources.

### 5.4 *Discussion of baseflow sensitivity analysis*

Significant variations in predicted baseflow can be generated by varying the selected time base component in the Institute of Hydrology method. In hydrogeological regimes similar to those found in Ireland, it is likely that the sensitivity to the time base parameter will be greatest in catchments supported by aquifers of a lower productivity or by aquifers overlain by thick low permeability strata. These systems tend to have more "flashy" flow regimes and will often have multiple subsurface groundwater pathways including, for example, peat, glacial tills and weathered rock. Lower productivity aquifers in Ireland will generally have thin unsaturated zones and there can be a distinct hydraulic separation between weathered rock and deeper fractures. As a consequence, hydrogeological regimes associated with these aquifers will tend to be more complex close to surface, with the potential for a wider range of subsurface pathways.

The sensitivity to the time base parameter will be greatest also in the winter months, when storms arrive much more frequently. In these periods, it is much more difficult, using simple automated techniques, to identify situations where groundwater flow and river flow are the same.

To derive a realistic baseflow estimation, it is necessary to a) understand conceptually the various subsurface pathways contributing flow to a river and b) to understand which pathway is of interest in terms of the baseflow output. Though the tests in this paper have focused on the IOH method, most automated baseflow separation methods employ empirical parameters, and the resulting baseflow estimate will be sensitive to variations in these parameters. Regardless of the technique chosen, it is likely that greatest care will be required when dealing with aquifers of a lower productivity, or where aquifers are overlain by thick low permeability strata.

Methods such as the Boughton two-parameter algorithm (Chapman, 1999) have recently been used in Ireland as part of a study into the links between groundwater recharge and groundwater vulnerability (Misstear and Brown, 2005). These analyses have been calibrated using summer hydrograph recessions, in order to ensure that the empirical parameters adequately reflect the real situation. The Boughton method, when applied to poorly productive aquifers overlain by thick low permeability tills, can yield realistically low values of BFI.

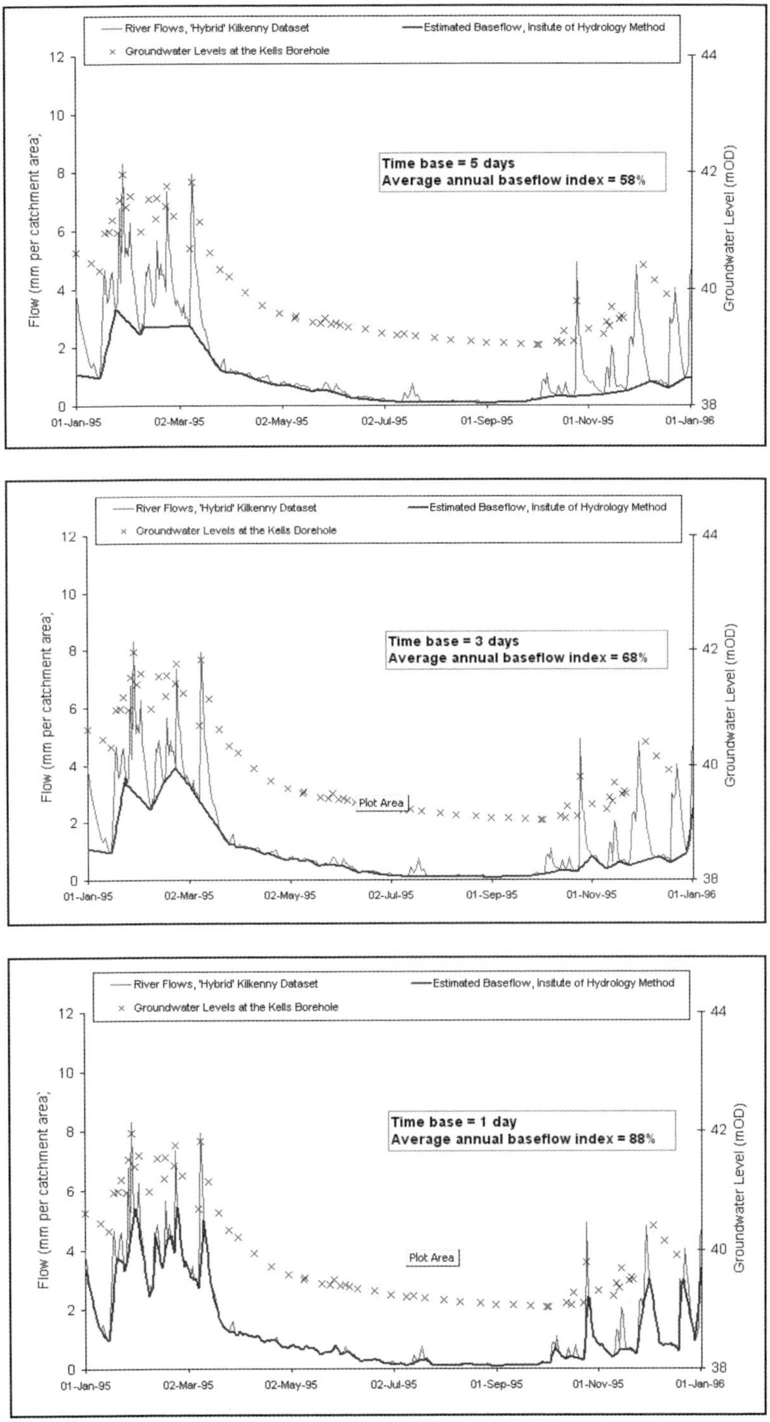


Figure 4. Influence of time base variation on baseflow estimates.

Based on the sensitivity analysis described in this paper, it is considered important that predictions adequately take account of winter baseflow. This can be achieved by a combination of borehole hydrograph analysis and assessment of short term winter storms. In low storage, rapid flow situations, borehole hydrographs can provide a means of assessing whether the separation method is adequately representing the response to recharge in the aquifer. Clearly, the borehole data will only be of value if they represent the recharge response in that portion of the subsurface pathway which is under consideration.

Winter recessions will often be difficult to isolate, particularly in higher rainfall areas where recessions will often be truncated by the arrival of new storms, and before releases from deep groundwater can become apparent. For long hydrograph records, the long-established concept of the master recession curve could be used to help identify key winter recessions. In this method, a characteristic recession slope for the whole record is derived from an examination of suitable recessions. For deep aquifer studies, the most suitable recessions will be the longest, but recessions can be selected to reflect shallower pathways, as appropriate. The resultant master slope will be defined by an exponential decay constant and, as such, the slope will be relatively shallow at lower flows in the summer months but relatively steep at higher flows in the summer months. This slope can then be used in a curve-fitting algorithm to help identify recessions across the whole record. These recessions will then form the foundation upon which to construct a baseflow separation.

In the example presented in Figure 5 below, the recessions have been joined using Nathan and McMahon's digital, filtering approach (1990), but any appropriate method could be adopted. In effect, the method is similar to the IOH (1989) method in that both methods identify key points on the hydrograph before joining these together into an estimation of baseflow. The difference is that the IOH method examines flow minima to approximate points where groundwater inputs comprise 100% surface water flows, while the slope fitting

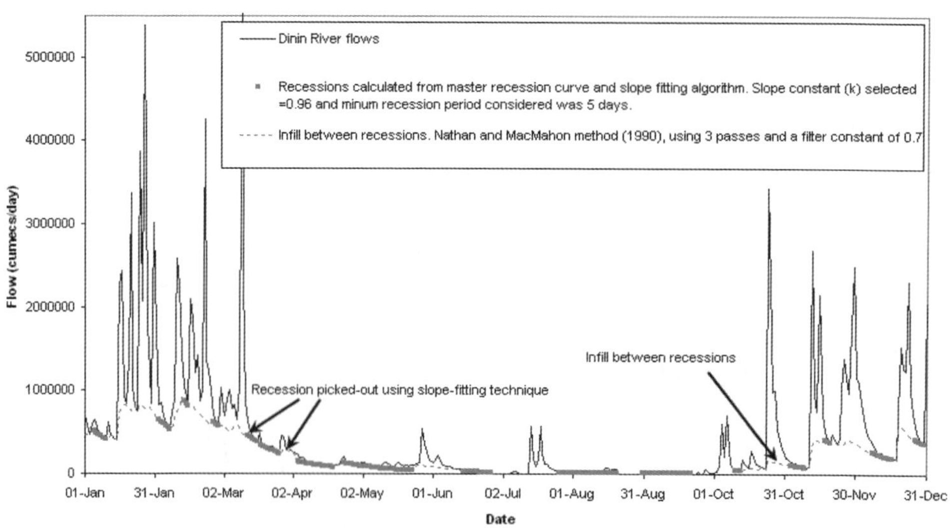


Figure 5. Example of the use of slope fitting techniques to select appropriate winter recessions, Dinin River catchment.

method examines slope minima to approximate periods where groundwater inputs comprise 100% surface water flows.

## 6 CONCLUSIONS

The recent work by the authors on soil moisture budgets, summarised in the first part of this paper, highlights the critical importance of the subsoil cover in determining recharge to the fractured aquifers of Ireland. The work shows that, in the Irish climate and vegetation conditions, recharge estimates are much more strongly influenced by the properties of the tills, than by the budgeting method adopted or by the detailed assumptions made about land cover or root constant. The till property of most importance is the hydraulic conductivity. The proportion of the effective precipitation determined from the soil moisture budget that leads to recharge – the recharge coefficient – can be as little as 2% where the aquifer is covered by thick, low permeability tills.

With regard to the estimation of recharge from river hydrograph analysis, the main points to note from the sensitivity analysis are:

1) Significant variations in predicted baseflow can be generated by varying the selected time base component (using the Institute if Hydrology method). In hydrogeological regimes similar to those found in Ireland, it is likely that the sensitivity to the time base parameter will be greatest:
   - in catchments supported by aquifers of a lower productivity or by aquifers overlain by thick low permeability strata;
   - in the winter months when storms arrive much more frequently and it is therefore much more difficult to isolate discrete recessions for calibrating baseflow estimations.
2) In order to derive a realistic answer it is necessary to have a clear conceptual model of the various subsurface pathways contributing flow to a river and also to know which pathway is of most interest in terms of the baseflow output. Most automated baseflow separation methods employ empirical parameters, and the resulting baseflow estimate will be sensitive to variations in these parameters. It is recommended that the parameters should be calibrated against borehole hydrographs (where suitable hydrographs are available). It is also recommended that the output should be calibrated against major summer recessions and short term winter recessions. The need to assess winter flows is not necessarily intuitive, as baseflow is normally associated with summer recessions. However, this is an important part of the assessment as baseflow predictions can be particularly variable in the winter months. A method to identify slope minima has been proposed to assist in this process.
3) Classic baseflow separation techniques are being superseded to an increasing degree by advances in numerical modelling techniques. Nevertheless, there is still merit in undertaking baseflow separation, particularly in the early stages of conceptual modelling or in situations such as diffuse pollution studies, where complex numerical modelling is either not feasible or not justified. Where methods are used inappropriately, or where the subsurface pathways leading to baseflow are not properly understood, the potential sensitivity of automated techniques to variations in their component parameters could generate significant errors. On the other hand, where there is an adequate understanding of the methods employed, and where there is a good initial conceptual

model, the sensitivity of baseflow estimates to the various parameters can actually be an advantage, allowing the assessment of a variety of baseflow scenarios which can be used to help improve the conceptualisation of the catchment characteristics.

## REFERENCES

Aslibekian O (1999) Regional assessment of groundwater recharge. GSI Groundwater Newsletter, 35: 8–11.

Aslyng HC (1965) Evaporation, evapotranspiration and water balance investigations in Copenhagen 1955–64. Acta Agriculture Scandinavica xv: 284–300.

Boorman DB, Hollis JM, Lilly A. (1995) Hydrology of soil types: a hydrologically-based classification the soils of the United Kingdom. Institute of Hydrology Report No. 126, 137p.

Bradbury CG, Rushton KR (1998) Estimating runoff-recharge in the South Lincolnshire Limestone. Journal of Hydrology 211: 86–99.

Cawley A (1994) Application of monthly water balance models to Irish catchments. In: Agmet Conference Proceedings, September 1994.

Chapman T (1999) A comparison of algorithms for stream flow recession and baseflow separation. Hydrological Processes 13: 701–714.

Daly EP (1994) Groundwater resources of the Nore River basin. Geological Survey of Ireland, RS 94/1

Drew D, Daly D (1993) Groundwater and karstification in mid-Galway, south Mayo and north Clare. Geological Survey of Ireland Report RS 93/3.

European Commission (2000) Directive 2000/60/EC of the European Parliament and of the Council of 23 October 2000 establishing a framework for Community action in the field of water resources. Official Journal of the European Community, L327.

Fitzsimons VP, Misstear BDR (2006) Estimating groundwater recharge through tills: a sensitivity analysis of soil moisture budgets and till properties in Ireland. Hydrogeology Journal 14: 548–561.

Food and Agriculture Organisation (1998) Crop Evapotranspiration: guidelines for computing crop water requirements. FAO Irrigation and Drainage Paper 56, Rome, Italy.

Grindley J (1970) Estimation and mapping of evaporation. IAHS Publication No. 1, 200–213.

Halford KJ, Mayer GC (2000) Problems associated with estimating ground water discharge and recharge from stream-discharge records. Ground Water 38: 331–342.

Hulme P, Rushton, KR, Fletcher S (2001) Estimating recharge in U.K. catchments. In: Impact of Human Activity on Groundwater Dynamics. International Association Hydrological Sciences Publ. 269: 33–42.

Institute of Hydrology (1989) Flow Regimes from Experimental and Network Data (FREND). NERC, Wallingford.

Lerner DN, Issar AS, Simmers I (1990) Groundwater recharge: A guide to understanding and estimating natural recharge. International Contributions to Hydrogeology 8, Heise, Germany.

Linsley RK, Kohler MA, Paulhus JLH (1975) Hydrology for Engineers. McGraw-Hill, New York.

Loughman E (2005) Groundwater and health issues. In: Proc. 25th Annual Seminar on "Groundwater in Ireland" (Irish National Committee of the International Association of Hydrogeologists), Tullamore, 19–20 April 2005, 101–108.

McCarthaigh M (1994) Summary details of water balances in Glyde, Dee, Finn and Blackwater (Monaghan) catchments for the period December 1975–November 1977. Environmental Protection Agency, Dublin.

Misstear BDR (2000) Groundwater recharge assessment: a key component of river basin management. In: Proc. National Hydrology Seminar on River Basin Management (Irish National Committees of the International Hydrology Programme and the International Committee for Irrigation and Drainage), Tullamore, 21 November 2000, 52–59.

Misstear BDR, Brown L (2005) Investigating the link between groundwater vulnerability and recharge. In: Proc. of the conference "Where waters meet" (Jointly convened by the New Zealand Hydrological Society, the International Association of Hydrogeologists, Australian Chapter, and the New Zealand Society of Soil Science), Auckland, 28 November to 2 December 2005.

Misstear BDR, Daly D (2000) Groundwater protection in a Celtic region: the Irish example. In: Robins NS & Misstear BDR. (eds) Groundwater in the Celtic Regions: Studies in Hard Rock and Quaternary Hydrogeology. Geological Society, London, Special Publications, 182, 53–65.

Nathan RJ, MacMahon TA (1990) Evaluation of automated techniques for baseflow and recession analysis. Water Resources Research 26:1465–1473.

Penman HL (1949) The dependence of transpiration of weather and soil conditions. Soil Sci., 1: 74–89.

Rushton KR (2003) Groundwater hydrology: conceptual and computational models. John Wiley & Sons.

Rushton KR (2005) Estimating recharge for British aquifers. Water and Environment Journal 19: 115–124.

Rushton KR, Ward C (1979) The estimation of groundwater recharge. J. Hydrol. 41: 345–361.

Serenath DCH, Rushton KR (1984) A routing technique for estimating groundwater recharge. Ground Water 22: 142–147.

# Groundwater quality in fractured rocks

CHAPTER 17

# Interpreting the evolution and stability of groundwaters in fractured rocks

Adrian Bath
*Intellisci, Willoughby on the Wolds, Loughborough, UK*

ABSTRACT: Groundwaters in fractured rocks have variable hydrochemical and isotopic compositions which are indicators of how these low permeability flow systems have evolved over short and long timescales in response to normal evolution and to episodic perturbations which might have affected hydrodynamic stability. Interpreting these geochemical signals (for example, salinity gradients and groundwater ages) is problematic because the effects of external perturbations on stability are dispersed and diluted by subsequent evolution. Hydrochemical and isotopic data from groundwaters in fractured rocks in Britain and Sweden illustrate these interpretations and the associated uncertainties. Understanding the evolution of such groundwaters is of general interest and is specifically relevant to considering the stability of these groundwater systems with respect to waste containment and contaminant movement.

## 1 INTRODUCTION

Hydrodynamic stability implies unchanging water flow, in terms of rate and direction, and may be a determining factor in hydrochemical stability. The most likely driving forces that would change water flow rates and/or directions are groundwater pressures at the system boundaries. One of the main ways of assessing hydrodynamic stability is by understanding groundwater movements in the past (i.e., palaeohydrogeology).

The main approaches to studying palaeohydrogeology and long-term groundwater stability are:

- Numerical modelling of hydrodynamics with parameters and boundary conditions that have been set according to expert judgement of past climate impacts (e.g., Hartley et al., 2005);
- Deductive interpretation and geochemical modelling of present-day groundwater chemical and isotopic compositions to deconvolute the "signatures" of past infiltration conditions and the mixing of water masses that characterises palaeohydrogeology;
- Characterisation and geochemical/isotopic analyses of recently-formed secondary minerals and their fluid inclusions to obtain information about the groundwaters from which they were precipitated (e.g., Bath et al., 2000; Degnan et al., 2005).

This paper is concerned with the second approach – interpretation of hydrochemical and isotopic data such as:

- Groundwater masses with different compositions whose spatial distribution can be interpreted as evidence of past flow and boundary conditions;
- Groundwater compositions that can be interpreted as mixtures of originally-distinct water sources;
- Isotopic compositions of groundwaters that can be interpreted to provide absolute or relative ages and thus evidence of the timescales of change and heterogeneity.

## 2 STABILITY OF GROUNDWATERS IN FRACTURED ROCKS

A *stable* groundwater system (which may or may not be at *steady state*) is not readily perturbed from its present state i.e. present flux and flow direction are relatively less responsive to potential changes in driving forces within the system and at the system boundaries. Stability is therefore not an absolute concept and does not mean that a groundwater system is inert. Greater groundwater stability is not necessarily directly correlated with lower permeability, since aquifers can also have inherently stable groundwater conditions.

Stability of a groundwater system must be considered in terms of the timeframes for perturbations of driving forces and for the consequent responses of the groundwater system. In a low permeability groundwater system, transient conditions are likely to prevail because the responses in terms of mass transfer of water lag behind the perturbations of driving forces which are pressure changes at the boundaries of the system. Putting past groundwater changes into an absolute or relative timeframe is one of the aims of geochemical and isotopic investigations in palaeohydrogeology. Three divisions of timescale are consistent with climatic periods and with information provided by isotopic interpretations: 0 to 100 years, 100 to $10^4$ years, and $10^4$ to $2 \times 10^6$ years. 0 to 100 years can be called the "modern" period which includes the occurrence of anthropogenic isotopic tracers due to nuclear activities in the last 50 years. 100 to $10^4$ years covers the Holocene period since the end of the last glaciation in northern Europe. $10^4$ to about $2 \times 10^6$ years covers the major part of the Quaternary period during which northern Europe experienced several ice age episodes that would have affected groundwater recharge, movements and hydrochemical evolution. In addition to the tectonic (i.e., crustal depression) and hydrological (i.e., melt water) effects of episodic cover by ice sheets, there were also considerably longer periods in the Quaternary when permafrost might have affected groundwater evolution.

## 3 GEOCHEMICAL INDICATORS OF HYDRODYNAMIC STABILITY

Geochemical indicators provide two basic types of information about palaeohydrogeology and hydrodynamic stability: (i) the physical processes of chemical mixing that are linked with water movements, and (ii) the timing and rates of water movements (Bath and Strömberg, 2004). The spatial scale of homogeneous water compositions in a groundwater system is determined by flow rates, mixing distances (i.e., dispersion and diffusion) and the timescale for which hydrodynamic conditions have been stable.

The hydrochemical and mineralogical parameters that are required for palaeohydrogeological interpretations about (i) the physical process of chemical mixing that is linked with

Table 1. Data for groundwaters that provide information about hydrochemical mixing, water movements and boundary conditions in the past.

| Data | Information |
|---|---|
| TDS | Total mineralisation (= salinity) is correlated with density which affects hydrodynamic stability via the effect on groundwater pressures. |
| Non-reactive solutes and isotopes: $Cl^-$, $Br^-$, $^{18}O/^{16}O$, $^2H/^1H$ | Non-reactive (i.e., conservative) solutes and isotopes are tracers of water components in groundwater mixtures. Component compositions may be identified by data analysis using linear mixing graphs or statistical analysis using Principal Components Analysis. Component compositions may be "signatures" of water sources, e.g. Br/Cl and $^{18}O/^{16}O$ signatures for sea water. |
| Other solutes: Na, K, Ca, Mg, $SO_4$, Sr, I, Li, B, etc. | Solutes that may have varying degrees of reactivity; expert judgement is required to assess possible alteration by water-rock reaction. |

Table 2. Data for groundwaters that provide information about the timing and rates of water and solute movements and mixing.

| Data | Information |
|---|---|
| $^{14}C$ (carbon-14) of DIC and DOC | Indicator of age of water since recharge or travel time along flow paths, showing the hydrodynamic structure of groundwater system |
| $^3H$ (tritium), $^{85}Kr$ and CFCs | Indicator of post-1950 recharge in water; indicates if modern water flows to depth in natural or perturbed system |
| $^{18}O/^{16}O$, $^2H/^1H$ | Qualitative indicators of distinct water sources that are also "signals" of climate at the time of recharge |
| $^{36}Cl/Cl$ of dissolved chloride | Ratio approaching that for secular equilibrium with respect to *in situ* production suggests hydrodynamic stability over timescale of ca. 1 Ma |
| $^{234}U/^{238}U$ | High values (>1) indicate radioactive decay series disequilibrium which might relate to relatively slow water movement |
| $^4He$ contents | Cumulative indicator of groundwater residence in contact with rock source of He. High contents are a qualitative indication of stability and a semi-quantitative estimate of water age |
| Ne, Ar, Kr, Xe contents | Dissolved noble gas data give semi-quantitative "signals" of temperature (i.e., climate) at the time of recharge |

water movements, and (ii) the timing and rates of water movements, are listed and described briefly in Tables 1 and 2.

Movements of groundwaters within the modern timescale can be detected with anthropogenic tracers, such as atmospheric tritium and krypton-85 ($^3H$ and $^{85}Kr$) and industrial chlorofluorocarbons (CFCs). Carbon-14 ($^{14}C$) dating of groundwaters gives useful information for ages up to about $3.5 \times 10^4$ years, i.e. the Holocene and late Pleistocene periods, and is the most prominent geochemical method for dating groundwater evolution through the last glacial period.

Stable oxygen ($^{18}O/^{16}O$) and hydrogen ($^2H/^1H$) isotopic ratios fingerprint water that originated as rainfall or snowfall at different temperatures, and therefore differentiate cold-climate recharge from Holocene and other interglacial recharge waters. The term "cold-climate recharge" describes water originating from any of the glacial periods during the Quaternary, prior to $10^4$ years ago of which the most recent was the late Pleistocene (Weichselian) glaciation. Dissolved atmospheric noble gas concentrations are also dependent on the temperature at recharge and are a complementary tool for tracing climatic conditions at time of recharge.

Chlorine-36 ($^{36}Cl$) and the uranium-thorium (U-Th) isotopic decay series provide the main geochemical tools for detecting groundwater evolution and stability in a timescale up to $2 \times 10^6$ a. However various complexities in interpretation of these isotopic data often decrease their quantitative reliability.

A stable groundwater system in low-permeability fractured rocks will mix by advective dispersion and diffusion so that old groundwater masses may no longer be distinguishable in terms of discrete sources and ages. In such a system, there is usually a pattern of vertical stratification of groundwater masses with the possibility of rapid increases in water age with increasing depth. Vertical segregation of water masses may be controlled by changes in permeability and by the decay of topographic head with increasing depth, and may be reinforced by density stratification where groundwaters have varying salinities.

## 4 ILLUSTRATIVE DATA FROM FRACTURED ROCKS IN BRITAIN AND SWEDEN

### 4.1 *Fractured metavolcanic basement rocks, northwest England*

Boreholes were drilled at Sellafield on the Irish Sea coast of northwest England to investigate groundwater conditions in the fractured metavolcanic basement rock, the Borrowdale Volcanic Group (BVG), which was being investigated as a potential host for a nuclear waste repository until the project was stopped in 1997. Sedimentary rocks, mainly sandstones of the Triassic Sherwood Sandstone Group (SSG), cover the basement over most of the area and are about 500 m thick in the centre of the area. Within the investigated area, the top surface of the basement dips from outcrop in the north-eastern part at 160 m above sea level to more than 1,500 m below sea level at the southwestern corner.

The regional hydraulic gradient is generally directed westwards through the area, from the high ground of inland hills towards the coast. The conceptual model has three main groundwater components which are related to hydrodynamics: a shallow fresh water aquifer which is topographically driven, a coastal and offshore basinal brine, and basement groundwaters which are mixtures of topographically-driven fresh water and deeper saline water, with the possibility of density variations being significant at depth. This area experienced glaciation and permafrost during the Quaternary up to 14 ka ago. This would have affected the groundwater system to some extent and geochemical data from water samples in deep boreholes were interpreted in terms of the perturbation due to glaciation and potential melt water intrusion and the degree of stability that is reflected by the stratification of waters with different origins, salinities and densities (Nirex, 1997; Bath et al., 2006).

Water samples were collected by pumping from 163 tests in discrete intervals in nineteen deep boreholes to a maximum depth of 1,950 m. Overall uncertainties in estimating *in situ* values for $^{3}H$, $^{14}C$ and $\delta^{13}C$ were dominated by the impacts of contamination by drilling fluids. The inorganic carbon system and specifically $^{13}C/^{12}C$ and $^{14}C$ were affected non-linearly by the drilling water additives. In a few cases of very low contamination, upper limits on $^{14}C$ have been inferred that provide minimum model ages.

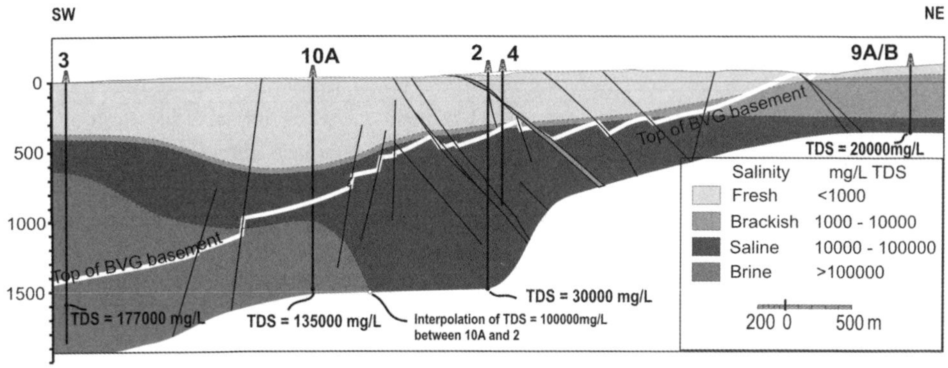


Figure 1. NE-SW section through the Sellafield area, showing some of the deep boreholes and contours of groundwater salinity (as mg/L TDS). The boundary at the base of sedimentary rocks and the top of the BVG basement rocks is shown; major faults are also shown schematically (adapted from Nirex, 1997).

The variations of salinity across the Sellafield area are illustrated in Figure 1. Groundwater compositions are: (i) fresh to brackish waters in relatively shallow groundwaters in the sedimentary aquifer strata (mostly down to at least 300 m depth), (ii) saline waters in BVG basement rocks at the eastern part of the area, and (iii) brines in the deep sedimentary and underlying basement formations at the western edge of the area where the thickness of sedimentary strata increases westwards under the coast.

The gradient with depth of chloride concentrations, $Cl^-$, in groundwaters in the BVG basement rocks in the eastern part of the Sellafield area is compared with the $Cl^-$ gradient in the deep sedimentary formations in western parts of the area in Figure 2. Data from outcropping basement crystalline rocks in southwest England and northern Scotland are also shown. The salinity gradients in basement rocks that outcrop or are overlain by relatively thin sedimentary cover are lower than the gradient in deep sedimentary formations and underlying basement at Sellafield which is similar for other Permo-Triassic basinal sedimentary sequences in the UK.

Salinity in the sedimentary sequence originates from the dissolution of evaporite minerals (mainly halite). The high salinity gradient in Figure 2 reflects equilibrium between the maintenance of high salinities at great depth due to dissolution of halite, upwards mixing of saline water by advective and diffusive mixing, and flushing of saline water by natural groundwater circulation. Therefore the regularity of the salinity gradient is an indication of, amongst other factors, the stability of the large scale regional hydrodynamic conditions. Integrated interpretation of geochemical indicators of stability, water ages and solute residence times suggests that brines in the basinal sedimentary sequence and in the underlying fractured basement rocks underneath the present-day coast at Sellafield are probably between 2 and 10 million years old and that movements of both sub-sea brine and onshore saline groundwater are directed towards the mixing zone and then upwards into shallow groundwaters underneath the coastal area.

Salinity in basement rocks at Sellafield is considered to have originated by downwards movement of dense brines from overlying sedimentary strata in the past, subsequent dilution during circulation of overlying fresh water, and chemical modification by water-rock

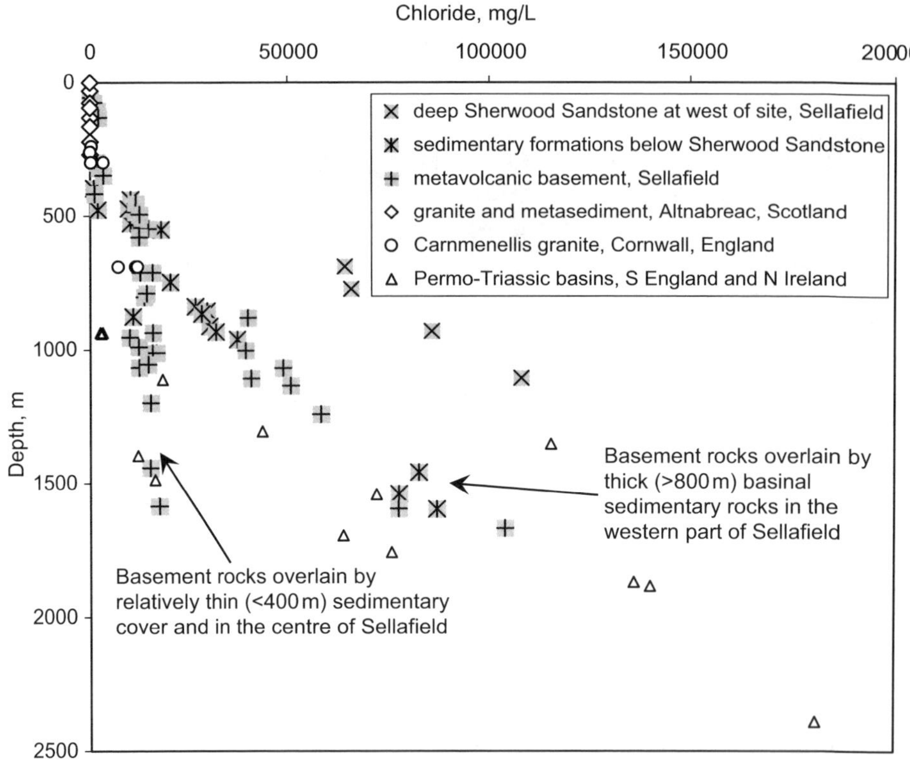


Figure 2. Chloride concentrations versus sample depth for groundwaters sampled in deep boreholes at Sellafield, northwest England; data from boreholes in crystalline rocks and in Permo-Triassic sedimentary basins are also shown. Data from Bath et al. (2006), Kay and Bath (1982), Downing and Gray (1986), Edmunds et al. (1987).

reaction (Bath et al., 2006). The lower salinity gradient in basement rocks in the centre of the Sellafield area, relative to that at comparable depths in basinal sedimentary rocks in the western part of the area, is indicative of a faster rate of groundwater circulation. The main factors which have influenced this apparent difference are the topographic hydrodynamic driving forces on the eastern boundaries of the system and the density-controlled stratification of groundwaters at the western boundary.

The interpretation of groundwater evolution and stability is supported by stable isotope ($^{18}O/^{16}O$ and $^{2}H/^{1}H$) and dissolved noble gases data. These show that all water types are of dominantly meteoric origins and indicate Quaternary ages for the basement saline groundwaters and pre-Quaternary ages for the basinal brine (Figure 3). Fresh-brackish groundwaters in the sedimentary formation in the centre of the investigated area have stable isotopic compositions similar to present-day recharge and are thus post-glacial. Relatively light $^{18}O/^{16}O$ compositions ($-7.5$ to $-7.0$ ‰ $\delta^{18}O$) for fresh-brackish groundwaters occur only down-gradient of the centre of the area at $>500$ m depth and in the up-gradient eastern part of the area at $>200$ m depth. These light isotope compositions indicate water that recharged in cold-climate conditions. Saline groundwaters in the basement formation also have relatively light isotopic compositions ($-8.1$ to $-6.8$ ‰ $\delta^{18}O$), indicating that they have a large component of cold-climate water mixed in with a smaller proportion of a

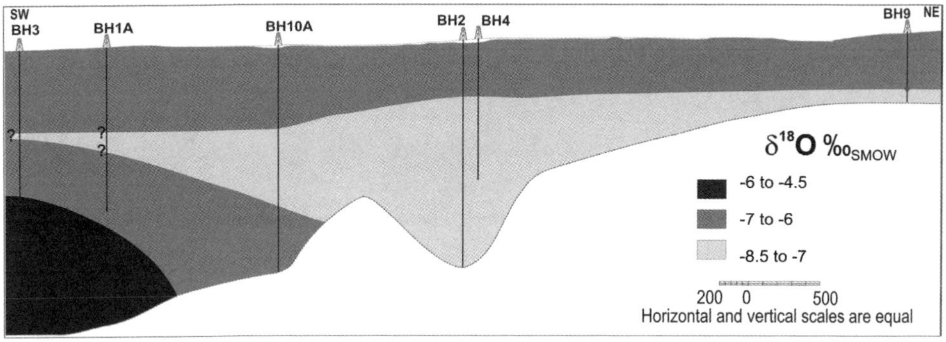


Figure 3. NE-SW section through the Sellafield area, showing contours of stable oxygen isotope compositions. The composition of present-day recharge is estimated to be between −6.7 and −6.0‰ $\delta^{18}O$. Depth scale and geological structure is shown in Figure 1 (adapted from Nirex, 1997).

much older saline water component. These interpretations are supported by noble gas concentrations which indicate average recharge temperatures of 4.8°C for saline groundwaters, in contrast to around 9.0°C for most fresh waters.

Cold-climate water is thought to contain a component of melt water from the ice sheet. It signifies that melt waters penetrated to these depths, i.e. to at least 1,500 m depth in the centre of the Sellafield area, and replaced, either partially or fully, the pre-existing groundwaters. It is likely that the penetration of melt water was promoted by high hydraulic gradients due to the head of the water column in the ice sheet. This would have happened during the several glacial periods in the Quaternary when the area was covered by ice, or perhaps in the periods of much shorter duration when the wet-based front of the ice sheet was over the Sellafield area. It is probable that pure melt waters would have had much lighter isotopic compositions than is seen in the present-day groundwaters, perhaps between −20 and −30 ‰ $\delta^{18}O$, so there has been substantial mixing of water from different sources, either at the time of melt-water intrusion and/or over the subsequent periods of normal hydraulic conditions.

The uniform stable isotope compositions of deep saline groundwaters in basement rocks in the eastern part of the Sellafield area indicate that pre-existing groundwaters were replaced, to the depths investigated, during the Pleistocene glacial period. $^{36}Cl/Cl$ ratios for these saline waters are at or close to secular equilibrium for *in situ* production of $^{36}Cl$ (Metcalfe et al., in press). This indicates that chloride ion has resided in these rocks (or rocks with similar U and Th contents and natural neutron flux) for at least 1.5 million years, in contrast to the water itself which has a late-Pleistocene age. The contrasting residence times of water and Cl suggest that meteoric water became saline after recharge, probably by mixing with relatively minor proportions of ancient saline groundwaters or brines. This process has probably varied since exposure of these rocks to meteoric water influx because of climate effects on the external boundaries. Water movements in the basement rocks have been sufficient to produce the fairly uniform salinity now seen down to at least 1,700 m depth in the centre of the area, but were also low enough, even during the periods of glacial melt water infiltration, not to flush completely the deep source of salinity.

Although a few samples from deeper fresh-brackish groundwaters in the centre of the area were obtained with low levels of contamination, there is still very poor knowledge of

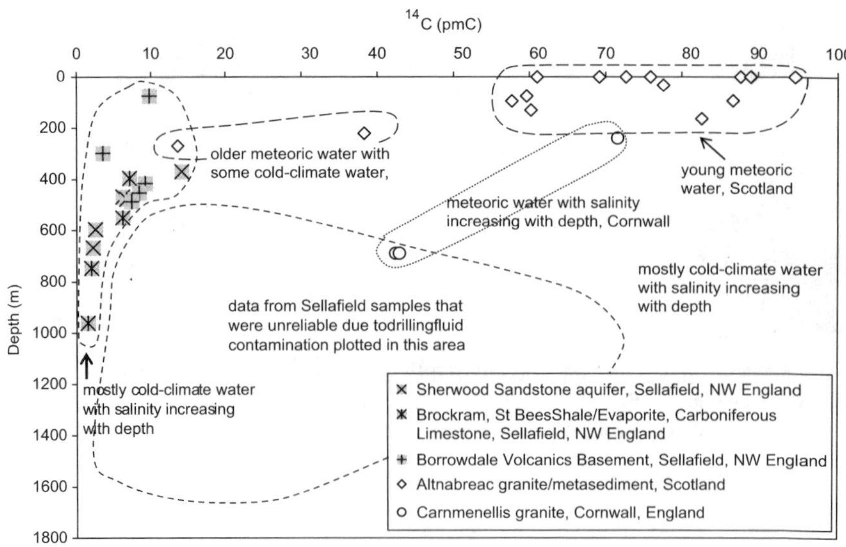


Figure 4. Carbon-14 versus sample depth for groundwaters sampled in deep boreholes at Sellafield, northwest England and from investigations of crystalline rocks in Scotland and southwest England. Data from Bath et al. (2006), Kay and Bath (1985), Edmunds et al. (1987).

*in situ* $^{14}$C contents and $\delta^{13}$C values for Sellafield groundwaters because of large uncertainties due to drilling fluid contamination of water samples (Bath et al., 2006). Data for the samples with lowest contamination are shown in Figure 4 in which data from groundwaters in outcropping crystalline rocks elsewhere in Britain are also shown. $^{14}$C values decrease sharply within the upper 200–300 metres, i.e. groundwaters become much older below this depth range. Model $^{14}$C ages calculated from the Sellafield data are between 0 and 30 ka which are minimum ages only because of the likelihood of extraneous $^{14}$C. These very uncertain interpreted age ranges add little to the qualitative cold-climate water ages that are inferred from $^{18}$O/$^{16}$O data.

$^{36}$Cl/Cl ratios in saline waters and brines across the brine-saline water interface in the western part of the area are not in secular equilibrium with the host sedimentary host formations (Metcalfe et al., in press). This suggests that the interface where mixing occurs of saline basement water with much older brine has moved between the different formations within the last 1.5 million years.

The depth to which modern groundwaters have penetrated in the shallow fresh-brackish groundwaters is uncertain. The high levels of sample contamination by drilling fluid raised the threshold for reliable detection of *in situ* tritium ($^{3}$H). Most of the less contaminated samples from underlying saline groundwaters which were analysed for $^{3}$H gave contamination-corrected values which are zero within the uncertainty range.

## 4.2 *Crystalline Shield rocks, Sweden*

Groundwaters in a small granitic intrusion at Stripa in central Sweden were investigated in an international cooperation project managed by SKB (the Swedish Nuclear Fuel and Waste Management Company) through 1980–90. Iron ore had been mined at Stripa for centuries, so the groundwater system around the mine has been disturbed by drawdown towards the pumped mine workings which are at about 300–400 m depth (Nordstrom et al., 1989). The spatial extent of hydraulic disturbance was estimated to extend to at least 3,000 m depth and several km laterally. The maximum sampled depth was 1,232 m.

Data from the Stripa mine study can be evaluated as geochemical indicators of hydrodynamic stability for which the palaeohydrogeology over the Quaternary timescale would have been affected by fresh water recharge and episodic glaciation. The site is far inland and would not have been influenced by palaeo-Baltic ("Littorina") sea water. Indications of stability are considered in two contexts: (a) the system that has been disturbed by mining, and (b) the natural system, prior to excavation of the mine. Geochemical indicators are potentially useful both for assessing evidence for response to past perturbations of the natural system, especially those due to glaciation, and also for investigating modern disturbance by the mine.

Groundwaters in the Äspö area at the Baltic coast were studied geochemically by SKB prior to excavation of the Hard Rock Laboratory (HRL) in undisturbed crystalline rock (Smellie and Laaksoharju, 1992; Laaksoharju et al., 1999a,b). Variations in the relative elevations of the rock surface and of sea water level during palaeo-Baltic stages might have affected groundwater stability in three ways: (i) changing hydraulic gradients between recharge zones and discharge zones, (ii) changing locations of meteoric water recharge and discharge as the shoreline moved, and (iii) changing pressure gradients due to density variation where sea water intrusion has varied spatially and compositionally over time. The adjacent Laxemar area, which is inland of Äspö and has therefore been less affected by sea water, has also been studied.

Tritium analyses of inflowing water into the Stripa mine suggest that perturbed flow from ground surface to the mine was faster than the 600 year travel time suggested by modelling (Nordstrom et al., 1989). $^3$H concentrations up to 42 TU were reported for some borehole samples from 300 m depth – suggesting transit time of water from the surface is no greater than 20 years with remarkably little dispersion or dilution by mixing (Moser et al., 1989). Other borehole samples contained insignificant $^3$H which indicates that connectivity of the fracture network is heterogeneous. Highly connected pathways which transmit water relatively rapidly when a high hydraulic gradient is imposed along them have similarly been identified from $^3$H contents of inflows in Shield mines in Canada (Clark et al., 1999; Douglas et al., 2000). When realistic analytical uncertainties are taken into account, all $^3$H data for water samples from below 50 m depth in the undisturbed system at Äspö are negligible. Reported $^3$H values in water samples from boreholes from the surface or from the HRL tunnel at depths to at least 70 m were up to 60 TU and therefore indicate significant drawdown of young recharge.

Dilution of brackish groundwaters due to progressive drawdown of fresher shallow groundwater after excavation has been spatially and temporally heterogeneous at both Stripa and Äspö. Fairly complex variations in $Cl^-$ concentrations are evident in deep borehole monitoring at Äspö (Rhén and Smellie, 2003) suggesting that draw-up of more saline groundwaters from greater depth has been superimposed on draw-down of more dilute groundwaters.

Groundwater samples from boreholes drilled from the Stripa mine were generally dilute (maximum 200 mg/L $Cl^-$) down to about 700 m depth (Figure 5). The maximum salinity of ~700 mg/L TDS was found at an intermediate depth, 810–910 m. $Cl^-$ concentrations were found to be rather heterogeneous, within the range 0–200 mg/L, over short distances,

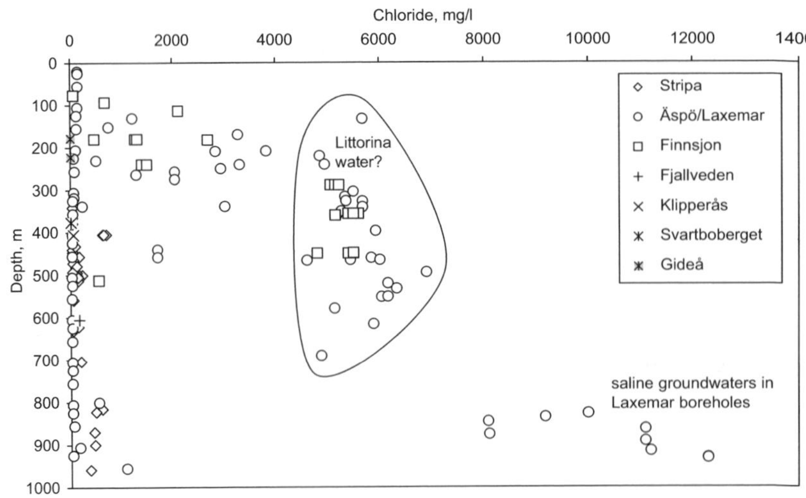


Figure 5. Chloride concentrations versus depth for groundwaters at <1000 m depth from sites in Sweden (note that deeper more saline groundwaters at Laxemar, ca. 47,000 mg/L $Cl^-$ at 1600 m depth, are not shown in this diagram). Stripa, Klipperås and Svartboberget are inland sites; Äspö and other sites may have had varying palaeohydrogeological influences of sea water. Littorina is a post-glacial stage of the palaeo-Baltic sea which was more saline than present Baltic water. Data from Smellie et al. (1985), Smellie and Wikberg (1991), Smellie and Laaksoharju (1992), Smellie et al. (1995), Laaksoharju et al. (1998).

showing that different flowpaths have been activated by draw-down towards the mine excavations. Monitoring of discrete packered intervals in one of the deeper boreholes showed that $Cl^-$ was increasing slightly with time, suggesting a continuing upconing of saline water. In contrast, $Cl^-$ decreased slightly in a short borehole from the mine workings, coupled with a rise of $^3H$, indicating flushing by drawn-down fresh water.

Variations of $Cl^-$ with depth in Äspö/Laxemar groundwaters are not associated with simple linear mixing of $\delta^{18}O$, indicating that these are mixtures of several end-members with varying $Cl^-$ and $\delta^{18}O$. However $\delta^{18}O$ is correlated with $Cl^-$ in the saline waters at depth at Laxemar, indicating that these are dominated by mixing of two components – glacial meteoric water and deep saline "Shield" water.

$\delta^{18}O$ values of $^3H$-containing groundwaters at Stripa that had penetrated to at least 300–400 m depth were $-10.5$ to $-12$ ‰ (Moser et al., 1989) which is typical of modern recharge. $\delta^{18}O$ values were found to be $-12$ to $-13.4$ ‰ in inflows at >400 m depth, i.e. lower by ~1–2 ‰, and also to have near-zero values of $^3H$. It can be concluded that low $\delta^{18}O$ values here indicate a cold-climate origin, i.e. >10,000 years old and probably glacial melt water, whilst $\delta^{18}O$ variations at shallower depths may simply be fluctuations in recent rainfall or could indicate pockets of water that are mixtures of components with different ages.

$\delta^{18}O$ and $\delta^2H$ of groundwaters at Äspö were found to have a wide range (e.g., $-7$ to $-16$ ‰ $\delta^{18}O$), indicating the presence of meteoric water from modern and cold-climate recharge, Baltic sea water and palaeo-Baltic precursors (e.g., Littorina), and ancient water from uncertain sources. Present-day Baltic Sea has around 3550–3800 mg/L Cl and $-5.9$ ‰ $\delta^{18}O$. The highest $\delta^{18}O$ values were found in fresh/brackish water at 0–200 m depth

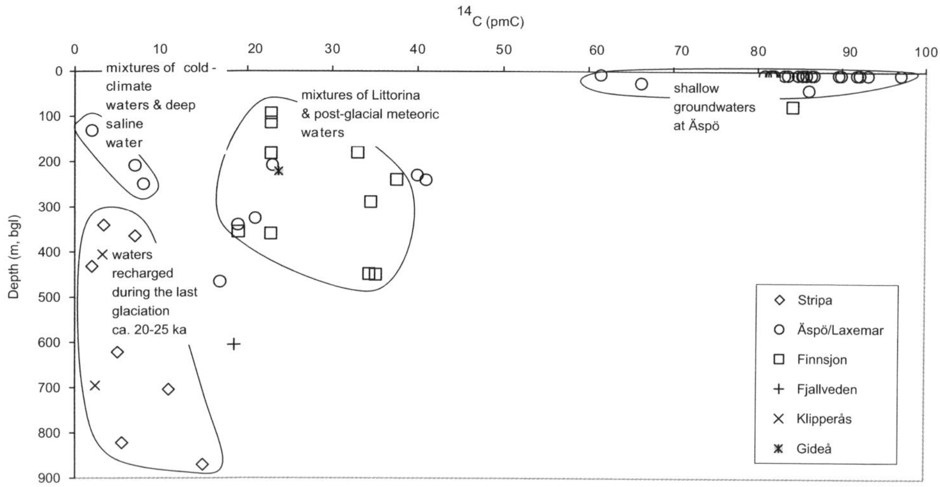


Figure 6. Carbon-14 data versus depth for groundwaters from sites in Sweden. Data from Smellie et al. (1985), Fritz et al. (1989), Smellie and Laaksoharju (1992), Laaksoharju et al. (1998), Tullborg and Gustafsson (1999).

(Laaksoharju et al., 1999b). The lowest values were found in localised pockets at 100–200 m depth in brackish waters which thus have a large component of cold-climate recharge, mixed with brackish water. That these pockets of water have not been replaced by more recent meteoric recharge implies compartmentalisation in the groundwater system and stability with respect to present-day boundary conditions.

Interpreted $^{14}C$ ages of deep groundwaters below 400–500 m depth at Stripa were reported to be 20,000–25,000 yrs, predating the end of the last glaciation, though the corrections using $\delta^{13}C$ to obtain these ages had large uncertainties (Figure 6; Fritz et al., 1989). These are slightly longer than the modelled hydrodynamic travel time of around 10,000 years from recharge. The ages, if correct and not much older, suggest that the meteoric water component to at least 800 m depth was renewed during the last glacial cycle and that the system has since been sufficiently stable to preserve these deep groundwaters.

The ages of the older meteoric water components at Äspö confirm the origin of the isotopically-light cold-climate waters. Pre-tunnelling Äspö data suggest that $^{14}C$ values in groundwaters deeper than 100 m are variably low (<40 pmC) and sometimes very low (<10 pmC) (Figure 6). One of the localised pockets of brackish water with light $\delta^{18}O$ (−15 ‰) has <8 pmC $^{14}C$. This indicates a pocket of glacial or cold-climate water mixed with brackish Littorina (palaeo-Baltic) water and is further evidence that the groundwater system is hydrodynamically compartmentalised.

$^{36}Cl$ contents in groundwaters at Stripa are derived mainly from *in situ* radiogenic production and the $^{36}Cl/Cl$ ratios at secular equilibrium were expected to be distinct in the granite and in surrounding metasediments because of their different U contents (Andrews et al., 1989). However $^{36}Cl$ data indicate that the $Cl^-$ has not been in the granite for long enough for secular equilibrium to be established and that flow times for movement of water into the granite formation are therefore less than 1.5 Ma, confirming that significant movement of these deep groundwaters has occurred within the Quaternary timescale (Andrews et al., 1989).

$^{36}Cl/Cl$ values for fresh/brackish water at 200–300 m depth at Äspö/Laxemar were 20–24 $\times$ $10^{-15}$ and were 40–43 $\times$ $10^{-15}$ in saline waters at 860 and 1,420 m depths (Louvat et al., 1999). The latter value in is close to that estimated for secular equilibrium and indicates, unsurprisingly, that the deep Shield saline water has been stable for at least 1.5 Ma.

These data provide only limited scope for comparison of groundwater evolution and stability between inland and coastal areas, because of the patchiness of representative data. The external changes that might have perturbed stable groundwater conditions in the past are glaciation (i.e., melt water, mechanical loading and permafrost) and varying sea water infiltration due to changes in palaeo-Baltic and isostatic conditions. Sea water infiltration is obviously only a potential influence in coastal areas. The distributions of palaeo-Baltic Littorina sea water in groundwaters at coastal sites reflect local palaeohydrogeological conditions. It is likely that inland areas have had longer durations of post-glacial fresh water infiltration than coastal areas, causing greater degrees of dilution and dispersion of pre-existing groundwaters and thus overprinting their hydrochemical and isotopic characteristics. Lower post-glacial hydraulic gradients relative to inland sites may account for the occurrence of more relict cold-climate water at coastal sites.

Some inferences about hydrodynamic stability can be made, but the strength of the evidence is rather variable. Firstly, it seems that glacial melt water penetrated many hundreds of metres and in some places to at least 1,000 m depth. However the low remaining proportions of melt water and of much older saline Shield water suggest that melt water flux did not fully displace pre-existing groundwaters at these depths. Secondly, where there has been post-glacial infiltration of palaeo-Baltic sea water, the density stratification or compartmentalisation effect coupled with low hydraulic gradient has reduced rates of subsequent fresh water circulation after shoreline recession. These tentative conclusions concerning degrees of relative stability or instability in various palaeohydrogeological environments remain as hypotheses that could be further tested with additional hydrochemical and isotopic investigations.

## 5 CONCLUDING SUMMARY

Geochemical data are valuable indicators of the evolution of groundwaters in low permeability fractured rocks over short and long timescales. Interpretations of stability focus on the palaeohydrogeological responses to external perturbations which have been superimposed on the normal evolution of the groundwater system. This is not a straightforward interpretation because the geochemical signals have been dispersed and diluted to varying degrees by subsequent processes.

Example sets of hydrochemical and isotopic data from groundwaters in British and Swedish fractured rocks illustrate the approach and problems of interpreting data in terms of palaeohydrogeological evolution and stability. In both cases, the greatest external perturbation in the longer timescale was glaciation and potential melt water infiltration. The other major influence on evolution and stability in many of the sites is the distribution of saline groundwaters with varying densities. Their marine, basinal or basement origins are characterised hydrochemically and isotopically. Ages of the various groundwater components are important for distinguishing the impacts of episodic perturbations from normal groundwater evolution.

There are various uncertainties in interpreting these geochemical indicators of evolution and stability in terms of the depths to which glacial melt waters have circulated and the degree to which they replace pre-existing groundwaters, and in terms of comparisons between inland and coastal groundwater systems. Uncertainties derive partly from the reliability of groundwater samples as being representative of *in situ* conditions, and partly from the non-uniqueness of interpretative models. Future investigations using these approaches should aim to improve sampling, to make conjunctive use of geochemical and isotopic indicators which have varying timescales and sensitivities, and to integrate these indicators with palaeohydrogeological modelling.

## ACKNOWLEDGEMENTS

Two reviewers are thanked for their constructive comments on the original manuscript which have resulted in significant improvements to this paper.

## REFERENCES

Andrews JN, Davis SN, Fabryka-Martin J, Fontes J-Ch, Lehmann BE, Loosli HH, Michelot J-L, Moser H, Smith B, Wolf M (1989) The in situ production of radioisotopes in rock matrices with particular reference to the Stripa granite. Geochim Cosmochim Acta 53:1803–1815.

Bath A, Strömberg B (2004) Geochemical indicators of groundwater stability. Proceedings of MRS 2003, Scientific Basis for Radioactive Waste Management XXVII, Kalmar, Sweden, June 2003. Mat Res Soc Symp Proc Vol 807. Materials Research Society, Boston, USA.

Bath A, Milodowski A, Ruotsalainen P, Tullborg E-L, Cortés Ruiz A, Aranyossy JF (2000) Evidence from mineralogy and geochemistry for the evolution of groundwater system during the Quaternary for use in radioactive waste repository safety assessment (EQUIP project). EC Report EUR 19613.

Bath A, Richards H, Metcalfe R, McCartney R, Degnan P, Littleboy A (2006) Geochemical Indicators of Deep Groundwater Movements at Sellafield, UK. J. Geochemical Exploration 90:24–44.

Clark I, Douglas M, Raven K, Bottomley D (1999) Tracing surface-to-depth flow and mixing at the Con Mine, Yellowknife, Canada: an analogue for the hydrogeology of radioactive waste repositories. Proc Intl Conf on Study of Environmental Change Using Isotope Techniques. IAEA-SM-361/42, 92–93. IAEA Vienna.

Degnan P, Bath A, Cortés A, Delgado J, Haszeldine S, Milodowski A, Puigdomenech I, Recreo F, Šilar J, Torres T, Tullborg, E-L. (2005) PADAMOT: Overview Report. EU 5th Framework Project "Palaeohydrogeological Data Analysis and Model Testing". UK Nirex Ltd., Harwell. 105 pp.

Douglas M, Clark ID, Raven K, Bottomley D (2000) Groundwater mixing dynamics at a Canadian Shield mine. J Hydrol, 235:88–103.

Downing RA, Gray DA (1986) Geothermal Energy – The Potential in the United Kingdom. Report of the British Geological Survey.

Edmunds WM, Kay RLF, Miles DL, Cook JM (1987) The origin of saline groundwaters in the Carnmenellis Granite (UK): further evidence from minor and trace elements. In: Saline Groundwaters and Gases in Crystalline Rocks (P Fritz and S K Frape, eds.). Geol. Assoc. Canada Spec. Paper 33:127–143.

Fritz P, Fontes J-Ch, Frape SK, Louvat D, Michelot J-L, Balderer W (1989) The isotope geochemistry of carbon in groundwater at Stripa. Geochim Cosmochim Acta 53:1765–1775.

Hartley LJ, Hoch AR, Hunter FMI, Jackson CP, Marsic N (2005) Regional hydrogeological simulations – Numerical modelling using CONNECTFLOW. Preliminary site description Simpevarp area – version 1.2. SKB Report R-05-12, Swedish Nuclear Fuel and Waste Management Co.

Kay RLF, Bath AH (1983) Groundwater geochemical studies at the Altnabreac research site. Rep. Inst. Geol. Sci. ENPU 82-12. British Geological Survey, Keyworth.

Laaksoharju M, Gurban I, Skårman C (1998) Summary hydrochemical conditions at Aberg, Beberg and Ceberg. SKB Report TR-98-03. Swedish Nuclear Fuel and Waste Management Co.

Laaksoharju M, Skårman C, Skårman E (1999a) Multivariate mixing and mass balance (M3) calculations, a new tool for decoding hydrogeochemical information. Appl Geochem 14:861–871.

Laaksoharju M, Tullborg E-L, Wikberg P, Wallin B, Smellie J (1999b) Hydrogeochemical conditions and evolution at the Äspö HRL, Sweden. Appl Geochem 14:835–839.

Louvat D, Michelot J-L, Aranyossy J-F (1999) Origin and residence time of salinity in the Äspö groundwater system. Appl Geochem 14:917–925.

Metcalfe R, Crawford MB, Bath AH, Richards HG (in press) Characteristics of deep groundwater flow in a basin marginal setting at Sellafield, northwest England: $^{36}Cl$ and halide evidence. Appl Geochem.

Moser H, Wolf M, Fritz P, Fontes J-Ch, Florkowski T, Payne, BR (1989) Deuterium, oxygen-18, and tritium in Stripa groundwater. Geochim Cosmochim Acta 53:1757–1763.

Nirex (1997) The Hydrochemistry of Sellafield: 1997 Update. Report SA/97/089. UK Nirex, Harwell.

Nordstrom DK, Olsson T, Carlsson L, Fritz P (1989) Introduction to the hydrochemical investigations within the International Stripa Project. Geochim Cosmochim Acta 53:1717–1726.

Rhén I, Smellie J (2003) Task force on modelling of groundwater flow and transport of solutes. Task 5 Summary Report. SKB Report TR-03-01.

Smellie J, Laaksoharju M (1992) The Äspö Hard Rock Laboratory: Final evaluation of the hydrogeochemical pre-investigations in relation to existing geologic and hydraulic conditions. SKB Report TR-92-31, Swedish Nuclear Fuel and Waste Management Co.

Smellie JAT, Wikberg P (1991) Hydrochemical investigations at Finnsjön, Sweden. J Hydrol 126:129–158.

Smellie JAT, Laaksoharju M, Wikberg P (1995) Äspö, SE Sweden: a natural groundwater flow model derived from hydrogeochemical observations. J Hydrol 172:147–169.

Smellie JAT, Larsson NA, Wikberg P, Carlsson L (1985) Hydrochemical Investigations in Crystalline Bedrock in Relation to the Existing Hydraulic Conditions. SKB Report TR-85-11.

Tullborg E-L, Gustafsson E (1999) $^{14}C$ in bicarbonate and dissolved organics – a useful tracer? Appl Geochem 14:927–938.

CHAPTER 18

# Universal controls on the evolution of groundwater chemistry in shallow crystalline rock aquifers: the evidence from empirical and theoretical studies

Bjørn Frengstad[1] and David Banks[2]
[1]*Norges geologiske undersøkelse, Trondheim, Norway*
[2]*Holymoor Consultancy, Brampton, Chesterfield, Derbyshire, UK*

ABSTRACT: Large empirical field studies of groundwater chemistry in crystalline rock aquifers have been carried out in Norway to ascertain how (and, indeed, whether) groundwater chemistry depends on lithology. A number of dissolved elements, such as radon, uranium and fluoride, exhibit a lithological dependence. However, the distribution of pH and several major ion components in groundwater can largely be explained in terms of feldspar hydrolysis and secondary calcite saturation / precipitation. We contend that the main trends in groundwater chemical composition in most crystalline silicate rock aquifers are *not* strongly dependent on lithology, but are dependent on five more universal factors; (i) the initial $PCO_2$ of the recharge water, (ii) the degree to which the aquifer geochemical system is "open" or "closed" with respect to $CO_2$, (iii) the availability and composition of hydrolysable silicate phases, (iv) the degree to which $CO_2$ has been consumed by plagioclase hydrolysis and (v) the extent to which feldspar hydrolysis has continued beyond the point of calcite saturation.

## 1 INTRODUCTION

Those great pioneers in the field of hydrogeochemistry, Garrels (1967) and Garrels and MacKenzie (1967) asserted that groundwater chemical composition in many crystalline silicate aquifers is dominated by plagioclase weathering. They formulated the bold and surprising hypothesis that groundwater composition is governed largely by the initial carbon dioxide content of the recharge water, by the degree to which plagioclase weathering has consumed available dissolved carbon dioxide by hydrolysis and by the composition of the plagioclase mineralogy.

A number of local and regional surveys of the hydrochemistry of crystalline rock groundwaters in Norway during the 1990's provided the opportunity to test Garrels and Mackenzie's hypotheses. Moreover, they afforded the chance to assess the health significance of certain dissolved parameters in groundwater from such lithologies: particularly fluoride, sodium, radon and uranium (Bjorvatn et al. 1992, Banks et al. 1995a,b, Sæther et al. 1995, Reimann et al. 1996, Morland et al. 1997, Banks et al. 1998a, Bårdsen et al. 1999).

The first part of this paper examines the data generated by the largest of these Norwegian regional groundwater surveys: the 1996–97 national survey of the hydrochemistry of crystalline rock groundwaters, carried out by the Geological Survey of Norway (NGU), in cooperation with the Norwegian Radiation Protection Authority (NRPA) and the Norwegian University of Science and Technology (NTNU), and reported by Banks et al. (1998a,b,c) and Frengstad et al. (2000, 2001). This part of the paper examines the distribution of solutes in Norwegian crystalline rock groundwaters and addresses the question of whether such groundwaters have an overall natural chemical quality suitable for human consumption.

The second part of the paper returns to Garrels' hypothesis. It attempts to ascertain whether the empirical datasets offer support for his simple hydrochemical evolutionary models and also to identify the main controls on groundwater chemistry in silicate lithologies.

## 2 GEOLOGICAL AND HYDROGEOLOGICAL SETTING

The mainland of Norway consists almost entirely of Precambrian and Palaeozoic crystalline bedrock. During the Quaternary, recurrent glaciations scoured away younger sedimentary rocks and weathered bedrock and left a dramatic topography of relatively unweathered, fresh rock, sometimes with a thin and discontinuous cover of glacial till and peat deposits. The terrain is incised with deep valleys and fjords, the former often partially infilled with glaciofluvial or alluvial sediments. Holocene uplift due to isostatic rebound has resulted in an upper postglacial marine limit between 0 and 221 m above the present sea level. Below this marine limit, marine clays may occur as a Quaternary cover, overlying the silicate-dominated bedrock.

Because arable soils are scarce in Norway, farmers have traditionally settled sparsely. Drilled wells in bedrock are therefore often the only economically viable drinking water supply in many parts of the country. More than 140,000 bedrock wells have been drilled in Norway and around 4,000 new boreholes are added to this total each year. The boreholes typically yield less than 1000 L/hr (median yield = 600 L/hr and median depth 56 m, based on a database of 12,757 boreholes: Morland 1997; Banks et al. 2005), and serve single households and farms, or groups thereof, in rural areas. It is estimated that 6% of the Norwegian population, or around 300,000 persons, are supplied by groundwater from crystalline rock wells. However, only one waterworks serving more than 1000 people is based on a crystalline rock aquifer.

## 3 THE 1996–97 SURVEY OF GROUNDWATER CHEMISTRY IN CRYSTALLINE ROCK AQUIFERS

During the 1996–97 national survey, around 2000 water samples were collected by local health authority representatives, following an open invitation to well-owners to have their drinking water sampled. Figure 1 shows the geographical distribution of the sampled wells and boreholes in South Norway, which in turn reflects the patterns of rural settlement (only a few samples were collected from Northern Norway due to the limited use of crystalline rock groundwater in this region). For the purpose of hydrochemical sampling, 500-ml polyethene flasks were filled at the well head or from a domestic tap via a sealed pressure tank. To minimise the risk of diverging sampling techniques it was decided not to filter the water samples. Separation of the sample into sub-aliquots and acid preservation was performed according to a strict protocol on arrival of each sample at NGU's laboratory. The

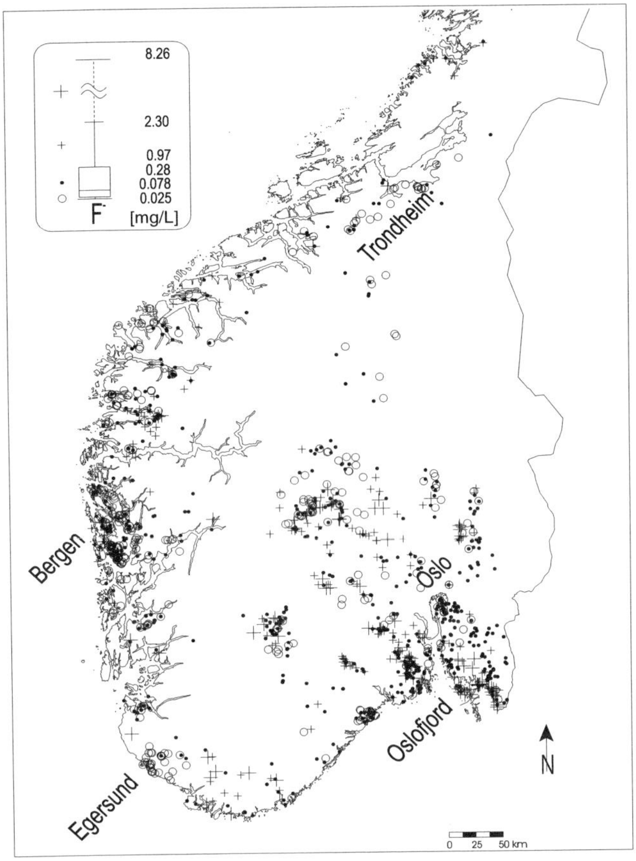


Figure 1. Map of South Norway showing the location of 1328 boreholes with coordinates and the distribution of fluoride in crystalline rock groundwater. High fluoride levels are especially found around the Oslo fjord, south of Bergen and in the southernmost part of Norway.

lack of field filtration implies that the samples thus represent the water as it is drunk. In fact, investigations by Reimann et al. (1999) and Frengstad (2002) have shown that filtration has very little effect on analyses of Norwegian bedrock groundwaters, probably due to their naturally very low particulate content. The samples were analysed at NGU's laboratory for 30 metals by Inductively Coupled Plasma Atomic Emission Spectrometry (ICP-AES), 7 anions by Ion Chromatography (IC), alkalinity by titration to a bicarbonate end- point and pH by electrode. Radon was field-sampled in 20 ml glass vials pre-filled with 10 ml Lumagel scintillation liquid and analysed after minimal delay by scintillation counting at NRPA's laboratory.

The dataset resulting from sample analysis was intensively quality controlled to remove all samples which were not *bona fide* crystalline rock groundwaters and to remove all samples which may have been subject to water treatment or which had an unacceptably high particulate content. This generated a final dataset of 1604 analyses, which will henceforth be referred to as "Rock_corr". A more lithologically representative subset of 476 samples was sent to the Federal Institute of Geosciences and Natural Resources in Hannover for ICP-MS analysis of 70 elements. Further details on the materials and methods, as well as on handling of data, can be found in Banks et al. (1998a,c), Frengstad et al. (2000) and Frengstad (2002). The results were interrogated by non-parametric statistical methods and presented graphically by the use of the programme code DAS (Dutter et al. 1992).

## 4 THE DISTRIBUTION OF HYDROCHEMICAL PARAMETERS IN CRYSTALLINE ROCK GROUNDWATER

A major report by Banks et al. (1998c) and a series of papers by Banks et al. (1998a,b) and Frengstad et al. (2000, 2001) document the distribution, according to various lithological and petrological groupings, of the various analysed hydrochemical parameters in the dataset resulting from the 1996–97 survey. In this paper, we merely present a small selection of parameters (Figures 2 and 3) that are of health significance or which are important for understanding hydrochemical evolution in crystalline rock aquifers. Most of the sampled groundwaters have low TDS with a median of 226 mg/L and less than 1% of the samples display TDS above 1000 mg/L. Table 1 compares the analytical results with Norwegian drinking water norms (which are typically based on European Union norms), or with other relevant norms.

Figure 2a shows the distribution of fluoride (determined by IC) in the 1604 groundwaters of "Rock_corr", both in the form of a cumulative frequency distribution and as boxplots. The figure also shows distributions within a selection of lithological sub-groupings (Table 2). The number codes of the groupings are based on the legend to bedrock map of Norway at scale 1:3,000,000 (Sigmond 1992). Although fluorine is an essential element for the healthy development of teeth and bones; excessive intake (often deemed to be >1.5 mg/L fluoride in drinking water) may harm the teeth during formation. Elevated fluoride concentrations were mainly found in groundwater derived from granites and leucocratic gneissic rocks (Figure 2a), with very low concentrations being reported in anorthosites. The element also shows some positive correlation with pH, which will be considered later in this paper. In all 16.1% of the 1604 samples exceeded the 1.5 mg/L norm (rising to over 50% in lithology 92: Precambrian granites).

Radon (Figure 2b) is a radioactive gas that may result in lung cancer through exposure during inhalation. Its solubility in water is high, but temperature-dependent. Radon can enter Norwegian dwellings in groundwater, especially given that most Norwegian domestic groundwater pumping systems are sealed and pressurised, allowing no opportunity for external degassing. The radon may then subsequently degas to the inside atmosphere via showers and washing machines, and may not easily escape if the house is well-insulated against the Scandinavian winter. As for fluoride, acidic lithologies such as granites and gneisses (as well as black shales) contain groundwater with the highest Rn concentrations. In fact, some 13.9% of 1601 analysed wells had radon concentrations exceeding the Norwegian guideline level of 500 Bq/L (rising to over 50% in lithology 92: Precambrian granites). Concentrations as high as 31,900 Bq/L have been recorded in the Precambrian Iddefjord Granite of south-east Norway (Banks et al. 2005). Fortunately, radon can be readily removed from water by aeration and storage prior to entry to the distribution system.

The solubility of uranium (Figure 2c) is low in reducing environments but, in most shallow oxidising groundwater environments, it is readily soluble. The element exhibits an extraordinary wide range of concentrations in the Norwegian groundwater dataset and spans more than six orders of magnitude. The toxicity of natural uranium is rather

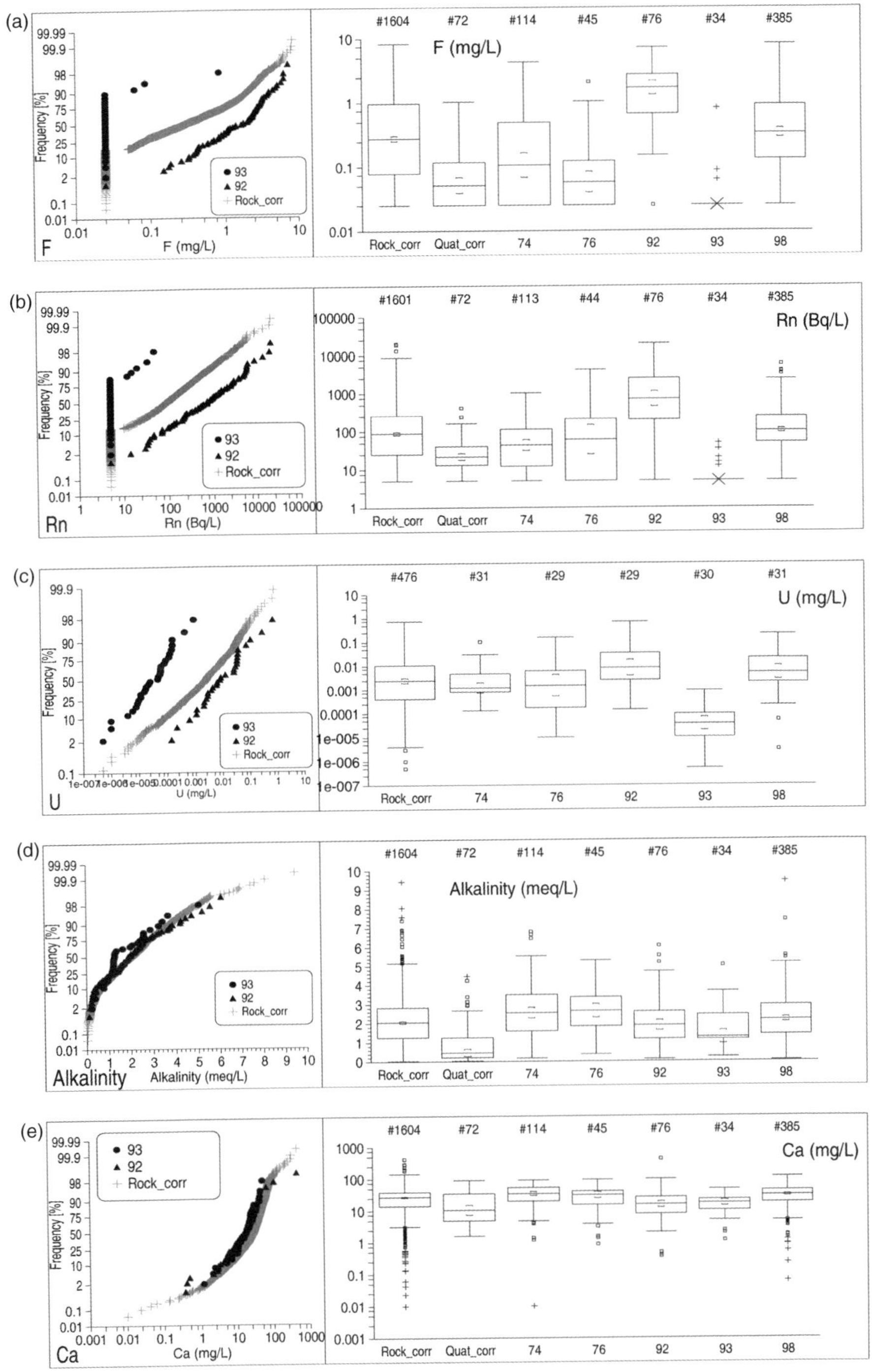


Figure 2. Cumulative frequency distributions (left) and boxplots (right), illustrating the distribution of (a) fluoride, (b) radon, (c) uranium, (d) alkalinity, and (e) calcium in crystalline rock groundwaters ("Rock_Corr", $n$ = 1604), various lithological subsets thereof (subsets 74, 76, 92, 93 and 98; see Table 2) and superficial Quaternary aquifers ("Quat_corr", $n$ = 72) in Norway. Note the probability scale on the y-axis of the cumulative frequency plots. Concentrations less than detection limit are plotted at a value of ½ x detection limit.

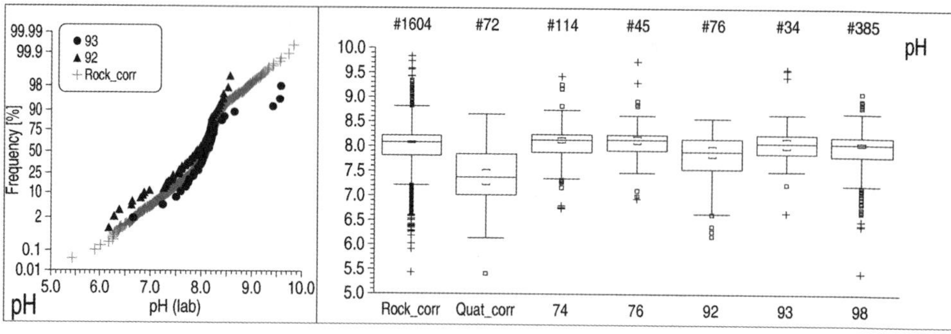


Figure 3. Cumulative frequency distribution (left) and boxplots (right), illustrating the distribution of pH (determined in the laboratory) in crystalline rock groundwaters ("Rock_Corr", $n = 1604$), various lithological subsets thereof (see Table 2) and superficial Quaternary aquifers ("Quat_corr", $n = 72$) in Norway. Note the flexure on the cumulative frequency distribution at approximately pH = 8.0–8.2 and the probability scale on the y-axis.

Table 1. Summary of proportion of groundwater samples exceeding Norwegian (when given) American or WHO drinking water norms for parameters of health-related or aesthetic significance (after Banks et al. 1998a and Frengstad et al. 2000).

| Parameter | Analytical technique | Norm | Foot-note | Violations |
|---|---|---|---|---|
| pH | pH electrode | <9.5 | a | 4/1604 (0.3%) |
| | | >6.5 | a | 18/1604 (1.1%) |
| Radon | Scintillation counting | <500 Bq/L | a | 222/1601 (13.9%) |
| Fluoride | IC | <1.5 mg/L | a | 258/1604 (16.1%) |
| Sodium | ICP-AES | <200 mg/L | a | 26/1604 (1.6%) |
| Iron | ICP-AES | <200 µg/L | a | 274/1604 (17.1%) |
| Manganese | ICP-AES | <50 µg/L | a | 445/1604 (27.7%) |
| Aluminium | ICP-AES | <200 µg/L | a | 70/1604 (4.4%) |
| Nitrate | IC | <44 mg/L | a | 13/1604 (0.8%) |
| Silver | ICP-MS | <10 µg/L | a | 0/476 |
| Arsenic | ICP-MS | <10 µg/L | a | 7/476 (1.5%) |
| Boron | ICP-MS | <1 mg/L | a | 0/476 |
| Beryllium | ICP-MS | <4 µg/L | b | 1/476 (0.2%) |
| Cadmium | ICP-MS | <5 µg/L | a | 1/476 (0.2%) |
| Chromium | ICP-MS | <50 µg/L | a | 0/476 |
| Copper | ICP-MS | <1 mg/L | a | 0/476 |
| Mercury | ICP-MS | <0.5 µg/L | a | 0/476 |
| Nickel | ICP-MS | <20 µg/L | a | 7/476 (1.5%) |
| Lead | ICP-MS | <20 µg/L | a | 1/476 (0.2%) |
| Antimony | ICP-MS | <10 µg/L | a | 0/476 |
| Selenium | ICP-MS | <10 µg/L | a | 1/476 (0.2%) |
| Thallium | ICP-MS | <2 µg/L | b | 0/476 |
| Uranium (US) | ICP-MS | <30 µg/L | c | 58/476 (12%) |
| Uranium (WHO) | ICP-MS | <15 µg/L | d | 100/476 (21%) |
| Zinc | ICP-MS | <300 µg/L | a | 14/476 (3%) |

a = Highest (or lowest) permitted concentration (Helse- og omsorgsdepartementet 2001)
b = American maximum allowable concentration (USEPA 2001)
c = American maximum allowable concentration (USEPA 2000)
d = World Health Organisation recommended guidance level (WHO 2004)

Table 2. Description of datasets depicted in Figures 2, 4 and 8.

| Dataset | Description |
|---|---|
| "Rock_corr" | Typically 1604 samples from all crystalline rock lithologies. The full, quality-controlled dataset resulting from the 1996–97 Norwegian Survey. |
| "Quat_corr" | 72 samples collected from wells in Quaternary sedimentary aquifers (described by Banks et al. 1998b). |
| "Egersund" | A limited set of samples collected by NGU staff in May 1997 from the Egersund anorthosite complex. Sampling was to a high, quality-controlled standard. Reported in full by Frengstad (2002) and Banks & Frengstad (2006). |
| 74 | Metasedimentary rocks of Cambro-Silurian age (Caledonian orogenic belt and Oslo-graben region). |
| 76 | Greenstones, greenschists, amphibolites and meta-andesites of Cambro-Silurian age. |
| 92 | Autochthonous Precambrian granites to tonalites. |
| 93 | Autochthonous Precambrian charnockites to anorthosites. |
| 98 | Precambrian gneiss, migmatite, foliated granites and amphibolite. |

The datasets 74, 76, 92, 93 and 98 are subsets of "Rock_corr" while "Quat_corr" and "Egersund" are independent datasets.

connected to its chemistry than to its radioactivity. No limits are set for uranium in drinking water by the EU although it is recognised that the element may cause kidney damage. The United States operates a maximum admissible concentration (MAC) of 30 μg/L (USEPA 2000) while WHO (2004) has recently suggested a guidance level of 15 μg/L. Compared to these norms, uranium would appear to pose a significant potential problem in crystalline rock boreholes in Norway.

It will have become apparent to the reader that lithologies 92 and 93 (Precambrian granites and anorthosites, respectively) offer hydrogeochemical end-members by which one can judge the degree of lithological dependence of concentrations of various parameters in groundwater. In fact, the following elements all show a tendency towards enrichment in granites and depletion in anorthosites compared to the dataset as a whole: Be, (Bi), Cd, Ce, F, Hf, (Hg), (In), La, (Li), (Mo), Nb, Pb, Rn, Ta, Th, (Ti), Tl, U, Y, Zr and rare earth elements.

The trace elements named above seem to show some degree of lithological dependence in groundwater. However, surprisingly little difference between various lithologies was observed for the distributions in groundwater of major ions such as alkalinity (reflecting bicarbonate and carbonate concentrations), calcium, magnesium and non-marine sodium (lithogenic sodium, i.e. sodium corrected for marine salts on the basis of chloride concentrations. Chloride itself exhibits a degree of inverse correlation with distance from the coast). Figures 2d,e show the distributions of calcium and alkalinity for example. Figure 3 exhibits the distribution of pH. The median pH of the "Rock_corr" dataset (n = 1604) is 8.07. It will be seen that, within each lithology, a considerable range in pH is observed. However, the range is very similar for every lithology, with a tendency for the median pH to be observed at around 8.0 to 8.2. The cumulative frequency distribution curve in Figure 3 exhibits a flexure at around this pH value and, indeed, around half of the samples fall within the pH interval of 8.0–8.3. In superficial Quaternary aquifers ("Quat_corr", Figure 3), the groundwater pH seems generally to be around one unit lower than pH in crystalline rock aquifers (Banks et al. 1998b).

## 5 THE RELATIONSHIP BETWEEN GROUNDWATER PH AND SOLUTE CONCENTRATIONS

The solubility of many elements in the groundwater is partly controlled by pH. The size of the "Rock_corr" dataset allowed us to study the correlation between element concentrations and pH. However, techniques involving xy-plots and regression analysis were difficult to interpret, due to clustering of pH values in the 8.0–8.3 interval. Therefore, the n = 1604 and n = 476 datasets were divided into five subsets of equal size ranked according to increasing pH values. The distributions of a given element's concentration within each pH range were then presented as boxplots. Figure 4 provides some examples of typical concentration dependence on pH for different elements (Frengstad et al. 2001). In brief, the following observations were made:

i. concentrations of heavy metals (Cd, Co, Cu, Ni, Pb, Tl, Zn) are negatively correlated with pH, most probably reflecting solubility constrains on hydroxides or carbonates with increasing pH.
ii. elements with amphoteric hydroxides, such as As, B, W, Al, Se, and Sn exhibit some signs of elevated concentrations at the highest pH levels

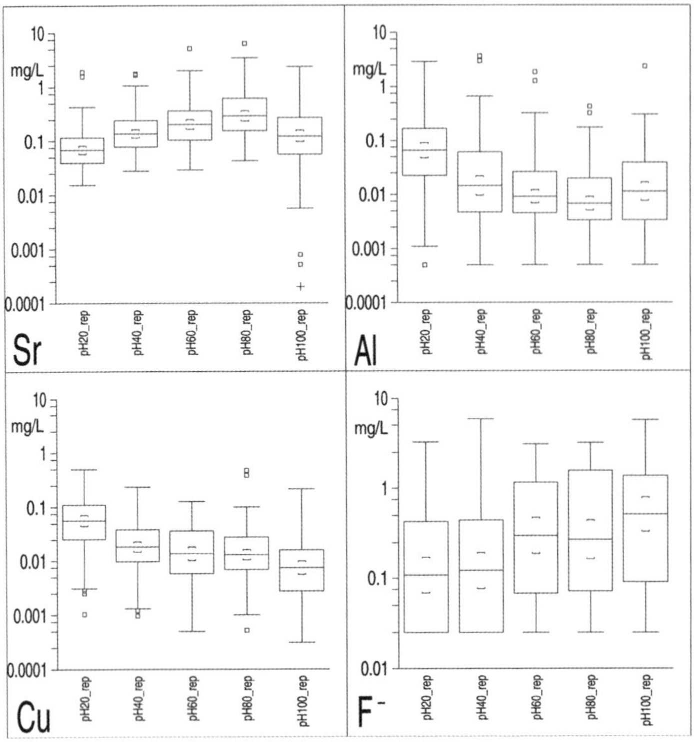


Figure 4. Examples of boxplots showing the distribution of element concentrations in crystalline rock groundwaters according to pH for the elements Sr, Al, Cu, F. The ($n$ = 476) data subset is divided into five percentile categories according to pH: 0–20% (pH 6.17–7.73); 20–40% (pH 7.74–8.01); 40–60% (pH 8.01–8.13); 60–80% (pH 8.14–8.22); 80–100% (pH 8.22–9.58).

iii. elements without obvious upper solubility constraints, but which are progressively accumulated via mineral-rock interaction, such as Na, Li and B, display positive correlations with pH
iv. fluoride also exhibit positive correlations with pH as the ion may be harboured in the crystal lattice of amphiboles and mica minerals and can be exchanged by $OH^-$ ions. Higher pH of the water (i.e., higher $OH^-$ concentration) will accelerate this process. There is also an indirect pH-dependence: high pH enables precipitation of calcite, hence, Ca-ions are removed from the water phase, thereby increasing the solubility of fluorite ($CaF_2$).
v. elements such as Ca, Sr, and Mg show initially positive correlations with pH, due to progressive hydrolysis of carbonate and silicate minerals, followed by a decrease at higher pH. This may reflect ion exchange or saturation and removal via precipitation of a carbonate phase (e.g., calcite for Ca).
vi. K and Si show no clear pH dependence because they attain saturation with respect to phases having a rather low solubility dependence on pH on an early stage.
viii. halides (chloride, bromide) and sulphate, with probably predominant atmospheric or marine sources, seem to exhibit concentrations that are rather independent of pH. The same applies for the inert gas Rn.

## 6 THE EVOLUTION OF GROUNDWATER CHEMISTRY IN SILICATE CRYSTALLINE ROCK AQUIFERS

From Figures 2 and 3 we note that the distributions of pH and major ions do not vary greatly between differing lithologies. We are minded of Garrels (1967) and Garrels and Mackenzie's (1967) observation that the evolution of groundwater composition in many crystalline rock lithologies is dominated by a single hydrochemical process: the hydrolysis of plagioclase by an aqueous solution of $CO_2$. This hypothesis can be justified by the modal abundance of plagioclase in many crystalline rocks and the fact that plagioclase reacts much faster than quartz and K-feldspar (White et al. 1998). Certainly, our Norwegian empirical data support the concept of similar hydrochemical evolutionary pathways taking place in all lithologies. The wide range of solute concentrations and pH values *within* each lithology could be interpreted as groundwaters simply having progressed to differing degrees along that evolutionary pathway (i.e., differing degrees of plagioclase hydrolysis, and consumption of $CO_2$). The clustering of pH values around 8.0 to 8.3 might suggest some particular buffering process taking place at that point in the groundwater's evolution.

We can thus formulate the following hypotheses for the early stage evolution of groundwaters in silicate crystalline rocks, from an initial recharge water with a given partial pressure of $CO_2$ ($PCO_2$)

i. if calcite is present, either in superficial materials or as an accessory mineral in the crystalline rock, its very rapid rate of reaction can lead to a groundwater dominated by calcium and bicarbonate (White et al. 1999, 2005):

$$H_2O + CO_2 + CaCO_3 = Ca^{2+} + 2HCO_3^- \tag{1}$$

The rapid rate of reaction can mean, however, that prolonged subaerial weathering effectively removes accessory calcite from the zone of groundwater circulation (Banks et al. 1998d).

ii. otherwise, the evolution of groundwater chemical composition will be dominated by plagioclase hydrolysis, with the ratio of Na/Ca reflecting that in the plagioclase mineralogy. Dissolved silica is released and a secondary clay mineral, such as kaolinite, will result:

$$2NaCaAl_3Si_5O_{16} + 6CO_2 + 9H_2O = 2Ca^{2+} + 2Na^+ + 6HCO_3^- + 4SiO_2 + 3Al_2Si_2O_5(OH) \quad (2)$$

Banks & Frengstad (2006) have examined two datasets; (i) the "Rock_corr" dataset from the national 1996/97 survey and (ii) a smaller dataset ("Egersund") derived from rigorous sampling of the Egersund Anorthosite aquifer of southwestern Norway, in the attempt to obtain samples from an aquifer with a modal dominance of plagioclase feldspar (probably approximating conditions as close to monomineralic plagioclase hydrolysis as it is possible to find in Nature). Both datasets are broadly consistent with the hypotheses outlined above. Moreover, the "Rock_corr" dataset exhibits very similar major ion evolutionary trends to the "Egersund" dataset, appearing to confirm the assertion that silicate hydrolysis (dominated by plagioclase) can be regarded as a dominant process in many crystalline rock lithologies.

When we examine pH-related trends in both datasets ("Rock_corr" and "Egersund" anorthosite), we find a distinct change between pH 8.0 and 8.5 (Figure 3). At around this pH, calcium ceases to accumulate in solution. Beyond this pH range, calcium appears to be removed from solution and sodium continues to accumulate. This process ultimately results in very high pH groundwaters (around pH 10) with a hydrochemistry dominated by sodium (as a cation) and alkalinity, with depleted concentrations of Ca (and also Mg). This can be seen in the sharply changing Na/Ca ratio in Figure 5.

Some researchers would be tempted to ascribe this process to ion exchange of $Ca^{2+}$ for $Na^+$, possibly on sodium-charged (following postglacial marine emergence) smectite fracture fillings, such as those described by Kocheise (1994). However, by judicious application of a hydrochemist's "Occam's Razor", Banks & Frengstad (2006) demonstrated that it is unnecessary to invoke cation exchange and that all the main hydrochemical trends observed can be explained purely by plagioclase hydrolysis. Frengstad & Banks (2000) and Banks & Frengstad (2006) showed, using a relatively simple PHREEQC model (Parkhurst 1995; Parkhurst & Appelo 1999), that pH would increase and bicarbonate and calcium initially accumulate in solution from plagioclase hydrolysis until, at a certain stage of groundwater evolution, calcite saturation would be achieved, removing calcium from solution and exerting a buffering effect on pH

$$Ca^{2+} + HCO_3^- = CaCO_3 + H^+ \quad (3)$$

Modelling suggested that, under typical closed $CO_2$ systems, this buffering effect occurs at around pH 8.0 to 8.5. Once calcium has effectively been removed from the solution, pH can continue to rise to relatively high values and result in a groundwater dominated by $Na^+$ and alkalinity, and depleted in $Ca^{2+}$ (Figure 6). The full equation for this process is:

$$4NaCaAl_3Si_5O_{16} + 8CO_2 + 14H_2O = 4CaCO_3 + 4Na^+ + 4HCO_3^- + 8SiO_2 + 6Al_2Si_2O_5(OH)_4 \quad (4)$$

or to a higher pH, with a greater degree of feldspar hydrolysis:

$$6NaCaAl_3Si_5O_{16} + 8CO_2 + 19H_2O = 6CaCO_3 + 6Na^+ + 2CO_3^{2-} + 2OH^- + 12SiO_2 + 9Al_2Si_2O_5(OH)_4 \quad (5)$$

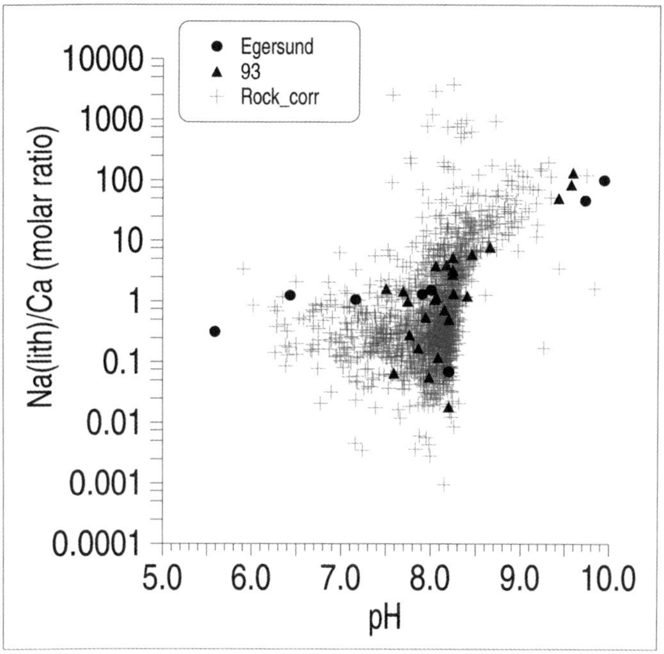


Figure 5. The molar ratio of Na/Ca plotted against pH for the Norwegian dataset "Rock_corr", its lithological subset 93 and the dataset ("Egersund") collected from a renewed field sampling campaign in the Egersund anorthosite aquifer in May 1997. Modified after Banks & Frengstad (2006).

Banks & Frengstad (2006) considered the impact of several factors on the evolution of the plagioclase-water-$CO_2$ system:

i. Plagioclase composition. It was found that a similar evolutionary pathway occurred for all plagioclase compositions. However, for calcium-rich plagioclases, calcium accumulates more "rapidly" in solution (relative to rate of plagioclase hydrolysis) and calcium depletion commences at an earlier stage.
ii. System openness or closure relative to $CO_2$. It was found that, while an open system could result in alkaline sodium bicarbonate waters via calcite precipitation, a very low $PCO_2$ (of $10^{-3}$ atm.) was required to achieve a pH as high as 8.2 and open systems could not realistically evolve the very high pH waters seen in real datasets.
iii. Initial $PCO_2$ concentration. Closed $CO_2$ systems could evolve the high pH, alkaline, Na-rich, Ca-poor waters seen in empirical datasets, although relatively high initial $PCO_2$ values of around $10^{-1.5}$ atm. were required to release the quantities of dissolved $Na^+$ approaching those seen in real waters (Figure 6).
iv. Alteration product. Defining calcic smectite (calcium montmorillonite in Garrels' (1967) parlance) as an alteration product rather than kaolinite, tended to slow the rate of calcium accumulation in solution (relative to plagioclase hydrolysis), but was found to be stoichiometrically inefficient to result in calcium *depletion* from solution. Calcite precipitation was still required for this purpose.

In summary, the major evolutionary trends observed in real datasets from anorthosites and from other crystalline silicate rock types in Norway can be explained by a model involving

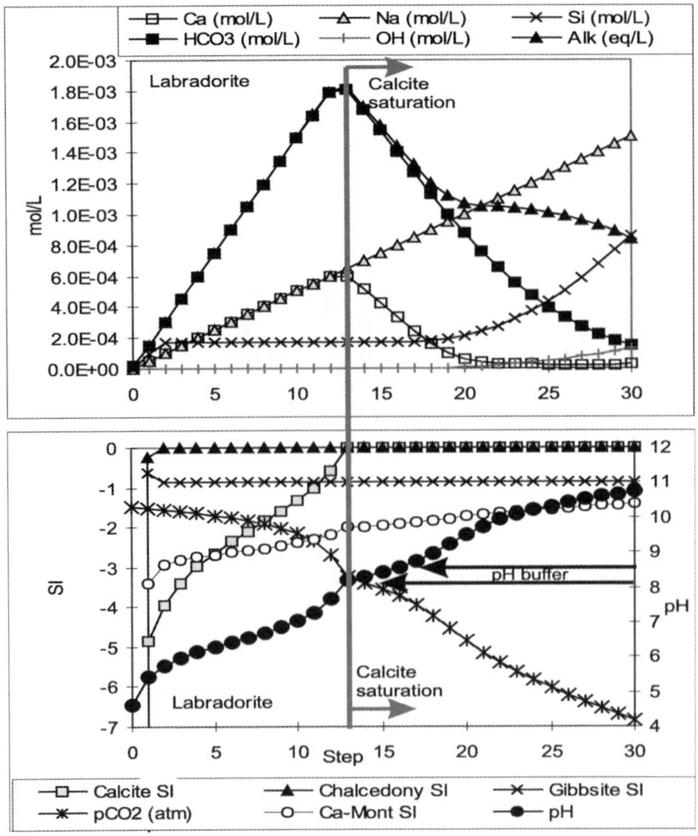


Figure 6. Modelled evolution of groundwater chemical composition, using PHREEQC, by dissolving $1.5 \times 10^{-3}$ moles of labradorite ($NaCaAl_3Si_5O_{16}$) in water of temperature 7°C and initial $PCO_2 = 10^{-1.5}$ atm. (closed $CO_2$ system) in 30 steps. Precipitation of calcite, kaolinite, gibbsite and chalcedony is permitted if oversaturated. The diagrams on the top show the evolution of total element concentrations, those on the bottom show pH and mineral saturation indices (SI). $\log_{10}(PCO_2)$ is also plotted on the SI axis. Note that alkalinity is defined as the equivalent per litre (eq/L) sum of $HCO_3^-$, $CO_3^{2-}$ and $OH^-$ and does not include other alkaline species formed at high pH such as Si-anions. Modified after Frengstad & Banks (2000) and Banks & Frengstad (2006).

plagioclase hydrolysis, coupled with calcite saturation and precipitation. Closed $CO_2$ systems with a relatively high initial $PCO_2$ of around $10^{-1.5}$ atm. were found to be necessary to generate both the observed pH values in excess of 8.5 and the quantities of dissolved solutes ($Na^+$, $Ca^{2+}$ and alkalinity) observed in real groundwaters. Banks & Frengstad (2006) report soil gas measurements from the Egersund area demonstrating that $PCO_2$ values as high as $10^{-1.5}$ atm. are *not* unrealistic, even for chilly Norwegian climates! In fact, even such high $PCO_2$ closed system models fail to generate quite the quantities of dissolved calcium and alkalinity observed in real waters. It would thus appear to be necessary to invoke open $CO_2$ weathering of plagioclase or calcite (e.g., in superficial deposits) to generate the observed concentrations of $Ca^{2+}$ and alkalinity, prior to system closure.

This is not to say, of course, that plagioclase hydrolysis is, in reality, the *only* process occurring in crystalline silicate aquifers. Other mineral phases may be important, for example:

- Accessory mineral phases may be responsible for the release of numerous trace elements, although the mobility of these may still ultimately be controlled by pH (e.g., many heavy metals), carbonate or calcium concentrations (e.g., fluoride). pH, carbonate and calcium concentrations will, in turn, be controlled by the extent of plagioclase hydrolysis. Note that, among those parameters exhibiting lithological dependence, uranium and radon are relatively soluble or chemically inert and do not experience clear pH-related solubility ceilings. Fluoride is a mobile anion that will accumulate in groundwaters if a source (apatite, fluorite, sheet silicates, amphiboles) is present. Its solubility may be constrained by the fluorite ($CaF_2$) solubility ceiling when calcium is present in the water. If calcite is removed from highly evolved silicate waters by calcite precipitation, however, this ceiling is "negated" and fluoride can accumulate to high concentrations in Ca-poor, high pH waters (Banks 1997, Figure 4).
- If calcite is present as an accessory mineral or in superficial deposits, the rapid carbonate hydrolysis reaction can dominate groundwater chemistry (White et al. 1999, 2005).

Clearly, there are also some silicate rock types where plagioclase feldspar may not dominate the weathering process (either kinetically or stoichiometrically). One example might be highly mafic or ultramafic rocks, where ferromagnesian silicates may dominate over feldspars. On hydrolysis, these would be expected to release bicarbonate alkalinity to solution, accompanied by other cations than calcium (magnesium, for example, in ultramafic rocks). For example, Barnes and O'Neil (1971), Barnes et al. (1978) and Drever (1988) describe the evolution of extremely high-pH (c. 11.7), Mg-poor, calcium hydroxide waters in magnesium silicate ultramafic rocks in the Western United States and in Oman, New Caledonia and Bosnia. They regard simple silicate weathering as the most likely explanation for this unusual water, where the geochemically (in the host rock) dominant major ion ($Mg^{2+}$) has been removed from the aqueous phase (by brucite, serpentinite and some degree of carbonate precipitation) and apparently replaced by another cation ($Ca^{2+}$) in high pH waters. This is analogous to the removal of $Ca^{2+}$ from highly evolved waters in calcium-rich anorthosites, by calcite precipitation, to be apparently replaced by sodium (Banks & Frengstad 2006).

## 7 CONCLUSION

A number of trace elements, occurring in groundwater in Norwegian crystalline rock lithologies, exhibit some degree of lithological dependence. Of particular note are radon, fluoride and uranium, all of which are significant in terms of human health and all of which tend to exhibit elevated concentrations in Norwegian granitic lithologies and low concentrations in, for example, anorthosites.

The main trends of major ion hydrochemical evolution of groundwaters in Norwegian silicate crystalline rock, however, appear to be similar in all lithologies. Within each lithology, a wide range of pH values and concentrations of major ions is observed. However, the range is similar in most lithologies, with a median pH of around 8.0 to 8.3 being typical. The major ion hydrochemical evolution in many crystalline silicate lithologies can largely be explained in terms of feldspar hydrolysis and secondary calcite saturation / precipitation.

We argue that the main trends in groundwater chemical composition in most crystalline silicate rock aquifers are *not* strongly dependent on lithology, but can be explained in terms of a more universal hydrolysable silicate (plagioclase) – $CO_2$ – $H_2O$ system and five variable factors:

i. the initial $PCO_2$ of the recharge water,
ii. the degree to which the aquifer geochemical system is "open" or "closed" with respect to $CO_2$,
iii. the availability and composition of hydrolysable silicate phases (dominated by plagioclase),
iv. the degree to which $CO_2$ has been consumed by plagioclase hydrolysis (determined by kinetic factors, grain size, fracture geometry, residence time)
v. the extent to which feldspar hydrolysis has continued beyond the point of calcite saturation.

## REFERENCES

Banks D (1997) Hydrogeochemistry of Millstone Grit and Coal Measures groundwaters, south Yorkshire and north Derbyshire, UK. Quart J Eng Geol 30:237–256.

Banks D, Reimann C, Røyset O, Skarphagen H & Sæther OM (1995a) Natural concentrations of major and trace elements in some Norwegian bedrock groundwaters. Appl Geochem 10:1–16.

Banks D, Røyset O, Strand T & Skarphagen H (1995b) Radioelement (U, Th, Rn) concentrations in Norwegian bedrock groundwaters. Env Geol 25:165–180.

Banks D, Frengstad B, Midtgård AaK, Krog JR, Strand T (1998a) The chemistry of Norwegian groundwaters. I. The distribution of radon, major and minor elements in 1604 crystalline bedrock groundwaters. Sci Tot Env 222:71–91.

Banks D, Midtgård AaK, Frengstad B, Krog JR, Strand T (1998b) The chemistry of Norwegian groundwaters. II. The chemistry of 72 groundwaters from Quaternary sedimentary aquifers. Sci Tot Env 222:93–105.

Banks D, Frengstad B, Krog JR, Midtgård AaK, Strand T & Lind B (1998c) Kjemisk kvalitet av grunnvann i fast fjell i Norge [*The chemical quality of groundwater in bedrock in Norway: in Norwegian*]. NGU Rep 98.058.

Banks D, Reimann C & Skarphagen H (1998d) The comparative hydrochemistry of two granitic island aquifers: The Isles of Scilly, UK and the Hvaler Islands, Norway. Sci Tot Env 209:169–183.

Banks D, Morland G & Frengstad B (2005) Use of non-parametric statistics as a tool for the hydraulic and hydrogeochemical characterization of hard rock aquifers. Scottish J of Geol 41:69–79.

Banks D & Frengstad B (2006) Evolution of groundwater chemical composition by plagioclase hydrolysis in Norwegian anorthosites. Geochim Cosmochim Acta (*in press*).

Barnes I & O'Neil JR (1971) The relationship between fluids in some fresh alpine-type ultramafics and possible modern serpentinization, western United States. Geol Soc Am Bull 80:1947–1960.

Barnes I, O'Neil JR & Trescases JJ (1978) Present day serpentinization in New Caledonia, Oman and Yugoslavia. Geochim Cosmochim Acta 42:144–145.

Bjorvatn K, Thorkildsen AH, Raadal M & Selvig KA (1992) Fluoridinnholdet i norsk drikkevann. Vann fra dype brønner kan skape helseproblemer. [*The fluoride content in Norwegian drinking water. Water from deep wells may cause health problems – in Norwegian*]. Norsk Tannlægeforenings tidsskrift 102:86–89.

Bårdsen A, Klock, KS & Bjorvatn K (1999) Dental fluorosis among persons exposed to high- and low-fluoride drinking water in western Norway. Community Dent and Oral Epidem 27:259–267.

Drever JI (1988) The Geochemistry of Natural Waters 2nd edn. Prentice Hall, Englewood Cliffs, New Jersey.

Dutter R, Leitner T, Reimann C & Wurzer F (1992) Grafische und geostatistiche Analyse am PC. Beiträge zur Umweltstatistik. [*Graphical and geostatistical analysis on PC. A contribution to environmental statistics – in German*]. Schriftenreihe der Technischen Univ Wien 29:78–88.

Frengstad B & Banks D (2000) Evolution of high-pH Na-$HCO_3$ groundwaters in anorthosites: silicate weathering or cation exchange? *In*: Sililo et al. (eds) Groundwater: Past Achievements and Future Challenges. Proc of XXXth IAH Congr, Cape Town. 493–498.

Frengstad B, Midtgård Skrede AaK, Banks D, Krog JR & Siewers U (2000) The Chemistry of Norwegian Groundwaters: III. The Distribution of Trace Elements in 476 Crystalline Bedrock Groundwaters, as Analysed by ICP-MS Techniques. Sci Tot Env 246:21–40.

Frengstad B, Banks D & Siewers U (2001) The Chemistry of Norwegian Groundwaters: IV. The pH-Dependence of Element Concentrations in Crystalline Bedrock Groundwaters. Sci Tot Env 277:101–117.

Frengstad B (2002) Groundwater quality of crystalline bedrock aquifers in Norway. Doktor Ingeniør Thesis 2002:53. Department of Geology and Mineral Resources Engineering, NTNU.

Garrels RM (1967) Genesis of some ground waters from igneous rocks. *In*: Abelson PH (ed.) Researches in Geochemistry. Wiley & Sons, New York, 405–420.

Garrels RM & Mackenzie FT (1967) Origin of the chemical compositions of some springs and lakes. In: Stumm W (ed.) Equilibrium Concepts in Natural Water Systems. Adv in Chem Series 67, Am Chem Soc, Washington D.C. 222–242.

Helse og omsorgsdepartementet (2001) Forskrift om vannforsyning og drikkevann [*Directive on water supply and drinking water – in Norwegian*]. FOR 2001-12-04-1372.

Kocheise R (1994) Svelleleire i undersjøiske tunneler [*Swelling clay in subsea tunnels – in Norwegian*]. Dr.ing. thesis 1994:124. Norges Tekniske Høgskole (NTH). Trondheim, Norway.

Morland G (1997) Petrology, Lithology, Bedrock Structures, Glaciation and Sea Level. Important Factors for groundwater Yield and Composition of Norwegian bedrock Boreholes? Doctor dissertation. Institut für Geowissenschaften. Montanuniversität Leoben.

Morland G, Reimann C, Strand T, Skarphagen H, Banks D, Bjorvatn K, Hall GEM & Siewers U (1997) The hydrogeochemistry of Norwegian bedrock groundwater – selected parameters (pH, F, Rn, U, Th, B, Na, Ca) in samples from Vestfold and Hordaland, Norway. NGU Bull 432:103–117.

Parkhurst D (1995) User's guide to PHREEQC – A computer program for speciation, reactive path, advective transport and inverse geochemical calculations. US Geol Surv, Lakewood, Colorado.

Parkhurst DL & Appelo, CAJ (1999) User's guide to PHREEQC (Version 2)–a computer program for speciation, batch-reaction, one-dimensional transport, and inverse geochemical calculations. U.S. Geol Surv Water-Res Inv Report 99–4259.

Reimann C, Hall GEM, Siewers U, Bjorvatn K, Morland G, Skarphagen H & Strand T (1996) Radon, fluoride and 62 elements as determined by ICP-MS in 145 Norwegian hard rock groundwater samples. Sci Tot Env 192:1–19.

Reimann C, Siewers U, Skarphagen H & Banks D (1999) Influence of filtration on concentrations and correlation of 62 elements analysed on crystalline bedrock groundwater samples by ICP-MS techniques. Sci Tot Env 234:155–173.

Sigmond EMO (1992) Berggrunnskart, Norge med havområder – Målestokk 1:3 millioner [*Bedrock map, Norway with marine areas – Scale 1:3 million*]. Norges geologiske undersøkelse.

Sæther OM, Reimann C, Hilmo BO & Taushani E (1995) Chemical composition of hard- and soft-rock groundwaters from central Norway with special consideration of fluoride and Norwegian drinking water limits. Env Geol 26/3:147–156.

USEPA (2000) National Primary Drinking Water Regulations; Radionuclides; Final Rule. Federal Register Dec 7, 2000, 65/236:76707–76753.

USEPA (2001) National Primary Drinking Water Standards. Office of Water (4606), EPA 816-F-01-007.

White AF, Blum AE, Schultz MS, Vivit DS, Stonestrom DA, Larsen M, Murphy SF & Eberl D (1998) Chemical weathering in a tropical watershed, Luquillo Mountains, Puerto Rico: I Long-term versus short-term weathering fluxes. Geochim Cosmochim Acta 62:209–226.

White AF, Bullen TD, Vivit DV, Schulz MS & Clow DW (1999) The role of disseminated calcite in the chemical weathering of granitoid rocks. Geochim Cosmochim Acta 63:1939–1953.

White AF, Schulz MS, Lowenstern JB, Vivit DV & Bullen TD (2005) The ubiquitous nature of accessory calcite in granitoid rocks: Implications for weathering, solute evolution, and petrogenesis. Geochim Cosmochim Acta 69:1455–1471.

WHO (2004) Guidelines for Drinking-water Quality (3rd Edition), section 12.122 Uranium. World Health Organisation, Geneva.

CHAPTER 19

# Factors influencing the microbiological quality of groundwater in Norwegian bedrock wells

Sylvi Gaut[1], Atle Dagestad[1], Bjørge Brattli[2] and Gaute Storrø[1]

*[1]Hydrogeology Section, Geological Survey of Norway, Trondheim, Norway*

*[2]Department of Geology and Mineral Resources Engineering, Norwegian University of Science and Technology, Trondheim, Norway*

ABSTRACT: The vulnerability of bedrock wells to microbiological contamination has been examined using microbiological data for 169 Norwegian waterworks using groundwater from bedrock. Inspections have been carried out at 49 of the 169 waterworks to identify possible causes to the recorded microbiological contamination. The microbiological water quality is correlated to (i) land use and contamination sources, (ii) type and thickness of surrounding superficial deposits, (iii) wellhead completion (including the well casing), and (iv) distance from wells to running water. Based on this study, wells are least vulnerable to microbiological contamination when the superficial deposits are >2.5 m thick and the wells are located >100 m from farmland and rivers/streams. It is recommended that the wellhead completion includes a well-house and a well casing of at least 5 m below and 0.5 m above ground level respectively. The gap between casing and bedrock should be sealed.

## 1 INTRODUCTION

In Norway the drinking water supply has mainly been based on surface water (lakes or rivers), and only about 15% of the population is supplied by groundwater (NGU 2002). About 2/3 of these are connected to waterworks, mostly small and medium sized (<1000 people), and 1/3 have private wells or springs. Many of the small waterworks, private households or holiday cottages are supplied by water from bedrock wells, and until the 1990s little emphasis was given to water quality in these wells. However, revision of the Norwegian drinking water regulations in the beginning of the 1990s, combined with Norwegian membership in the European Economic Agreement (EEA), put focus on groundwater through the Program for Improved Water Supply (PROVA). This revealed clear evidence of water quality problems in bedrock wells, which made it necessary to increase knowledge about the factors influencing the groundwater quality in these wells.

Several regional investigations of the hydrogeochemistry of groundwater in Norway have been carried out (e.g., Hongve et al. 1994; Banks et al. 1995a; Banks et al. 1995b; Reimann et al. 1996; Morland 1997; Frengstad 2002), whereas examination of the vulnerability of bedrock wells to microbiological contamination has not been studied. Therefore a study was initiated by the Geological Survey of Norway (NGU) in 1998 to investigate the

Table 1. The Norwegian standards for drinking water quality of 1995 and 2002. Only microbiological parameters relevant to this article are shown. In the revised regulations of 2002 the guidance level is removed, HPC at (37°C) is no longer measured and *E.coli* is analysed instead of FC*.

| Norwegian standard for drinking water quality | Type of analysis | Guidance level (number) | Maximum allowable concentration (number) |
|---|---|---|---|
| 1995 | HPC[1] (22°C)/ml | 100 | – |
| | HPC (37°C)/ml | 10 | – |
| | TC[2] (37°C)/100 ml | 0 | 0 |
| | FC[3] (44°C)/100 ml | 0 | 0 |
| 1 January 2002 | HPC (22°C)/ml | – | Not given. If the number exceeds 100, investigations must be initiated |
| | TC (37°C)/100 ml | – | 0 |
| | *E. coli*[4]/100 ml | – | 0 |

**E.coli* and FC are set as equal and used as one parameter in this study because *E.coli* is regarded as the most common member of the fecal coliform group detected in groundwater (Hellesnes 1979; Østensvik 1998).
[1]HPC = Heterotrophic plate counts, [2]TC = Total coliforms
[3]FC = Fecal coliforms, [4]*E. coli* = *Escherichia coli*

vulnerability of groundwater wells in bedrock to microbiological contamination (Gaut 2005). Microbiological analyses of raw water and tapwater from 169 waterworks in Norway have been examined for the period 1996–98. The data showed that only 24% of the waterworks met the Norwegian Standard for Drinking Water Quality (NSDW) of 1995 (Table 1) with respect to microbiological parameters. In order to assess factors influencing the microbiological quality, field inspections were subsequently carried out at 49 of these 169 waterworks (Figure 1). Prior to this work it was assumed that a correlation existed between microbiological contamination and the following factors:

- Land use and contamination sources
- Superficial deposits (type, thickness and extent)
- Design and protection of the well
- Distance from rivers/streams

## 2 DATA COLLECTION AND METHODS

The first part of the study included collection of monthly microbiological analyses from waterworks using groundwater derived from bedrock wells. A list of such waterworks was put together based on information from databases at the Norwegian Institute of Public Health (NIPH) and NGU. This was regarded as the most comprehensive list of waterworks based on groundwater from bedrock wells supplying more than 100 persons or 20 households.

Microbiological analyses in the period 1996–98 were done by the laboratories of the Norwegian Food Control Authority (SNT), and they were in 1998 contacted and requested to provide microbiological analyses for the waterworks in this study. Initially reported microbiological water quality from 195 waterworks in Norway, was collected from the laboratories. However, quality control of the dataset, which ensured that (i) all waterworks

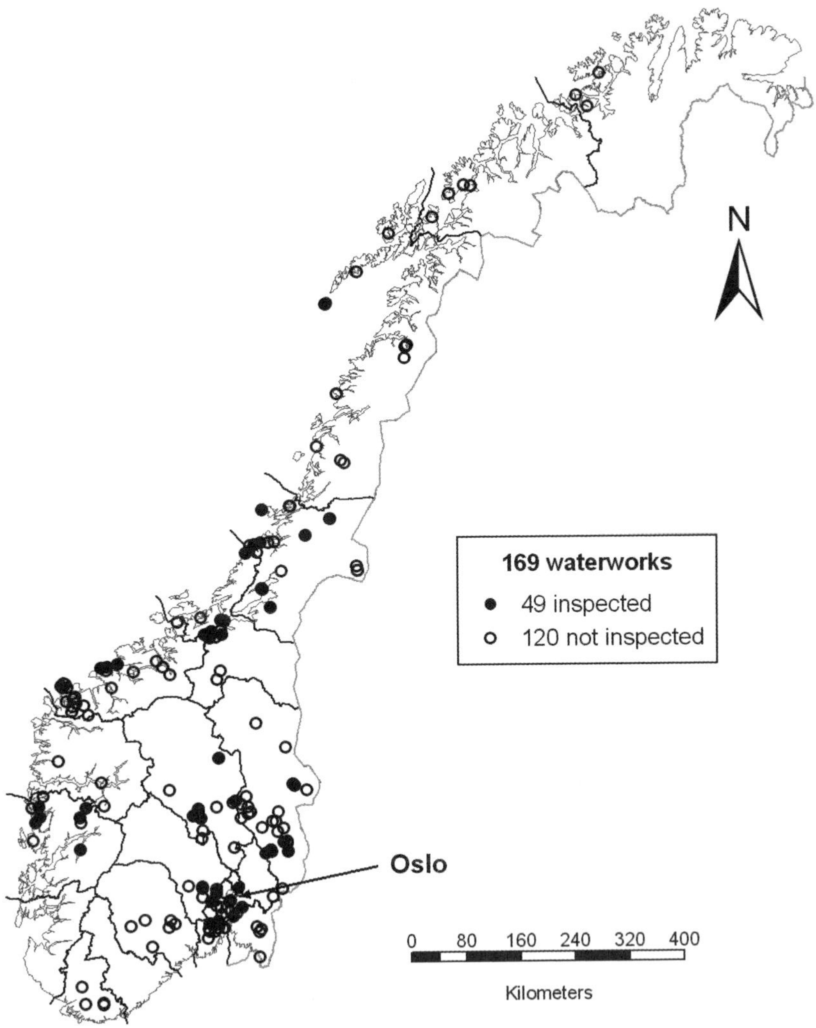


Figure 1. Geographical distribution of the 169 waterworks in this study. In all 49 of the 169 waterworks (filled circles) were inspected in the summer/autumn of 2000 and 2001.

were based on groundwater from bedrock wells, (ii) sampling frequency was at least 4 times a year and (iii) the water samples represented the period 1996–98, lowered the number of waterworks in this study to 169 (Gaut 2005).

Many of the 169 waterworks supplied water that did not meet the NSDW of 1995 (Sosial- og helsedepartementet 1995) regarding the number of total coliforms (TC) and/or fecal coliforms (FC) or heterotrophic plate count (HPC) at 22°C and/or 37°C (Table 1). Based on these results, 49 of the 169 waterworks were selected for further investigations, and field inspections were carried out at these waterworks during the summer/autumn of 2000 and 2001 (Figure 1). The 49 waterworks consisted of 135 wells in total, and all were examined with regard to the following parameters: Land-use and possible sources of contamination in the vicinity of the well, thickness and extension of superficial deposits in the well area, location relative to the occurrence of marine sediments, design and protection of the well and distance from rivers/streams.

To verify possible causes of microbiological contamination, field observations from each well have been compared with reported microbiological quality. Unfortunately, several of the waterworks did not collect water samples directly from the supply well. Instead, they collected raw water in the vicinity of the treatment plant or pressure tank, or tapwater from the distribution line. Thus, the reported water quality represented an integrated value in the cases where the waterwork was supplied from more than one well. As a consequence, well specific microbiological quality of groundwater existed for only 63 of the 135 inspected wells, and it is data from these 63 wells that are presented in this paper.

Because the project proceeded for several years, microbiological analyses from the period 1999–2003 were added to the dataset and the wells were grouped according to reported microbiological water quality using the revised NSDW of 2002 (Helse- og omsorgsdepartementet 2001) (Table 1). Consequently data on HPC at 37°C were discarded:

- Good microbiological quality (all reported water samples meet the revised NSDW). In tables and figures these wells are classified as "None".
- HPC at 22°C is periodically reported to exceed 100 colony forming units (cfu)/ml. Coliforms are not detected in these wells. In tables and figures these wells are classified as "$HPC_{only}$".
- Coliforms (TC and/or FC) are periodically detected. In tables and figures these wells are classified as "Coliforms". In some of the wells, HPC at 22°C is also reported to exceed 100 cfu/ml.

To evaluate statistical significance (95% confidence interval) boxplots (Tukey 1977) or student t-tests (Swan et al. 1995) were used. The latter was used on data presented in the tables. Generally only statistically significant differences are commented in this paper.

## 3 RESULTS AND DISCUSSION

Of the total of 169 waterworks, 76% did not meet the NSDW of 1995 (Table 1) for one or more water samples analysed in the period 1996–98. About 1/3 of these waterworks exceeded only the guidance level for either HPC at 22°C or both HPC at 22°C and 37°C. The rest had problems with the presence of coliforms (TC and/or FC) or both coliforms and HPC. About 40% of the latter group detected coliforms in $\geqslant 1/4$ of the samples. As presented by Gaut (2005) seasonal changes were observed and the microbiological contamination could be related to high water infiltration during snowmelt and autumn precipitation.

To identify possible causes to the microbiological contamination field inspections were carried out at 49 of the 169 waterworks and a total of 135 wells were inspected. Reported microbiological quality for 63 of the 135 inspected wells was then compared with parameters collected at the field inspections as described in "data collections and methods", and the results are presented and discussed in this section.

### 3.1 *Land use and potential sources of contamination*

Land use and contamination sources are found to be closely related. The land use around the 63 wells was divided into three main groups; Farmland, outlying fields and built-up areas or

Table 2. Number of wells with microbiological quality exceeding the NSDW of 2002 related to type of land use. Total number of wells is 63. For 45 of the 63 wells only one type of land use is observed (bold).

| | Total number of wells | Wells exceeding the NSDW (%) | | |
|---|---|---|---|---|
| | | None | $HPC_{only}$ | Coliforms |
| Farmland (>100 m from the well) | 10 | 40 | 20 | 40 |
| Farmland (<100 m from the well), incl. the 11 wells with farmland as the only land use | 19 | 26.5 | 26.5 | 47 |
| **Farmland only < 100 m from the well)** | **11** | **9** | **27** | **64** |
| **Outlying fields only** | **24** | **42** | **33** | **25** |
| **Built-up areas or scattered houses only** | **10** | **30** | **40** | **30** |
| Built-up areas or scattered houses incl. wells >100 m from farmland | 18 | 33.3 | 33.3 | 33.3 |

scattered houses. Farmland includes arable land, pasture or production of grass. Table 2 shows that for 11 wells farmland was the only land use within 100 m, built-up areas or scattered houses (including cottage development areas) were the only land use close to 10 wells, and as many as 24 wells were situated in outlying fields where no buildings or farmland existed.

During field inspection of the 63 wells, possible microbiological contamination sources in the catchment area were registered. The most common sources were (Table 3):

- Farming <100 m from the well (incl. grazing sheep in outlying fields)
- Septic tanks, sewage infiltration systems and sewer leakages
- Wildlife (moose and deer) in the well area
- Surface runoff towards the well and accumulation of surface water (pools/ponds) close to or in contact with the wellhead

Droppings from wildlife like moose and deer may constitute a source of contamination for all wells, however 10 wells were situated in areas where this type of wildlife was reported to be more extensive.

Compared to outlying fields, wells situated in the vicinity of farmland or built-up areas were more often contaminated with coliforms (Table 2), and wells should not be located within 100 m of a farming area. However, in Norway sheep and cattle graze in outlying fields and this makes the wells vulnerable to microbiological contamination if the animals are allowed access to the wellhead. During the field inspections sheep were observed at the well site for five of the 24 wells. Of these wells, 2 have reported coliforms and 3 have reported HPC exceeding 100 cfu/ml.

Manure, either by manure spreading or grazing animals, may have contaminated wells in the vicinity of farmland. Further examination of the dataset indicates that wells furthest from the grazing land are less likely to be contaminated, but exceptions are found. Similar results were found by Goss et al (1998) who discovered that the number of wells with microbiological contamination decreased with increasing distance from feedlot or exercise yards. The correlation was more pronounced for dug or bored wells than drilled wells.

Table 3. Number of wells with microbiological quality exceeding the NSDW of 2002 related to potential sources of contamination registered in the catchment area. Total number of wells is 63. In all 12 wells are registered with two potential contamination sources and 2 wells with three potential contamination sources.

| | Total number of wells | Wells exceeding the NSDW (%) | | |
|---|---|---|---|---|
| | | None | $HPC_{only}$ | Coliforms |
| Wells with no obvious contamination source | 16 | 63 | 31 | 6 |
| Farming <100 m from the well incl. grazing sheep in outlying fields | 24 | 21 | 33 | 46 |
| Septic tanks, sewage infiltration systems and reported possible sewer leakages | 8 | 25 | 12.5 | 62.5 |
| Wildlife (e.g., moose and deer) in the well area | 10 | 40 | 10 | 50 |
| Possible contamination from surface water accumulated close to or in contact with the wellhead. No other contamination source is registered | 10 | 40 | 50 | 10 |
| Possible contamination from surface water accumulated close to or in contact with the wellhead. Other possible contamination sources are also registered | 10 | 10 | 0 | 90 |

Wells situated close to built-up areas or scattered houses may be contaminated from sewage systems, septic tanks or pit latrines (Daly 1985; Daly et al. 1993; Macler and Merkle 2000). This is also confirmed for 2 wells located <50 m from a septic tank in this study, and improper sewage treatment by infiltration is stated as the reason for microbiological contamination at one waterwork. Improper sewage treatment is also reported by Gaut & Tranum (2003) as one of the reasons of poor microbiological quality for several wells in a small community outside Oslo.

## 3.2 *The superficial deposits*

The thickness and extent of the superficial deposits were evaluated for a radius of 20 m around each well using field observations in combination with Quaternary geology maps. The superficial deposits were then classified in two categories; (1) medium to thick, or (2) thin or discontinuous. In category 1 the superficial deposits have average thickness of at least 0.5 m and no bedrock exposed, whereas in category 2 bedrock is exposed, although the thickness of the superficial deposits locally can be more than 0.5 m. Data presented in Table 4 suggest that wells located in areas with medium to thick superficial deposits are less susceptible to microbiological contamination, which is also found by Conboy & Goss (2000).

The thickness of the superficial deposits at the well site was recorded in the well log for 48 of the 63 wells. In order to evaluate the correlation between the thickness of the superficial deposits and the microbiological water quality, the 48 wells were divided in two

Table 4. Number of wells with water exceeding the NSDW of 2002 in relation to extent and thickness of superficial deposits (63 wells), depth to bedrock at the well point (48 wells), and well location above or below the marine limit (63 wells).

| | Total number of wells | Wells exceeding the NSDW (%) | | |
|---|---|---|---|---|
| | | None | $HPC_{only}$ | Coliforms |
| Medium to thick superficial deposits (category 1) | 40 | 40 | 35 | 25 |
| Thin or discontinues superficial deposits (category 2) | 23 | 26 | 31 | 43 |
| Depth to bedrock from surface at the well point >5 m | 14 | 43 | 36 | 21 |
| Depth to bedrock from surface at the well point ≤5 m | 34 | 24 | 38 | 38 |
| No data | 15 | – | – | – |
| Depth to bedrock from surface at the well point >2.5 m | 22 | 45 | 32 | 23 |
| Depth to bedrock from surface at the well point ≤2.5 | 26 | 15 | 42.5 | 42.5 |
| No data | 15 | – | – | – |
| Well location above marine limit (a.m.l.) | 21 | 19 | 48 | 33 |
| Well location below marine limit (b.m.l.) | 42 | 43 | 21 | 36 |

groups according to recorded depth from surface to bedrock at the well point (Table 4). In the table the subdivision is done for two depths; more or less than 5 m and more or less than 2.5 m. However, only when comparing wells with thickness of deposits >2.5 m and ≤2.5 m a statistical significant difference in microbiological water quality was found, showing that more wells reported good microbiological quality in the first group. It is therefore suggested that the superficial deposits at the well site should be at least 2.5 m thick with no bedrock exposed in the well area.

Sediment type, i.e. grain and pore size, is also regarded as an important factor controlling migration of water and microorganisms in soil (Gerba and Keswick 1981; Robertson and Edberg 1997). Marine sediments are found in Norway at elevations up to about 220 m above sea level ("marine limit"). Statistically less wells situated below the marine limit supply water exceeding the NSDW of 2002 (Table 4). This is because the sediments below the marine limit are dominated by marine deposits with high clay content and low permeability, which reduce the infiltration rate and give good protection against microorganisms.

### 3.3 *Design and protection of the well*

Improper design or protection of the groundwater well may cause microbiological contamination, as described by among others Conboy & Goss (1999), Daly (2000) and Korkka-Niemi (2001). Based on the dataset, the possibility to avoid microbiological contamination increases with increasing casing length (Table 5, Figure 2). No wells with casing length less than 2.5 m have reported good microbiological water quality (Table 5) and the casing length should therefore be at least 2.5 m. Generally the well casing is drilled into bedrock

Table 5. Number of wells exceeding the NSDW of 2002 regarding microbiological parameters. The table presents length of well casing (48 wells), location of top of casing above or below ground level (63 wells) and observed leakages during inspection with downhole video camera (18 wells).

| | Total number of wells | Wells exceeding the NSDW (%) | | |
|---|---|---|---|---|
| | | None | $HPC_{only}$ | Coliforms |
| Length of well casing > 5 m | 23 | 43.5 | 39 | 17.5 |
| Length of well casing >2.5 but ≤ 5 m | 16 | 31 | 31 | 38 |
| Length of well casing >0 but ≤2.5 m | 9 | 0 | 22 | 78 |
| Top of well casing is above ground level | 42 | 40 | 29 | 31 |
| Top of well casing is below ground level | 21 | 24 | 33 | 43 |
| Observed leakage between bottom of well casing and bedrock | 8 | 37.5 | 37.5 | 25 |
| No visible leakage between bottom of well casing and bedrock | 10 | 10 | 30 | 60 |

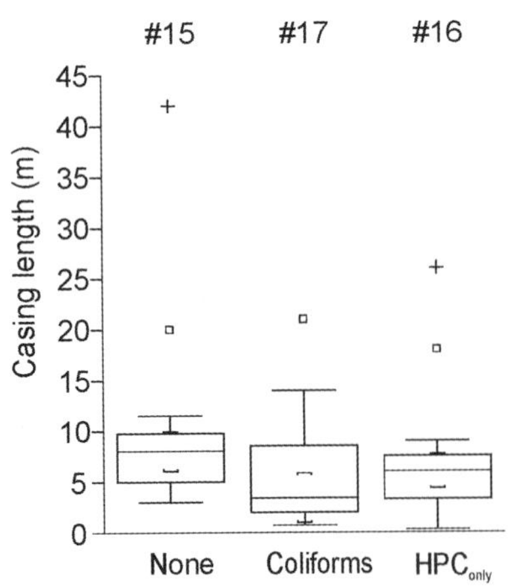


Figure 2. Boxplots presenting the correlation between wells with groundwater exceeding the NSDW of 2002 and the length of the well casing for 48 wells. None = no wells exceed the NSDW.

and consequently the casing length is influenced by the thickness of the superficial deposits at the well site (Gaut 2005). According to Figure 2, wells with detected coliforms had a statistically significant shorter casing length (<5 m) than wells where no microbiological problems were reported. Correlation between casing length and depth to bedrock shows that most of the wells with depth to bedrock <2.5 m and casing length < 5 m periodically supply water with either coliforms or HPC exceeding the NSDW of 2002 (Gaut 2005). Based on the observations above, and the indication that wells with the casing protruding above ground level are less likely to have microbiological contamination (Table 5), it is suggested that the casing should have a minimum length of 5 m below the ground and protrude 0.5 m above ground level.

Figure 3. Examples of wellhead completions. a) Well-house, b) concrete well-protection (manhole), c) concrete well-protection with well-cover (small "house") instead of a concrete lid and d) no protection except a proper cap on top of the casing.

However, the dataset shows that also wells with casing length longer than 5 m occasionally supply water that did not meet the NSDW of 2002. For 18 of the 63 wells, a downhole video camera was used to investigate the presence or status of sealing and indications of water leakage between casing and bedrock at the bottom of the well casing (Table 5). Water leakages at the bottom of the casing were observed for 8 wells, whereas 10 wells had neither visible leakage at the day of inspection nor any indications of previous leakages. No statistical differences in reported microbiological water quality exist between wells with or without leakages. A reason for this can be that most wells have a casing too short to seal off shallow fractures, which may lead to inflow of groundwater with short residence time in the subsurface. Short casings ($<3$ m) and water inflow in the uppermost 10 m of the borehole are also found for several other wells inspected by NGU in 2004/2005 (Storrø et al. 2006).

An interesting observation is that 5 of the 8 wells with visible leakages were reported to be sealed with bentonite or cement-based suspensions, which shows that it is important that the sealing is correctly performed to avoid leakages.

Inspection of the well sites in this study revealed a multitude of wellhead completions (Figure 3). Wells were protected by well-houses (Figure 3a), concrete well-protections (manhole) (Figure 3b) or a combination of both (Figure 3c). There were also wells where the only protection was a solid cap on top of the casing (Figure 3d).

Number of wells with water samples reported to exceed the NSDW of 2002 in relation to wellhead completion are presented in Table 6. Based on the microbiological analyses and the field observations, wells should be protected by a well-house (Figure 3a). The main reasons for this are that a well-house is (i) above ground level, (ii) most likely to be constructed properly and (iii) less likely to be destroyed or neglected compared to a concrete well-protection.

Table 6. Number of wells with water samples reported to exceed the NSDW of 2002 in relation to existence of well-house and/or concrete well-protection. Examples of the wellhead completions are shown in Figure 3. Total number of wells is 58.

| | Total number of wells | Wells exceeding the NSDW (%) | | |
|---|---|---|---|---|
| | | None | $HPC_{only}$ | Coliforms |
| 1. Well-house with concrete floor | 12 | 42 | 25 | 33 |
| 2. Concrete well-protection, no well-house or well-cover | 30 | 23 | 27 | 50 |
| 3. Concrete well-protection in combination with well-cover or well-house | 16 | 50 | 44 | 6 |
| 1+ 3. Well-house with concrete floor and concrete well-protection in combination with well-cover or well-house | 28 | 46 | 36 | 18 |

However, data in this study indicate that thick superficial deposits at the well site and sufficient distance from potential contamination sources are more important for good microbiological water quality than a perfect wellhead completion. This is so, because the 5 wells protected by a well-house reporting good microbiological quality (42% of group 1), were situated in areas with no known contamination sources and generally thicker superficial deposits than the 7 wells with water quality problems (Table 6). Additionally, the 16 wells located in a concrete well-protection combined with a well-house or well-cover (Figure 3c) were all located in areas with medium to thick superficial deposits and in outlying fields. Even though there were flaws in the concrete well-protections, which include incomplete sealing between the concrete floor and the casing, only one well periodically reported coliforms.

Surface water having direct access to the well can contaminate the groundwater (Daly 2000). In all, 30 of the 63 inspected wells (Table 6) were protected by a plain concrete well-protection (Figure 3b), and at least 12 of them had visible flaws in the construction. Mainly due to leakage between the concrete ring and the concrete floor and/or between the well casing and the concrete floor, water had accumulated inside these manholes. For the wells located in areas with medium to thick superficial deposits, the water level inside the manhole was influenced by the groundwater level at the well site. In cases where the well casing did not protrude ground level the water periodically overtopped the casing, allowing direct inflow of possible contaminated water to the well. Even though no known contamination sources like septic tanks or farmland were observed in the well area, several of these wells reported HPC exceeding the NSDW of 2002. This is because HPC at 22°C represents microorganisms that are part of the natural microbiota in soil and water. It is therefore important to hinder surface water to accumulate close to wells even at sites where no obvious source of coliforms exists. To support this recommendation, results presented in Table 3 show that wells located in areas with potential contamination sources (e.g., farming and septic tanks) and with surface water accumulated close to or in contact with the wellhead, had problems with both HPC and coliforms in the groundwater. This indicates that wells supplying water with periodically high values of HPC are vulnerable to contamination by coliforms if coliforms are present in the well area.

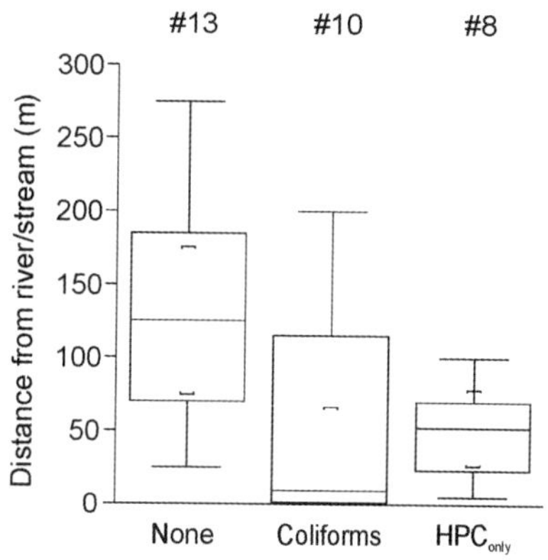


Figure 4. Boxplots presenting correlation between wells with groundwater exceeding the NSDW of 2002 and distance from river/stream. Number of wells is 31. None = no wells exceed the NSDW.

## 3.4 *Well location related to distance from rivers/streams*

Malard et al (1994) showed that infiltration of river water into an aquifer can cause microbiological pollution. In this study, 31 of the 63 wells were situated within a distance of 300 m of a river/stream and using these wells a statistically significant correlation was found between distance to the watercourse and well site and the microbiological water quality (Figure 4). Most contaminated wells were located less than 75 m from the river/stream and based on these observations, new wells should be located at least 100 m from such watercourses to avoid coliforms and high HPC in the groundwater.

# 4 CONCLUSIONS

Based on the observations in this study the following conclusions can be made:

1. The microbiological quality of groundwater from bedrock wells is correlated to land use and contamination sources and type and thickness of the superficial deposits in the well area, in addition to the design and protection of the well.
2. Groundwater wells should be placed well apart from any known contamination source, especially septic tank systems and farmland with manure spreading or grazing animals. Recommended minimum distance is 100 m. However, the best location is in outlying fields with no grazing livestock.
3. In order to achieve sufficient residence time for water infiltrated from rivers or streams, a safety distance of 75–125 m to these watercourses is recommended.
4. Bedrock wells located in areas with substantial extent and thickness of superficial deposits are less vulnerable to microbiological contamination of the groundwater than wells situated directly on bedrock or where the superficial deposits are thin or discontinous. Based on the dataset the thickness of the deposits should be at least 2.5 m to ensure attenuation of possible pathogenic microorganisms. Wells located below the marine limit are in general better protected than those situated above the marine limit.

5. Proper wellhead completion (including the well casing) hinders contamination of the groundwater through the wellhead and inflow (leakage) of possibly microbiologically contaminated water at the bottom of the casing. Recommended wellhead completion includes a well-house and a casing of at least 5 m below and 0.5 m above ground level respectively. The gap between casing and bedrock should be sealed. Groundwater inflow at shallower depth than 10 m should be avoided.

## ACKNOWLEDGEMENTS

The authors wish to thank the managers of various waterworks who allowed data from the waterworks to be used, the persons at the 49 waterworks that were visited, and those who contributed with further information when needed. Thanks also to Stephen Lippard, NTNU for editing the English text and NGU for financing the study.

## REFERENCES

Banks D, Røyset O, Strand T, Skarphagen H (1995a) Radioelement (U, Th, Rn) concentrations in Norwegian bedrock groundwaters. Environmental Geology 25:165–180.

Banks D, Reimann C, Røyset O, Skarphagen H, Sæther O (1995b) Natural concentrations of major and trace elements in some Norwegian bedrock groundwaters. Applied Geochemistry 10:1–16.

Conboy MJ, Goss MJ (2000) Natural protection of groundwater against bacteria of fecal origin. Journal of Contaminant Hydrology 43:1–24.

Conboy MJ, Goss MJ (1999) Contamination of rural drinking water wells by fecal origin bacteria – Survey findings. Water Quality Research Journal of Canada 34(2):281–303.

Daly D (2000) Practical Approaches to Preventing Pollution of Wells. The GSI Groundwater Newsletter 38:9–15 (http://www.gsi.ie/workgsi/groundwater/groundwaterfra.htm).

Daly D (1985) Groundwater quality and pollution. It affects you. It depends on you. Geological Survey of Ireland, Information Circular 85/1, 25 pp.

Daly D, Thorn R, Henry H (1993) Septic tank systems and groundwater in Ireland. Report series RS 93/1 (Groundwater), Geological Survey of Ireland, Department of Transport, Energy and Communications, 30 pp.

Frengstad B (2002) Groundwater quality of crystalline bedrock aquifers in Norway. Dr. ing. thesis 2002:53, Norwegian University of Science and Technology, Norway.

Gaut A, Tranum I (2003) Sørbråten og Solemskogen. Gjennomgang av drikkevannskvalitet og vurdering av muligheten for en utvidet, tilfredsstillende vannforsyning. [Sørbråten and Solemskogen. Existing drinking water quality and evaluation of the possibility of an extended and satisfactory water supply]. Rapport 132681-1, Statkraft Grøner AS, 13 pp.

Gaut S (2005) Factors influencing microbiological quality of groundwater from potable water supply wells in Norwegian crystalline bedrock aquifers. Doctoral thesis 2005:99, Norwegian University of Science and Technology, Norway.

Gerba CP, Keswick BH (1981) Survival and transport of enteric viruses and bacteria in ground water. In: van Duijvenbooden W et al. (eds) Quality of Groundwater. Proceedings of an International Symposium, Noordwijkerhout, The Netherlands, 23–27 March 1981. Studies in Environmental Science 17:511–515.

Goss MJ, Barry DAJ, Rudolph DL (1998) Contamination in Ontario farmstead domestic wells and its association with agriculture: 1. Results from drinking water wells. Journal of Contaminant Hydrology 32:267–293.

Hellesnes I (1979) Indikatorer med hygienisk betydning i vann. [Hygenic indicators for water]. VANN 14(1B):57–75.

Helse- og omsorgsdepartementet (2001) Forskrift 4. desember 2001 nr 1372 om vannforsyning og drikkevann (Drikkevannsforskriften) [Directive on water supply and drinking water]. (http://www.lovdata.no/for/sf/ho/ho-20011204-1372.html. Cited 23. March 2006).

Hongve D, Weideborg M, Andruchow E, Hansen R (1994) Landsoversikt – drikkevannskvalitet. Spormetaller i vann fra norske vannverk [National overview – drinking water quality. Trace metals in water from Norwegian waterworks]. VANN 92, Statens Institutt for Folkehelse, Oslo, 110 pp.

Korkka-Niemi, K., 2001: Cumulative geological, regional and site-specific factors affecting groundwater quality in domestic wells in Finland. Monographs of the Boreal Environment Research, 20, Finnish Environment Institute, 98 pp.

Macler BA, Merkle JC (2000) Current knowledge on groundwater microbial pathogens and their control. Hydrogeology Journal 8:29–40.

Malard F, Reygrobellet J, Soulié M (1994) Transport and Retention of Fecal Bacteria at Sewage-Polluted Fractured Rock Sites. J. Environ. Qual. 23:1352–1363.

Morland G (1997) Petrology, lithology, bedrock structures, glaciation and sea level: Important factors for groundwater yield and composition of Norwegian bedrock boreholes? NGU Report 97.122 I, Geological Survey of Norway, 274 pp.

NGU (2002) Geology for society: Groundwater. http://www.ngu.no/index.asp?ilangid=1. Cited January 2005.

Østensvik Ø (1998) Fekale indikatorbakterier i drikkevann [Fecal indicator bacteria in drinking water]. Norsk veterinær tidsskrift 110(10):606–614.

Reimann C, Hall GEM, Siewers U, Bjorvatn K, Morland G, Skarphagen H, Strand T (1996) Radon, fluoride and 62 elements as determined by ICP-MS in 145 Norwegian hard rock groundwater samples. The Science of the Total Environment 192:1–19.

Robertson JB, Edberg SC (1997) Natural Protection of Spring and Well Drinking Water Against Surface Microbial Contamination. I. Hydrogeological Parameters. Critical Reviews in Microbiology 23(2):143–178.

Sosial- og helsedepartementet (1995) Forskrift om vannforsyning og drikkevann m.m. (Drikkevannsforskriften) [Directive on water supply and drinking water etc.]. I-9/95, Sosial- og helsedepartementet, 40 pp.

Storrø G, Gaut S, Sivertsvik F, Gundersen P, Sørdal T, Berg T (2006) Kvalitet av borebrønner i fjell – inspeksjon av brønnutforming [Quality of bedrock wells – inspection of well constructions]. NGU Rapport 2006.031, Norges geologiske undersøkelse.

Swan ARH, Sandilands MMcCabe P (1995) Introduction to Geological Data Analysis. Blackwell Science Ltd, 446 pp.

Tukey JW (1977) Exploratory Data Analysis. Addison-Wesley, 506 pp.

CHAPTER 20

# Evaluating mineral water quality trends of Pedras Salgadas (Portugal)

Carla Lourenço[1] and Luís Ribeiro[2]

[1]*INETI – National Institute of Engineering, Technology and Innovation (ex-IGM), Estrada da Portela Zambujal, Apartado Amadora, Portugal*

[2]*CVRM – Geosystems Center, Instituto Superior Técnico, Av. Rovisco Pais Lisboa, Portugal*

ABSTRACT: The mineral waters of Pedras Salgadas of north of Portugal are located in fractured rock formations associated with different types of granites. Their genesis is related to the great Penacova-Régua-Verin fault. These waters are hypo-thermal, naturally carbonated with values of free $CO_2$ between 500 mg/l and 5800 mg/l. TDS range from 384 mg/l to 5415 mg/l and the hydrochemical facies is bicarbonate-sodium. In order to detect the trends and the rate of change of some main physical-chemical parameters, a seasonal Mann-Kendall test was applied to a sequence of monthly values, observed in 5 wells, from 1995 to 2005. To estimate the rates of change per unit time, a robust estimator of the trend slope was also calculated by Thiel and Sen regression method.

## 1 INTRODUCTION

Portugal is one of the richest E.U. countries with respect to mineral waters. The carbonated waters are a very particular group of natural mineral waters and the Vidago-Pedras Salgadas hydro-mineral field is one of the most important Portuguese concessions that has contributed in 2005 with 33,200,000 euros to the Portuguese economy. There is a growing need to apply adequate tools that permit a proper quantitative/qualitative evaluation of these waters; one of the main objectives to analyse the temporal evolution of the main physicochemical parameters and the magnitude of their trends.

## 2 STRUCTURAL GEOLOGY OF VIDAGO/PEDRAS SALGADAS REGION

The existing carbonate waters in mainland Portugal are located in the northern area of the Hesperic Massif in the middle Galiza-Subzone (see Figure 1). Their geographic location is correlated with regional fault systems such as the Verin-Régua-Penacova Fault, Vilariça Fault, and Rio Minho Fault. They emerge on the granites and schist, usually on the intersection between the great regional faults and their conjugates, because these sites offer the best conditions for the rising of the fluids from the deep zones of the crust exist.

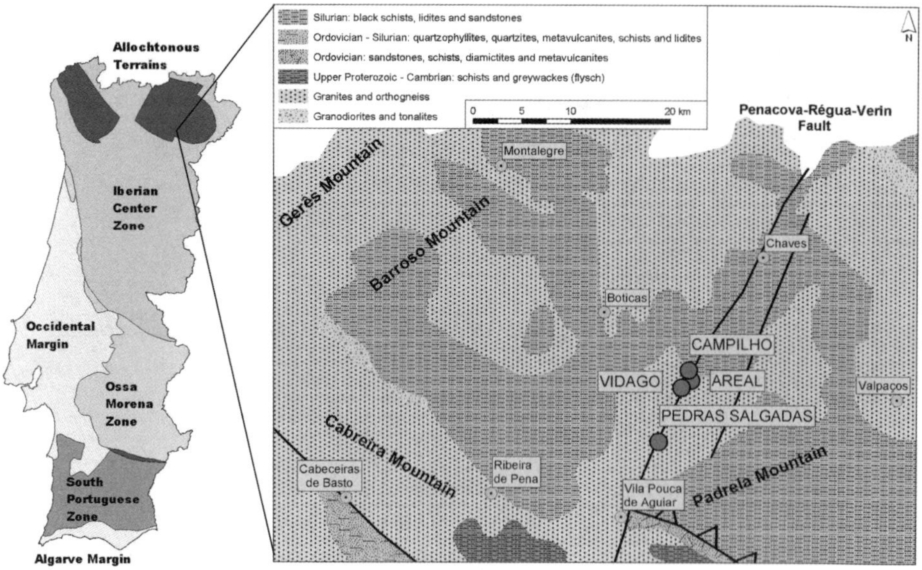


Figure 1. Location of Vidago-Pedras Salgadas hydro-mineral field in Portugal. (source:INETI)

The carbonate waters of Pedras Salgadas are associated with the great fault Penacova-Régua-Verin (with a NNE-SSW direction), which is long and deep and affects the entire continental crust. The springs occur mainly in the W branch of the fault, and sometimes in their conjugate WNW-ESSE, emerging on areas where the post-tectonic granitoid rocks are dominant.

The structural features that greatly influence the hydro-mineral trajectories are discussed in Ribeiro (1992); however, according to Carvalho (1993), the recharge areas and the conditions of in-depth circulation are not completely understood. The tectonic model of Ribeiro (1992) points out the recharge areas in Serra de Padrela and in E border of the tectonic depression. Notwithstanding, the isotopic study made by Palma (1993) suggests "the recharge area of the mineral aquifer is situated south of the emergencies themselves and might be connected with the surface of Alvão or the surface of Padrela". Same suggestion was made by Costa (1992) that considers that the mineral water acquires the principal's characteristics through a deep circulation in tectonic accidents SSW-NNE / S-N, being the differences of chemical patterns resultant of a set of reactions that, in the slow ascension of the water during the phase of discharge of the hydro-mineral aquifer, develop with different intensities; according to this author the alterations that has been observed, most of all derived from the exploitation regime of exploration of the carbonated waters, point in this direction.

## 3 WELL MONITORING

In Pedras Salgadas concession exists 4 carbonated water wells and 3 natural springs legalised as natural mineral water. Notwithstanding, the extraction of mineral water is only made through 4 wells (AC 12A, AC 13, AC 17 and AC 25) with a depth that varies between 70 m (well AC12) and 185 m (well AC25). Well AC12A was build in 1998, replacing

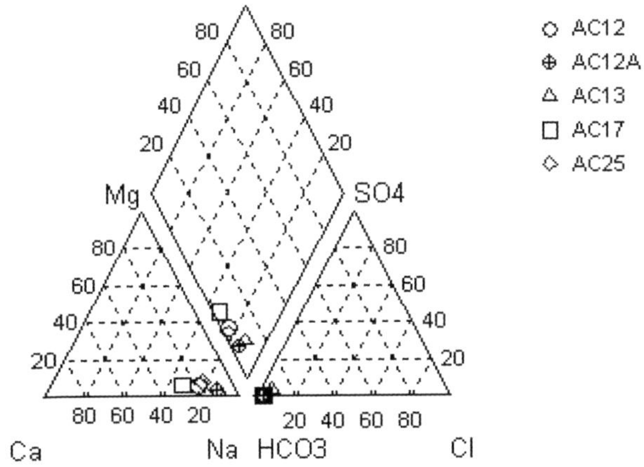


Figure 2. Piper diagram of carbonated waters of Pedras Salgadas.

AC12, whose water revealed some physical-chemical instability, which was an aspect that was verified in Ribeiro and Lourenço (1999). The natural springs mentioned above are part of a study of regular monitoring made by the concessionaire with the objective to evaluate the available mineral resource.

A monitoring program was carried out by the former Portuguese Geological-Mining Institute, since 1986 that consisted of 3 to 4 simple physical-chemical analyses per year and one complete analysis every 4 or 5 years. For this trend analysis, observations ranging from 1995 to 2005 were used. Each observation consists on information of the following variables: $F^-$, $CO_2$, $SiO_2$, $Cl^-$, $HCO_3^-$, $SO_4^{2-}$, $Na^+$, $Mg^{2+}$ and $Ca^{2+}$.

## 4 EXPLORATORY AND VISUALIZATION DATA ANALYSIS

Mineral waters of Pedras Salgadas are hypo-thermal, naturally carbonated, with values below 20°C (at surface) and with a rich content in $Na^+$ (Figure 2). Analysing Figures 3 to 6 we concluded that median contents of $HCO_3^-$ are from 403 mg/l to 3175 mg/l, with a maximum value of 4087 mg/l detected in AC25 (Figure 3) and median concentrations of $Na^+$ between 148 mg/l (AC13) and 931.50 mg/l (AC25).

The highest $F^-$ concentrations were found in well AC13 (median equal to 4.5 mg/l) and the lowest values of $CO_2$ were observed in the same well (median equal to 712.5 mg/l) in contrast to AC 25 where a median of 4955 mg/l was determined.

In short, we can say that well AC25 depicts water with the most mineralized properties, in opposition to the hydrochemical facies observed in AC13. The physical-chemical parameters with the greatest range are $CO_2$ and $HCO_3^-$, followed by Na. AC25 is the well that presents the largest variation of these parameters.

For a first visualisation of the temporal evolution of each parameter, the correspondent time series were plotted for each parameter. For instance Figure 7 shows in general a downward trend of the parameter $HCO_3^-$ in the 5 cases, with special evidence in AC17 and AC25 wells.

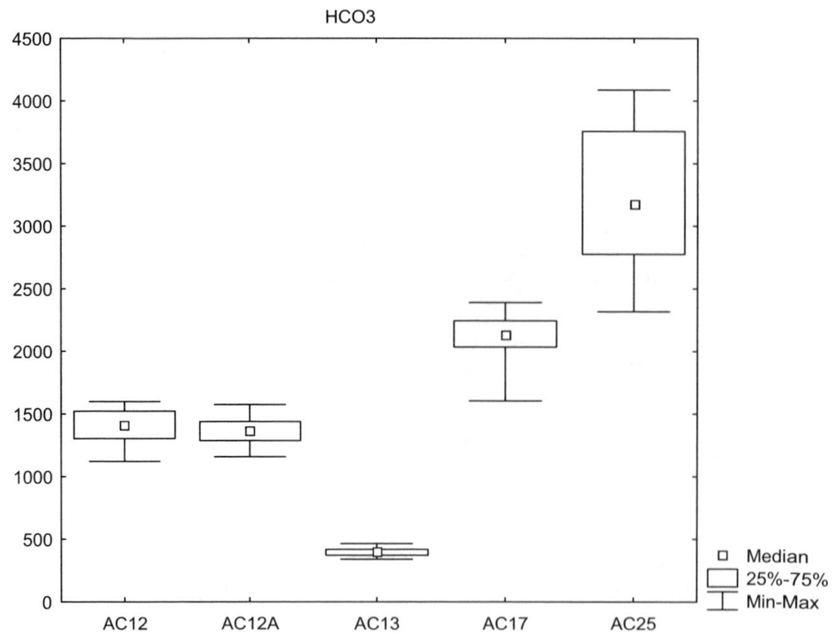


Figure 3. Multiple box-plots of $HCO_3^-$ (mg/l) calculated for the 5 wells.

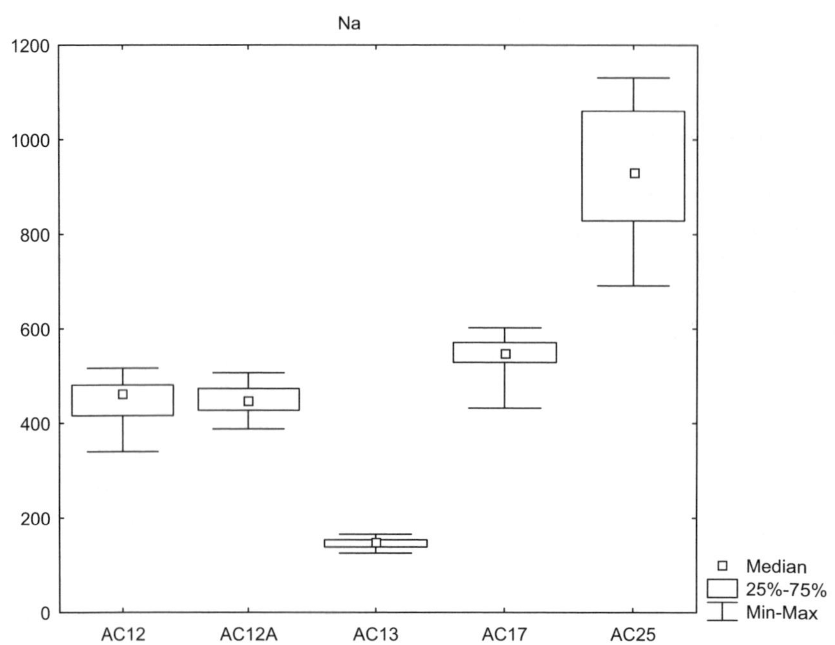


Figure 4. Multiple box-plots of $Na^+$ (mg/l) calculated for the 5 wells.

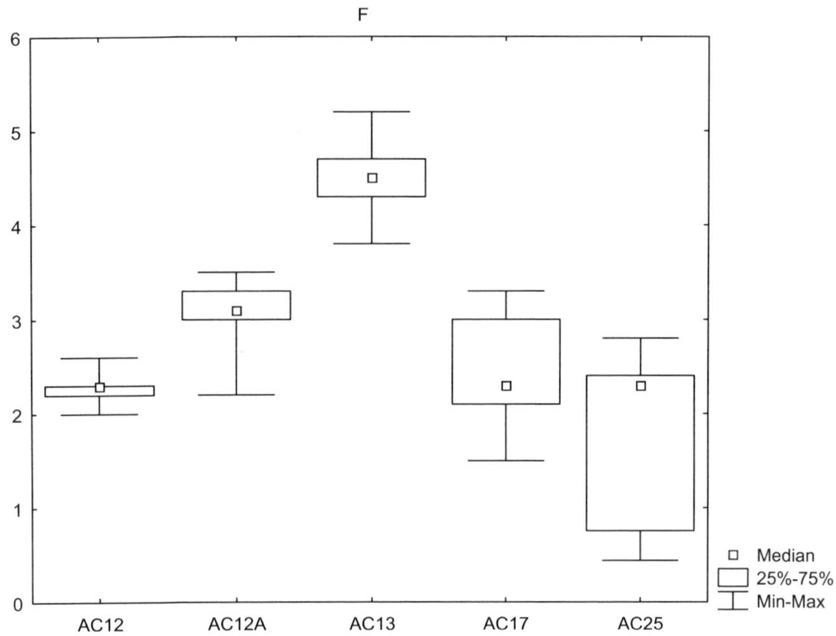


Figure 5. Multiple box-plots of $F^-$ (mg/l) calculated for the 5 wells.

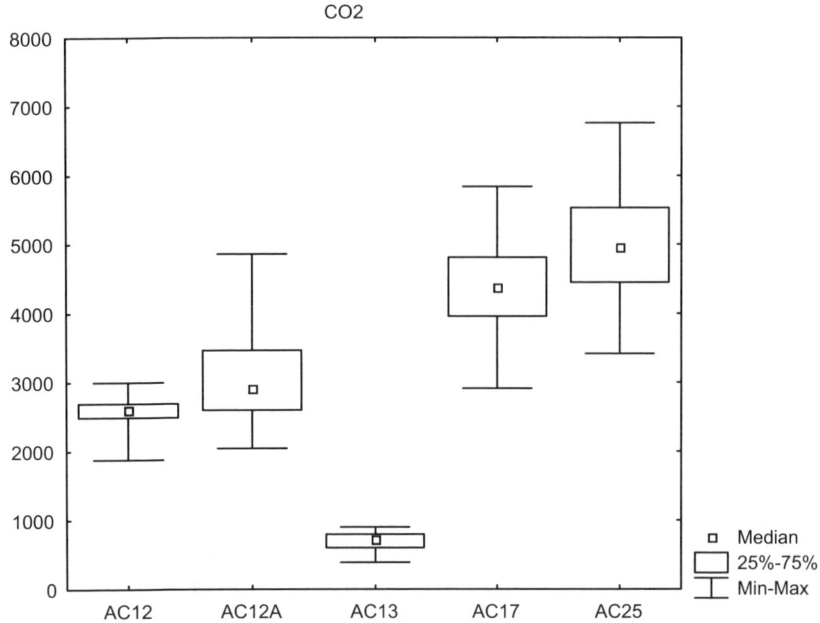


Figure 6. Multiple box-plots of $CO_2$ (mg/l) calculated for the 5 wells (mg/l).

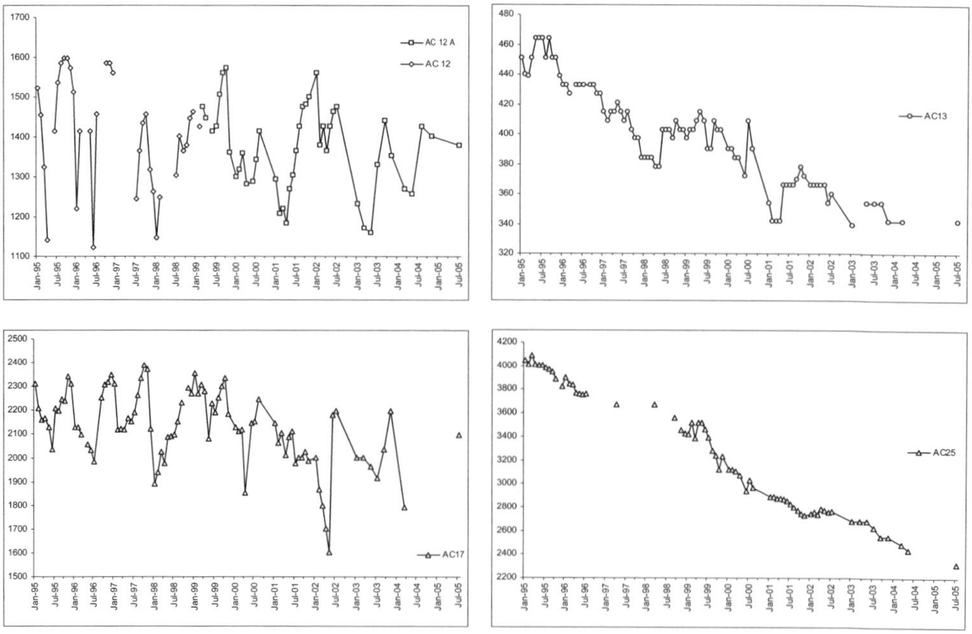


Figure 7. Temporal evolution of $HCO_3^-$ (mg/l) in AC12, AC12A, AC13, AC17 and AC25 wells.

## 5 TREND ANALYSIS

### 5.1 *Methodology*

The Mann-Kendall test is a non-parametric statistical method mainly used to detect and assess the trends of the water quality parameters (Hirsch et al., 1982). The selection of a non-parametric approach is particularly suitable to analyse data exhibiting one or more of the following properties: free distributions, missing values, presence of outliers, seasonality, or censored data. The test is applied to a sequence of observation ordered by time $q_1$, $q_2$, …, $q_n$, where a parameter S is build according to the following expression:

$$S = \sum_{k<j} f(q_j - q_k) \tag{1}$$

Where:

f is a function defined by: $f(\theta) = 1$ (if $\theta > 0$); $f(\theta) = 0$ (if $\theta = 0$) and $f(\theta) = -1$ (if $\theta < 0$)

A null hypothesis $H_0$ of no trend is tested and rejected or confirmed according to the statistical distribution. To measure the strength of the upward or the downward trend a statistic p is calculated. In general trends are considered statistically significant when $p<0.1$. For this study a seasonal Mann-Kendall test procedure is used and an estimation of the correspondent slope is performed by a robust regression suggested by Thiel and Sen.

Using this technique, the slope of the trend is given by the median (B) calculated on the basis of all the values $d_{jk}$ following:

$$d_{jk} = (q_j - q_k) / (j - k) \tag{2}$$

Table 1. Water quality trends in Pedras Salgadas.

| Wells Parameters | $F^2$ | $CO_2$ | $SiO_2$ | $Cl^2$ | $HCO_3^2$ | $SO_4^{22}$ | $Na^1$ | $Mg^{21}$ | $Ca^{21}$ |
|---|---|---|---|---|---|---|---|---|---|
| **AC 12 (till 1998)** | 0.00 ↔ 0.46 | −4.86 ↓ 0.21 | −0.12 ↓ 0.01 | −0.1 ↓ 0.00 | −5.96 ↓ 0.00 | 0.01 ↑ 0.07 | −1.37 ↓ 0.21 | −0.05 ↔ 0.12 | −0.41 ↓ 0.00 |
| **AC12A (from 1998)** | 0.00 ↔ 0.86 | 7.71 ↔ 0.2 | 0.05 ↑ 0.06 | −0.02 ↓ 0.04 | −1.42 ↔ 0.15 | −0.02 ↓ 0.04 | −0.41 ↓ 0.02 | −0.02 ↓ 0.03 | −0.08 ↓ 0.17 |
| **AC13** | 0.004 ↑ 0.00 | −3.02 ↓ 0.00 | −0.03 ↓ 0.00 | −0.01 ↓ 0.01 | −1 ↓ 0.00 | 0.08 ↑ 0.00 | −0.28 ↓ 0.00 | −0.02 ↓ 0.00 | −0.05 ↓ 0.00 |
| **AC17** | 0.01 ↑ 0.00 | −7.57 ↓ 0.00 | −0.02 ↔ 0.17 | −0.04 ↓ 0.00 | −1.63 ↓ 0.00 | 0.00 ↔ 0.42 | −0.3 ↓ 0.00 | −0.05 ↓ 0.00 | −0.29 ↓ 0.00 |
| **AC25** | 0.02 ↑ 0.00 | 3.52 ↔ 0.29 | −0.04 ↓ 0.00 | −0.15 ↓ 0.00 | −14.93 ↓ 0.00 | 0.04 ↑ 0.00 | −3.62 ↓ 0.00 | −0.3 ↓ 0.00 | −1.02 ↓ 0.00 |

legend:0.01: slope (per month)
↑ = Upward trend ↓ = Downward trend ↔ = No trend Statistic p = 0.00

for all the pairs $q_K$ with $1 \leqslant k < j \leqslant n$

The trend slope is expressed by change per unit time and is related with S by:

If $S>0$ then $B \geqslant 0$;

If $S < 0$ otherwise.

The test can be applied with other techniques in a joint methodology in order to achieve a particular objective (Ribeiro and Macedo, 1995).

## 5.2 *Results and discussion*

The non-parametric test was applied to the mineral waters of Pedras Salgadas, to the period from January 1995 to July 2005 in a monthly basis. Observations of the physical-chemical parameters listed above were used for this purpose. Table 1 displays the results of the detected upward, downward or no trends, the estimated slope (change per month) and the statistical p. From the analysis of this table the following features may be highlighted:

(a) In the majority of the wells, significant downward trends ($p < 0.1$) were detected in the following water quality descriptors: $Cl^-$, $Na^+$, $Mg^{2+}$, $HCO_3^-$ and $Ca^{2+}$.
(b) In some wells, however, significant upward trends were also observed in $F^-$ (wells AC13, AC17 and AC25) and $SO_4^{2-}$ (AC12, AC13 and AC25).

Concerning the magnitude of the estimated slopes, we emphasize the decrease of 14.93 mg/l per month of $HCO_3^-$ calculated in AC25. However such high value should be interpreted according to the relative high median values obtained in this well when compared with the others.

A stability of the temporal evolution of these parameters should be expected in this hydro-mineral field and not the occurrence of significant upward and downward trends detected in all parameters and in the majority of the wells. This situation maybe caused by some exogenous influences such as the alteration of the well pumping rates. The consequences are visible in the behaviour of $CO_2$ and in the major and trace composition, especially $HCO_3^-$.

## CONCLUSIONS

The temporal evolution of 9 water quality parameters monitored in 5 wells of Pedras Salgadas carbonate waters, showed significant upward and downward trends, although the estimated slopes are, in general, not important considering the high mineralization of these waters. The high variability in these trends detected in some parameters is closely related to the pumping system, with indirect influences in the fluctuations of the $CO_2$.

Lourenço (2000) revealed that, on the contrary of the wells, the natural springs of Pedras Salgadas hydro-mineral field do not show significant upward and downward trends. This fact may indicate that some of the variations verified in the wells are a consequence of the system pump, that cause variations in $CO_2$, and that consequently may imply a set of reactions that in turn can influences modifications in the majority and residuals component of these waters, mainly in $HCO_3^-$.

This is a dual phase hydrogeological system, gas and water, and the pumping process favours the separation of the two phases. Generally the system has to reach a new equilibrium, and such fact can delay some time.

Further studies should be conducted in order to understand the influence of the alteration of pumping rates in the water quality trends detected in one of the most important hydro-mineral fields of Portugal.

## REFERENCES

Carvalho, M., 1993: Definição dos Perímetros de Protecção das Águas Minerais das Bacias de Pedras Salgadas e Vidago. Sondagens e Fundações A.CAVACO. final report, 25p.

Costa, A., 1992: Estudo Hidrogeológico das Águas Minerais de Fte. Romana e Sabroso e das Zonas de Protecção, 44p.

Hirsch, R. M., Slack, J. R. and Smith, R. A., 1982: Techniques of Trend Analysis for Monthly Water Quality Data. Water Resources Research., vol. 18, (1), pp.107–121.

Lourenço, M. C., 2000: Modelação Estatística das Águas Gasocarbónicas de Vidago e Pedras Salgadas. MSc thesis in Georesources, Instituto Superior Técnico, Lisbon, 145p.

Palma, F., 1993: Estudo Isotópico das Águas Gaso-carbónicas das Bacias de Pedras Salgadas e Vidago, report, 17p.

Ribeiro, A., 1992: Controle Estrutural e Geomorfológico das Nascentes de Águas Minerais da Região de Vidago-Pedras Salgadas, DGFCUL/UTAD, 105 p.

Ribeiro L. and Lourenço C., 1999 : A Study of Trend Analysis on Mineral Waters in the North of Portugal ; *in* Fendeková and Fendek (eds.) Hydrogeology and Land Use Management, 717–720 Slovak Association of Hydrogeologists Publ.

Ribeiro, L. and Macedo, M.E., 1995 – Application of Multivariate Statistics, Trend- and Cluster Analysis Groundwater Quality in Tejo and Sado Aquifer' *in* K. Kovar & J. Krásný (eds.) GQ95 – Proc. of the International Conference on Groundwater Quality: Remediation and Protection –, Praga, República Checa, IAHS publication n° 225, 39–47.

CHAPTER 21

# Questions and answers about the evolution of $CO_2$-rich thermomineral waters from Hercynian granitic rocks (N-Portugal): a review

José M. Marques[1], Mário Andrade[2], Paula M. Carreira[2], Rui C. Graça[1] and Luís Aires-Barros[1]

[1] *Instituto Superior Técnico. Centro de Petrologia e Geoquímica. Av. Rovisco Pais, Lisboa, Portugal*

[2] *Instituto Tecnológico e Nuclear (ITN). Estrada Nacional Sacavém, Portugal*

ABSTRACT: This paper reviews geochemical and isotopic studies carried out on hot and cold $HCO_3/Na/CO_2$-rich mineral waters issuing along a major regional NNE-trending fault (North Portugal). $\delta^2H$ and $\delta^{18}O$ values are similar to those of the local meteoric waters, indicating a meteoric origin for the mineral waters. $\delta^{13}C$ values from $CO_2$ gas and total dissolved inorganic carbon, and previously reported $^3He/^4He$ values, indicate that the carbon in these $CO_2$-rich mineral waters is mainly derived from a deep-seated (upper-mantle) source. The low $^{14}C$ activity measured in some of the cold $CO_2$-rich mineral waters is incompatible with the systematic presence of $^3H$ in those waters, indicating that total carbon in the recharge waters is being masked by large quantities of deep-seated ($^{14}C$-free) $CO_2$. Differences in the $^{87}Sr/^{86}Sr$ values presented by the thermomineral waters seem to be related with the existence of water-rock interaction with different granitic rocks.

## 1 INTRODUCTION

During the last years, the geohydrology of the Vilarelho da Raia – Pedras Salgadas region (Figure 1) has been studied in detail (e.g., Aires-Barros et al., 1995, 1998; Andrade 2003; Marques et al., 1998a,b, 2000a,b, 2001, 2003). Geochemical and isotopic ($^2H$, $^{13}C$, $^{18}O$, $^{87}Sr$, $^3H$, $^{14}C$) data has been discussed in order to update the geohydrological conceptual model of the Chaves low-temperature geothermal field. The results obtained have also been used to understand the relations between these hot (Chaves, 76°C) and cold (Vilarelho da Raia, Vidago, Pedras Salgadas, 17°C) $HCO_3$-Na-$CO_2$-rich mineral waters issuing at the northern part of the Portuguese mainland, associated with fractured Hercynian granitic rocks. These $CO_2$-rich mineral waters flow from natural springs and boreholes located either in granitic outcrops or in the peribatholitic boundaries concordant to the main NNE-SSW fault trend (Figure 1), the so-called Verin – Chaves – Régua – Penacova trending fault.

The aim of this paper is to review the hydrogeological investigations performed at Vilarelho da Raia – Pedras Salgadas region using both geochemical and isotopic approaches, providing additional information to answer the most common questions associated with these complex hydrogeological systems, namely: (i) what is the origin of the thermomineral waters? (ii)

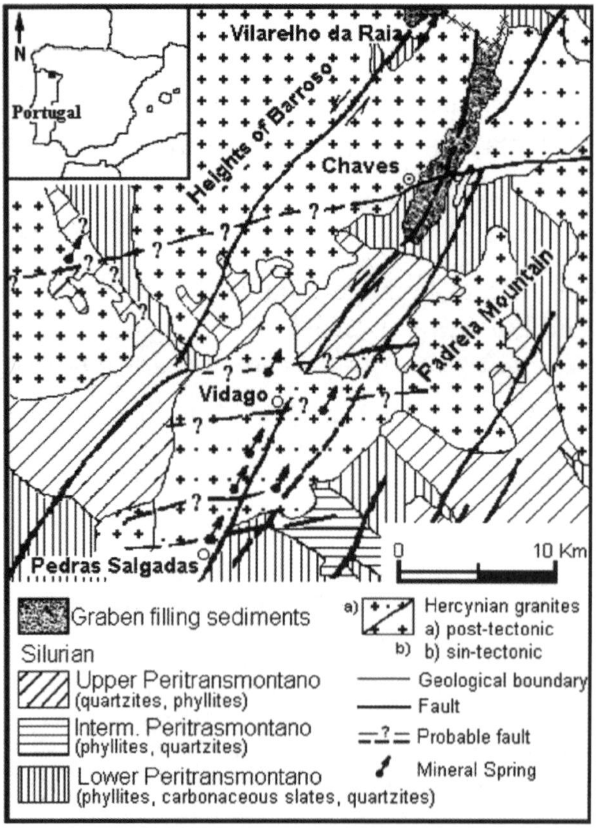


Figure 1. Schematic geology of the Vilarelho da Raia-Pedras Salgadas region. After Sousa Oliveira (1995).

what is the nature of $CO_2$? (iii) which are the main water-rock interactions processes occurring at depth? Understanding the origin of carbon dioxide Earth degassing and associated $CO_2$-mineral springs is the key to improve knowledge on important hydrogeological processes (e.g., to determine whether or not these hot and cold carbon dioxide mineral waters could be considered surface manifestations of large-scale underground flowpaths). Furthermore, increasing knowledge on the local geohydrology is extremely important to achieve the sustainable use of this "invisible" georesource, since most of the studied $CO_2$-rich mineral waters are used both as a source of bottled water and a recreational resource (spa facilities, tourism, etc.).

## 2 GEOLOGICAL BACKGROUND

The region under research is located in the tectonic unit of Middle Galicia/Trás-os-Montes sub-zone of the Central-Iberian Zone of the Hesperic Massif. According to several authors (Baptista et al., 1993 and Sousa Oliveira and Portugal Ferreira, 1996), who described the geology of Vilarelho da Raia – Pedras Salgadas, the main geological formations are: (i) Hercynian granites (syn-tectonic – 310 Ma and post-tectonic – 290 Ma) and (ii) Silurian metasediments of the Upper, Intermediate and Lower Peritransmontano Group, that consists on a sequence of quartzites, phyllites and carbonaceous slates. The most recent formations are Miocene-Pleistocene graben filling sediments with variable thickness, showing their maximum development along the central axis of Chaves Depression, which is a graben whose axis is oriented NNE-SSW (Figure 1). It is bounded on its eastern side by the edge of the Padrela Mountain escarpment with a 400 m throw. The western block is formed by several grabens from the Heights of Barroso towards the Chaves Depression. The Vidago/Pedras Salgadas area is also mainly composed of post-tectonic Hercynian granites with some outcrops of metamorphic rocks of Silurian age, as well as Cenozoic cover deposits (Baptista et al., 1993). Chaves thermal waters emerge within a wide graben, whereas the cold mineral waters (Vidago/Pedras Salgadas) are found in areas where the NNE-SSW megalineament does not exhibit such an important morphological structure.

Tectonically, the region is mainly controlled by the NNE-SSW hydrothermally active fault playing an important role in the circulation of the thermomineral waters. The 200 km long NNE-SSW megalineament, which reaches a depth of 30 km in the study area (Baptista et al., 1993) could play an important role in $CO_2$ extraction and migration from its deep source (upper mantle) to the surface. Sousa Oliveira and Portugal Ferreira (1996) pointed out the fact that the mineral waters emerge in places where the following subvertical fracture systems intersect: (1) N-S to NNE-SSW, (2) ENE-WSW, (3) NNW-SSE to NW-SE, and (4) WNW-ESE to W-E.

## 3 FIELD INVESTIGATIONS AND METHODS

Water samples for geochemical and isotopic analysis were collected from rain, shallow cold dilute groundwaters (spring waters) and from the Vilarelho da Raia, Chaves, Vidago and Pedras Salgadas $CO_2$-rich thermomineral waters. Rock samples and mineral separates from local geological outcrops were also analysed for Sr concentrations and $^{87}Sr/^{86}Sr$ ratios. Temperature (°C), pH and electrical conductivity (μS/cm) of the waters were determined "*in situ*". Total alkalinity was measured a few hours after collection. The following methods were applied for chemical analyses performed at the Laboratório de Mineralogia e Petrologia of Instituto Superior Técnico (LAMPIST): atomic absorption spectrometry for Ca and Mg; emission spectrometry for Na, K and Li; colorimetric methods for $SiO_2$ and F; ion chromatography for $SO_4$, $NO_3$ and Cl; potentiometry for alkalinity, here referred to as $HCO_3$. Dry residuum (DR) was calculated following the U.S. Geological Survey procedure, which calls for drying at 180°C for 1 hour (Hem, 1970). The data on free $CO_2$ content of the thermomineral waters was kindly supplied by the Águas de Carvalhelhos Company, the Municipality of Chaves, and Vidago – Melgaço & Pedras Salgadas Company. The free $CO_2$ data was obtained through the carbonate alkalinity, determined by the back acid-base titration method. The free $CO_2$ of the samples was estimated by means of a graphical technique, taking into consideration the carbonate alkalinity, pH and ionic strength. A detailed description of this method is given by Ellis and Mahon (1977).

$\delta^2H$ and $\delta^{18}O$ measurements (*vs* V-SMOW, Vienna Standard Mean Ocean Water) were performed by mass spectrometry (SIRA 10–VG ISOGAS) at the Instituto Tecnológico e Nuclear (ITN – Portugal) following the analytical methods of Epstein and Mayeda (1953) and Friedman (1953), with an accuracy of $\pm 1.0$ ‰ for $\delta^2H$ and $\pm 0.1$ ‰ for $\delta^{18}O$. The $^3H$ water content (reported in Tritium Units, TU) was also determined at ITN, using electrolytic enrichment followed by liquid scintillation counting method (standard deviation varies between $\pm 0.9$ and $\pm 1.3$ TU, depending on tritium activity). $^{13}C$ values were measured, at ITN, on the TDIC (Total Dissolved Inorganic Carbon) of groundwater, precipitated in the field as $BaCO_3$ at a pH environment higher than 9.0. The values are reported in ‰ to V-PDB (Vienna Peede Belemnite) standard, with an accuracy of ±0.1 ‰. The $^{14}C$ content was measured at the Physics Department of Utrecht University through accelerator mass spectrometry as described by Van der Borg et al. (1984, 1987). $^{14}C$ is given in pmC (percentage of modern Carbon). The Sr concentrations and Sr isotope ratios ($^{87}Sr/^{86}Sr$) were performed at the Geochron Laboratories-USA. The Sr concentrations were determined by isotope mass dilution spectrometry and $^{87}Sr/^{86}Sr$ ratios have been performed by mass spectrometry. All Sr data were normalised to $^{87}Sr/^{86}Sr = 0.1194$ (long term reproducibility of NBS-987 at MIT: 0.710247 ± 0.000014 (2 sigma s.d.).

## 4 GEOCHEMICAL MODEL – QUESTIONS AND ANSWERS

### 4.1 *What is the origin of the mineral waters?*

The tectonic and geomorphologic features (Chaves, Vidago and Pedras Salgadas grabens), and the geological environment (dominated by granitic rocks) of the region under research, seem to be responsible for the occurrence of a lot of hot and cold $Na/HCO_3/CO_2$-rich mineral waters (gas bubbles can be directly observed in the springs). From the geochemical point of view, the waters from Vilarelho da Raia/Chaves and Vidago/Pedras Salgadas areas form two separate groups. Recent hydrogeochemical data (Andrade, 2003) confirms the previous thermomineral waters classification (Aires-Barros et al., 1998; Marques et al., 1998a,b) where two main groups of $CO_2$-rich mineral waters have been proposed:

Group I: Chaves spring and borehole waters – are hot waters with temperatures ranging from 48°C to 76°C, dry residuum (DR) of about 1600 mg/L to 1850 mg/L, and free $CO_2$ between 350 mg/L and 1100 mg/L. The pH values are close to 7. The associated gas phase issued from the $CO_2$-rich springs is practically pure $CO_2$ (other minor components are $O_2 = 0.05\%$, $Ar = 0.02\%$, $N_2 = 0.28\%$, $CH_4 = 0.009\%$, $C_2H_6 = 0.005\%$, $H_2 = 0.005\%$ and $He = 0.01\%$ – data from Almeida, 1982). In this group is also included the low temperature (≈17°C) Vilarelho da Raia cold spring and borehole waters, with similar chemical composition comparatively to Chaves hot waters (DR values are between 1790–2260 mg/L and free $CO_2$ is of about 790 mg/L).

Group II: Vidago and Pedras Salgadas spring and borehole waters – present low temperature (≈17°C) and pH values (pH ≈6) and are distinguished from Group I thermomineral waters by higher Ca, Mg and free $CO_2$ content ($CO_2$ up to 2500 mg/L). Some of Vidago borehole waters show considerable higher mineralization (DR ≈4300 mg/L).

The rather constant ratios of ionic species such as $HCO_3$, Na, and Sr plotted against a conservative element such as Cl (Figure 2) should be faced as geochemical signatures of different degrees of water-rock interaction within a similar geological environment. The trend enhanced in Figure 2 seems to indicate that most Cl in the mineral waters could be derived from granitic rocks by leaching. In fact, the mineral waters displaying the highest $HCO_3$ and Na concentrations are also those presenting the highest Cl values. As stated by Edmunds et al. (1985), acid hydrolysis of plagioclase and biotite could be the main source of salinity in groundwaters percolating through granitic rocks. However, this trend is not so obvious for some Pedras Salgadas mineral waters. Some samples show a rather wide variation in the Sr, Na and $HCO_3$ ions, and a rather constant chloride concentration. This pattern seems to be ascribed to the fact that the recharge areas of such group of mineral

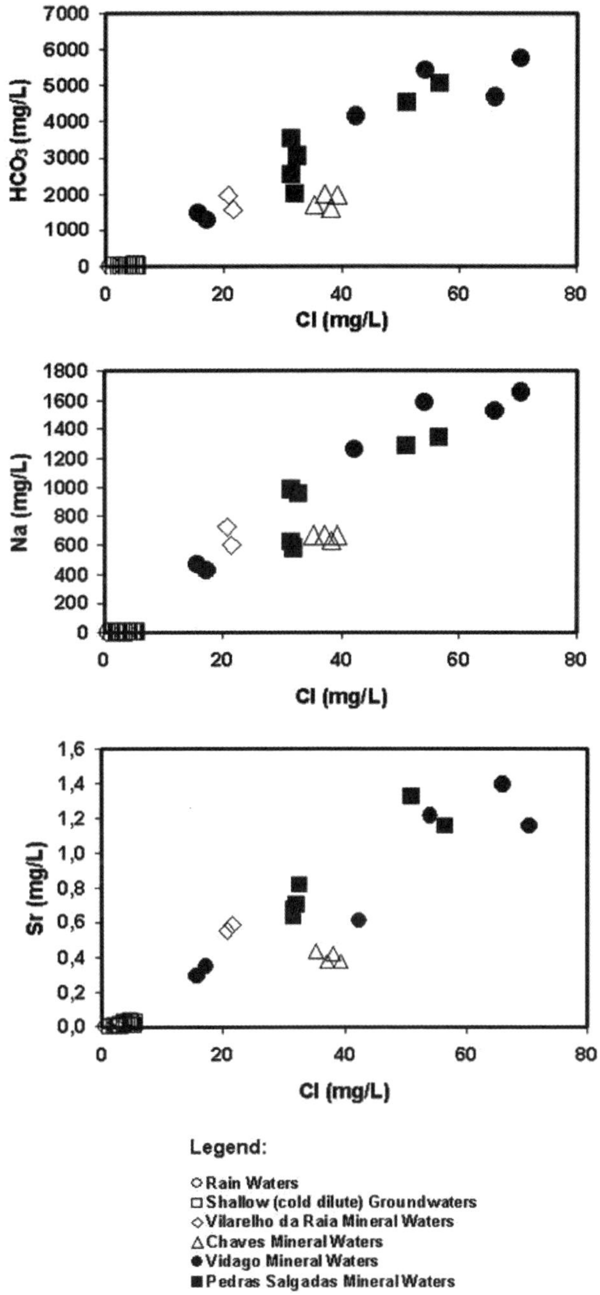


Figure 2. Cl *vs* $HCO_3$, Na and Sr for the waters from Vilarelho da Raia – Pedras Salgadas region. Modified from Andrade (2003).

waters should be located within the contact between the phyllites and granitic rocks, where preferential leaching of phyllites can occur.

The quartz and chalcedony oversaturation (see Aires-Barros et al., 1995) of Chaves hot mineral waters (encouraging silica deposition rather than dissolution of rock-forming quartz) should be faced as a geochemical signature that fluid inclusions are not the main source of salinity in these mineral waters.

The reason why some of the most mineralized waters in the area are cold waters is related with the fact that in $CO_2$-rich thermomineral waters carbon dioxide (more than temperature) is one of the most important species which influences the chemical and physical characteristics of the fluids (Criaud and Fouillac 1986; Greber 1994). In these type of thermomineral waters, water-rock interaction is enhanced by low temperatures since the solubility of $CO_2$ in water increases with decreasing temperature. The fact that water-rock interaction is mainly governed by $CO_2$ rather than by high temperatures is indicated by the good correlation between Cl (a tracer of water-rock interaction) and Na (dominant cation). As stated by Stumm and Morgan (1981) the solubility of albite increases considerably with rising partial pressure of $CO_2$. In the diagrams of Figure 2 the data from Chaves hot mineral waters is plotted in a cluster (presenting same geochemical signatures), which is a good indication of the existence of a common reservoir for Chaves hot mineral waters. On the other hand, Vidago and Pedras Salgadas cold $CO_2$-rich mineral waters (Group II waters) presenting distinct geochemical signatures seem to be ascribed to different water-gas-rock interactions and consequently different underground flowpaths. Since this group of waters are cold, they should be associated to shallow underground flowpaths, favouring a high $CO_2$ content, which could explain the high mineralization of some of these cold mineral waters. The calculated values of $P_{CO2}$, using the computer program HIDSPEC (a hydrogeochemical model that calculates the speciation of natural waters, in: Carvalho and Almeida, 1989), are very high (between 0.16 atm in Chaves and 1.20 atm in Vidago mineral waters) suggesting a deep source for the $CO_2$ (Aires-Barros et al., 1998). The relatively low Sr (and $HCO_3$ values) found in the Chaves hot waters may be caused by boiling whereas the relatively high Sr and $HCO_3$ at Vidago and Pedras Salgadas cold mineral waters strongly demonstrate the increasing solubility of divalent ions (Sr, Ca and Mg) with decreasing temperature (Marques et al., 2001). As stated by Goff et al. (1991), although the concentrations of the conservative species increase during flashing or boiling, the Sr concentration may decrease due to loss of $CO_2$ during flashing by the reaction:

$$Sr^{2+} + 2HCO_3^- \rightarrow SrCO_3 + CO_2 + H_2O$$

Chemical geothermometry is one of the most important geohydrologic tools in the exploration of thermomineral water resources. The main objective of geothermometric interpretation is to use the chemistry of hot springs in order to estimate chemical and physical properties of the reservoir fluid. This methodology depends upon the temperature dependence of the concentrations of certain species, the chemical equilibrium between minerals and water and various chemical reactions. Chemical geothermometers should be applied with caution to the $CO_2$-rich thermomineral waters since chemical modifications associated with cooling in upflow zones can be large and major and trace-element concentrations in natural spring waters could be mainly controlled by secondary reactions (Michard, 1990). Nevertheless, concerning the studied $CO_2$-rich mineral waters, the simultaneous evaluation of the results of the $SiO_2$ and $K^2/Mg$ geothermometers have been performed by Aires-Barros et al. (1998). The data from Chaves $CO_2$-rich mineral waters (the most representative of the deep fluids in this area) indicate equilibrium temperatures around 120°C, which are in agreement with the issue temperatures of Chaves mineral waters (76°C). Considering the mean geothermal gradient of 30°C/km (Duque et al.,

1998), we can estimate a maximum depth of about 3.5 km reached by the Chaves water system. This value was obtained considering that:

$$\text{depth} = (T_r - T_a)/gg$$

where $T_r$ is the reservoir temperature (120°C), $T_a$ the mean annual temperature (15°C) and gg the geothermal gradient (30°C/km). Accepting that water mineralization is more controlled by the availability of $CO_2$ rather than by the temperature (Greber, 1994), the cold $CO_2$-rich mineral waters from Vilarelho da Raia, Vidago and Pedras Salgadas should be faced as different stages of water-rock interaction processes involving local circulation of shallow groundwaters.

Isotope geochemistry has greatly contributed to the present understanding of Vilarelho da Raia, Chaves, Vidago and Pedras Salgadas thermomineral waters. In this paper we will review the use of isotopes ($^{18}O$, $^2H$ and $^3H$) to address questions associated to the studied thermomineral waters, in particular to recharge, flow systems and mixing problems, emphasising that investigations that uses stable isotope data integrated with chemical and other relevant data (such as lithological or morphostructural characteristics) usually produce important results. $\delta^{18}O$ and $\delta^2H$ values of the hot and cold $CO_2$-rich mineral waters (Figure 3) lie on or close to the Global Meteoric Water Line (GMWL: $\delta^2H = 8\ \delta^{18}O + 10$) indicating that they are meteoric waters which have been directly infiltrated into the bedrock (with no signs of surface evaporation). Based on $\delta^{18}O$ and $\delta^2H$ values of cold dilute spring water samples collected at different altitudes in the Vilarelho da Raia – Pedras Salgadas region, the local meteoric water line (LMWL: $\delta^2H = 7.01 \pm 0.74\ \delta^{18}O + 4.95 \pm 3.9$) was calculated (Figure 4). Data presented in Figures 3 and 4 indicates that both thermomineral and shallow cold dilute groundwaters present a similar range in the isotopic ($\delta^2H$ and $\delta^{18}O$) composition. This pattern indicates that both mineral waters and cold dilute spring waters have been recharged under similar climatic conditions, and that even the hot $CO_2$-rich mineral waters have no signs of water-rock interaction at elevated temperatures (consistent with the results of the chemical geothermometry). In fact, no significant $^{18}O$-shift was observed in the hot and

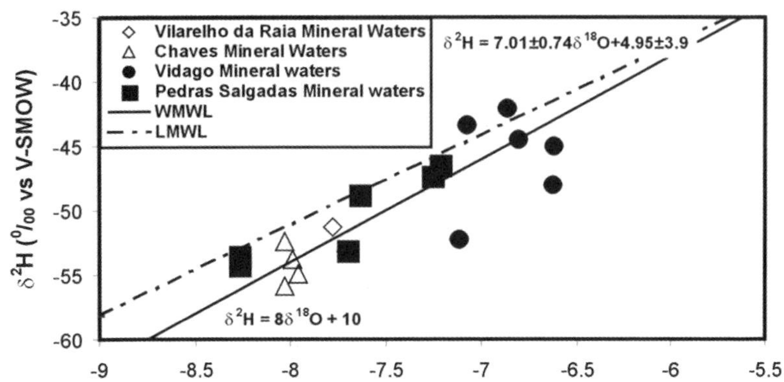


Figure 3. Plot of $\delta^{18}O$ *vs* $\delta^2H$ (‰ *vs* V-SMOW) for the studied $CO_2$-rich mineral waters. Modified from Andrade (2003).

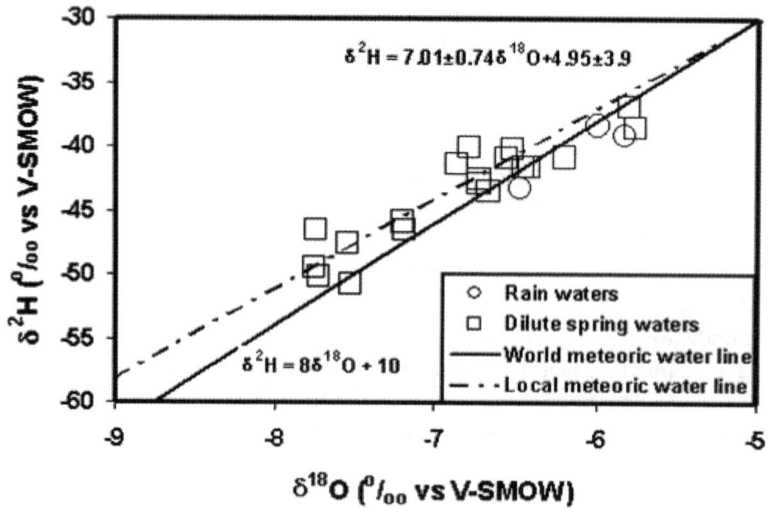


Figure 4. Plot of $\delta^{18}O$ *vs* $\delta^2H$ (‰ *vs* V-SMOW) for the cold dilute spring waters of the region. Modified from Andrade (2003).

cold $CO_2$-rich mineral waters indicating that, within this group of waters, water-rock interaction takes place at temperatures bellow 150°C, at depth. It should be stated that important $^{18}O$-shifts are common in "high-temperature" geothermal systems (e.g., The Geysers/USA, Wairakei/New-Zealand, Larderello/Italy) due to water-rock interaction at temperatures higher than 150°C (see IAEA, 1981).

Since the isotopic composition of meteoric groundwater generally matches the mean composition of precipitation over the recharge area (IAEA, 1981), we have used locally derived shallow groundwaters (cold dilute spring waters) to characterise the altitude dependence of the isotopic composition of the $CO_2$-rich thermomineral waters (Figure 5). All the selected spring waters show $^3H$ concentrations up to 8.5 TU, ascribed to local recharge. The results obtained indicate that in the studied region there is a decrease in $\delta\ ^{18}O$ of −0.23 ‰ per 100 m rise in altitude. According to IAEA (1981) the isotope gradients vary between −0.15 and −0.5 ‰ $\delta^{18}O$/100 m, pointing to an average rate of depletion of −0.26 ‰ (calculations based on the isotopic composition of precipitation water samples). Using these values we have achieve an approximate mean recharge elevation for the studied $CO_2$-rich mineral waters, namely: Chaves and Vilarelho da Raia 1000 m a.s.l; Vidago 500–700 m a.s.l.; Pedras Salgadas 800–1200 m a.s.l. According to the geomorphology of the region, the recharge of the studied mineral waters seems to occur preferentially on the E block of the Chaves Depression (Padrela Mountain).

The systematic (Aires-Barros et al., 1995, 1998; Marques et al., 2000b; Andrade 2003) presence of tritium (2 to 4.5 TU) measured in some of the Vidago and Pedras Salgadas cold $CO_2$-rich mineral waters (AC16 and AC17 boreholes, respectively) is difficult to attribute to a mixing process with local shallow (cold dilute) groundwaters (Figure 6).

On one hand, the Cl concentration of Vidago AC16 $CO_2$-rich mineral waters could be faced as a mixing signature, which is not consistent with the calculated $P_{CO2}$ values (around 1.20 atm, see Aires-Barros et al., 1998). On the other hand, Pedras Salgadas AC25 $CO_2$-rich mineral waters present similar Cl contents to the AC17 $CO_2$-rich mineral waters

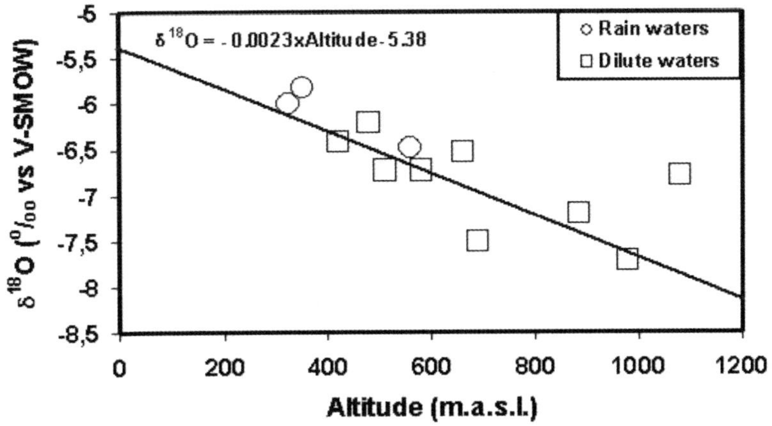


Figure 5. Plot of altitude *vs* $\delta^{18}O$ (‰ *vs* V-SMOW) for the shallow groundwaters (cold dilute spring waters) of the region. Modified from Andrade (2003).

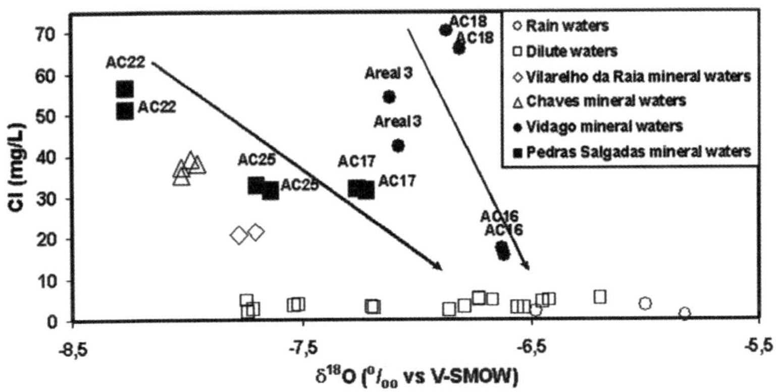


Figure 6. Cl *vs* $\delta^{18}O$ for the waters from Vilarelho da Raia – Pedras Salgadas region. Adapted from Andrade (2003).

(see Figure 6), but no $^3H$ concentration (Annex 1). So, the systematic $^3H$ concentrations found in Vidago AC16 and Pedras Salgadas AC17 cold mineral waters seems to be ascribed to shallow (and short) underground flow paths, being the water mineralization strongly controlled by the $CO_2$ content. The $CO_2$ is, most probably, initially transported as a separate gas phase that is subsequently mixed, in a shallow environment, with the circulating waters in a fractured system.

## 4.2 *What is the source of $CO_2$ in the mineral waters?*

The carbon dioxide present in thermomineral waters can be attributed to two main origins: organic and inorganic (Panichi and Tongiorgi, 1975). Concerning the organic source, $CO_2$ can be produced from decay of organic matter with mean $\delta^{13}C$ values around −26 ‰ and −22 ‰. Among the inorganic sources, carbon in thermomineral waters may originate from: (i) deep-seated (upper mantle) carbon showing $\delta^{13}C$ values ranging between $-8$ ‰ and $-1$ ‰, (ii) dissolution of limestones which have mean $\delta^{13}C$ values close to 0 ‰, or (iii) metamorphism of carbonates producing $CO_2$ with slightly positive $\delta^{13}C$ values (Truesdell and Hulston, 1980).

Carbon-13 determinations carried out on total dissolved inorganic carbon (TDIC) of the $CO_2$-rich mineral waters gave values lying in the range of $-6$ ‰ to $-1$ ‰ *vs* V-PDB (Marques et al., 2000b) indicating a deep-seated (upper mantle) origin for the $CO_2$. These results corroborate the $\delta^{13}C$ values measured on $CO_2$ gas samples ($\delta^{13}C(CO_2) = -5.72$ ‰ *vs* PDB) referred by Almeida (1982), for the case of Chaves thermomineral waters. Nevertheless, it should be noted that the use of measurements on free $CO_2$ may lead to wrong conclusions if the carbon remaining in water is not considered (R. Kreulen 1996 – personnel communication). Another possible interpretation of our $\delta^{13}C$ data could be a simple mixing of magmatic $CO_2$ and $CO_2$ derived from limestones. In fact, some calcareous lenses have been reported by Brink (1960) in Vila Real area (at the south of our study area). They have a width of about 30 m and a length of about 1.5 km, and occur in a zone of low-grade metamorphism that mainly comprises the chlorite-sericite schists and phyllites. Schermerhorn (1956) correlates the limestones intercalated in the Beira schists (the so-called Ante-Ordovician schisto-graywacke complex) of northern Portugal (Trás-os-Montes, Douro region) with those occurring NW Spain (Galicia). However, the $^{87}Sr/^{86}Sr$ data (forward discussed) suggests that that the presence of carbonate rock levels at depth as a possible carbon contributor for the mineral waters should be excluded. So, an alternative interpretation is that the $\delta^{13}C$ values of the $CO_2$ (of magmatic origin) could be shifted to less negative values as the result of fractionation at low temperature. Fractionation could take place during the exsolution of dissolved $CO_2$ (Aires-Barros et al., 1998).

Pérez et al. (1996) measured and discussed the $^3He/^4He$ and $^4He/^{20}Ne$ from terrestrial fluids in the Iberian Peninsula. According to them, the isotopic signature of helium in a fluid sample from Cabreiroa, located at the same NNE-trending fault of the studied $CO_2$-rich thermomineral waters, is significantly higher than typical crustal helium ($^3He/^4He$ value of 0.69). Pérez et al. (1996) estimated that, in the fluid sample from Cabreiroa, the helium's fractions from mantle, crustal and atmospheric reservoirs were 8.35%, 91.62% and 0.02%, respectively. The relatively high $^3He/^4He$ found in Cabreiroa sample corroborates an important mantle-degassing component.

## 4.3 *What is the age of the mineral waters?*

The use of radioactive isotopes in dating groundwater play an important role in assessing the dynamics of the groundwater systems, in the characterisation of the water resources and in planning its exploitation. Dating confined aquifers (old groundwater) remains one of the mains concerns of the isotope hydrology. Radioactive isotopes ($^3H$, $^{14}C$) together with stable isotopes ($^2H$, $^{13}C$ and $^{18}O$) are fundamental tools in the identification of old groundwaters. Among the radioactive isotopes with a half-life higher than $10^3$ years, carbon-14 (half-life = 5730 years) represents the most important tool in groundwater dating. This isotope is present in the atmosphere, soils, aquifer matrix, etc. The different source of carbon incoming to the groundwater system makes dating complicated, because the geochemical reactions that can be involved.

In the case of Chaves thermomineral waters the issue temperature (76°C) indicates the existence of a geothermal system associated with a relatively deep circulation and consequently a large residence time, supported by the absence of $^{3}H$ and a $^{14}C$ activity of 4.3 pmC (Aires-Barros et al., 1998). Nevertheless, admitting a residence time large enough for all tritium decay, one have to consider an income of carbon, free of $^{14}C$, to the water system. This assumption is based on the $\delta^{13}C$ signatures ($-2.8$ ‰ *vs* PDB) presented by Chaves thermomineral waters (Aires-Barros et al., 1998). In contrast, the systematic presence of tritium (2 to 4.5 TU) in Vidago (AC16) and Pedras Salgadas (AC17) cold $CO_2$-rich mineral waters could be attributed to a shallow (and short) flow path. In these mineral waters the $^{3}H$ content is not in agreement with the $^{14}C$ values $\approx$7.9 to 9.9 pmC (Aires-Barros et al., 1998). Using these $^{14}C$ values, the apparent $^{14}C$ groundwater ages estimated by a piston model (Salem et al., 1980) are close to 9 ka BP. These results indicate that total carbon in the recharge waters is being masked by larger quantities of $CO_2$ ($^{14}C$-free) introduced from deep-seated (upper mantle) sources.

## 4.4 *What are the main water-rock interaction processes occurring at depth?*

Sr isotopes, commonly applied to the resolution of petrological problems, can used as an important tool in geohydrology (e.g., Stettler and Allègre, 1978; McNutt et al., 1990). The $^{87}Sr/^{86}Sr$ ratios are very good hydrogeochemical tracers because the high atomic weight of strontium avoids isotopic fractionation by any natural process. The measured differences in the $^{87}Sr/^{86}Sr$ ratios in waters can be ascribed to the mixing of Sr derived from different rock sources with different isotopic compositions. Water-rock interaction causes rocks with different chemical characteristics and ages to release Sr into water. So, the $^{87}Sr/^{86}Sr$ ratios in waters depend on the Rb/Sr ratios and the age of the percolated rocks (Faure, 1986).

The $^{87}Sr/^{86}Sr$ values of the following suite of mineral waters (Vilarelho da Raia/Chaves – Vidago – Pedras Salgadas) do not show a progressive enrichment in radiogenic strontium (Figure 7): Vilarelho da Raia/Chaves ($^{87}Sr/^{86}Sr$ = 0.727154 to 0.728035), Vidago ($^{87}Sr/^{86}Sr$ = 0.720622 to 0.72428), and Pedras Salgadas ($^{87}Sr/^{86}Sr$ = 0.716713 to 0.717572). This behaviour can be ascribed to the inexistence of an hydraulically related flow path

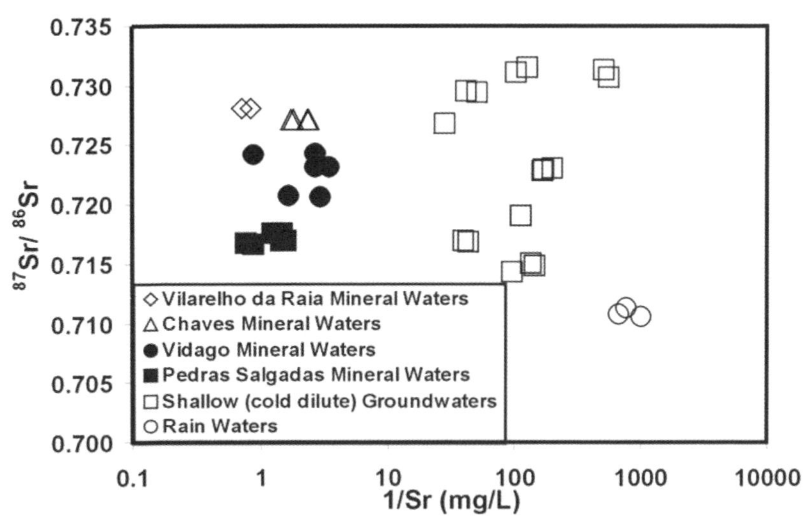


Figure 7. 1/Sr *vs* $^{87}Sr/^{86}Sr$ for the waters from Vilarelho da Raia – Pedras Salgadas region. Modified from Marques et al. (2001).

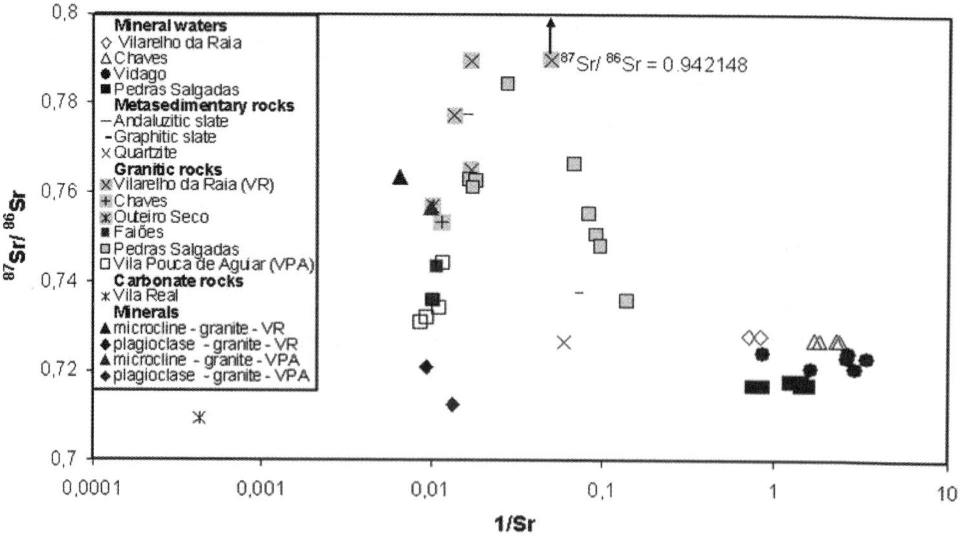


Figure 8. Plot of 1/Sr (mg/L) *vs* $^{87}Sr/^{86}Sr$ for the rocks, minerals and $CO_2$-rich mineral waters. Adapted from Andrade 2003.

from Vilarelho da Raia towards Pedras Salgadas, along the Verin – Chaves – Régua – Penacova trending fault. This trend confirms the idea that the three group of thermomineral waters could be the result of similar hydrogeological systems but not the same system, being the different Sr isotopic signatures related with the existence of water-rock interaction with different granitic rocks. As stated by Marques et al. (2001), the scattering of the Sr data of local rain waters, shallow cold dilute groundwaters and thermomineral waters can be understood through the existence of three end-members ((a) Vilarelho da Raia/Chaves, (b) Vidago and (c) Pedras Salgadas) of a concentration trend, from the rain waters towards the thermomineral waters (Figure 7).

The large difference observed between the $^{87}Sr/^{86}Sr$ ratios in the shallow cold dilute groundwaters can be explained by the difference in the nature and degree of weathering of the percolated rocks. The $^{87}Sr/^{86}Sr$ ratio of the waters is usually mainly constrained by the relative leaching stabilities of the different minerals. As indicated by Stettler and Allègre (1978), plagioclases and minerals such as biotite, pyroxene, and hornblende provide most of dissolved ions. On the other hand, potassium feldspars and silica are more slowly attacked. In view of the average Sr distribution among these minerals and characteristic $^{87}Sr/^{86}Sr$ ratios found in the dominant rock types of the region, one can presume that the isotopic ratio in the water should be roughly similar to the ratio of the percolated rock. In our case, the slight drop of the $^{87}Sr/^{86}Sr$ ratio in waters could be interpreted as the result of the reduced contribution from K-feldspar which usually bears Sr with relatively high $^{87}Sr/^{86}Sr$ ratios.

A comparison of $^{87}Sr/^{86}Sr$ values among the mineral waters and the rocks/minerals from Vilarelho da Raia – Pedras Salgadas area is shown in Figure 8. The Sr isotopic signatures indicate (i) that no equilibrium has been reached between the mineral waters and the whole rocks and (ii) that the Sr isotope values were obtained from equilibrium being reached between those fluids and particular granite minerals. In fact, the mean Sr isotopic ratio of the mineral waters ($^{87}Sr/^{86}Sr_{mean}$ = 0.722419) is similar to the Sr isotopic ratios of the plagioclases of the granitic rocks presented by Aires-Barros et al. (1998): Vilarelho da Raia-plag. $^{87}Sr/^{86}Sr = 0.72087$ and Vidago-plag. $^{87}Sr/^{86}Sr = 0.71261$, suggesting that the plagioclase dissolution is the main water-rock interaction process.

The hypothesis of the existence of water-limestone interaction to explain the origin of $CO_2$ should be considered rather limited, once the composition of the Sr and the $^{87}Sr/^{86}Sr$ ratio of limestone ($^{87}Sr/^{86}Sr = 0.709485$) is rather different from the mineral waters. This interpretation was also based on the fact that the $CO_2$-rich mineral waters displaying the less negative $\delta^{13}C$ value (Vidago AC18: $-1.0$ ‰ *vs* V-PDB) show a high $^{87}Sr/^{86}Sr$ ratio ($^{87}Sr/^{86}Sr = 0.724280$).

## 5 CONCLUDING REMARKS

This study has demonstrated the usefulness of coupled geochemical and isotopic studies on a suite of $CO_2$-rich thermomineral waters issuing in the Northern part of Portuguese mainland. $CO_2$-rich mineral waters from Vilarelho da Raia – Pedras Salgadas region represent locally meteoric waters recharged at different altitudes. Isotopic signatures of C present in these mineral water systems indicate a deep-seated (upper-mantle) origin for the $CO_2$. The most probable explanation by which the carbon dioxide could be transported from its deep source to the surface involves migration as a separate gas phase being incorporated in the infiltrated meteoric waters (at considerable depth in the case of the hot $CO_2$-rich mineral waters and at shallow levels in the case of cold $CO_2$-rich mineral waters). Other solutes (such as Na and $HCO_3$) are originated from the local granitic rocks, being the dissolution favoured by the high $CO_2$ content of the circulating waters. The Sr isotopes and Sr concentrations in the thermomineral waters and rocks have provided a clear picture on the influence of varying rock types on the thermomineral waters signatures. The Sr-isotope data presented in this study strongly suggest that Vilarelho da Raia, Chaves, Vidago and Pedras Salgadas $CO_2$-rich thermomineral waters should be faced as surface manifestations of different underground flowpaths.

## ACKNOWLEDGMENTS

We would like to thank the Águas de Carvalhelhos Enterprise, the Municipality of Chaves and the Vidago, Melgaço and Pedras Salgadas Enterprise for the help in the fieldwork campaigns. This work was funded by the PRAXIS R&D Project "FLUMIRE" under the Contract No. C/CTE/11004/98. An early draft of this manuscript was critically read by Peter Kralj and we gratefully acknowledge his contribution.

## REFERENCES

Aires-Barros L. Marques JM, Graça RC (1995) Elemental and isotopic geochemistry in the hydrothermal area of Chaves / Vila Pouca de Aguiar (Northern Portugal). Environmental Geology 25 (4): 232–238

Aires-Barros L. Marques JM, Graça RC, Matias MJ, van Der Weijden CH, Kreulen R, Eggenkamp H GM (1998) Hot and cold $CO_2$-rich mineral waters in Chaves geothermal area (northern Portugal). Geothermics 27 (1): 89–107

Almeida FM (1982) Novos dados geotermométricos sobre águas de Chaves e de S. Pedro do Sul [New geotermometric data on Chaves and S. Pedro do Sul waters]. Comunicações Serviços Geológicos de Portugal 68 (2): 179–190

Andrade MPL (2003) A geoquímica isotópica e as águas termominerais. Contribuição dos isótopos do Sr ($^{87}Sr/^{86}Sr$) e do Cl ($^{37}Cl/^{35}Cl$) na elaboração de modelos de circulação. O caso de algumas águas gaso-carbónicas do N de Portugal [Isotopic geochemistry and thermomineral waters. Contribution of Sr ($^{87}Sr/^{86}Sr$) and Cl ($^{37}Cl/^{35}Cl$) isotopes to the elaboration of circulation models. The case of some $CO_2$-rich waters from N Portugal]. MSc Thesis. Technical University of Lisbon. Instituto Superior Técnico

Baptista J, Coke C, Dias R, Ribeiro A (1993) Tectónica e geomorfologia da região de Pedras Salgadas/ Vidago e as nascentes minerais associadas [Tectonics and geomorfology of Pedras Salgadas region and associated mineral springs]. Comunicações da XII Reunião de Geologia do Oeste Peninsular I: 125–139

Brink AH (1960) Petrology and ore geology of the Vila Real – Sabrosa – Vila Pouca de Aguiar region, northern Portugal. Comunicações dos Serviços Geológicos de Portugal XLIII

Carvalho MR, Almeida C (1989) HIDSPEC, um programa de especiação e cálculo de equilíbrios água/rocha [HIDSPEC, a speciation and water/rock equilibrium calculation program]. Revista da Universidade de Aveiro 4 (2): 1–22

Criaud A, Fouillac C (1986) Étude des eaux thermominérales carbogazeuses du Massif Central Français. II. Comportment de quelques métaux en trace, de l'arsenic, de l'antimoine et du germanium. Geochimica et Cosmochimica Acta 50: 1573–1582

Duque R, Monteiro Santos FA, Mendes-Victor LA (1998) Heat flow and deep temperatures in the Chaves Geothermal system, northern Portugal. Geothermics 27 (1): 75–87

Edmunds WM, Kay RLF, McCartney RA (1985) Origin of saline groundwaters in the Carnmenellis granite (Cornwall, England): natural Processes and reaction during hot dry rock reservoir circulation. Chemical Geology 49: 287–301

Hem JD (1970) Study and Interpretation of the Chemical Characteristics of Natural Water. 2nd Eds. Geological Survey Water. United States Department of the Interior.

Ellis AJ, Mahon WAJ (1977) Chemistry and geothermal systems. Energy Science and Engineering: resources, technology, management. An International Series. Academic Press

Epstein S, Mayeda T (1953) Variation of $^{18}O$ content of waters from natural sources. Geochimica et Cosmochimica Acta 4: 213–24

Faure G (1986) Principles of Isotope Geology. 2nd Edition. John Wiley & Sons

Friedman I (1953) Deuterium content of natural waters and other substances. Geochimica et Cosmochimica Acta 4: 89–103.

Goff F, Wollenberg HA, Brookins, DC, Kristler RW (1991) A Sr-isotopic comparison between thermal waters, rocks, and hydrothermal calcites, Long Valley caldera, California. Journal Volcanology and Geothermal Research 48: 265–281

Greber E (1994) Deep circulation of $CO_2$-rich palaeowaters in a seismically active zone (Kuzuluk/ Adaparazi, northwestern Turkey). Geothermics 23 (2): 151–174

IAEA [International Atomic Energy Agency] (1981) Stable isotopes hydrology. Deuterium and oxygen-18 in water cycle. IAEA, Vienna, Technical Reports Series 210

Marques JM, Aires-Barros L, Graça RC, Matias MJ, Basto MJ (1998a) Fluid chemistry and water-rock interaction in a $CO_2$-rich geothermal area, Northern Portugal. In: Arehart GB, Hulstron JR (eds) Proceedings of the 9th International Symposium on Water-Rock Interaction – WRI-9 / Taupo, New Zealand, A.A. Balkema, Rotterdam: 637–640.

Marques JM, Aires-Barros L, Graça RC, Matias MJ, Basto MJ (2000a) Water/rock interaction in a $CO^2$-rich geothermal area (Northern Portugal): an $^{18}O/^{16}O$ and $^{2}H/^{1}H$ isotope study. Geothermal Resources Council Transactions 24: 253–258

Marques JM, Andrade M, Aires-Barros L, Graça RC, Eggenkamp HGM, Antunes da Silva M (2001) $^{87}Sr/^{86}Sr$ and $^{37}Cl/^{35}Cl$ signatures of $CO_2$-rich mineral waters (N-Portugal): preliminary results. In: Seiler K-P, Wohnlich S (eds) New approaches characterizing groundwater flow. A.A. Balkema: 1025–1029

Marques JM, Andrade M, Carreira PM, Graça RC, Aires-Barros L (2003) Evolution of $CO_2$-rich mineral waters from Hercynian granitic rocks (N-Portugal): questions and answers. In: Krásny J, Hrkal Z, Bruthans J (eds) Proceedings of the International Conference on Groundwater in Fractured Rocks, Prague: 217–218

Marques JM, Carreira PM, Aires-Barros L, Graça RC (1998b) About the origin of $CO_2$ in some $HCO_3$/Na/$CO_2$-rich Portuguese mineral waters. Geothermal Resources Council Transactions 22: 113–117

Marques JM, Carreira PM, Aires-Barros L, Graça RC (2000b) Nature and role of $CO_2$ in some hot and cold $HCO_3$/Na/$CO_2$-rich Portuguese mineral waters: a review and reinterpretation. Environmental Geology 40 (1): 53–63

McNutt RH, Frape SK, Jones MG, MacDonald IA (1990) The $^{87}Sr/^{86}Sr$ values of Canadian Shield brines and fracture minerals with applications to groundwater mixing, fracture History, and geochronology. Geochimica et Cosmochimica Acta 54: 205–215

Michard G (1990) Behaviour of major elements and some trace elements (Li, Rb, Cs, Sr, Fe, Mn, W, F) in deep hot waters from granitic areas. Chemical Geology 89: 117–134

Panichi C, Tongiorgi E (1975) Carbon isotopic composition of $CO_2$ from springs, fumaroles, moffetes and travertines of Central – Southern Italy: a preliminary prospection method of geothermal area. Proceedings of the Second United Nations Symposium on the Development and Use of Geothermal Resources, S. Francisco, C.A: 815–825

Pérez NM, Nakai S, Wakita H, Albert-Bertrán JF, Redondo R (1996) Preliminary results on $^3He/^4He$ isotopic ratios in terrestrial fluids from Iberian peninsula: seismoctectonic and neotectonic implications. Geogaceta 20 (4): 830–833

Salem O, Visser JM, Dray M, Gonfiantini R (1980) Groundwater flow patterns in the western Libyan Arab Jamahiriya evaluated from isotopic data. In: International Atomic Energy Agency (eds) Arid-Zone Hydrology: investigation with isotope techniques. Proc. Advisory Group Meeting Vienna (1978): 165–179

Schermerhorn LJG (1956) The age of the Beira Schists (Portugal). Boletim da Sociedade Geológica de Portugal XII (1/2): 77–100

Sousa Oliveira A (1995) Hidrogeologia da Região de Pedras Salgadas [Hydrogeology of Pedras Salgadas Region]. MSc Thesis. Universidade de Trás-os-Montes e Alto Douro. Vila Real, Portugal

Sousa Oliveira A, Portugal Ferreira MR (1996) A estruturação do sistema graben – horst cruzado da região de Pedras Salgadas – Vidago (Norte de Portugal): enquadramento das emergências hidrominerais associadas [Structure of the cross graben-horst system in the Pedras Salgadas – Vidago region (North of Portugal): frame of the associated hydromineral springs.]. Proceedings do 3° Congresso da Água / VII SILUBESA III, Lisboa: 123–130

Stettler A, Allègre CJ (1978) $^{87}Rb$-$^{87}Sr$ studies of waters in a geothermal area, the Cantal, France. Earth and Planetary Science Letters 38: 364–372

Stumm W, Morgan JJ (1981) Aquatic Chemistry – An Introduction Emphasizing Chemical Equilibria in Natural Waters (2nd edn), Wiley-Interscience, New-York

Truesdell AH, Hulston JR (1980) Isotopic evidence on environments of geothermal systems. In: Fritz P, Fontes JCh (eds) Handbook of Environmental Isotope Geochemistry, 1, The Terrestrial Environment: 179–226

Van der Borg K, Alderliesten C, Haitjema H, Hut G, Van Zwol NA (1984) The Utrecht accelerator facility for precision dating with radionuclides. Nuclear Instruments and Methods in Physics Research B5: 150–154

Van der Borg K, Alderliesten C, Houston CM, de Jong, AFM, Van Zwol NA (1987) Accelerator mass spectrometry with $^{14}C$ and $^{10}Be$ in Utrecht. Nuclear Instruments and Methods in Physics Research B29:143–145

CHAPTER 22

# Origin of salinity in a fractured bedrock aquifer in the Central West Region of NSW, Australia

Karina Morgan and Jerzy Jankowski

*UNSW Groundwater Centre, School of Biological, Earth and Environmental Sciences, The University of New South Wales, Sydney, N.S.W. 2052, Australia*

ABSTRACT: The origin of salts in Na(Mg)-Cl-rich groundwaters contained within the deep fractured bedrock aquifer of the Oakdale Formation in the Spicers Creek catchment was determined using hydrogeochemical and isotopic data. Stable isotope data showed that these groundwaters were depleted relative to SMOW and plotted along the LMWL. This data implied that groundwaters were meteoric in origin and had not undergone evaporation prior to recharge. Therefore, an increase in salinity in the abovementioned groundwaters was not controlled by evaporation, diffusion of connate seawater or interaction with hydrothermal fluids. Observed $^{87}Sr/^{86}Sr$ isotope ratios in Na(Mg)-Cl-rich groundwaters resulted from isotopic exchange with Sr-rich plagioclase contained within the bedrock. Evolution of these groundwaters from the fractured bedrock aquifer contained within the Oakdale Formation occurred during prolonged and extensive water-rock interaction.

## 1 INTRODUCTION

The Spicers Creek catchment is located approximately 400 km west of Sydney and is contained within the larger Macquarie River Catchment that is located in the Central West Region of New South Wales, Australia. The catchment covers an area of approximately 500 km$^2$, with the Spicers Creek forming the main surface water system. Soils are fertile and support a healthy dryland agricultural industry of wheat and canola products. The Spicers Creek catchment is an example of a catchment experiencing fluctuating water tables, waterlogging and salinisation of surface water leading to a decline in agricultural productivity. Identification of the source of salt in this catchment is an important factor for the implementation of land management options to halt and/or reverse land degradation processes.

Groundwaters contained within the fractured bedrock aquifers located in the Spicers Creek catchment have varying salinities and groundwater chemistries that are dependent on aquifer lithology. The origin of salts contained within these fractured bedrock aquifers was identified by using various hydrogeochemical techniques. The presence of salt in the system appears to be the primary influence on hydrogeochemical processes. From initial hydrogeochemical and isotopic observation, the intermediate and deep groundwater systems appear to have two distinctive groundwater types; the Na(Mg)-Cl-rich and Na-$HCO_3$-rich groundwaters. Other groundwaters appear to represent mixing between these end-member groundwater groups and rainfall recharge to the system (Morgan, 2005).

## 2 HYDROGEOLOGY

Spicers Creek catchment is geologically and hydrogeologically complex. Schofield (1998) identified three main aquifer systems in the catchment. The first aquifer is the deep regional aquifer system contained within the fractured basement units of the Oakdale and Gleneski Formations. The second is the intermediate aquifer contained within the Mesozoic sediments and the third is the shallow unconsolidated aquifer system that consists of recent colluvial/alluvial sediments that form a thin veneer over the landscape. These units have been described extensively in Morgan (2005).

The main focus of this study is the deep fractured bedrock units throughout the catchment. The deep regional system is contained within Ordovician to Devonian fractured volcanic rocks. They include the Oakdale Formation in the west and the Gleneski Formation in the eastern part of the catchment. Groundwater flow within the fractured rock aquifers is generally controlled by tectonic fracture networks and shear zones (Meakin and Morgan, 1999). Major fracture zones in the Oakdale Formation have hydraulic conductivities that are several orders of magnitude greater than the non-fractured sections of the aquifer (Morgan et al., 2005).

## 3 HYDROCHEMICAL CATEGORISATION OF DEEP GROUNDWATERS

Groundwater chemistry in the Spicers Creek catchment is heterogeneous and is discussed according to the aquifer type that the groundwater samples were abstracted from. The Oakdale Formation is predominantly a submarine mafic volcanic unit (Morgan et al., 1999a). Groundwater EC of the Oakdale Formation range from 4,460 $\mu S\,cm^{-1}$ to 14,510 $\mu S\,cm^{-1}$. $Na^+$ dominates the cation composition with an average concentration of 6,850 $mg\,L^{-1}$ and $Cl^-$ dominates the anion composition with an average concentration of 4,720 $mg\,L^{-1}$. Sulphate concentrations are also high with concentrations greater than 1,200 $mg\,L^{-1}$. Water types range from Na-Cl-rich groundwaters to Mg-Na-Cl-rich and to mixed Na-Mg-Cl-$HCO_3$-rich waters. Groundwaters are neutral with an average pH of 7.1, reducing with an average Eh of −197 mV and tepid with an average temperature of 21°C. Oakdale Formation groundwaters contain elevated concentrations of $Sr^{2+}$, As and V. Most Oakdale Formation groundwaters have a strong $Cl^-$ dominance with varying concentrations of $Mg^{2+}$ and $Na^+$. The Oakdale Formation groundwaters represent the endmember Na(Mg)-Cl-rich groundwaters in the Spicers Creek catchment.

The Gleneski Formation is predominantly a submarine felsic volcanic unit (Morgan et al., 1999b). The Gleneski Formation groundwaters are generally less saline than Oakdale Formation groundwaters, with salinities ranging from 955 $\mu S\,cm^{-1}$ to 6,650 $\mu S\,cm^{-1}$, with an average of 3,640 $\mu S\,cm^{-1}$. They possess different chemical composition than those of the Oakdale Formation and are varied in cation and anion chemical composition. Groundwater facies range from Mg-Na-(Ca)-Cl-$HCO_3$ to Na-(Ca-Mg)-Cl-$HCO_3$ to Na-$HCO_3$-Cl-rich. They contain less $Na^+$ (average 330 $mg\,L^{-1}$), $Cl^-$ (average 1,980 $mg\,L^{-1}$), $Sr^{2+}$ (average 1,800 $\mu g\,L^{-1}$) and $SO_4^{2-}$ (average 67 $mg\,L^{-1}$), than Oakdale Formation waters. Gleneski Formation groundwaters have high $Mg^{2+}$ concentrations. These groundwaters have higher pH values, and they are reducing with an average Eh of −149 mV. The anion composition of Gleneski Formation groundwaters generally comprises 40 to 60% of $HCO_3^-$. Gleneski Formation groundwaters contain the lower concentrations of trace elements $Sr^{2+}$, B and $Li^+$ concentrations than the other aquifers in Spicers Creek catchment.

Groundwaters from the intermediate system that overlie the deep regional system are varied in groundwater chemistry and salinity because they have been sampled from various lithological units, which consist of terrestrial and marine sediments of Mesozoic age. Groundwater ECs range from 3,070 $\mu S\,cm^{-1}$ to 21,250 $\mu S\,cm^{-1}$ with an average of 8,400 $\mu S\,cm^{-1}$. Intermediate groundwaters range from highly reducing (−364 mV) to oxidised (+55 mV), with an average value of −130 mV. Intermediate groundwaters are acidic (pH 5.28) to alkaline (pH 8.50), with an average of 6.80 and have an average temperature of 21°C. Intermediate groundwaters contain the highest ionic variation in the catchment ranging from saline Na-Cl-rich groundwaters to mixed Na-(Mg-Ca)-Cl-$HCO_3$-rich, to relatively fresh Na-$HCO_3$-Cl-rich groundwaters. They contain high $Na^+$ concentrations (>4,500 $mg\,L^{-1}$) and high $Cl^-$ concentrations (>6,900 $mg\,L^{-1}$). Intermediate groundwaters have elevated $HCO_3^-$ and $SO_4^{2-}$ concentrations. They have the highest $Mg^{2+}$ (655 $mg\,L^{-1}$) and $Ca^{2+}$ (488 $mg\,L^{-1}$) concentrations in the catchment. Trace elements are also elevated in the intermediate groundwaters with notable concentration of B, $Sr^{2+}$, $Li^+$, As, Mn, Ni, Zn and Ag.

Na-$HCO_3$-rich groundwaters are part of the intermediate aquifer systems and the least saline bedrock groundwaters within the catchment, with an average EC value of 4,590 $\mu S\,cm^{-1}$. Na-$HCO_3$-rich groundwaters have a slightly higher average temperature (22°C) and are reducing with an average Eh of −162 mV and DO of 0.2 $mg\,L^{-1}$. Na-$HCO_3$-rich groundwaters have the lowest pH, with an average of 6.43 and are the most homogenous in hydrochemical character. These groundwaters are composed of mainly $Na^+$ with an average concentration of 1,077 $mg\,L^{-1}$, and $HCO_3^-$ with an average of 3,460 $mg\,L^{-1}$. These groundwaters also contain elevated $CO_2$ concentrations, with an average concentration of 1,190 $mg\,L^{-1}$ and elevated $K^+$ concentrations. These groundwaters have low $Cl^-$, $SO_4^{2-}$, $Mg^{2+}$ and $Ca^{2+}$ concentrations. Na-$HCO_3$-rich groundwaters contain elevated trace element concentrations such as B, Ba, Cr, Ga, $Li^+$ and Rb.

The general hydrochemical character of deep groundwaters in the Spicers Creek catchment indicate that groundwater chemistry is dependent on two main factors; the aquifer lithology and mixing between two end-member groundwaters present in the catchment. These end-members include the Na-$HCO_3$-rich groundwaters from the intermediate groundwater system and the Na(Mg)-Cl-rich groundwaters associated with the Oakdale Formation.

## 4 ORIGIN OF Na(Mg)-Cl-RICH GROUNDWATERS

Isotopic data was used to confirm the origin of water Na(Mg)-Cl-rich groundwaters and their salinity in the Spicers Creek catchment. $\delta^{18}O$ isotopic values of Na(Mg)-Cl-rich groundwaters range from −5.28‰ to −4.4‰, with an average of −4.96‰. $\delta^2H$ values range from −36.2‰ to −31.8‰, with an average of −33.06‰. Mixed groundwaters have $\delta^{18}O$ values ranging from −7.51‰ to −4.32‰, with an average of −5.29‰. $\delta^2H$ values range from −45.3‰ to −27‰, with an average of −34.26‰. Variations in these values

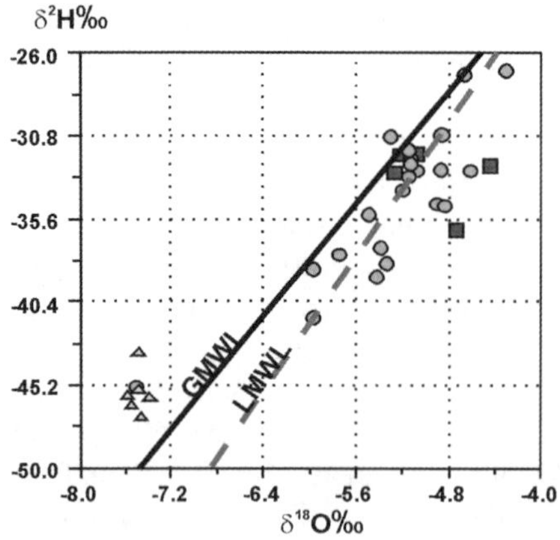


Figure 1. $\delta^2$H vs. $\delta^{18}$O for deep and intermediate groundwaters in Spicers Creek catchment.

occur due to the mixing of Na-$HCO_3$-rich groundwaters that are depleted in $\delta^{18}$O and $\delta^2$H and the deeper Na(Mg)-Cl-rich groundwaters that have more enriched $\delta^{18}$O and $\delta^2$H values.

The relationship between $^{18}$O and $^2$H was identified by plotting $\delta$ $^{18}$O versus $\delta$ $^2$H against the Global Meteoric Water Line (GMWL $\delta^2H‰ = 8.13\delta^{18}O + 10.8$) (Rozanski et al., 1993) and the Local Meteoric Water Line (LMWL $\delta$ $^2H‰ = 9.74\delta^{18}O + 16.63$) (Schofield, 1998). $\delta^{18}$O and $\delta^2$H values indicate that deep groundwaters in the Spicers Creek catchment are depleted relative to SMOW, with most values extrapolating towards the meteoric water lines (Figure 1). This trend implies groundwaters are derived from local rainfall. The relatively depleted isotopic signatures of $\delta^{18}$O and $\delta^2$H in Na(Mg)-Cl-rich groundwaters do not indicate the presence of connate marine or hydrothermal fluids. Connate seawater would result in enriched $\delta^{18}$O and $\delta^2$H values close to SMOW and hydrothermal fluids evolved from rock-water interactions at elevated temperatures would have enriched oxygen-18 values in the resultant water (Richter et al., 1993).

Due to the elevated salinities of Na(Mg)-Cl-rich groundwaters the possibility of evaporative concentration must be considered. This is assessed by analysing the slope of $\delta^{18}$O and $\delta^2$H data for Na(Mg)-Cl-rich groundwaters. This graph is likely to yield information on evaporation processes occurring prior to groundwater recharge (Gat, 1981; Clark and Fritz, 1997). Regression analysis of $\delta^{18}$O and $\delta^2$H data for Na(Mg)-Cl-rich groundwaters indicates they have a slope of $-1.85$, which implies that these groundwaters have not undergone evaporation prior to recharge. If groundwaters have undergone evaporation prior to recharge they would plot along a slope of approximately 4 on a $\delta^{18}$O versus $\delta^2$H bivariate plot.

To further show that evaporation is not a likely process affecting the Na(Mg)-Cl-rich groundwaters, the relationship between $\delta^{18}$O and $Cl^-$ was assessed (Figure 2). As $Cl^-$ concentrations increase in the groundwater sample, an increase in heavy $^{18}$O isotope would also be observed if evaporation had occurred (Turner et al., 1987). The correlation between $\delta^{18}$O and $Cl^-$ is weak ($r^2 = 0.410$) for Na(Mg)-Cl-rich groundwaters and mixed groundwaters show no correlation ($r^2 = 0.001$).

Stable isotope data implies Na(Mg)-Cl-rich groundwaters are of meteoric origin and have not undergone evaporation prior to recharge. Therefore, evaporation or mixing with

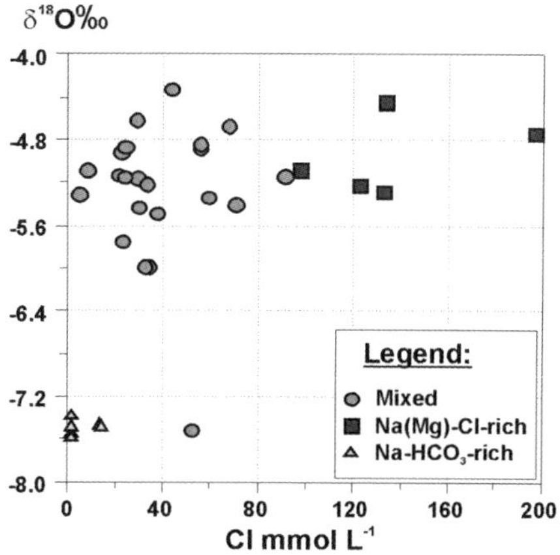


Figure 2. $\delta^{18}O$ vs. Cl for deep and intermediate groundwaters in Spicers Creek catchment.

connate seawater or hydrothermal fluid can be discounted as the primary mechanism leading to the increasing salinity in the Na(Mg)-Cl-rich groundwaters.

## 5 GEOCHEMICAL EVOLUTION OF Na(Mg)-Cl-RICH GROUNDWATERS

Na(Mg)-Cl-rich groundwaters evolve from the Oakdale Formation aquifer, which is composed of coarse epiclastic, rocks, lavas and shallow intrusives of felsic to intermediate origin (Ashley, 2001). Average mineral abundances were estimated from thin section. The rock samples were found to be composed of approximately 45% plagioclase, 15% K-feldspar, 10% quartz, 10% chlorite, 10% biotite, 5% sericite, 1% carbonate, 1% haematite, with minor epidote and trace minerals. Pyrite was also observed in the Oakdale Formation during drilling (MIM, 2002). Based on the observed lithology the following geochemical reactions are likely to influence the observed chemistry of in the Oakdale Formation aquifer. They include: the incongruent weathering of plagioclase, albite, anorthite, K-feldspar, chlorite and biotite; oxidation of pyrite; and the congruent dissolution of calcite and/or dolomite.

The geochemical evolution of Na(Mg)-Cl-rich groundwaters were assessed by using bivariate plots and ionic ratio calculations based on methods employed by Sami (1992) and Richter et al. (1993) and considering the above mentioned mineralogy of the Oakdale Formation. The relationship between $Na^+$ versus $Cl^-$ for Na(Mg)-Cl-rich groundwaters shows that they plot on or close to the 1:1 halite dissolution line, with a slight excess of $Cl^-$ relative to $Na^+$ (Figure 3). As groundwaters evolve or become more saline they move closer to unity, where Na/Cl ratios approach 1 with increasing groundwater salinities (Figure 4). This is the opposite process that was observed by Salama et al. (1993) and Acworth and Jankowski (2001) where these authors noted a decrease in Na/Cl ratios with increasing salinity.

Ca + $Mg/HCO_3$ ratios increase as the groundwater salinity increase, which implies $Mg^{2+}$ and $Ca^{2+}$, are added to Na(Mg)-Cl-rich groundwaters at a greater rate than $HCO_3^-$ (Table 1). Most potential lithological sources of $Ca^{2+}$ and $Mg^{2+}$ have a $(Ca+Mg)/HCO_3$

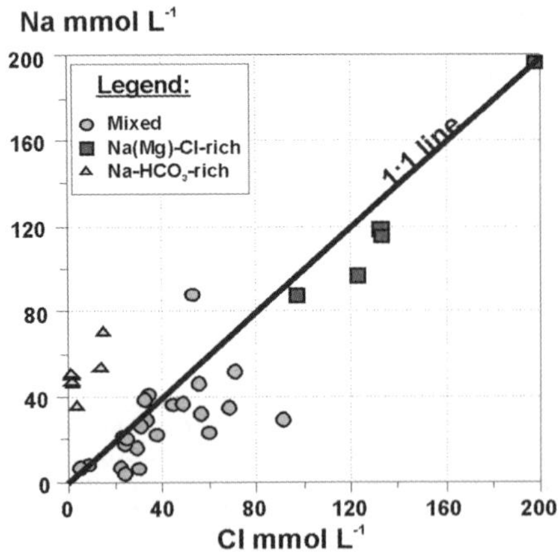


Figure 3. $Na^+$ vs. $Cl^-$ for deep and intermediate groundwaters in Spicers Creek catchment.

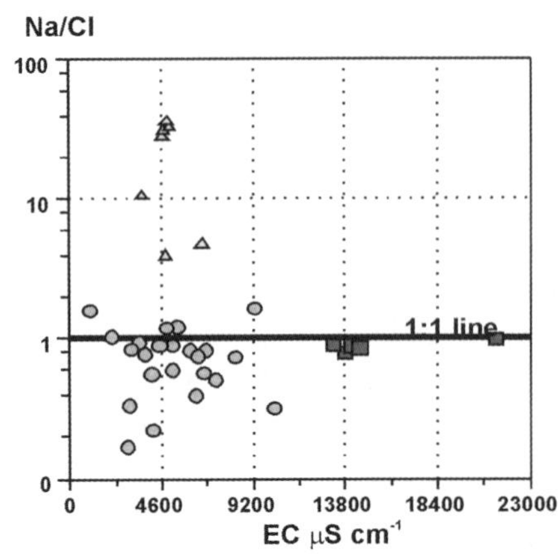


Figure 4. Na/Cl mmol $L^{-1}$ vs. EC for deep and intermediate groundwaters in Spicers Creek catchment.

Table 1. Ion ratios for deep and intermediate groundwaters in Spicers Creek catchment.

| Ion ratios | Na(Mg)-Cl-rich | Mixed | Na-$HCO_3$-rich |
|---|---|---|---|
| Ca+Mg/$HCO_3$ | ~1 | 0.3 to 11 | ~0.1 |
| Ca+Mg/$SO_4$ | 1.6 to 4.5 | 2 to 500 | ~ 600 |

ratio of ~0.5 (Mahlknecht, et al., 2004). At low salinities or in the Na-$HCO_3$-rich groundwaters the ratios are <0.5. The ratio increases with salinity, implying $Mg^{2+}$ and $Ca^{2+}$ are contributed at a greater rate than $HCO_3^-$, as salinity increases. If $Mg^{2+}$ and $Ca^{2+}$ concentrations are only derived from the weathering of carbonates, the Ca+Mg/$HCO_3$ ratios

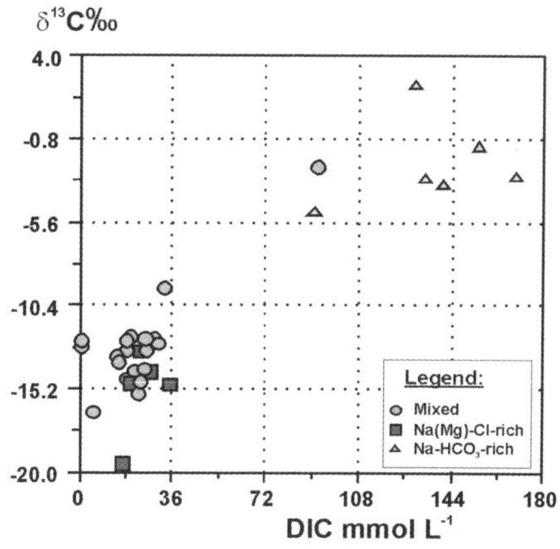


Figure 5. $\delta^{13}C_{DIC}$ vs. DIC for deep and intermediate groundwaters in Spicers Creek catchment.

would equal 0.5 (Mahlknecht, et al., 2004). This suggests that $Ca^{2+}$ and $Mg^{2+}$ is contributed from sources other than carbonate dissolution and the source of these ions is likely to be the weathering of aluminosilicate minerals in this system.

Consideration of $SO_4^{2-}$ ratios will identify if halite and gypsum dissolution are the main contributors of salts to the Na(Mg)-Cl-rich groundwaters. According to Richter and Kreiter (1987) when molar ratios of (Ca + Mg)/$SO_4$ and of Na/Cl that are close to unity they suggest that halite and gypsum are not the only sources of dissolved constituents to the groundwater. Salt water derived from deep-basin brines are characterised by Na/Cl ratios of less that 1 and (Ca + Mg)/$SO_4$ molar ratios greater than 1 (Richter and Kreitler, 1987; Richter et al., 1993). Molar ratios of Ca + Mg/$SO_4$ in the Na(Mg)-Cl-rich groundwaters are greater than 1, ranging from 1.6 to 4.5, implying that the observed salinity of these groundwaters is not solely from halite and gypsum dissolution. Sulphate and $Ca^{2+}$ therefore appear to be contributed from additional sources or geochemical process.

$\delta^{13}C_{DIC}$ values for Na(Mg)-Cl-rich groundwaters show that deep groundwater in the Spicers Creek catchment is relatively depleted, with values ranging from −19.50‰ to −13.04‰, with an average of −15.31‰. Mixed groundwaters have signatures ranging from −16.52‰ to −2.37‰, with an average of −12.56‰. The variance experienced in the mixed groundwaters reflects the mixing of depleted Na(Mg)-Cl-rich groundwaters and more $\delta^{13}C_{DIC}$-enriched Na-$HCO_3$-rich groundwaters. The $\delta^{13}C_{DIC}$ values observed in Na(Mg)-Cl-rich groundwaters are indicative of groundwaters that have evolved from a non-marine aquifer system. In a marine carbonate system, heavier $\delta^{13}C$ values would be expected with increasing groundwater residence time (Edmunds et al., 2003). The relationship between $\delta^{13}C_{DIC}$ and DIC in Na(Mg)-Cl-rich groundwaters shows $\delta^{13}C_{DIC}$ values are not related to the influx of DIC into the system (Figure 5).

As $Cl^-$ concentrations increase the $\delta^{13}C_{DIC}$ values become slightly more depleted, implying further depletion occurs as carbonate precipitation occurs from Na(Mg)-Cl-rich groundwaters (Figure 6). Therefore, terrestrial sourced carbonate precipitates are likely to be influencing the $Ca^{2+}$ and $HCO_3^-$ concentrations in Na(Mg)-Cl-rich groundwaters.

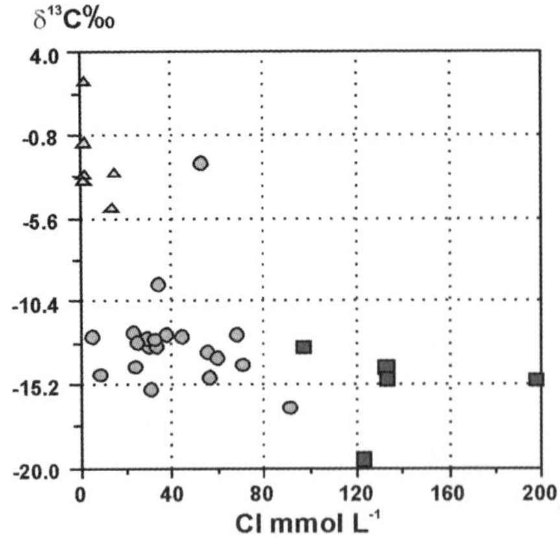


Figure 6. $\delta^{13}C_{DIC}$ vs. $Cl^-$ for deep and intermediate groundwaters in Spicers Creek catchment.

## 6 $^{87}Sr/^{86}Sr$ ISOTOPIC EVIDENCE FOR THE ORIGIN OF SALINITY

Radiogenic isotopes of strontium are extremely useful for identifying the source(s) of strontium, hence the source of salt to the groundwater system. $^{87}Sr/^{86}Sr$ isotopic ratio data provided further indication of water-rock interaction and how it plays a major role in influencing $Sr^{2+}$ geochemistry in the Na(Mg)-Cl-rich groundwaters. $^{87}Sr/^{86}Sr$ isotopic ratios of dissolved strontium are maintained in groundwater and are controlled by water-rock interaction and therefore represent the isotopic composition of the strontium bearing minerals contained in the host rock (Faure, 1986; Jorgensen and Banoeng-Yakubo, 2001). $^{87}Sr/^{86}Sr$ ratios are not significantly altered by the radiogenic decay of $^{87}Rb$ to $^{87}Sr$ after deposition (Faure, 1986). Rubidium is concentrated primarily in mica, K-feldspar and clay minerals whereas strontium occurs in plagioclase, feldspar, apatite and carbonate minerals (Faure, 1986). The $^{87}Sr/^{86}Sr$ isotopic ratios of any Rb-containing minerals will increase with time while that of Rb-free minerals will not (McNutt et al., 1990). $^{87}Sr$ enrichment occurs from the dissolution of older Rb-bearing rock units (McNutt, 2000). These isotopic ratios are source-dependent and produce information on geochemical processes occurring in the groundwater system. Generally, hydrologically young waters exhibit less radiogenic strontium ratios than do older waters (Collerson et al., 1988).

A large number of sources and sinks of strontium are present in the groundwaters and aquifer systems of the Spicers Creek catchment. Sources of strontium include aerosols, the dissolution of Sr-rich minerals within the unsaturated zone and saturated zone of the system. Strontium sinks may include the precipitation of strontium-rich minerals, ion exchange reactions and the formation of clay minerals.

$^{87}Sr/^{86}Sr$ isotopic ratios for Na(Mg)-Cl-rich groundwaters range from 0.7054 to 0.7060, with an average of 0.7056. Mixed groundwaters range from 0.7057 to 0.7093, with an average of 0.7070. $^{87}Sr/^{86}Sr$ isotopic ratios are relatively un-radiogenic, which was not expected, considering that Cartwright et al. (2002) showed that the whole rock $^{87}Sr/^{86}Sr$ isotope ratios of Ordovician turbidites of the Lachlan Fold Belt, to have radiogenic signatures that range from 0.7626 to 0.7763. These results are compared with other studies of $^{87}Sr/^{86}Sr$ isotope ratios in deep fractured rock environments, which indicate radiogenic signatures greater than 0.715 are generally encountered (McNutt et al., 1990; Banner et al., 1989).

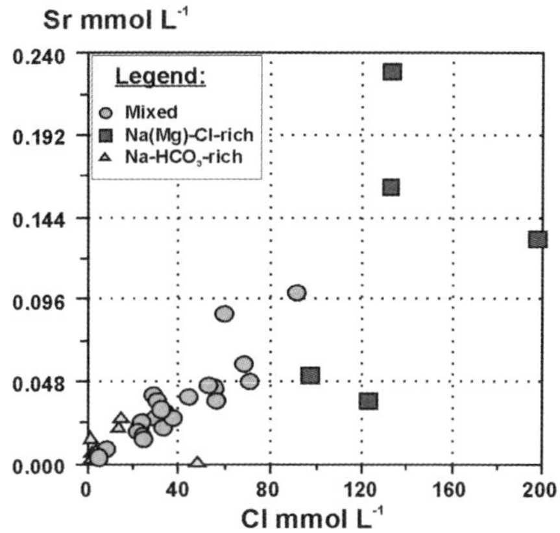


Figure 7. $Sr^{2+}$ vs. $Cl^-$ for deep and intermediate groundwaters in Spicers Creek catchment.

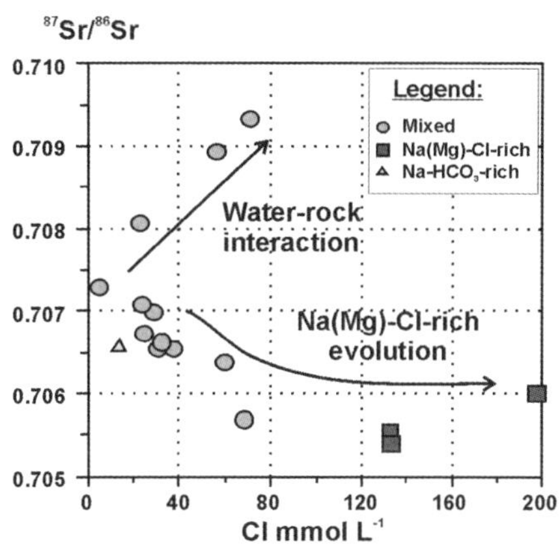


Figure 8. $^{87}Sr/^{86}Sr$ vs. $Cl^-$ for deep and intermediate groundwaters in Spicers Creek catchment.

A plot of $Sr^{2+}$ versus $Cl^-$ shows that $Sr^{2+}$ increases as $Cl^-$ concentration increases (Figure 7); this is especially true for Na(Mg)-Cl-rich groundwaters. This trend shows that the source of $Cl^-$ and $Sr^{2+}$ is likely to be similar because both ions are elevated in the Na(Mg)-Cl-rich groundwaters, particularly those Na(Mg)-Cl-rich groundwaters that have the highest concentrations of $Cl^-$ and $Sr^{2+}$ and plot in the top right hand corner of Figure 7.

The relationship between $^{87}Sr/^{86}Sr$ isotopic ratios versus $Cl^-$ and $Sr^{2+}$ shows the evolution of groundwaters from two different lithologies (Figures 8 and 9). The first occurs due to water-rock interaction, which causes $^{87}Sr/^{86}Sr$ isotopic ratios to become more radiogenic with increasing $Sr^{2+}$ and $Cl^-$ concentrations. Shallow groundwaters recently recharged groundwaters (>5 m bgs) in the area where found to have $^{87}Sr/^{86}Sr$ isotopic ratios signatures close to that of marine $Sr^{2+}$. Subsequently with time the $^{87}Sr/^{86}Sr$ isotope ratios increase as alkali feldspars and mica minerals contained within the Gleneski Formation and Intermediate aquifer became involved in geochemical reactions. The second trend is the evolution of Na(Mg)-Cl-rich groundwaters in the Oakdale Formation aquifer. As $Cl^-$ and $Sr^{2+}$

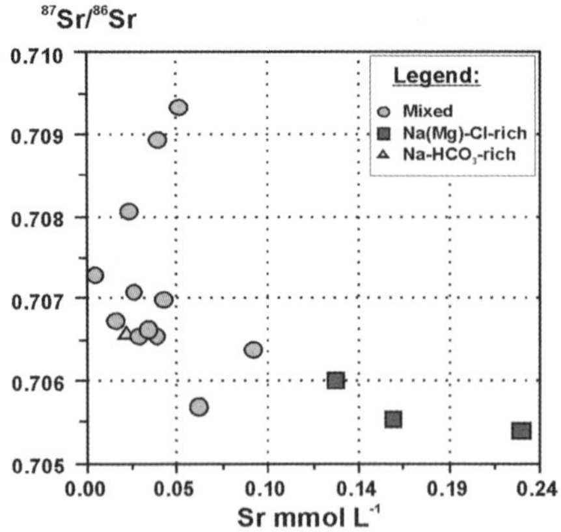


Figure 9. $^{87}Sr/^{86}Sr$ vs. $Sr^{2+}$ for deep and intermediate groundwaters in Spicers Creek catchment.

concentrations increase the $^{87}Sr/^{86}Sr$ isotopic ratios become less radiogenic. Considering the mineralogy of the aquifer the most likely process leading to these values is the addition of non-radiogenic $Sr^{2+}$ from a non-radiogenic plagioclase source.

Whole rock analysis of Canadian Shield pluton units by Franklyn et al. (1991) showed $^{87}Sr/^{86}Sr$ isotopic ratios values of between 0.706 to 0.717. They then analysed $^{87}Sr/^{86}Sr$ isotope ratios of individual minerals associated with these rocks and found biotite has radiogenic values ranging from 0.782 to 0.894, plagioclases were low ranging from 0.703 to 0.715 and secondary mineral phases such as epidote, which were present as alteration minerals coating fracture walls, had ratios ranging from 0.7039 to 0.7055 (Franklyn et al., 1991). These results indicate the importance of evaluating the role different mineral phases on the observed $^{87}Sr/^{86}Sr$ isotope signature in groundwaters (McNutt et al., 1990).

Under low temperature conditions, plagioclase is the first major susceptible phase to react with water, readily losing $Sr^{2+}$ to the water (Franklyn et al., 1991). The Oakdale Formation aquifer is composed of approximately 45% plagioclase. Therefore, the addition of non-radiogenic $Sr^{2+}$ from the plagioclase phase is most likely leading to the observed un-radiogenic signature found in the Na(Mg)-Cl-rich groundwaters. It appears that $^{87}Sr/^{86}Sr$ isotope ratios observed in Na(Mg)-Cl-rich groundwaters are dominated by isotopic exchange with plagioclase. Franklyn et al. (1991) also observed this to be the case in the Canadian Shield rocks. McNutt et al. (1990) also found plagioclase to be the main source of $Sr^{2+}$ within Canadian Shield groundwaters. They also noted that with time the $^{87}Sr/^{86}Sr$ isotope ratios increased as biotite weathering occurred because biotite has a high Rb/Sr ratio and dissolution of these minerals leads to a radiogenic input of $Sr^{2+}$. The observed $^{87}Sr/^{86}Sr$ isotope ratios showed that weathering of biotite has a minor influence on the observed groundwater chemistry of the Na(Mg)-Cl-rich groundwaters investigated in this study.

The observed trend of decreasing radiogenic $Sr^{2+}$ and the addition of non-radiogenic $Sr^{2+}$ to the more geochemically evolved Na(Mg)-Cl-rich groundwaters was also observed by Armstrong et al. (1998) in Milk River Aquifer pore waters where they found the $^{87}Sr/^{86}Sr$ ratios became less radiogenic with increasing distance from the recharge area. The $^{87}Sr/^{86}Sr$ ratios of the aquifer units such as the sandstone and shale whole rocks,

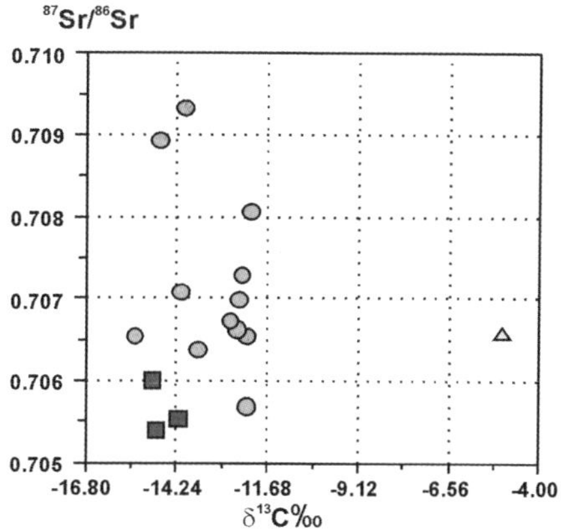


Figure 10. $^{87}Sr/^{86}Sr$ vs. $\delta^{13}C_{DIC}$ for deep and intermediate groundwaters in Spicers Creek catchment.

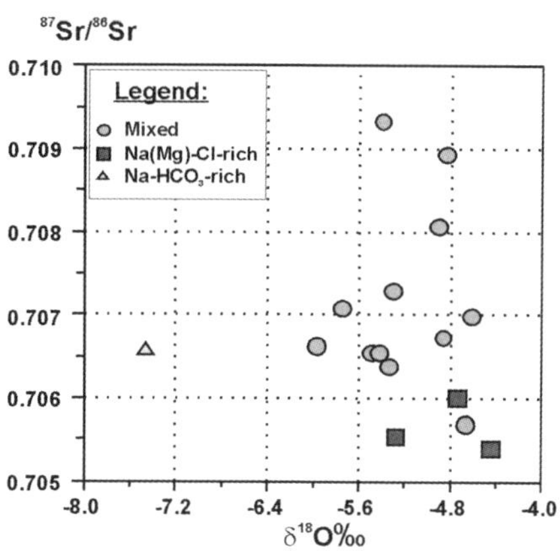


Figure 11. $^{87}Sr/^{86}Sr$ vs. $\delta^{18}O$ for deep and intermediate groundwaters in Spicers Creek catchment.

leachates and feldspars were found to be too radiogenic to account for low $^{87}Sr/^{86}Sr$ ratios measured in the evolved pore waters (Armstrong et al., 1998).

Bivariate plots of $^{87}Sr/^{86}Sr$ isotopic ratios versus $\delta^{13}C_{DIC}$ and $\delta^{18}O$ further indicate that these Na(Mg)-Cl-rich groundwaters are geochemically distinct (Figures 10 and 11). These groundwaters are less radiogenic than other groundwaters in the catchment. They are more depleted in $\delta^{13}C_{DIC}$ and also slightly more enriched in $\delta^{18}O$. Therefore, Na(Mg)-Cl-rich groundwater samples plot in a cluster in the bottom left hand corner of Figure 10 and bottom right hand corner of Figure 11 showing that these waters are isotopically different to the mixed and Na-$HCO_3$-rich groundwaters. It appears that as Na(Mg)-Cl-rich groundwaters become more geochemically evolved the observed $\delta^{13}C_{DIC}$ values become more depleted which is most likely occurring due to $\delta^{13}C_{DIC}$ fractionation from calcite precipitation. This is supported with carbonate saturation indices that indicate that all Na(Mg)-Cl-rich groundwater samples are supersaturated with carbonate minerals implying that carbonate minerals are likely to be precipitating out of these waters. Subsequently, it is appears that as plagioclase weathers in the Oakdale Formation aquifer and $Ca^{2+}$ and $HCO_3^-$ are released into the groundwaters increasing the carbonate saturation indices in the groundwater leading to carbonate precipitation.

## 7 CONCLUSIONS

The origin of salts contained within the deep fractured bedrock groundwaters of the Oakdale Formation was determined using hydrogeochemical and isotopic evidence. Two distinctive end-member groundwaters where identified within the deep fractured bedrock systems. These were the brackish Na-$HCO_3$-rich and saline Na(Mg)-Cl-rich endmembers.

This research discovered the origin and geochemical evolutionary pathways of the Na(Mg)-Cl-rich groundwaters. The Na(Mg)-Cl-rich groundwaters were found to be meteoric in origin and have not undergone significant evaporation prior to recharge. They have Na/Cl ratios of ~1 and elevated concentrations of $Na^+$, $Cl^-$, $SO_4^{2-}$, As, $Sr^{2+}$ and V. $^{87}Sr/^{86}Sr$ isotopic ratios were used to identify the source of elevated concentrations of $Sr^{2+}$ in the Na(Mg)-Cl-rich groundwaters. Isotopic ratios showed that the elevated $Sr^{2+}$ was from the weathering of plagioclase phases contained within the Oakdale Formation aquifer. These results also showed that Na(Mg)-Cl-rich groundwaters inherited less radiogenic signatures as $Cl^-$ and $Sr^{2+}$ concentrations increase. The evolution of Na(Mg)-Cl-rich groundwaters from the Oakdale Formation has occurred due to long residence time and extensive geochemical reactions. The significance of these findings shows that the majority of salt contained within deep fractured aquifers has been derived from extensive water-rock interaction processes within the Oakdale Formation at depth.

## REFERENCES

Acworth I, Jankowski J (2001) Salt source for dryland salinity – evidence from an upland catchment on the Southern Tablelands of New South Wales. Australian Journal of Soil Research (39): 39–59.

Armstrong SC, Sturchio C, James Hendry M (1998) Strontium isotopic evidence on the chemical evolution of pore waters in the Milk River Aquifer, Alberta, Canada. Applied Geochemistry (13): 463–475.

Ashley PM (2001) Petrographic report on seven drill core samples from the Yarindury prospect, north of Wellington, NSW. Earth Sciences University of New England Armidale. Order B01903. for MIM Exploration Pty Ltd.

Banner J, Wasserburg G, Dobson P, Carpenter A , Moore C (1989) Isotopic and trace element constraints on the origin and evolution of saline groundwaters from central Missouri. Geochimica et Cosmochimica Acta (53): 383–398.

Cartwright I, Weaver T, Tweed S, Ahearne D, Cooper M, Czapnik K, Tranter J (2002) Stable isotope geochemistry of cold $CO_2$-bearing mineral spring waters, Daylesford, Victoria, Australia: sources of gas and water and links with waning volcanism. Chemical Geology (185):71–91.

Clark I, Fritz P (1997) Environmental Isotopes in Hydrogeology. CRC Press LLC, Bocca Raton, New York, 328pp.

Collerson KD, Ullman WJ, Torgersen T (1988) Groundwaters with unradiogenic $^{87}Sr/^{86}Sr$ ratios in the Great Artesian Basin, Australia. Geology (16): 59–63.

Edmunds WM, Guendouz AH, Mamou A, Moulla A, Shand P, Zouari K (2003) Groundwater evolution in the Continental Intercalaire aquifer of southern Algeria and Tunisia: trace element and isotopic indicators. Applied Geochemistry (18): 805–822.

Faure G (1986) Principles of Isotope Geology. Wiley, New York, 589pp.

Franklyn MT, McNutt RH, Kamineni DC, Gascoyne M, Frape SK (1991) Groundwater $^{87}Sr/^{86}Sr$ values in the Eye–Dashwa Lakes pluton, Canada: Evidence for plagioclase – water reaction. Chemical geology (86): 111–122.

Gat, JR (1981) Chapter 10 Groundwater. In: Gat, J. R. and Gonfiantini, R. 1981. Stable Isotope Hydrology. Deuterium and Oxygen-18 in the Water Cycle, IAEA, Technical Report Series 210: 223–241.

Jorgensen NO, Banoeng-Yakubo BK (2001) Environmental isotopes ($^{18}O$, $^{2}H$, and $^{87}Sr/^{86}Sr$) as a tool in groundwater investigations in the Keta Basin, Ghana. Hydrogeology Journal (9): 190–201.

McNutt RH, Frape SK, Fritz P, Jones MG, Macdonald IM (1990) The $^{87}Sr/^{86}Sr$ values of Canadian Shield brines and fracture minerals with applications to groundwater mixing, fracture history, and geochronology. Geochimica et Cosmochimica Acta (54): 205–215.

McNutt RH (2000) Chapter 8: Strontium Isotopes. In: Cook, PG, Herczeg, AL (eds), Environmental Tracers in Subsurface Hydrology. Kluwer Academic Publishers: Australia.

Mahlknecht J, Schneider JF, Merkel BJ, Leon IN, Bernasconi SM (2004) Groundwater recharge in a sedimentary basin in semi-arid Mexico. Hydrogeology Journal (12): 511–530.

Meakin NS, Morgan EJ (1999) Dubbo Geological sheet 1:250,000, S1/55-4 2nd ed. Explanatory Notes. Facer RA, Stewart JR (eds). Geological Survey of New South Wales, Sydney. 504.

MIM Exploration Pty Ltd (2002) Exploration Data. EL5623 Yarindury and EL5758 North Comobella.

Morgan K (2005) Evaluation of salinisation processes in the Spicers Creek catchment, Central West Region of New South Wales. PhD Thesis, University of New South Wales, Sydney (unpublished).

Morgan K, Jankowski J, Taylor G (2005) Structural controls on groundwater flow and groundwater salinity within a dryland affected catchment, Central West region, New South Wales. Hydrological Processes. In press.

Morgan EJ, Colquhoun GP, Watkins JJ, Raymond OL, Warren AYE, Meakin S, Henderson GAM, Scott MM, Percival IG (1999a) Middle to Late Ordovician (To ?Early Silurian). In Dubbo Geological sheet 1:250,000, S1/55-4 2nd edn. Explanatory Notes. Facer RA, Stewart JR (eds). Geological Survey of New South Wales, Sydney: 29–60.

Morgan EJ, Barron LM, Colquhoun GP, Henderson GAM, Watkins JJ, Pogson DJ, Scott MM (1999b) Silurian – Mumbil Group. In Dubbo Geological sheet 1:250,000, S1/55-4 2nd edn. Explanatory Notes. Facer RA, Stewart JR (eds). Geological Survey of New South Wales, Sydney: 77–99.

Nordstrom D, Ball J, Donahoe R, Whittmore D (1989) Groundwater chemistry and water-rock interactions at Stripa. Geochimica et Cosmochimica Acta (53): 1727–1740.

Richter BC, Kreitler CW (1987) Source of ground water salinization in parts of West Texas. GWMR Fall 1987: 75–84.

Richter BC, Kreitler CW, Bledsoe BE (1993) Geochemical Techniques for Identifying Sources of Ground-Water Salinisation. By C.K. Smoley. CRC Press, Inc. Florida. 258pp.

Rozanski K, Araguas-Araguas L, Gonfiantini R (1993) Isotopic patterns in modern global precipitation. In Continental Isotope Indicators of Climate. American Geophysical Union Monograph. Cited by Clark and Fritz, 1997.

Salama RB, Farrington P, Bartle GA, Watson GD (1993) The chemical evolution of groundwater in a first – order catchment and the process of salt accumulation in the soil profile. Journal of Hydrology (143): 233–258.

Sami K (1992) Recharge mechanisms and geochemical processes in a semi-arid sedimentary basin, Eastern Cape, South Africa. Journal of Hydrology (139): 27–48.

Schofield S (1998) The Geology, Hydrogeology and Hydrogeochemistry of the Ballimore Region, Central New South Wales. PhD Thesis, University of New South Wales, Sydney (unpublished).

Turner JV, Arad A, Johnston CD (1987) Environmental isotope hydrology of salinized experimental catchments. Journal of Hydrology (94): 89–107.

CHAPTER 23

# Isotope studies of a groundwater-flow system in granite, Middle Hungary

László Palcsu[1]*, Éva Svingor[1], Mihály Molnár[1], Zsuzsanna Szántó[1], István Futó[1], László Rinyu[1], István Horváth[2], György Tóth[2] and István Fórizs[3]

[1] *Laboratory of Environmental Studies, Institute of Nuclear Research of the Hungarian Academy of Sciences, Debrecen, Hungary*
[2] *Geological Institute of Hungary, Budapest, Hungary*
[3] *Laboratory for Geochemical Research, Research Centre of Earth Sciences of the Hungarian Academy of Sciences, Budapest, Hungary*
* *now at the Institut für Umweltphysik, University of Heidelberg, In Neuenheimer Feld 229, Heidelberg 69115, Germany, palcsu@iup.uni-heidelberg.de*

ABSTRACT: A large research project is going on in Hungary to find the most suitable territory for a low and intermediate level radioactive waste repository. A granite massif located near Bátaapáti, between Mecsek Mountains and Szekszárdi Hills seems to be the best geological formation for this purpose. This paper deals with the isotope studies conducted to define the groundwater-flow system. The granite rock of Carboniferous age is covered by 30–70 metres thick Holocene and Pleistocene sediments. The sediment is mostly aquitard with several infiltration zones. The peripheral part of the granite body is more fractured than the inner part. The properties of the flow system were investigated by isotope measurements of groundwater samples from boreholes. Tritium content, δ-values and radiocarbon ages show that Holocene waters appear in the upper and peripheral part of the granite body, and glacial, pre-glacial waters are present in the deeper part of the bulk. The vertical water movement is only a few $cm \cdot yr^{-1}$. Based on isotope examination of groundwater, we conclude that this field is suitable for repository from hydrogeological point of view. We also find that the cooling due to climate change during the last ice age was $7.9 \pm 1.5°C$.

## 1 INTRODUCTION

Radioactive waste disposal is a great challenge in the world. In Hungary, a project has been going on for many years to find the most suitable site for a low and intermediate level radioactive waste repository. The initial screening showed that a granite complex located near the village of Bátaapáti (46°13′N, 18°36′E), between Mecsek Mountains and Szekszárdi Hills in Middle Hungary, seems to be the best geological formation for this purpose (see asterix in Figure 1). The site is called "Üveghuta" (see Üh-identifiers in Table 1).

One of the most important issues to be clarified in the preliminary safety assessment is that how the water flows in the granite rock. On the other hand, other questions have to be answered: how old is the groundwater in the granite, how does the water age change with the depth, or is there any groundwater–rock interaction within the aquifer. Based on the isotope data of groundwater samples the cooling during the Last Glacial Maximum (LGM) can be investigated. The groundwater flow system is studied using tritium, $^{14}C$, $\delta^{13}C$, $\delta D$, $\delta^{18}O$, dissolved salts, noble gases, etc. measurements. In the present study, we have focused only to the isotope evaluations, the more detailed hydrogeology and water modelling are published elsewhere (*Horváth* et al., *1997*).

## 2 METHODS

Tritium, radiocarbon, stable isotope ratios were measured in all water samples, and dissolved helium isotopes in a few samples were analysed. The tritium was determined by the helium-3 ingrowth method using a noble gas mass spectrometer (VG 5400, Fisions Instruments) in the Institute of Nuclear Research (INR), Debrecen, Hungary. The detection limit of this method is about 0.005 TU (tritium unit: 1 TU corresponds to the $^3H/^1H$ ratio of $10^{-18}$). The accuracy of the measurements is 2–4% in the range of 1–20 TU (*Palcsu* et al., *2002*; *Palcsu, 2003*). Stable isotope ratios ($\delta D$ and $\delta^{18}O$) of water were measured by a Finnigan Mat delta S mass spectrometer in the Laboratory for Geochemical

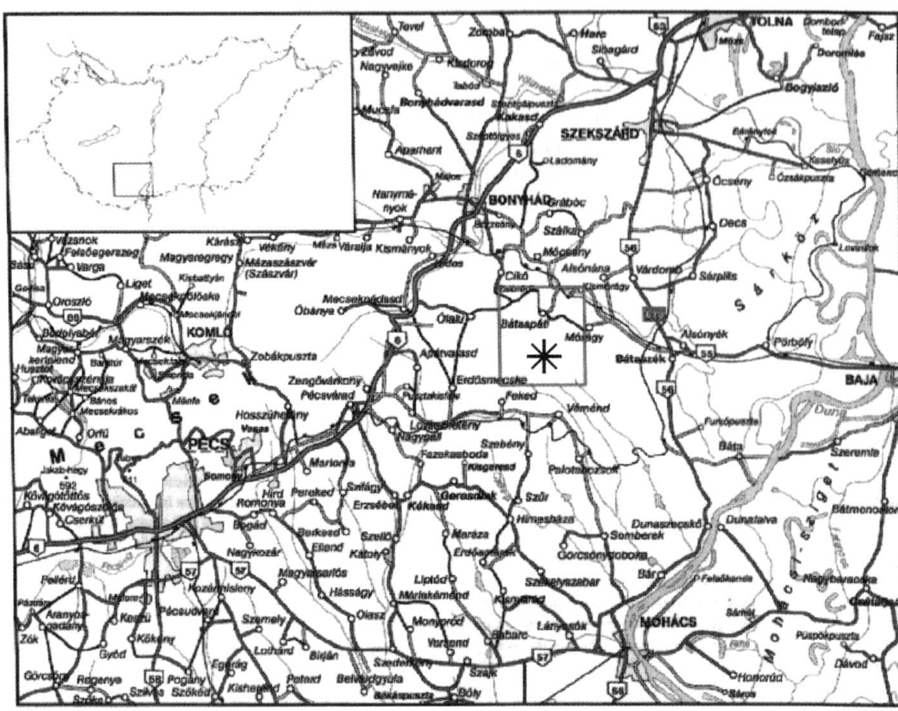


Figure 1. The location of the studied area represented by the asterix, Middle Hungary.

Research, Budapest, Hungary, while the $\delta^{13}C$ of dissolved inorganic carbon were measured by a McKinney-Nier-type stable isotope ratio mass spectrometer (*Hertelendi* et al., *1987*) in the INR. Dissolved noble gas measurements, in practice helium, were carried out with the above mentioned noble gas mass spectrometer. The radiocarbon activity of the dissolved inorganic carbon was determined by gas proportional counting. The gas proportional system consisting of 9 proportional counters was designed and built in the INR. (*Csongor and Hertelendi, 1986*). The samples were converted to high purity $CO_2$ gas and measured for 3–4 days. The overall precision of the system for modern carbon samples is better than 4 [‰] after a counting period of seven days (*Hertelendi* et al., *1989*). The radiocarbon ages of the water samples were calculated according to the Ingerson-Pearson model (*Ingerson and Pearson, 1964*) and had an overall error of about 3000 years.

## 3 GEOLOGICAL BACKGROUND

The granite is Carboniferous and covered by 30–70 metres of Holocene sediment and Pleistocene loess sediment (*Horváth* et al., *1997*). The sediment is generally in low permeability, but there are zones where deep infiltration is possible. The peripheral part of the granite, in 60–80 m thickness, is fractured, which definitely implies that the granite was at the surface about 2–3 million years ago (*Balla* et al., *2000*; *Horváth* et al., *1997*), whereas the inside part of the granite massif is unaltered. The flow velocity in the compact granite is very low, about $10^{-9} ms^{-1}$, but in the fractured granite it might be higher (*Tóth* et al., *2003*). The properties of the flow system were investigated by isotope measurements of groundwater samples from the boreholes drilled down to 350–400 metres into the granite. The water samples come from different screen depth (see Table 1). A and B represent shallow monitoring wells very close to the given boreholes of the same number. The monitoring wells provide water samples from the first shallow aquifer. Several water samples could be taken from a certain borehole using the packer technique.

## 4 RESULTS AND DISCUSSION

In order to determine the young component of the water samples, tritium concentrations were measured. All samples, mostly in the shallow ones, contained tritium (Table 1, Figure 2 and 3). It is natural in case of water tables close to the surface, but in the deeper layers the most probable reason of presence of young water component can be the remains of the drilling water or water infiltrated during the drilling. Otherwise, there are studies which discussed on presence of tritium at relatively greater depths. For instance, *Sheldon and Solomon* (*2001*) observed above-background tritium concentrations at 60 m depth in fractured dolostone from Smithville, Ontario, *Bradbury and Muldoon* (*1992*) measured tritium at 70 m depth in fractured dolomite in Wisconsin, and *Busenberg and Plummer* (*1996*) found tritium to over 100 m depth in schists in New Hampshire. In our study, tritium content decreases with the depth and sharply goes under a very low limit around at 80 m of depth where the border of the Pleistocene-Holocene sediment and the granite massife is located. Young water containing tritium is not present in the deeper granite formation but occurs in significant amounts in the upper layers of the granite and in the sediment cover.

Table 1. Data of the water samples from the examined boreholes in the research area (pMC: % modern carbon, BP: before present, TU: tritium unit,–: not measured.

| Sample identifier | Date of sampling | Screen (m) | $\delta^{13}C_{VPDB}$ (‰) | $^{14}C$ (pMC) | Radiocarbon age (yr BP) | $\delta D_{VSMOW}$ (‰) | $\delta^{18}O_{VSMOW}$ ([‰]) | Tritium (TU) |
|---|---|---|---|---|---|---|---|---|
| Üh-25A | 14/06/2002 | 66.8–68.7 | – | – | – | – | – | 0.040 ± 0.007 |
| Üh-25 | 06/09/2002 | 192.5–203.2 | −11.9 | 10.6 | 12460 | −82.0 | −11.39 | 0.030 ± 0.010 |
| ” | 17/10/2002 | 249.6–260.0 | −12.2 | 4.8 | 19170 | −93.2 | −12.84 | – |
| ” | 04/11/2002 | 289.8–300.7 | −12.9 | 3.9 | 21360 | −97.0 | −13.23 | 0.017 ± 0.007 |
| Üh-26A | 14/06/2002 | 62.0–73.0 | – | – | – | – | – | 0.300 ± 0.060 |
| Üh-26 | 17/07/2002 | 211.8–229.0 | −10.8 | 4.1 | 19500 | −97.7 | −13.36 | 0.010 ± 0.008 |
| ” | 01/09/2002 | 304.2–324.3 | −11.6 | 9.2 | 13380 | −92.1 | −12.58 | – |
| ” | 26/08/2002 | 389.8–400.6 | −11.5 | 3.3 | 21780 | −90.0 | −12.5 | 0.080 ± 0.020 |
| Üh-27 | 25/07/2002 | 242.9–253.5 | −13.4 | 23.0 | 7000 | −74.4 | −10.47 | 0.009 ± 0.006 |
| ” | 02/08/2002 | 273.7–284.3 | −13.9 | 29.2 | 5300 | −78.7 | −11.15 | 0.032 ± 0.009 |
| ” | 17/08/2002 | 380.3–391.1 | −13.8 | 10.0 | 14100 | −87.0 | −11.94 | – |
| Üh-28A | 24/07/2002 | 51.4–52.7 | – | – | – | – | – | 0.050 ± 0.010 |
| Üh-28 | 05/10/2002 | 149.2–158.1 | −12.7 | 17.7 | 8740 | −70.7 | −10.18 | 0.032 ± 0.008 |
| ” | 29/09/2002 | 246.9–271.5 | −11.2 | 3.6 | 21270 | −89.6 | −12.4 | 0.013 ± 0.007 |
| Üh-6 | 14/06/2002 | 71.7–79.7 | – | – | – | – | – | 0.059 ± 0.008 |
| Üh-29 | 21/08/2002 | 0.0–5.0 | – | – | – | – | – | 8.2 ± 0.2 |
| ” | 14/09/2002 | 54.2–65.7 | −14.7 | 92.9 | 0 | −68.7 | −9.86 | 9.1 ± 0.2 |
| ” | 19/09/2002 | 85.3–104.8 | −12.6 | 39.9 | 1920 | −79.3 | −11.33 | – |
| ” | 08/10/2002 | 179.2–190.8 | −11.4 | 3.9 | 20330 | −89.6 | −12.43 | – |
| Üh-30 | 06/08/2002 | 0.0–19.9 | −14.2 | 78.5 | – | −70.2 | −9.83 | 1.5 ± 0.05 |
| ” | 27/08/2002 | 50.1–78.9 | −13.8 | 55.2 | – | −76.6 | −10.85 | 0.130 ± 0.020 |
| ” | 18/09/2002 | 119.7–130.3 | −12.4 | 6.0 | 17440 | −96.1 | −13.28 | 0.051 ± 0.009 |
| ” | 22/09/2002 | 148.2–160.3 | −12.7 | 0.5 | 38210 | −99.2 | −13.58 | – |
| Üh-31A | 22/06/2002 | 31.8–36.6 | −14.7 | 85.4 | 0 | −70.7 | −9.98 | 5.5 ± 0.2 |
| Üh-31B | 30/06/2002 | 0.0–18.2 | −15.5 | 100.4 | 0 | −68.1 | −9.72 | – |
| ” | 10/07/2002 | 34.7–40.0 | – | – | – | – | – | 3.1 ± 0.1 |
| Üh-32A | 29/06/2002 | 26.9–50.0 | −13.1 | 31.9 | 4100 | −78.0 | −10.87 | 0.080 ± 0.010 |
| Üh-33 | 15/10/2002 | 35.6–44.0 | −13.7 | 74.5 | 0 | −69.6 | −9.79 | 0.520 ± 0.020 |
| Üh-35 | 14/11/2002 | 30.0–44.2 | – | – | – | – | – | 0.330 ± 0.020 |
| Mo-7A | 17/07/2002 | 37.5–50.0 | −14.8 | 94.6 | – | – | – | 5.3 ± 0.2 |
| Mo-7B | 19/09/2002 | 32.6–42.6 | −14.5 | 98.7 | – | – | – | 5.9 ± 0.1 |

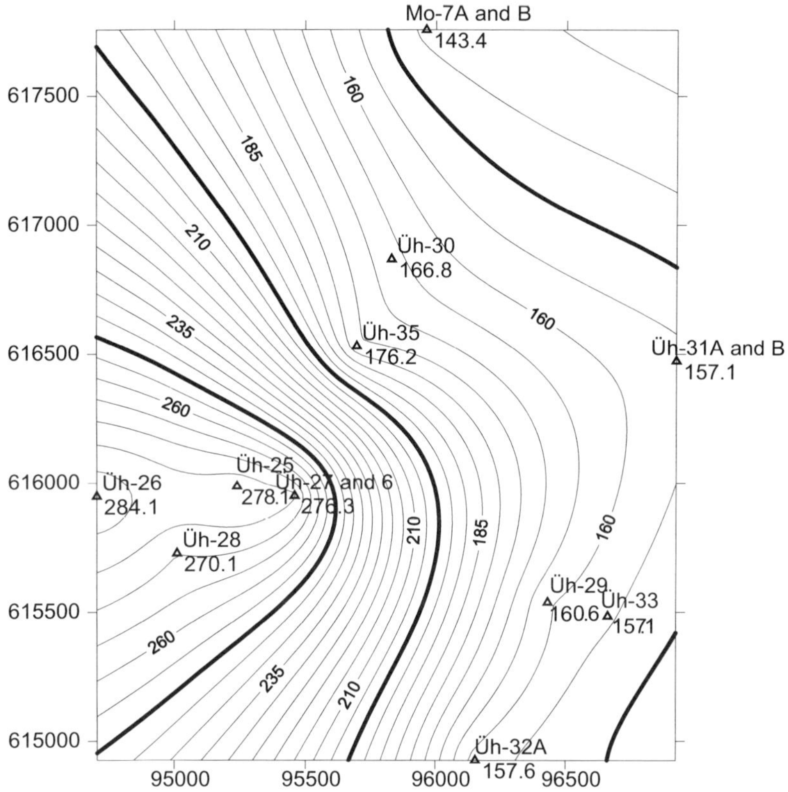


Figure 2. The locations of the wells sampled.

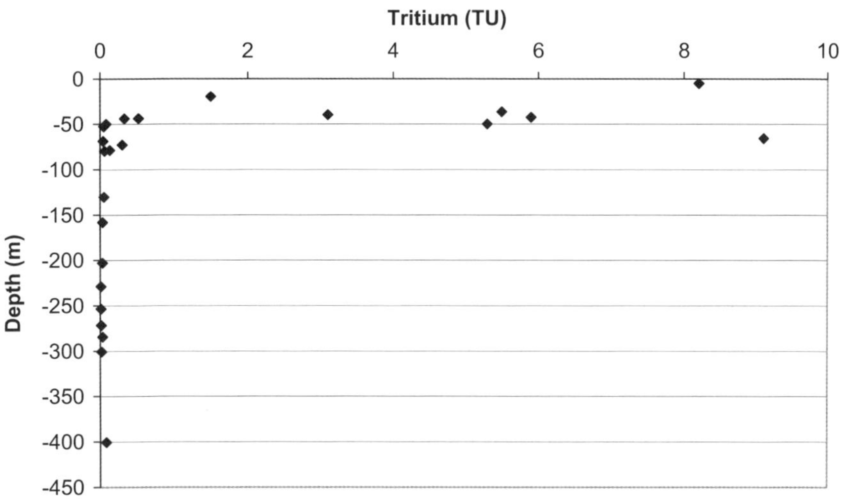


Figure 3. Tritium concentration of groundwater samples plotted against sampling depth.

There are unexpectedly high values in higher depth: 5–9 TU water could be recharged from the 1960's, and the 9 TU value may be the rest of the bomb peak of the tritium when the tritium concentration in rain was at least two orders of magnitude higher than recently.

Radiocarbon ages of water samples were determined following the method described by *Ingerson and Pearson* (*1964*) who suggested calculating the radiocarbon age using concentration of $^{14}C$ and stable isotope ratio of $^{13}C$ in the dissolved inorganic carbon of water. This method does not need any hydrocarbonate and major ion chemistry data, unlike other methods like, for example, of Fontes and Garnier (*1979*) do. The field studied in this paper contains a groundwater flow-system where the groundwater loses its salinity along the flow path (*Balla* et al., *2000*; *Tóth* et al., *2003*). The changing of the salt and ion contents in the groundwater did not allow us to use ion chemistry in radiocarbon age determination.

It can be seen in Table 1 that the water becomes older with increasing depth according to the radiocarbon ages, which increased up to 21–22 kyr BP. These waters are thought to infiltrate in the last ice age. A borehole (Üh-30) provided also very old water, the age of the water from 150–160 m in well Üh-30 was the oldest one (38.2 kyr BP) implying that its recharge happened before the LGM. All water ages are increasing with the depth which is not surprising because the hydraulic heads are also decreasing with the depth (*Horváth* et al., *1997*), therefore, the groundwater flows downwards. A very simple estimation of the vertical groundwater velocity can be done using the radiocarbon ages. Assuming that the water flows down with constant velocity, which must be not true, the ratio between depth and age should mean the vertical component of the water velocity. In this calculation (Figure 4), the vertical velocity of the groundwater is $0.4$–$5.5\,cm \cdot yr^{-1}$ with an average of $1.2\,cm \cdot yr^{-1}$. In earlier studies of this site, using groundwater modelling based on hydraulic head data (*Horváth* et al., *1997*), $10\,cm \cdot yr^{-1}$ of the overall groundwater velocity in this water system was obtained. This is in good agreement with our findings based only on the connection between the age and the depth of the groundwater.

These suggest that the perpendicular water motion is very slow which is very important, if a radioactive waste repository was planned to be built. Establishing of a final disposal of low and intermediate level radioactive waste requires at least a retardation of 600 years, which is 20 times of the half life of $^{137}Cs$ that is a reference isotope with a relatively high half life of 30 years for low and intermediate level radioactive waste. In 20 half lives this isotope completely decays. As we have found, the groundwater movement downwards in the granite at depths of hundreds of meters is very slow. The discharge area of the groundwater flow system from the upper granite along the flow path is at least a few hundreds of meter far away (*Horváth* et al., *1997*; *Balla* et al., *2000*), while from the inner side of the ground it can be much more farther. The water velocity we have got is enough so that groundwater goes just 7.2–60 m in 600 years, which means that a potentially escaped radioactive isotopes are not able to reach the surface.

There was a limited possibility to take noble gas samples dissolved in groundwater. Dissolved helium measurements were done on three samples taken into copper tube sampler (*Schlosser* et al., *1988*). Table 2 shows the helium concentrations and isotope ratios ($^{3}He/^{4}He = R$ divided by the atmospheric ratio $R_a$ which is $1.3841 \cdot 10^{-6}$) data of the samples which came from the deepest water layer. These helium data are correlated with the radiocarbon ages: the oldest water (Üh-30 with 38.2 kyr BP) contained the largest amount of helium, while the least helium was in the water sample from a shallow borehole (Üh-35, filtered between 30.0 and 44.2 m). The radiogenic component of helium ($^{4}He_{rad}$), which arises from the alpha-decay of the elements of uranium and thorium series, was also determined

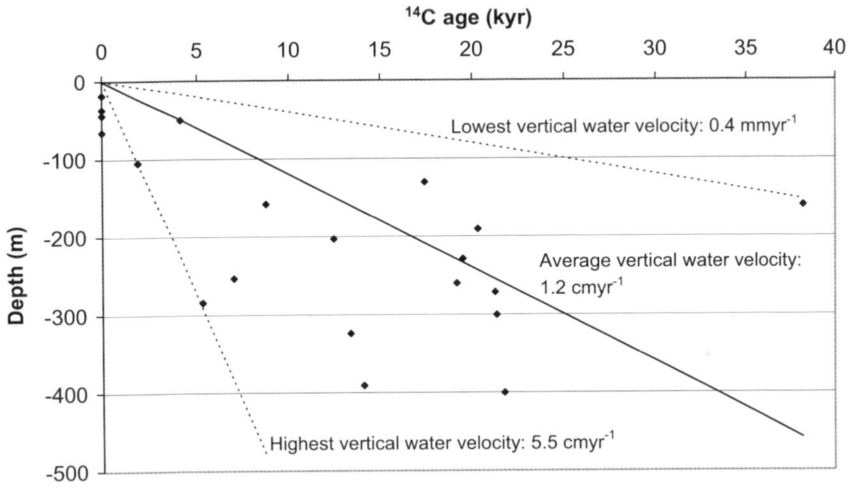


Figure 4. Radiocarbon ages versus depth, and the vertical component of the groundwater velocity.

Table 2. Helium data of water samples of three boreholes from the deepest layer (cm$^3$STP: cubic centimetre at standard temperature and pressure).

| Sample identifier | $^{14}$C age (yr BP) | Helium (cm$^3$STP$_{He}$/g$_{H2O}$) | $^3$He/$^4$He | R/R$_a$ |
|---|---|---|---|---|
| Üh-27 | 14100 | $5.0 \cdot 10^{-6}$ | $3.59 \cdot 10^{-8}$ | 0.026 |
| Üh-30 | 38210 | $3.7 \cdot 10^{-5}$ | $1.94 \cdot 10^{-8}$ | 0.014 |
| Üh-35 | 300 | $8.7 \cdot 10^{-7}$ | $7.58 \cdot 10^{-7}$ | 0.548 |

(see Eq. 1) using the measured helium value ($^4He_{meas}$), the atmospheric helium components from the solubility equilibrium ($^4He_{eq}$) and excess air ($^4He_{ex}$), and the mantle helium ($^4He_{mantle}$). Previous studies demonstrated that radiogenic helium diffused to the porewater from the hostrock, and accumulated in the groundwater. This "in-situ" accumulation rate ($A_{He}$) depends on the uranium-thorium content, density and porosity of the rock. It can be calculated according to the Eq. 2 (*Stute* et al., *1992*):

$$^4He_{rad} = {}^4He_{meas} - {}^4He_{eq} - {}^4He_{ex} - {}^4He_{mantle} \tag{1}$$

$$A_{He} = \Lambda_{He} \cdot \rho_r \cdot \rho_w{}^{-1} \cdot (C_U \cdot P_U + C_{Th} \cdot P_{Th}) \cdot n_{eff}{}^{-1} \tag{2}$$

The release factor $\Lambda_{He}$ is the fraction of produced $^4$He released from the mineral into the water and it is usually taken to be 1 (*Andrews and Lee, 1979*). The density of the rock ($\rho_r$) is taken as 2860 kgm$^{-3}$ for granite, $\rho_w$ is the density of the water. The $^4$He production rates from U and Th decay are $P_U = 1.19 \cdot 10^{-13}$ cm$^3$ STPμg$_U{}^{-1}$ yr$^{-1}$ and $P_{Th} = 2.88 \cdot 10^{-14}$ cm$^3$ STPμg$_{Th}{}^{-1}$ yr$^{-1}$. The $n_{eff}$ is the effective porosity of the rock. In our study, the real accumulation rate was calculated from the connection of groundwater age and the radiogenic helium component, despite the fact that we had only three data points. We found that the real accumulation rate was $9 \cdot 10^{-10}$ cm$^3$ STPg$^{-1}$ yr$^{-1}$, while the calculated in-situ accumulation rate was $7.9 \cdot 10^{-11}$ cm$^3$ STPg$^{-1}$ yr$^{-1}$ according to the 5 μgg$^{-1}$ of uranium and 30 μgg$^{-1}$ of thorium content, and 0.05 effective porosity. The real accumulation rate exceeds the

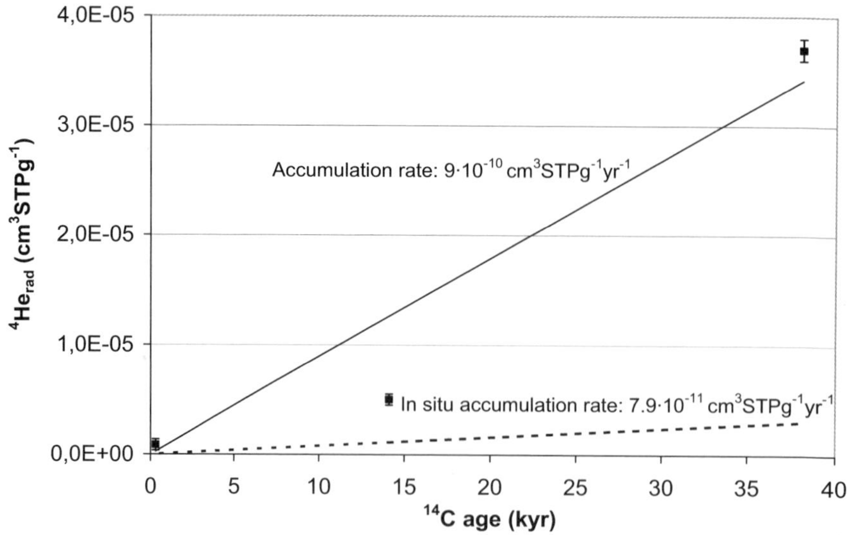


Figure 5. Radiogenic helium component of helium content of three water samples, the helium accumulation rate calculated from the $^4He_{rad}$ values and the radiocarbon ages, and the in situ helium accumulation rate calculated from the uranium and thorium content of the granite body (note that there are unfortunately three data points).

in-situ rate by an order of magnitude (Figure 5). This finding is quite common in hydrogeological studies: the real helium accumulation rate is usually higher than the in-situ one. The reason can be the release of stored helium from fresh sediment rocks or a helium flux from adjacent layers or even the whole underlying crust (*Torgersen and Clarke, 1985*; *Solomon* et al., *1996*; *Aeschbach-Hertig* et al., *2002*). Significant release of stored old helium from the granite is not expected because of the Carboniferous age of the rock, therefore the excess terrigenic helium can come from the deeper part of the aquifer by diffusion. However, the obtained helium accumulation rate could be employed for age determination of very old water from this area in later studies based on helium measurements where the radiocarbon dating is not applicable because of the very low $^{14}C$ content.

The stable isotope ratios of the water samples fit the Global Meteoric Water Line (GMWL) (*Dansgaard, 1964*) extremely well ($\delta D = (8.03 \pm 0.13) \cdot \delta^{18}O + (9.93 \pm 1.49)$‰, Figure 6). The correlation proves the meteoric origin of these waters. It also shows that isotope exchange between the hostrock and the groundwater might play only a minor or even negligible role. On the other hand, the waters become more depleted in deuterium and oxygen-18 with increasing depth (−70‰ to −99‰, and −10‰ to −14‰, respectively). This isotope signature provides information on the water evolution and history in the past. A good correlation exists between the surface temperature and the isotope composition of precipitation, whereby higher temperatures correspond to more positive $\delta^{18}O$ and $\delta D$ values (*Danskgaard, 1964*).

The changing of the isotope ratios in the samples implied that the older groundwater in the deeper layer infiltrated under a colder climate. The radiocarbon ages as well as the decrease of hydrogen and oxygen isotope ratios with increasing depth allow us to suppose that the water in the deepest layer infiltrated before the Holocene, namely during the Last Glacial Maximum. In the $\delta^{18}O-{}^{14}C$-age plot (Figure 7), the different water samples can be distinguished with respect to their $\delta^{18}O$ values and radiocarbon ages. We found that the average

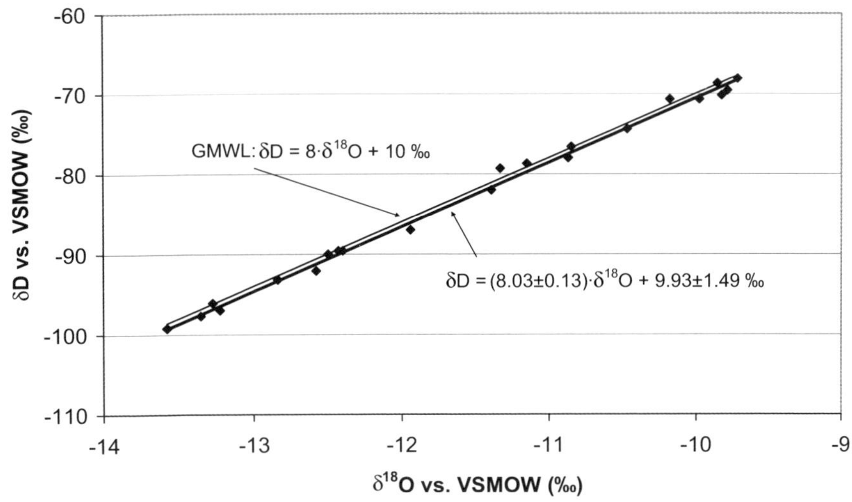


Figure 6. The $\delta D$–$\delta^{18}O$ relationship in groundwater samples from the studied area (thick line) and for the Global Meteoric Water Line (GMWL, thin line).

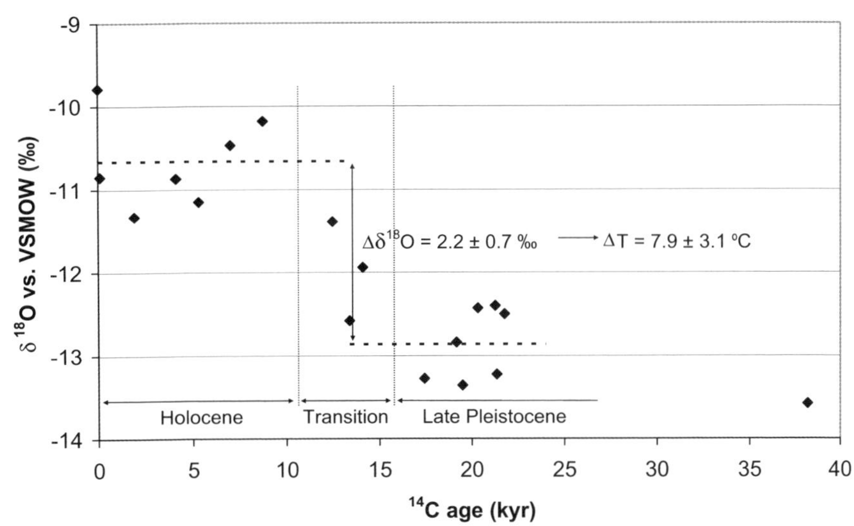


Figure 7. The changing of $\delta^{18}O$ during the transition of Holocene and Pleistocene can be used to calculate the climate cooling in the Last Glacial Maximum compared to the Holocene (see text).

$\delta^{18}O$ value for the Holocene groundwater was $-10.6 \pm 0.5$‰, while for the late Pleistocene it was $-12.8 \pm 0.4$‰, thus the difference was $2.2 \pm 0.7$‰. Using the difference between the $\delta^{18}O$ of Holocene waters and late Pleistocene waters, the temperature change can be determined if the $\delta^{18}O$–surface temperature relation is known. Huneau et al found in Southern France that the $\Delta\delta^{18}O/\Delta T$ relation is 0.22‰/°C (*Huneau* et al., *2000*), while in the Black

Forrest, Germany, it was found to be 0.38 ‰/°C (*Clarke and Fritz, 1997*). Palcsu examined the isotopic composition of recent precipitation in Hungary, and found that the $\Delta\delta^{18}O/\Delta T$ relation was 0.28‰/°C (*Palcsu, 2003*), which was between the values found in France and Germany. On the basis of these relations and the $\Delta\delta^{18}O$ change, the temperature change might be 10.0 ± 3.2°C, 5.8 ± 1.8°C and 7.9 ± 2.5°C, respectively.

The average temperature change is 7.9 ± 1.5 °C, which agrees with the finding of an earlier study in the Great Hungarian Plain about 200 km far away from our studied area. Stute and Deák (1989) investigated the dissolved noble gases in groundwater and determined noble gas temperatures (NGT's). They found that the difference in NGT's for Holocene and ice-age waters was about 8.5°C. This corresponds to our findings, and confirms that the $\Delta\delta^{18}O/\Delta T$ relation obtained from in situ isotopic investigation of recent precipitation gives the best estimation for groundwater in a given area.

## 5 CONCLUSION

Holocene water can be found in the upper and peripheral part of the granite body, but in the deeper part of the bulk glacial and pre-glacial waters appear. On the basis of isotope data we can assume that the water does flow very slowly in the inner part of the granite bulk and so the field is suitable for low and intermediate level radioactive waste repository from hydrogeological point of view. On the other hand, we can conclude that the cooling during the last ice age was 7.9 ± 1.5°C at this site.

## REFERENCES


Aeschbach-Hertig W, Stute M, Clarke JF, Reuter RF, Schlosser P (2002) A paleotemperature record derived from dissolved noble gases in groundwater of the Aquia Aquifer (Maryland, USA). Geochimica et Cosmochimica Acta 66(5):797–817.

Andrews JN and Lee DJ (1979) Inert gases in groundwater from the Bunter Sandstone of England as indicators of age and palaeoclimatic trends. Journal of Hydrology 41:233–252.

Balla Z, Horváth I, Rotár-Szalkai Á, Tóth Gy (2000) Hydrogeological Characteristics of the Üveghuta Site. Annual Report of the Geological Institute of Hungary.

Bradbury KR and Muldoon MA (1992) Hydrogeology and groundwater monitoring of fractured dolomite in the Upper Door priority watershed Door County, Wisconsin, Wisconsin Geological and Natural History Survey Open File Report 1992–2.

Busenberg E and Plummer LN (1996) Concentrations of chloro-fluorocarbons and other gases in ground water at Mirror Lake, New Hampshire. In: Morganwalp DW, Aronson DA (ed) US Geological Survey Toxic Substances Program Proceedings of the Technical Meeting, Colorado Springs, Colorado. US Geological Survey, Water Resources Investigations Report, 94-4015:151–158.

Clark I and Fritz P (1997) Environmental isotopes in hydrogeology, Lewis Publishers, New York.

Csongor É and Hertelendi E (1986) Low-level counting facility for $^{14}C$ dating. Nuclear Instruments and Methods in Physics Research "B" 17:493–498.

Dansgaard W (1964) Stable isotopes in precipitation. Tellus 16:436–468.

Fontes JC and Garnier JM (1979) Determination of the initial $^{14}C$ activity of the total dissolved carbon: A review of the existing models and a new approach. Water Resources Research 15(2):399–413.

Ingerson E and Pearson FJ (1964) Estimation of age and rate of motion of groundwater by the $^{14}C$-method. In Recent Researches in the Fields of Atmosphere, Hydrosphere and Nuclear Geochemistry, Nagoya University, Nagoya, Japan, 263–283.

Hertelendi E, Gál J, Paál A, Fekete S, Györffi M, Gál I, Kertész Zs, Nagy S (1987) Stable isotope mass spectrometer. Fourth Working Meeting Isotopes in Nature. Ed.: Wand U, Strauch G Akademie der Wissenschaften der DDR Zentralinstitut für Isotopen und Strahlenforschung, Leipzig 323–328.

Hertelendi E, Csongor É, Záborszky L, Molnár J, Gál J, Györffi M, Nagy S (1989) A counter system for high-precision $^{14}C$ dating. Radiocarbon 31:399–406.
Horváth I, Deák J, Hertelendi E, Szőcs T (1997) Hydrogeochemical investigations in the Tolna Hills area. Geological Institute of Hungary, Annual Report,1996/II.:271–280.
Huneau F, Aeschbach-Hertig W, Kipfer R, Blavoux B (2000) Isotopic evidences of the last-glacial/holocene climatic transition from groundwaters preserved in the Valreas Miocene aquifer (Southeastern France). EGS 25th General Assembly, Nice, EGS.
Palcsu L, Molnár M, Szántó Zs, Svingor É, Futó I (2002) Metal container instead of glass bulb in tritium measurement by helium-3 ingrowth method. Fusion Science and Technology 41(3):532–535.
Palcsu L (2003) A nemesgáz-tömegspektrometria hidrológiai és atomerőművi alkalmazásai (Noble gas mass spectrometry in hydrology and nuclear industry). PhD Theses, University of Debrecen, Hungary [in Hungarian].
Schlosser P, Stute M, Dörr C, Sonntag C, Münnich KO (1988) Tritium/$^{3}$He-dating of shallow groundwater. Earth and Planetary Science Letters 89:353–362.
Sheldon AL and Solomon DK (2001) Dissolved gases and tritium in fractured dolostone: implications for groundwater recharge and helium distribution. Draft Report. Smithville Phase IV Bedrock Remediation Program.
Solomon DK, Hunt A, Poreda RJ (1996) Source of radiogenic helium 4 in shallow aquifers: Implications for dating young groundwater. Water Resource Research 32:1805–1813.
Stute M and Deák J (1989) Environmental isotope study ($^{14}C$, $^{13}C$, $^{18}O$, D, noble gases) on deep groundwater circulation systems in Hungary with reference to paleoclimate. Radiocarbon 31(3):902–918.
Stute M, Sonntag C, Deák J, Schlosser P (1992) Helium in deep circulating groundwater in the Great Hungarian Plain: Flow dynamics and crustal and mantle helium fluxes. Geochimica et Cosmochimica Acta 56:2051–2067.
Torgersen T and Clarke WB (1985) Helium accumulation in groundwater, I: An evaluation of sources and the continental flux of crustal $^{4}$He in the Great Artesian Basin, Australia. Geochimica et Cosmochimica Acta 49:1211–1218.
Tóth Gy, Horváth I, Marsó K, Muráti J, Nagy P, Rótárné-Szalkai Á, Szőcs T (2003) Integrált vízföldtani értelmezés, I–II. kötet. Kis és közepes radioaktivitású atomerőművi hulladékok végleges elhelyezése. Kézirat. (Integrated hydrogeological interpretation. Volume I–II. Final disposal of low and intermediate low radioactive waste of nuclear power plant.) BA 03 123 [in Hungarian].

CHAPTER 24

# Structure conditional thermal springs in Central Brazil

Uwe Tröger
*Technical University of Berlin, Institut of Hydrogeology, Berlin, Germany*

ABSTRACT: In Central Brazil thermal water springs occur which are not linked to any thermal anomaly or have volcanic evidence. All springs are situated on fractures or faults and rise from aquifers consisting of quartzite of Meso-Neoproteozoic formations which are localised in the Tocantins Fault Belt. The most prominent are the Rio Quente springs at Pousada west of Serra de Caldas, an oval dome structure in Goias. Long term research have shown that the monolithic structure was deformed strongly and a quartzite karstification provides an enlargement of the fractures and a bigger and bigger underground reservoir. Deep wells in Caldas Novas confirm an extension of the fractures in depth. The karstification can be observed in all thermal aquifers in Central Brasil in the quartzite Paranoá Formation. Isotopic and hydrochemical analysis confirm the deep flow and heating process.

## 1 INTRODUCTION

The thermal springs in Brazil are a tourist attraction but, due to the high pumping rates of the warm water from many deep wells in case of Caldas Novas Goias, the reservoirs are also in danger to be overexploited. In Caldas Novas (picture 1) the drawdown of the thermal water aquifer reached over 50 m. Even after controlling groundwater abstraction the water table is now 30 m lower than before wells were drilled. At the same time of pumping over 1 $m^3/s$ in the city of Caldas Novas the flow of the Rio Quente springs diminished from 1,7 $m^3/s$ to 1,2 $m^3/s$. Little understanding of the flow in the complex fractured system and the behaviour of the thermal water reservoir arose to a bad management of the important spa. Technical University of Berlin and University of Brasilia carried out the existing research as well as hydrogeologists from Caldas Novas Goias. They also were active in the other thermal springs. The major focus of this article will be the Serra de Caldas in Goias.

## 2 LOCALIZATION AND CLIMATE

The Serra the Caldas Dome (SCD) with its springs is situated in the state of Goias in the center of Brazil (Figure 1). The thermal water occurs in springs east and west of SCD and is as groundwater mainly exploited from wells in the city of Caldas Novas. The locality of one of the biggest thermal springs in the world, Pousada do Rio Quente west of the SCD, is situated 350 km south of Brasilia the capital of the country. The morphology exposes a remarkable dome with a N-S extension of 14,5 km and E-W extension of 9,4 km. The difference in altitude is approximately 250 m between the top of the dome and the surrounding area. From the top and from far the dome has the shape of a strato volcano (Figure 2).

The climate in the whole region is subtropical. Heavy tropical rainfall up to 2500 mm/y (mean value for Caldas Novas, Goias 1600 mm/year) dominates the summer and hard droughts with humidity down to 15% are registered during the winter. The variation of rainfall is from 1100 mm/a to 2300 mm/a during the last decade. The average temperature in Caldas Novas is 23°C. The real evapotranspiration is up to 850 mm/y. A good amount of rainfall can infiltrate even under bad conditions as registered some years ago. Other thermal water springs are situated over 500 km north of the capital Brasilia in the state of Tocantins (Figure 1).

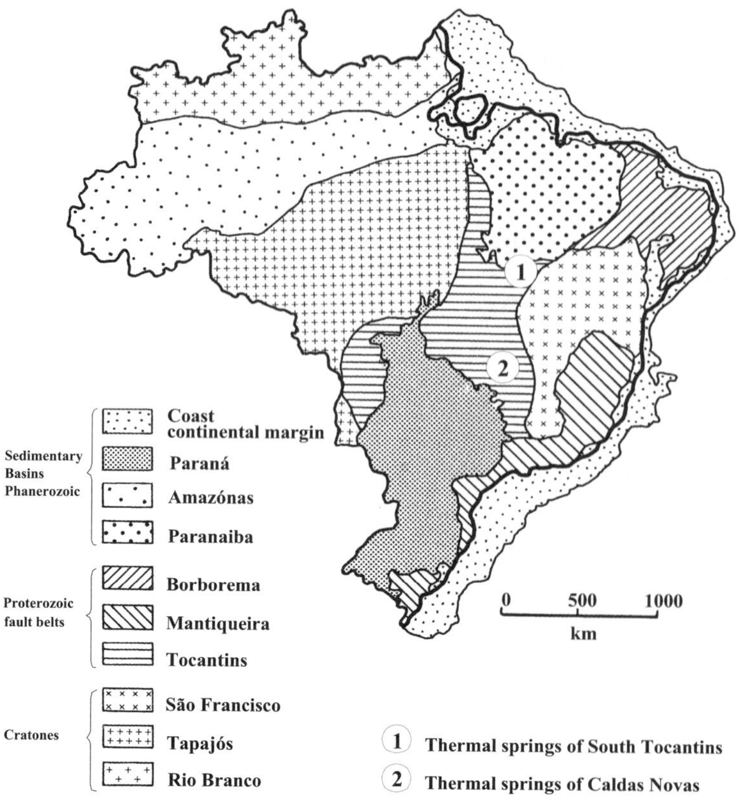


Figure 1. Geological features of Brazil and spring localization.

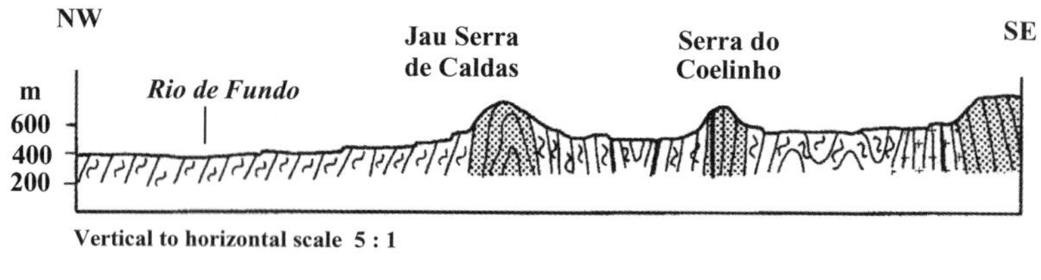


Figure 2. Cross section of the Jau Serra de Caldas in Tocantins.

The name Serra de Caldas is all places conjoint. Thus the name implies all thermal springs in Central Brazil are situated at cliffs or foots of hills or small mountains. The morphology of the other spring areas is not as impressive as SCD.

The precipitation in south Tocantins state is little higher than in south Goias state and the mean temperature is up to 5°C higher.

## 3 GENERAL GEOLOGY

The geological map of Brazil (Figure 1) shows the situation of all thermal springs in the Tocantins Fault belt which extents over some Brazilian states. The orogenic main phases happened 750–600 Ma ago during the Brasiliano cycle. The complex geology and tectonic was published by Silva et al. (2004). The basin predominantly consists of marine sediments. Ripple marks and cross bedding show shallow sedimentation conditions. Pharneozoic laterites with up to 50 m thickness cover the quartzite in some areas.

## 4 GEOLOGY OF TOCANTINS SPRINGS

The springs in Tocantins occur in fractured quartzite, metapellites, and metasediments which are older than the quartzite of SCD. The cross section of the area shows the anticline of the Serra de Caldas close to Jau (Figure 2). The folding of the consolidated rocks resulted in many faults and fractures similar to Parana spring area shown in Figure 4. At the southern end many small springs occur in the fractured quartzite banks. Microfissured schist surround them (Tröger et al. 2004).

The metasediments of the Serra da Mesa unit (Mesoproterozoic) close to Parana (Figure 3) are exposed in large steeply dipping monoclines extending for tenth of kilometres. Granites and gneisses surround outcropping metamorphic sedimentary rocks (Marini, 1984).

The springs of Mansão das Águas Mornas (Araí, Group) close to the village Paraná are located at the end of a long ridge.

Schists are intercalating with thick quartzite banks and in some units marbles were found on the basis (Dardenne, 1981). The quartzite banks itself are over- and underlain by schistic siltstones (Figure 3). The springs 1 and 2 in Figure 4 are different in temperature and occur in a deep cut of the ridge.

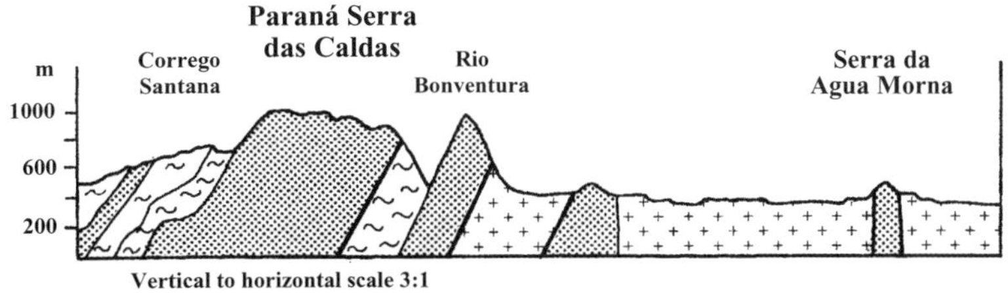


Figure 3. Cross section of the Parana Serra de Caldas.

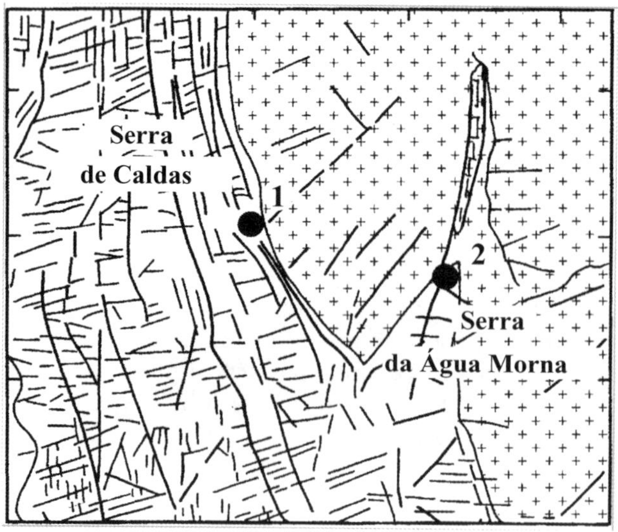


Figure 4. Extreme fractured and faulted quartzite and less effected basement. Monocline pitches to the south.

## 5 STRUCTURES OF TOCANTINS SPRINGS

Normal confinement of quartzite rock formations in the Mesoproterozoic and Neoproterozoic series are the consequence of basin inversion during the Brasiliano cycle. All thermal water springs are located on tectonic features which result from normal folding. The axis of the ridges are striking NNE and faults are striking mainly south-east to north-west and long extending faults are striking north-south. In the Parana Serra de Caldas the N striking ridges of quartzite are crossed by many faults striking east-west or north-south (Figure 4).

## 6 HYDROGEOLOGY OF THE TOCANTINS SPRINGS

The temperature of the water of the springs in the state of Tocantins ranges from 38° to 40°C. The electrical conductivity is in the range of rain water and does not reach values higher than 40 μS/cm. This low value demonstrates that the water is only in contact with mainly inert rocks consisting of quartz and some clay slates. The main ions are analysed in small quantities or are not abundant. The pH is acid and in the range of 5.2 to 5.7 which shows the absence of carbonate rocks. Groundwater that circulates in the basement shows the same hydrochemical composition.

The geological units have a long extension and the springs are not located at the lowest point of the mountain ridge but they occur where fractures are observed on air photos in both examples. The geological units are highly fractured over the whole outcrop. The mechanism of the springs can only be understood if the model of a preferential pathway is accepted. Drillings close to the springs that reached the basement did not produce water of

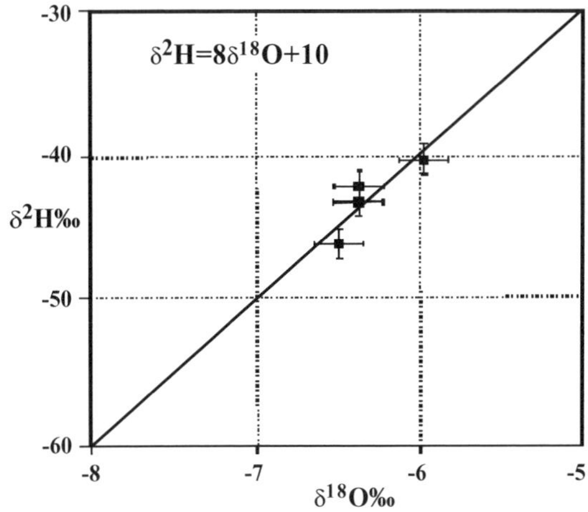


Figure 5. $\delta^{18}O/\delta^{2}H$ diagram of samples from the springs in Tocantins.

the same temperature. Also the yield was very low because no good permeability structures were encountered. Thus, it is necessary to differentiate between a deep water pathway in which the thermal gradient influences the temperature of the spring water. The upward flow is only possible through extended fractures. Smaller and shallow fracture networks bear less and cold water in small springs which occur temporary during the rainy season.

The isotopic composition of the water shows a local recharge of these aquifers. The stable isotopes $^{2}H$ and $\delta^{18}O$ indicate that during the recharge process climatic conditions are similar to the recent ones. The water has no signature of evaporation processes as Figure 5 shows. The sample position is close to the meteoric water line. The $^{13}C$ isotope ratios of −19.6 and −20.3‰ are rather light. Taking into account the original $^{13}C$ signature of the atmosphere, the subsequent fractionation within the plants' metabolic processes and the hydrochemical processes that ensue, the $^{13}C$ content mainly reflects the influence of the Cerrado plants (bushes, shrubs and small trees). The impact of carbonate dissolution that would shift the values towards a less depleted isotopic signature is obviously not possible because of absence of any carbonate rock.

The groundwater residence time has been analysed based on tritium and $^{14}C$ data. The samples have 90–95% modern $^{14}C$, taking into account a minor reservoir effect (carbon uptake during recharge), the groundwater can be considered recent in terms of $^{14}C$ dating. Some of the samples contain tritium and indicated mixing with post-bomb rainfall (less than 50 years), some samples are tritium free and indicate groundwater with residence times of more than a few decades. The oldest water from Jaú Serra das Caldas spring with 90% $^{14}C$ contains no tritium. The infiltration is fast and must be far from the spring because the water is free of tritium. This and the elevated water temperature indicated a fast deep circulation. The recharge to the fractures must be of regional scale as there is no detectable tritium-bearing component.

The spring in the Parana region (Figure 4) contain 0,6 Tritium Units (TU). Obviously this groundwater must mix with a minor part of younger water as there is some tritium in the samples. The groundwater recharge for this spring can be localized north and south of the

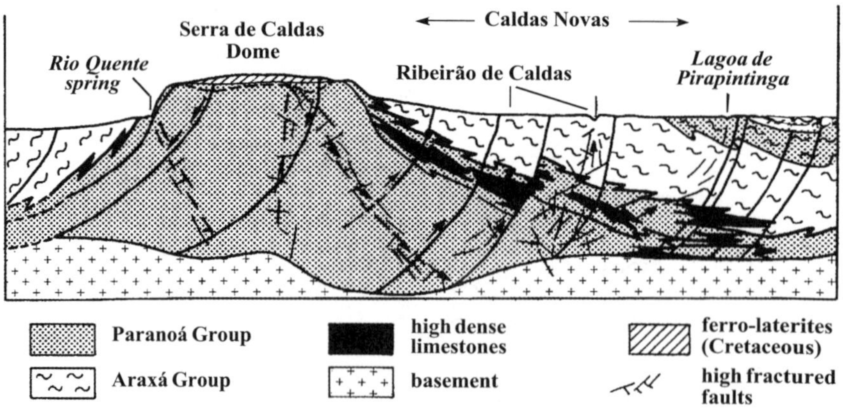


Figure 6. Cross section of Serra de Caldas Dome with the indication ↗ of water flow direction.

spring but can also be in the west of the overturned anticline (Figure 3). $^{14}C$ indicate a smaller system than in the example of Jaú where no tritium was detected. The outcrop area of the quartzite can be calculated to more than 1000 km$^2$. The $SiO_2$ content is only a little lower than in Jau (~20 mg/L) and indicates deep water temperatures of 50°–60° C. This value shows suggests a deep flow from the west flank of the quartzite ridges. Also it is clear that the groundwater follows preferential pathways along fractures.

## 7 GEOLOGY OF SERRA DE CALDAS DOME

Campos et al. (2000) carried out a detailed stratigraphic map. Horizontal lying greenschist and metamorphic quartzite of the Paranoá Group with a low radial dip form the oval dome. Schist and thin to thick banks of quartzite built up the younger Araxá Group exposed in the surrounding (Figure 6 & 7). Between both sediment bodies an unconformity is notified. Drillings show a mylonite of some meters thickness.

Cretaceous ferro-laterites with a high content of manganese which was mined in a few places build the top of the SCD. It is difficult to estimate the exposure time of the quartzite of SCD but probably only a few meters of the overthrusting schist covered the massive and were eroded rapidly. The dome consists mainly of quartzite horizontal layered beds of the Paranoá sequence. On top of this sequence marble or high dense calcite built up a strata which is not always detected. The younger strata in the surrounding of the dome are schist with intercalating thin quartzite bands of the Araxá Group (Figure 6).

## 8 STRUCTURES OF SERRA DE CALDAS DOME

The structure of the SCD is in contrast to the examples from Tocantins much more complex. The form of the eroded plateau as shown (Figure 9 and picture 1) reminds a strata volcano. The axis strikes NNW and the elliptic form is approximately 17 km long and 10 km wide. Drake Jr. (1980) explained the Serra as a window of younger into older sediments. Campus et al. gave the stratigraphic correct interpretation. Only recent research reconstructed in detail the nature of the strange structural conditioned dome (D'el Rey et al., 2004).

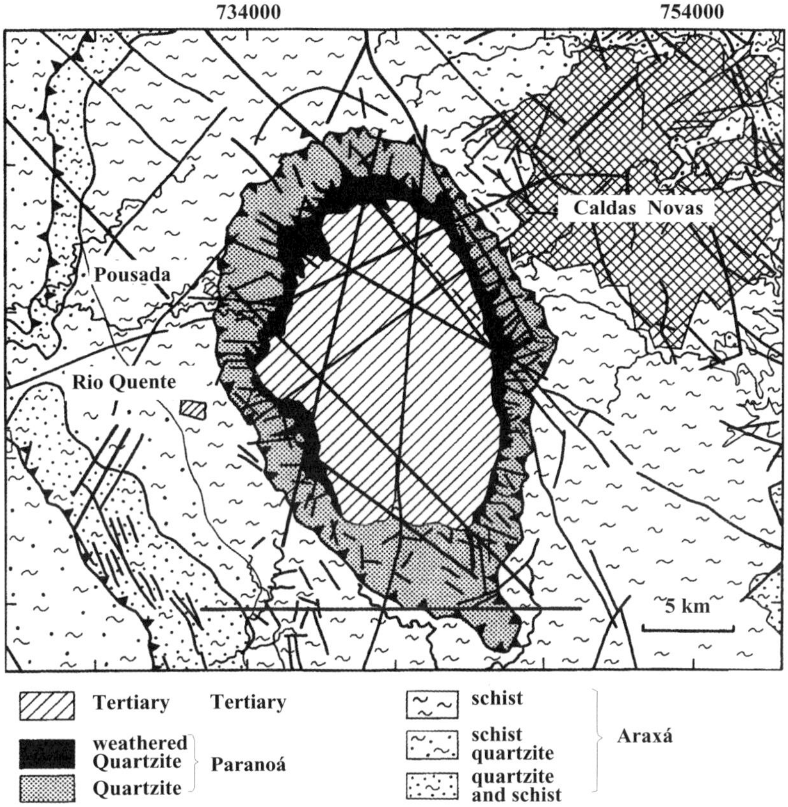


Figure 7. Structure and geological map of the Serra de Caldas Dome and the surrounding.

Haralyi (1978) presumed an oval sedimentary body up to a depth of more than 2000 m after carrying out gravimetric measurement.

During the Brasiliano tectonic cycles three events of ductile progressive deformation took place (750–600 Ma). The Tocantins Fault Belt (Figure 1) came under stress due to the confinement of the Amazon and São Francisco plates. The inversion of the basin ended in an overthrusting what from a nappe of the Araxá Group remains. The ramps of the movement can be observed at places on the west-side of the Serra de Caldas D'el Rey et al., 2004). The maximum subhorizontal compression $\sigma_1$ is oriented 290–300°/110–120°. This stress deformed the Araxá schist and quartzite banks more than the Paranoá massive Quartzite. Finally the stress deformed the dome and a duplex (Figure 8) was structure at depth was possible. Structures can be observed on the satellite image (picture 1) and its interpretation not completely shown in Figure 7.

Gravimetry measurements of Haraly (1978) have shown such forms. Many faults and fractures are the consequence of the duplex structure as well as mylonites in the thrust planes and sliding faults.

A model (essential result in Figure 8) for the evolution of the Serra de Caldas was elaborated from D'el Rey et al., (2004). Originally the Paranoá Group was 200 m thick. The strata were dipping to WNW. The nappe of the Araxá Group (schist signature) pushed and the internal ramp in the Paranoá Group was developed. The dome was rising up due to a

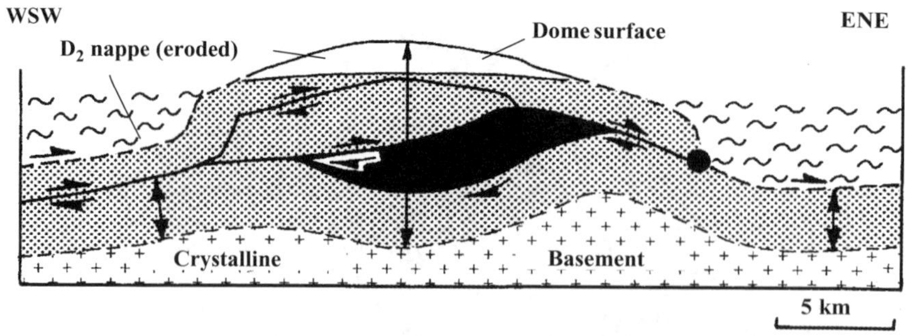


Figure 8. Model of the evolution of the SCD (D'el Rey et al., 2004).

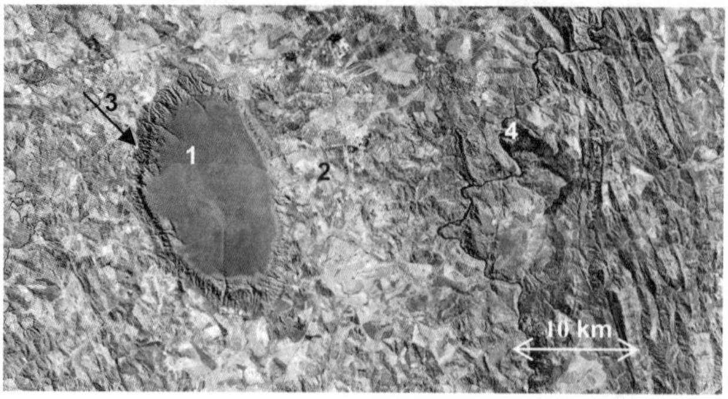


Picture 1. Satellite image of SCD (1), Caldas Novas (2) Rio Quente springs (3), and Lagoa de Pirapitinga (4).

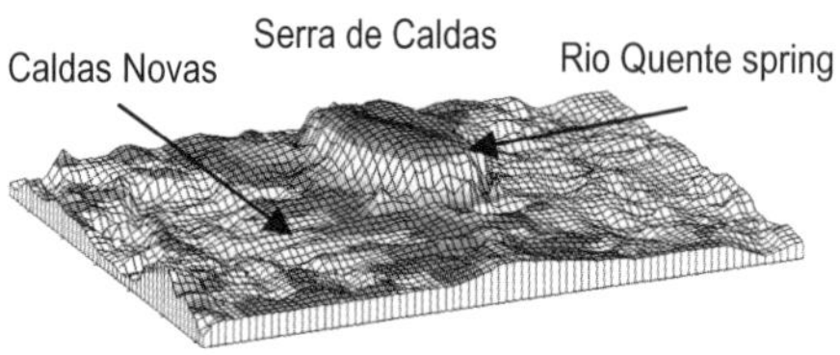


Figure 9. Shape of the dome structure with localization of Caldas Novas and Rio Quente spring.

continuos stress. Due to various ramps which also multiplied the thickness of the Paranoá Group to over 1700 m the dome could develop. The reason for the origin of the Dome might be the irregularity of the basement. This evolution is important for many fractures and the continuous water pathway to the bottom of the sedimentary series. Also for this reason almost all faults and fractures dip to the west. The east dipping brittle zones are the effect of the stretching during the overthrusting.

## 9 QUARTZITE KARSTIFICATION

The karstification of quartzite plays an important role in the subtropical climate of South America. White (1996) reports karst-similar features from Venezuela.

The recharge and the groundwater flow is only possible because the fractures have a good permeability which is a result of a dissolution of the quartz. The rain water has a mean temperature of 23°C and thus the quartz dissolute faster than in cooler climate where this process can not be observed. No soil cover did exist during some hundred million years. Fractures in the exposed quartzite enlarged during the time and the infiltration into the fractures can be observed till today in areas where the quartzite banks have no soil cover. The brittle deformation of the quartzite banks left a network of faults and fractures wide open. Cretaceous or Tertiary sediments cover the quartzite outcrops in some places of Central Brazil. Nevertheless this material reduces the fast infiltration in the opened fractures but has only the function of a filter. It is important to mention that the localization of SCD was always in the tropics since Triasic time.

The thermal water occurs in depth and as deeper the water flows as more widen the fractures are. Fractures of 20 cm width were observed in deep wells in Caldas Novas already at 300 m below surface. Estimates of the total space from dissolution in the dome of the Serra de Caldas Paranoá Group quartzite from Reinhold (2005) resulted 2,5 $km^3$. As the outcrops occur in other places than in Goias mostly in steep dip of the banks fractures are not as extended from dissolution as in the dome.

## 10 HYDROGEOLOGY OF THE SCD

The SCD has an extraordinary situation because of the unique tectonic features. The sub-horizontal to horizontal lying massive low metamorphosed quartzite banks were exposed after erosion of the highly fractured and folded Araxá slates and thin quartzite banks. The overthrusting of the Araxá group and the building of the duplex structure caused west-dipping fractures and faults as well as pull down fractures dipping east following the model of D'el-Rey (2004, Figure 8).

Crenulation fractures arose along the thrust plane. Water could always infiltrate and reached already a certain temperature in the Dome. The consequence was an extension of the fractures in depth along an unknown time due to the dissolution of the quartz. The vertical permeability is much higher than the horizontal permeability. Video logs show wide opened vertical fractures in the depth. Only some of the horizontal fractures are sufficiently permeable. Therefore the groundwater follows the west dipping fractures down and rises up in the mylonized east dipping pull down fractures in the western region of the Dome and the other way around in the east.

The Tertiary cover of the SCD is very permeable. It could be observed that after a heavy rainfall of 50 mm/h the water disappeared after a period of minutes. Between the weathered zone and the Tertiary soil exists a clay condensation horizon where part of the infiltrating water is drained to the rim of the massive. There occur some permanent springs with little run off (Pietzner, 2001).

Before 1975 the thermal water occurred in three spring areas: Lagoa de Pirapitinga and along the Riberão de Caldas (Caldas River) east of the Dome and the Pousada do Rio Quente west of the Dome (Figure 10). At the last spot the springs have had a run off of more than 1,66 $m^3$/s. The other springs have had a flow of approximately of 0,1 $m^3$/s. Since the first wells were drilled in Caldas Novas immediately the springs along the Caldas River felt dry. The water table in the SCD dropped locally between 1994 and 2000 almost 100 m (Figure 11) as a result of the increasing pumping in Caldas Novas. Before 1994 the draw down in the city was dramatic because of an uncontrolled situation.

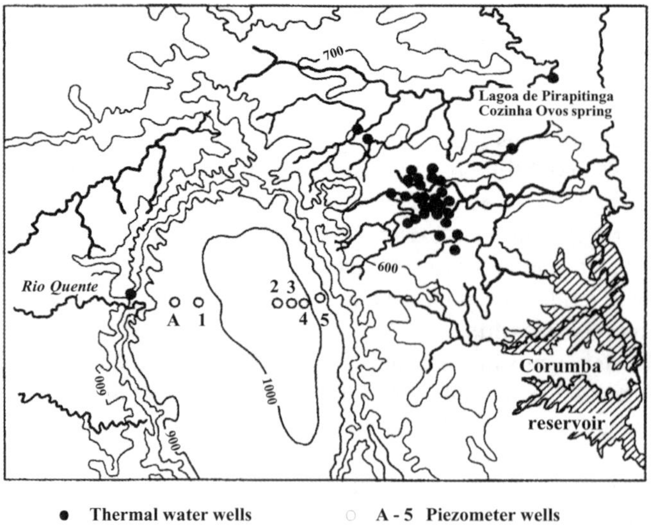


Figure 10. Localization of piezometer wells (see also below) and thermal water wells in Caldas Novas. The springs near the Pirapitinga river are circled oval.

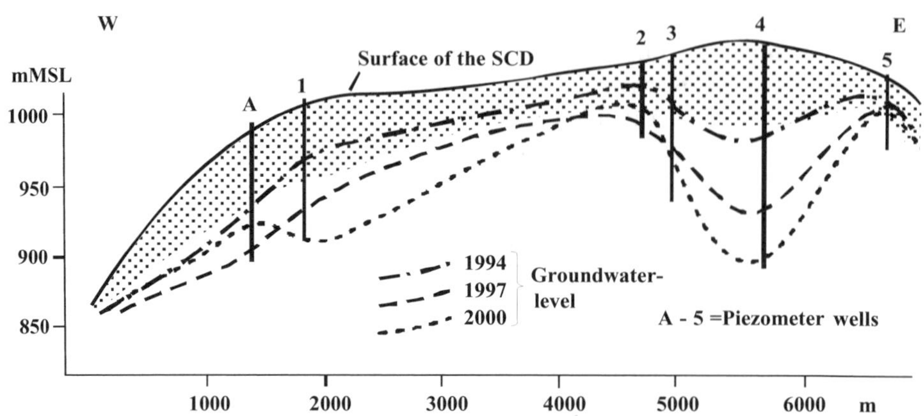


Figure 11. The lowering of the water table from 1994 to 2000 on the SCD top follows structures (see also Figure 6).

The recharge of the thermal water takes place on top of the SCD. The pressure head in the Serra aquifer was measured 950 m MSL. The springs of the Rio Quente occur in 700 m MSL. The springs and the pumped thermal water from wells in Caldas Novas yield more water than the calculated and measured recharge of the Serra de Caldas (Pietzner, 2001). The evidence of an additional recharge in the Araxá Group is given with the hydro-geochemical interpretation below. The pumping rate from the wells in Caldas Novas reaches almost 1 $m^3/s$. The wells which reach the Paranoá aquifer are different in their chemical composition from the ones which reach only fractures in the Araxá Group.

In most drillings very dense limestone (probably marble) was detected on top of the Paranoá quartzite. Fractures are generally less observed and of smaller aperture. It is possible that the groundwater flows with very low velocity and enriches in calcium. Karst features in the marble are not reported from the drilling companies. The chemical fingerprint of calcite is abundant in all thermal water samples taken from deeper wells in the Araxá slates.

The Araxá slates are covered with a thin Cambisol or Latosol soil. The clay content is very low and infiltration is very good. The strong deformation of the slates left many fractures which are important to recharge an upper aquifer. The water reaches a little bit higher temperature than the rain water.

## 11 HYDRAULICS OF SCD

Three main faults connect the deep thermal Paranoá aquifer with the springs placed in Pousada, along the Ribeirão de Caldas (no more active), and Lake Pirapitinga. The pressure head of the aquifer reaches the morphologic lowest areas in the region. The flow of the springs along the Ribeirão de Caldas was reduced after the wells in Caldas Novas started to pump larger amounts of water. On an area of 20 km$^2$ over 200 wells were drilled in the last 30 years. The water table in the wells dropped down dramatically and started to rise to the actual level after government control was implemented (Figure 10).

The run off quantity from springs in Lake Pirapitinga was not registered and observations were subjective. The flow reduction in the Rio Quente was significant but nevertheless it is still much more than expected. In the mean time the head on top of the SCN dropped much more. The first drilled observation wells felt dry in a depth of 50 m and the water table still is falling (Figure 11).

Fractures play an important role in this scheme. The first abstraction wells were shallow and the temperature of the water of little thermal influence. Some wells had a high and other a small draw down during pumping depending on the communicating fractures. Most of the deep wells of over 600 m depth cross wide open fractures and have a draw down of some meters during pumping 25 m$^3$/h. Shallow wells with a yield of 15 m$^3$/h have a draw down of over 100 m. Figure 12 shows the specific yield in relation to the depth (Fach 2002). Collapsed shallow wells are substituted with deep wells of up to 850 m depth. However, there is tendency of increasing temperatures in all wells while maintaining the abstraction. Continuous temperature measurements from officials show the same tendency. Reinhold (2005) could show from chemical analysis (see Hydrochemistry) that the fraction of groundwater from the Paranoá quartzite increases in the same time. The flow path and the mixture of the groundwater in the SCD is shown schematic in Figure 13. The low temperature in the springs of rio Quente can be explained with the mixture of water from different depth.

New drilled wells cleared new pathways for the thermal groundwater. A well situated in the valley of the Ribeirão de Caldas reached the thermal aquifer in 850 m depth and after sealing maintains a constant head of 1 bar. The temperature maintains in the same way at 59°C for the last four years.

Taking the specific yield in account the deep wells are more efficient. Shallow wells are less efficient even if some have a high yield. The explanation for this effect is the difference of the fractures in the different type of rock. Faults only have a better hydraulic conductivity in the Araxá slates. Normal fractures have a low hydraulic conductivity. However, the fractures in the Araxá slates are very important for the local recharge of the aquifer. It is

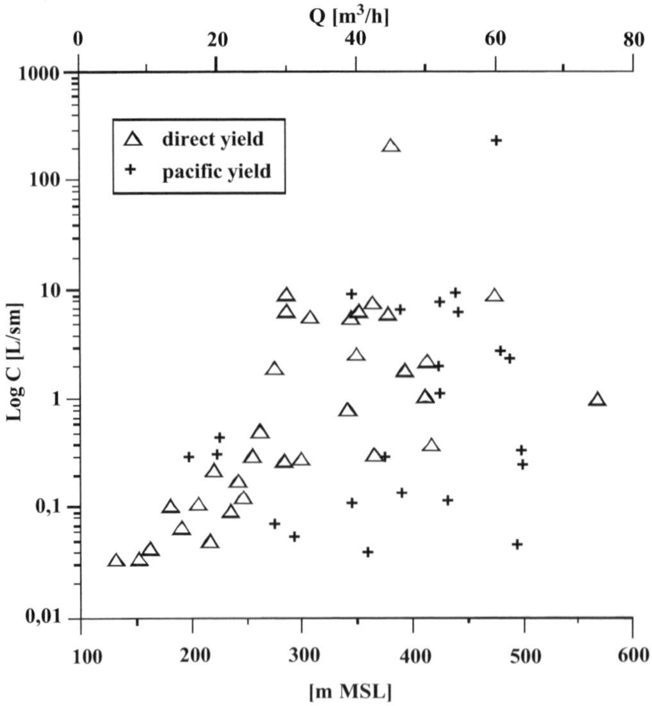


Figure 12. Specific yield C in relation to depth. Triangles show the direct yield.

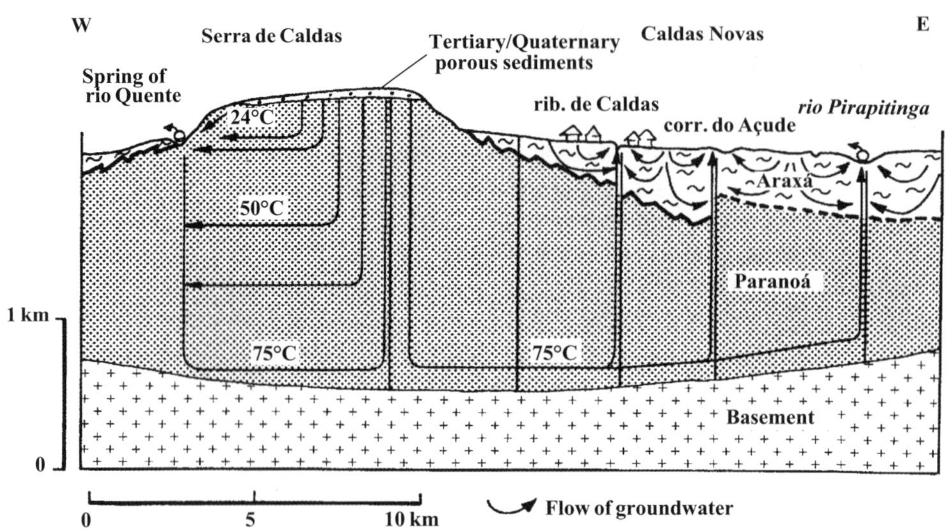


Figure 13. Schematic flow diagram of thermal groundwater in the SCD region showing a cold water cycle and the hot water flow with the mixture of both in the lower Araxa schist section.

necessary to take in account that the recharge decreased during the last decade because the urban area increased some hundred percent. Therefore the sealed surface increased in the same way and in some wells the water table is much lower than before.

It is also important to remark that some wells have a better yield even if not deep. The overthrusting has in consequence a block slip faulting. The blocks are dividing the aquifer and depending where the wells are located the more shallow or deeper they reach the Paranoá quartzite. A high yield from shallow wells (up to 250 m depth) does not mean a higher temperature. The good hydraulic conductivity of the faults of the Araxá Group allows a deep infiltration of the dense cold water.

## 12 HYDROCHEMISTRY

All thermal water wells in the surrounding of the SCN were sampled two times in 2000 and 2004. The chemical composition of the groundwater is a key to the understanding of the flow. Three characteristic types of groundwater can be identified (Table 1). The thermal water of the Paranoá quartzite has a mean electric conductivity (EC) significant lower after a four year interval of the measurements. Also important is the decrease of the EC in the deep Araxá 2 aquifer. In the same time the temperature rose up 0,5°C. Remarkable is the increase of the temperature in the deep Araxá groundwater. This water is mixed water of uprising thermal water from Paranoá and locally recharged water from Araxá 1 (see Figure 6 and Figure 13).

In the four year interval the temperature of the shallow aquifer decreased 0,5°C. It can be estimated that the recharge decreased too, even if there is no evidence but the sealed area of the city occupies 250% more space. The temperature of the Araxá 2 groundwater finally increased 1°C in the same period. At the same time the temperature in the springs of the Rio Quente and Lake Pirapitinga (Cozinha Ovos) maintained their temperature. Even if there is a loss of runoff from these springs.

The Piper diagram (Figure 17) shows the similarity of the character of the groundwater (Reinhold 2005) from Araxá 1 and Paranoá. Chloride and sulphate are almost absent and the bicarbonate is the dominate ion. The bicarbonate is not in equilibrium with the cations (Zschocke 2000). The composition of the water from Araxá 1 is various and therefore more spread in the Piper diagram. It is obvious that the shallow aquifer is characterized by the soil and the weathered zone. The different rock types of the Araxá group produce different types of soil. As this area has been leached during the time the water is low mineralised (Table 1). The deeper the water flows into the fractures of the Araxá 2 series the more the water is mineralised. The lithology does not change with depth and a further mineralization is not to expect.

Table 1. Electric conductivity [μS/cm], temperature [°C] and pH of the groundwater from pumping wells in comparison of 2000 and 2004. Measurements were carried out in a flow through cell after 12 h to 24 h pumping and constant values. Monthly official measurements show the same trend.

| Year | Mean [mmhos/cm = μS/cm] | | Mean Temperature [° C] | | Mean pH | |
|---|---|---|---|---|---|---|
| | 2000 | 2004 | 2000 | 2004 | 2000 | 2004 |
| Paranoá quartzite | 62 | 47 | 55,6 | 56,2 | 6,1 | 6,0 |
| Araxá 1 (shallow) | 61 | 63 | 32,5 | 32,0 | 6,7 | 6,4 |
| Araxá 2 (deep) | 179 | 138 | 41,0 | 41,5 | 7,4 | 7,4 |

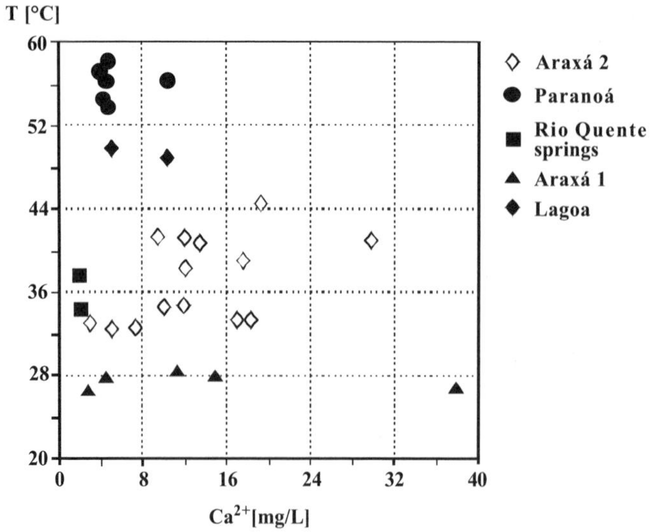


Figure 14. T/$Ca^{2+}$ relation for different aquifer.

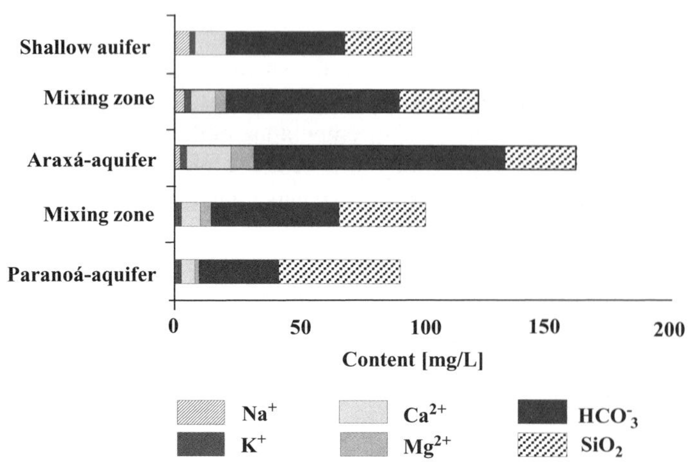


Figure 15. Typical distribution of the principal ions in the pseudo-aquifers of the SCD groundwater (analysis from Zschocke 2000).

To understand the process it is necessary to follow the groundwater flow (Fig. 13). The thermal water is recharged on top of the SCD. It flows along the fractures and faults to the karstified part of the quartzite in a depth of 2000 m and more. Following the gradient to the springs the water must flow upward through fractures and faults. The cold dense water which is recharged in the wider surrounding of the wells of Caldas Novas flows in opposite to the thermal water movement downward and mixes in the Araxá 2 fractured network zone with water that is rising from the Paranoá thermal aquifer. Evidence of this process is given from the chemical composition of the groundwater. The low mineralised thermal water from the Paranoá quartzite (Figure 15) dissolves marbles in the upper Paranoa

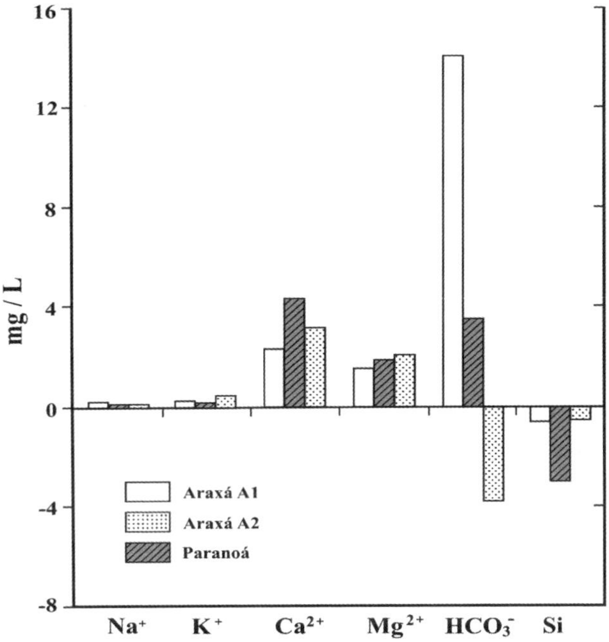


Figure 16. Variation in ion content of 87 wells. Minus means a higher content in 2004 comparing with 2000. The influence of deep water is shown with the increase of silica aquifer.

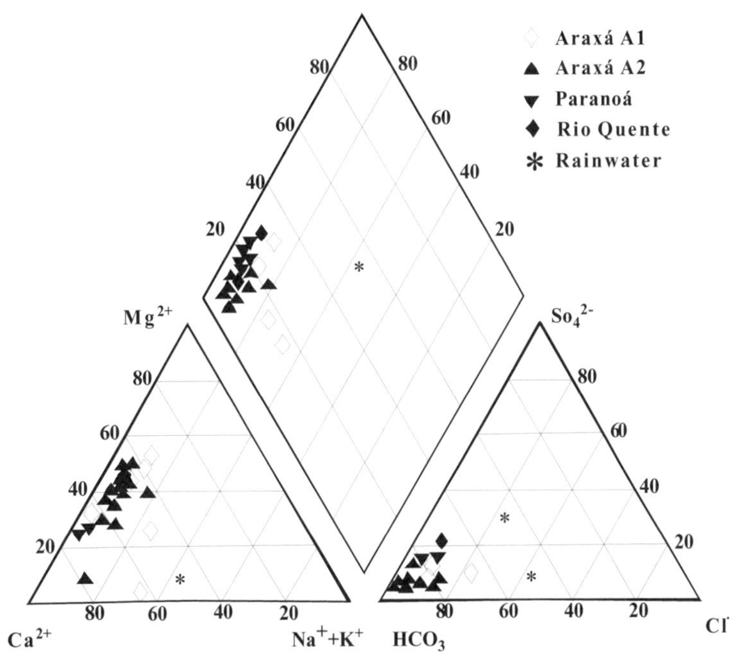


Figure 17. Piper diagram of 91 samples: 87 wells and 3 springs as well as rainwater.

group and triples its content of calcite and magnesium (Figure 15). The mixing with the normal groundwater from the Araxa 1 recharge can be detected. The deeper groundwater of the Araxá 2 is more mineralised than the water of the upper part of the Araxa 1 cold groundwater (Figure 15). The chemical fingerprint can be observed also in the scattered calcium diagram in Figure 14.

Figure 16 shows the variation of the ion content in the wells in Caldas Novas. In 2004 Calcium is less abundant and silica more than in 2000. The groundwater from the semi-confined deeper aquifer in the future is migrating more intensive along the fractures into the upper aquifer which is obviously an effect of overexploitation.

## 13 CONCLUSIONS

Thermal water springs are wide spread in Central Brazil. They occur in quartzite rock formation from the Upper Middle Proterozoic to the lower Upper Proterozoic and are independent from hot spots or volcanic activities. They all are exposed for a long period and the fractures are widen by silica dissolution (quartzite karstification). The intensive tectonics built up a functional network of fractures serving the warm water transport. The folding of the mountain ridges enables a deep flow of groundwater and therefore the thermal effect. In case of the Serra de Caldas Dome the complex structure and the overthrusting as well as the duplex formation formed a fracture network which reaches over 2000 m depth. The thermal water reaches the highest temperatures in Brazil. Due to the overexploitation the run off of the Rio Quente decreases and the draw down in the wells is continuously. A good management is necessary for all springs if the spas should exist in the future.

## ACKNOWLEDGEMENT

The author thanks his students for their excellent collaboration in Caldas Novas. Their master's theses were of an enormous value for the understanding of the hydrogeology of the area and without their help the understanding of the groundwater flow would be still very incomplete.

## REFERENCES

Albuquerque, C. (1998) Caldas Novas: Ecológia. Kelps, Caldas Novas/Goiás.

Campos, E.C. & Costa, J.F.G. (1980) Projeto estudo hidrogeológico da Região de Caldas Novas. Vol. I. MME/DNPM/CPRM. Goiânia. P.34–47.

Campus, J.E.G., Tröger, U. & Haesbaert, F.F. (2005) Águas Quentes de Caldas Novas, Goiás Notável ocorrência de águas termais sem associação com magmatismo, www.unb.br/ig/sigep/sitio113/sitio113.pdf.

CPRM, SMET, Metano, UnB (1999) Geologia e Recursos Minerais do Estado de Goiás e Distrito Federal. Texto Explicativo dos Mapas Geológico e de Recursos Minerais, Escala 1:500.000, Ministério de Minas e Energia – CPRM, Goiânia.

Dardenne, M.A. (1981) Os Grupos Paranoá e Bambuí na Faixa Dobrada Brasília. Simpósio sobre e Craton do São Francisco e suas Faixas Marginais, Salvador, *Anais* da Soc. Bras. Geol. Salvador, SBG, 1981, p. 147–157.

D'el-Rey Silva, L.J.H., Klein, P.B.W. & Walde, D. (2004) The Caldas Novas Dome , Central Brasil: Structural evolution and implications for the evolution of the Neoproterozoic Brasilia Belt, Jour. Of South America Earth Sciences, 17, 153–169.

Drake Jr., A.A. (1980) Tectonic studies in the Brazilian Shield. The Serra de Caldas Window, Goiás. Washington. USGS DNAE DNPM CPRM. Geological Survey Professional Paper 1119-A/B, p.1–11.

Fach, A. (2002) Fliessverhalten im Kluftgefüge von Caldas Novas, unpublished MSc Thesis, Technische Universität Berlin, 66 pp appendix 32 pp.

Furnas Centrais Electricas SA (1996) Encarregon de Sondotécnia SA de estudar anomalis termais da região de Caldas Novas – Vol 1 – Relatório final textos, Cor-151 Re, 133 pp.

Haralyi, N.L.E. (1978) Carta Gravimétrica do Oeste de Minas Gerais, Sudeste de Goiás e Norte de São Paulo. Phd Sci. Geol., Instituto de Geociências da Universidade de São Paulo, unpublished., p. 125–133.

Marini, O.J., (1984) As faixas de dobramentos Brasília, Uruaçu e Paraguai-Araguaia e o Maciço Mediano de Goiás. In Schobbenhaus Filho, C. et alii. *Geologia do Brasil*. Brasília: DNPN, p. 251–303.

Pietzner, O. (2001) Grundwasserneubildung und Grundwas-serabfluss im Porenwasserleiter der Serra de Caldas, Goiás, Brasilien. unpublished MSc Thesis, Technische Universität Berlin, 122 pp.

Reinhold, E. (2005) Hydrogeochemische Korrelation von Grundwasserproben des Thermalgrundwasserleiters von Caldas Novas, Goiás – Brasilien, unpublished MSc Thesis, Technische Universität Berlin, 77 pp, VI appendix 42 pp.

Tröger, U., Costa, J.F.G. & Haesbaert, F. F. (2003) The Thermal Aquifer and the Springs of the Serra de Caldas Area – Goias – Brazil, Int. Conf. Groundwater in Fractured Rocks, Prag, p. 225–226.

Tröger, U., Peixote Filho, S. & Lacerda, H. (2004) Thermal Water Springs in Central Brazil – Hydrogeology, Chemical Composition, and Isotope Studies, XXXIII IAH Congress ISBN 970-32-1749-4, Zacatecas, Mexico.

UHE CORUMBÀ (1999) Reservatório e Termalismo , Etapa IV, Regiao de Caldas Novas e Área de Influência de Reservatório, Monitoramento, Vol. T.021.98-RO, Rio de Janeiro.

White, W.B. (1960) Quartzite karst in southeastern Venezuela. In: Report omn the Angel Falls Expedition. The Netherworld New, USA, 8/3, p. 29–31.

Zschocke, A. (2000) Zur Hydrochemie der Grundwässer von Caldas Novas – Zentralbrasilien –, Implikationen für das Thermalgrundwasser, MSc Dissertation, Technische Universität Berlin, 67 pp.

# Investigation and interpretation methods in fractured environment

CHAPTER 25

# Characterization of fractured porous media: Aquifer Analogue Approach

Ralf Brauchler[1], Martin Sauter[1], Christopher I. McDermott[2], Carsten Leven[3], Matthias Weede[4], Peter Dietrich[3], Rudolf Liedl[5], Heinz Hötzl[4], and Georg Teutsch[3]

[1] *Geoscience Center, University of Göttingen, Göttingen, Germany*
[2] *Center for Applied Geoscience, University of Tübingen, Tübingen, Germany*
[3] *Center for Environmental Research (UFZ), Leipzig, Germany*
[4] *University of Karlsruhe, Department of Applied Geology, Karlsruhe, Germany*
[5] *Department of Geoscience, University of Dresden, Dresden, Germany*

ABSTRACT: After a short overview on characterization techniques, the concept of the Aquifer Analogue Approach to fractured porous media is presented. In contrast to consolidated rocks of low permeability (e.g., granite), fractured porous materials show significant storage and matrix conductivity. To investigate these characteristics, flow and transport of fractured porous sandstone materials were examined on different scales. On the laboratory scale tomographical and single well experimental set-ups based on gas flow measuring techniques were developed and applied for the characterization of different size fractured sandstone samples ranging from approx. 0.02 $m^3$ to 1 $m^3$. These methods distinguished flow and transport occurring in the porous matrix and in the fracture network. For the derivation of actual hydraulic parameters, a travel time based tomographic inversion approach was developed. These techniques were implemented to investigate a larger fractured sandstone block (approx. 200 $m^3$) with pneumatic and hydraulic flow and transport experiments.

## 1 INTRODUCTION

The characterization of flow and transport parameters of fractured porous aquifers using conventional aquifer testing methods can be considered as a challenge. Generally, specially adapted spatially high resolution investigation techniques are required to resolve the large contrast in hydraulic properties between highly permeable fractures and the much less permeable rock matrix, which induces complex flow and transport patterns within the rock material (e.g., Karasaki et al., 2000; McDermott et al., 1998). The availability of spatially and temporally high resolution data sets are a prerequisite for the prediction of the large variability in discharge and the high vulnerability to contamination, which plays a major role in developing innovative investigation techniques for sustainable water resources management for fractured porous aquifers. Recent monographs provide comprehensive overviews of experimental and modeling techniques in the field of fractured rocks (Singhal and Gupta, 1999; Evans et al., 2002; National Research Council, 1996, 2001; Dietrich et al., 2005; Faybishenko et al., 2005).

In contrast to igneous fractured rocks where flow and transport in the matrix plays a minor role, for fractured *porous* media, the contributions of both fracture network and matrix must be considered. This paper, besides a brief review of special characteristics of igneous fractured rocks and investigation techniques generally applied, concentrates on the research of fractured porous media using the Aquifer Analog Approach. The Fractured Rock Aquifer Analogue Approach is a concept that attempts to reconcile *directly* available information on geometric properties of the fracture network in the outcrop with integral data or data recorded in the immediate vicinity of a borehole that are generally available from hydraulic testing. Therefore, common techniques for the determination of structural and hydraulic parameters are briefly reviewed followed by analogue investigation techniques utilizing tomographical and single well methods conducted on laboratory samples as well as transport experiments on the field block scale are used to evaluate fracture-controls on flow in a fractured sandstone.

Generally, the first step in the characterization of fractured rocks is the *direct determination* of structural parameters of the aquifer such as fracture density, orientation, aperture distribution, and fillings. The scanline method of Priest (1993) is one example of a practical method to record fracture geometries at easily accessible outcrop walls. For aquifers, which are usually not directly accessible except through boreholes, the fracture networks have to be recorded with *indirect methods*, including borehole-geophysical techniques, flow meter measurements, temperature measurements, image logging and fluid logging (Morin et al., 1988; Silliman, 1989; Paillet et al., 1987; Tsang et al., 1990; Williams and Johnson, 2004; and Deltombe and Schepers, 2004). Detailed reviews about the potential of geophysical logs to characterize fractured rocks are given by Keys (1979); Kobr et al. (1996); and National Research Council (1996). The derivation of hydraulic properties based on structural information may, however, encounter problems of ambiguity, because the characterization of hydraulically active zones depends strongly on the geology of subsets of fractures contributing to large scale flow and transport. Studies performed during the last decades have shown that commonly only 10% or less of all fractures play an important role on a scale important for contaminant flow (e.g., National Research Council, 1996; Olsson, 1992). This coincides with the observations, of a lack of correlation between the intensity of fracturing and hydraulic conductivity (Neuman 1997; Chen et al., 2000; Ando et al., 2003). To investigate flow in fracture networks on the field scale, different types of *hydraulic tests* were developed including hydraulic single well tests, interference tests arranged in a tomographical configuration, and tracer tests.

*Single Well Test:* Packer, drillstem and slug tests are employed to obtain hydraulic and geometric parameters at well-defined fracture intervals. Evaluation procedures for pumping and injection tests are generally based on a conceptual model assuming flow from a homogeneous matrix towards a single (equivalent) fracture, directly connected to the borehole. A review of interpretation methods adopted from the petroleum-engineering field for individual conceptual models and well/fracture configurations is presented in Gringarten (1982) who discussed the effects of well storage and well skin and their influence on test interpretation. More comprehensive summaries are presented by Matthews and Russel (1967); Earlougher (1977); and Streltsova (1988). The application of the Laplace inversion technique in the field of groundwater (Moench and Ogata, 1981) consider more complex problems with elaborate conceptual models in the field. There are numerous analytical solutions for more complex set-ups, partly based on double-porosity approaches (e.g., Naceur and Economides, 1988; Karasaki et al., 1988; Barker, 1988). Barker (1988) presented the concept of a fractional flow dimension for the analysis of hydraulic tests in fractured systems. The flow geometry towards the well can be derived from the fractional dimension. Heterogeneity, however, can be the reason for ambiguity in the interpretation. Chang and Yortsos (1990) introduced solutions for fracture networks which can be described by a fractal dimension. Hsieh (2000) presents a review of hydraulic testing and evaluation procedures in fractured and tight formations as well as illustrations of problems involved in rationalizing of local-scale measurements. Beckie and Harvey (2002) have shown that the estimated transmissivities from slug test interpretation are unbiased and less sensitive to near well heterogeneities. Based on single well pneumatic packer tests in six boreholes drilled into unsaturated fractured tuff at the Apache Leap Research Site, Chen et al. (2000); Vesselinov et al. (2001b) have reconstructed a three-dimensional representation of the air permeability using kriging techniques.

*Interference Tests*: Recently developed tomographic surveys have a potential for high resolution nondestructive determination of hydraulic properties of fractured aquifers. The procedure consists of performing a series of hydraulic or pneumatic short term interference tests in which the position of the stressed interval in the pumping well varied between tests. The sequence of tests produce a pattern of crossing streamlines in the region between pumping and observation interval similar to the crossed ray paths used in seismic or radar tomography. McDermott et al. (2003a) showed that it is possible to define oriented flow and transport tensors in a laboratory sample of unsaturated fractured sandstone using a tomographical experimental set-up. Based on a similar experimental set-up and laboratory samples Brauchler et al. (2003); Brauchler (2005) presented a hydraulic travel time based tomographic approach for the direct determination of a three-dimensional pneumatic diffusivity distribution of a fractured sandstone sample. On the field scale Vesselinov et al. (2001a,b); Chen et al. (2000) have evaluated numerous pneumatic cross-hole interference tests at the Apache Leap Test Site using a geostatistical inversion technique, yielding three-dimensional reconstruction of air permeabilities at a spatial resolution of 1 $m^3$. The theory of the pilot inverse method was applied to a set of geologic facies data and hydraulic data sets by Lavenue and De Marsily, 2001. They successfully determined the heterogeneity of a fractured dolomitic aquifer by generating one hundred three-dimensional conductivity fields and calibrated them with the pressure heads of several interference pumping tests by assigning different geostatistical hydraulic properties to the fractured and unfractured parts of the aquifer. Meier et al. (2001) interpreted cross-hole pumping tests within a subvertical shear zone in granite at the Grimsel Test Site using a geostatistical inverse approach. 40 different transmissivity fields were determined which equally honor the hydraulic data but only a single transmissivity field was in line with simulated tracer test data. A more computationally efficient technique based on the inversion of peak arrival times and amplitudes of transient head data was proposed by Vasco et al. (2000). By evaluating a pair of interference tests, they were able to reconstruct an image of a shallow fracture in a limestone.

*Tracer Test:* Transport in fracture networks can be investigated with tracer experiments which require substantial effort for a differentiated measuring of tracer breakthrough. In Grimsel and STRIPA (Abelin et al., 1987), sorbing and non-sorbing tracers were collected throughout the entire ceiling of former mine workings to obtain individual tracer-breakthrough curves. The interpretation demonstrated that only a few fractures dominate tracer transport and that matrix diffusion (Skagius and Neretnieks, 1986) and channeling (Tsang and Tsang, 1989; Tsang et al., 1991) play an important role in the distribution of the concentration measured. The development of specialized packer equipment by Novakowski and Lapcevic (1994) contributed largely to the advancement of the field testing of tracer experiments. The test and interpretation techniques which are currently generally available as well as the problems encountered are summarized by NRC (1996). Some experience in the investigation of single-fracture flow processes was also obtained from field experiments (e.g., Novakowski and Lapcevic, 1994; Himmelsbach, 1993; and Kelley et al., 1987). Recently, fractured crystalline basement rocks have receive attention as potential aquifers. Particularly in the African Shield and the Scandinavian countries, these aquifers can be very important locally (e.g., Krasny, 2002; Olofsson, 2002; and Gudmundsson et al., 2002). Several recent conferences were devoted to this topic, summarized in the proceedings of Rohr-Torp and Roberts (2002) and Krasny and Mls (1996).

Most of the above mentioned studies have been concerned fractured crystalline (igneous and metamorphic) rocks and the findings cannot necessarily be applied directly to fractured systems with a porous matrix. In addition, the complexity of the flow and transport processes increases considerably for non-igneous systems, especially as a result of matrix heterogeneities, e.g. due to different sedimentological structures, and the increased conductivity and storage of the matrix. In addition, the extent of required knowledge on properties of porous fractured aquifers differs from the level at which such aquifers can actually be investigated with field methods. Most investigation techniques have limitations with respect to the spatial resolution needed to characterize such systems that is typically an inaccessible saturated aquifer. It is believed that the Aquifer Analogue Approach can provide some insight into the specifics of the problem and point in the direction of potential solutions.

## 2 AQUIFER ANALOGUE APPROACH

The research concept of the "Fractured-Rock Aquifer Analogue Approach" is based on the outcrop analogue method, in which well defined samples from sandstone outcrops (quarries) are used as a realistic representation of those sections of the (aquifer) system where access is limited to (a few) boreholes. Outcrop analogue studies have mainly been used in the petroleum industry for reservoir characterization (Flint and Bryant, 1993). Based on a detailed sedimentary analysis of outcrops, assumed to represent the characteristics of the reservoir rocks, deductions are made about their hydraulic properties. The advantages of analogue studies are the accessibility of the material in a context larger than the borehole and the opportunity to obtain detailed small scale measurements in two and sometimes three dimensions. For fractured systems, the aquifer analogue approach has rarely been employed and then only in two dimensions (Kiraly 1969; Billaux et al., 1989). In the mine adit of Fanay-Augères, Billaux et al. (1989) attempted to generate fracture networks statistically based on a 2D survey of fracture traces. However, this approach suffers from the problem of ambiguity (i.e., there is an infinite number of possibilities that can satisfy the 2D observations). Applying various statistical and pattern-recognition techniques, La Pointe (2000) showed some relationship between increased flow rates measured by well tests and geological data (structure, lithology) and derived a regional hydrogeological model on the basis of eight boreholes. Similar ambiguities apply here as well. In the frame of the research initiative "Aquifer Analogue Approach for Fractured Porous Aquifers" of the Deutsche Forschungsgemeinschaft (DFG, German Research Foundation), the flow and transport properties of fractured porous sandstone materials were determined on different scales and appropriate measuring techniques were developed.

The choice of a particular outcrop for the analogue investigation is critical, because the samples chosen determine how representative the obtained results are for the porous fractured aquifer system. Attention is also directed to possible experimental artifacts (e.g., widening of the fracture apertures due to formation unloading). In this application of the approach, the focus of investigation is on fractured aquifers of the shallow subsurface so that such alterations in aperture are regarded to be of minor importance. In this part of the investigation the main focus is on basic experimental techniques for the determination of effects of the heterogeneous system on the employed measurements and on the identification of the governing processes (i.e., possible deviations from natural systems are of minor importance). The development of analogue investigation techniques in the laboratory (samples range from $0.02\,m^3$ to approx. $1\,m^3$) are described, comprising the physical description of the fracture geometry as well as pneumatic tomographic and single well experiments. These techniques are then implemented to investigate a larger fractured sandstone block (approx. $10 \times 8 \times 2.20\,m^3$).

## 2.1 *Laboratory experimental design*

Criteria for the design of laboratory experimental techniques required: (a) access to various positions along the surface and (b) rapid performance of numerous flow and transport experiments. To fulfill these criteria, a specially adapted tomographical set-up in combination with gas flow technique was developed, enabling to perform rapidly a large number of flow and transport experiments in all spatial direction.

*Sample recovery*

Laboratory samples were recovered from the upper part of the Triassic Stubensandstein Formation, which is quarried in southern Germany (Mc Dermott et al., 2003b). This arkosic sandstone, which is ideally suited for the experiments because of the high matrix porosity and the dense regularly spaced fracture network. For the study of the fractured porous system bench scale samples (30 cm diameter) and large (block, approx. $1\,m^3$) scale laboratory samples were recovered. The greatest problem with the recovery of fractured material is that during sampling or preparation the fractures, natural planes of weakness open up and the sample disintegrates before reaching the laboratory. The problem is not new was demonstrated by Wichter and Gudehus (1976) in their recovery of large scale fractured samples. A number of methods have more recently been suggested involving resin injection (Frieg et al., 1998), but the danger always exists that the resin impregnates the rock mass and alters the characteristics of the material being sampled. A further disadvantage is the sometimes complex preparation of the resin itself under field conditions.

In order to overcome these problems, a sheath of 3 mm thick PVC plastic that had been cut and glued together with an internal radius of some 5 mm less than the external radius of the *in situ* drilled sample. In the field the sheath was heated to a temperature of 70°C causing it to expand radially, which allowed it to be placed over the *in situ* sample. The sheath then cooled and contracted around the sample exerting a slight stabilizing pressure on the sample and remained transparent. The sample can then be removed, cut to the required dimensions and transported to the laboratory without significant damage to the natural fracture network. An example of the samples collected using this approach are illustrated in Figure 1.

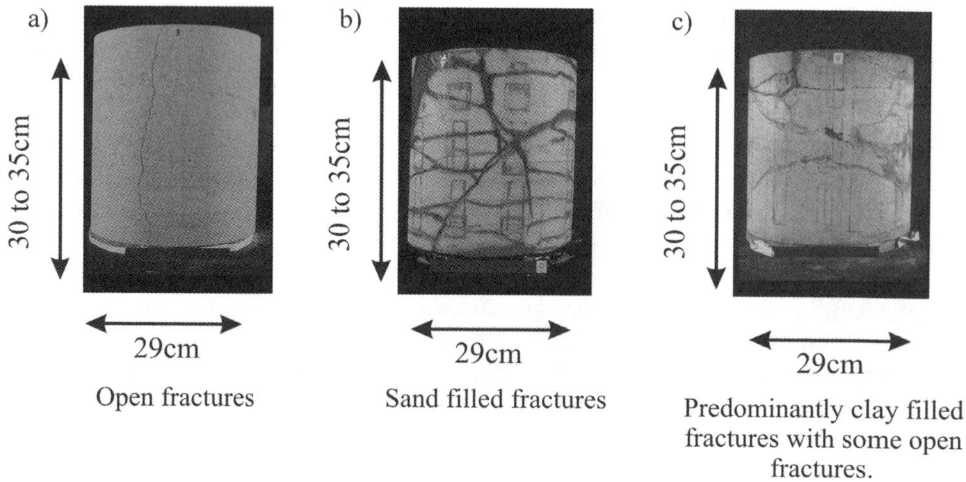


Figure 1. Example of highly fractured samples recovered from the field. Reprinted from Engineering Geology, 69, McDermott CI, Sinclair B, Sauter M, Recovery of undisturbed highly fractured bench scale samples (diameter 30 cm × 35 cm) for laboratory investigation, pp. 161–170, 2003, with permission from Elsevier.

For the block size ($1\,m^3$) the required mechanical protective membrane was chosen to provide the necessary hydraulic seal for latter experimentation. Here it was necessary to characterize the sample prior to the application of the mechanical seal. To satisfy all these requirements the block was coated with epoxy resin to a thickness of at least 5 mm.

## 2.2 *Tomographical experiments performed on laboratory cylinders*

The concept of three-dimensional tomographical investigations for the determination of the internal structure of a solid body based upon the distribution of measurable parameters is well established, and offers a method whereby more detail over the formation of the integral signal from several discrete elements can be derived. In principal the tomographic approach involves the determination of the parameter distribution within the body investigated by combining several different spatially orientated measurements. The combination and inverse interpretation of these measurements allow a three-dimensional image of the parameter distribution to be defined. In practical experimental terms to apply the tomographical investigation concept, it is necessary to make numerous point to point or surface to surface measurements of flow or both flow and transport across a sample.

To allow the collection of such a large number of measurements, gas flow techniques were applied enabling rapid measurement of flow (pressure and flow rate) and transport (tracer concentration) signals. Employing gas flow techniques there is no need to saturate the samples prior to the investigation, and very small flow rates could easily be measured rapidly with a high degree of accuracy. In accordance with the tomographical concepts and the sample geometry, a special experimental cell was developed which allowed access to pre-selected positions along the surface of the sample ports (Figure 2).

To investigate the flow and transport parameters, a stable linear flow field is established across the sample from the input port/ports to the output port/ports. The flow across the sample

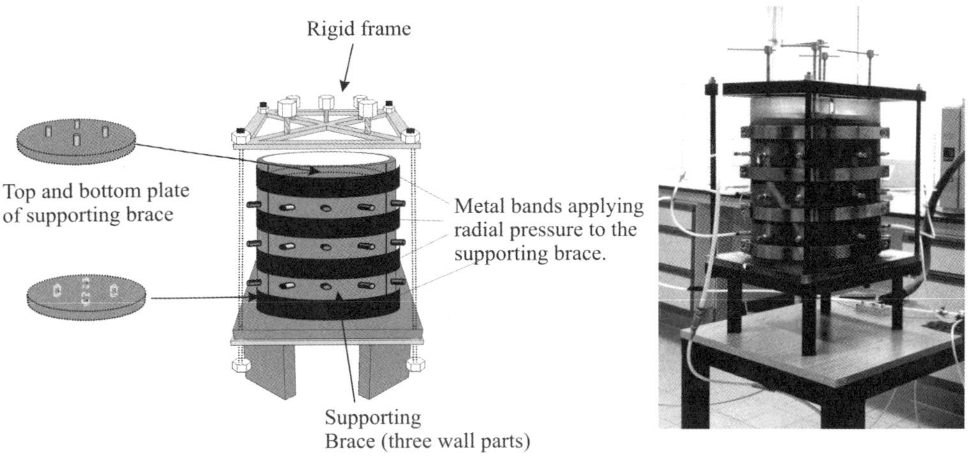


Figure 2. Experimental set-up for the tomographical investigation of bench scale samples. Reprinted from Journal of Hydrology, 278 (1–4), McDermott CI, Sauter M, Liedl R, New experimental techniques for pneumatic tomographical determination of the flow and transport parameters of highly fractured porous rock samples, pp. 51–63, 2003, with permission from Elsevier.

is recorded using a flow meter. Gas tracer (Helium) is then injected via a flow-through loop into the flow field and the breakthrough of the gas tracer at the output port/ports is recorded using a mass spectrometer. The exact mass of the tracer introduced into the system is known and the location and rate of recovery are measured. Spatially orientated point to point measurements of the flow and transport characteristics could then be carried out from port to port. Examples of flow rate tensors derived from the experimental results for the highly fractured sandstone sample illustrated in Figure 1c are presented in Figure 3. Here, three different levels of the sample are investigated via point to point measurements.

Typical results of transport experiments are illustrated in Figure 4. Where fractures and matrix contribute to the transport then interesting double and even triple peaks have been observed.

For the derivation of actual hydraulic parameters a travel time based tomographic inversion approach based on point to point flow rate measurements was developed. Note we use the terminology hydraulic instead of pneumatic tomography because the practical application of this method is the high resolution characterization of fractured aquifers. As an alternative to solving the standard transient groundwater flow equation by analyzing transient head data as a function of time (e.g., Gottlieb and Dietrich, 1995; Yeh and Liu, 2000; Vesselinov et al., 2001 a,b; Bohling et al., 2002; Zhu and Yeh, 2005), this approach is based upon the inversion of travel times of pressure changes. The approach follows the procedure of seismic ray tomography. In seismic ray tomography the propagation of a seismic signal is described by a line integral relating the travel time t with the spatial velocity distribution v of an investigated area:

$$t = \int_{x_1}^{x_2} \frac{ds}{\nu(s)} \tag{1}$$

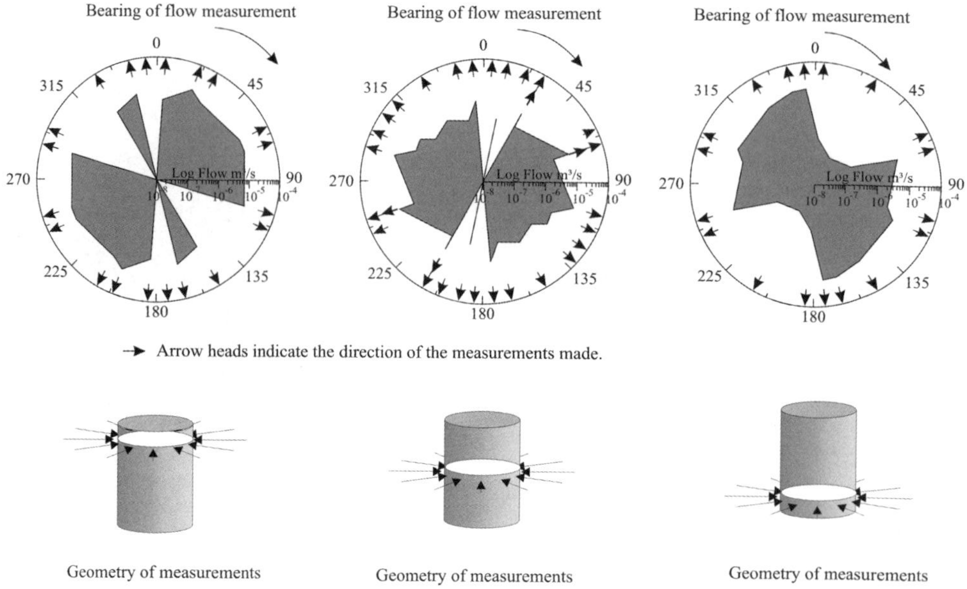


Figure 3. Example of flow rate tensors derived from a highly fractured sample, Reprinted from Journal of Hydrology, 278 (1–4), McDermott CI, Sauter M, Liedl R, New experimental techniques for pneumatic tomographical determination of the flow and transport parameters of highly fractured porous rock samples, pp. 51–63, 2003, with permission from Elsevier.

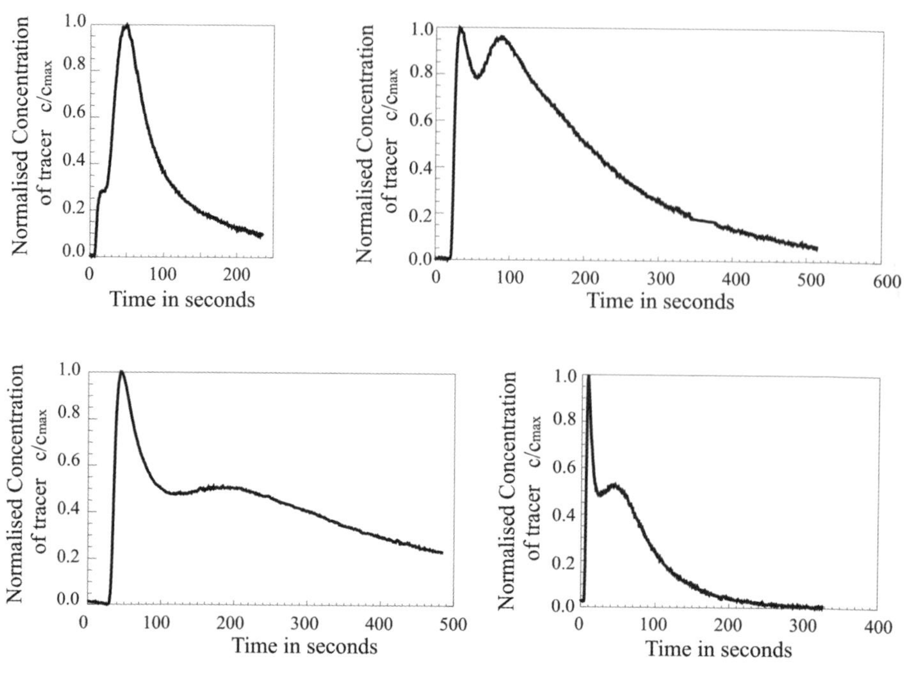


Figure 4. Typical experimental transport results where fracture and matrix, or double fractures are involved. Reprinted from Journal of Hydrology, 278 (1–4), McDermott CI, Sauter M, Liedl R, New experimental techniques for pneumatic tomographical determination of the flow and transport parameters of highly fractured porous rock samples, pp. 51–63, 2003, with permission from Elsevier.

Using equation (1), Kulkarni et al. (2000) and Vasco et al. (2000) derived a line integral relating the arrival time of a "hydraulic signal" to the reciprocal value of hydraulic diffusivity:

$$\sqrt{t_{peak}(x_2)} = \frac{1}{\sqrt{6}} \int_{x_1}^{x_2} \frac{ds}{\sqrt{D(s)}} \qquad (2)$$

where $t_{peak}$ is the travel time of the peak of a Dirac signal from the point x1 (source) to the observation point x2 (receiver) and D the hydraulic diffusivity. In our approach we utilize the similarity between the equations (1) and (2), which allows the use of the same inversion software. The inversion results presented in the following were conducted with the commercial software package GeoTom 3D, which is based on the tomography program 3DTOM (Jackson and Tweeton, 1996). The program uses a least square solution of a linear inverse problem. The algorithm is called SIRT (Simultaneous Iterative Reconstruction Technique) and belongs to the group of series expansion methods. These methods allow curved ray paths and streamlines trajectories, respectively, through the target area and are therefore well suited for applications in seismic and hydraulic tomography. For an explicit discussion of the series expansion methods and different ray bending algorithms we refer to Jackson and Tweeton (1996); Um and Thurber (1987); Dines and Lytle (1979); Gilbert (1972); and Lo and Inderwiesen (1957).

The proposed inversion technique is based on the relationship between the peak travel time of a recorded transient pressure curve with a Dirac source at the origin and the diffusivity of the geological medium. This situation is not very satisfying because much of the information is not used and therefore lost for the interpretation. We have therefore developed a transformation factor which allows the application of our approach to several travel times characteristic for each signal. For technical reasons a Heaviside source is easiest to implement especially if adequate signal strengths for long distances are required. In order to avoid the differentiation of each recorded curve a conversion factor relating Heaviside to Dirac sources is developed (Brauchler et al., 2003).

The presented methodology was applied to data from a set of point to point flow rate measurements conducted on the fractured sandstone cylinder illustrated in Figure 1a. The flow rate measurements were performed with a similar experimental set-up described above. At the injection port, which can be described as a point source a Heaviside signal was applied. Compressed air was injected into the unsaturated sandstone block and the flow rate recorded at the measuring port. All other ports are open to the atmosphere. The flow rate was recorded until a steady state flow field was established between the injection and the measuring port. It took approximately 20 seconds to obtain steady state conditions. The pressure difference between the injection port and the measuring port was kept constant during the experiment by applying 0.5 bar air pressure at the injection port. In total 487 measurements were conducted.

In Figure 5, two inversion results are illustrated based on early travel times (the arrival time associated with 1% of the maximum flow rate, Figure 5a–c) and late travel times (the arrival time associated with 40% of the maximum flow rate, Figure 5d–f) by means of three depicted slices of different depths of the cylinder showing the gas diffusivity distribution.

The comparison of both diffusivity reconstructions reveals that the inversion based on early travel times shows a better agreement between the reconstructed high diffusivity zone and the mapped fracture than the inversion based on late travel times. This result is

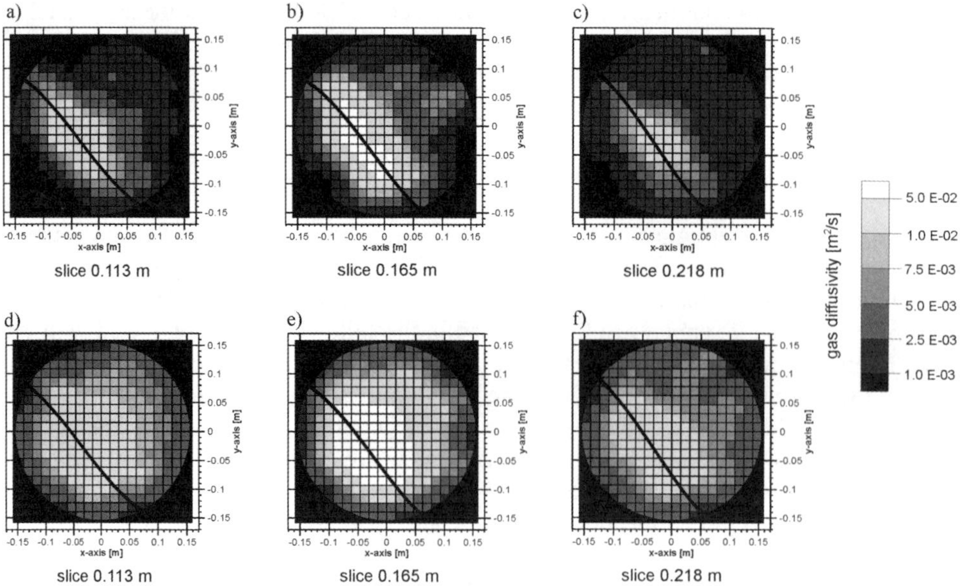


Figure 5. Inversion results for three depicted slices of different depths of the cylinder. The black line indicates the deducted trace of the fracture, derived from a surface mapping. (a)–(c) Inversion results associated with 1% of the maximum flow rate. (d)–(f) Inversion results associated with 40% of the maximum flow rate (after Brauchler et al., 2003).

consistent with the fact that the signal propagation of a pressure pulse follows Fermat's principle. Thus, early time signals, characterizing the initial part of a signal, follow the fastest pathways between source and receiver. The fastest pathways are usually identical to preferential flow paths, which in this case is the fracture. Whereas late travel times, characterizing the late part of a signal, reflect more or less integral behavior, since the pressure difference between source and receiver then affects the entire system.

## 2.3 *Single well methods performed on a laboratory block*

In Figure 6a, a schematic scheme of the large scale laboratory block is shown that was recovered for the study of the fractured porous system. In order to avoid disintegration of the sample during preparation due to the numerous fractures, it was only possible to cut the block to lengths of approximately $0.9 \times 0.9 \times 0.8\,\text{m}^3$ resulting in somewhat irregular edges. After recovering and sealing the block with an air-tight resin coating, the block sample is prepared for a variety of pneumatic test series, e.g. a radial flow field under fully controlled boundary conditions. A specific multi-purpose measuring device was developed to account for the specific task of simultaneously monitoring flow and pressure measurements at a large number of locations (Figure 6b, Leven, 2002). For the single well set-up, a vertical borehole was drilled through the center of the sandstone block.

In the following, a series of pneumatic tests is presented, where a steady-state flow field with constant injection pressure is imposed on the fractured porous sandstone block. For this test series, compressed air is injected through a vertical borehole at constant pressure, as shown in Figure 6c. All ports on the block surface with exception of those at top and

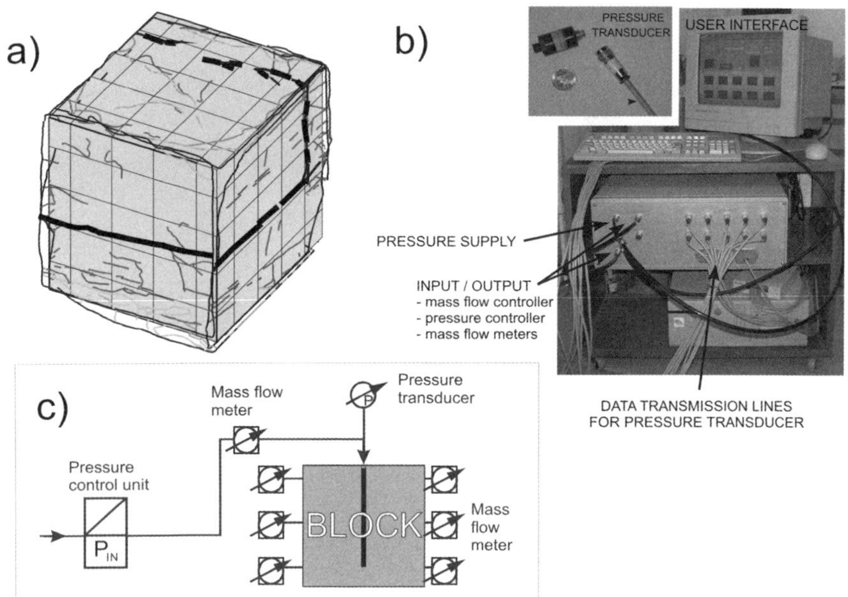


Figure 6. (a) Schematic of the fractured sandstone block ($0.9 \times 0.9 \times 0.8\,\text{m}^3$). The solid black lines mark dominant fractures in the block sample (after Leven et al., 2004). (b) Set-up of the multi-purpose measuring device and illustration of the LabVIEW® based user interface, which allows a variety of distinct hydraulic measuring configurations (Leven, 2002). (c) Illustration of an operating state of the multi-purpose measuring device for performing pneumatic tests with a steady state flow field with constant injection pressure.

bottom faces were open to the atmosphere (i.e., the injected gas was allowed to escape through the external ports, where discharge of the injected gas was measured). The injection pressure and the injected flow rate remained constant during the experiment.

Figure 7a illustrates a frequency distribution of detected flow rates normalized to the radial distance of the individual ports to the central borehole. It is obvious that the data are bimodally distributed, showing two distinct lognormal distributions. The lognormal distribution with smaller normalized flow rates is characteristic for the response of the porous matrix. A lognormal distribution of hydraulic conductivity in porous media is commonly adopted while assuming a linear relationship between flow rate and hydraulic conductivity (Leven et al., 2004). Because the fracture apertures are lognormally distributed (not shown, Leven et al., 2004), a lognormal distribution for the detected flow rates can be expected on the basis of an idealization of the cubic law which relates the flow through a fracture system to the cube of the fracture width (e.g., Streltsova, 1988), which is reflected by the lognormal distribution for higher normalized flow rates.

Horizontal sections of the steady-state flow field in the fractured sandstone block are illustrated in Figure 7b. The heterogeneous nature of the fractured porous system due to preferential flow paths within the fracture network becomes evident. The effect of the fracture network on the induced flow field by increased flow rates is clearly apparent in both plots of Figure 7b. The main component of the fracture network is a fissure with large apertures extending from side I to III on the right parallel to side II. Other peaks in the flow field pattern indicate regions where the flow is also influenced by the presence and orientation of other individual fractures.

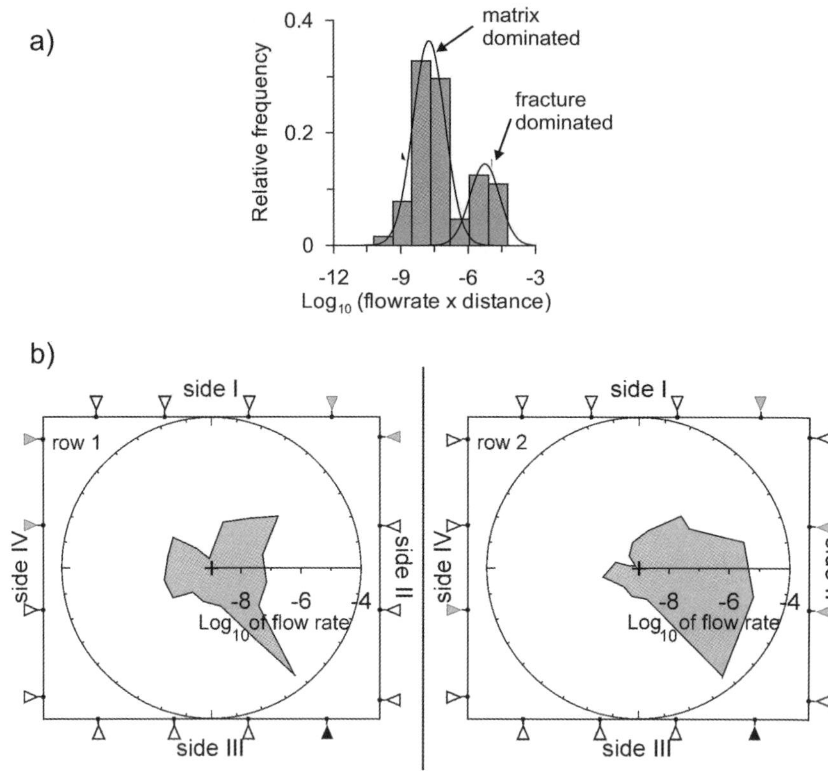


Figure 7. (a) Relative frequency of measured flow rates at the outlets on the block surface for pneumatic steady state flow test with constant injection pressure showing a bimodal distribution of flow rates at the outlets normalized to the radial distance to the central borehole. The curves illustrate two distinct lognormal Gaussian distributions, characteristic for the porous matrix (left) and the fracture network (right). (b) Sections parallel to horizontally aligned outlets on the block surface illustrating distribution of the flow fractions within the block under steady-state conditions (Symbols: △ – matrix dominated connection, ▲ – matrix connection in direct vicinity of fracture, ▲ – direct fracture-fracture connection). The centre of each plot matches the position of the vertical borehole (flow rate standardized to the radial distance from the central borehole in $[m^3/s] \times [m]$) (after Leven et al., 2004).

From the experiments presented above it can be concluded that in a fracture porous system flow is concentrated in only a few highly conductive fractures and that the precise characterization of the fracture network geometry is a prerequisite for the identification of the particularities of the system behaviour (Guglielmi and Mudry, 2001; Leven et al., 2004). In other words, as stated by Streltsova (1988), "it is fracture continuity and interconnectivity, however, not the characteristics of individual fractures that are responsible for the particulars of a fractured reservoir".

## 2.4 *Pneumatic and hydraulic transport experiments on the field block scale*

The insights gained from the laboratory experiments were employed for the design of a test set-up that allows to perform flow and transport experiments under natural field conditions. Therefore, a block of ca. 200 $m^3$ volume in the same Triassic sandstone as the laboratory blocks is carved into an axisymmetric shape. (Thüringer 2002; Dietrich et al., 2005, Figure 8a). For

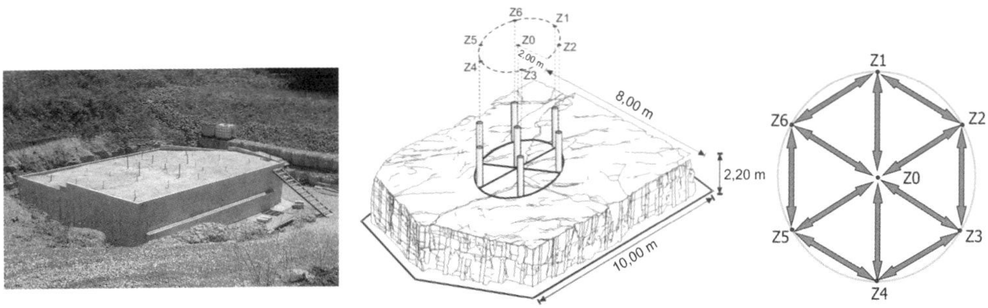


Figure 8. (a) Sealed sandstone block. (b) Fracture traces and experimental configuration of boreholes. (c) Borehole set-up in the centre of the field block.

controlled boundary conditions, the side walls and surface of the sandstone are covered by a double-layer coating of water and gas tight concrete. Basal layers of clay and silt are regarded as a natural seal of the field block. Prior to the sealing process, a detailed statistical analysis of local fracture system parameters has been applied by Witthüser and Himmelsbach (1998), (Figure 8b). Facilities were installed to allow experiments for unsaturated (original state) and saturated conditions in order to compare flow and transport results between pneumatic and hydraulic tests for comparable and controlled boundary conditions.

Boreholes were drilled through the sandstone block into the underlying clay (Figure 8c). The configuration of the boreholes was optimized to avoid the influence of the outer sealed block boundaries during flow and transport experiments. The pneumatic and the hydraulic tracer experiments were designed as symmetric dipole fields. For comparison, the same gradient was applied for both test set-ups (hydraulic and pneumatic). Tracer was injected after a steady-state flow field was established. For the pneumatic test helium and for the hydraulic tests sodium chloride was used. Recovery of the tracer was recorded with a mass spectrometer or electric conductivity meter, respectively.

Helium (unsaturated conditions) and sodium chloride (saturated conditions) breakthrough curves are illustrated in Figure 9, normalized to the respective maximum concentration for the connection Z0–Z1. Both breakthrough curves show next to the main peak two further peaks (marked with arrows), interpreted as different preferential pathways. In contrast to the breakthrough curve for saturated conditions, the peaks in the helium breakthrough curve are considerably sharper. The similarity between the two breakthrough curves indicates that the tracer test under saturated conditions appears to be controlled by the same transport paths as tracer test under unsaturated conditions. Differences can be assigned to the higher dynamic viscosity of water compared to gas which in turn leads to a lower penetration of the tracer into minor flow paths and matrix and lower flow velocities.

For the simulation of the tracer experiments a discrete model approach is used. The field block is described by a three-dimensional numerical hybrid model (Kolditz, 1997) consisting of approx. 3,40,000 triangular finite elements. The positions of the fractures are chosen in accordance with the photogrammetric interpretation of fracture traces of the block surface (Figure 8b). Porosity was determined by Hg-porosimetry and pycnometry in the laboratory and the molecular diffusion coefficients of $Na^+$ and $Cl^-$ in water was taken from Reeves 1979. The model was calibrated by varying the fracture apertures and bulk matrix permeability. The best match could be achieved for a fracture aperture of $1.0 \times 10^{-3}$ m.

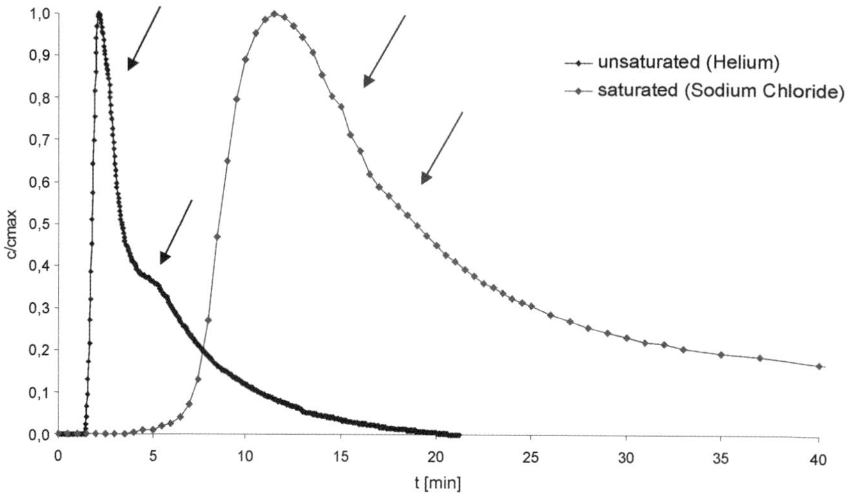


Figure 9. Tracer breakthrough curves for saturated (sodium chloride) and unsaturated (Helium) conditions.

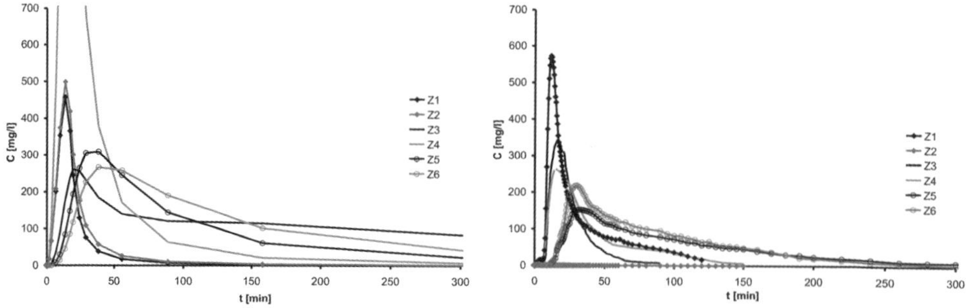


Figure 10. Comparison between simulated (left) and measured tracer breakthrough curves (right) of tracer tests (saturated conditions) with injection at Z0 and extraction in the outer boreholes.

This calibrated aperture is significantly higher than the fracture aperture determined on sandstone core samples (33–88 μm, Baraka-Lokmane 2002). The difference can be explained by the presence of numerous microfractures in the matrix and channeling effects inside the natural fractures (Weede and Hötzl 2005), which are not accounted for by the model. The calibrated hydraulic conductivity of the matrix is $1.0 \times 10^{-7}$ m/s, which is a reasonable value for the sandstone. The deviation from the measured average value in the laboratory ($1 \times 10^{-7}$–$5 \times 10^{-9}$ m/s) can be explained by heterogeneities within the matrix as well as the sampling bias for laboratory samples, favoring lower conductivities. It was not the intention to match every measured breakthrough curve exactly with the simulated model but rather to mimic the overall system response with the model so a homogeneous matrix was assumed.

Figure 10 shows breakthrough curves of a test set-up with injection in the central borehole and recordings in the outer ring boreholes. Simulated and measured concentration show good agreement for Z0–Z1, Z0–Z3, Z0–Z5, and Z0–Z6. The simulated concentration of Z4 is far too high. A possible explanation might be that the fracture connection between the injection and extraction point derived from the fracture mapping is not continuous. Note, the experimental curve of connection Z0–Z2 is not illustrated because of experimental problems.

## 3 CONCLUSIONS

This study has shown that the Aquifer Analogue Approach is well suited to provide insights into the complex flow and transport pattern of fractured porous media. New experimental techniques were developed and applied to laboratory samples ranging in size from approx. 0.02 $m^3$ to 1 $m^3$ as well as to a 200 $m^3$ field block.

On the laboratory scale pneumatic flow and transport experiments are described using tomographical and single well set-ups. From the tomographical experiments performed on a sandstone cylinder (approx. 0.02 $m^3$) three dimensional flow rate tensors can be derived illustrating the influence of the fracture network on the flow pattern. For the derivation of actual hydraulic parameters a travel time based tomographic approach is used. The procedure which was developed in analogy to seismic tomography allows the inversion of different travel times of transient pressure signals. The inversion results are encouraging, and thus applications of this approach to field problems appear feasible. The single well experiments were performed on a larger laboratory block (approx. 1 $m^3$). The experimental series with an induced steady-state radial flow and open external ports provide information on the heterogeneity in terms of preferential flow paths within the fractured block. The influence of the matrix was less significant.

In a final step the experimental set-up and insights from the laboratory measurements were applied to characterize a block (approx. 200 $m^3$) under natural field conditions. A measurement set-up was installed to compare pneumatically (unsaturated conditions) and hydraulically (saturated conditions) performed flow and transport experiments for comparable boundary conditions. The comparison between the two test set-ups indicates that the transport measurements under saturated conditions were controlled by the same transport paths. Numerical investigations based on a discrete model approach show a strong dependence of the simulation results from the photogrammetric interpretation of the main flow and transport paths of the field block.

## ACKNOWLEDGEMENTS

The investigations were performed with the financial support of the German Research Foundation to the "Fractured-Rock Aquifer-Analogue: Experiments and Modeling" project.

## REFERENCES

Abelin H, Birgersson L, Gidlund J, Neretnieks I, Tunbrandt S (1987) Results from some tracer experiments in crystalline rocks in Sweden. Academic Press Inc, 28: 363–379.

Ando K, Kostner A, Neuman SP (2003) Stochastic continuum model of flow and transport in a crystalline rock mass: Fanay Augères, France revisited. Hydrogeology Journal, 11(5): 521–535.

Baraka-Lokmane S. (2002): Determination of hydraulic conductivities from discrete geometrical characterisation of fractured sandstone cores. Tübinger Geowissenschaftliche Arbeiten (TGA) C 51, Geowissenschaftliche Fakultät der Eberhard-Karls-Universität; Tübingen.

Barker, J (1988) A generalized radial-flow model for pumping tests in fractured rock. Water Resources Research, 24(10): 1796–1804.

Beckie R, Harvey CF (2002) What does a slug test measure: An investigation of instrument response and the effect of heterogeneity, Water Resources Research 38(12), 1290, DOI:10.1029/2001WR001072.
Billaux D, Chiles J, Hestir K. Long, J (1989) Three-dimensional statistical modeling of a fractured rock mass – an example from the Fanay-Augere mine. International Journal of Rock Mechanics and Mining Sciences and Geomechanics Abstracts, 26, 281–299.
Bohling GC, Zhan X, Butler JJJr, Zheng L (2002) Steady shape analysis of tomographic pumping tests for characterization of aquifer heterogenous. Water Resources Research, 38(12), 1324, doi: 10.1029/2001WR001176.
Brauchler R, Liedl R, Dietrich P (2003) A travel time based hydraulic tomographic approach, Water Resources Research, 39(12), 1370, doi:10.1029/2003WR002262.
Brauchler R (2005) Characterization of fracture porous media using multivariate statistics and hydraulic travel time tomography, Tübinger Geowisenschaftliche Arbeiten, (TGA), Reihe C, (87): 78 pp.
Chang J and Yortsos Y (1990) Pressure transient analysis of fractal reservoirs. SPE Formation Evaluation, SPE.
Chen G, Illman WA, Thompson DL, Vesselinov VV, Neuman SP (2000) Geostatistical, type curve and inverse analysis of pneumatic injection tests in unsaturated fractured tuffs at the Appache Leap Research Site near Superior, Arizona. In Faybishenko B et al. (eds) Dynamics of flow and transport in fractured rocks. Am Geophys Union Monograph Series:73–98.
Deltombe JL, Schepers R (2004) New developments in real time processing of full waveform acoustic televiewer data, Journal of Applied Geophysics, 55: 161–168.
Dietrich P, Helmig R, Sauter M, Hötzl H, Köngeter J, Teutsch G (2005): Flow and Transport in Fractured Porous Media. Springer; Berlin, Heidelberg: 419 pp.
Dines KA, Lytle RJ (1979) Computerized Geophysical Tomography, Proc. IEEE, 67(7): 1065–1073
Earlougher, R. (1977). Advances in well test analysis. Society of Petroleum Engineers, Dallas, Texas
Evans DD, Nicholson TJ, Rasmussen TC (eds) (2001) Flow and transport through unsaturated fractured rock. Geophys Monogr 42 nd edn, Am Geophys Union, Washington, DC: 196 pp.
Faybishenko B, Witherspoon PA, Gale J (eds) (2005) Dynamic of fluids in fractured rock. Geophys Monogr 162, Am Geophys Union, Washington, DC: 250p.
Flint S, Bryant IE (1993) The Geological Modelling of Hydrocarbon Reservoirs and Outcrop Analogues, Volume 15 of Spec. Publs. Int. Ass. Sediment.
Frieg B, Alexander WR, Dollinger H, Bühler C, Haag P, Möri A, Ota K (1998): In situ resin impregnation for investigating radionuclide retardation in fractured repository host rocks. Journal of Contaminant Hydrology, 35: 115–130.
Gilbert P (1972) Iterative methods for three-dimensional reconstruction of an object from projections. J. theor. Biol. 36: 105–117.
Gringarten, A (1982). Flow-test evaluation of fractured reservoirs. In: T.N. Narasimhan, Recent trends in hydrogeology, GSA, Special Paper, 189.
Gottlieb J, Dietrich P (1995) Identification of the permeability distribution in soil by hydraulic tomography.- Inverse Problems, 11: 353–360.
Gudmundsson A, Gjesdal O, Brenner S, Fjeldskaar I (2003) Effects of linking up of discontinuities on fracture growth and groundwater transport. Hydrogeo. J., 11: 84–99.
Guglielmi Y, Mudry J (2001) Quantitative measurements of channel-block hydraulic interactions by experimental saturation of a large, natural, fissured rock mass. Ground Water 39(5): 696–701.
Himmelsbach T (1993) Untersuchungen zum Wasser- und Stofftransportverhaltenvon von Störungszonen im Grundgebirge Albtalgranit, Schwarzwald). Ph.D. thesis, Reihe Angew. Geol. Karlsruhe, 23.
Hsieh P (2000) A brief survey of hydraulic tests in fractured rocks. In: Dynamics of Fluids in Fractured Rock. Geoph. Monograph, 122.
Karasaki K, Long J, Witherspoon, P (1988) Analytical models of slug tests. Water Resources Research, 24:115–126.
Karasaki K, Freifeld B, Cohen A, Grossenbacher K, Cook P, Vasco D (2000) A multidisciplinary fractured rock characterization study at Raymond field site, Raymond, CA. Journal of Hydrology 236: 17–34.
Kelley V, Pickens J, Reeves M. Beauheim R (1987) Double-porosity tracer-test analysis for interpretation of the fracture characteristics of a dolomite formation. Proc. Solving Ground Water Problems with models, Denver, Colorado.

Keys WS (1979) Borehole geophysics in igneous and metamorphic rocks, Transaction of the SPWLA 20th Annual Logging Symposium, Houston, Texas: P1-P26.

Kiraly L (1969) Statistical analysis of fractures (orientation and density) Geol. Rundschau, 59.

Kobr M, Mares S, Lukes J (1996) Contribution of geophysical well-logging techniques to evaluation of fractured rocks, Acta Universitatis Carolinae, 40: 257–268.

Kolditz O. (1997): Strömung, Stoff- und Wärmetransport in Kluftgestein.Gebrüder Borntraeger; Berlin, Stuttgart: 262 pp.

Krasny, J (2002) Quantitative hardrock hydrogeology in a regional scale. Proc. of Nordic Workshop 2001, NGU, 439.

Krasny J, Mls J (1996) First workshop on "Hardrock hydrogeology of the Bohemian Massif", 1994. Acta Universitatis Carolinae Geologica, 40(2).

Kulkarni KN, Datta-Gupta A, Vasco DW (2001) A Streamline Approach to Integrating Transient Pressure Data into High Resolution Reservoir Models. SPE Journal, 6(3).

Jackson MJ, Tweeton DR (1996) 3DTOM: Three-dimensional Geophysical Tomography. United States Department of the Interior, Bureau of Mines, Report of Investigation 9617: 84 pp.

La Pointe P. Hudson J (1985) Characterization and Interpretation of Rock Mass Joint Patterns. Special paper 199, Geological Society of America.

Lavenue M, de Marsily G (2001) Three dimensional interference test interpretation ion a fractured aquifer using pilot point inverse method. Water Resources Research 37(11): 2659–2675.

Leven C, Sauter M, Teutsch G, Dietrich P (2004) Investigations of the effects of fractured porous media on hydraulic tests-an experimental study at laboratory scale using single well methods. Journal of Hydrology 297: 95–108.

Lo T, Inderwiesen PL (1957) Fundamentals of seismic tomography. Geophysical monograph series. (6): 178 pp.

Matthews C, Russel, D (1967). Pressure build-up and flow tests in wells. Society of Petroleum Engineers, Dallas, Texas.

McDermott CI, Sauter M, Liedl R (1998) Investigating fractured-porous systems-The Aquifer Analogue Approach. In Rossmanith (ed.), Mechanics of Jointed and Faulted Rock, Balkema, Rotterdam: 607–612.

McDermott CI, Sauter M, Liedl R (2003a) New experimental techniques for pneumatic tomographical determination of the flow and transport parameters of highly fractured porous rock samples. Journal of Hydrology, 278(1–4): 51–63.

McDermott CI, Sinclair B, Sauter M. (2003b) Recovery of undisturbed highly fractured bench scale samples (diameter 30 cm × 35 cm) for laboratory investigation. Engineering Geology, 69(122): 161–170.

Meier PM, Medina A, Carrera J (2001) Geostatistical inversion of cross-hole pumping tests for identifying preferential flow channels within shear zones. Ground Water 39(1): 10–17.

Moench A, Ogata, A (1981) A numerical inversion of the LaPlace transform solution to radial dispersion in a porous medium. Water Resources Research, 17(1):250–252.

Morin R, Hess A, Paillet F (1988) Determining the distribution of hydraulic conductivity in a fractured limestone aquifer by simultaneous injection and geophysical logging. Ground Water, 26: 587–595.

Naceur B, Economides M (1988). Production from naturally fissured reservoirs intercepted by a vertical hydraulic fracture. Proc. – Society of Petroleum Engineers of AIME, California Regional Meeting.

National Research Council (1996) Rock fractures and fluid flow: contemporary understanding and applications. U.S. National Committee for Rock Mechanics, National Academy Press, Washington, D.C: 551 pp.

National Research Council (2001) Conceptual models of flow and transport in the fractured vadose zone. U.S. National Committee for Rock Mechanics, National Academy Press, Washington, D.C: 374 pp.

Neuman SP (1979) Stochastic approach to subsurface flow and transport: A view to the future. In: Dagan G, Neuman SP (eds) Subsurface flow and transport: A stochastic approach. Cambridge University Press, Cambridge UK: 231–241.

Novakowski K, Lapcevic P (1994) Field measurement of radial solute transport in fractured rock. Water Resources Research, 30: 37–44.

Olofsson B (2002) Estimating groundwater resources in hardrock areas – a water balance approach. Proc. of Nordic Workshop 2001, NGU, 439.

Olsson O (1992) Site characterization and validation, final report, Stripa Proj. 92-22, Conterra AB, Uppsala, Sweden.

Paillet F, Hess, A, Cheng, C, Hardin, E (1987). Characterisation of fracture permeability with high resolution vertical flow measurements during borehole pumping. Ground Water, 25: 28–40.

Priest, SE (1993) Discontinuity Analysis for Rock Engineering. Chapman & Hall, London.

Reeves MJ. (1979): Recharge and pollution of the English Chalk: Some possible mechanisms. – Engineering Geology 14: 231–240.

Rohr-Torp E, Roberts D (2002) Hardrock hydrogeology – proceedings of a Nordic workshop. Oslo August 2001, NGU, 439.

Silliman, SE (1989) An interpretation of difference between aperture estimates derived from hydraulic and tracer tests in a single fracture. Water Resources Research, 25(10): 2275–2283.

Singhal BBS, Gupta RP (1999) Applied hydrogeology of fractured rocks. Kluwer Academic Publishers, Dordrecht, The Netherlands, 400 pp.

Skagius K, Neretnieks I (1986) Porosities and diffusivities of some nonsorbing species in crystalline rocks. Water Resources Research, 22: 389–398.

Streltsova T (1988) Well testing in heterogeneous formations. Exxon monographs, John Wiley & Sons: 413 pp.

Thüringer, C (2002) Untersuchungen zum Gastransport im ungesättigten, geklüftet-porösen Festgestein. Schriftenreihe der Angewandten Geologie, 70: 128 pp.

Tsang YW, Tsang CF (1989) Flow channelling in a single fractures a two dimensional strongly variable heterogeneous medium. Water Resources Research, 25(9): 2076–2080.

Tsang C, Hufschmied P, Hale F (1990) Determination of fracture in- flow parameters with a borehole fluid conductivity logging method. Water Resources Research, 26(4): 561–578.

Tsang C, Tsang Y, Hale F (1991) Tracer Transport in Fractures: Analysis of field data based on a variable–aperture channel model. Water Resources Research, 27(12): 3095–3106.

Um J, Thurber C (1987) A fast algorithm for two – point seismic ray tracing. Bulletin of the Seismological Society of America, 77(3): 972–986.

Vasco DW, Keers H, Karasaki K (2000) Estimation of reservoir properties using transient pressure data: An asymptotic approach, Water Resources Research, 36(12): 3447–3465.

Vesselinov VV, Neumann SP, Illmann WA (2001a) Three-dimensional numerical inversion of pneumatic cross-hole tests in unsaturated fractured tuff, 1. Methodology. Water Resources Research 37(12): 3001–3017.

Vesselinov VV, Neumann SP, Illmann WA (2001b) Three-dimensional numerical inversion of pneumatic cross-hole tests in unsaturated fractured tuff, 2. Equivalent parameters, high-resolution stochastic imaging and scale effects. Water Resources Research 37(12): 3019–3041.

Weede M, Hötzl H. (2005) Strömung und Transport in einer natürlichen Einzelkluft in poröser Matrix – Experimente und Modellierung. Grundwasser, 10(3): 137–145.

Wichter L, Gudehus, G (1976) Ein Verfahren zur Entnahme und Prüfungen von geklüfteten Großbohrkernen. Proc. 2 Nat. Taguung über Felsmechanik, Aachen.

Williams JH, Johnson CD (2004) Acoustic and optical borehole-wall imaging for fracture rock aquifers, Journal of Applied Geophysics, 55: 150–160.

Witthüser K, Himmelsbach Th. (1998): Determination of fracture parameters for a stochastic fracture system generation. Grundwasser, 3(3): 103–106.

Yeh TCJ, Liu S (2000) Hydraulic tomography: Development of a new aquifer test method. Water Resources Research 36(8): 2095–2105.

Zhu J, Yeh TCJ (2005) Characterization of aquifer heterogeneity using transient hydraulic tomography. Water Resources Research, 41 W07028. doi:10.1029/2004WR003790.

CHAPTER 26

# Application of the oriented core drilling/sampling method to an environmental site investigation

James P.G. Bunker,
*Environ International Corporation, Los Angeles, California, USA*

ABSTRACT: Collection of oriented core is an effective tool where detailed structural geologic analysis is necessary to identify and quantify contaminant migration pathways and structurally complex sedimentary bedrock sites. These subsurface features include fractures, faults, and variable bedding thickness, lithology, and continuity. This paper focuses on the application of the method to a specific environmental investigation, and summarizes the methodology and procedures, and advantages/disadvantages. The case history involves a hydrocarbon remedial investigation in Los Angeles, California (Meredith/Boli & Associates, 1994, 1997, 1997a). Thinly interbedded, fractured sedimentary rock of Miocene-age Puente Formation underlies the site. Structural and lithologic information obtained from oriented core provided an understanding of migration pathways within permeable sandstone laminae interbedded with impermeable fine-grained strata. These hydrocarbon-bearing strata in turn were interconnected by two near-vertical fracture sets prominent within the contaminant plume. These fractures permitted the hydrocarbons to migrate via a laminae-fracture-laminae sequence across impermeable bedding thus providing the necessary transport mechanism for hydrocarbons movement.

## 1 INTRODUCTION

The collection and analysis of oriented soil and/or bedrock core is an underutilized technique in the environmental consulting industry. Many scientists and engineers are unaware of the availability of this sampling technique and its effectiveness for collecting undisturbed samples during site characterization. This method can be highly beneficial where detailed structural geologic analysis is necessary to identify and quantify contaminant migration pathways in the vadose and/or saturated zones. The oriented core sampling technique is more widely employed in petroleum exploration where it is used to evaluate primary sedimentary structures and fractures, in mineral exploration and development to determine fracture and mineral-bearing vein orientations, and in engineering geology for dam site, tunnel, and bridge evaluations. This paper focuses on the application of the oriented core method to a specific environmental investigation, and summarizes the overall methodology and procedures, and advantages/disadvantages of the method.

In the environmental industry, the oriented core method appears to have important applications for sedimentary bedrock sites with complex structure and stratigraphy. These complex subsurface features could include sedimentary rock strata which are heavily fractured or faulted and/or possess variable bedding thickness, lithology, and continuity. This method also is applicable to fractured crystalline rock settings, although other methods of determining fracture orientations (e.g., an acoustic borehole televiewer® or similar wireline geophysical tool) may yield the structural data necessary for most environmental investigations. These alternative wireline methods, however, typically are less useful in a soft sedimentary rock environment where mudcake on the borehole sidewall can severely inhibit the tool resolution.

The acquisition of bedrock core is, in itself, a considerable benefit for any subsurface investigation because it provides the consultant with a continuous and undisturbed rock record of the drilled subsurface. The collection of continuous oriented core provides an even greater advantage for an environmental or geotechnical investigation, because it allows the consultant to view and analyze the formation structure and subsequently to prepare a detailed subsurface geologic model. Subsurface data obtained from previous or subsequent investigations, using less expensive conventional drilling techniques, then can be input into this model with a high level of confidence.

Several disadvantages exist when using the oriented core drilling/sampling method. This method can only provide useful data if the drilling contractor produces good core recovery, which is often not the case in weakly indurated, sedimentary rock. The case history discussed below provides an excellent example of the importance of good core recovery. The drilling contractor for the initial investigation at the site produced approximately 60 percent core recovery that provided only adequate structural information. However, a follow-on investigation, which relied on a different and more experienced contractor, produced a 96 percent core recovery that yielded a greater amount and quality of data for subsequent subsurface modeling. The superior results of the follow-on investigation points out the need to obtain the services of a knowledgeable and skillfully trained drilling company. Due to the substantial amount of time required for preparation, mobilization, core retrieval, and core analysis, the oriented core method is relatively expensive when compared to more commonly used sampling techniques such as split-barrel, "push"-type sampling, and pitcher barrel, etc. However, non-oriented core collection methods also are more costly than conventional sampling methods. The installation of several properly positioned oriented coreholes may, however, provide superior overall subsurface information and may in the long run, prove more cost effective than a greater number of less expensive conventionally installed borings. Another notable disadvantage is that bedding/structure orientation data between the ground surface and approximately 20 feet (6.1 m) below grade can not be collected due to the magnetic influence of metallic aboveground equipment (e.g., drill rig, pipe, support vehicles, etc.).

## 2 SITE BACKGROUND

The work performed in this case study was completed from 1994 through 1998 (Meredith/Boli & Associates, 1994, 1997, 1997a). Prior to that time, multiple borings and groundwater monitor wells had been installed during previous site investigations. Previous conventional hollow-stem auger drilling and sampling indicated the presence of geologic features such as fractures, and thin permeable laminae interbedded with relatively thick sequences of mudstone and siltstone. Although no undisturbed, oriented samples had been collected, the available subsurface data together with a detailed regional geologic analysis suggested that bedding plane orientation and fractures were influencing contaminant migration. Based on this analysis, it became apparent that the collection and analysis of continuous, oriented core could provide useful in-situ structural data with which the influence of geologic structures on contaminant migration could be evaluated. An oriented core drilling/sampling program initially was designed and implemented in 1994 to assess the diesel fuel release. This initial assessment was followed by a 1995–1998 study of the gasoline release in a different part of the site.

The site lies on the southern flank of the Elysian Park Hills, which together with other low hills that extend eastward to the Puente Hills, demarcate much of the northern margin of the approximately 277-square mile (717 $km^2$) Central Basin of the Los Angeles Coastal Plain (Figure 1). Both large- and small-scale faults and folds of varying scale and amplitude typify this tectonically active region. The Elysian Park Hills constitute a moderate topographic rise formed by uplifted beds of primarily Miocene sedimentary rocks. Several thousand feet of the Miocene-age Puente Formation underlie the site region. The Puente Formation primarily comprises interbedded siltstone, sandstone, and mudstone that have a mean bedding orientation of N79°W, 29°SW (Lamar, 1970; Dibblee, 1989) in the site vicinity.

The Central Basin is the most prominent groundwater basin in the Los Angeles metropolitan area. Published hydrogeologic studies for the Central Basin suggest that subsurface flow from these bedrock highland areas is negligible. The results of the investigations discussed in this paper suggest that although the fractured and thinly bedded sedimentary rocks are water bearing, they possess only limited transmissivity, and do not contribute significantly to subsurface flow into the Central Basin.

## 3 ORIENTED CORING METHODS AND PROCEDURES

Based on information supplied by DDS (1994), the Christensen Hügel® oriented coring system yields continuously scribed formation core along with borehole azimuth and inclination data. The system can be employed for boreholes drilled vertically, on angle, or horizontally. Core barrel sizes of both NXB (75.69-mm hole size and 47.62-mm core size) and HXB (92.71-mm hole size and 60.96-mm core size) can be used. System components include a downhole multi-shot magnetic survey instrument, modified Christensen quad-latch wireline double-tube core barrel, the Hügel scribing shoe positioned immediately above the core bit, and a 10-foot (3 m) segment of non-magnetic (beryllium-copper alloy) drill rod (Figure 2).

The collection of oriented core requires drilling by the rotary wash drilling method. Oriented core was collected from seven borings during the site investigations. The borehole locations were selected to provide a broad coverage of the investigation areas and to provide structural information both parallel to and perpendicular to the regional strike of bedding.

The seven investigation boreholes were cored continuously using a modified 92.71-mm diameter Christensen® diamond surface-set core bit and the Christensen Hügel® oriented coring system. During coring, the core was scribed continuously by three differentially spaced tungsten carbide knives in the scribing shoe. A down-hole multi-shot magnetic survey instrument enclosed in the upper inner tube above the lower inner tube (core tube) provided a continuous film record during each coring run. Once the survey instrument was sealed/enclosed in the upper inner tube, the core can be retrieved consistent with standard wireline core barrel operation. The instrument and reference scribe were aligned prior to coring, and orientation

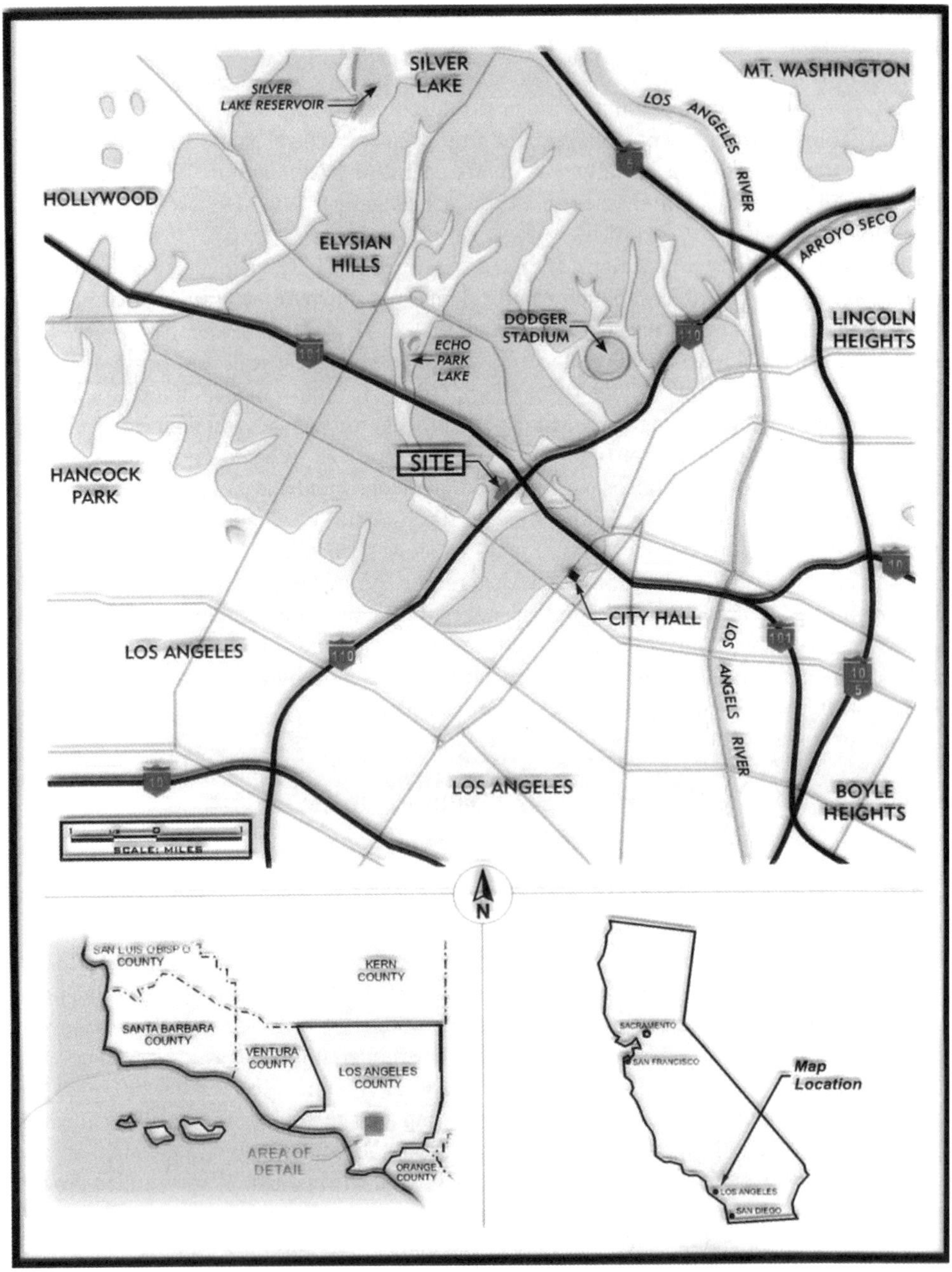


Figure 1. Site location map (1 cm approximately equals 1 km).

data including azimuth and inclination were recorded on film exposed at 2-minute intervals. The film record was indexed to specific depth intervals during the run. Following completion of the coring activities, the film was developed and the orientation data were determined from a film reader.

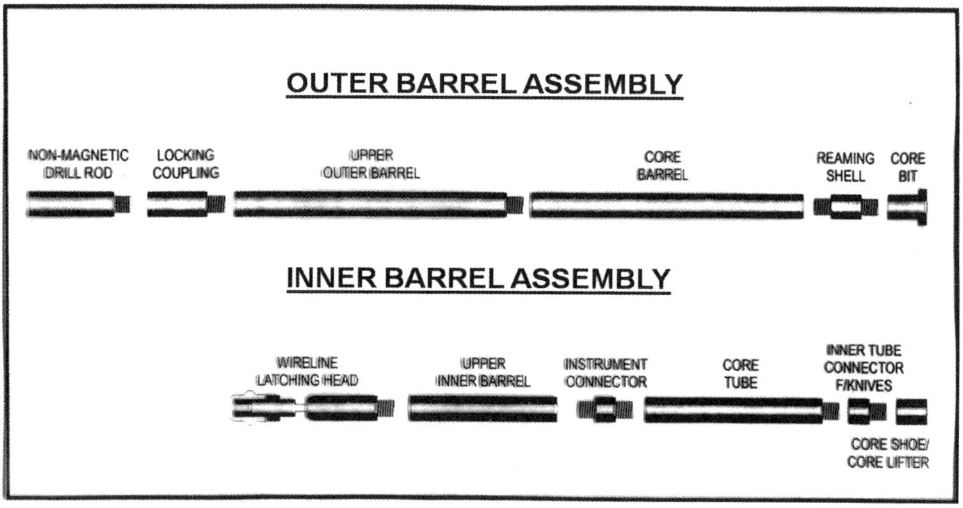


Figure 2. Components of the Christensen Hugel oriented core system (modified from directional drilling services).

## 4 ORIENTED CORE ANALYSIS

Once retrieved from the borehole, the core was extruded from the inner core barrel, placed on core trays, and cleaned of mud-cake rind. An initial core log describing pertinent core information and geology was completed simultaneously with the coring operation. A detailed lithologic and structural core log was prepared subsequent to collection of all cores using the depth/orientation data from the downhole directional survey. Core sections with well preserved scribe lines and structural features of interest were placed in a mechanical goniometer and positioned such that the reference scribe was properly aligned as determined from the film reader. A measuring arm on the goniometer then was rotated and visually aligned with the planar feature of interest (i.e., bedding plane, fracture, etc.) whereupon the strike and dip of the feature was recorded. After completion of core analysis, the core was slabbed and photographed to document the core sample results. An example of the photographed core features such as variable bedding thickness and lithology, and both near-vertical and bedding plane fractures, is shown in Figure 3.

A total of 89 measurements was made in this fashion, consisting of 64 bedding planes, 2 faults, and 23 fractures. To help identify and quantify fracture sets, and to evaluate variations in bedding orientation, the data were subjected to computer-aided analysis (i.e., Rockware Utilities Stereo® software application) including a statistical evaluation and graphical representation as equal-area stereonets.

## 5 INVESTIGATION RESULTS

The structural and lithologic information obtained from the oriented core was integral to developing a conceptual model for gasoline and diesel fuel contaminant transport at the site. The conceptual model was based on a dual-porosity transport system, whereby gasoline migration occurred both within isolated permeable beds and through interconnected

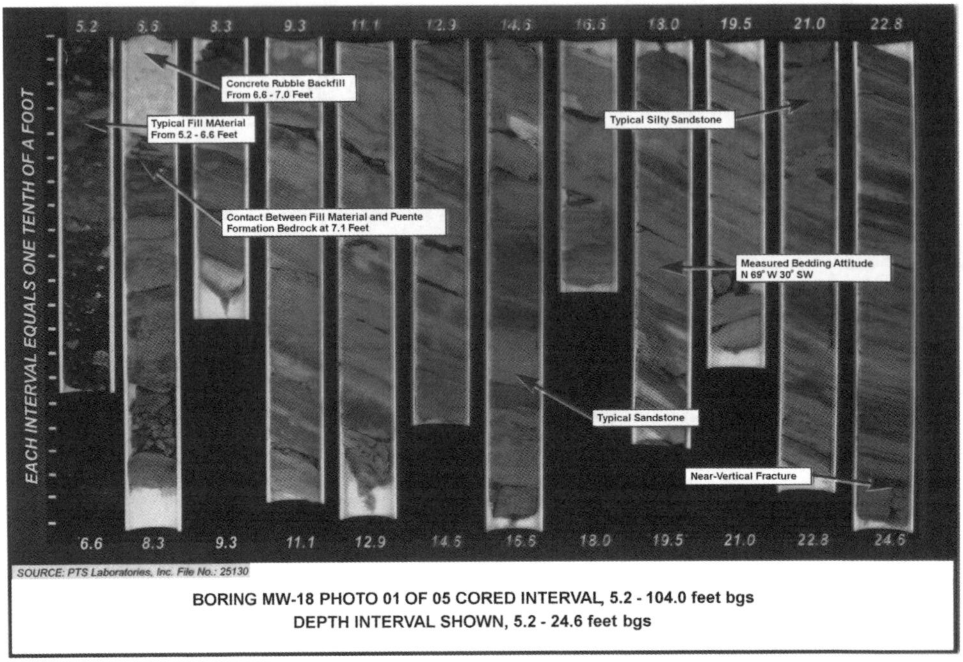


Figure 3. Slabbed Core Photograph showing typical thinly interbedded silty sandstone bedding orientation, and near vertical fractures.

fractures of varying geometry and extent (Meredith/Boli & Associates, 1997a). Near vertical, crosscutting fractures appeared to facilitate hydraulic interconnection between these otherwise isolated permeable beds. These findings were consistent with estimates of hydraulic conductivity (Figure 4), which also suggest that most of the fuel hydrocarbon transport at the site occurs within permeable laminae and, to a lesser extent, fractures.

Data generated from the seven oriented coreholes were used to provide "anchor points" for detailed 2- and 3-dimensional stratigraphic evaluations. The site stratigraphy typically consists of an upper coarser-grained, lithofacies predominantly composed of finely interbedded sandstone, siltstone, and silty mudstone in the south part of the site. This lithofacies is underlain by either a relatively thick and more massively bedded sequence of silty mudstone/clayey siltstone, mudstone, and siltstone in the central part of the site, or a lower variably bedded lithofacies dominated by sandy siltstone/silty sandstone in the north part of the site (Figure 5). Bedding continuity is highly variable; in most cases, individual beds were shown to be continuous over distances no greater than 50 to 100 feet (15 to 30 m). Individual bed thickness typically varies from less than 1 foot to more than 5 feet. (0.3 to 1.5 m) Bedding orientation is relatively uniform across the site with a mean orientation striking N79°W and dipping 25°SW; this mean orientation is consistent with the regional bedding orientation.

The site hydrostratigraphy, which consists of thin permeable sandy lamina interbedded with impermeable mudstone and siltstone, is conducive to the formation of small-scale isolated groundwater flow and storage zones. The bedrock and fractures are capable of storing

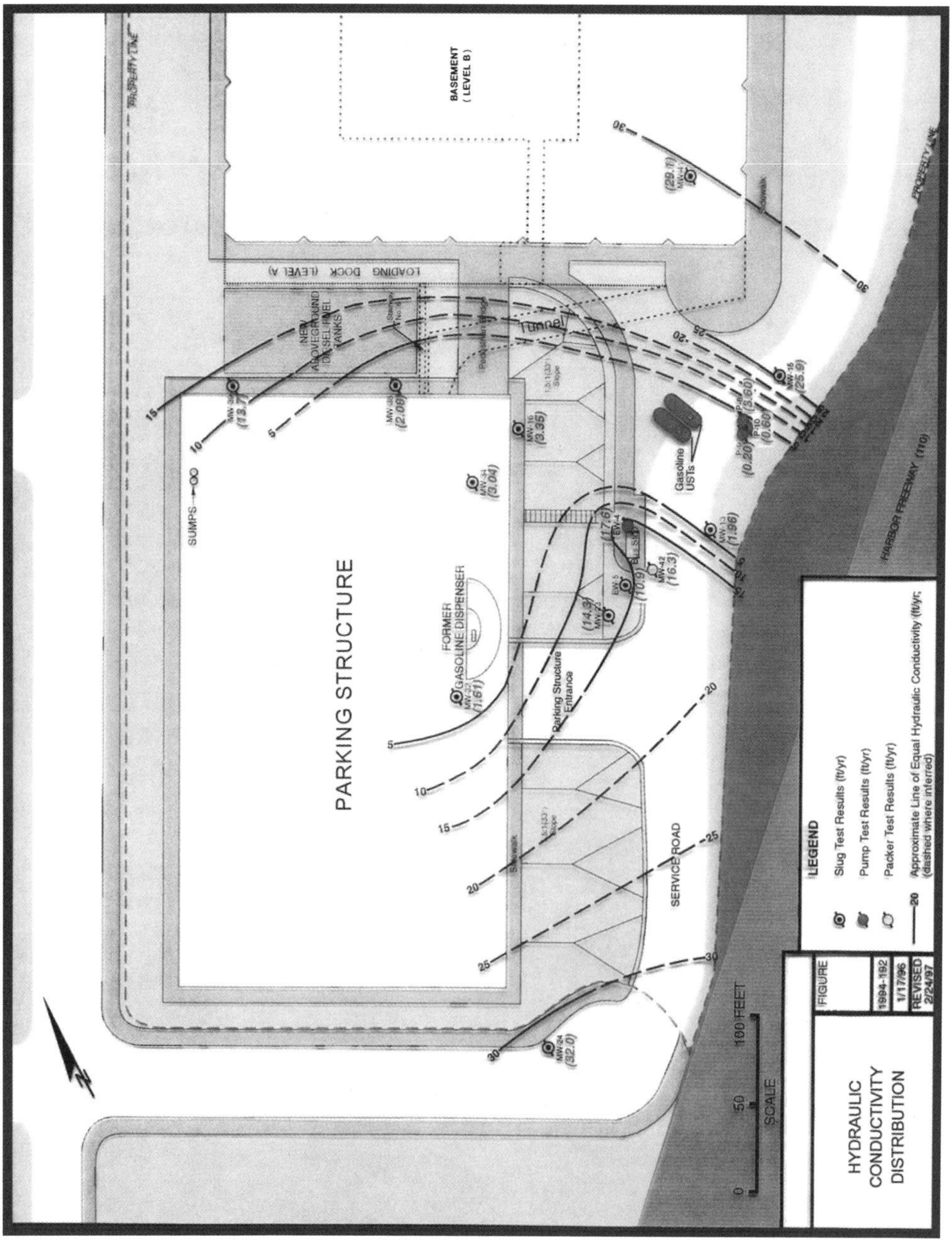


Figure 4. Hydraulic conductivity distribution (1cm = 10m).

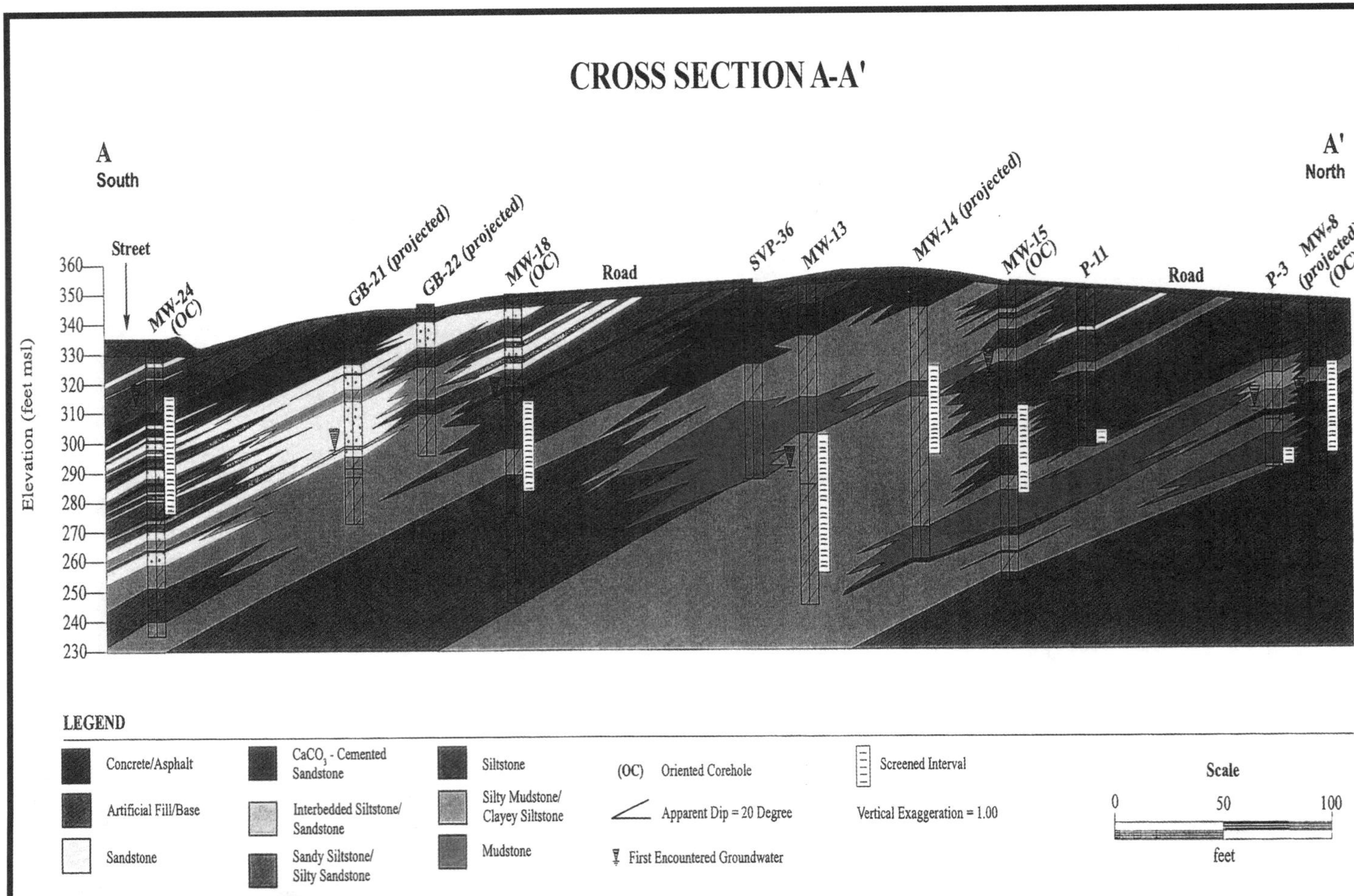


Figure 5. Diagrammatic Cross Section A-A′, illustrating the complex site stratigraphy. Oriented core data was instrumental in the development of a 3-dimensional subsurface model for the site.

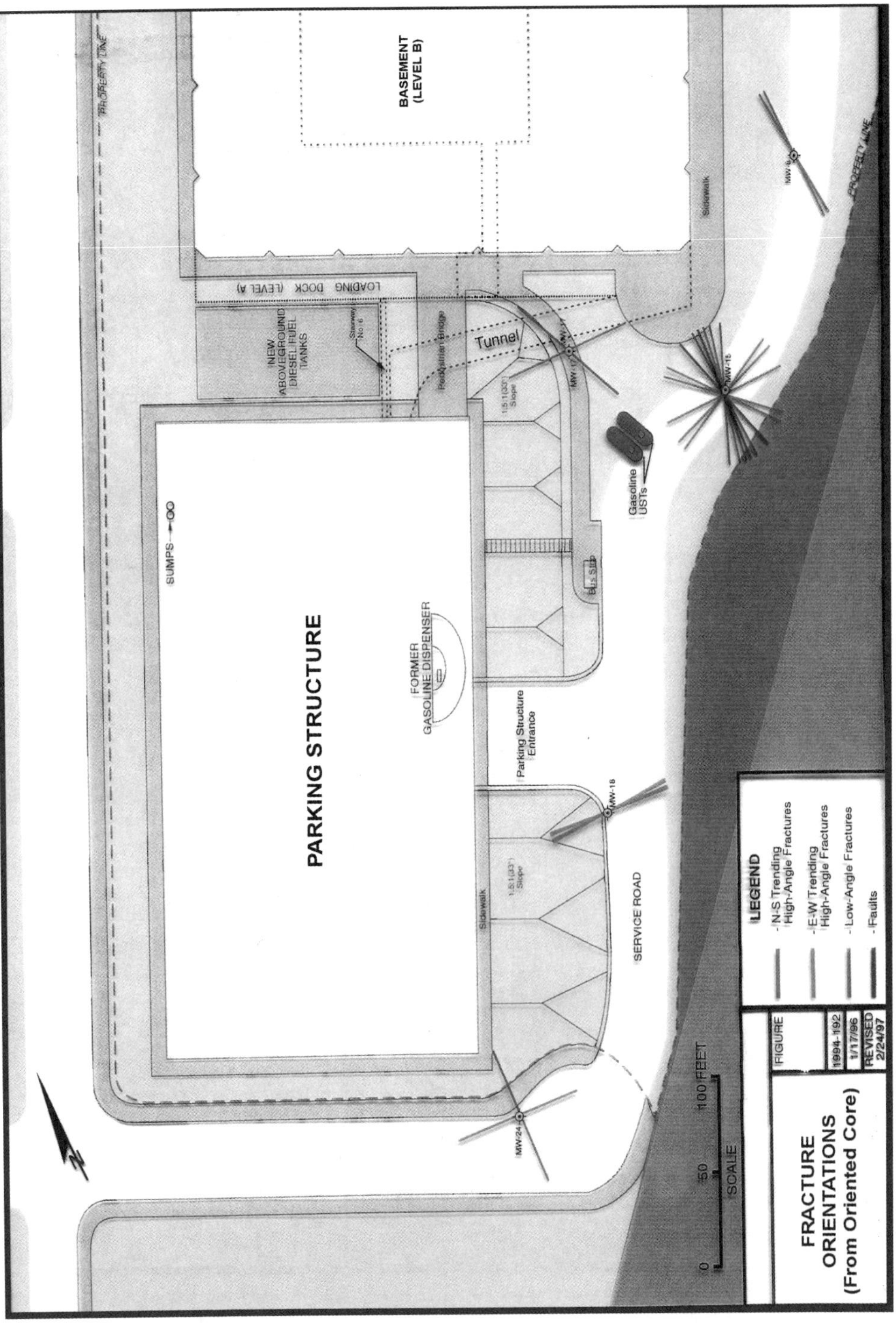


Figure 6. Fracture orientations measured from oriented core (1 cm = 10 m).

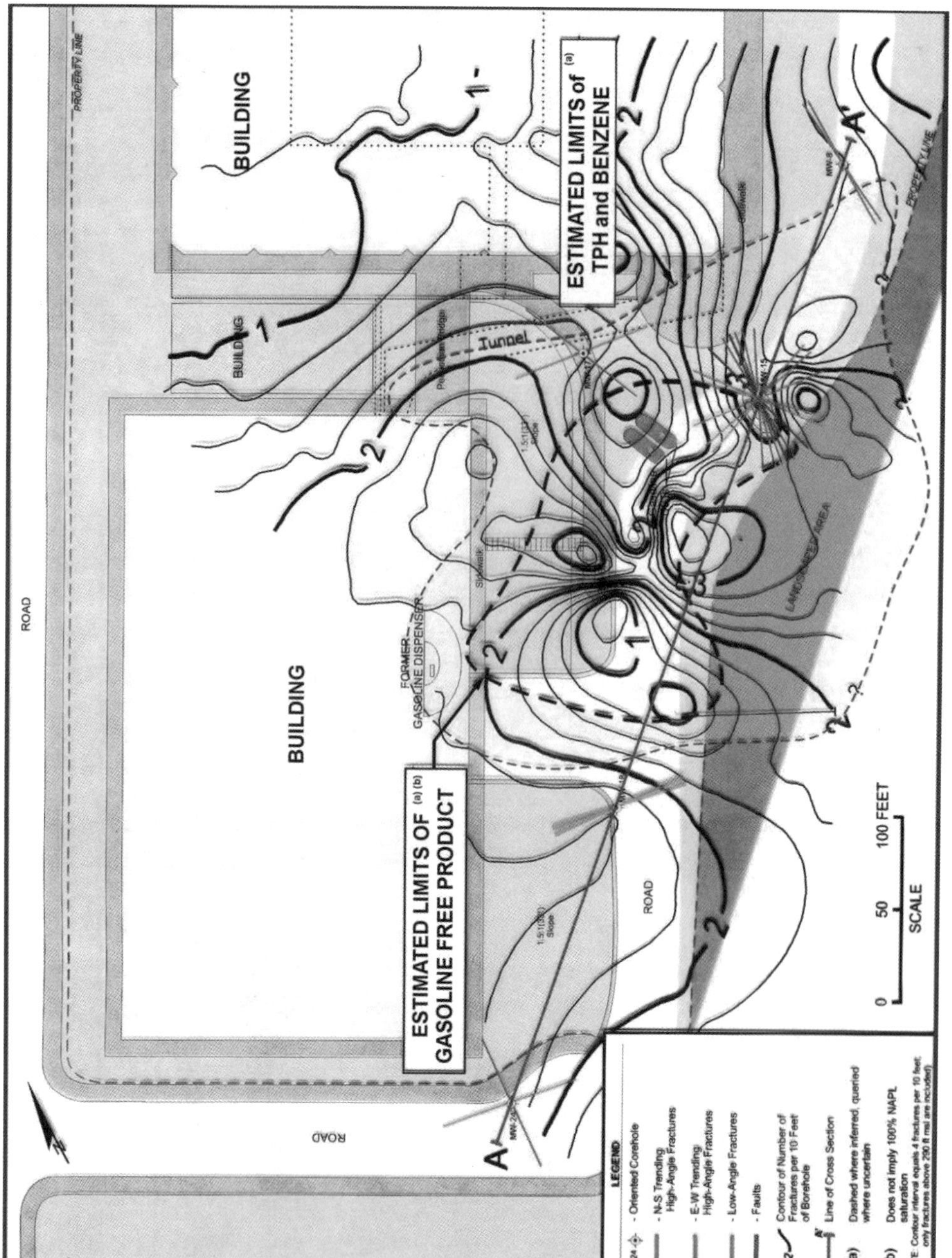


Figure 7. Map showing fracture density and orientation. Gasoline contamination is most prevalent in areas where north-south trending vertical fractures are abundant (1 cm = 1 m).

appreciable quantities of groundwater due to their generally higher porosity, but transmit water slowly and in small quantities. Groundwater at the site flows generally southward in a down-dip direction. This relationship suggests that higher permeability strata within the formation play a significant role in controlling groundwater movement. These conditions favor groundwater recharge of permeable beds that dip beneath less permeable confining beds, thereby creating local pockets of higher potentiometric pressure. Pressure gradients are relieved and groundwater equilibrates where high-angle fractures intersect these locally confined beds.

Numerous fractures were observed during the logging and analysis of bedrock samples; in some places, the fracture density exceeded 4 fractures per linear foot (0.1 per linear m) (Figure 6). Visible fracture aperture ranged widely from no visible aperture to apertures exceeding 10 mm. The majority of bedrock fractures occurs within the vadose zone and uppermost saturated zone at depths less than 40 feet (12 m) below ground surface. Significant fracturing above 20 feet (6 m) is present, but oriented data above this depth could not be obtained due to the magnetic influence of metallic aboveground equipment.

The strong similarities in the depths where gasoline-bearing fractures occur, as well as the close correspondence of these fractures to the water table in the center of the investigation area, demonstrate the local ability of fractures to store gasoline and suggest that fractures are serving as local migration pathways. High-angle fractures are particularly important because they cut across and interconnect otherwise hydraulically isolated beds. Figure 7 shows the orientation of the fracture sets and small-scale faults in plan view as well as the contoured fracture density in the study area. The figure illustrates the relationship between fracture orientation and density to gasoline contaminant migration; the area east from the gasoline underground storage tanks, is dominated by abundant north-south trending high-angle fractures and low-angle bedding plane fractures.

## 6 CONCLUSION

The oriented core method is an extremely useful tool for evaluating structurally complex sedimentary rock environments where conventional environmental assessment techniques may fail to obtain the data necessary to solve critical contaminant transport and migration problems. In the case of this investigation, the structural and lithologic information obtained from oriented bedrock core was essential in understanding and developing a conceptual model for gasoline and diesel fuel contaminant migration. This knowledge in turn was utilized to design and install a remediation system at the site. This method, as discussed herein, has strong application potential in both the environmental and engineering geology practices.

## ACKNOWLEDGEMENTS

Thanks to Arlen Ruen (Ruen Drilling; Clark Fork, Idaho), and Larry Kunkel (PTS Laboratories; Santa Fe Springs, California) for exceptional drilling and core retrieval, and logistical/laboratory support on this project.

## REFERENCES

CBC WELNAV (Christensen Boyles Corporation), 1993, Oriented Cores, 11 p.

Dibblee, T. W., 1989, Geologic Map of the Los Angeles Quadrangle, Los Angeles County, California, Dibblee Geological Foundation, Santa Barbara, California, 1 sheet, scale 1:24,000.

Directional Drilling Services (formerly CBC WELNAV), 1993, Oriented Core: Orientation Core Barrel Operation Manual, Directional Drilling Services, a Division of Layne Christensen, Salt Lake City, Utah.

Lamar, D.L., 1970, Geology of the Elysian Park-Repetto Hills Area, Los Angeles County, California, CDMG Special Report 101, San Francisco, California, 45 pp.

Meredith/Boli & Associates, Inc., 1997a, Gasoline UST System Feasibility Study/Corrective Action Plan Bank of America NT&SA, Los Angeles Data Center.

Meredith/Boli & Associates, Inc., 1997, Gasoline UST System Site Assessment and Interim Remedial Measures, Bank of America, Los Angeles Data Center, Los Angeles, California.

Meredith/Boli & Associates, Inc., 1994, Supplemental Investigation, Feasibility Study, and Corrective Action Plan for Bank of America, Los Angeles Data Center, Los Angeles, California.

CHAPTER 27

# Probabilistic delineation of groundwater protection zones in fractured-rock aquifers

Júlio F. Carneiro
*Centro de Geofísica de Évora, Col. Luís Verney, Évora Codex, Portugal*

ABSTRACT: Groundwater protection zones in fractured-rock aquifers are delineated with a hybrid method that combines discrete fracture network (DFN) models and equivalent continuum models. The method, known as the Statistical Continuum Method (SCM), relies on statistics of movement of particles in a local-scale DFN followed by the use of those statistics to mimic movement in a catchment-scale continuum model. Because the method is based in a Monte-Carlo procedure, it is possible to delineate probability contours for protection zones. The method is implemented in a catchment-scale synthetic case study that uses MODFLOW as the continuum groundwater flow model. Reverse particle tracking is conducted and the locations of the particles at time t are used to create probability contours of the t-isochrone protection zone. The computational needs for upscaling from local to regional scale are not heavy and probability contours of protection zones can be easily computed.

## 1 INTRODUCTION

Common practice of groundwater protection zone delineation in fractured-rock aquifers often ignores the specific features of solute transport in fractured rocks and resorts to the same analytical or numerical techniques used for continuous porous media aquifers. Fractured-rock aquifers are thus envisaged as Equivalent Porous Media (EPM) with heterogeneity and discontinuity being largely ignored.

The effect of ignoring the influence of fractures is usually not quantified. This results in hardly defensible protection zones, especially considering that field tests have indicated that transport of solutes in fractured rocks can be extremely complex. It is not unusual for tracer tests in fractured rocks to fail or not to comply with the physical models (Fick's laws) that describe solute transport in continuous media. Specifically, it is commonly accepted that a "scale effect" may result in parameters, such as dispersivity, increasing with transport distance, and consequently showing non-Fickian behaviour. Furthermore, several studies have shown that a Representative Elementary Volume (REV) – a necessary condition for assuming equivalence to porous medium – may be defined only for very large scales or not defined at all (Bear, et al. 1993, de Marsily 1986, Painter 1999).

Where the EPM assumption proves unacceptable, alternative ways of delineating protection zones must be sought, bearing in mind that the characteristic features of solute transport in fractured rocks are the relevant factors that will determine the protection zones shape and size. Thus, it is promising to use Discrete Fracture Network (DFN) models to delineate protection zones because they are seen as reliable tools to simulate solute transport in fractured rocks (e.g., Cacas, et al. 1990, Dershowitz and Eiben 1999, Endo, et al. 1984, Outters and Shuttle 2000).

Despite the considerable amount of research on DFN models, few studies related to protection zone delineation are known. Bradbury and Muldoon (1994) were the first to apply DFN models to delineation of protection zones. They used fracture statistics to generate a two-dimensional fracture network representation of the aquifer. A Monte Carlo procedure using particle tracking completed the delineation of probabilistic capture zones. They concluded that the EPM approach underestimates the capture zones, not providing adequate resource protection.

Robinson and Barker (1999) suggest an hybrid approach that couples the use of the 3-D DFN models with stochastic continuum models that apply effective hydraulic parameters resulting from the DFN simulations. The method enabled them to delineate probabilistic protection zones at the catchment scale in a fractured rock aquifer. Rayne et al. (2001) used a 2-D DFN code to assess the influence of flow in the fractured unsaturated zone on the size of a capture zone. They show that, for the case study, the capture zone should be widened by tens of metres to account for the effect of fractures. Chevalier et al. (2001) used a 3-D DFN code to delineate protection zones in a fractured environment; they concluded that fracture aperture and orientation are key elements in determining size and shape of capture zones.

DFN models are usually restricted to local scale studies due to the computational requirements. Given that protection zones are usually defined at the catchment scale, alternatives must be sought to upscale DFN transport features to regional scale models. The challenge is to couple the rigour of solute transport in DFN models with the versatility of the continuum approach in modelling large areas.

Several techniques for upscaling the flow properties have been proposed (e.g., Hestir and Long 1990, Long, et al. 1982, Robinson and Barker 1999, Robinson 1984, Zimmerman and Bodvarsson 1996). Most of these are based on the use of a DFN model at a small-scale to find "effective" hydraulic parameters that are then applied in a large-scale continuous model. These approaches have proved their value in upscaling flow, and a similar approach is used in this study to solve the groundwater flow problem at the catchment-scale. However, it is questionable that such approaches are reliable to upscale solute transport (Endo, et al. 1984, Guérin and Billaux 1994, Herbert and Lanyon 1995).

This study addresses the applicability of a different solute transport upscaling technique, the **Statistical Continuum Method**, to delineate probabilistic protection zones.

## 2 OVERVIEW OF THE STATISTICAL CONTINUUM METHOD

The Statistical Continuum Method (SCM) (Schwartz and Smith 1988) uses particle tracking in which physical transport is simulated in terms of velocity and velocity variation. Solute transport is modelled by collecting statistics on particle movement in a subdomain using a DFN model and then using these statistics in a continuum model to conduct particle tracking in a larger scale domain. The DFN sub-domain is a small but representative piece of a much larger continuum (Figure 1a).

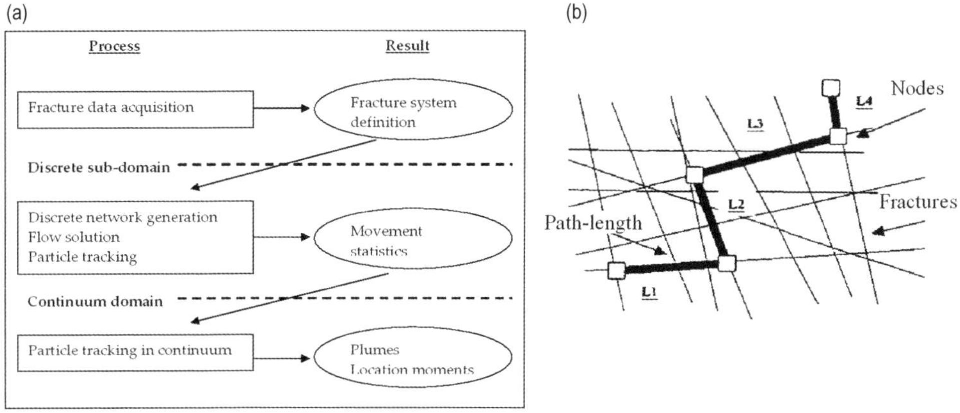


Figure 1. (a) Flow-chart for the Statistical Continuum Method; (b) Path-length ($L_i$) and movement direction. In this network there are 8 possible movement directions, since the particles can travel locally against the overall hydraulic gradient. After Parney (1999).

The statistics collected are the frequency that a particle chooses to move in each direction and frequency distributions of fracture lengths between intersections – path-length (Figure 1b) – and of velocity.

This methodology, albeit still experimental and requiring validation against field situations, has several advantages (National Research Council 1996); it (1) upscales particle movement without the need to account explicitly for every fracture that would occur at a regional scale; (2) makes no assumption about Fickian behaviour on solute transport; particle tracking is conducted in the DFN domain; and (3) the continuum model mimics the particles movements without considering dispersion tensors.

Smith et al. (1990), Parney and Smith (1995) and Parney (1999) conducted research towards assessing the reliability of the method to determine the adequate statistical distributions that describe particle movement, and its applicability to 3-D fractured environments. These authors restricted the analysis to two fracture sets and to uniform flow conditions. The results were encouraging and seemed to confirm the SCM as a reliable tool for modelling solute transport in fractured rocks at the catchment scale. Therefore, it may be a suitable method to delineate groundwater protection zones in fractured rock aquifers, as long as the following is achieved:

(a) the method is generalised so that it can account for any number of fracture sets;
(b) the SCM is linked to a standard groundwater modelling code, so that it can recognise the flow solution and conduct particle tracking according to the statistics collected in the DFN subdomain;
(c) particles location at time t are used to compute probabilistic protection zones.

This study is limited to 2-D environments. Reliable 2-D DFN modelling only occurs in well-connected fracture networks. Since the method is applied to model solute transport at the catchment scale, a 2-D approach is assumed adequate, once connectivity is guaranteed, given that the vertical flow and transport are less important than the horizontal because those are constrained by aquifer thickness.

## 3 DELINEATING PROTECTION ZONES WITH THE STATISTICAL CONTINUUM METHOD

Delineation of groundwater protection zones with the SCM was conducted in a synthetic case study. A confined fractured aquifer has north and south impermeable boundaries, while the east and west boundaries are constant head. Three wells pumping at constant rate create a complex flow pattern (Figure 2a). The fracture characteristics of the aquifer are considered as homogeneous throughout the whole domain (Table 1). A DFN subdomain is used to collect the statistics for several orientations of hydraulic gradient. The subdomain has a rectangular area of 75 m × 75 m, with constant head boundaries at X = 0 m and X = 75 m, and impermeable boundaries at Y = 0 m and Y = 75 m (Figure 2b).

### 3.1 *Statistics of movement in the discrete fracture sub-domain*

The statistics of movement are gathered using a DFN code in which particle tracking is conducted for several hydraulic gradient orientations. The code chosen to simulate transport in the small scale domain was SDF (Rouleau 1988), which is a 2-D DFN model that generates

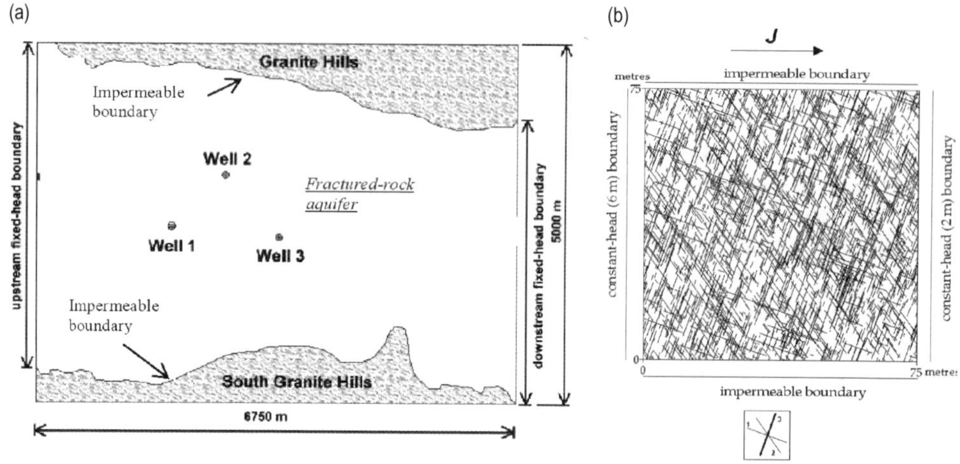


Figure 2. Synthetic case study. (a) Layout of the aquifer (based on Chiang and Kinzelbach 2001); (b) Sample realisation of the discrete subdomain (75 m × 75 m). In the smaller square, thickness of fractures is proportional to fracture aperture. Set 3 is responsible for a high anisotropy of the fracture system. J denotes hydraulic gradient.

Table 1. Fracture statistics for the base case network.

| Fracture set | Fracture orientation | Fracture density ($m^{-1}$) | Fracture length Mean (m) | Fracture length Stand. Dev. (m) | Fracture aperture (m)* | Transmissivity ($m^2$/d)** |
|---|---|---|---|---|---|---|
| Set 1 | N110°, 15°NE | 0.6 | 4.0 | 2.2 | $5 \times 10^{-5}$ | $6.7 \times 10^{-3}$ |
| Set 2 | N145°, 65°E | 0.6 | 2.7 | 2.2 | $5 \times 10^{-5}$ | $6.7 \times 10^{-3}$ |
| Set 3 | N20°, 55°W | 1.2 | 3.3 | 2.2 | $1 \times 10^{-4}$ | $5.3 \times 10^{-2}$ |

* Fractures are considered as smooth parallel plates of uniform aperture.
** Transmissivity of individual fractures, computed according to the cubic law for plane fractures.

random fracture networks, solves the groundwater flow problem, and conducts particle tracking to simulate solute transport by advection. A Monte Carlo procedure is implemented, so that the statistics are collected in multiple realisations of the fracture network. The statistics collected for the 6 possible movement directions are:

(a) **directional choice** – i.e. frequency each direction is chosen by the ensemble of particles;
(b) mean and variance of **path-length** – the distance a particle travels along a fracture between the fracture intersection in which it enters and the intersection through which it exits (Figure 1b);
(c) mean and variance of **velocity** – normalization by the hydraulic gradient is done, which converts velocity into the ratio $K/n_e$, where K is hydraulic conductivity and $n_e$ is kinematic porosity. Here "kinematic porosity" (or advective porosity) is the ratio between the volume of circulating water and the total volume of rock (de Marsily 1986), and is assumed to have tensorial properties, its value varying with flow direction (Neuman 1990). For simplicity, the ratio $K/n_e$ will be designated hereafter as "velocity".

The hydraulic gradient orientation varies in 15° increments and again multiple realisations of the DFN are performed, in order to build a matrix of statistics for directional choice, path-length and velocity against hydraulic gradient orientation.

Directional choice is described by a discrete uniform distribution and a two-parameter gamma distribution is used to fit path-length and velocity retrieved from the DFN model

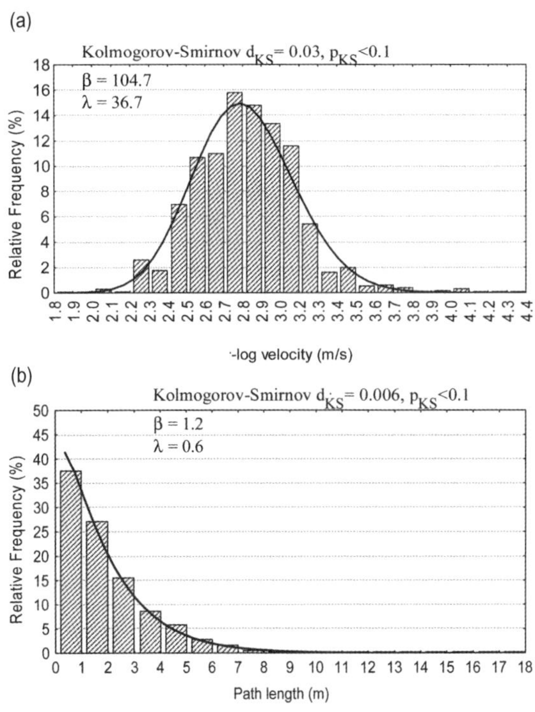


Figure 3. Gamma distribution fit to the values of: (a) –log velocity; (b) path-length. Fracture set 1, down-gradient. Other fracture sets fit to the same distribution type. Sample size: 4691. $p_{ks}$ and $d_{ks}$ denote the level of significance and maximum difference, respectively, for the Kolmogorov-Smirnov test.

(Figure 3). Notice that the Kolmogorov-Smirnov test indicates that the hypothesis of the data being distributed according to a gamma distribution is acceptable with probability above 0.9 (for a detailed discussion of probability distributions and the SCM see Parney 1999).

The two-parameter gamma distribution is defined as:

$$p_x(x) = \frac{\lambda e^{-\lambda x}(\lambda x)^{\beta-1}}{\Gamma(\beta)} \qquad -\infty < x < \infty \tag{1}$$

where $\Gamma(\beta)$ is the gamma function. The distribution is dependent on the shape ($\beta$) and scale parameters ($\lambda$) which are estimated by (Weisstein 1999):

$$\beta = \frac{\overline{\mu}^2}{\sigma^2}, \ \lambda = \frac{\overline{\mu}}{\sigma^2} \tag{2}$$

where $\overline{\mu}$ and $\sigma^2$ are mean and variance, respectively, of path-length or of velocity. As $\beta$ approaches infinity, the gamma distribution converges to the Gaussian distribution.

To ensure stabilisation of the movement statistics with number of realisations, the maximum number of particles (200) supported by the DFN code was used in each realisation and the sampling area (75 m $\times$ 75 m) was as large as admissible for the prescribed fracture density. All parameters stabilised quickly, except for the standard deviation of velocity, which with 32 realisations still varies up to 7% with respect to 16 realisations. More realisations should be conducted but due to computer limitations, the number of realisations was kept at 32 for each hydraulic gradient orientation. A sensitivity analysis demonstrated that the results of the SCM are almost indistinguishable when considering 8, 16 or 32 realisations.

### 3.2 *Correlation between path-length and velocity*

Partial distributions of velocity for several ranges of path-length show that, although gamma distributions can always fit the velocity data, the shape and scale parameters tend to increase with path-length (Figure 4a). That is, it is necessary to vary the gamma distribution parameters with path-length.

Figure 4b is a typical scatterplot of path-length $\times$ velocity, which closely resembles the scatterplots of Parney and Smith (1995) that led them to suggest the adoption of a path-length $\times$ velocity correlation. This path-length and velocity dependence is explained by the DFN code relying on the relative proportion of flow rate in each of the intersecting fracture segments to compute the probability that the particles may travel along each of those segments. Higher speeds coincide with higher flow rates, so once a particle enters a high velocity fracture it is likely to continue along it, imposing large path-lengths.

Similar plots for several different fracture sets indicate that the best fit is provided by:

$$\textit{mean velocity} = c_0 - c_1 \ln(\textit{path-length}) \tag{3}$$

where $c_0$ and $c_1$ are constants retrieved from the movement statistics. The correlation coefficient is usually small because of large dispersion for small path-lengths. Velocity is not

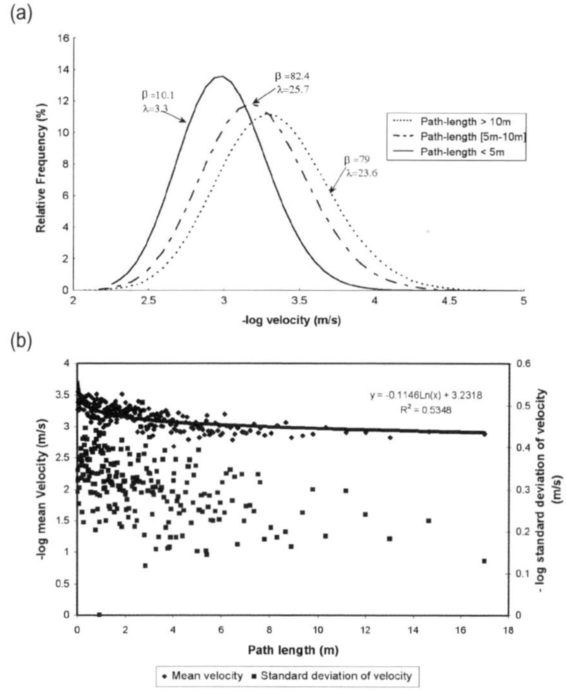


Figure 4. Correlation of velocity and path-length. (a) gamma distribution fit to velocity data for several path-length ranges; (b) plot of mean and standard deviation of velocity versus path-length. Also shown the best-fit regression between mean velocity and path-length. Fracture set 3, down-gradient, sample size: 9502. Other fracture sets show the same behaviour.

evaluated from equation (3), which is used to find a tendency for variation of mean velocity with path-length. Velocity is always generated from a gamma distribution using the mean and variance of velocity, with the shape and scale parameters of the distribution varying with path-length. Thus, the distribution used for each movement direction is not constant and can assume a whole range of gamma distributions, such as in Figure 4a.

## 4 UPSCALING TO THE CATCHMENT SCALE CONTINUUM DOMAIN

To implement the SCM procedure at a catchment scale, it is necessary to know the hydraulic gradient orientation and value at every location. Thus, the groundwater flow problem has to be solved using a continuum model, such as MODFLOW (McDonald and Harbaugh 1988). Once the flow solution is found, a code designated SCPATH computes the hydraulic gradient and conducts particle tracking according to the following procedure:

(a) estimate the hydraulic gradient at the location where particles are to be released;
(b) from the matrix of the movement statistics against hydraulic gradient orientation, a uniform distribution is used to randomly generate the direction that the particle will take;

(c) a gamma distribution is used to randomly generate path-length, with the scale and shape parameters being evaluated from equation (2);
(d) equation (3) is used to generate the mean velocity, after which a gamma distribution with shape and scale according to equation (2) is used to generate the velocity of the particle;
(e) the product of velocity (or more accurately $K/n_e$) by the hydraulic gradient value returns advective velocity;
(f) the particle is moved to the end of the path-length and travel time is calculated;
(g) steps b) through f) are repeated until the end of the time-step or until the particle changes cell. When either of these occurs, the hydraulic gradient is estimated at the new particle location and steps b) through f) are repeated. The number of time steps is key to the accuracy of the procedure, since it relates to the number of hydraulic gradient readings;
(h) steps a) through g) are repeated until the end of the prescribed travel time and a new particle is then selected to move.

The code SCPATH includes an option to conduct "standard" particle tracking in the continuum domain. It uses the same hydraulic parameters on which the MODFLOW flow solution was computed, but it includes anisotropy in kinematic porosity. This option is assumed realistic, since in fractured rocks kinematic porosity is certain to vary with flow direction.

### *Comparison of solute transport according to the SCM and DFN models*[1]

A comparison was made between solute transport in the DFN sub-domain and solute transport simulated using the SCM. No rotation of the hydraulic gradient was conducted because the purpose was to assess the reliability of the SCM under uniform hydraulic gradient conditions. The statistics were used in a continuum domain of same size and boundary conditions as the DFN sub-domain (Figure 2b). Plumes of particles were back-tracked, both in the discrete domain and in the continuum domain. In the SCM procedure 1000 particles were tracked in one single realisation while in the discrete domain five realisations of 200 particles each were used. It is not expected that five realisations of the DFN fit perfectly the results of the SCM, but the main tendencies of movement should be preserved.

The main tendency is of coincidence between movements of the two clouds of particles (Figure 5), but it is also obvious that longitudinal dispersion seems to be underestimated in the SCM.

The SCM and DFN results can be compared by plotting the first (mean) and second (variance) moments about the mean particle location. The mean x and y coordinates are the 1st moments, while the variance of those coordinates are the 2nd moments about the mean. The slope of the 1st moments is equivalent to the advective velocity in x and y direction, and the slope of the 2nd moments is related to the dispersion coefficient, although it may not coincide because the plumes may not be symmetrical.

The fit of the 1st moments in x and y is quite good (Figure 6a), showing that velocity in the x and y directions is very similar to that occurring in the DFN. At late times the 1st moments of x and y diverge from a straight line due to the particles reaching the limit of the discrete domain. As for the 2nd moments, dispersion in the y direction shows a good fit until

[1] This section aims to assess the reliability of the SCM for the case study (Figure 2, Table 1). For a systematic analysis of reliability of the Statistical Continuum Method refer to Parney (1999).

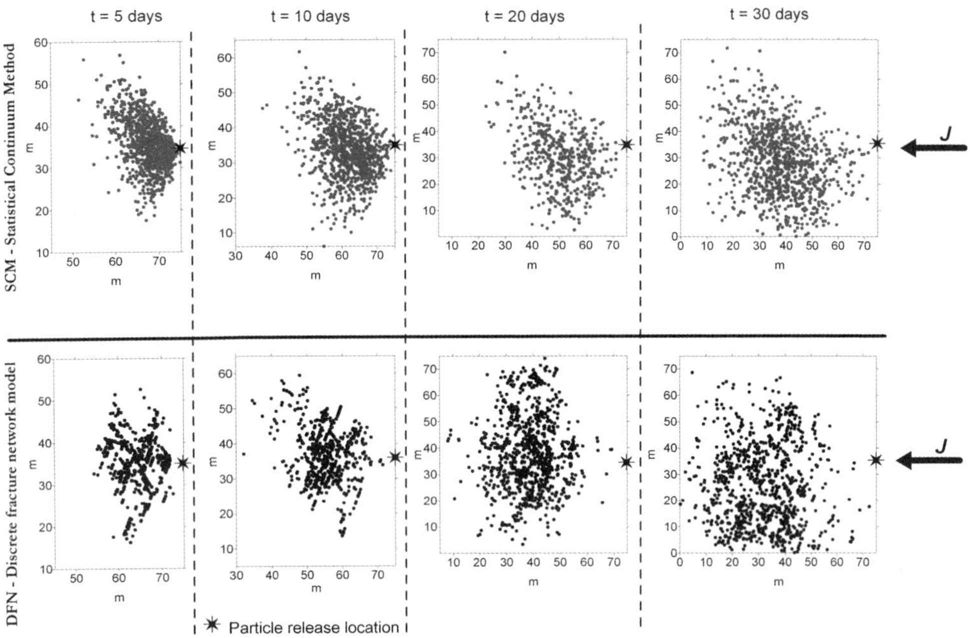


Figure 5. Plumes of particles tracked using SCM and DFN models. 1000 particles instantaneously released at $X_0 = 75\,\mathrm{m}$, $Y_0 = 37.5\,\mathrm{m}$. Uniform hydraulic gradient (J) from right to left. Notice that scale varies.

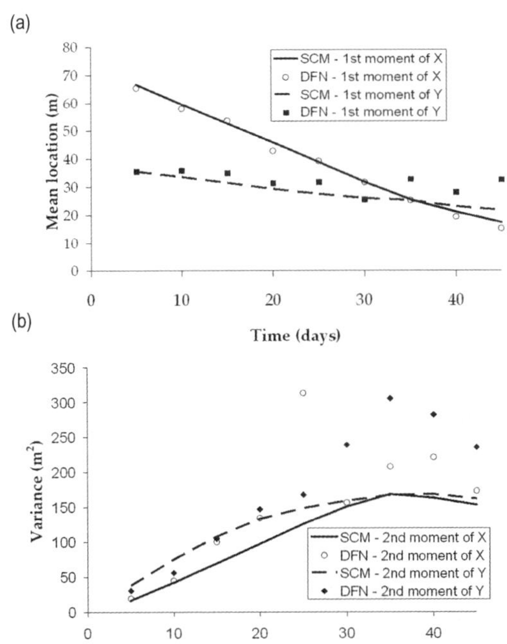


Figure 6. Spatial moments of x and y. Particles released at $X_0 = 75\,\mathrm{m}$, $Y_0 = 37.5\,\mathrm{m}$.

particles start leaving the fracture domain (Figure 6b). After 30 days, dispersion in they direction in the DFN domain is larger than in the SCM. The DFN model identifies the lateral boundaries as impermeable, and so the particles are deflected back into the domain, while the SCM does not identify the boundary as impermeable and consequently the particles leave the domain, decreasing lateral dispersion.

Comparison of dispersion in the x direction is not as good as in the y direction. Both in the DFN and in the SCM particles were released at location $X_0 = 75\,m$ and $Y_0 = 37.5\,m$. The SCM uses precisely this location and generates the first element starting from that point. In the discrete domain, particles are shifted to the closest node of the network. Therefore, in the DFN model there will be an apparent dispersion even for $t = 0$, which can explain the less good fit between 2nd moments of x and y Nevertheless, the overall tendency of dispersion is well represented in the SCM.

## 5 DELINEATION OF GROUNDWATER CAPTURE ZONES

To apply the SCM at the catchment scale, discretization of the aquifer shown in Figure 2a must be such that the finite difference cells are larger than the discrete sub-domain ($75\,m \times 75\,m$) where the statistics were calculated. The fracture pattern of the aquifer is homogeneous in the whole domain, thus the statistics of movement are applicable to all cells. If the fracture pattern varied throughout the aquifer, different statistics of movement should be applied in the relevant cells.

The effective hydraulic parameters to input into MODFLOW to establish the flow solution must be consistent with the fracture network model. Two approaches were used. The main practical difference between the two approaches is that the second method allows us to account for anisotropy in K and $n_e$.

- "field" data, meaning the use of data that one would gather in a real situation: hydraulic conductivity, K, from a pumping test and kinematic porosity, $n_e$ from a tracer test. A pumping test was simulated in the DFN for a single realisation of the base case network. Although being specific for isotropic and homogeneous medium, which is not the case, Thiem's equation was used to compute the value of K, to increase the similarity with a real situation in which Thiem's would likely be used. A mean, isotropic, value of K was adopted. Tracer tests were also simulated in the same DFN realisation with particles being dropped at one boundary and the time of arrival at the other boundary being used to compute the mean $n_e$ according to $n_e = qt/l$ (Guérin and Billaux 1994), where q is specific discharge, t is travel time and l is the travel distance. Mean values of $K = 4 \times 10^{-3}\,m/d$ and $n_e = 6 \times 10^{-5}$ were found;
- "integral simulation", meaning that K and $n_e$ are computed for several hydraulic gradient orientations. A uniform flow field is imposed on the domain and K is computed using Darcy's law and the flow rate crossing the domain. It is possible to define an ellipsoid of $1/\sqrt{K(\Theta)}$ and an anisotropy factor in $K_x/K_y$. Kinematic porosity was also computed for every hydraulic gradient orientation and an anisotropy factor was estimated. The minimum value of K occurred for a hydraulic gradient at 0° to the x-axis ($K_x = 10^{-3}\,m/d$) and the maximum value occurred for hydraulic gradient oriented at 90° ($K_y = 5.8 \times 10^{-3}\,m/d$). The corresponding values of effective porosity found were

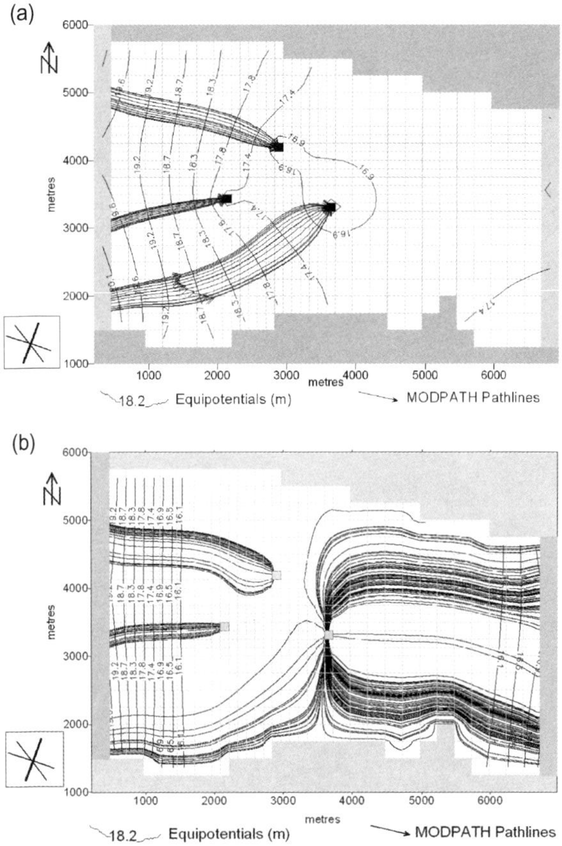


Figure 7. Capture zones of the three wells, according to an EPM approach, i.e. ignoring the solute transport features of the fractured media. (a) Flow solution computed with $K_x = K_y$; (b) Flow solution computed with $K_y/K_x = 5.8$. 1000 particles were backtracked from each well. The small squares show the orientation of the fracture sets that compose the aquifer.

$(n_e)_x = 6 \times 10^{-5}$ and $(n_e)_y = 9 \times 10^{-5}$, for hydraulic gradients at 0° and 90° with the x-axis, respectively.

Figure 7 shows the flow solution found with MODFLOW, both for the situation considering isotropic K (Figure 7a) and for the situation considering anisotropy in K (Figure. 7b). Also shown are the capture zones for the equivalent porous medium approach (i.e., ignoring the solute transport features of the fractured media). For the isotropic case the particle tracking program MODPATH (Pollock 1989) was used, but in the anisotropic case the capture zones were found with the SCPATH option including anisotropy in $n_e$.

Notice that the capture zones depicted in Figures 7a and 7b are strikingly different not only due to the anisotropy in $n_e$, but also because the flow solutions were not calibrated to any field data; such calibration would force the flow solutions to be more alike and consequently the capture zones would also be more similar.

Figure 8 compares the capture zones delineated according to the SCM procedure with the capture zones shown in Figure 7 which were delineated using an EPM approach. If the

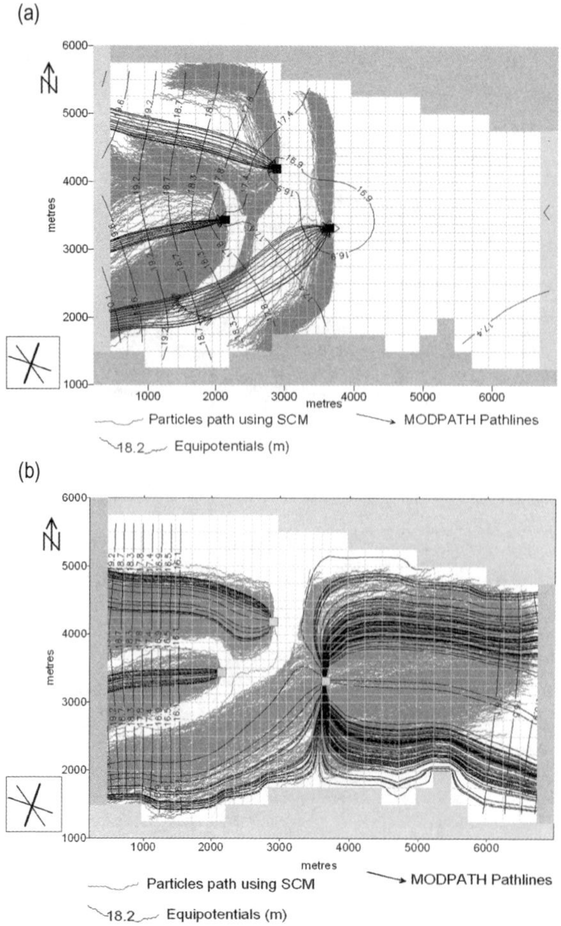


Figure 8. Comparison between capture zones delineated by the SCM approach and by the EPM approach. (a) Flow solution computed with $K_x = K_y$; (b) Flow solution computed with $K_y/K_x = 5.8$. 1000 particles were backtracked from each well. In SCM the tracking of particles was limited to 10,000 days, which explains why some of the paths do not reach the limit of the capture zone.

flow solution obtained with an isotropic K is used (Figure 8a), the capture zones are completely different when delineated with the SCM procedure and when delineated assuming EPM conditions. This is due to the highly anisotropic behaviour and discontinuous character of the fracture system. Dispersion is higher and the main flow direction is imposed by the most permeable fracture set (number 3), oriented at N20°E. The results of the SCM seem to be consistent with the imposed flow field and with the anisotropy of the fracture system, which is in favour of the reliability of the method.

Figure 8b compares the groundwater capture zones delineated using the flow solution that considers anisotropy in K. In this case the continuous porous media capture zones were computed considering also anisotropy in $n_e$. As for the capture zone delineated according to the SCM, they indirectly encompass both the anisotropy in K and the anisotropy in $n_e$, and are consistent with the hydraulic gradient and anisotropy of the fracture network. The capture zones retrieved by the EPM and the SCM approaches are now more similar, the main differences resulting from the dispersion imposed by the fracture network.

Notice that Figure 8b shows a better resemblance between the SCM solution and the continuous porous medium approach due to the adoption of anisotropy in $n_e$. However, in most common applications the discrepancy between the two approaches would be more akin to the one shown in Figure 8a, since the standard continuous models do not consider directional kinematic porosity.

## 6 PROBABILITY CONTOURS OF GROUNDWATER PROTECTION ZONES

Solute transport modelling in fracture networks is closely associated with the concept of randomness. Even if advection is the single driving force, particle locations vary between realisations of the network and even in each single realisation. The result of the modelling process is a plume of particles regardless the particles being released at the same location. Such occurs both in the DFN model and in the SCM. The spread of the particles can be interpreted as a dispersive effect or as the probability of a particle being at a certain location at time t. That is, the t isochrone no longer is a single surface line, but rather is distributed over a wide area according to some probability function. Thus, probabilistic protection zones must result from analysis of particles' location through time.

The probability of any finite difference cell being part of an isochrone t is computed using a procedure based on a methodology followed by van Leeuwen (2000):

- whenever each particle i reaches a cell j before time t an integer $Z_i = 0$ is assigned to the cell. Whenever the travel time to each cell is above t, an integer $Z_i = 1$ is assigned;
- the sum of $Z_i$ values at each cell j is divided by the total number of particles n reaching the cell to find the probability P(j, t) that particles crossing that cell will take longer than time t to reach the well, according to $P(j, t) = \sum_{i=1}^{n} Z_i / n$.

The result is the probability that any particle passing out of that cell will take longer than time t to reach the well. For instance, the 75% contour for the 50-day isochrone indicates that 75% of the particles crossing that contour will take longer than 50 days to reach the well, thus defining the 0.75 probability of the well being totally protected.

Figure 9a shows the probability contours for the isotropic K flow solution. There is considerable difference between the protection zones delineated with MODPATH and the protection zones delineated with SCM, both in orientation and size. The protection zones delineated by the SCM are marked by particles travelling primarily along fractures that are more permeable (fracture set 3).

Figure 9b shows the groundwater protection zone delineated when considering anisotropy in K and $n_e$. The uncertainty contours are consistent both with the piezometric surface and with the fracture anisotropy, with lateral dispersion being very high despite the steep hydraulic gradient parallel to x. In Figures 9a and 9b it is demonstrated that the distance between the 10% contour and 100% contour can be of up to 500 metres, for a total travel distance of 1000 to 2000 metres. This can be regarded as a measure of the uncertainty associated with the protection zone, which in this case is high and cannot be ignored.

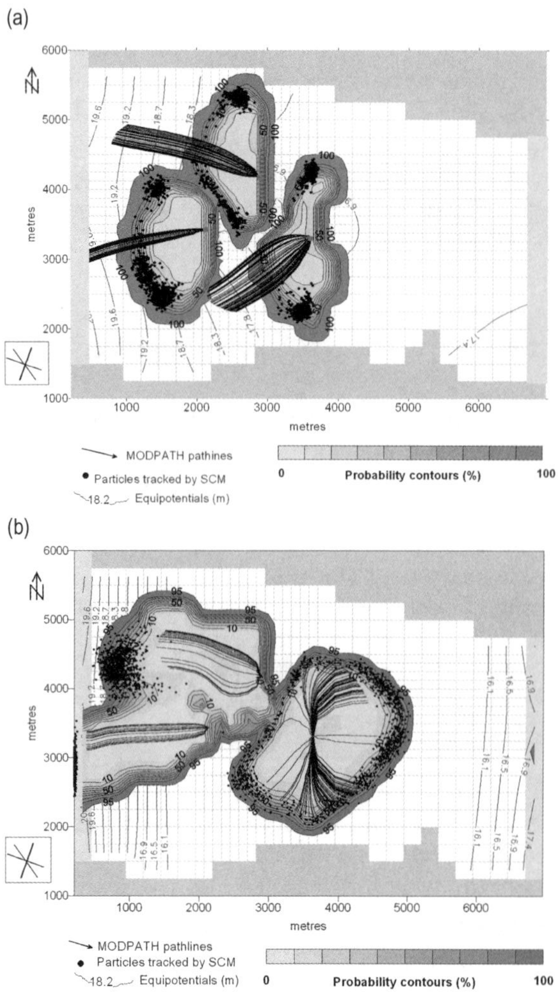


Figure 9. Ten-year protection zone. (a) Flow solution computed with $K_x = K_y$; (b) Flow solution computed with $K_y/K_x = 5.8$. Contours represent the probabilistic capture zone computed with the output from the SCM. MODPATH path-lines realise the protection zone as computed with a continuous porous media model.

## 7 CONCLUSIONS

The specific features of solute transport in fractured rocks environments may result in considerable deviations in the shape and size of groundwater protection zones that one would establish using the EPM assumption. Thus, whenever possible, models that are able to represent the transport behaviour in fractured rocks should be used to delineate protection zones. The uncertainty inherent to transport modelling in fractured rocks should be translated into probabilistic protection zones.

Protection zones in a case study representing a fractured-rock aquifer were delineated using a hybrid approach which combines DFN models and equivalent continuum models. The method, known as the Statistical Continuum Method (SCM), relies on statistics of movement of particles in a local-scale DFN followed by the use of those statistics to mimic movement in a catchment-scale continuum model. Probabilistic protection zones result from the procedure.

The effect of discontinuity and heterogeneity of the fracture network imposes capture and protection zones strikingly different from those computed using the EPM assumption (i.e., MODPATH). A better approximation would result if the continuous porous medium model includes anisotropy in effective porosity. However, even in this circumstance, the protection zones delineated under the EPM assumption and by the technique applied in this study would not be similar, due to the dispersive effect inherent to transport in fracture networks. This dispersive effect is reflected in uncertainty about the protection zone size and shape, and should be addressed through delineation of probability contours.

In the simulations, the fracture sets have constant orientation and aperture. Simulations were conducted where those parameters were left to vary. Variations in aperture enhances the accuracy of the SCM, which results from directional choices becoming more clear due to the increased differences between flow rate in the several fracture sets. When fracture orientation was the varying parameter, the fit between the 1st moments was equally good, but the fit of the 2nd moment was worse.

The SCM provides a valid approach to simulate solute transport in fractured rocks at the catchment scale, where DFN models may not be practical. The methodology is easy to implement and can be made to work with standard finite difference flow models. The computational requirements for upscaling from local to regional scale are not heavy and probability contours of protection zones can be easily computed.

The data requirements are, for the most, of common gathering in fractured rocks studies, where information about fracture sets orientation, length and density are collected in a more or less routine manner. A major difficulty presents itself when hydraulic properties are to be assigned to each individual fracture set. The transmissivity of each fracture set (or its aperture, if a correlation between transmissivity and aperture is to be used) and the percentage of hydraulically connected fractures are difficult to assess, and are one of the most active branches of research dealing with fractured-rocks hydrogeology (for a discussion on this issue see Robinson and Barker 1999). Packer tests may be of practical use in this respect. These difficulties hamper the use of the SCM in a straight forward manner for delineation of protection zones. Still, and although requiring more research, the methodology is applicable for assessment of the influence that fracture networks may have on the shape and size of protection zones, particularly in comparison to the EPM based techniques.

The SCM has difficulties in dealing with networks close to the percolation threshold (i.e., networks in which the connection between fractures is at the limit to ensure the existence of a continuous path between opposite faces of a rock mass and that flow can occur) and with networks of fractures with very different scale lengths (Parney 1999). According to National Research Council (1996), large discrete fractures, fracture zones or faults, must be modelled explicitly whatever the modelling approach is. Possibly this is also the case for the Statistical Continuum Method.

## ACKNOWLEDGEMENTS

Prof. John A. Barker, from the University College London, is gratefully acknowledged for his advice and supervision. Thanks are also due to Prof. Ken Bradbury, from the Wisconsin Geological and Natural History Survey, for providing the code SDF. This study was partially supported by FCT – Fundação para a Ciência e Tecnologia – under the Third Community Support Framework.

## REFERENCES

Bear J, Tsang C-F, de Marsily G (1993) Flow and contaminant transport in fractured rock. Academic Press San Diego.

Bradbury KR, Muldoon MA (1994) Effects of fracture density and anisotropy on wellhead protection area delineation in fractured aquifers. Applied hydrogeology 17–23.

Cacas MC, Ledoux E, de Marsily G, Barbreau A, Calmels P, Gaillard B, Margritta R (1990) Modeling fracture flow with a stochastic discrete fracture network – calibration and validation.2. The transport model. Water Resources Research 3:491–500.

Chevalier S, Bues MA, Tournebize J, Banton O (2001) Stochastic delineation of wellhead protection area in fractured aquifers and parametric sensitivity study. Stochastic Environmental Research and Risk Assessment 3:205–227.

de Marsily G (1986) Quantitative hydrogeology: groundwater hydrology for engineers. Academic Press Orlando, FL.

Dershowitz B, Eiben T (1999) SR-97 Alternative models project. Discrete fracture network modelling for perfomance assessment of Aberg. SKB Technical report 99/43. Stockholm.

Endo HK, Long J, Wilson CK, Witherspoon P (1984) A model for investigating mechanical transport in fractured media. Water Resources Research 10:1390–1400.

Guérin F, Billaux DM (1994) On the relationship between connectivity and the continuum approximation in fracture flow and transport modelling. Applied hydrogeology 3:24–31.

Herbert AW, Lanyon GW (1995) Fracture networks and the use of the REV concept. AEA Technology report NSS/R358 Harwell.

Hestir K, Long JCS (1990) Analytical expressions for the permeability of random two-dimensional poisson fracture networks based on regular lattice percolation and equivalent media theories. Journal of Geophysical Research 21565–21581.

Long JCS, Remer JS, Wilson CR, Witherspoon PA (1982) Porous media equivalents for networks of discontinuous fractures. Water Resources Research 3:645–658.

McDonald MG, Harbaugh AW (1988) A modular three dimensional finite difference groundwater flow model. USGS report USGS Techniques of water resource investigations TWRI Book 6-chapter A1 Washington D.C.

National Research Council (1996) Rock fractures and fluid flow: contemporary understanding and applications. National Academy Press Washington, D.C.

Neuman SP (1990) Universal scaling of hydraulic conductivities and dispersivities in geologic media. Water Resources Research 8:1749–1754.

Outters N, Shuttle D (2000) Sensitivity analysis of a discrete fracture network model performance assessment of Aberg. SKB report R-00-48 Stockholm.

Painter S (1999) Long-range spatial dependence in fractured rock – empirical evidence and implications for tracer transport. SKB report 34.

Parney R, Smith L (1995) Fluid velocity and path length in fractured media. Geophysical Research Letters 11:1437–1440.

Parney RW (1999) Statistical continuum modelling of mass transport through fractured media, in two and three dimensions. PhD, Univ. of British Columbia, Canada.

Pollock DW (1989) Documentation of computer programs to compute and display pathlines using results from the USGS modular 3D finite difference groundwater model. USGS open file report 89–381.

Rayne T, Bradbury KR, Muldoon MA (2001) Delineation of capture zones for municipal wells in fractured dolomite, Sturgeon bay, Wisconsin, EUA. Hydrogeology Journal 5:432–450.

Robinson JN, Barker JA (1999) A fractured/fissured approach to groundwater protection zones. Draft Project Record. Jackson Environmental Institute. University College London report W6D(96)2 London.

Robinson PC (1984) Connectivity, flow and transport in network models of fractured media. PhD, Oxford University.

Rouleau A (1988) A numerical simulator for flow and transport in stochastic discrete fracture networks. National Hydrology Research Institute report Saskatoon, Saskatchewan.

Schwartz FW, Smith L (1988) A continuum approach for modelling mass transport in fractured media. Water Resources Research 8:1360–1372.

Smith L, Clemo T, Robertson M (1990) New approaches to the simulation of the field-scale solute transport in fractured rocks. Paper presented at the 5th Canadian/American Conference on Hidrogeology, Dublin (Ohio).

van Leeuwen M (2000) Stochastic determination of well capture zones conditioned on transmissivity data. PhD, Imperial College London.

Weisstein EW (1999) CRC concise encyclopedia of mathematics. CRC Press Boca Raton, Fla.

Zimmerman RW, Bodvarsson GS (1996) Effective transmissivity of two-dimensional fracture networks. Int J Rock Mech and Min Sci and Geomech Abs 4:433–438.

CHAPTER 28

# Interconnected flow measurement between boreholes using a borehole flowmeter

Lasse Koskinen[1] and Pekka Rouhiainen[2]
[1]*VTT Technical Research Centre of Finland, VTT Processes, Finland*
[2]*PRG-Tec Oy, Espoo, Finland*

ABSTRACT: This paper introduces a characterization method for groundwater flow in fractured crystalline rocks that combines a sophisticated experimental set-up and advanced numerical flow modeling methodology. The experimental part included an interference test consisting of a few separate pumping tests that were carried out in a group of five boreholes in gneissic rock. In addition to them, the pumping responses were also observed in another four shallow boreholes. A sensitive flowmeter was used to measure steady-state pumping responses. "Background" flow rates (i.e., flow rates without pumping in any borehole) were first measured. Thereafter, one borehole was pumped at a time and steady-state flow responses were measured in all boreholes, including the pumped borehole. The boreholes were open during all the tests and flow rates were measured with two meters intervals. The numerical flow modeling part makes use of a number of assumptions which are discussed together with some preliminary results.

## 1 INTRODUCTION

Characterization of groundwater flow properties of fractured crystalline hard rocks in countries like Finland and Sweden is motivated, in particular, by the plans of disposing of nuclear waste, including spent fuel in the deep rock repositories. Also in this context, this has proven to be a challenging task, while, generally speaking, characterization means have developed in two fronts: field testing and modeling. In this paper we introduce an investigation method that is able to produce large amounts of measurement data connected to the hydrogeologic properties within a block of rock embraced by multiple boreholes. The method is based on the use of the Posiva flowmeter, which is an advanced and sensitive instrument to measure groundwater flow in open boreholes in fractured crystalline rocks. Due to the complexity of a three-dimensional fracture system and the large body of measurement data, the characterization of the rock properties embraced by the boreholes calls for numerical tools. In this study we will apply the Connectflow software (Hartley et al., 2001) which was designed to compute parameters corresponding to the Posiva flowmeter measurements in the open boreholes.

## 2 THE SITE

Olkiluoto is an island (approx. 10 km$^2$), separated from the mainland by a narrow strait, on the coast of the Baltic Sea. The Olkiluoto nuclear power plant with two reactors in operation and a third one under construction, and the VLJ repository for low and intermediate waste are located in the western part of the island. The repository for spent fuel will be constructed in the central part of the island (Figure 1). The suitability of Olkiluoto for a location of a spent fuel repository has been investigated over fifteen years by means of ground- and air-based methods and from shallow and deep (300–1000 metres) boreholes.

Olkiluoto Island is quite flat, and only 4% of it is represented by outcrops of bedrock. The highest point of Olkiluoto is about 18 m above sea level. The outcrops are mainly granitic or grey gneisses, but investigation trenches have shown that migmatitic mica gneisses are the most abundant rock types (Posiva, 2003). On the site scale, the rock also is characterized with a few tens of identified fracture zones between which the rock mass is sparsely fractured.

## 3 THE FLOWMETER

Unlike traditional types of borehole flowmeters, the Difference flowmeter method measures the flow rate into or out of limited sections of the borehole instead of measuring the

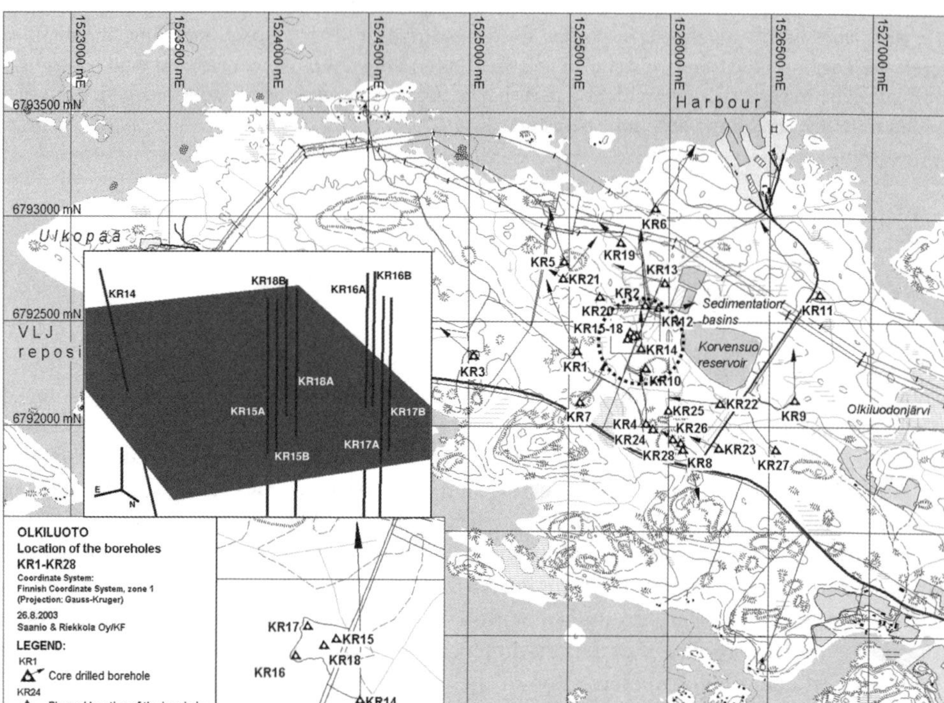


Figure 1. Deep boreholes and the location of the underground rock characterisation facility ONKALO at Olkiluoto. The size of the grid squares is 500 × 500 m$^2$. The inserted figure shows the boreholes employed in the interference test in a 3D sketch together with the plane of an imaginary "fracture" to aid the eye. The dashed circle indicates the area where boreholes KR14 to KR18A, B are located.

total cumulative flow rate along the borehole. The advantage of measuring the flow rate in isolated sections is a better detection of the incremental changes of flow along the borehole, which are generally very small and can easily be missed using traditional types of flowmeters (Rouhiainen, 2001).

Rubber disks at both ends of the downhole tool are used to isolate the flow in the test section from that in the rest of the borehole, see Figure 2. The flow along the borehole outside the isolated test section passes through the test section by means of a bypass pipe and is discharged at the upper end of the downhole tool (Öhberg & Rouhiainen, 2000). Flow is measured using the thermal pulse and/or thermal dilution methods. Measured values are transferred in digital form to the PC computer. The winch for the 1500 m long cable is installed on a trailer.

Flow can be measured in many ways and logging speed varies according to chosen system. A typical logging speed with a 0.1 m point interval is 10 m/hour.

Besides incremental changes of flow, the downhole tool of the Difference flowmeter can be used to measure:

- The electric conductivity (EC) of the borehole water and fracture-specific water. The electrode for the EC measurements is placed on the top of the flow sensor, Figure 2.
- The single point resistance (SPR) of the borehole wall (grounding resistance), The electrode of the single point resistance tool is located in between the uppermost rubber disks, see Figure 2. This method is used for high resolution depth/length determination of fractures and geological structures.
- The prevailing water pressure profile in the borehole (not used in this study). The pressure sensor is located inside the electronics tube and connected via another tube to the borehole water.

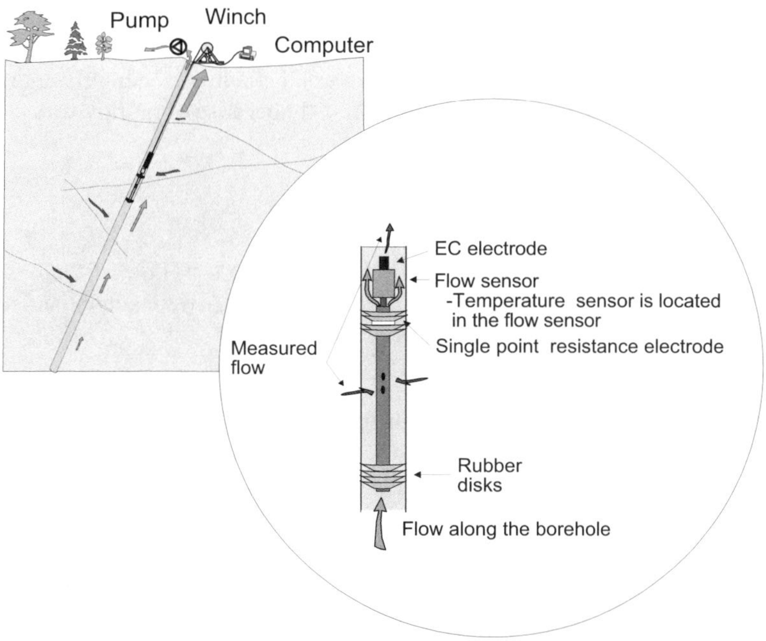


Figure 2. A schematic visualization of the downhole equipment used in the difference flowmeter.

Table 1. Equipment specifications.

| | |
|---|---|
| Type of instrument | Posiva Flow Log/Difference Flowmeter |
| Borehole diameters | 56 mm, 66 mm and 76–77 mm |
| Length of test section | A variable length flow guide is used |
| Method of flow measurement | Thermal pulse and/or thermal dilution |
| Additional measurements | Temperature, single point resistance, electric conductivity of water, caliper, water pressure |
| Winch | Mount Sopris Wna 10, 0.55 kW, 220 V/50 Hz 220 V/50 Hz. Steel wire cable 1500 m, four conductors, Gerhard -Owen cable head |
| Length determination | Based on the marked cable and on the digital length counter |
| Logging computer | PC, Windows 2000 |
| Total power consumption | 1.5–2.5 kW depending on the pumps |

Table 2. Range and accuracy of sensors.

| Sensor | Range | Accuracy |
|---|---|---|
| Flow | 6–300 000 ml/h | +/− 10% curr.value |
| Temperature (middle thermistor) | 0–50°C | 0.1°C |
| Temperature difference (between outer thermistors) | −2–+2°C | 0.0001°C |
| Electric conductivity of water (EC) | 0.02–11 S/m | +/− 5% curr.value |
| Single point resistance | 5–500 000 Ω | +/− 10% curr.value |
| Groundwater level sensor | 0–0.1 Mpa | +/− 1% fullscale |
| Absolute pressure sensor | 0–20 MPa | +/− 0.01% fullscale |

- Temperature of the borehole water. The temperature sensor is placed in the flow sensor, Figure 2.

The lowest flow rate (6 ml/h) can be measured in good conditions using the thermal pulse method. In practice this is not always possible, mostly because of muddy water.

## 4 BOREHOLES IN THE INTERFERENCE TEST

Altogether nine boreholes were utilized in the interference test. Five of them were deep, reaching deeper than 125 m. Four of them, KR15A to KR18A were accompanied by shallow boreholes, named with a respective suffix "B" (Table 3).

## 5 MEASUREMENTS IN THE INTERFERENCE TEST

The interference test was carried out from 4 January 2001 till 21 March 2002 (Rouhiainen & Pöllänen, 2003). Each of the deep boreholes – KR14, KR15A, KR16A, KR17A, and KR18A – was pumped at a time such that the pumping followed a recovery period lasting for a few days (Table 4). The pumping drawdown in was about 10 m except for KR14 in which a large pumping rate (25 l/min) was obtained with a clearly weaker pumping drawdown (6 m). The whole duration of the experiment was about three months. A steady state condition with respect to water levels in the boreholes was attempted; however, the pumping rates

Table 3. Summary of borehole specifications. The ground surface entry point of the boreholes is located about 9 m above sea level. The diameter of the boreholes is 76 mm.

| Borehole | Length [m] | Azimuth | Inclination | Casing [m] |
|---|---|---|---|---|
| KR14 | 514 | 0° | 70° | 9.5 |
| KR15A | 234 | Vertical | | 40 |
| KR15B | 45 | Vertical | | 4.5 |
| KR16A | 170 | Vertical | | 40 |
| KR16B | 45 | Vertical | | 4.5 |
| KR17A | 157 | Vertical | | 40 |
| KR17B | 45 | Vertical | | 4.1 |
| KR18A | 125 | Vertical | | 40 |
| KR18B | 46 | Vertical | | 6.5 |

Table 4. Summary of basic data on the interference test.

| | KR14 | KR15A | KR16A | KR17A | KR18A |
|---|---|---|---|---|---|
| Start date | 12 Mar. | 4 Jan. | 21 Jan. | 4 Feb. | 25 Feb. |
| End date | 21 Mar. | 18 Jan. | 1 Feb. | 19 Feb. | 7 Mar. |
| Duration of pumping [days] | 9 | 12 | 11 | 15 | 10 |
| Recovery time [days] | 14 | 3 | 3 | 6 | 5 |
| Pumping rate [l/min] | 25 | 6.7 | 5.9 | 7.7 | 5.3 |
| Pumping drawdown [m] | 6 | 10 | 10 | 11 | 10 |
| Remark | Last pumping borehole | First pumping borehole | | | A change in the pumping rate from 6.8 to 5.3 l/min |

in KR17A and KR18A were not quite steady but gradually lowering to the values given in Table 4.

## 6 OBSERVED FRACTURING AND TRANSMISSIVITY

Fracturing in the boreholes has been mapped both in the borehole core logs and televiewer imagery (Wild et al., 2002; Julkunen et al., 2003). An example of a borehole image is shown in Figure 3. Based on the observations, the fracturing in the bedrock is much more intensive in the upper 50 m of the boreholes. This is reflected in Table 5 such that the number of transmissive features in the shallow "B" boreholes (with depths down to about 40 m) is roughly the same as those in much deeper "A" boreholes (in which there is the 40-m top casing). Currently these data have provided information about fracture orientations, which seem to support a predisposition to horizontal to subhorizontal directions.

Before the actual interference test each borehole was scanned with the flowmeter for transmissive fracture intersections (Pöllänen & Rouhiainen, 2001, 2002). The results from these single-borehole measurements are summarized in Table 5, which covers transmissivities higher than $4 \times 10^{-8}\,m^2/s$. They roughly obey a log-normal distribution, with a mean of about $10^{-6}\,m^2/s$ and standard deviation in the log-space of 3.

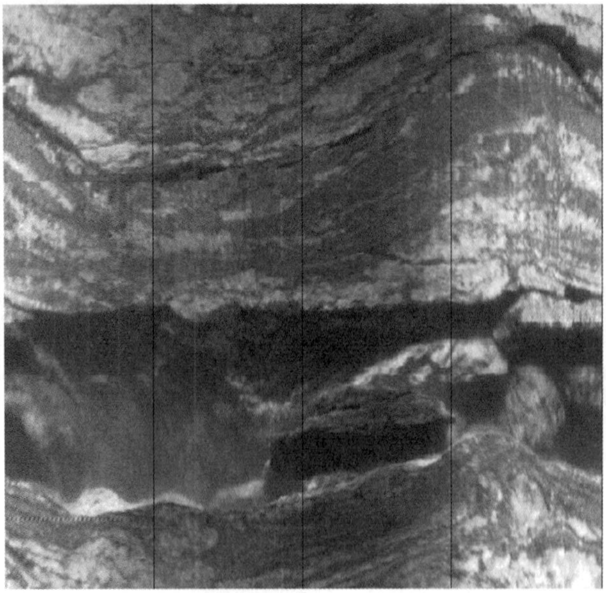

Figure 3. A 7-cm opening in the televiewer imagery from borehole KR14 at a depth of about 50 m (along the borehole). This crack is connected to the high measured transmissivity of $4 \times 10^{-5}$ m$^2$/s (see Table 5). The total height of the figure is about 20 cm. The image covers the whole 360° of borehole wall of the diameter of 76 mm.

Table 5. Determined transmissivities higher than $4 \times 10^{-8}$ m$^2$/s of the fractures intersecting the boreholes. The number of all fractures, i.e., the whole population also consisting in less transmissive fractures, is much higher than those in this table. The first column of the table is the order of the intersection in each borehole from the top. The data is given by a pair of numbers such that first is the depth (along the borehole) and the second is the transmissivity in unit of $10^{-6}$ m/s. For example, at a depth of 50 m there is a fracture having a transmissivity of $4 \times 10^{-5}$ m$^2$/s. The depth values are measured along the boreholes but these deviate from the actual, linear depth only for borehole K14 as it is not due vertical but has a dip of 70 degrees. Measurements for borehole KR17B did not succeed. Naturally, the 40-metre top casings in the "A" boreholes make a detection of any transmissive fracture above that depth impossible.

| | KR14 | KR15A | KR16A | KR17A | KR18A |
|---|---|---|---|---|---|
| I | 14, 2 | 42, 2 | 43, 2 | 45, 0.4 | 51, 4 |
| II | 42, 0.8 | 61, 4 | 49, 6 | 51, 2 | 56, 0.8 |
| III | 50, 40 | 73, 0.4 | 57, 0.6 | 68, 6 | 63, 0.04 |
| IV | 80, 0.8 | – | 71, 0.04 | 128, 0.6 | 78, 0.08 |
| V | – | – | 79, 0.2 | 132, 0.06 | – |
| VI | – | – | 87, 0.04 | – | – |
| VII | – | – | 152, 0.2 | – | – |
| | | KR15B | KR16B | KR17B | KR18B |
| I | | 12, 1.4 | 18, 1.4 | – | 10, 0.6 |
| II | | 15, 1.6 | 24, 0.6 | – | 16, 2 |
| III | | 22, 14 | 30, 0.4 | – | 22, 6 |
| IV | | 34, 4 | 38, 0.4 | – | 32, 18 |
| V | | 42, 4 | 42, 1.0 | – | |

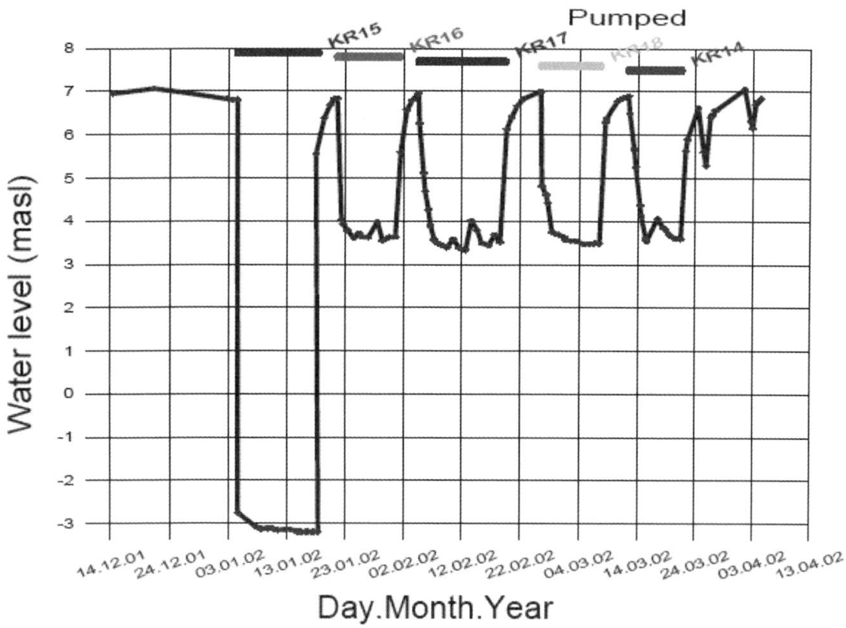


Figure 4. Water level in borehole KR15 during the interference experiment. The first stage of pumping took place in this borehole as is reflected the high change in the water level in it. The experiment ended with a recovery stage after the borehole KR14 pumping in March 2002.

Table 6. Measured steady-state drawdowns. The pumping borehole is in the first column. For example, pumping of borehole KR15A brought about a drawdown of 1.0 m in borehole KR14 and 1.3 m in borehole KR18B.

| | KR14 | KR15A, B | KR16A, B | KR17A, B | KR18A, B |
|---|---|---|---|---|---|
| KR14 | 6.0 | 3.4, 4.4 | 3.0, 3.0 | 3.0, 0.2 | 3.0, 5.6 |
| KR15A | 1.0 | 10, 2.0 | 3.5, 2.4 | 3.0, 1.0 | 3.2, 1.3 |
| KR16A | 0.9 | 3.3, 1.0 | 10, 1.7 | 3.4, 0.4 | 3.7, 0.9 |
| KR17A | 1.0 | 3.5, 2.0 | 3.9, 1.5 | 11, 0.2 | 3.7, 0.9 |
| KR18A | 0.8 | 3.5, 0.9 | 3.5, 1.5 | 3.2, 0.2 | 10, 0.8 |

## 7 RESPONSES IN THE INTERFERENCE TEST

The hydraulic head response, i.e. the drawdown was determined as the drop in the water level in the (open) boreholes. An example is shown in Figure 4, which shows that reasonably steady conditions were achieved during the pumping. It is also seen, that the recovery after each pumping phase took a few days. The drawdowns from all pumping phases are compiled in Table 6.

It is striking from Table 6 that the drawdown responses are so high. In particular it is curious that the responses in boreholes KR14 and KR18B are almost identical – while the distance of the boreholes from each other is about 80 m. The drawdowns in the two boreholes are very similar to the extent that even the pumping in borehole KR14 brought about almost the same drawdown in KR18B as the pumping drawdown itself. Only a very good hydrogeologic connection could possibly be behind such a strong pumping response. Because of this reasoning, we incorporated such a horizontal zone at a depth of a few tens of metres to see its effect in comparison with the case without such pre-determined fracture. As a very high transmissivity is to be associated with this fracture one would expect it to be visible in televiewer imagery as is depicted in Figure 3 showing a crack of the width of up to about 7-cm in borehole KR14 at the depth of 50 m (borehole depth). Besides this knowledge, the location of the pre-determined fracture was decided upon the basis of some other conclusions drawn from flow rate measurements as is explained below.

Presumably, the pumping drawdowns are transmitted from the pumping borehole to the observation boreholes by a fracture or fractures that have a high transmissivity. From the Posiva flowmeter readings we are able to identify those spots in the boreholes. Such data are compiled in Tables 7 and 8. These data clearly show the concentration of the response on a few places in the boreholes only. The numbers in the tables reveal some interesting patterns in the hydrogeologic connections. Let us, for example, take a look at flow rates measured in boreholes KR14 and KR15B (Table 7). We see that when borehole KR14 is pumped the largest water flow – by far – into this borehole takes place at a depth of 50 m. This is completely in line with the high transmissivity measured in the same place. However, when borehole KR15B is pumped we see that there is not an outflow detected in this spot but an inflow. An outflow spot is located some 30 m deeper. As a strong outflow in an observation borehole is necessarily associated with a good hydraulic connection between it and the pumping borehole, we conclude that the highly transmissive fracture that intersects borehole KR14 at the depth of 50 m can *not* form a hydrogeologic connection between two boreholes. Instead this connection is formed by a fracture that intersects borehole KR14 at the depth of 80 m and by a fracture that intersects borehole KR15A at a depth of 42 m. We do not know and can not know based on the available data whether this connection is formed by one or more fractures.

One issue related to quality of measurements is the balance of the water flowing into and out of the *observation* boreholes. In a perfect case the water balance, the "net flow" in Table 7 in the observation boreholes, should naturally vanish. It is seen that this condition is fairly well met in borehole KR16 and KR18B (excluding the pumping in KR14 for the latter), and also in borehole KR18A except when borehole KR15A was pumped. Otherwise the boreholes show less than perfect "conservation of volumetric flow". In particular, it is remarkable that the net flow is positive (except in borehole KR18A in which also mild negative net flows are obtained) indicating that more water is entering the borehole than is leaving it. Of course this cannot happen and an explanation should be found. One possibility could be leaking top casings, which are steel tubes. As the topmost horizon of the Olkiluoto bedrock is highly fractured with a predisposition in the horizontal direction, it is possible that in the case of a leaky top casing a hydraulic connection concealed by it is able to remove water from the observation borehole without becoming detected by the flowmeter. A similar thing can also happen in the pumping borehole as the horizontal orientation of such a hydraulic connection prevents it from being detected in the pumping borehole either. Actually this hypothesis is also supported by that fact that in the pumping boreholes, the measured total flow rate and applied pumping flow rate (Table 4) disagree somewhat – the former is smaller – in particular for borehole KR14 (that has the shallowest top casing).

Table 7. Compilation of the Posiva flowmeter measurements in the interference experiment for the "A" boreholes. The first column indicates the location of transmissive feature from the top of the borehole. The second column is the transmissivity (Table 5) at the intersection of the transmissive feature and the borehole. A positive flow rate means water flows into the borehole and negative means the opposite. The flow rates are expressed in ml/hour. "NP" stands for No Pumping, i.e., the flow rate without the influence of pumping in any borehole. The response to the pumping is then determined by comparing the values in this column with the other columns. For example (see the marked cells), in borehole KR14, the flow at the depth of 80 m is 0.9 l/hour out of the borehole. When borehole KR15A was pumped this flow was increased to about 32 l/hour out of the borehole.

| | $T/10^{-6}$ [$m^2$/s] | NP | KR14 | KR15A | KR16A | KR17A | KR18A |
|---|---|---|---|---|---|---|---|
| **KR14** | | | | | | | |
| I (14 m) | 2 | 3,200 | 62,600 | 12,100 | 12,400 | 13,800 | 12,900 |
| II (28 m) | 2 | 3,100 | 79,700 | 4,800 | 4,800 | 2,900 | 3,800 |
| III (42 m) | 0.8 | 0 | 14,300 | 300 | 300 | 100 | 200 |
| IV (50 m) | 40 | −2,400 | 783,600 | 18,000 | 8,800 | 4,100 | 10,500 |
| V (80 m) | 0.8 | −900 | 17,500 | −31,800 | −20,800 | −18,300 | −20,900 |
| Net flow | – | 3,000 | 957,700[1] | 3,400 | 5,500 | 2,600 | 6,500 |
| **KR15A** | | | | | | | |
| I (42 m) | 2 | 3,200 | −12,600 | 95,000 | 16,800 | 15,600 | 14,900 |
| II (61 m) | 4 | −3,170 | 14,400 | 224,200 | −12,300 | −10,600 | −12,500 |
| III (73 m) | 0.4 | 0 | 100 | 3000 | 0 | −800 | 0 |
| Net flow | | 38 | 1,900 | 342,200[2] | 4,500 | 4,200 | 2,400 |
| **KR16A** | | | | | | | |
| I (43 m) | 0.2 | 100 | −300 | 900 | 10,900 | 500 | 600 |
| II (49 m) | 6 | 1,100 | −2,000 | −1,800 | 231,000 | −5,300 | −300 |
| III (57 m) | 0.6 | −100 | −300 | −2,300 | 26,500 | 700 | −1,100 |
| IV (71 m) | 0.04 | 0 | 100 | 0 | 3,400 | 0 | −700 |
| V (79 m) | 0.2 | −200 | 800 | 1,000 | 7,300 | 1,400 | −600 |
| VI (87 m) | 0.04 | 0 | 300 | 300 | 1,700 | 600 | 400 |
| VII (152 m) | 0.2 | −500 | 1,100 | 1,600 | 6,700 | 2,100 | 1900 |
| Net flow | – | 500 | −300 | −300 | 287,500[3] | 0 | 200 |
| **KR17A** | | | | | | | |
| I (45 m) | 0.4 | 200 | −200 | −100 | −300 | 26,300 | −100 |
| II (51 m) | 2 | 3,800 | −1,800 | −9,000 | −20,100 | 119,600 | −17,100 |
| III (68 m) | 6 | 1,200 | −5,500 | 6,900 | 16,200 | 222,000 | 12,300 |
| IV (128 m) | 0.6 | −4,900 | 7,600 | 4,300 | 9,800 | 53,700 | 10,200 |
| V (132 m) | 0.06 | 0 | 0 | 0 | 200 | 1,900 | 200 |
| Net flow | – | 300 | 100 | 2,100 | 5,800 | 423,500[4] | 5,500 |
| **KR18A** | | | | | | | |
| I (51 m) | 4 | 200 | 1,200 | 11,000 | −900 | −4,200 | 238,200 |
| II (56 m) | 0.8 | 200 | −2,600 | −11,300 | 500 | 1,500 | 49,800 |
| III (63 m) | 0.04 | 0 | −200 | −1,100 | 0 | 0 | 2,900 |
| IV (78 m) | 0.8 | −600 | 1,800 | 3,200 | 0 | 2,400 | 31,800 |
| Net flow | – | −200 | 200 | 1,800 | −400 | −300 | 322,700[5] |

(1) The actual pumping rate from Table 4 is 1.6 times the total flow log measurements.
(2) The actual pumping rate from Table 4 is 1.2 times the total flow log measurements.
(3) The actual pumping rate from Table 4 is 1.2 times the total flow log measurements.
(4) The actual pumping rate from Table 4 is 1.1 times the total flow log measurements.
(5) The actual pumping rate from Table 4 matches to the total flow log measurements.

Table 8. Compilation of the Posiva flowmeter measurements in the interference experiment for the shallow, "B", boreholes. The first column indicates the location of transmissive features from the top of the borehole. The second column is the transmissivity determined at the intersection of a transmissive feature and the borehole. A positive flow rate means water flows into the borehole and negative means the opposite. The flow rates are expressed in ml/hour. "NP" stands for No Pumping, i.e., the flow rate without the influence of pumping in any borehole. The response to the pumping is then determined by comparing the values in this column with the other columns. While this summary flow measurements in the "B" boreholes is given here, these were not taken into account in the preliminary study described by this paper (except for the clear hydrogeologic connection between boreholes KR14 and KR18B whose existence is corroborated by the strong observed responses in the latter to the pumping former borehole as discussed in the text). However, the "B" boreholes were taken into account as highly conductive one-dimensional hydrogeologic features.

| | $T/10^{-6}$ [$m^2/s$] | NP | KR14 | KR15A | KR16A | KR17A | KR18A |
|---|---|---|---|---|---|---|---|
| **KR15B** | | | | | | | |
| I (12 m) | 1.4 | 1,000 | 10,800 | 5,700 | 3,000 | 3,800 | 3,200 |
| II (15 m) | 1.6 | 400 | 3,600 | 4,800 | 1,800 | 2,800 | 2,400 |
| III (22 m) | 17 | 2,500 | −13,300 | 31,900 | 4,400 | 14,800 | 11,900 |
| IV (34 m) | 4 | −800 | −15,100 | 0 | −5,000 | −5,600 | −3,700 |
| V (42 m) | 4 | −3,200 | 15,800 | −38,200 | −2,200 | −2,100 | −11,300 |
| Net flow | – | −100 | 1,800 | 4,200 | 2,000 | 13,700 | 2,500 |
| **KR16B** | | | | | | | |
| I (18 m) | 1.4 | 1,100 | 2,900 | 7,100 | 6,300 | 5,000 | 4,100 |
| II (24 m) | 0.6 | 700 | −4,100 | 4,700 | 4,200 | 3,600 | 2,300 |
| III (30 m) | 0.4 | 100 | −1,000 | −600 | −1,800 | −2,400 | −2,100 |
| IV (38 m) | 0.4 | −200 | 100 | −1,700 | −2,600 | −2,800 | −2,300 |
| V (42 m) | 1.0 | −800 | 100 | −3,400 | −3,400 | −300 | −600 |
| Net flow | – | 900 | 2,000 | 6,100 | 2,700 | 3,100 | 1,400 |
| **KR17B** | | | | | | | |
| I | ?? | | | | | | |
| II | ?? | | | | | | |
| III | ?? | | | | | | |
| Net flow | – | | | | | | |
| **KR18B** | | | | | | | |
| I (10 m) | 0.6 | 200 | 4,000 | 600 | 700 | 600 | 600 |
| II (16 m) | 2 | 1,300 | 14,800 | 1,700 | 2,000 | 1,900 | 1,600 |
| III (22 m) | 6 | 1,400 | 21,300 | 1,900 | 2,000 | 1,300 | 1,000 |
| IV (32 m) | 18 | −2,800 | −33,200 | −3,700 | −4,100 | −3,300 | −2,900 |
| Net flow | | 100 | 6,900 | 500 | 600 | 500 | 300 |

## 8 APPROACH TO THE ANALYSIS

The analysis of the interference test aims at understanding the observed interference test responses from the basis of the bedrock fracturing at Olkiluoto. The approach to the analysis – carried out in two phases – is based on stochastic fracture network flow simulations with the Connectflow software (Hartley et al., 2001). The stochastic realizations also covered the case for one deterministic fracture connecting boreholes KR14 and KR18B. The existence of such fracture was recognized upon the basis of the strong response in the latter to the pumping in the former and the very similar drawdowns in the both boreholes in general. The statistical parameters for fracture orientations were adopted from studies by Poteri & Laitinen (1999) and Poteri (2001), which make use of the data from deeper than 200 m in deep boreholes. They inferred orientations that show some predisposition to horizontal fracture directions. For the current study, the fracture intensity was fixed to roughly yield the number of observed *flowing* fractures in the boreholes (i.e., the number roughly corresponding to the number of the fractures in Table 5). It is noteworthy that this fracture intensity is much smaller that the fracture intensity determined by Poteri & Laitinen (1999) and Poteri (2001). This is remarkable in the sense that they only used borehole data from deeper that 200 m to determine the fracture intensity, but without a possibility of making use of direct observation of water flow in the fractures. In other words, the rock mass that is supposed to be much less fractured than the topmost 50 m, for example, actually has a far larger fracture intensity that the intensity of fractures bearing a dominant part of water flow in the topmost 50 m of the bedrock at Olkiluoto. For the fracture size a power law distribution was specified such that it yielded fractures large enough to span over the distances separating the boreholes but with an exponent (3.0) that led to a quickly decreasing frequency of larger fractures. Among them was a coupling between fracture size and transmissivity similar to that in studies by SKB (2004). This defines a simple and direct dependency between the fracture extent, L, expressed in metres, and fracture transmissivity:

$$T = aL^b \ [m^2/s].$$

In the numerical groundwater flow simulations in this study, we used $a = 5 \times 10^{-7}$; $b = 3$; and $a = 7 \times 10^{-7}$; and $b = 1.8$. These are meant to serve as first trials; no exhaustive probing for "proper" values has been done. This choice of the form of the dependency may be criticized for being too simplistic – there surely are large fractures that would not show such a remarkable *effective* transmissivity; and smaller ones with a clearly higher transmissivity than that representing the simple dependency. However, while this may be true, one still might expect a positive correlation between the fracture size and fracture transmissivity. Furthermore, albeit formally the simple relationship has no constraint on the values of the transmissivity – however large a value would be produced – in our case, the upper limit of the transmissivity is implicitly imposed by setting an upper limit to the fracture size (100 m).

The starting point of the first phase is to assign the pumping drawdown (rather than pumping flow rate) as the pumping boundary condition in a pumping borehole – say, borehole KR14. Then the drawdowns in the observation boreholes are calculated and compared with the corresponding measured values. If the agreement is deemed fair (formally, an absolute value of the sum of differences between the observed and computed drawdowns in the observation boreholes was required to be less than a criterion value) the realization is exposed to boundary conditions modeling the pumping in the "next" borehole (e.g., borehole KR15A) and so on. For simplicity the measurements for the "B" boreholes were not taken into account in this preliminary study – except for the knowledge that supported the existence of a very good hydrogeologic connection between boreholes KR14 and KR18B. The "B" boreholes, nevertheless, were included in the modeling as one-dimensional highly conductive features. Finally, in the case of a "good" realization, i.e. a realization that shows reasonable coincidence with the drawdown measurements for each separate pumping test, is tested for the observed flow rates in the second phase of the analysis.

## 9 DISCUSSION ON THE PRELIMINARY RESULTS

Based on the analysis of the observations in connection to the interference test, it is evident that the connected network – the group of fractures forming the connections transmitting the pumping responses – is much sparser than that based on the observer total bedrock fracturing. Furthermore numerical modeling supports the idea that the connected network is essentially formed by fractures whose size is of the order of the distance of the boreholes from each other, and at the same time, it seems favorable to assume a correlation between the fracture size and fracture transmissivity. This probably is a generic result. As the observed responses – the responses which must be transmitted by good hydrogeologic connections – over the applied separations have to be realized in numerical simulations one most straightforward way of doing this is the simplest possible connection formed by a minimum number of connecting fractures. Clearly, the easiest way of realizing this is with one large highly transmissive fracture that spans between the boreholes. In our study, the direct relation between fracture transmissivity and fracture size ensured that such a fracture was assigned with a high transmissivity. On the other hand, supposedly a similar kind of connection could, in principle, be realized with a "chain" of smaller conductive fractures, but a simple and straightforward stochastic generation of realizations would produce these "chains" very infrequently.

One interesting aspect explored is that of the impact of the deterministic feature. This fracture was motivated by the evident good hydrogeologic connection between boreholes KR14 and KR18B. Our findings from numerical simulations seem to strongly support the existence of such a fracture – it greatly increases the likelihood of occurrence of a realization for which the calculated response is much closer to the observations than the realizations without such a fracture. Interestingly, the existence of such a horizontal feature has got independent corroboration from analyses on other field investigations at the Olkiluoto site.

## 10 OUTLOOK

This tentative study could clearly be expanded in a few directions. The first and most straightforward step is to make use of the knowledge of the orientations of the fractures in the boreholes employed in the interference test as this information has now become available. Furthermore, the body of data currently collected from the boreholes would offer a basis to extensively condition the stochastic realization. The location, orientation and transmissivity of each fracture intersecting any of the boreholes have been determined now. Conditioning each borehole with its observed fracturing is possible but calls for some further development of the modeling tool.

In addition, a technique to realize a positive correlation between fracture size and transmissivity while maintaining a lognormal structure of the transmissivity statistics should be developed. Formally this would probably be easily achieved if a one-to-one correspondence of the fracture size and transmissivity is to be required but it would be more difficult if some (stochastic) variability were to be allowed. There would be rather little actual data to support determining the correlation, though, but proper sensitivity analyses could be carried out.

Ultimately, there is the difficult issue of the internal heterogeneity of the fractures at Olkiluoto. It may be argued that for the drawdown, this heterogeneity is not so important provided that an effective value for the fractures can be determined. However, we would expect the Posiva flowmeter measurements to depend on the internal heterogeneity much more sensitively. Unfortunately, any attempt to take the internal heterogeneity into account is made difficult by the demands it imposes on the computer resources but also – and more importantly – there is scant *in-situ* information to offer a solid basis to quantify this heterogeneity. While the lack of hard data can formally be dealt with through proper sensitivity and uncertainty analyses, this would also enlarge the burden on the computer resources.

It would be of course very interesting to test the ability of the inferred fracture zone model to predict the outcome of some other field test carried out within the same block of rock. Actually such a test of different character has been carried out at the site. This test represents a traditional pressure drawdown test with multiple double packers. Further investigation of this test along with the lines discussed here is yet to be done.

## REFERENCES


Hartley, LJ, Cliffe, KA, Herbert, AW, Shepley, MG and Wilcock, PM. 2001. *Groundwater flow modelling with a combined discrete fracture network and continuum approach using the code CONNECTFLOW*, AEA Technology ReportAEA/ENV/R/0520.

Julkunen, A, Kallio, L, Lehtonen, T and Välitalo, J. 2003. Optical imaging in boreholes KR7, KR10, KR11, KR15–19 and KR15B–KR18B at Olkiluoto site, 2002. Posiva Oy, Eurajoki. Working Report 2003-50.

Posiva 2003. *Baseline conditions at Olkiluoto.* Posiva Oy, Report POSIVA 2003-02.

Poteri, A and Laitinen, M. 1999. *Site-to-canister scale flow and transport in Hästholmen, Kivetty,* Olkiluoto and Romuvaara. Posiva Oy, Report POSIVA 99-15.

Poteri, A. 2001. *Estimation of the orientation distributions for fractures at Hästholmen, Kivetty,* Olkiluoto and Romuvaara. Posiva Oy, Working Report 2001–10.

Pöllänen, J and Rouhiainen, P. 2001. *Difference flow and electric conductivity measurements at the Olkiluoto site in* Eurajoki, *boreholes KR13 and KR14*. Posiva Oy, Working Report 2001–42.

Pöllänen, J and Rouhiainen, P. 2002. *Difference flow and electric conductivity measurements at the Olkiluoto site in Eurajoki, boreholes KR15–KR18 and KR15B–KR18B*. Posiva Oy Working Report 2002–29.

Rouhiainen, P and Pöllänen, J. 2003. *Hydraulic crosshole interference tests at the Olkiluoto site in Eurajoki, boreholes KR14–KR18 and KR15B–KR18B*. Posiva Oy, Working Report 2003–30.

Rouhiainen, P. 2001. Posiva groundwater flow measuring techniques. In: *Proceedings of the XXXI International Association of Hydrogeologists Congress Munich/Germany/*10–14 September 2001. New Approaches Characterizing Groundwater Flow Volume 2, ISBN 90 2651 850 1.

SKB 2004. Preliminary *site description Forsmark area – version 1.1*. Swedish Nuclear Fuel and Waste Management Co (SKB), Report R-04-15.

Wild, P, Siddans, A and Kennaugh, K. 2002. *Optical Televiewer survey and processing in Olkiluoto site, Finland 2001. Boreholes KR9, KR12, KR13 and KR14*. Posiva Oy, Working Report 2002–02.

Öhberg, A and Rouhiainen, P. 2000. *Posiva groundwater flow measuring techniques.*, Posiva Oy. Report POSIVA 2001–12.

CHAPTER 29

# Characterizing flow in natural fracture networks: comparison of the discrete and continuous descriptions

Tanguy Le Borgne*, Olivier Bour, Jean Raynald De Dreuzy and Philippe Davy

*Géosciences Rennes, Université de Rennes 1, campus de Beaulieu, Rennes cedex, France*
** Now at Laboratoire de Tectonophysique, Université de Montpellier 2, Montpellier, France*

ABSTRACT: We illustrate alternative approaches to characterize and model flow in fractured rocks at site scale using field data. Approaches include (*i*) explicitly characterizing the main fracture flow paths with cross borehole flowmeter tests, and (*ii*) modelling the system response to large scale pumping test by a uniform diffusion equation that takes into account scaling of flow properties. The application of these approaches at the Plœmeur fractured crystalline aquifer shows that cross-borehole flowmeter tests have very interesting possibilities for aquifer flow path imagery. However, an exhaustive imagery of the fracture network is not possible. On the other hand, the interpretation of pumping test experiments shows that a mean flow model may be defined for most of the site. Such representation that allows characterizing the mean flow properties arising from the multiscale heterogeneity is complementary to local flow measurements.

## 1 INTRODUCTION

After several decades of investigations on the characterization and modeling of flow in natural fracture networks (*National Research Council*, 1996), the prediction of flow through fractured aquifers remains very difficult. The main reason for that is the presence of heterogeneity at all scales, ranging from the millimeter to the kilometer scale (Segall and Pollard, 1983; Okubo and Aki, 1987; Villemin and Sunwoo, 1987; Barton and Hsieh, 1989; Hirata, 1989; Scholz and Cowie, 1990; Olsson, 1992; Davy, 1993; Ouillon and Sornette, 1996; Odling, 1997; Bonnet et al., 2001; Bour et al., 2002). The consequence of the presence of heterogeneity at virtually all scales is that it is extremely difficult to build large scale aquifer models that can include all the relevant heterogeneity (*National Research Council*, 1996). The alternative is to use upscaled models to represent the average flow properties, without explicitly representing the heterogeneity. However, the absence of characteristic length scale in fracture networks suggests that upscaled models should include scaling of flow properties. Such discrete and continuous descriptions appear to be the two main possible approaches to characterize flow in fractured and heterogeneous aquifers. The objective of this paper is to discuss what are the advantages and limits of these two approaches through a field example based on the Plœmeur crystalline aquifer in France.

The Plœmeur site is characterized by a well connected fracture network over scales of several hundreds of meters. It is instrumented with pumping and observation wells with a wide range of separations. The site is used for water supply and the question of the prediction of flow at large scale is directly posed. Below we recapitulate the main characteristics of the Plœmeur aquifer and attempt to characterize explicitly the main fracture flow paths at site scale. Because most boreholes are equipped with a slotted casing, packers could not be used. Thus, the characterization relies on the use of single and cross borehole flowmeter tests for borehole separations ranging from 7 to 150 meters. Finally, we present an alternative approach that consists of modelling the system response to large scale pumping test by a uniform diffusion equation that explicitly takes into account scaling of flow properties.

## 2 CASE STUDY: THE PLŒMEUR AQUIFER

The Plœmeur aquifer is located on the south coast of Brittany in crystalline bedrock terrain characterized by igneous and metamorphic rocks (Figure 1). Compared to other bedrock aquifers in Brittany, this aquifer is outstandingly productive. Structural analysis reported by Touchard (1999) shows that the site is located at the intersection of two regional tectonic features: a contact between Late Hercynian granite and micaschist dipping about 30° to the north, and a dextral normal fault zone striking north 20° and dipping 70° to the east. The regional structural contact, which is characterized by enclaves of micaschist and granite dykes, may reach 100 meters in thickness. Ductile deformation within the micaschist increases towards the contact. This deformation, attributed to the granite emplacement, has been reactivated by late brittle deformation. Cores and drill cuttings show very heterogeneous lithological characteristics with an alternation of micaschist enclaves and granitic dykes such as aplites and pegmatites. The degree of fracturing is highly variable. Thus, the contact zone where most flow occurs is located in a heterogeneous fractured zone with various lithologies. A number of geophysical experiments (seismic, electric imaging, borehole geophysics) have been performed on the site. However, because of the structural heterogeneity of the site and because the imaging of lithological properties variations are not sufficient

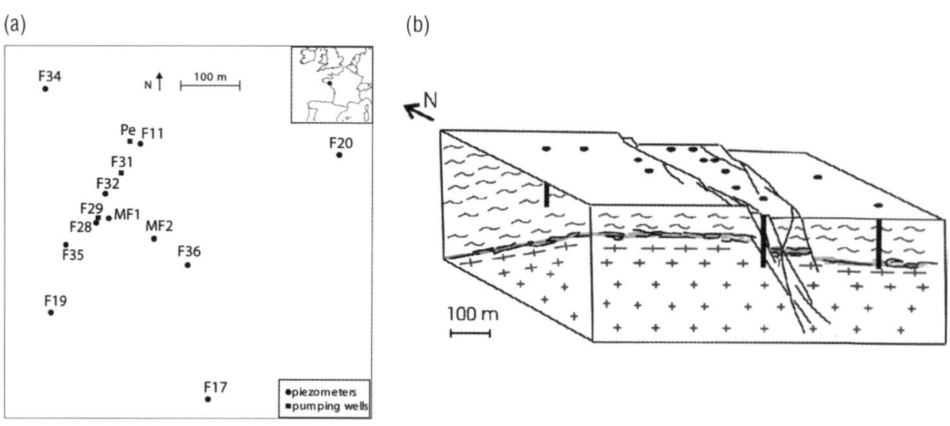


Figure 1. (a) Map of the Plœmeur well field showing the location of boreholes used for this study. (b) Schematic 3D diagram of the Plœmeur aquifer geology showing the regional contact dipping toward the North and the N° 20 dextral normal fault along which many boreholes are drilled.

to image flow paths, the geophysical investigations have been relatively unsuccessful in identifying deep aquifer-scale flow structures at the Plœmeur site.

Over forty boreholes have been drilled around the Plœmeur site. In the present study, we restrict our analysis to those that are the closest to the three pumping wells (Pe, F31, F29) within an area of approximately 600 meters in diameter (Figure 1). These boreholes react relatively rapidly to the pumping variations even for large separations. Most boreholes are about 100 meters deep and intersect a series of producing zones with variable inflows. In some cases the boreholes encounter a highly productive zone of a few meters in thickness, and located at a typical depth of about 70 m. Monitoring wells were completed with a 115 mm inner diameter casing that is slotted over the thickness of the productive zones. The length over which boreholes are slotted varies from 40 meters to 100 meters for the deepest borehole.

## 3 DISCRETE DESCRIPTION

In this section, we illustrate to what extent it is possible to characterize the geometry and distribution of hydraulic property of flow paths in fractured or heterogeneous aquifers. The main issues are that (*i*) the number of observation boreholes is limited; (*ii*) flow paths are three dimensional so that the properties observed along boreholes may not be simply interpolated between boreholes. Note that, in heterogeneous medium such as the Plœmeur site, where lithological variations are complex, classical geophysical investigations methods may be relatively unsuccessful in identifying deep aquifer-scale flow structures. In such context, the best methods to characterize the connectivity, geometry and hydraulic properties of flow paths are based on the use of straddle packers (Day-Lewis et al., 2000; Nakao et al., 2000). However, like at Plœmeur, long-screened observation boreholes are often the only measurement points available so that packers cannot be used. Thus, in this case, cross borehole flowmeter tests are the only alternative technique to characterize aquifer-scale flow structures (Paillet, 1998; Williams and Paillet, 2002; Le Borgne et al., 2006[a,b]).

While borehole flow logs have been used to estimate the distribution of permeability along boreholes (Hess, 1986; Molz et al., 1989; Paillet et al., 1987), cross-borehole flowmeter tests have been proposed as a means to characterize flow paths between boreholes (Paillet, 1998). Such experiments are based on the idea that changing the pumping conditions in a given aquifer will modify the hydraulic head distribution in large-scale flow paths to produce measurable changes in the vertical flow profiles in observation boreholes. We have proposed a methodology for the inversion of flow measurements to derive the hydraulic properties of inter-borehole flow-paths (Le Borgne et al., 2006[a]).

At the Plœmeur site, an extensive set of flowmeter experiments have been performed. The experiments include both single and cross borehole flowmeter tests. Borehole flows were first measured in ambient conditions in 6 boreholes using a heat-pulse flowmeter. For each borehole, a second borehole flow profile was obtained while pumping at about 30 litres per minute and measuring the vertical distribution of flow after drawdown had stabilized. Figure 2a shows an example of single borehole flowmeter test for borehole F11. Four inflow points are identified by inspection of the pairs of ambient and pumping flow profiles. Ambient up flow of about 2 to 4 L/min is measured in borehole F11. This ambient flow is due to the boundary conditions, and in particular to the operating well field conditions, but also to heterogeneous drawdown propagation in the different flow zones. The ambient flow gives a first indication of the presence of multiple flow paths that are isolated from each

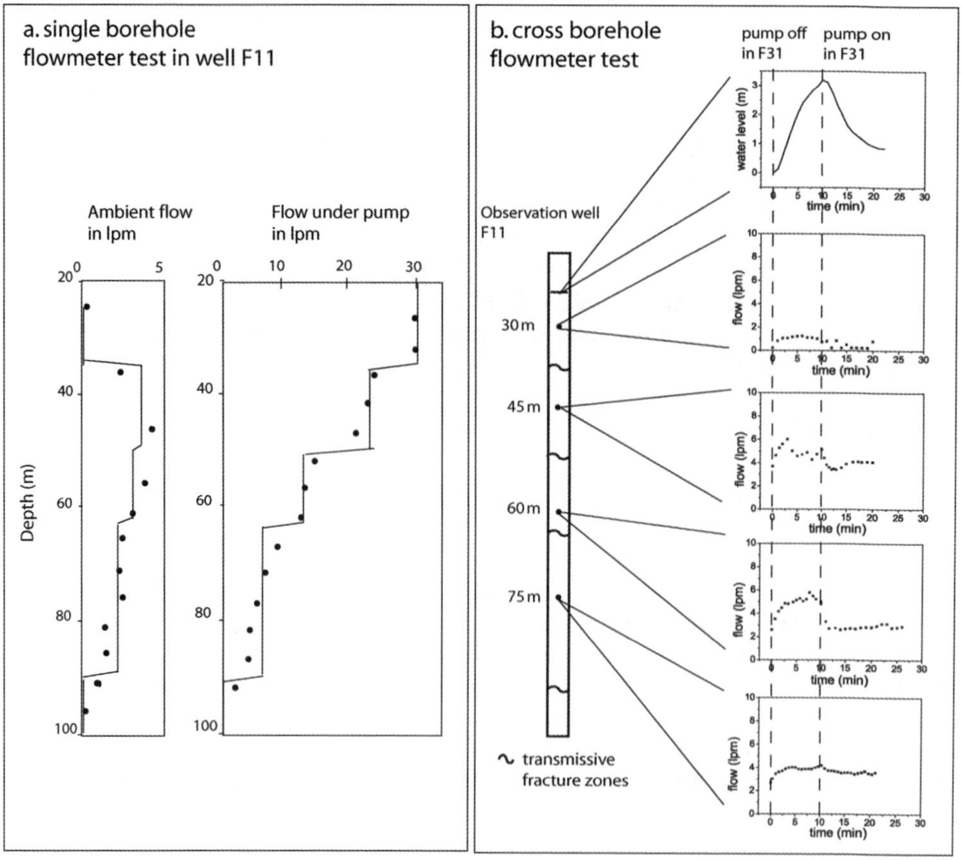


Figure 2. (a) Single flowmeter tests for well F11; the solid lines represent the borehole flow model fit that are used to estimate flow zone transmissivities. (b) Transient cross-borehole flowmeter test data at stations 30 m, 45 m, 60 m and 75 m and water level variation in borehole F11 when varying the pumping rate in borehole F31.

other or partially connected to each other. Flow modeling techniques (Paillet, 2000) applied to the single borehole flowmeter data sets yielded estimates for zone transmissivity ranging from less than $10^{-5} m^2/s$ to more than $10^{-3} m^2/s$, and hydraulic head differences between zones ranging from a few centimetres to more than a meter.

After the inflow points were identified and the transmissivity of each such point was estimated, the influence of various nearby pumping wells was investigated by turning off individual production pumps and checking for measurable changes in flow within each observation borehole. Such cross-borehole tests were conducted for all pairs of observation and production wells where measurable changes were identified under the influence of changes in distribution of well field production. Figure 2b shows an example of cross-borehole experiment where the pump in well F31 was turned off and then turned on after 10 minutes while flow was measured in borehole F11 at different depths. The inter-borehole distance between the observation borehole F11 and the pumped borehole F31 is 62 meters. The observed well water level increased to a maximum of about 3 meters and decreased back towards the level measured at the start of the experiment when the pump was turned on in

F31. The measured borehole flows at locations between and above the flow zones in borehole F11 are shown as the discrete data points in Figure 2b. For the 60-meter measurements, the flow response is a sharp increase in upflow followed by a return to ambient flow when the pump is turned on in F31. This implies that the consequence of turning off the pump in F31 is an increase in hydraulic head in a flow zone below 60 meters in F11, showing the existence of a connection between these two wells below 60 meters. However, the flow responses measured at the other depths are more complex, suggesting that there may be multiple connections between the two wells involving in particular some delayed responses.

The interpretation of cross borehole flowmeter data to define the main hydraulic connections between pumping well and observation well requires solving two inverse problems. The first inverse problem consists of estimating the hydraulic head variations that drive the transient borehole flow observed in the cross-borehole flowmeter experiments. The second inverse problem is related to estimating the geometry and hydraulic properties of large-scale flow paths in the region between pumping and observation wells that are compatible with the head variations deduced from the first problem (Le Borgne et al., 2006[a]). For the example of Figure 2b, we performed such an interpretation and deduced that all flow zones in F11 are connected to F31 except for the 50 meters zone, which is isolated. The connecting zones at different depths have different hydraulic properties, which imply multiple delayed hydraulic head responses in F11 when the pump in F31 is turned off.

The results of the single and cross borehole flowmeter experiments at the Plœmeur site have been synthesized with other available data, including driller air-lift borehole logs as well as large-scale pumping tests, to define a conceptual model of fracture network geometry and connectivity (Le Borgne et al., 2006[b]). Despite the vertical heterogeneity, the highest connectivity zones are generally found between $-35$ meters above sea level and $-70$ meters above sea level whatever the distance to the main pumping well. The flowmeter experiments suggest the presence of a sub-horizontal high connectivity zone located between $-35$ and $-45$ meters, which intersects most wells of the site. However, the depth of the main connected zones in boreholes F31, F32 and F35, which is comprised between $-60$ and $-70$ meters, indicate that this high connectivity zone may locally be slightly shifted or that the flow system may be locally more complex, with several fast connectivity zones.

Flowmeter experiments yield valuable insights into the large scale aquifer flow structures and help provide a useful aquifer conceptual model at first order. They also provide very important information for sampling the groundwater chemistry or performing tracer tests in such heterogeneous aquifers. At the Plœmeur site, these experiments showed that the site is characterized by a large range of hydraulic properties and that high transmissivity zones are connected over long spatial ranges. These results are relatively consistent with the geological model presented in Figure 1b. At first order, the most permeable zone is sub-horizontal and may correspond to the contact between granite and micaschist that is slightly dipping towards the north. Secondary structures, that may have shifted locally this contact, may be linked to late faulting like the N 20° structure (Figure 1). Furthermore, the cross-borehole flowmeter data may be used to estimate the hydraulic properties of the flow zones connecting the pumping and observation wells (Le Borgne et al., 2006[b]). Inversion of cross-borehole flowmeter tests data from Plœmeur yielded transmissivity estimates ranging from less than $10^{-4} m^2/s$ to nearly $10^{-2} m^2/s$, and storage coefficient estimates ranging from$10^{-5}$ to more than $10^{-3}$ for borehole separations ranging from 7 to 150 meters.

Two issues need to be stressed: (*i*) the number of observation boreholes is limited so that the constraints on flow paths geometry remains relatively weak; or at least undersampled (*ii*) the flowmeter experiments allow characterizing mainly the fast connection zones but the network of secondary fractures, which may be fundamental for long term hydraulic response, is poorly characterized. Most boreholes are characterized by both fast and slow connections between each other. Thus, a single flow zone cannot describe flow in the subsurface that certainly involves a complex fracture network. Hence, given the available characterization techniques, it is difficult to build large scale aquifer models that include explicitly all the relevant information. An alternative is to use upscaled equivalent models to represent the average properties of the system without explicitly representing the heterogeneity. In the following section, we present the application of such a continuous approach that consists of modelling the system response to large scale pumping test by a uniform diffusion equation that explicitly takes into account scaling of flow properties.

## 4 CONTINUOUS DESCRIPTION

We replace an exhaustive description of heterogeneous media with a continuous description to model the mean flow properties of the natural medium. Using extensive pumping test data, we explored whether or not a mean equivalent flow model can be defined to describe hydraulic head variations on a large range of spatial and temporal scales (Le Borgne et al., 2004). The data consist of pumping tests performed in the Plœmeur aquifer by monitoring drawdown time-series at several observation piezometers located at distances ranging from 2 m to 400 m from the pumping well. The pumping test durations varied from 5 days to 80 days with sampling rates as short as one minute giving a scale range of 4 to 5 orders of magnitude in time.

### 4.1 *Homogeneous porous media representation*

The simplest model is the classical Theis analysis that corresponds to a homogeneous porous media representation. Figure 3 shows two examples of the best fit that can be obtained with the Theis model for drawdown measurements at observation wells during short term and long term pumping tests. In both cases, the fits that can be obtained with this model do not closely match the evolution of drawdown, especially for later times. It is particularly clear for the long-term dataset, for which the Theis model does not match at all the late time evolution of drawdown. This observation appears to be valid for all piezometers, regardless of the distance from the pumping well (Le Borgne et al., 2004).

The inconsistency of the Theis model with observations may arise from the multiscale nature of fractured media that is likely to prevent the existence of a characteristic scale above which the application of a homogeneous porous media representation may be justified. Furthermore, parameter estimates from Theis model present a relatively large variability depending on the observation borehole (Figure 4). Although transmissivity estimates do not vary so much compared to the possible range of variation, storage coefficient estimates vary over nearly two orders of magnitude. Such variability may be due either to large-scale heterogeneity or to a possible scale dependence of the hydraulic properties which is expected to occur in such heterogeneous medium (Clauser, 1992; Sanchez-Vila et al., 1996; Illman, 2006). This suggests that the parameters depend on the observation time and on the distance between the pumping and observation well. Analyzing the variability of the parameters derived from the Theis model as a function of scale may be useful for providing some

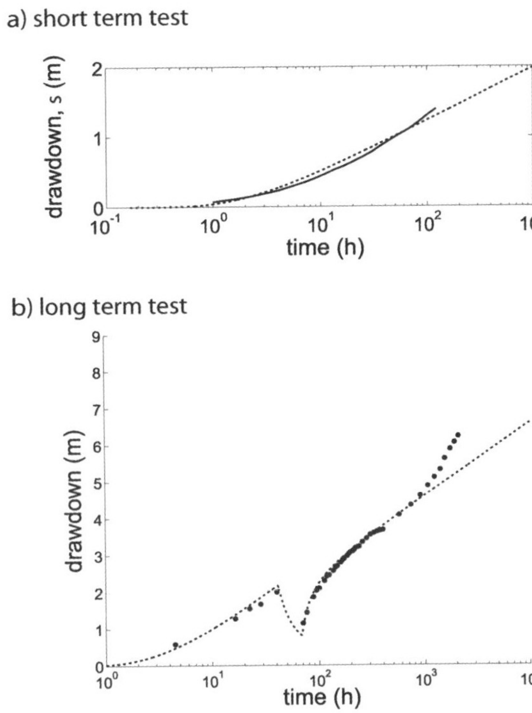


Figure 3. Examples of Theis model best fits to (a) short term pumping test data (5 days duration, pumping well: F31, observation well: F36), (b) long term pumping test data (3 months duration, pumping well: Pe, observation well: F19). During the long term test, the pump stop one day after the start of the experiment has also been taken into account in the model.

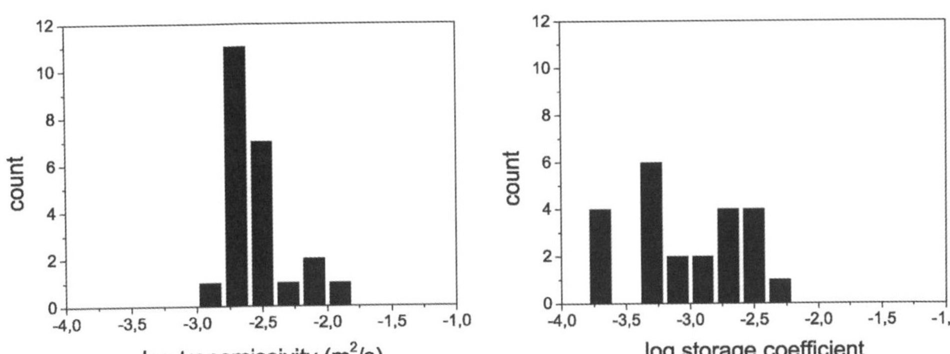


Figure 4. Estimates of transmissivity and storage coefficient from the pumping tests analysis at the different observation wells.

insights about the medium heterogeneity at the Plœmeur site. However, it provides neither a relevant conceptual model of flow nor a predictive tool for modeling pressure diffusion in the medium. Scale effects should therefore be addressed by using flow models that integrate scaling properties.

### 4.2 *Representation of the flow properties arising from multiscale heterogeneity*

Recently, new models have been developed to integrate scaling properties that are particularly relevant for fracture networks and highly heterogeneous media. First, the generalized radial flow model of Barker (1988) recognizes that flow may not fill the three dimensional space entirely. Hence, this model assumes that flow may be characterized by any flow dimension, n, with $1 \leqslant n \leqslant 3$. Barker's model is generalized in the sense that classical mono-, bi- and tri-dimensional flow models are recovered for the integral dimensions $n = 1, 2, 3$.

The model of Acuna and Yortsos (1995) is based on the scaling properties of some fracture networks, such as the absence of a characteristic length scale (Bonnet et al., 2001; Bour et al., 2002). It is a generalization of Barker's model and accounts for the possible anomalously slow diffusion that has been shown to occur in media with fractal properties (Chang and Yortsos, 1990). The topological properties of fractal structures, in particular the existence of the dead ends and bottle necks without characteristic size, imply that the mean radius of diffusion is evolving slower that the square root of time. The Chang and Yortsos model is exact for some deterministic fractal structures including the Sierpinski gasket (O'Shaughnessy and Procaccia, 1985) and introduces scalings for the storativity and transmissivity: $S \sim r^{D-d}$ and $T \sim r^{D-d-(2-dw)}$, respectively, where D is the fractal dimension and $d_w$ is the anomalous diffusion exponent and d is the embedding dimension. The generalized applied diffusion equation (O'Shaughnessy and Procaccia, 1985; Chang and Yortsos, 1990) can be written as:

$$S\frac{\partial h}{\partial t} = \frac{T}{r^{D-1}}\frac{\partial}{\partial r}\left(r^{D-dw+1}\frac{\partial h}{\partial r}\right). \qquad (1)$$

Barker's model is recovered for $d_w = 2$, case for which diffusion is called "normal" as opposed to the case $d_w > 2$. For the latter case, diffusion is said to be anomalously slow (Halvin and Ben-Avraham, 1987). Such anomalously slow diffusion occurs in particular in percolation networks at threshold (Stauffer and Aharony, 1992). In two-dimensional percolation networks, $d_w = 2.86$ (Stauffer and Aharony, 1992).

The full validation of these scale dependent models requires the observation of drawdowns at different scales on a sufficiently large range of spatial and temporal scales. We proposed a methodology for the assessment of these scale dependent models that is based on the fact that diffusion equations of Theis' model, Barker's model and Acuna and Yortsos' model can be written in the same general formulation (equation 1) (Le Borgne et al., 2004). We recall that Barker's solution reduces to Theis solution when $n = 2$, as well as Acuna and Yortsos (1995) model does when $D = 2$ and $dw = 2$. To obtain a general analytical solution for the three models, we express the analytic solution for drawdown $s(r, t) = h(r, t = 0) - h(r, t)$, in a general formulation that depends on three independent parameters:

$$s(r, t) = h_0(r)\Gamma\left(\frac{n}{2} - 1, \frac{t_c(r)}{t}\right). \qquad (2)$$

Where $\Gamma$ is the (complementary) incomplete gamma function.

The parameter n is the hydraulic dimension, related to D and $d_w$; the two other parameters, $h_0$ and $t_c$, describe the characteristic amplitude and the characteristic time of drawdown evolution. Note that $t_c$ can be interpreted as the characteristic time after the start of pumping for the drawdown cone to reach a given observation well. We derived the expressions of $h_0$, $t_c$ and n in (2) for the three models by reformulating the different transient flow equations (Le Borgne et al., 2004). The evaluation of the relevance of these models consists of comparing the drawdown observed in the field to their solution given by (2). Since the three parameters $h_0$, $t_c$ and n of (2) are independent, they can be obtained by a single fit to the data. The hydraulic dimension n is determined by the late-time shape of the curve. $t_c$ and $h_0$ are time and amplitude factors. Theis', Barker' and Acuna and Yortsos' models can be distinguished by the value of n and by the scaling of $h_0(r)$ and $t_c(r)$. For Theis model we expect $n = 2$ and $t_c \sim r^2$. For Barker's model, we may have $n \neq 2$ but the characteristic time should always scales as $t_c \sim r^2$. For Acuna and Yorstos' model $n \neq 2$ and $t_c \sim r^{dw}$ with $d_w > 2$. The discrimination between Barker's and Acuna and Yortsos' models requires thus the existence in the field of piezometers located at different distances from the pumping well that span the widest possible scale range to investigate the variations with scale of n and $t_c$.

To estimate the hydraulic dimension n, we fit the general analytical solution for hydraulic variations (equation 1) to the pumping test data. Figure 5 shows two examples of drawdown curves for a short and long term pumping tests. For both examples, a model with $n = 1.6$ fits well the variation of drawdown with time contrary to the Theis model that corresponds to $n = 2$ (Figure 3). For the different observation wells and the different

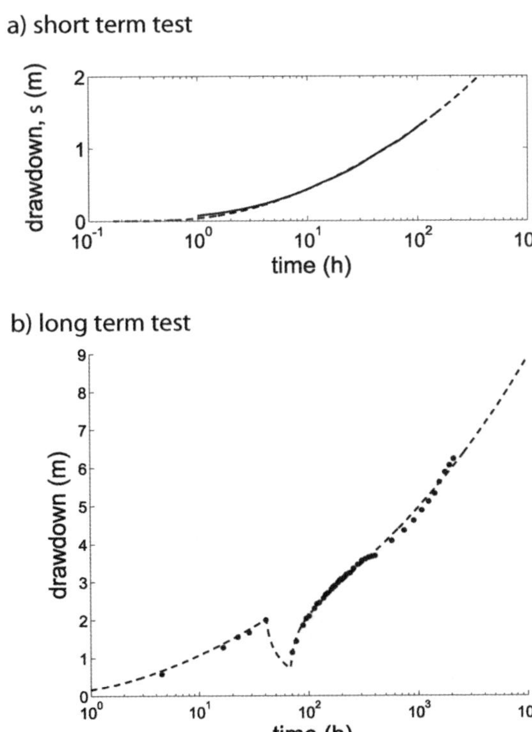


Figure 5. Examples of fractional flow model best fits with $n = 1.6$ to (a) short term pumping test data (5 days duration, pumping well: F31, observation well: F36), (b) long term pumping test data (3 months duration, pumping well: Pe, observation well: F19).

pumping tests, the best estimates of the hydraulic dimension, n, are systematically lower than 2 and mostly comprised in the range [1.4, 1.7]. Short, medium and long-term pumping tests therefore provide consistent results. It thus appears that a mean fractional flow model with a flow dimension of about 1.6 is relatively consistent over the whole dataset that covers 2 orders of magnitude in space and 4 orders of magnitude in time.

The full validation of these scale-dependent models requires the modeling of drawdown at observation well at different distances on a sufficiently large range time scales. This implies a consistent scaling of the characteristic time and amplitude $t_c$ and $h_0$ as a function of the observation borehole separation distance with the pumping well. The detailed analysis of a pumping test with pumping well F31 (Le Borgne et al., 2004) shows that, except for the two piezometers located east of a main fracture zone (F20 and F36 in Figure 1), all derived values of $t_c$ follow a general trend such that $t_c \sim r^{dw}$ with $d_w = 2.8$. For these boreholes, the scaling of the characteristic amplitude $h_0$ is consistent with such exponent. The derived exponent implies that hydraulic head diffusion is anomalously slow; this characteristic that is taken into account only in the model of Acuna and Yortsos (1995), based on diffusion in fractals. Hence, despite local heterogeneities that are intrinsic to any fractured crystalline aquifers, it appears possible to define a mean flow model at the scale of the site, at least for the major fault zone. Note that even in such approach, some large scale geological features must be explicitly considered. In the case of the Plœmeur site, the differentiation of two distinct fracture systems is supported by geological arguments: all the piezometers following the anomalously slow diffusion trend intersect the normal fault zone except one that is slightly west of the zone. The two outliers correspond to piezometers that intersect only the zone of contact between micaschists and granite (Figure 1).

In summary, the definition of a global flow model that is consistent with the properties of natural fractured media allows (*i*) the possibility to make reliable long term predictions, (*ii*) to assess some fundamental properties of fracture networks related to scale effects in hydraulic properties. Therefore, it represents an interesting and complementary alternative to the discrete description. However, research is needed to understand the link between the properties of the structure supporting flow and the exponents that can be derived from pumping tests: the fractal dimension D and the anomalous diffusion exponent $d_w$. De Dreuzy et al. (2004) have proposed some insights about this issue by relating these exponents to the properties of multi-fractal media using numerical simulations.

## 5 CONCLUSION

This paper illustrates through a field example the different approaches that may be used to characterize flow at site scale in fractured or heterogeneous aquifers. This study is based on the Plœmeur fractured crystalline aquifer where observation boreholes are available at different separation distances. The characterization relies on the use of single and cross borehole flowmeter tests for borehole separations ranging from 7 to 150 meters (Le Borgne et al., 2006[b]). These experiments provided valuable insights into the large scale aquifer flow structures and conceptual model consistent with the geological heterogeneity. However, two key issues are: (*i*) the number of observation boreholes is limited so that the constraints on flow paths geometry are relatively weak; (*ii*) the flowmeter experiments characterize mainly the fast connection zones. The network of secondary fractures, which may be fundamental for long term hydraulic response, is yet poorly characterized.

As an alternative to the discrete description, we investigated whether pressure diffusion at aquifer scale may be modeled by a uniform diffusion equation (Le Borgne et al., 2004). Such upscaled models should represent the average properties of the system, without explicitly representing the heterogeneity. We tested the classical Theis model and found inconsistencies with observed drawdown curves that led to a relatively large parameter variability depending on the observation well. The Theis does not provide neither a relevant conceptual model of flow nor a predictive tool for modeling pressure diffusion at this site.

We tested models based on scaling properties relevant for natural fracture networks. Such models include additional scaling parameters that describe on average the flow properties arising from the multiscale heterogeneity. Interpretation of drawdown time-series, recorded in piezometers located at distances ranging from 2 to 400 meters from the pumping well and of durations ranging from 5 to 90 days, show temporal evolutions of drawdown that are well modeled by fractional flow models (Barker, 1988; Acuna and Yortsos, 1995). Estimates of flow dimensions are consistent across the whole site and lie in the range [1.4–1.7]. To investigate the nature of diffusion, we defined a methodology based on the evolution of the characteristic time or amplitude of hydraulic head variations with the distance from the pumping well. The derived exponents imply that hydraulic head diffusion is anomalously slow, a characteristic that is taken into account only in the model of Acuna and Yortsos (1995), based on the theory of diffusion in fractals. Hence, despite local heterogeneities intrinsic to any fractured crystalline aquifers, it is possible to define a mean flow model at the scale of the site, at least for one of the major fault zone. Such description accounts for the effect of the whole fracture network, including small and large fractures as well as the distribution of hydraulic properties in fractured zones. The continuous approach appears well suited to predict the average long term evolution of hydraulic heads. It also allows assessing some fundamental properties of natural fracture networks related to scale effects in hydraulic properties. Therefore, it represents an interesting complementary approach to the discrete description.

## ACKNOWLEDGMENTS

We gratefully acknowledge the Plœmeur council staff for their cooperation during field work. This work was supported by the European project "SALTRANS" (contract EVK1-CT-2000-00062). We also wish to thank the French CNRS for their financial support, in particular through the INSU project "ORE H+" and through the support of a specific project on flow measurements in heterogeneous media.

## REFERENCES

Acuna, J.A. and Y.C. Yortsos. (1995) Application of fractal geometry to the study of networks of fractures and their pressure transient, Water Resour. Res., 31(3), 527–540.

Barenblatt, G.I., I.P. Zheltov and I.N. Kochina. (1960) Basic concept in the theory of seepage of homogeneous liquids in fissured rocks, J. Appl. Math., 24, 1286–1303.

Barker, J.A. (1988) A generalized radial flow model for hydraulic tests in fractured rock, Water Resour. Res., 24(10), 1796–1804.

Barton, C.C. and P.A. Hsieh. (1989) Physical and Hydrological flow Properties of Fractures, Field Trip Guide, AGU, Washington, D.C., T385.

Bonnet, E., O. Bour, N. Odling, I. Main, B. Berkowitz, P. Davy and P. Cowie. (2001) Scaling of Fracture Systems in Geological Media, Rev. of Geophys., 39(3), 347–383.

Bour, O., P. Davy, C. Darcel and N. Odling. (2002) A statistical scaling model for fracture network geometry, with validation on a multiscale mapping of a joint network (Hornelen Basin, Norway), J. of Geophys. Res., 107(B6), 2113, doi: 10.1029/2001JB000176.

Bourdarot, G. (1999) Well Testing: interpretation methods, in Collection des Cours de l'ENSPM, 350 p., edited by editions TECHNIP, IFP, Paris, France.

Chang, J. and Y.C. Yortsos. (1990) Pressure-transient analysis of fractal reservoirs, SPE Form. Eval., 5, 631.

Clauser, C. (1992) Permeability of crystalline rock, Eos Trans. AGU, 73(21), 237–238.

Davy, P. (1993) On the frequency-length distribution of the San Andreas fault system, J. Geophys. Res., 98, 12141–12151.

Day-Lewis, F.D., Hsieh and P.A., Gorelick, S.M. (2000) Identifying fracture-zone geometry using simulated annealing and hydraulic connection data, Water Resour. Res., 36 (7), 1707–1721.

de Dreuzy, J.R., P. Davy, J. Erhel and J. de Brémond d'Ars (2004). Anomalous diffusion exponents in continuous two-dimensional multifractal media. Physical Review E, 70.

Halvin, S. and D. Ben-Avraham. (1987) Diffusion in disordered media, Adv. in Phys., 36(6), 695–798.

Hess, A.E. (1986) Identifying hydraulically conductive fractures with a slow velocity borehole flowmeter. Canadian Geotechnical Journal, 23: 69–78.

Hirata, T. (1989) Fractal dimension of fault system in Japan: Fractal structure in rock fracture geometry at various scales, Pure Appl. Geophys., 121, 157–170.

Illman, W.A., in press. Strong evidence of directional permeability scale effect in fractured rock. J. of Hydrol.

Kiraly, L. (1975) Rapport sur l'état actuel des connaissances dans le domaine des caractères physiques des roches karstiques, in Hydrogeology of karstic terrains p., edited by A. Burger and L. Dubertret, International association of hydrogeologists, Paris.

Le Borgne, T., O. Bour, J.-R. de Dreuzy, P. Davy and F. Touchard. (2004) Equivalent mean flow models for fractured aquifers: Insights from a pumping tests scaling interpretation. Water Resources Research: 10.1029/2003WR002436.

Le Borgne, T., F. Paillet, L, and O. Bour. (2006[a]) Cross borehole flowmeter tests for transient heads in heterogeneous aquifers. Ground Water Vol. 44, No. 3 p. 444–452.

Le Borgne, T., O. Bour, F.L. Paillet, and J-P. Caudal. (2006[b]). Assessment of preferential flow path connectivity and hydraulic properties at single-borehole and cross-borehole scales in a fractured aquifer. J. of Hydrol., Vol. 328, No 1–2, p. 347–359.

Molz, F.J., R.H. Morin, A.E. Hess, J.G. Melville, and O. Güven. (1989) The impeller meter for measuring aquifer permeability variations: evaluation and comparison with other tests. Water Resources Research, 25(7): 1677–1683.

Nakao, S., J. Najita, and K. Karasaki. (2000) Hydraulic well testing inversion for modelling fluid flow in fractured rocks using simulated annealing: a case study at Raymond field site, California. Journal of Applied Geophysics, 45(3): 203–223.

Odling, N. (1997) Scaling and connectivity of joint systems in sandstones from western Norway, J. of Struct. Geol., 19(10), 1257–1271.

Okubo, P.G. and K. Aki. (1987) Fractal geometry in the San Andreas fault system, J. Geophys. Res., 92, 345–355.

Olsson, O. (1992) Site characterization and validation, in Final Report, Stripa Project 92-22, Conterra AB, Uppsala, Sweden.

O'Shaughnessy, B. and I. Procaccia. (1985) Diffusion on fractals, Phys. rev. A, 32(5).

Ouillon, G. and D. Sornette. (1996) Unbiased multifractal analysis : Application to fault patterns, Geophys. Res. Lett., 23(23), 3409–3412.

Paillet, F.L. (1998) Flow modelling and permeability estimations using borehole flow logs in heterogeneous fractured formations. Water Resources Research, 34(5): 997–1010.

Paillet, F.L. (2000) A field technique for estimating aquifer parameters using flow log data. Ground Water, 38(4): 510–521.

Paillet, F.L., A.E. Hess, C.H. Cheng, and E. Hardin. (1987) Characterization of fracture permeability with high-resolution vertical flow measurements during borehole pumping. Ground Water, 25(1): 28–40.

Sanchez-Vila, X., J. Carrera, and J.P. Girardi. (1996) Scale effects in transmissivity. Journal of Hydrology, 183(1–2): 1–22.

Scholz, C.H. and P. Cowie. (1990) Determination of total strain from faulting using slip measurements, Nature, 346, 837–838.

Segall, P. and D. Pollard. (1983) Joint formation in granitic rock of the Sierra Nevada, Geol. Soc. of Am. Bull., 94, 563–575.

Stauffer, D. and A. Aharony. (1992) Introduction to percolation theory, second edition, 181 p., Taylor and Francis, Bristol.

Theis, C.V. (1935) The relation between the lowering of the piezometric surface and the rate and duration of discharge of a well using groundwater storage, Eos Trans. AGU, 16, 519–524.

Touchard, F. (1999) Caractérisation hydrologique d'un aquifère en socle fracturé, Ph.D. thesis, in Mémoires de Géosciences Rennes, n° 87, 271 p., Rennes, France.

Villemin, T. and C. Sunwoo. (1987) Distribution logarithmique self similaire des rejets et longueurs de failles: exemple du bassin Houiller Lorrain, C. R. Acad. Sci. , Ser. II, 305, 1309–1312.

Williams, J.H. and F.L. Paillet. (2002) Using flowmeter pulse tests to define hydraulic connections in the subsurface: a fractured shale example. J. of Hydrol., 265, 100–117, 8/30.

CHAPTER 30

# New analysis of pump tests in fractured aquifers

Gerard Lods and Philippe Gouze
*Laboratoire de Tectonophysique, UMR, CNRS – Université de Montpellier, ISTEEM-MSE, Montpellier Cedex, France*

ABSTRACT: Identifying the hydraulic characteristics and transport properties of fractured reservoirs requires the development of models accounting: (1) for medium heterogeneity (i.e., the presence of major conductive fractures that delimit matrix blocks) and (2) for spatial organization of the major conductive fractures, which dominate flow at the pump test scale. These two main features are tackled by combining generalized radial flow (flow with fractal dimension) with the theories of double-porosity and leakance. This nD model, with n not necessarily integer, extends usefully the domain of application of the usual 1D/2D/3D double-porosity/leakance models for a large range of connectivity levels of the fracture networks. These models account for the impact of the network structure on the overall hydrodynamic behaviour, even though they do not handle the geometry and properties of the fractures individually or by directional families. The accuracy of the coupled transient behaviours analysis is augmented by modelling pumping well storage and skin effects. The combination of all these features allows matching a wide range of pumping test curves with a small number of independent parameters. Several examples of pumping test, covering a large range of hydrogeological situations in sedimentary and crystalline rocks, are analysed in order to demonstrate how the different models can be used to improve predictions.

## 1 INTRODUCTION

Identifying the hydraulic characteristics and transport properties of fractured reservoirs requires the development of models accounting (1) for medium heterogeneity, i.e. the presence of major conductive fractures that delimit capacitive matrix blocks, and (2) for geometrical arrangement and the spatial distribution of the major connected fractures, which dominates the flow at the scale of the pump tests. Flow around wells in fractured media has been studied intensively since the straightforward 2D line source or Theis's model (1935), which considers an equivalent porous medium and assumes confined, homogeneous, isotropic, infinite, and initially at rest. The leakance concept (Hantush, 1956) extends the Theis approach to semi-confined aquifers and produces behaviours intermediate between those for confined and unconfined aquifers. The double porosity concept was proposed by Barenblatt et al. (1960) to model fractured aquifer. In this model the system is represented by two overlapping continua: the main fractures network that controls flow at the scale of the pump tests, and the matrix blocks, drained by the fractures, that may contain poorly connected, weakly open, dead end or isolated fractures. Inter-continua mass exchange can be either stationary (Warren and Root, 1963) or transient (Boulton and Streltsova, 1977). The later was extended by the parameterization of the fracture skin effect (Moench, 1984), embodying the hydraulic impedance of the skins. When this impedance becomes important, Moench's solution reduces to the stationary exchange. In the following, we consider transient exchanges with fracture skin effect, generalizing exchanges between blocks and fractures. Fractional flow dimensions find their usefulness when integral dimensions (1, 2, 3) solutions are not satisfying. They allow accounting for a scale dependence (fractality) of hydrodynamic properties (Le Borgne et al., 2004). Barker's solution (1988) for one continuum was extended to leakance and double porosity systems by Hamm and Bidaux (1994, 1996). In sufficiently fractured media, the flow around PWs (pumping wells) is dominated by the intercepted discontinuities network. The flow dimension constitutes a parameter, reflecting the dominating geometry of the flow in the network. For example, this parameter is 1 for a parallel flow (subvertical fault intercepted by the well, or one channel fracture), 2 for cylindrical flow (fully penetrating well in a homogeneous porous aquifer), and 3 for spherical flow (packer tests with one intercepted fracture in a dense and extended network). For all the combinations of these geometries, generalization to non integral dimensions is necessary. PW skin effect (Everdingen and Hurst, 1953), which is potentially important, can be taken into account. It includes laminar and turbulent head losses occurring in the hole, in its equipment (strainer, gravel pack) and in the vicinity of the well (mud invasion and turbulence). Once the well has been developed, or when the fractures are well open at the well face, a relative head gain, called negative skin effect, can be observed. Additionally, PW storage effect (Papadopoulos and Cooper, 1967), due to the difference in storage and conductivity between the zone occupied by the well and the formation, can be estimated. It is particularly significant for large diameter wells and when the rock formation is weakly conductive. In models considering an infinitesimal well, such as Theis, this effect cannot be accounted for, although in reality a part of the measured flow rate contributes to its drainage or filling, for abstraction or injection respectively, producing a drawdown delay. The models presented below couple leakance, double porosity, PW storage and skin effects, with fractal dimension flow. They improve our ability in tackling all together the medium heterogeneity, the geometrical arrangement of the major connected fractures network, and the PW storage and skin effects.

## 2 THE MODELS

### 2.1 *Theory*

The aquifer is assumed confined, isotropic, radially infinite, and initially at rest. Each continuum (e.g., the "fractures" continuum, the "matrix" continuum and the aquitard continuum) is homogeneous (Figure 1) . Under the leakance assumption, the drawdown in the network $h_f$ (m) obeys the equation (Hamm and Bidaux, 1994):

$$S_{sf}\frac{\partial h_f}{\partial t} = \frac{K_f}{r^{n-1}}\frac{\partial}{\partial r}\left(r^{n-1}\frac{\partial h_f}{\partial r}\right) + K_f\frac{h_f}{B^2} \tag{1}$$

where $K_f$ ($m \cdot s^{-1}$) and $S_{sf}$ ($m^{-1}$) are the hydraulic conductivity and specific storativity of the fractures network, respectively, n is the flow dimension, r(m) is the radial distance from the PW, t(s) is the elapsed time since the test beginning, and B(m) is the leakance factor, which characterizes the exchanges with the aquitard. Under the double porosity

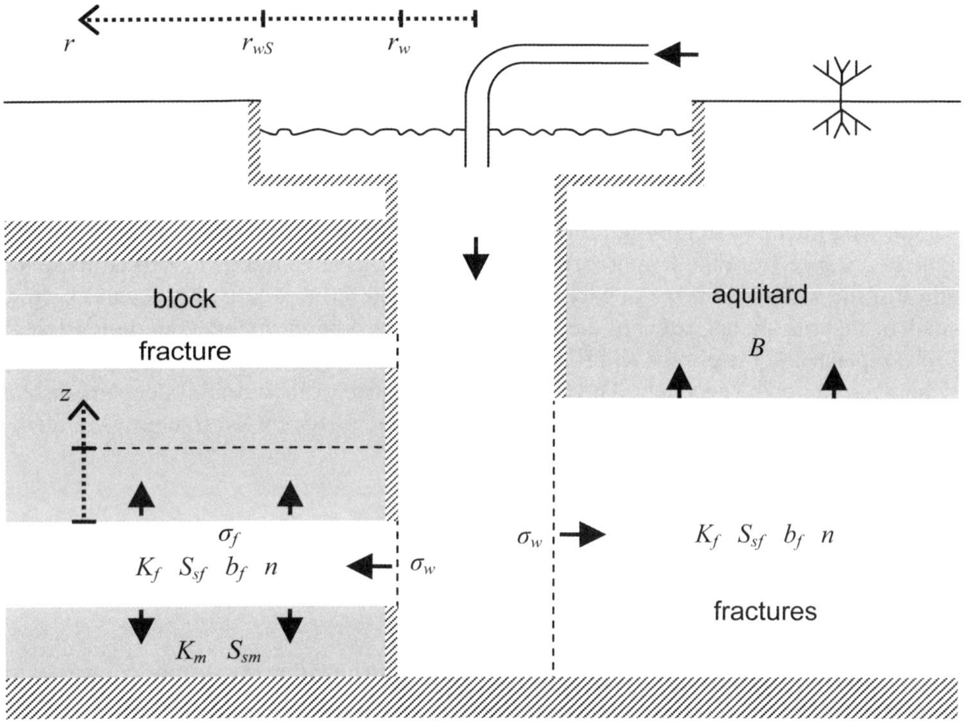


Figure 1. Conceptual scheme (injection) of the (left) double porosity model with transient exchange and fracture skin, and (right) of the leakance model, combined with fractal dimension flow and pumping well storage and skin effects. Hatching symbol indicates no flow boundaries (confining layers and well casing) and hypothetical negligible flow boundaries (matrix/well and aquitard/well interfaces).

assumption, drawdowns in the network and in the blocks, $h_m$ (m), obey the equations (Hamm and Bidaux, 1996):

$$S_{sf}\,\frac{\partial h_f}{\partial t} = \frac{K_f}{r^{n-1}}\,\frac{\partial}{\partial r}\left(r^{n-1}\,\frac{\partial h_f}{\partial r}\right) + \frac{K_m}{b_m}\left(\frac{\partial h_m}{\partial z}\right)_{z=0} \tag{2}$$

$$S_{sm}\,\frac{\partial h_m}{\partial t} = K_m\,\frac{\partial^2 h_m}{\partial z^2} \tag{3}$$

where $K_m$ ($m \cdot s^{-1}$) and $S_{sm}$ ($m^{-1}$) are the hydraulic conductivity and specific storativity of the blocks, respectively, z (m) is the coordinate in the block on the draining path (z = 0 at the interface), $b_m$ (m) is the mean length of the draining paths, or blocks mean half thickness ($\partial h_m/\partial z = 0$ for $z = b_m$). In place of $S_{sm}$, $K_m$, $b_m$ and z, it is usual to consider the dimensionless parameters

$$\omega = \frac{S_{sf}}{S_{sf} + S_{sm}} \tag{4}$$

$$\lambda = \frac{K_m r_w^2}{K_f b_m^2} \tag{5}$$

$$z_D = z/b_m \tag{6}$$

denoting the fractures storativity ratio, the interporosity flow coefficient, and the dimensionless distance from the fracture, respectively. The exchange radius $r_w$ (m) is the inner radius of the well in the zone of exchange with the aquifer. Similarly, $K_f$, $S_{sf}$ and $b_f$ (m), which is the ortho-radial extent of the flow front, i.e. the aquifer thickness in dimension 2, can be replaced by the groups $K_f\, b_f^{3-n}$ and $S_{sf} b_f^{3-n}$, denoting the generalized transmissivity and storativity, respectively. The hydraulic impedance of the mineral deposits on the fracture walls is modelled as a singular head loss characterized by the fracture skin factor $\sigma_f\,(-)$ (Moench, 1984):

$$h_m(z = 0) = h_f + b_m \sigma_f \left( \frac{\partial h_m}{\partial_z} \right)_{z=0} \tag{7}$$

The exchanges between the PW and the matrix and the aquitard are minor compared to its exchange with the fracture continuum. The PW balance equation is (Barker, 1988):

$$S_w \frac{\partial h_w}{\partial t} = Q + K_f b_f^{3-n} \alpha_n r_w^{n-1} \left( \frac{\partial h_f}{\partial r} \right)_{r=r_w} \tag{8}$$

where $h_w$ (m) is the well drawdown. The $\alpha_n$ coefficient is equal to $2\pi^{n/2}/\Gamma\,(n/2)$ where $\Gamma$ is the gamma function, $Q\,(m^3 \cdot s^{-1})$ is the test flow rate, positive for injections, $S_w\,(m^2)$ is the PW capacity, which for a test without packer, corresponds to its free surface area. The PW capacitive radius, $r_{wS}$ (m), is the mean radius of that area ($S_w = \pi r_{wS}^2$). For packer tests, $S_w$ depends on the deformability of the exchange chamber (packer jacket, tubing, …). The PW skin effect is represented by a singular head loss, through the PW skin factor $\sigma_w\,(-)$:

$$h_w = h_f(r_w) - r_w \sigma_w \left( \frac{\partial h_f}{\partial r} \right)_{r=r_w} \tag{9}$$

A negative value of $\sigma_w$ indicates a good fractures opening at the well-rock interface. In the presence of non linear skin effect, due to turbulent flow, the skin factor increases with the flow rate. To take into account the occurrence of turbulent flow, the value of the critical rate $Q_c\,(m^3 \cdot s^{-1}; Q_c > 0)$, delimiting the laminar and the turbulent flow, can be introduced in the usual model, such that the Q-derivative of $\sigma_w$ is continuous (Lods, 2000):

$$\sigma_w = \sigma_{wl} + \sigma_{wt}\, \theta(|Q| - Q_c)\, (|Q| - Q_c)^p / |Q| \tag{10}$$

where $\sigma_{wl}$ (−) and $\sigma_{wt}$ ($m^{3(1-p)} \cdot s^{p-1}$; $\sigma_{wt} \geqslant 0$) are the linear (laminar) and non-linear (turbulent) skin factors, respectively, p ($p > 1$) is the non-linear head loss dimensionless exponent (Jacob, 1947; Rorabaugh, 1953), θ is the Heaviside distribution. The PW storage effect vanishes when the left member in (8) is much smaller than the last term of the right member. In the presence of skin effects, Bourdarot (1996) indicates two quantitative criteria used in the petroleum engineering to estimate the storage effect duration $t_{wS}$ (s) for a PW tapping a single continuum aquifer, Ramey's criterion:

$$t_{wS} > \frac{2.725 \times 10^{-5} \rho g S_w}{K_f b_f} (200 + 12\sigma_w) \tag{11}$$

and Chen and Brigham's criterion:

$$t_{wS} > \frac{2.725 \times 10^{-5} \rho g S_w}{K_f b_f} 170 \exp(0.14\sigma_w) \tag{12}$$

where ρ ($kg \cdot m^3$) is the water mass density and g ($m \cdot s^{-2}$) is the gravity acceleration. The differential term in (9) represents the skin effect $h_{\sigma w}$ (m). Once the PW storage effect is negligible, $h_{\sigma w}$ can be rewritten by applying Darcy's law at the well-rock interface (8) as follows:

$$h_{\sigma w}(Q) = \sigma_w(Q)\, Q/(K_f b_f^{3-n} \alpha_n r_w^{n-2}) \tag{13}$$

Transient flow rates tests, such as step drawdown tests, harmonic tests (periodic flow rate perturbation), and slug tests (short flow impulse), as well as flow rate fluctuations due to pump regime of backpressure fluctuations, can be modelled by applying the superposition principle. In order to model potential non-linearity due to the PW skin effect, this principle can be applied as follows (Lods and Gouze, 2004). For a flow rate history $\{(t_i, Q_i), i = 1, n\}$, where the rate $Q_i$ is applied between times $t_i$ and $t_{i+1}$, the superimposed drawdown $h_{sup}$ at time t ($t_i \leqslant t < t_{i+1}$) is:

$$h_{sup}(t) = \sum_{j=1}^{i-1} [h(t - t_j, Q_j) - h(t - t_{j+1}, Q_j)] + h(t - t_i, Q_i) \tag{14}$$

where h(t, Q) is the drawdown at time t, produced by a flow pulse Q beginning at time zero. Hence, according with equation (13), the superimposed PW skin effect when wellbore storage effect is negligible is $h_{\sigma w}(Q_i)$.

### 2.2 *Analytical solutions*

Dimensionless drawdown solutions in the PW and in the aquifer are obtained analytically in the Laplace domain. By introducing the dimensionless variables:

$$t_D = 4K_f t/[(S_{sf} + S_{sm}) r_w^2] \tag{15}$$

$$S_{wD} = S_w/[\pi^{n/2} b_f^{3-n} r_w^n (S_{sf} + S_{sm})] \tag{16}$$

$$r_D = r/r_w \tag{17}$$

$$B_D = B/r_w \tag{18}$$

$$h_{iD} = h_i 4\pi^{n/2} K_f b_f^{3-n}/(Q r_w^{2-n}),\ i = w, f, m \tag{19}$$

where $S_{sm} = 0$ for the leakance model, and applying Laplace transform with respect to $t_D$, one obtains (Hamm and Bidaux, 1994, 1996):

$$\bar{h}_{wD} = \left[ p \left( pS_{wD} + \frac{1}{2\Gamma(n/2)(K^{\nu}_{\nu-1}(\sigma) + \sigma_w)} \right) \right]^{-1} \tag{20}$$

$$\bar{h}_{fD} = \frac{2\Gamma(n/2)}{p(1 + 2\Gamma(n/2)pS_{wD}(K^{\nu}_{\nu-1}(\sigma) + \sigma_w))} \frac{r_D^{\nu} K_\nu(\sigma r_D)}{\sigma K_{\nu-1}(\sigma)} \tag{21}$$

$$\bar{h}_{mD} = \frac{\bar{h}_{fD}}{1 + \sigma_f \eta \tanh(\eta)} [\cosh(\eta z_D) - \tanh(\eta)\sinh(\eta z_D)] \tag{22}$$

$$\langle \bar{h}_{mD} \rangle = \zeta \bar{h}_{fD} \tag{23}$$

where overline denotes Laplace transform, p the Laplace variable. $K_\nu$ the modified Bessel function of second kind and order $\nu$, $\langle h_m \rangle$ is the z-average of $h_m$, and:

$$\sigma^2 = 4p + 1/B_D \quad \text{for the leakance model} \tag{24}$$

$$\sigma^2 = 4p[\omega + \zeta(1 - \omega)] \quad \text{for the double porosity model} \tag{25}$$

$$\zeta = f(\eta)/[1 + \sigma_f \eta^2 f(\eta)] \tag{26}$$

$$\eta^2 = 4p(1 - \omega)/\lambda \tag{27}$$

$$f(x) = \tanh(x)/x \tag{28}$$

$$v = 1 - n/2 \tag{29}$$

$$K^{\nu}_{\nu-1}(x) = K_\nu(x)/xK_{\nu-1}(x) \tag{30}$$

Solutions (20-23) are implemented and numerically inverted in the time domain using Stehfest's algorithm (1970), in the WTFM software (Lods and Gouze, 2004, ftp://saphir.dstu.univ-montp2.fr/TPHY/lods/wtfm ), which offers a user-friendly interface.

The generalized line source solution, which is the modified form of the Theis's model (2D flow) for fractal dimension flow, assumes an infinitesimal well radius and no skin effect. For a single porosity, the Laplace transform can be inverted analytically (Barker, 1988):

$$h_f = \frac{Qr^{2-n}}{4\pi^{n/2}K_f b_f^{3-n}}\Gamma(-\nu, r_D^2/t_D), \qquad \nu < 1 \tag{31}$$

where $\Gamma(-\nu, u)$ is the complementary incomplete gamma function, which for $n = 2$, reduces to the integral exponential. The asymptotic form of (31) allows investigating the flow dimension:

$$h_f = \frac{Qr^{2-n}}{4\pi^{n/2}K_f b_f^{3-n}\nu}\left[\left(\frac{4K_f t}{S_{sf}}\right)^{\nu} - \Gamma(1-\nu)r^{2\nu}\right], \qquad \nu \neq 0 \tag{32}$$

It is an extension of the logarithmic approximation (Cooper and Jacob, 1946). Asymptotically, when $n < 2$, the time dependent term dominates, and the late time slope of the logarithmic derivative is $\nu$ on bi-logarithmic plot (Figure 2), whereas when $n > 2$, the time dependent term tends to zero and a steady state can be achieved.

## 3 MODELS FEATURES

The models behaviours are presented with synthetic examples sharing the parameters presented in Tab. 1, of PZ (piezometer) responses with radial distance from PW of 1 m and a flow rate of 30 L/mn, illustrated in Figure 2. The leakance effect (1PorLeak) tends to stabilize the heads. The double porosity effect (2Por models) can be appreciated as follows. For a PZ intercepting the fractures network (2PorFiss), at the beginning of the test, essentially only the network is drained, and the double porosity response is identical to that of a single porosity (1Por) with conductivity $K_f$ and specific storativity $S_{sf}$. Then progressively the second porosity is drained and the response reaches that of a single porosity (1PorStot) with conductivity $K_f$ and specific storativity $S_{sf} + S_{sm}$. For a PZ in the matrix (2PorMatr), the response is delayed, and reaches that of the 1PorStot model, at the same time. Finally, for the PZ of the double porosity model, the use of the matrix mean drawdown (2PorMatrMean) allows reducing the parameters number by eliminating z. The logarithmic derivative ($\partial h/\partial \ln(t)$), more sensitive than its primitive, constitutes another fitting curve, and is helpful for discriminating the model type (Bourdet et al., 1983). The numerically computed derivatives of the 1Por and 2Por models are plotted on the right plot of Figure 2. PW storage produces an initial linear increase, followed as time increases, by a decrease of the drawdown derivative. For a PZ intercepting the fractures network, the double porosity effect displays a U-shaped perturbation, corresponding to the progressive

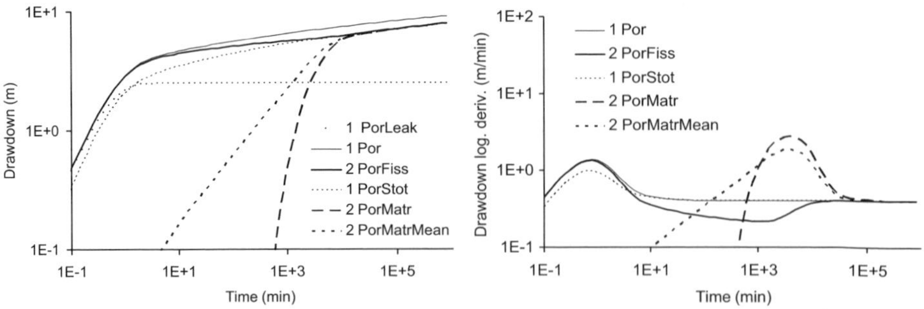


Figure 2. Synthetic PZ responses built with the parameters presented in Table. 1, showing the effects of leakance (1PorLeak), and double porosity: piezometer intercepting the fractures network alone (1Por), the fractures network embedded in the matrix (2PorFiss), the fractures network alone with storativity increased with that of the matrix (1PorStot), piezometer in the matrix (2PorMatr), and mean drawdown in the matrix (2PorMatrMean).

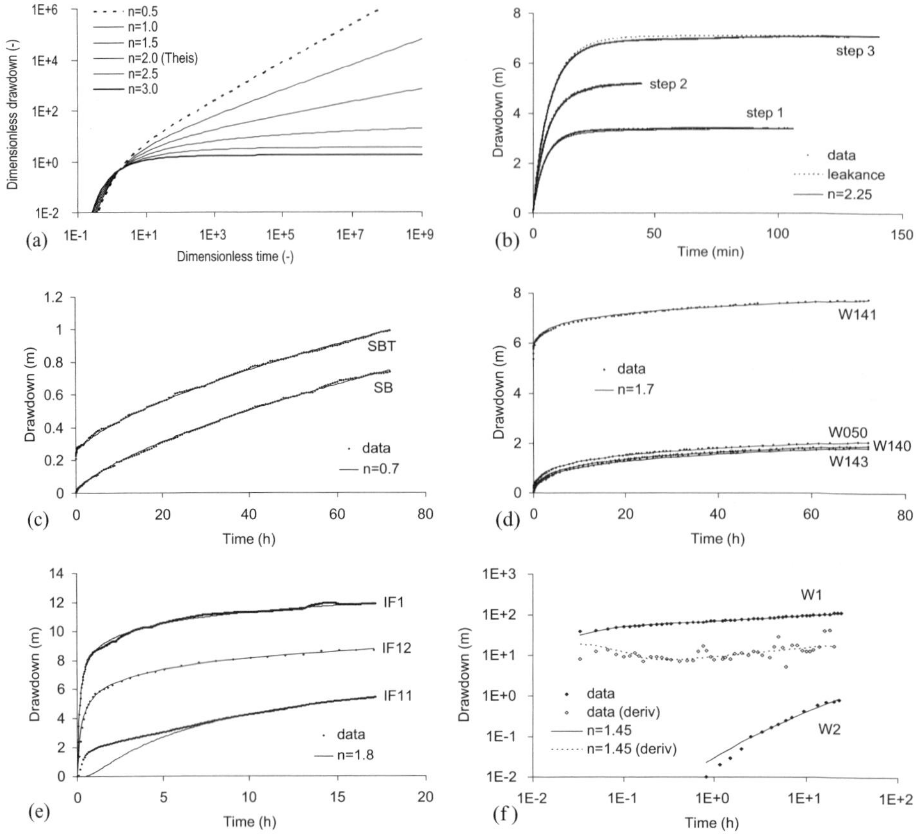


Figure 3. (a) Dimensionless generalized line source solution : influence of the flow dimension n. (b) Case study 1: step drawdown test in the fractured limestone aquifer. (c) Case study 2: pump test in the fractured harzburgite aquifer. (d) Case study 3: pump test in the fractured gabbro aquifer. (e) Case study 4: pump test in the fractured granite aquifer. (f) Case study 5: pump test in the fractured granite aquifer.

Table 1. Parameters set for the synthetic examples (SE) and the 5 field case studies (CS 1 to CS 5) discussed in the text.

| | SE | CS 1 | CS 1 | CS 2 | CS 3 | CS 4 | CS 5 |
|---|---|---|---|---|---|---|---|
| Model type | | 1 Por. Leak. | 1 Por. Fract. | 2 Por. Fract. | 1 Por. Fract. | 2 Por. Fract. | 2 Por. Fract. |
| $r_w$(m) | 0.05 | 0.055 | 0.055 | 0.1 | 0.108 | 0.15 | 0.1 |
| $r_{ws}$(m)$^{(*)}$ | 0.03 | 0.052/ 0.054 | 0.052/ 0.0557 | 0.1 | 0.12 | 0.15 | 0.0757 |
| $\sigma_w$ (−)$^{(*)}$ | 0 | 3.2/5.8 | 1.88/3.4 | 6230 | 59.9 | −5.38 | 230 |
| $K_f\, b_f^{3-n}$ ($m^{4-n} \cdot s^{-1}$) | 1E−4 | 4.15E−5 | 4.15E−5 | 2.90E+1 | 2.78E−2 | 1.62E−4 | 3.25E−3 |
| $S_{sf}\, b_f^{3-n}$ ($m^{2-n}$) | 1E−6 | 3.50E−4 | 3.50E−4 | 2.46E−2 | 1.03E−1 | 5.00E−6 | 1.16E−4 |
| n (−) | 2 | 2 | 2.25 | 0.7 | 1.7 | 1.8 | 1.45 |
| $\lambda$ (−) | 1E−9 | – | – | 4.87E−5 | – | 1.39E−8 | 1.79E−11 |
| $\omega$ (−) | 5E−2 | – | – | 2.59E−4 | – | 5.55E−4 | 9.10E−2 |
| $\sigma_f$ (−) | 0 | – | – | 5 | – | 0 | 0 |
| $z_D$ (−) | 1 | – | – | – | – | 0.016 | – |
| B (m) | 20 | 11 | – | – | – | – | – |

* Minimum and maximum values are reported.

drainage of the matrix. It can be superimposed with PW storage effect. The later time part of the curve allows estimating the flow dimension by using the asymptotic form of the line source solutions. Figure 3a illustrates the flow dimension influence.

## 4 EXAMPLES OF APPLICATION

Case study 1 shows an application of both high dimensions and leakance flow emphasizing their similar behaviour. The L1 hole, drilled down to 100 m, in the fractured limestones of the Montpellier University, was tested at 3 flow rates, 6.14, 8.75 and 9.89 L/min, with recovery period after each of the pumping steps. Drawdown curves (Figure 3b) suggest a leakage effect, a high dimension flow, or the presence of a recharge boundary. With identical generalized transmissivity and storativity (Tab. 1), one obtains the same overall quality level of the data fitting, whether a simple leakance model or a flow of dimension 2.25 is considered. The similar behaviour is due to the fact that a flow with $n > 2$ can account for by a vertical exchange with the aquitard. Step 1 is slightly better fitted by the simple leakance, while the others are slightly better fitted with a flow of dimension 2.25. The slight differences in well capacities can be explained by weak variations in pump discharge and/or of the well capacitive radius. The determination of flow boundary is not possible, because it would require the knowledge of storage properties, that could have been obtained from the data recorded in a distant PZ. The parameters of the PW skin have been fitted with $p = 2$. Hence, with 3 steps and at least 2 steps above the critical rate, the 3 parameters $\sigma_{wl}$, $\sigma_{wt}$, $Q_c$ are unique. For both models, one obtains a critical rate between those of steps 1 and 2. The non linear head loss seems to be due to turbulence within the formation in the vicinity of the well and not in the well itself, because it is open, and displays in-well Reynolds numbers less than 1700.

Case study 2 is an application of low-dimensional flow (Figure 3c). The SB and SBT holes are drilled, 26.2 m apart, down to about 40 m in a harzburgite massif of the ophiolitic complex of Oman (Dewandel, 2002). Comparing the weakness of the recorded drawdowns with the flow rate applied in SBT ($83.6\,m^3 \cdot h^{-1}$), indicates a highly permeable medium, but the drawdown curves do not stabilize. Modelling such behaviour using cylindrical flow would require the introduction of tight flow boundaries that are not observed on the site. With fractal dimension flow, this type of behaviour can be interpreted using low dimension flow. The best fit is obtained for a dimension of 0.7 (Tab. 1), assuming that the PZ intercepts the fractures network of a double porosity medium. The very low dimension indicates a possible anisotropy and/or the presence of multiple impervious boundaries. On the other hand, the high transmissivity and storativity both indicate the presence of highly connected fractures network in an unconfined aquifer. It should be noted, however, that confined solutions can be successfully applied to unconfined aquifers provided that the drawdown is small relative to aquifer thickness. The very high value for the PW skin factor can be explained by turbulent head losses (related to high in-well Reynolds numbers, $>10^5$) and more importantly by a geometrical effect related to the PW shape for low flow dimensions. The skin factor produces the same effect as the correction coefficient used in case of partial penetration (Hantush, 1961). For example, the analysis of a pump test in a homogeneous confined aquifer, laterally bounded by two tight parallel boundaries, using a dimension of 1, gives a high skin factor reflecting the additional head loss caused by the flow convergence toward the well. The high value for $\lambda$ indicates an important block permeability, which could suggest that the double porosity approach is useless. However the value of $\sigma_f$ indicates high head loss at the matrix-fracture interface, possibly due to mineral coating.

Case study 3 shows a relevant application of fractal dimension flow in fractured gabbros of the ophiolitic complex of Oman (Figure 3d). The site comprises a PW (W141), pumped at $75.6\,m^3 \cdot h^{-1}$, and three PZs, (W050, W140 and W143), at distances of 10.1 m, 14.6 m, and 20 m from the PW, respectively (Dewandel, 2002). These holes are drilled down to about 40 m. With a dimension of 2, none of the curves can be fitted accurately, even with double porosity, whereas a dimension of 1.7 allows fitting with a single porosity, simultaneously three of the four curves with exactly the same parameters (Tab. 1). A tortuosity coefficient slightly larger than 1 (i.e., PZ distance increases with a rate of 1.1), is necessary to fit the remaining PZ, which is not surprising in fractured media, and denotes probably a higher flow path heterogeneity in that direction. The high transmissivity and storativity indicate the presence of high conductivity fractures in an unconfined aquifer. The large PW skin factor can be explained by a PW geometrical effect in dimension 1.7 as described above, and turbulence in the PW, where the in-well Reynolds number overflows $10^5$.

Case study 4 is an application to a pump test ($4.5\,m^3 \cdot h^{-1}$) in fractured granites in India showing atypical results (Maréchal et al., 2003). The two PZ, IF11 and IF12, located at the same distance, 22 m, from the PW IF1, react at the same time, but with different behaviours (Figure 3e). The dissimilarity cannot be explained by property anisotropies because the PZs are located symmetrically apart from the PW. The fit was performed by considering IF11 and IF12 in the matrix and in the fracture respectively. The best fit is obtained for a dimension of 1.8 (Table 1). In this case, IF1 and IF12 are very well fitted at all times, whereas only the last ten hours of IF11 can be completely fitted. That can be explained by the de-saturation, during the test, of a sub-domain of the fractures network that was initially saturated. The negative PW skin factor indicates the interception by the PW of particularly open fractures.

Case study 5 shows an excellent simultaneous fit using the double porosity approach, of the drawdown at the PW ($26.25\,m^3 \cdot h^{-1}$) and at the PZ, with the former fitted with the mean matrix drawdown (Figure 3f, Table 1). The rock formation, located in South Korea, is made of fractured granites. Depth of the PW is 600 m, whereas that of the PZ, located at a distance of 240 m, is unknown (Lim et al., 1990). With a cylindrical flow, the data recorded at the PZ can be well fitted, but not the PW response. With a flow dimension of 1.45, both curves are fitted. The important PW skin factor can be explained by a PW geometrical effect in dimension 1.45 as described above, and turbulence in the well, where the Reynolds number exceeds $4 \cdot 10^4$.

For the last three examples, the flow dimensions obtained (1.7, 1.8 and 1.45) are often observed in the field. This could be explained by the anomalously slow diffusion occurring namely on percolation networks, due to dead ends and bottlenecks which slow down diffusion (Stauffer and Ahorony, 1992), producing flow dimensions less than that of the basic 2D flow.

## 5 CONCLUSIONS

These set of generalized models, based on averaged properties, improves our ability to assess medium heterogeneity in distinctly different situations. These models are particularly appropriate for the analysis of packer tests, which are likely to produce flows with high dimension. Specifically, they allow fitting accurately the initial times which are affected by PW storage, and characterizing the PW skin effect which determines the operating rate for abstraction wells. The selected examples cover a large range of behaviours and demonstrate the models ability to differentiate the parameters that control the drawdown trends. In many cases, these generalized models give a very powerful analysis tool at the scale of the pump tests. In particular, the combination of the double continuum approach with fractal dimension flow offers a wide range of behaviours with a limited number of parameters, allowing an accurate match of the data, and, as a consequence, improved predictions.

## ACKNOWLEDGEMENTS

This research is partially funded by the European Commission-RDG-(EKVT1-CT-00091 ALIANCE).

## REFERENCES

Barenblatt G.R., Zheltov I.P., Kochina I.N. (1960) Basic concepts in the theory of seapage of homogeneous liquids in fractured rocks. J. Appl. Math. Mech., Engl Transl. 24(5):1286–1303.

Barker J.A. (1988) A generalized radial flow model for hydraulic tests in fractured rock. Water Resources Research 24(10):1796–1804.

Boulton N.S., Streltsova T.D. (1977) Unsteady flow to a pumped well in a fissured water-bearing formation. J. of Hydrology 35:257–269.

Bourdarot G. (1996) Essais de puits: méthodes d'interprétation. Ed. Technip, Paris, 352 pp.
Bourdet D., Ayoub J.A., Whittle T.M., Pirard Y.M., Kniazeff V. (1983) Interpreting well tests in fractured reservoirs. World Oil 1975, 77–87.
Cooper H., Jacob C. (1946) A generalized graphical method for evaluating formation constants and summarizing well field history. Trans. Amer. Geophys. Union 27, 526–534.
Dewandel B. (2002) Structure and hydrogeological functioning of a discontinuous aquifer: the Oman ophiolite. Mémoires Géosciences Montpellier 26, ISBN-2-912411-25-4, 328 pp.
Everdingen A.F. van, Hurst W. (1953) The skin effect and its influence on the productive capacity of the well. Trans. Am. Inst. Min. Metall. Pet. Eng. 198:171–176.
Hamm S.-Y., Bidaux P. (1994) Transient flow with fractal geometry and leakage : theory and application. C.R. Académie des Sciences, Paris, 318, série II, n° 2:227–233.
Hamm S.-Y., Bidaux P. (1996) Dual-porosity fractal models for transient flow analysis in fractured rocks. Water Resources Research 32(9):2733–2745.
Hantush M., (1956) Analysis of data from pumping tests in leaky aquifers. Trans. Amer. Geophys. Union 37(6), 702–714.
Hantush M. (1961) Aquifer tests on partially penetrating wells. J. Hydraul. Div., Proc. Amer. Soc. Civil Eng. 87(HY5):171–195.
Jacob C.E. (1947) Drawdown test to determine effective radius of an artesian well. Trans. Proc. Amer. Soc. Civil Eng. 112, pap. 2321:1047–1064.
Le Borgne T., Bour O., de Dreuzy J.R., Davy P., Touchard F. (2004) Equivalent mean flow models for fractured aquifers: insights from a pumping tests scaling interpretation. Water Resources Research, 40, W03512, doi: 10.1029/2003WR002436.
Lim J.U., Kim Y.K., Kim J.K., Sung K.S., Kim S.Y. (1990) Investigation report of geothermal potential, Shynbook-myeon, Pocheon area, Rep. 43, Korea Inst. of Energy and Resources, 54 pp.
Lods G. (2000) Pump tests with flow logging in discontinuous heterogeneous aquifers, application in carbonate medium. Mémoires Géosciences Montpellier 20, ISBN-2-912411-19-X, 230 pp.
Lods G., Gouze Ph. (2004) WTFM, software for well test analysis in fractured media combining fractional flow with double porosity and leakance approaches. Computers & Geosciences, 30:937–947.
Maréchal J.C., Dewandel B., Subrahmanyam K., Torri R. (2003) Review of specific methods for the evaluation of hydraulic properties in fractured hard-rock aquifers. Current Science of India, August 2003, 85:511–516.
Moench A.F. (1984) Double porosity models for a fractured groundwater reservoir with fracture skin. Water Resources Research 20(7):831–846.
Papadopoulos I.S., Cooper H.H.Jr. (1967) Drawdown in a well of large diameter. Water Resources Research 3:241–244.
Rorabaugh M.J. (1953) Graphical and theoretical analysis of step-drawdown test of artesian well. Proc. Amer. Soc. Civil Eng. 79(362), 23pp.
Stauffer D., Ahorony A. (1992) Introduction to percolation theory. Taylor and Francis, Philadelphia, Pa, 2nd ed., 181pp.
Stehfest H. (1970) Numerical inversion of Laplace transforms. Comm. ACM 13(1):47–49.
Theis C.V. (1935) The relation between the lowering of the piezometric surface and the rate and duration of discharge of a well using ground water storage. Trans. Amer. Geophys. Union 16:519–526.
Warren J.E., Root P.J. (1963) The behaviour of naturally fractured reservoirs. SPE J 3(2):245–255.

CHAPTER 31

# Estimating groundwater flow rates in fractured metasediments: Clare Valley, South Australia

Andrew Love[1], Craig T Simmons[2], Peter Cook[3], Glenn A Harrington[4]
Andrew Herczeg[3] and Todd Halihan[5]
[1]*Research and Innovation Group DWLBC, St Adelaide SA,*
[2]*Flinders University of South Australia, Adelaide*
[3]*CSIRO Land and Water, Private Bag 2, Glen Osmond SA*
[4]*DWLBC, Mt Gambier SA*
[5]*Oklahoma State University, School of Geology, Stillwater, OK*

ABSTRACT: Different techniques were used to estimate vertical and horizontal flow rates in fractured rock aquifers of the Clare Valley, South Australia. Horizontal flow was estimated from well dilution tests, cubic law and a comparison of unpurged and purged radon concentrations. Vertical flow was estimated by the vertical distribution of environmental tracers to determine groundwater ages combined with a parallel plate model. A flow net analysis indicated that horizontal flow estimated from radon is consistent with recharge rates calculated from vertical groundwater age profiles. Horizontal flow calculated from well dilution tests and the cubic law both on a local and regional scale are not consistent with estimates of recharge. This is attributed to fractures being well connected on a local scale but poorly connected on a regional scale. The geological setting and fracture style in the Clare Valley supports the hypotheses of discontinuous regional horizontal fracturing resulting in a reduction of connectivity with increasing scale. Groundwater flow rates estimated from radon displayed large temporal variability varying by a factor of 3 through out the year. This variability is larger than can be accounted for by seasonal variations in the regional water table. We attribute this to a decrease in fracture connectivity with depth, where the highly weathered fracture zone at the top of the water table becomes disconnected from the regional flow system when the water table declines.

## 1 INTRODUCTION

Whilst many previous studies (see NRC, 1996 for comprehensive review) have investigated fractured rock aquifers with contaminant transport as a primary driver, this paper examines the movement of water and solutes in a fractured meta-sedimentary aquifer directed towards water resources analyses. The goal for sustainable water resource management requires estimates of volumetric flow rates and, not estimates of flow rates through individual fractures. The key characteristic of fractured rock aquifers is extreme spatial variability in hydraulic conductivity. As a result of the complex physical framework of fractured formations, the identification and connection of fractures are difficult to characterise, which in turn makes estimating average volumetric flow rates technically challenging. For example, traditional hydraulic approaches that rely on average hydraulic properties to estimate flow in porous media are not directly transferable to water resource investigations in fractured rock aquifers. In addition, small-scale (fracture scale) approaches often used in contaminant studies may not be directly transferable as they estimate groundwater flow through individual fractures and not volumetric flow through the fractured formation. On a regional scale groundwater flow and transport in fractured rock are controlled by the connectivity of the fracture network. This raises an important but unresolved question of how to estimate large-scale properties from small-scale field measurements because the scale of our field measurements is not commonly commensurate with the scale of the problem.

The study site is the Clare Valley where a major viticultural industry thrives. Its water supply is sourced from fractured metasediments of Proterozoic age. The primary objective of the study is to provide water resource mangers with volumetric flow rates to assist in the development of a water allocation policy. The use of environmental tracers for estimating horizontal flow and groundwater recharge rates has been discussed in a number of recent publications (Cook et al. 1999, Cook and Simmons 2000, Love et al. 2002, Love 2003 and Cook et al. 2005). In this manuscript we compare the results from the different techniques and discuss the implication for groundwater flow and connectivity within the Clare Valley. The primary objectives of this paper are to:

1. Compare estimates of horizontal groundwater flow rates calculated from hydraulic data (cubic law) with other recently developed methods by relating estimates of horizontal flow and recharge to those of regional flow by using a simple flow net analysis.
2. To determine what inferences can be made from our measurements about scale and fracture connectivity to improve conceptual understanding of the system.

### 1.1 *Groundwater flow rates*

In a porous media aquifer, groundwater flow rates are determined using Darcy's Law from estimates of hydraulic conductivity and hydraulic gradient. Alternatively if groundwater ages can be determined using environmental tracers ($^{14}C$, $^{36}Cl$ and CFC's) then estimates of flow rates can be determined (Cook and Herczeg 1999). However in fractured rock aquifers neither method is straightforward. Hydraulic conductivity is highly variable and estimates of flow using Darcy's Law may be unreliable. Similarly, matrix diffusion complicates environmental tracer interpretation because the apparent age of the tracer we may not reflect the age of the water.

### 1.2 *Scale and parameter estimation*

Consider of hydraulic conductivity versus the scale of the measurement. At very small scales the values of hydraulic conductivity will vary between the hydraulic conductivity of the matrix and that of a fracture, resulting in large variability. As the scale of the test increases, then the variability of hydraulic conductivity will be reduced. At a certain scale the measurements of hydraulic conductivity may approach a constant value equal to the aquifer hydraulic conductivity. This scale is referred to as a representative elementary volume (REV). Bear et al. (1993) defined this as the scale beyond which there are no significant statistical variations in the value of the hydraulic property of interest.

In fractured rock systems a REV may not exist (NRC, 1996; Cook, 2003). NRC notes that there are now two additional views on scales and continuity in fractured rock aquifers. The first is that the hydraulic conductivity increases with the scale of measurement. This scenario is based on the premise that as the scale of the test increases there is a greater probability of intersecting a large fracture. Thus, the hydraulic conductivity of the aquifer and fracture connectivity of the system increases. This may correspond to an aquifer system with a large number of small fractures but a small number of large fractures. The second is that hydraulic conductivity decreases with increasing scale of measurement and fracture connectivity is reduced. This scenario is a consequence of fractures having finite lengths. An example of such a system could be a fracture network dominated by vertical fractures with discontinuous horizontal fractures that do not connect the vertical fractures on a regional scale. In this situation it is unlikely that a REV exists.

## 2 STUDY SITE

The study site is the Wendouree Winery in the Clare Valley located approximately 100 km north of Adelaide South Australia (33°50′ S; 138°37′E). Groundwater occurs in a fractured carbonaceous dolomite of low matrix porosity and permeability. The aquifers at Wendouree are best described as a purely fractured media where there is minimal matrix porosity and effectively zero matrix permeability with all of the groundwater flow occurring in the facture network (Love et al. 2002, Cook 2003). Precipitation is 650 mm $yr^{-1}$ and potential evapotranspiration (ET) is approximately 1,975 mm $yr^{-1}$ (Love 2003). The major geological feature in the region is the Hill River Syncline. Most of the irrigation wells occur on the western limb of the syncline where the geological units are dipping vertically. The field site at Wendouree Winery is intensively instrumented with a set of 10 vertically nested piezometers up to 100 m depth as well as 7 wells with open completions up to 120 m. The local hydraulic gradient (dh/dx) varies from 0.0003 to 0.0007 throughout the year. The hydraulic gradient at the nested piezometer is 0.0005 vertically downwards. A regional groundwater divide occurs 2.5 kms from the site, which corresponds to a regional dh/dx of 0.026.

### *Geological structure and fracturing*

At Clare, fractures appear to be relatively evenly distributed on the well scale and in exposed outcrop. However, fracture spacings would vary in their spatial distribution due to variations in regional geology and stress regimes over large spatial scales. The ridge and valley geomorphologic style at Clare is a result of regional compressional tectonics (during the Delamerian Orogeny (Preiss 1987) and later tectonic activity during the Tertiary) that has resulted in a suite of rolling anticlines and synclines corresponding with the topography. In such a setting, the nature of the fracturing would vary spatially resulting in fracture clusters of different character depending upon their location in a geological context. For example, the hinge zones of anticlines (ridge lines) would correspond to a zone of maximum extension, where fractures may have large apertures but be widely spaced. Conversely the hinge zone of a syncline (valley floor) would correspond to a zone of maximum compression, where fractures are closely spaced but with a high probability of being closed. Bore yields in the Clare region confirm this pattern with higher yields occurring on ridges and along the western limb of the Hill River Syncline and lower yields within the hinge zone of the syncline. At Wendouree the bedding plane is near vertical (i.e., limb of an eroded inclined syncline) in this extensional setting fractures are likely to be near vertical (as observed in core and outcrop) with minimal cross cutting horizontal fractures. This would suggest that any horizontal fractures are unlikely to be regionally extensive.

Fracture data were collected from an outcrop of the Saddleworth Formation 2 kms NE of Wendouree. The outcrop is a vertical wall which represents a fractured bedding plane of ~50 m in length and 6 m in height. Fracture spacing data were collected along vertical and horizontal scan lines. The mean fracture spacing for the horizontal scan line was 2B = 0.16 m (which is used to estimate vertical fracture spacings (Halihan et al. 1999). While the mean fracture spacing measured from core at Wendouree was 2B = 0.33 m, this represents a vertical scan line and is used to determine horizontal fracture spacing (Love 2003).

## 3 ESTIMATES OF FLOW RATES

We estimated horizontal and vertical flow rates at the well scale.

### 3.1 *Horizontal flow*

Horizontal flew rates were estimated using hydraulic methods, well dilution tests, and radon tracer data.

#### 3.1.1 *Hydraulic methods*

Single well aquifer hydraulic tests can be a useful initial approach to estimate equivalent porous media properties of discrete fracture flow (Lapcevic et al. 1999). We equate our measured values of bulk hydraulic conductivity from pumping tests to discrete fracture properties from the following:

$$K_b = \frac{\rho g (2b)^3}{12\mu(2B)} \tag{1}$$

where, $K_b$ is the bulk hydraulic conductivity over the test interval [$LT^{-1}$] (where $K_b$ = transmissivity/test interval), ρ is the fluid density, g is the acceleration due to gravity, 2b is the fracture aperture, μ is the kinematic viscosity and 2B is equal the fracture spacing (which is equal to the test interval if we assume only one fracture) [L]. Alternatively by rearranging equation (1) we can estimate an equivalent fracture aperture ($2b_{eq}$) from the pumping test:

$$2b_{eq} = \left(\frac{12\mu(2B)K_b}{\rho g}\right)^{1/3} \tag{2}$$

If we assume that all groundwater flow occurs in fractures and there is no flow in the matrix, then the effective porosity for groundwater flow is the ratio of fracture aperture to fracture spacing (i.e., 2b/2B). Hence the groundwater velocity through the fractures, $v_{ave}$ [$LT^{-1}$] can be determined by combining equation (1) with Darcy' Law to give.

$$v_{ave} = \frac{\rho g}{12\mu}(2b_{eq})^2 \frac{dh}{dx} \tag{3}$$

Where in addition to the previously defined parameter dh/dx is the hydraulic gradient [−]. In this way the product of $2b_{eq}$ and $v_{ave}$ multiplied by the number of fractures in a particular interval can determine estimates of horizontal groundwater flow rates.

Single well constant discharge tests were performed on the nested piezometers at Wendouree (Love et al. 2002). Bulk hydraulic conductivity ($K_b$) obtained from the nested piezometers at Wendouree ranged from 7–10 m $day^{-1}$ in the upper 40 m to an average of ~0.1 m $day^{-1}$ at depths >40 m. The aquifer pumping tests are influenced by both vertical and horizontal hydraulic conductivity. The presence of vertical and sub vertical fractures in the core and at outcrop indicates the importance of vertical flow at the site. For our analyses here, we assume that values of $K_b$ are the same in both the horizontal and vertical direction.

Below, piezometer values of $K_b$ are converted to estimates of equivalent hydraulic apertures and average velocities through the fractures. The average value of $K_b$ obtained from the nested piezometers is approximately 8 m $day^{-1}$ in the top 40 m. Assuming a horizontal fracture spacing (2B) of 0.33 m (from core mapping) and assuming that all of the fractures are identical, then this would convert to an average aperture of 330 μm and an average velocity through each fracture of 40 m $day^{-1}$ (where dh/dx = 0.005). For depths below 40 m and where $K_b < 0.1$ m $day^{-1}$, using 2B = 0.33 m values of $2b_{eq} < 78$ μm and $v_{ave} < 2$ m $day^{-1}$ are obtained.

Estimates of equivalent porous media (EPM) flow were determined from variations in the local hydraulic gradient at Wendouree (dh/dx = 0.003 to 0.007) to vary from 1 to 2.4 $m^2$ $day^{-1}$ which corresponded to seasonal variations in water table (expressed as flow per unit width). Furthermore regional flow rates were also estimated from values of $K_b$ from Wendouree pumping tests in the upper 40 m and assuming a regional hydraulic gradient of dh/dx = 0.026 from a water table elevation map to be 8.5 $m^2$ $day^{-1}$.

### 3.1.2 *Point and well dilution tests*

The point dilution method is based on injecting a tracer into an interval of a well isolated by packers and monitoring the concentration of the tracer with time as it is flushed by regional groundwater flow. The groundwater flow rate is related to the time it takes for the tracer to be removed from the test interval. The faster the tracer is removed the faster the horizontal flow. For a perfectly mixed well the changes in concentration within the well can be expressed as:

$$C(t) = C_\infty + (C_\infty - C_o) \exp\left[\frac{-2qt}{\pi r}\right] \tag{4}$$

where, C = concentration in the test interval $[ML^{-3}]$, t = time measured after injection of the tracer, $C_\infty$ is the concentration in the well prior to injection of the tracer (assumed to be equal to the concentration in the aquifer in the vicinity of the well), $C_o$ = concentration of the tracer in the well immediately after injection, q is the horizontal flow rate (or Darcy velocity) through the test interval (assumed to be constant throughout the duration of the test) $[LT^{-1}]$. The important assumption of the test is that the tracer concentration ($C_o$) is completely mixed through the entire packed-off volume for the entire test.

We have developed a variation on this technique, the Well Dilution method, which is an adaptation of the point dilution tests but covers the entire well. The well column is disturbed by pumping water from the bottom of the well and re-injecting it into the top of the well (so that no water is removed from the well). The method requires a contrast in concentration between water in the well, $C_\infty$ and a tracer $C_0$. For aquifers where groundwater EC increases with depth no added tracer is required. For cases where there is no stratification of EC in the well, a concentrated tracer such as NaCl is slowly added into the injection line for the whole time the well is pumped. The pump is quickly removed and EC is monitored at different times as it returns to its pre mixed concentration. Unlike standard point dilution tests this method seeks to examine changes in flow rate with depth, rather than to mix packed off zones and measure a mean flow rate for the interval. Horizontal flow into the well can be estimated by monitoring the time required for the mixed profile $C_0$ to return back to its original concentration $C_\infty$. When $C_0$ approaches $C_\infty$ rapidly; high horizontal flow can be inferred. Conversely low horizontal flow is inferred when $C_0$ returns slowly to $C_\infty$. Under conditions of horizontal flow, the changes in concentration with time at each depth should be given by equation (4). If vertical flow is important then quantification of horizontal flow rates is more difficult and requires numerical modelling.

Well Dilution experiments were performed in October 1998 and May 2000, chosen to represent seasonal variation in the water table, 4 m below the ground surface in October and 8 m in May (Figure1). In these experiments we did not add tracer but took advantage of the natural EC stratification in the well. For both experiments the well was disturbed by pumping relative saline water from ~90 m and re-introducing the water at the top of the well. The experiments resulted in an inversion of the original EC profile with more saline water stratified over lower salinity water. EC was monitored in vertical profile at different time periods from 15 minutes (immediately after removing the pump) until 15 days as it

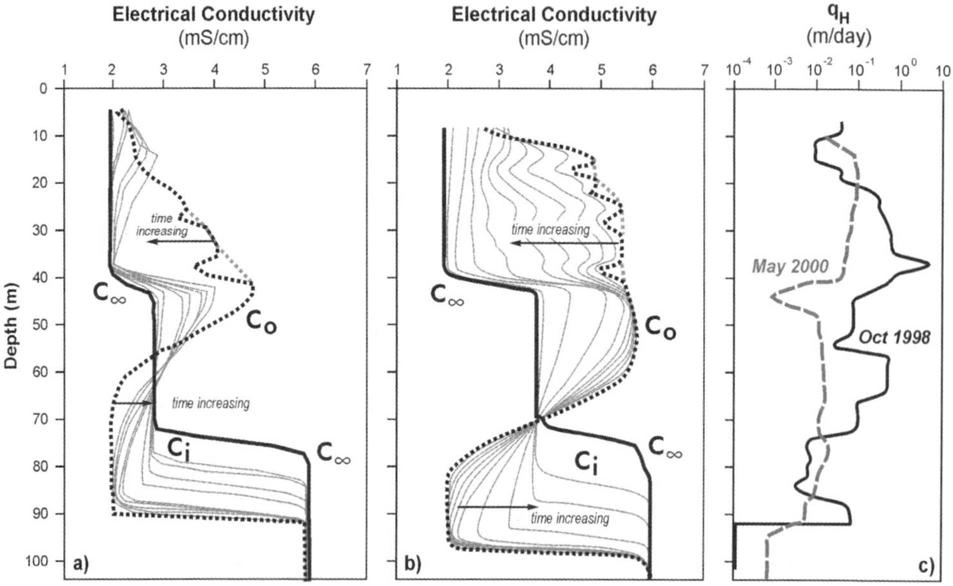


Figure 1. Results of Well Dilution experiments. (a) October 1998, (b) May 2000. The initial EC profile before mixing is denoted $C_\infty$, (solid thick line) which is assumed to reflect the EC profile in the aquifer. The initial profile immediately after mixing is denoted $C_0$ (dashed thick line) and the transient EC profiles (CI, dashed thin lines) as they evolve at different times after mixing. Arrows denote increasing time. (c) Results of the CMC modelling for horizontal flow through the well.

returned to its pre-mixed profile. In both experiments it was not critical to obtain a perfectly mixed well profile, rather what is important is to obtain a disturbed concentration profile ($C_o$) that is significantly different to the ambient concentration profile ($C_\infty$).

The tests can be examined qualitatively by examining the time taken for $C_o \rightarrow C_\infty$. In order to obtain semi-quantitative estimates of flow, a compartmental mixing cell (CMC) was developed to separate the effects of horizontal and vertical flow. The well is divided into a one-dimensional vertical domain that is discretized into a number of cells. Each cell receives inputs of mass (water and solute) both horizontally from the aquifer and vertically from the well. Further details of the modelling approach including issues of non-uniqueness are presented elsewhere (Love 2003).

A CMC model was run for different time periods in order to simulate the transient behaviour of the EC profiles. The total flow through the well was calculated from the final CMC model calibration by summing the product of the flow rate q in m $day^{-1}$ and the cell size. For October 1998 the value is $30\,m^2\,day^{-1}$ and $11.7\,m^2\,day^{-1}$ for May 2000. A correction needs to be made for hydrodynamic flow distortion where the flow rate through an open well is twice of that through the aquifer (Gasper 1972). This corresponds to horizontal flow through the aquifer of 15 and $5.9\,m^2\,day^{-1}$ for October 1998 and May 2000 respectively. Horizontal flow in October 1998 is approximately 2.5 times greater than May 2000. For October 1998 70 % of this flow occurs above 40 m while for May 2000 85 % of the flow occurs in the top 40 m. This clearly indicates a very active flow system in the upper 40 m with a significant reduction in horizontal flow below 40 m.

### 3.1.3 *Horizontal groundwater flow rates from Radon*

$^{222}Rn$ is produced from in situ decay of $^{238}U$ in the aquifer. Its concentration depends on the amount of uranium and thorium in the host rock as well as the pore space geometry. Because of its short half-life of 3.83 days, $^{222}Rn$ has been used to locate zones of active groundwater in flow into wells (Cook et al. 1999, Hamada 1999). If the aquifer mineralogy is fully homogeneous, then the concentration of radon in the well reflects groundwater flow rates. The locations of active groundwater into the well correspond to high $^{222}Rn$ concentrations. If there was low flow from the aquifer to the well, we expect radon concentrations in the unpurged well to be low due to radiogenic decay.

Cook et al. (1999) developed a new method to estimate horizontal flow rates by comparing unpurged to purged $^{222}Rn$ concentrations from the piezometer nest. If horizontal flow is faster than radon can decay, then concentrations of radon in the piezometers should approach that of the aquifer. Thus the ratio of radon concentrations in unpurged piezometers to radon concentrations in the aquifer ($C/C_o$) will be a function of radioactive decay and flow rate.

The method is a variation of a point dilution test, where instead of injecting a tracer into the well, an environmental tracer already in the groundwater is used. The changes in concentration within the well can be expressed as:

$$V\frac{dC}{dt} = CoQ - CQ - \lambda CV \tag{5}$$

Equation 5 assumes that all groundwater inflow into the well is horizontal throughout the well column, with no vertical flow in the well. $C_o$ and C are the concentrations of a particular solute in the aquifer and well respectively. Q is the volumetric flow rate through the well [$L^3 T^{-1}$], V is the volume of the well [$L^3$] and $\lambda$ is the decay constant [$T^{-1}$] for $^{222}$Rn ($\lambda = 0.18$ day$^{-1}$, we assume steady state conditions where dc/dt = 0. Substituting $V = \pi r^2 h$, where r is the radius of the well and h is the height of the interval tested, the groundwater flow rate, q, is given by equation (6):

$$q = \frac{C}{C - Co} \frac{\lambda \pi r}{2} \tag{6}$$

To calculate the flow rate, q, in $LT^{-1}$ the well volume is divided by the cross sectional area (Q = 2qrh). At high flow rates decay of radon during flow through the well is minimal therefore, the technique is not very accurate for high flow rates (Cook et al. 1999). Flow rates calculated through the well require a correction for hydrodynamic distortion to determine the flow rate in the aquifer. Because the well acts as an open conduit with high hydraulic conductivity, groundwater flow will converge towards the well. These results in flow rates through the well being greater than flow rates through the aquifer. Corrections for flow distortion in the well will vary with well geometry and hydraulic conductivity of gravel pack and piezometers [Gasper, 1987; Drost, 1968]. For the configuration of the nested piezometers at Wendouree the flow through the piezometer is approximately 4 times greater than the flow through the well.

Unpurged radon was sampled on three occasions (January 2001, December 2001 & June 2002) and purged radon concentrations were sampled twice in January 2001 and December 2001 from the piezometer nest at Wendouree (Figure 2). Variations in unpurged concentrations are a function of aquifer mineralogy and flow rates. The highest unpurged radon concentrations were measured in the upper 40 m ranging from 15 to 41 Bq $L^{-1}$.

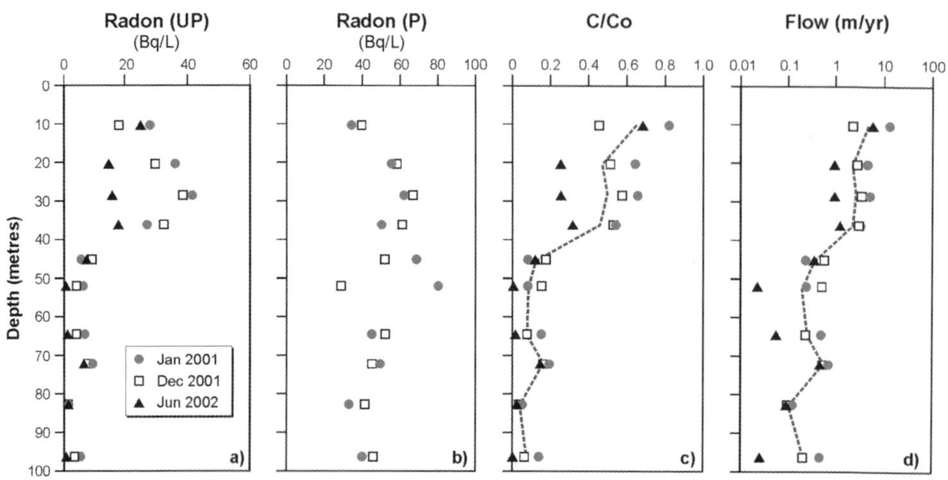


Figure 2. Radon profiles from the piezometer nest at Wendouree; a) Unpurged (UP) $^{222}$Rn (Bq $L^{-1}$): b) Purged (P) $^{222}$Rn (Bq $L^{-1}$): c) Unpurged to purged radon concentrations ($C/C_o$) the dashed line represents average values of $C/C_o$: d) Groundwater Flow rate (m year$^{-1}$) through the piezometer, where the dashed line represents the average flow rate. Sampling $^{222}$Rn (UP) occurred in January 2001, December 2001 and June 2002, while Purged (P) occurred in January 2001 & December 2001. All data are plotted at the mid point of the piezometer completion interval.

Below 40 m radon concentrations are $<10$ Bq $L^{-1}$. Radon concentrations are generally highest in January 01 and December 01 and decrease in June 02, which corresponds to periods of high and low water tables respectively. There is very little variation in purged radon concentrations between the two different sampling times (with the exception of 41497-1) and these purged values have been averaged to estimate horizontal flow rates for the different sampling times.

The ratio of $C/C_o$ (unpurged to purged radon concentrations) and the calculated horizontal flow rate from equation (6) for the nested piezometers are shown in Figure 2. The flow rates calculated from the piezometer data indicate relatively high horizontal flow in the upper section of the well and a significant reduction in horizontal flow with depth. The ratio of $C/C_o$ shows a large variability for the different sampling times with the largest values of $C/C_o$ occurring in January 2001 and December 2001 and reduced values of $C/C_o$ in June 2002. The higher $C/C_o$ ratios in January 2001 and December 2001 occur when the water table is close to the land surface (2.5 m) while the lower ratios of June 2002 correspond to when the water table is deeper (5.2 m below ground level).

The total volumetric flow rates through the well were calculated by a linear interpolation from the flow rates in the piezometers to be 0.68, 0.31 and 0.2 $m^2$ $day^{-1}$ for January 2001, December 2001 and June 2002 respectively. These flow rates through the well were converted to aquifer flow rates by correcting for flow distortion to range between .05 to 0.18 $m^2$ $day^{-1}$. The average annual flow rate through the aquifer was determined from average values of C and $C_o$ to be 30 $m^2$ $year^{-1}$. The largest percentage of this flow (88%) occurs in the upper 40 m.

### 3.2 *Vertical flow*

There are very few, if any, reliable methods for estimating vertical flow rates in fractured rock aquifers. Below we describe a new method for estimating vertical flow rates and recharge rates based on the vertical distribution of groundwater ages and the parallel plate model. For the parallel plate model we need to have knowledge of various aquifer parameters including fracture aperture (2b) that we obtained from aquifer pumping tests fracture spacing (from fracture mapping), and estimates of the matrix diffusion coefficient. We assume that groundwater flow occurs through vertical planar, planar parallel fractures with uniform matrix properties. For the case of a constant source conservative tracer, subject to radioactive decay, concentrations within the fractures can be related to vertical flow within the fractures $V_w$ by (Neretnieks 1981).

$$V_w = \left[1 + \frac{\theta_m D^{1/2}}{b\lambda^{1/2}} \tan h(BD^{-1/2}\lambda^{1/2})\right] / \left[\frac{\delta t_a}{\delta z}\right] \tag{7}$$

Where $V_w$ is the water velocity in the fracture $[LT^{-1}]$, $\theta_m$ is the matrix porosity $[-]$, D is the diffusion coefficient within the matrix $[L^2T^{-1}]$, b is the half aperture [L], B is the half fracture spacing [L], $\lambda$ is the $^{14}C$ decay constant $[T^{-1}]$ ($\lambda = 1.21 \times 10^{-4}$ $yr^{-1}$ for $^{14}C$) and $\delta ta/\delta z$ is the $^{14}C$ age gradient $[TL^{-1}]$. The above equation can also be applied to other radioactive tracers by substituting the appropriate decay constant and using the age gradient of that tracer. Cook and Simmons; (2000) substituted the decay constant for $^{14}C$ with the exponential growth rate for CFC-12 ($k = 0.06$ $yr^{-1}$) to enable vertical flow rates to be calculated from CFC-12 age gradients. The mean volumetric flow rate through the fracture, $Q_v$, is given by

$$Q_v = V_w \frac{b}{B} \tag{8}$$

Environmental tracer data ($^{14}C$, CFC-12 & $^3H$) at Wendouree indicate rapid vertical circulation in the top 40 m (Figure 3). The degree of vertical connection decreases with depth in the aquifer system with discontinuities in $^{14}C$ activities below the fractures at 38 and 52 m indicating reduced vertical flow. The sub-horizontal fractures at these depths intersect vertical flow and divert it horizontally.

In the upper flow system the CFC-12 age gradient is 0.92 yr $m^{-1}$ (23 year/25 m). Assuming an average bulk hydraulic conductivity, $K_b = 8\,m\,day^{-1}$ from the pumping test data at the piezometers and $B = 0.08\,m$ (where $2B = 0.16\,m$ is the mean vertical fracture spacing from the nearby Saddleworth outcrop) yields $b = 1.315 \times 10^{-4}\,m$ (from equation 3). Assuming $D = 10^{-4}\,m^2\,yr^{-1}$, $\theta_m = 0.02$ and substituting these values into equation (7) one obtains $V_w = 7.6\,m\,yr^{-1}$, $\theta_f = 1.64 \times 10^{-4}$ and $Q_v = 12.4\,mm\,yr^{-1}$. The value of $Q_v \sim 12.4\,mm\,yr^{-1}$ also represents the recharge rate at the site, assuming a parallel plate model with one-dimensional vertical flow.

Before applying a correction for retardation of $^{14}C$ due to matrix diffusion (equation 7) the radiocarbon data needs to be adjusted for any geochemical interactions that may modify the initial activity ($A_o$) of $^{14}C$. The results of this approach are detailed in Love 2003 and are not repeated here. Regardless of which geochemical correction scheme is chosen, all of the $^{14}C$ correction models give negative groundwater ages for upper four piezometers indicating that this groundwater has a thermonuclear $^{14}C$ component. This is consistent with thermonuclear $^3H$ and the presence of post 1960 CFC-12 concentrations. Hence no

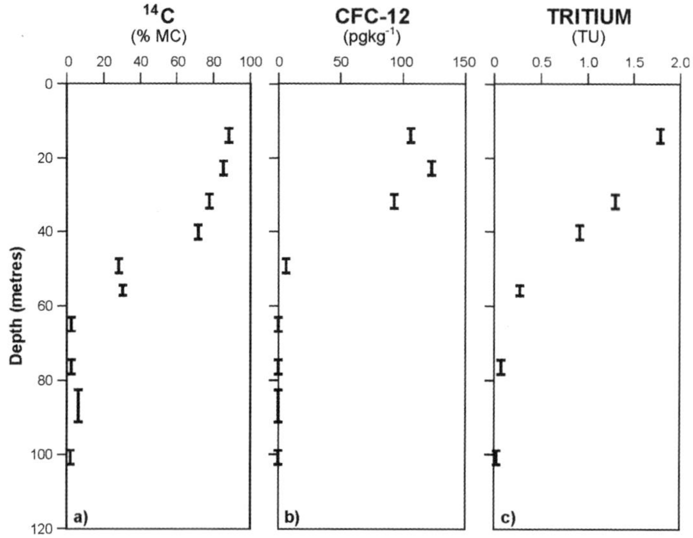


Figure 3. Environmental tracer data from piezometer nest at Wendouree, a) $^{14}C$ (%MC); b) CFC-12 (pg $kg^{-1}$); c) Tritium (TU), The depths are shown below ground level. The vertical bar represents the length of the piezometer interval.

$^{14}C$ age gradient can be determined in the top 38 m and instead an approximate $^{14}C$ age gradient of ~1 yr $m^{-1}$ is used that reflects the time since thermonuclear testing in the early 1960's (i.e., 40 years in ~40 m to reflect the $^{14}C$ "bomb" recharge pulse). For the upper flow system the following is used: B = 0.08 m (from the nearby outcrop which reflects the vertical half fracture spacing) yields b = $1.31 \times 10^{-4}$ m, $\theta_f = 1.64 \times 10^{-3}$, $V_w$ = 13.1 m $yr^{-1}$) and an estimated recharge value $Q_v$ = 21.6 mm $yr^{-1}$.

## 4 REGIONAL ESTIMATES OF GROUNDWATER FLOW

The methods used to estimate horizontal flow rates were all performed on individual wells. Single wells are often the only means available for fracture characterisation and estimating groundwater flow rates at depth. Data from single well tests are often used to scale up aquifer properties from the near well scale to larger spatial scales. However, it is not clear how representative scaling up is because of potential drilling artefacts, small spatial sampling scales and the assumption of regional continuity of aquifer properties may be invalid. Disturbance of the aquifer due to drilling can result in either an increase or a decrease in hydraulic conductivity near the well. The restricted sampling size may mean that flow rates can only be extrapolated over larger spatial scales by considering the aquifer as a continuum. In a fractured rock aquifer where flow is controlled by the inter connection of the fractured network, this assumption may be incorrect.

In order to assess whether or not estimates of horizontal flow at Wendouree can provide reasonable estimates of regional groundwater flow, a flow net calculation was performed. Estimates of volumetric horizontal flow rates ($q_H$) calculated previously from hydraulic, radon and well dilution methods can be related to those of regional flow from:

$$R = \frac{(q_{H_0} w - q_{H_1} w)}{L} \tag{9}$$

Where, R = recharge rate [$LT^{-1}$], $q_{H_0}$ and $q_{H_1}$ are the outflow and inflow rate [$L^2 T^{-1}$] respectively and w is the width of the flow tube (assumed to be of unit width) and L is the length to the flow divide [L]. From a water table contour map (Love 2003) it is assumed that the groundwater divide is at the same location as the surface water and topographic divide, and at this location $q_{H_1}$ = 0. Results of the analysis are presented in Table 1.

Table 1. Results of the flow net analysis to estimate recharge on a regional scale. For radon the range in $q_{H_0}$ is from June 2002 and January 2001, Well Dilution from May 2000 and October 1998 (for both the range of $q_{H_0}$ corresponds ~ to periods of low and high water tables). For EPM the range in $q_{H_0}$ corresponds to seasonal variation in the horizontal hydraulic gradient measured at the Wendouree field site (dh/dx = .003 to 0.007). Regional EPM horizontal flow was calculated assuming a regional hydraulic gradient of 65 m in 2.5 kms dh/dx = 0.026 from the water table elevation map. The distance to the flow divide (L) was assumed to be 2.5 kms.

| | $q_{H_0}$ ($m^2$ $day^{-1}$) | Recharge rate (mm $yr^{-1}$) |
|---|---|---|
| Radon | 0.05–0.18 | 7.3–26.3 |
| EPM | 1.0–2.4 | 146–350 |
| Well Dilution | 5.9–11.8 | 861–1723 |
| Regional EPM | 8.5 | 1241 |

The mean recharge rate from radon is in the range estimated from CFC-12 and $^{14}C$ age gradients. Recharge rates for both hydraulic approaches and the well dilution method are clearly unrealistic with values in some cases approaching or exceeding annual rainfall.

The above approach assumed the groundwater flow was related to a continuum of interconnected fractures on a regional scale. Although natural fractures occur over a wide range of spatial scales their interrelationships is not well understood. Fracture connectivity is a function of fracture spacing, orientation, and length. There is evidence to suggest that fractures are not well connected at the Wendouree field site. Field observations of fractures at the nearby outcrop indicate that many horizontal fractures are isolated with finite lengths. A large-scale pump test at a well at Wendouree indicated that for late time data, hydraulic conductivity decreased by a factor of two. Values of $K_b$ calculated from the piezometers were for pumping tests ranging from 80 to 150 minutes duration and are more likely representative of values close to the well (i.e., the size and number of fractures that intersect the well) and not for larger spatial scales.

## 5 DISCUSSION

Fracture connectivity at Clare appears to be strongly influenced by decrease in fracture connectivity with increasing scale, and seasonal variations in the water table.

### 5.1 *Decrease in fracture connectivity with increasing scale*

Whether or not a REV (with respect to either hydraulics or solutes) exits at Clare cannot be explicitly determined. It is possible that the sampling size of our tests are too small to determine the presence or not of a REV. However, there is evidence to suggest that with increasing scale that hydraulic conductivity and fracture connectivity decreases. Furthermore, the calculated values of groundwater flow from the hydraulic and well dilution methods are unrealistically high to represent regional flow. The most plausible model for fracture connectivity for Clare is one where with increasing scale the hydraulic conductivity and fracture connectivity of the system decreases. This model has similarities to the fractured rock system at Mirror Lake [Hsieh, 2000] where the fractured network was proposed to occur in a series of highly conductive clusters that are connected by fractures of a lower transmissivity.

At Clare although fractures appear to be ubiquitous on the well scale and exposed outcrop, one would not necessarily expect the fracture pattern to vary smoothly over larger spatial scales. This is because at large scales the distribution of fractures would largely be controlled by regional geological stresses. Although the relationship between geological structure and fracture distribution has not been well studied here, one can still make some broad scale inferences. The ridge and valley geomorphologic style at Clare is a result of regional compressional tectonics (during the Delamerian Orogeny and later tectonic activity during the Tertiary) that has resulted in a suite of rolling anticlines and synclines. In such a setting the nature of the fracturing would vary and result in fracture clusters of different character depending upon their relative location in a geological context. For example, the top of anticlinal structures (ridges) would correspond to a zone of maximum extension, where fractures may have large apertures but be widely spaced. Conversely the hinge of a syncline (valley floor) would correspond to a zone of maximum compression, where fractures are closely spaced but with a high probability of being closed. The overall fracture pattern at Clare is likely to be similar in style to Mirror Lake where clusters of transmissive and poorly transmissive fractures occur in different zones (as a result of varying tectonic stress, i.e. compression, extensional or intermediate between these two members). At Wendouree the bedding plane is near vertical (i.e., limb of an eroded inclined syncline) in this setting fractures are likely to be near vertical (as observed in core and outcrop) with minimal cross cutting horizontal fractures. This would suggest that any horizontal fractures are unlikely to be regionally extensive. The relationship between structural style and fracture patterns would require further work to elaborate on the above hypothesis.

## 5.2 *Seasonal variations in the water table*

As well as a conceptual model of fracture connectivity decreasing with scale, seasonal variation in the water table also appears to have an important influence on fracture connectivity. Both the well dilution and radon methods indicated large seasonal variation in groundwater flow with flow being ~1/3 lower at times when the water table was deeper. This suggests that the fractures near the surface (weathered zone, upper 10–15 m) are not connected to the flow system at times of the year when the water table is deep (May–June). When the water table rises the near surface fractures become re-connected to the flow system and horizontal flow increases. (Figure 4) This model of fracture connectivity has many similarities to the model proposed by Moore, (1996) at Oak Ridge.

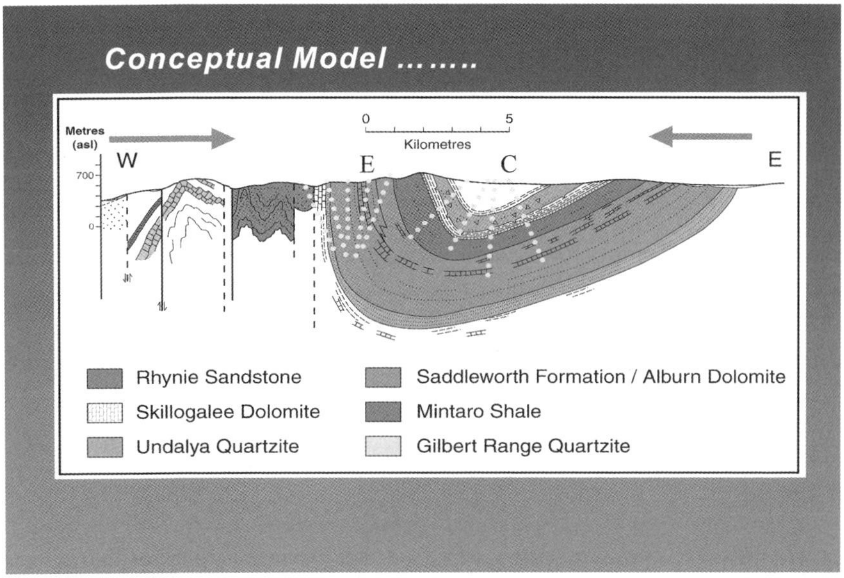


Figure 4. Schematic representation of the Clare Valley hydrogeological units. The main geological feature in the area is the Hill Valley Syncline which plunges ~10° to the N. The arrows represent the dominant direction of compressional tectonics in the region (i.e., E-W). EX represents zone of maximum extension while, C represents a zone of maximum compression. The dotted lines represent parallel to diverging fracture sets on the limb of the syncline. In such a setting we would expect minimal horizontal fracturing. This represents the zone of high bore yields where the majority of the irrigation activity occurs. Converging fracture sets occur in the hinge of the syncline that corresponds to very low bore yields. The fracture pattern results in a series of high and low transmissive fracture clusters.

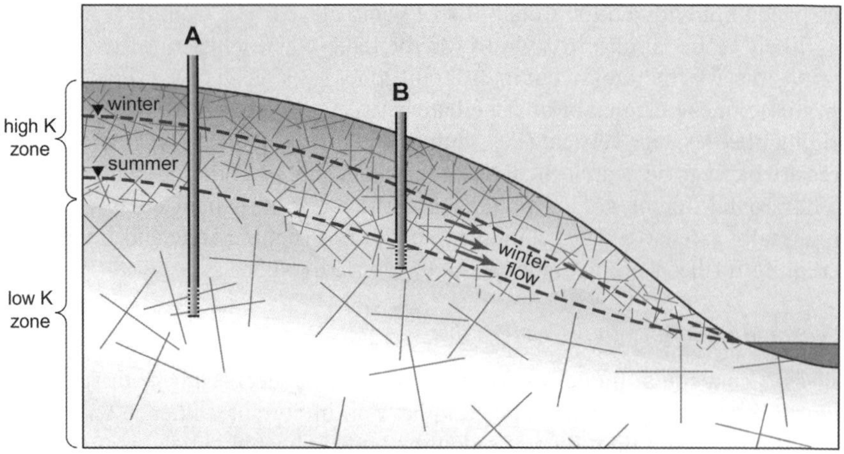


Figure 5. Schematic representation of flow within the Clare Valley, The intensity of fractures decreases with depth. On a regional scale it is postulated that hydraulic connections only occur through the shallow highly fractured zone. The well intercepts a number of fractures when the water table is close to the surface, when the water table falls the connection to the regional groundwater system is disconnected so that flow though these fractures decreases rapidly even though the regional hydraulic gradient is still relatively high (Figure reproduced from Cook 2003).

## 6 CONCLUSIONS

Groundwater flow rates using the Darcy's Law equivalent, the Cubic Law for fractured rock aquifers, are unrealistically high. It is suggested that the fracture network in the Clare Valley is well connected on a local scale but poorly connected on a regional scale. Groundwater recharge rates estimated from groundwater "ages" are consistent with horizontal flow rates determined from radon concentrations. Accurate estimates of groundwater flow rate to fractured rock aquifers remain a significant and yet unresolved issues for water resource managers. The main issue is parameterising Darcy's Law at the correct scale; a key to solving this problem is to obtain a greater understanding of the fracture network on a regional scale.

## REFERENCES

Bear J, C-Fu, Tsang G. de Marsily (eds), (1993) Flow and contaminant transport in fractured rock, Academic Press, Inc, ISBN 0-12-083980-6, 560 pp.

Cook PG, Love AJ, Dighton JC, (1999) Inferring groundwater flow in fractured rock from dissolved radon. Ground Water, 37(4): 606–610.

Cook PG. Herczeg AL, (eds) (1999) Environmental Tracers in Subsurface Hydrology. Kluwer, Boston, 529 pp.

Cook PG. Simmons CT (2000) Using environmental tracers to constrain flow parameters in fractured rock aquifers; Clare Valley, South Australia. In: B. Faybishenko, P.A. Witherspoon and S.M. Benson (eds) Dynamics of Fluids in Fractured Rock. Geophysical Monograph, 1222 American Geophysical Union, 337–347.

Cook PG. (2003) A guide to regional groundwater flow in fractured rock aquifers Contributions from M. Williams, C. Simmons, A. Love, T. Halihan, G. Herbert and G. Heinson, ISBN 1 74008 233 8 Seaview Press, pp 108, 2003.

Cook PG. Love AJ. Robinson NI. Simmons CT (2005) Groundwater Ages in fractured rock aquifers. Journal of Hydrology 308:284–301.

Drost W, Klotz D, Koch A, Moser H, Neumaier F, Rauert W (1968) Point dilution methods of investigating ground water flow by means of radioactive isotopes. Water Resour Res 4(1):125–46.

Gasper E, Oncescu M (1972) Radioactive Tracers in Hydrology, 342pp., Elsevier, Amsterdam,..

Halihan T, Love AJ, Cook PG. (1999) Use of outcrop, well, and creek data to develop a regional conceptual model of the fractured rock aquifer of the Clare Valley, South Australia, Abstracts with Program, Geological Society of America Annual General Meeting, Denver, CO.

Hamada 1999. Estimation of groundwater flow rate using the decay of $^{222}$Rn in a well. Journal of Environmental Radioactivity, 47:1–13.

National Research Council (NRC) (1996) Rock fractures and fluid flow: Contemporary understanding and applications. National Acadamy Press, Washington DC, 551 pp.

Hsieh PA. (2000) A brief survey of hydraulic tests in fractured rock. In: Faybishenko B., Witherspoon P.A. and Benson S.M. (eds) Dynamics of fluid in fractured rock. Geophys. Monogr., 122, American Geophysical Union, 59–66, 2000.

Lapcevic PA, Novakoski KS, Sudicky EA (1999) Groundwater flow and solute transport in fractured media. In: The handbook of Groundwater Engineering. Ed. Jacques W Delleur, CRC Press, USA. Ch17, 1–38.

Love AJ, Cook PG, Harrington GA, Simmons CT. (2002) Groundwater flow in the Clare Valley DWR02.03.2002.

Love AJ (2003) Groundwater flow and solute transport dynamics in a fractured meta-sedimentary aquifer. PhD Dissertation FUSA.

Love AJ, Cook PG, Harrington GA, Simmons CT. (2002) Groundwater flow in the Clare Valley. Department for Water Resources, South Australia. Report DWR02.03.0002, 43 pp.

Love AJ, Cook PG, Halihan T, Simmons CT. (1999) Estimating groundwater flow rates in a fractured rock aquifer, Clare Valley, South Australia. Water 99. The Institution of Engineers Australia, 25th Hydrology and Water Resources Symposium. Brisbane, Australia, July 6–8, 1999.

Moore GK. (1996) Quantification of ground-water flow in fractured rock, Oak Ridge, Tennessee, Ground Water., 35(3):478–482, 1996.

Neretnieks I. (1981) Age dating of groundwater in fissured rock: influence of water volume in micropores. Water Resour. Res., 17(2):421–422.

Preiss, WV. (1987) Compiler, The Adelaide Geosyncline – late Proterozic stratigraphy, sedimentation, palaeontology and tectonics. Bull geol. Surv .S, Aust., 53.

CHAPTER 32

# Fracture networks in Jurassic carbonate rock of the Algarve Basin (South Portugal): Implications for aquifer behaviour related to the recent stress field

Inga Moeck[1], Michael Dussel[2], Uwe Tröger[3] and Heinz Schandelmeier[4]

*[1] GeoForschungsZentrum Potsdam, Dep. Geoengineering, Telegrafenberg, Potsdam, Germany*
*[2] GEOTOP, Hönower Strasse 35, Berlin, Germany*
*[3] TU Berlin, Inst. of Hydrogeology, Ack 1-2, Ackerstrasse 76, Berlin, Germany*
*[4] TU Berlin, Inst. of Exploration Geology, BH2, Ernst Reuter Platz 1, Berlin, Germany*

ABSTRACT: The understanding of fault and fracture systems is essential for fracture characterisation of aquifers. Where data from boreholes indicate a strong dependency of fault permeability on stress state, a multidisciplinary approach is introduced to specify the relation between permeability of faults and stress field. The study combines hydrogeological, structural, and electromagnetic analyses in Jurassic carbonate rocks of the Central Algarve Basin (South Portugal). Fault slip data are used to calculate different the paleostress and current stress fields. Tracer tests provide the main groundwater flow direction that is supported by electromagnetic measurements. The results indicate that recently stressed shear fractures are the most hydraulically conductive structures in this system. This contrasts previous models in which extensional fractures are considered as most transmissive. The discrimination of potentially active shear fractures are thought to play an important role in aquifer behaviour of tectonically active regions.

## 1 INTRODUCTION

Apart from layer-parallel conduits, subvertical and vertical brittle faults and fractures provide the most permeable pathways for groundwater migration. Some workers (e.g., Long et al. 1991) have demonstrated that relatively few fractures serve as primary conduits for groundwater flow in hard rocks and that only selected fractures and faults in karstified carbonate rocks develop significant conduits. This study specifies which fractures play an important role in aquifer behaviour of the Central Algarve Basin. The improved understanding of fracture systems may help to explain why only particular fractures and faults are more karstified or more permeable than others. Fault and fracture kinematics generally reflect a specific state of stress in a rock mass so that the problem of fracture permeability was approached by some workers (e.g., Larsson 1972, Barton et al. 1995, Banks et al. 1996) by discriminating significant faults and fractures. The fundamental "hydrotectonic model" of Larsson (1972) postulates that tensile fractures are more transmissive than shear fractures because the former are not closed by a significant component of normal stress.

However, tensile fractures, which are formed normal to the least principal stress-direction, represent only one particular set of fractures of a total fracture population that may be characterized by a great variety in spatial orientation. A set of pre-existing fractures within the reference frame of the current stress field contains tensile fractures, hybrid tensile fractures, hybrid shear fractures, shear fractures and fractures that bear no apparent relationship with the current stress field. To address the problem of hydraulic conductivity of faults on a regional scale we apply geophysical techniques (audio magnetotelluric and very low frequency electromagnetic) that are frequently used for groundwater exploration. In combination with structural analyses, we investigate the relationship between the kinematic and dynamic significance of faults and fractures and their potential hydraulic conductivity with respect to the current stress field in the Central Algarve Basin of South Portugal.

## 2 GEOLOGIC SETTING

The test area is in one of the major aquifer systems of the Central Algarve, which consists of intensely karstified carbonate Liassic rocks. The landscape is characterized by typical karstic morphology with dolines, poljes, canyons, and structural depressions. The Algarve Basin is one of the regions within the Iberian plate (Figure 1) that have only been slightly inverted and deformed during Alpine tectonic activities (Ribeiro et al. 1990).

It is situated near the seismically active Azores-Gibraltar mega-shearzone that is the active plate boundary between Europe and Africa. Due to the limited impact of Cenozoic inversion tectonics, several pre-Mesozoic tectonic phases are recognised in the rocks of the Algarve Basin.

The study area is situated in the north-central Algarve, some 50 km NW of Faro (Figures 1 and 2). The basin, which began to open in the Middle Triassic due to transtensional strike-slip movement along the nearby Maghreb-Gibraltar shearzone, has a typical half-graben geometry. Basin fill ranges from Triassic siliciclastics to lower Liassic clays, evaporites, and volcanics to Upper Jurassic carbonates to Lower Cretaceous marls. Between the Upper Cretaceous and the Quaternary, the central part of the basin evolved under terrestrial conditions, which was interrupted by a short period of marine incursion during the upper Miocene (Almeida 1985).

The current fault pattern in the basin, deduced from LANDSAT-TM interpretation, is characterised by NW-SE to NNW-SSE and NE-SW to NNE-SSW strike-slip faults cut by E-W to ENE-WSW oblique reverse faults. The conjugate set of NW-SE and NE-SW oriented faults were generated through reactivation of pre-existing Paleozoic fault systems that underly the basin sediments, whereas the E-W to ENE-WSW striking faults were generated during the Triassic rifting phase (Moeck 2005). From the Upper Cretaceous onwards the basin has been subjected to transpressional and compressional stress fields that variously reactivated and modified the pre-existing faults.

## 3 METHODS

To investigate faults and fractures, their hydraulic conductivity and their relationship with the current stress field, it is necessary to understand the surface and subsurface conditions of the

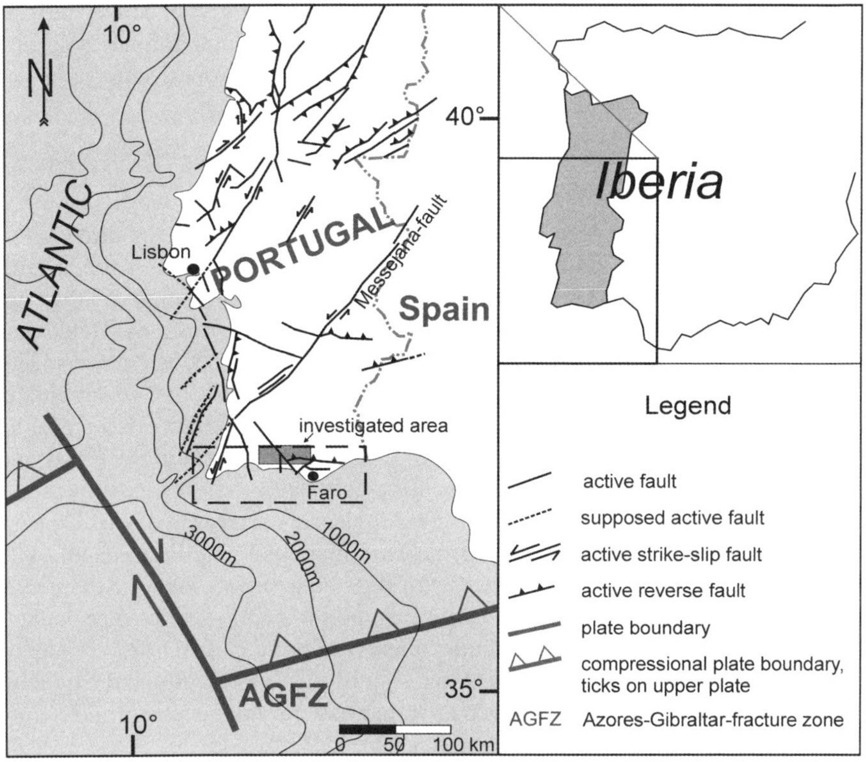


Figure 1. Generalised neotectonic framework of Portugal and adjacent areas in the Atlantic. The AGFZ represents the current plate boundary between the African and the Iberian plate. Dashed box: Algarve Basin, grey box: studied area (modified after Srivastava et al. 1990, and Cabral 1995).

study area. Our interdisciplinary approach involve hydrogeological, tectonic, geophysical and 3D-modelling analyses. Surface conditions are investigated by means of fault-slip analyses to discriminate different sets of fractures with respect to the related stress fields. The knowledge of the paleostress fields is necessary to evaluate their impact on the pre-existing fault pattern and hydraulically conductive structures and to isolate the effects of the current stress field. Geophysical measurements (AMT and VLF-EM) reveal the subsurface conditions with respect to the directions of hydraulically conductive structures of the study area and finally the geophysical results are tested at one location by a bacteriophage tracer test.

### 3.1 *Hydrogeological measurements*

In the Liassic aquifer seven short-time pumping tests and two tracer tests were conducted to obtain data about permeability as related to tectonic structures. Additionally, tracer test results of the "Projeto dos Regadios da Província do Algarve" (Tröger 1987) were also considered to get a more general overview of the hydraulic behaviour of this carbonate aquifer. In the Upper Jurassic formations south of the Algibre fault (Figure 2), a tracer test result were compared with electromagnetic measurements. The general groundwater flow of the aquifer system is deduced from measurements of the piezometric level across the test area. Piezometric levels indicate a general groundwater flow direction although this is not an exact method for determining groundwater flow in karstic areas. 70 groundwater samples were taken from springs and wells with depths of up to 280 m.

## 3.2 *Structural geological analyses*

Fault kinematic analyses are based on the orientation of fault planes, the direction of slip along slickenside planes as indicated by striae orientation, and the sense of slip, determined according to the criteria of Means (1981) and Petit (1987). Commonly, relative ages of slickensides are determined from cross-cutting relationships of striations. Heterogeneous fault slip data sets were separated and decomposed into homogeneous data sets using the P/T-method of Sperner et al. (1993), Hoeppner diagrams (Meschede 1994), and fluctuation diagrams (Etchecopar et al. 1981). The decomposed homogeneous data sets were compared with stylolite populations of different ages. All data were statistically processed with the computer program TectonicsFP 1.5 of Reiter and Acs (1999).

The geometrical constraints of the mapped fault system are represented in a 3D geological model, developed from surface data (geological map and satellite image). The 3D earthVision® modelling procedure started with digitizing surface data of the geological map and related six cross sections. The data preparation was followed by the calculation of fault and horizon grids applying a minimum tension technique. A 3D fault model of 107 faults and 134 fault blocks was processed in a step by step calculation and reconciliation of nine sub-fault trees. This model served as framework for the horizon gridding process according to geologic and specific intersection rules (Moeck et al. 2005).

## 3.3 *Electromagnetic measurements*

AMT (audio magnetotelluric) and continuously registering VLF-EM (very low frequency electromagnetic) measurements were conducted to detect directions of maximum apparent

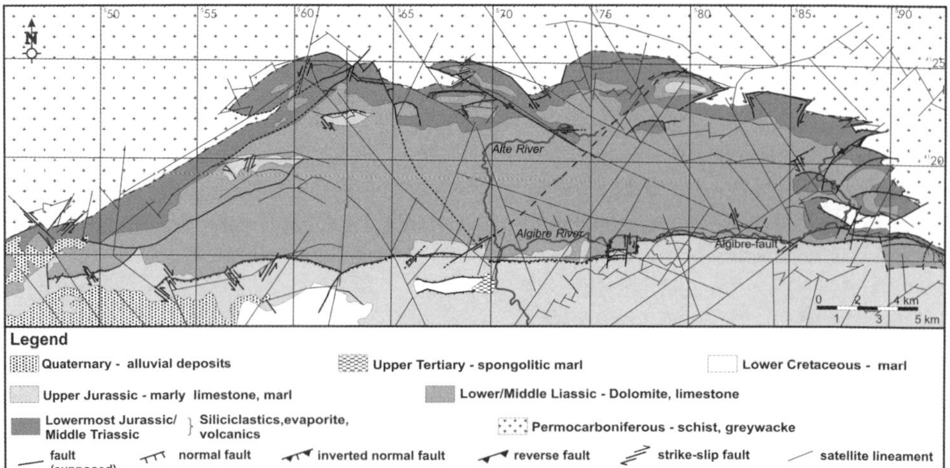


Figure 2. Geological map of the central Algarve Basin with major fault zones and satellite lineaments.

resistivities in the aquifer. The applied electromagnetic devices were developed by the University of Neuchâtel (Prof. I. Müller). AMT-measurements were carried out at 20 locations to obtain information on deeper structures and preferential groundwater pathways. Each site was measured using ten different frequencies (35–12,700 Hz). Apparent resistivities and phase differences were recorded in six directions at each measuring site because fractured and/or karstified aquifers are known to generally have a high factor of anisotropy. The electrical field was measured at each site with the line of two electrodes in 0°, 30°, 60°, 90°, 120° and 150° direction . At two sites the lines were laid in 20°, 50°, 80°, 110°, 140° and 170° (see Figure 7, more details in Dussel, 2005). The magnetic field was measured with one coil and a mean error of approximately ±15° was assigned to each of the measured directions. The direction in which the highest apparent resistivity occurs is considered to represent the most conductive structure (Koll and Müller 1989, Fischer 1985). The highest frequency and the frequency in which the highest factor of anisotropy ($\rho_{max}/\rho_{min}$) occurs, are chosen to deduce preferential directions of both infiltration and maximum groundwater flow in the Liassic aquifer. VLF-EM data were generated by a continuously registering device. The VLF-EM profiles were measured in parallel order to provide orientation of hydraulically conductive structures.

## 4 RESULTS

Most of the karstified structures strike along 20° to 30° direction. Subordinate karstification in the eastern part of the Liassic aquifer is along 90° to 110° striking structures (Figure 3). In the central part of the aquifer, the Paderne plain, these structures represent strike-slip faults that are important for groundwater flow. In the north-eastern part of the aquifer groundwater flow is directed from NE to SW (Figure 4).

In the central part of the aquifer, groundwater flow direction changes to ESE-WNW subparallel to the Algibre fault related to 110° striking structures. Groundwater flows also

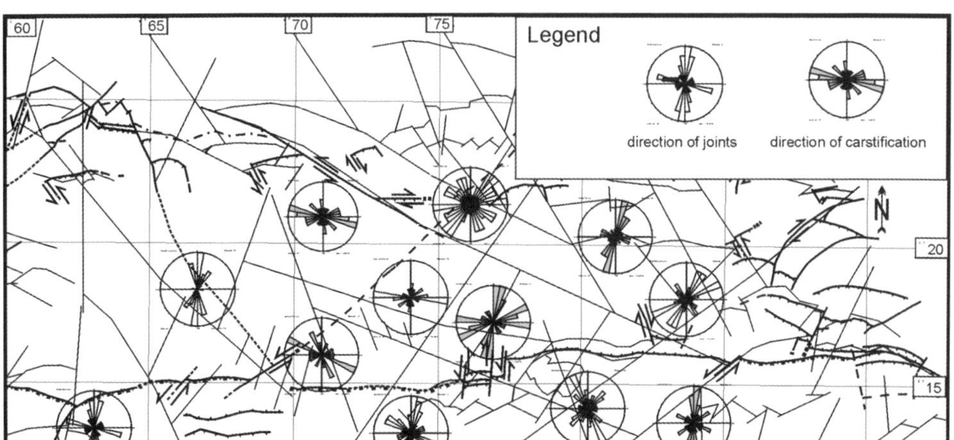


Figure 3. Orientation of fractures and karstification in the eastern part of the study area (Tröger et al. 2001).

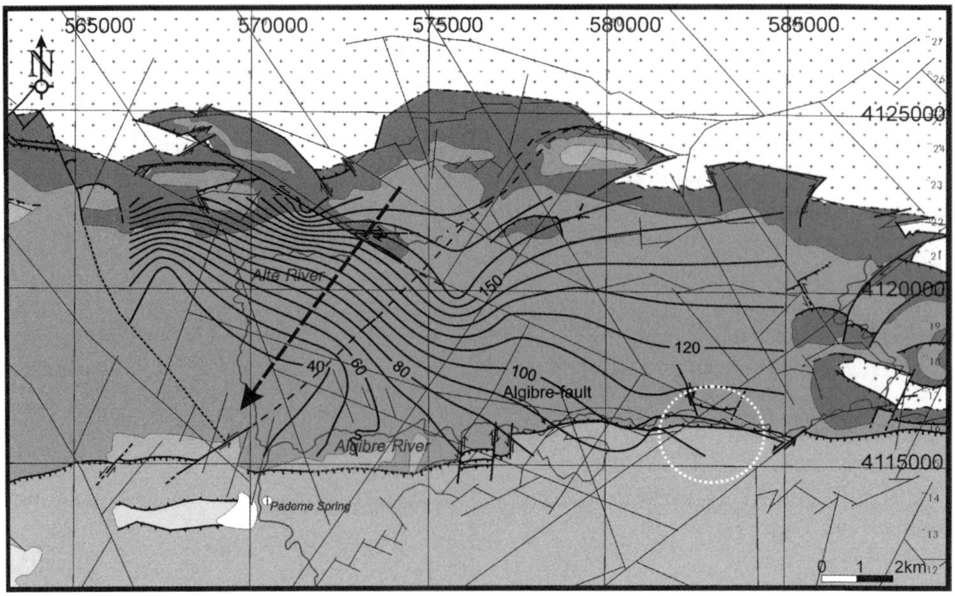


Figure 4. Groundwater contour map of October 1998, based on 53 piezometric measurements (elevation in m amsl) dashed arrow indicates main groundwater flow; dashed circle shows the combined tracer and AMT test (Dussel et al. 2000).

along N-S striking transcurrent faults, that intersect the E-W trending Algibre fault. Here surface water drainage and groundwater flow corresponds to each other because the Algibre river is subparallel to the fault direction. Geophysical measurements (VLF-EM and AMT) and a tracer test reveal groundwater flow along a 20° striking structure which can be tracked into the Liassic rock north of the Algibre fault (Figure 4 and Carvalho-Dill et al. 1999). In the fluvial bed of the Algibre river, a spring is located on a 110° striking satellite lineament; this is another indication of fracture dominated flow.

Six pumping tests reveal transmissivities between 50 to 1,700 $m^2$/day. and short time pumping tests suggest a relation between permeability, satellite lineaments, and faults (Table 1). In comparison, five pumping tests during the "Projecto dos Regadios da Província do Algarve" (Tröger 1987) showed locally high transmissivities without any draw-down ($Q_{min}$ = 2.4 L/s and $Q_{max}$ = 11 L/s).

Interpretation of tracer tests (Tröger 1987, Dussel 2005) is generally difficult in the Liassic aquifer because of the changing draw-downs from irrigation wells. Agricultural consumption of groundwater gives rise to derivation of the natural groundwater flow. As a result of these tests many different flow directions with highest flow velocities were registered in different regions (Tröger 1987, Dussel 2005), dependent on regional fracture patterns, hydraulic gradient and groundwater consumption by agriculture. Phreatic groundwater flow and supposed subterranean rivers ("fluviokarst") are responsible for groundwater velocities that range from 3 to 3,500 m/day. In the western part of the Liassic aquifer, a tracer test (Tröger 1987) indicated groundwater flow along the E-W striking Algibre fault in the area of Sobrado (Figure 5). Highest groundwater velocities are in 135° SE (16 m/day) and 210° SW (11.2 m/day). These results correspond very well with the high transmissivity at

Table 1. Results of pumping tests in boreholes below the carbonate aquifer ($T_w$: transmissivity according interpretation of recovery data) (after Dussel 2005).

| No | Lithology | T ($m^2/d$) | Remarks |
|---|---|---|---|
| P25 | dolomite | 12 | subrosion zone, chloride affected |
| P26 | dolomite, limestone | 225 | located on the E-W striking Algibre-fault, which is acting as recent shear zone |
| P27 | dolomite | 4,493 | without any relation to a known tectonic structure |
| P28 | dolomite | without draw down | located on an intersection of 20° and 135° striking satellite lineaments, cavern bored |
| P29* | sandstone | 3 | located on E-W striking fault |
| P30* | sandstone | 3 | located on a 120° striking satellite lineation |
| P31 | dolomite | $T_1 = 1{,}280$<br>$T_2 = 6.6, T_w = 4.6$<br>$(S = 3.7 \cdot 10^{-5})$ | base of the carbonatic aquifer |

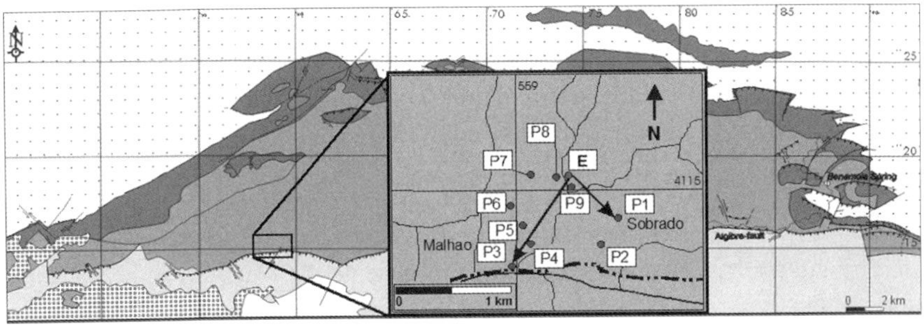


Figure 5. Tracer test near Sobrado indicate flow in 135° and 210° orientation. E is injection site, P1–P9 are monitoring sites. Legend is same as for Figure 1 (modified after Tröger et al. 2001).

the intersection point of two satellite lineaments oriented in 20° and 135° in central part of the aquifer.

Two tracer tests in 2000 showed groundwater velocities of approximately 500 m/day in the high water zone, about 40 m/day in the high phreatic zone, and nearly 3 m/day in the deep phreatic zone (Dussel 2005). Groundwater flow in these two tracer tests in the eastern part of the aquifer was mainly directed NNE-SSW (190° and 200° respectively).

Most of the water samples indicate a normal earth alkaline, mainly hydrogencarbonatic water type. High concentrations of sodium, calcium, chloride and sulphate occur in the vicinity of E-W-striking structures characterised by uplifted gypsum in the hanging wall. The lowest water temperature of 70 water samples with an average of 19.2° C was measured at the intersection of two satellite lineaments with orientations in 120°N and 30°N. The physical-chemical parameters of the water samples are related to the nearby geologic structures.

Six different stress fields are discriminated – two extensional stress fields with (I) $s_3 = 27°$ and (II) with $s_3 = 124°$ and four inversion stress fields with (III) $s_1 = 99°$ to (IV) $s_1 = 27°$ to (V) $s_1 = 179°$ and to (VI) $s_1 = 148°$ (Table 2 and Moeck 2005). The latest

Table 2. Paleostress fields (AA-C) and current stress field (D), listed from the oldest at the top and the current at bottom.

| stress field | σ1 Az./Dip | σ2 Az./Dip | σ3 Az./Dip | stress ratio *R* | stress regime |
|---|---|---|---|---|---|
| AA | 246/71 | 120/14 | 027/19 | 0,4225 | normal faulting |
| BB | 127/85 | 033/05 | 304/05 | 0.1955 | normal faulting – radial extension |
| A | 279/29 | 068/57 | 176/11 | 0.1995 | transpression |
| B | 207/03 | 121/63 | 295/25 | 0.3954 | strike slip – transpression |
| C | 179/03 | 287/28 | 107/52 | 0.2221 | transpression |
| D | 146/13 | 239/24 | 336/67 | 0.2928 | transpression |

stress-field is documented by an abundance of stylolites, slickensides (i.e., striae on fault planes that are used as indicator of slip direction) and tension gashes. The P-axis of the latest stress-field is oriented in NW-SE direction (146°) and reflects a transpressional stress regime (Table 2 stress field D). The direction of the maximum stress and the stress regime corresponds to the current stress field in southern Portugal, as presented by various authors (Moreira 1985, Zoback 1992, Ribeiro et al. 1996, Jimenez-Munt 2000). Consequently, we consider the latest stress field D as the current stress field in the Algarve region.

The orientation of the two latter stress fields (Miocene to recent) is particularly important to discriminate between critically stressed and non-critically stressed faults and their significance with respect to their hydraulic conductivity. The hydraulic structures were generated preferably during the most important phase of karstification in the carbonate rocks in lower Miocene.

According to Koll and Müller (1989), the direction of the highest apparent resistivity measured by the applied electrode line order represents the orientation of the maximum hydraulic conductivity. In our study the AMT investigations show the highest apparent resistivity and lowest phase difference (f = 3880 Hz) in mainly 0°–30° (N-NNE) and subordinately in 90°–120° (E-ESE) (Figure 6). The groundwater flow is in the same direction in 0°–30° determined by tracer tests in the central and eastern part of the study area (Carvalho-Dill et al. 1999). Thus, our results agree with the indications of Koll and Müller (1989). In the early to late Jurassic carbonate rocks of the basin fill the maximum groundwater flux is controlled by 0° to 30°-trending structures (N to NNE). At other points located near a prominent 34°-striking lineament identified on the LANDSAT-TM scene, AMT-profiles also revealed 0° to 30°-trending conductive structures (N to NNE). At points located near the Paderne spring in Quaternary cover rocks, 0° to 30° (N to NNE) and 90°–120° directions (E to ESE) host major conductive structures, whereas in Liassic aquifer rocks underneath the Quaternary cover the highest apparent resistivity is in 0°-direction (N) (Figure 6). Nearby dolines are aligned along 110° (ESE) trend.

The results of the electromagnetic investigations show distinct orientations of high conductive structures aligned mainly 0°–34° (N-NNE) and subordinately in 90°–120° (E-ESE).

## 5 DISCUSSION

The combined use of stress tensor orientations coupled with geophysical and hydrogeological data lead to a good estimate of groundwater flux in fractured rock within the aquifer system of the Central Algarve Basin.

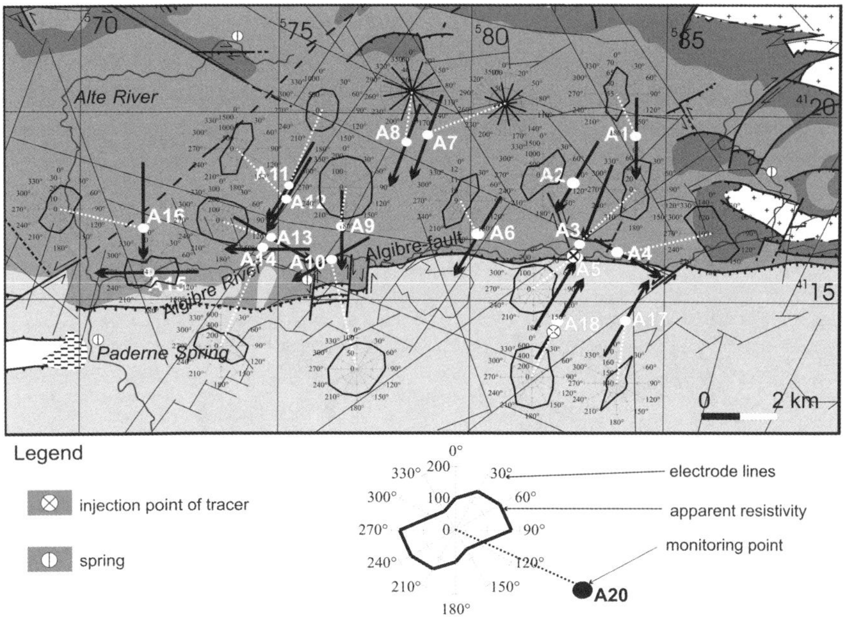


Figure 6. AMT (audio magnetotelluric) measuring sites, results, and interpreted hydraulic conductivity, symbolized by arrows.

In his fundamental hydrotectonic model, Larsson (1972) stated that tensional fractures in Precambrian basement areas of Sweden would be more transmissive and open than shear fractures, because the latter are held closed by a component of normal stress. However, Larsson's model has some weak arguments. For instance, first, the theory assumes that shear or tension fractures generated in Precambrian times will retain their shear or tensional nature to the present; second, it does consider that tensional fractures are possibly less interconnected than shear fractures, which generally show a complex interconnected fracture grid; and, third, tensional fractures may be filled in and sealed with later mineralization that dramatically reduces the permeability.

To specify the relationship between computed paleostress fields and the geophysical measurements, fault and fracture orientations were classified with respect to their kinematic significance within the various stress fields (Figure 7). It should be noted, that the maximum conductive structures, oriented in 0°–30° and subordinately 90°–120°, as shown by the electromagnetic measurements, correspond with the orientation of conjugate shear fractures within the current stress field.

In contrast, extensional fractures and the conjugate shear fracture systems of the older stress fields seem to bear no major relationship with the hydraulically most conductive structure orientation (Figure 8).

This conceptual fracture model agrees with the study of Himmelsbach (1993), who described the draining character of fault zones in the Albtal granite of the southern Black Forest in Germany. In this study tracer tests showed a N-S trending strike slip zone as the dominant hydraulically conductive structure. This fault zone is critically stressed related to the current local stress field characterized by a 155° oriented maximum horizontal stress.

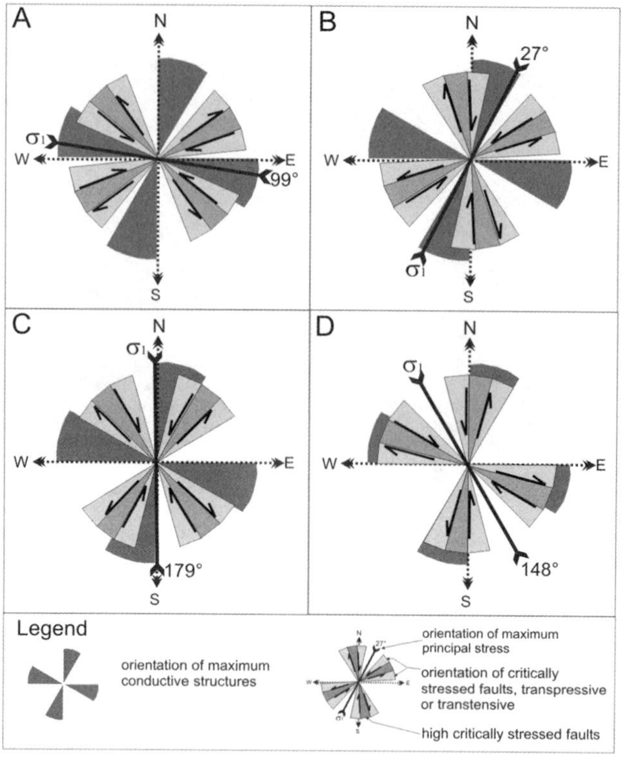


Figure 7. Kinematic significance within stress fields from Table 2 and compared with main hydraulic directions (Moeck 2005).

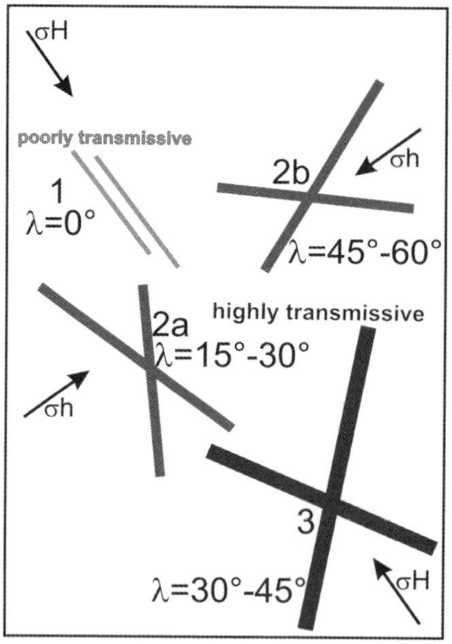


Figure 8. Conceptual fracture model: Relation of fracture permeability, recent stress field and fault kinematics. λ = angle between σH and fracture plane. 1 tensional fractures, 2 hybrid shear fractures: 2a transtensional, 2b transpressional, 3 shear fractures.

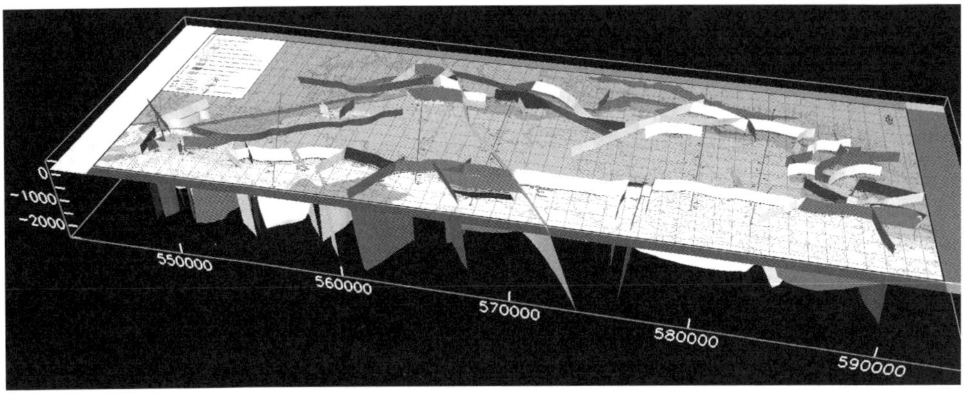


Figure 9. 3D fault model of the studied area. The conceptual fracture model of Figure 9 is applied on the fracture system of the 3D model and shows the transmissive (grey) and non-transmissive (white) faults.

Ufrecht (2001) found highly mineralised water rising from the shear zones in Middle Triassic limestones near Stuttgart (South Germany). Ufrecht investigated hydrochemical parameter of the Neckar river and determined hydrochemical anomalies of springs which are related to a recent strike slip zone crossing the Neckar river.

The high permeability along shear fractures might reflect the typical dense fracture network, that is built up by conjugate shear fractures (Riedel shears). Riedel shears range from macro to micro scale (Hancock 1994) and can be considered as increased porosity along the main fault or fracture zone. The study area is situated in a region with high horizontal stresses, indicated by the maximum main stress in the horizontal. Investigation in other lithological and tectonic provinces with a different stress field might show other results in terms of dependency of fracture permeability and stress field. Therefore more scientific work is needed in this field of tectonically influenced fracture permeability. Our results indicate, however, that the hydraulic conductivity of structures are significantly influenced by the stress field and that the hydraulically most conductive fractures are critically stressed faults with respect to the current stress field.

The conceptual model of fracture characteristics is applied in the 3D geological model, which is based on the field map and satellite image. The 3D model shows white faults as non-critically stressed and dark grey as critically stressed faults with respect to the maximum horizontal stress sH = 148° (Figure 9). Analysing the area of the combined tracer VLF-EM test, the 3D model reveals a 25° striking 72° westward dipping critically stressed fault north of the test area. The fault might extend to the test area, acting as groundwater pathway as indicated by the tracer test. This example shows that 3D geological models can help to define the geometrical constraints of fault strike, dip, fault characteristics and aquifer behaviour.

## 6 CONCLUSION

Combined with tectonic analyses, remote sensing, and hydrogeological investigations, AMT and continuously registering VLF-EM methods are useful tools to investigate groundwater paths in fractured rock.

Our results reveal a significant relation between hydraulically conductive structures and fault and fracture kinematics of the current stress field. Critically stressed faults and fractures with respect to the current stress field correspond to the most conductive structures, oriented in 0°–30° and (subordinate) 90°–120° directions. Thus the hydraulically highly conductive structures are critically stressed faults, that are characterised by a component of high shear stress. This contrasts with conventional models, in which extensional faults oriented parallel to the maximum principal stress vector are considered to be the preferential groundwater conduits. We conclude, that (i) the hydraulic conductivity depends on the current stress field and (ii) critically stressed faults influence the hydraulic conductivity to a larger extent than extensional faults.

Critically stressed faults are specified by a shear angle of $\lambda = 30°–45°$ to the maximum main stress s1 direction. In the Algarve Basin, the maximum main stress is the maximum horizontal stress s1 = sH. Knowledge of the regional stress field might allow an prediction of fracture permeability. We note, however, that our results are from a tectonically active region, which is situated near a plate boundary. The dependence of hydraulic gradient on stress field may be less pronounced in tectonically less active regions. More studies of fracture characterisation in similar regions and in regions of less maximum horizontal stress is needed to define the relation between tectonic stress, fracture properties and aquifer behaviour.

## ACKNOWLEDGEMENTS

This research was supported by the "Deutsche Forschungsgemeinschaft" DFG (Tr 274/2–1,2). We thank for A. Champell and D. Bruhn for worthwhile reviewing. We thank Dynamic Graphics Ldt. Paris for support of 3D model development.

## REFERENCES

Almeida C (1985) Hidrogeologia do Algarve Central. PhD thesis, Faculdade de Ciências da Universidade de Lisboa.

Banks D, Odling NE, Skarphagen H, Rohr-Torp E (1996) Permeability and stress in crystalline rocks. Terra Nova 8: 223–235.

Barton CA, Zoback MD, Moos D (1995) Fluid flow along potentially active faults in crystalline rock. Geology 23: 683–686.

Cabral J (1995) Neotectonica em Portugal continental. Memorias do Instituto Geológico e Mineiro, 31: pp. 265, Universidade da Lisboa/Portugal.

Carvalho Dill A, Dussel M, Reis E, Baptista R, Coimbra R, Reis M (1999) The combined use of electromagnetic methods and tracers to detect preferential groundwater pathways. In: Actualidad de las técnicas geofisicas aplicadas en hidrogeología, Alarcón MO, López Geta JA (eds). Inst Tecnológico GeoMinero de España.

Dussel M (2005) Hydrotektonik und Grundwasserdynamik im rezenten Spannungsfeld am Beispiel der verkarsteten Gesteine im Zentralalgarve (Portugal), http://edocs.tu-berlin.de/diss/2005/dussel_michael.pdf.

Dussel M, Moeck I, Müller I, Schandelmeier H (2000) Preferential groundwater pathways along critically stressed faults in carbonate rocks: (I) Evidence from hydrogeologic and electromagnetic (AMT, RF-EM) data. EGS Abstracts, 2, 25th General Assembly, Nice, France.

Etchecopar A, Vasseur G, Daignieres M (1981) An inverse problem in microtectonics for determination of stress tensors from fault striation analyses. J Struct Geol 3: 51–65.

Fischer G (1985) Some remarks on the behaviour of magnetotelluric phase. Geophys Prospecting 33: 716–722.

Hancock PL (1994) From joints to paleostress. Roure, F. (ed) *In*: Peri-Tethyan Plattforms, Paris, p. 145–158.

Himmelsbach T (1993) Untersuchungen zum Wasser- und Stofftransportverhalten von Störungszonen im Grundgebirge (Albtalgranit, Südschwarzwald). Dissertation. Schriftenreihe Angewandte Geologie der Universität Karlsruhe, 23, 238 pp.

Koll J, Müller I (1989) Elektromagnetische Very Low Frequency-Resistivity (VLF-R) Prospektion zur Erkundung von Grundwasserleitern im paläozoischen Mittelgebirge am Beispiel des Oberharzes. Steirische Beiträge zur Hydrogeologie 40: 103–122.

Jimenez-Munt I (2000) Neotectonica of the Azores-Gibraltar plate-boundary. EGS, 25th General Assembly, Nice 2000.

Larsson I (1972) Groundwater in granite rocks and tectonic models. Nordic Hydrol 3: 111–129.

Long JCS, Karasaki K, Davey A, Peterson J, Landsfeld M, Kemeny J, Martel S (1991) An inverse approach to the construction of fracture hydrology models conditioned by geophysical data. Int J of Rock Mechanics, Mineral Science & Geomechanics Abstracts 28: 121–142.

Means WD (1981) The concept of steady state foliations. Tectonophysics 78: 179–199.

Meschede M (1994) Methoden der Strukturgeologie: Ein Leitfaden zur Aufnahme und Auswertung strukturgeologischer Daten im Gelaende und im Labor. Enke, Stuttgart, Federal Republic of Germany. 175 pp.

Moeck I (2005) Hydrotektonik von Grundwasserleitern: Rekonstruktion von Spannungsfeldern und 3D Modellierung einer geologischen Karte des Zentral-Algarve (Südportugal), http://opus.kobv.de/tuberlin/volltexte/2005/1057/pdf/moeck_inga.pdf.

Moeck I, Schandelmeier H., Dussel M, Luxey P (2005) Aquifer characterization: 2D geological maps as portal to 3D conceptual geological models. Extended Abstracts CD ROM, p. G003, EAGE 67th Conference & Exhibition, 13–16 June 2005, Madrid, Spain, ISBN 9073781981.

Moreira VS (1985) Seismotectonics of Portugal and its adjacent area in the Atlantic. Tectonophysics 117: 85–96.

Petit JP (1987) Criteria for the sense of movement on fault surfaces in brittle rocks. J Struct Geol 9: 597–608.

Reiter F, Acs P (1999) TectonicsFP 1.5: A computer program for structural geology. Universität Innsbruck.

Ribeiro A, Baptista R, Cabral J, Matias L (1996) Stress pattern in Portugal mainland and the adjacent Atlantic region, West Iberia. Tectonics 15/3: 641–659.

Ribeiro A, Kullberg M C, Kullberg J C, Manupella G, Phipps S (1990) A Review of Alpine tectonics in Portugal. Forland detachment in basement and cover rocks. Tectonophysics 184: 357–366.

Sperner B, Ott R, Ratschbacher L (1993) Fault-striae analysis: a Turbo Pascal program package for graphical presentation and reduced stress tensor calculation. Computer & Geosciences 19: 1361–1388.

Srivastava SP, Roest WR, Kovacs LC, Oakey G, Levesque S, Verhoef J Macnab R (1990) Motion of Iberia since the Late Jurassic; results from detailed aeromagnetic measurements in the Newfoundland Basin. Boillot, G. & Fontbote Joseph, M. (ed) *In*: Alpine evolution of Iberia and its continental margins.: Tectonophysics 184 (3–4). Elsevier, Amsterdam, Netherlands, p. 229–260.

Tröger U, Dussel M, Moeck I, Schandelmeier H (2001) A new approach to the reconnaissance of groundwater flow in hard rocks hydrotectonic. Seiler Klaus, P. & Wohnlich, S. (ed) *In*: New approaches characterizing groundwater flow; proceedings. A.A. Balkema. Lisse, Netherlands. 2001.

Tröger U (1987) Vorkommen und Nutzung von Grundwasser im Mittelabschnitt des Algarve – Portugal. Habilitationsschrift, TU Berlin.

Ufrecht W (2001) Vulnerabilität und Schutzmassnahmen im Quellgebiet der Stuttgarter Mineral- und Heilwässer. – Zeitschrift für Angewandte Geologie, 47/1, 47–54.

Zoback ML (1992) First- and second-order patterns of stress in the lithosphere: the world stress map project. J geophys Res 97/B8: 11,703–11,728.

# Anthropogenic impacts on fractured environment

CHAPTER 33

# Using specific capacity to assign vulnerability to diffuse pollution in fractured aquifers in Scotland

Mark Betson[1] and Nick Robins[2]
[1]*Westcott House, Jesus Lane, Cambridge*
[2]*British Geological Survey, Wallingford*

ABSTRACT: A stochastic approach, applied to field scale pumping test data from Scotland, is used to estimate a quantitative indicator of effective hydraulic conductivity ($K_{ef}$), where $K_{ef}$ is considered a quantitative indicator of aquifer vulnerability to diffuse pollution. The unit specific capacity (USC) is defined as a representative correlate to aquifer permeability. A survey of pumping test data in fractured aquifers in Scotland yielded a set of USC values with significant subsets for Devonian and Permian aquifers. The geometric mean of these subsets indicated USC values of 0.19 m/d for the Devonian and 2.47 m/d for the Permian aquifer rocks. The data set for the Devonian also permitted $K_{ef}$ to be evaluated as 0.20 m/d.

## 1 INTRODUCTION

Descriptions of aquifer vulnerability in the British Isles have normally related to the presence and thickness of surficial strata (NRA 1992, Palmer et al. 1995, Environment Agency 1998, SEPA 1997 and DoE for Northern Ireland 1994, 1999) or subsoil thickness and permeability (DoELG et al. 1999) over potential aquifers. Thus, the permeability of the vadose zone is the most important vulnerability indicator despite its being a variable dependent on moisture conditions. In Scotland and Northern Ireland the surficial and subsurface geology were originally classified as highly permeable, moderately permeable and weakly permeable and combined with soil classes to form groundwater vulnerability classes (Lewis et al. 2000). This system was broadly similar to that previously adopted in England and Wales only the labels Major, Minor and None Aquifer were somewhat misleadingly used.

The Water Framework Directive (European Community 1996) upsets the conventional aquifer vulnerability philosophy by designating a groundwater body (the basic groundwater management unit) as a distinct volume of groundwater within an aquifer or aquifers capable of supplying at least $10\,m^3d^{-1}$ of drinking water as an average. As even low permeability systems contain some groundwater, new approaches are being sort to classify vulnerability classes for all potential aquifers. One such methodology is the removal of recharge or aquifer productivity from the assessment and to focus only on transport (permeability or permeability indices) and attenuation processes of pollution migration (Ó Dochartaigh et al. 2005). When applied to fractured aquifer systems this reverses the vulnerability assessment from one of weakly permeable, low storage and, therefore, low vulnerability, to rapid transport, poor attenuation and high vulnerability. There remains the problem of determining the overall permeability of fractured systems.

Tellam and Lloyd (1981) demonstrated the variability in hydraulic properties that occur in such rocks and reported field hydraulic conductivities for Palaeozoic mudrocks ranging from $10^{-6}$ m d$^{-1}$ to $10^{-1}$ m d$^{-1}$. Robins and Misstear (2000) divided fractured basement rocks into two distinct categories (Table 1) each with discrete hydrogeological characteristics. Any scheme of vulnerability assessment that lumps all secondary permeable rocks into one class can cause an overstatement of vulnerability where a risk averse policy is pursued. The heterogeneous nature of these strata is significant, even for the regionally important Carboniferous Limestone. This is illustrated by data for 225 boreholes in the west of Ireland in which there is a large range of yields skewed towards the lowest values (Table 2).

The variety of fractured rocks and unsaturated depths indicate that a range in vulnerability is possible. Morris et al. (2003), for example, state that extreme vulnerabilities are associated with highly fractured aquifers with shallow water tables that offer little chance for contamination attenuation. In the case of poorly permeable fractured rocks the opportunity for attenuation is increased and the "aquifer" is less vulnerable.

The vulnerability of aquifer units is in part, at least, a function of their permeability (Foster 1998). It is difficult to make general characterisations of the permeability for the fractured and karstic flow regimes predominating in the deeper Scottish aquifers. Any attempt, to ascribe a generalised bulk hydraulic conductivity for these aquifers assumes they statistically exhibit a degree of stationarity in permeability between particular scales of measurement (Dagan 1986, Bear 1972). Such a heterogeneous aquifer may be described in terms of the probability density functions of its properties, including permeability, distributed according to random space functions. If the probability density function

Table 1. Principal aquifer characteristics in upland hard rock areas.

| Class | Geology | Properties | Transmissivity ($m^2 d^{-1}$) | Storativity | Borehole yield (l/s) |
|---|---|---|---|---|---|
| Regionally important aquifers | Carboniferous Limestone, Devonian, some volcanic rocks | Anisotropic; fracture flow dominant; regional and local flow paths. | 100–4000 | 0.01–0.20 | 5–40 |
| Locally important aquifers | Precambrian and Lower Palaeozoic; some Upper Palaeozoic; some volcanic rocks | Anisotropic, secondary porosity dominant, local flow paths. | 20–50 | <0.05 | 1–5 |

Table 2. Summary of performance characteristics of 225 boreholes in Carboniferous Limestone in the west of Ireland (after Drew and Daly 1993).

| Variable | Maximum | Minimum | Mean |
|---|---|---|---|
| Depth (m) | 177 | 3 | 57 |
| Yield (l/s) | 76 | 0 | 2.4 |
| Specific capacity (l/s m$^{-1}$) | 7.6 | 0 | 0.8 |

exhibits stationarity at a given scale, then a stochastic approach can be used to estimate a generalised value for the property (Dagan 1986). Provided the analysis is undertaken at a coarse scale (i.e., catchment scale and greater), this method can be applied to fractured but poorly permeable rocks. In their review of calculation methods of equivalent permeabilities Renard and de Marsily (1997) consider stochastic methods the most suitable for calculation of the effective hydraulic conductivity where there is only an approximate knowledge of the aquifer system. The rule of geometric averaging demonstrated by Matheron (1967) for a uniform flow field in two dimensions is one of the few solutions to yield an exact result, which is extended to D dimensional space by Landau and Lifshitz (1960) with a formula that is a first-order approximation of the effective permeability ($K_{ef}$),

$$K_{ef} = \mu_a^{(D-1)/D} \mu_a^{1/D} \tag{1}$$

Where $\mu_a$ is the arithmetic mean. In three dimensions in the case of a log-normal permeability distribution this formula becomes,

$$K_{ef} = \mu_g \exp\left[\sigma^2_{\ln k}\left(\frac{1}{2} - \frac{1}{D}\right)\right] \tag{2}$$

Where $\mu_g$ is the geometric mean and $\sigma^2_{\ln k}$ is the variance of the permeability logarithm.

Renard and de Marsily (1997) list a number of successful tests of the above formula, including examples where the linear integral scale of the media is greater than the modelled unit of aquifer to which the permeability is being applied. Nœtinger et al. (2005) explore aspects of the formula with a view to its use in stochastic and upscaling methods in hydrogeology; they conclude that the stochastic method is an "elegant way to embed both averaging, upscaling and inverse problems". It is also noteworthy that the formula has been used successfully with sandstone and limestone aquifers at a range of scales (Nœtinger and Jacquin 1991).

The nature of this averaging process does, however, mean that it should be considered inadequate for estimating the local risk from point source pollution (at the scale or smaller than a modelled grid unit) where the effective permeability indicated by a bulk hydraulic conductivity could completely underestimate the value for coincident preferential flow pathways (e.g., fracture zones) and point sources. Other techniques, such as the use of tracer studies that are more directionally sensitive may be more appropriate under these circumstances.

The aim of this study is to suggest that a stochastic assessment of permeability is a sensible measure of vulnerability to diffuse sources of pollution; where the scale of spatial averaging is commensurate with the scale of distribution of the pollutant and data sources used to assess its impact.

## 2 METHOD

The scale for vulnerability mapping of diffuse pollution depends on the resolution of available data such as climate, land use, soil physical properties, and agricultural practice

(Anthony et al. 1996). In the U.K., a mapping grid of 1 $km^2$ has been used in constructing a tool to support government policy development on diffuse nitrate pollution (Lord and Anthony 2000).

The term effective hydraulic conductivity, $K_{ef}$, can be applied in a stochastic context for the hydraulic conductivity of a medium that is assumed statistically homogeneous over a series of scales (Renard and de Marsily 1997). If the fracture geometry (fracture spacing, length, and aperture) are such that the spatial change in hydraulic conductivity follows a log-normal distribution, $K_{ef}$, is equal to the geometric mean of the range of conductivities. It is assumed that the geometric mean of a parameter directly correlated to hydraulic conductivity is also representative of the medium.

For a grid unit size of 1 $km^2$, the appropriate length scale of measurement is $10^2$–$10^3$ m (approximately the same as the unit cell dimensions for model grid). For aquifer yield in fractured media, $K_{ef}$ is most appropriately defined by large-scale flow and pressure head measurements (e.g., pumping tests), although this may not reflect aspects of contaminant transport.

Pumping tests data in Scotland are scarce, although there is an adequate number with which to conduct stochastic analyses. The locations of each data point are correlated with the associated geology type from British Geological Survey (BGS) 1:625,000 digital maps. From common parameters in the pumping test data, specific capacity, SC, was selected as the most available indicator of aquifer permeability. Mathematically, however, it is inconsistent to use SC as a measure of permeability because it is proportional to transmissivity (e.g., Logan 1964, Razack and Huntley 1991) and therefore dimensionally incorrect.

Numerous authors (e.g., Ahmed and de Marsily 1987, Kupfersberger and Blöschl 1995) have investigated the correlation of SC to aquifer transmissivity, where transmissivity offers a better evaluation of aquifer hydraulic properties. Razack and Lasm (2005) collected the results of those studies they have been able to identify that relate the parameters for fractured aquifers. Their findings are tabulated in Table 3 where T and SC are in units of $m^2/d$. The relationship is assumed to follow a power law of the form,

$$T = A \cdot SC^B \tag{3}$$

Where A and B are the empirically derived parameters.

Using the effective hydraulic conductivity as a bulk measure of aquifer permeability it is possible to write the following dimensionally correct function using the unit specific capacity (*USC*),

$$K_{ef} = \sqrt[n]{\frac{T_1}{h_{sat_1}} \cdot \frac{T_2}{h_{sat_2}} \cdot \frac{T_3}{h_{sat_3}} \cdots \frac{T_n}{h_{sat_n}}} = A \cdot \sqrt[n]{\frac{SC_1^B}{h_{sat_1}} \cdot \frac{SC_2^B}{h_{sat_2}} \cdot \frac{SC_3^B}{h_{sat_3}} \cdots \cdot \frac{SC_n^B}{h_{sat_n}}} \tag{4}$$

Where n is the number of measurements and $h_{sat}$ is the saturated aquifer thickness. Assuming a fully penetrating borehole, the USC is defined as

$$USC = \frac{SC}{h_{sat}} \tag{5}$$

Table 3. Empirically derived parameters for equation 3 from a selection of studies.

| Reference | Aquifer | A | B |
|---|---|---|---|
| Huntley et al. (1992) | Peninsular Ranges batholith | 0.12 | 1.18 |
| Mace (1997) | Calcareous fractured aquifer | 0.76 | 1.08 |
| Fabbri (1997) | Calcareous fractured aquifer | 0.85 | 1.07 |
| Jalludin and Razack (2004) | Fractured basalts | 2.99 | 0.94 |
| Hamm et al. (2005) | Volcanic aquifers | 0.99 | 0.89 |
| Razack and Lasm (2005) | Highly fractured and metamorphic aquifer | 0.33 | 1.3 |

However, boreholes are commonly partially penetrating , therefore, we redefine USC as

$$USC = \frac{SC}{h_{bhsat}} \tag{6}$$

where $h_{bhsat}$ is the saturated depth of the borehole in the aquifer. This redefinition adds errors with piezometric head measurements at or near the pumped borehole (Kruseman and de Ridder 1990) affecting the estimate of SC, but this is the only practical method for using the available data where records of aquifer penetration are not always available.

From Equation 4, given the empirical constants of A and B, a bulk hydraulic conductivity value for an aquifer may be estimated from the geometric mean of all the USC values available from pump tests. The exact solution of Matheron (1967) to the stochastic problem varies locally, however, and is progressively less sensitive to local variations in permeability as the difference between the scale of local variation and the scale of measurement increase (Nœtinger et al. 2005). Therefore, if included in any risk evaluation, the effective hydraulic conductivity so determined would indicate transport potential within the aquifer but should not be considered a precise prediction of transport.

Despite this caveat, the method presents a quantitative assessment of vulnerability by using the commonly available parameter of specific capacity.

## 3 DATA

A key task in this work is collation and formatting of the pumping test data available for Scotland based on published reports and information archived by the BGS. In Betson and Robins (2003), we used 75 pump tests that were subsequently refined to exclude those without sufficient information to calculate SC or USC values. Unfortunately, this left a much reduced data set that only permitted the assessment to be made for the Devonian Old Red Sandstone and Permian aquifers (Figure 1). Following the original study, additional data were recovered for the Old Red Sandstone and new pump test data added. These additional data permitted a better assessment of the USC values for the Old Red Sandstone and permitted the calculation of the empirical constants A and B from Equation 3. Table 4 gives the original number of results available for SC and USC values and the geology type the results are associated with.

The characteristics of the range in SC values were compared to those for Devonian aquifers in England and Wales (Morris et al. 1998, Table 5). The comparison shows the ranges of SC results obtained.

Figure 1. The 1:625,000 solid geology map of Scotland overlain with the sites of the pump tests (triangles) and those pump tests with SC values (circles).

Table 4. The available data from Scottish pumping tests that provides specific capacity and unit specific capacity values.

| Geology Type | No. SC Values | No. USC Values |
|---|---|---|
| Devonian Old Red Sandstone | 27 | 11 |
| Permian Basal Breccias, Sandstones and Mudstones | 16 | 10 |
| Igneous | 4 | 4 |
| Carboniferous Limestone | 4 | 3 |
| Undifferentiated Permian and Triassic Sandstone | 5 | 2 |
| Namurian Milstone Grit | 1 | 1 |
| Westphalian Coal Measures | 1 | 1 |
| Lower Lias | 1 | 0 |

Table 5. Comparison of the characteristics of the Devonian aquifers specific capacity values from this study for Scotland and values for England and Wales (Morris et al. 1998).

| | Devonian Scotland | Devonian, South-East England | Devonian, Wales and Welsh Borders |
|---|---|---|---|
| Number of records | 27 | 170 | 148 |
| SC Max | 855 | 9101 | 1226 |
| SC Min | 0.432 | 0.1 | 0.000001 |
| SC Geo. Mean | 25.36 | 9.7 | 8.29 |
| SC Arith. Mean | 79.77 | 118 | 39.3 |

## 4 RESULTS

Following the above methodology, the geometric mean was calculated for the two most significant sets of USC data available on the Permian and the Devonian; the results are presented in Table 6. The new information recovered for the Devonian tests included transmissivity values that permitted the calculation of the empirical constants from the plot of ln T against ln SC (Figure 2).

The linear regression in Figure 2 indicates values for A and B of ($e^{-0.255}$) and 1.117, respectively. Given the value of the geometric mean for the Devonian in Table 6, according to equation 4, the value of $K_{ef}$ can now be calculated as 0.20 m/d, which assumes in the method that the hydraulic conductivity distribution is lognormal. This assumption was tested for the two data sets by plotting the cumulative probability based on a lognormal distribution against the USC values obtained (Figure 3) and the natural log of the USC values arranged in order of lowest to highest (as in Table 6). A linear fit to the data indicates

Table 6. The USC values in m/d ($m^3$/d/m/m) for the Devonian and the Permian basal breccias listed from the lowest to the highest and the geometric mean of the each data set given at the bottom.

| Permian | Devonian |
|---|---|
| 0.62 | 0.01 |
| 1.00 | 0.01 |
| 1.23 | 0.04 |
| 1.82 | 0.07 |
| 2.16 | 0.09 |
| 2.84 | 0.18 |
| 3.74 | 0.27 |
| 3.90 | 0.39 |
| 5.68 | 1.13 |
| 12.04 | 1.84 |
| | 13.06 |
| ***Geometric Mean*** | |
| 2.47 | 0.19 |

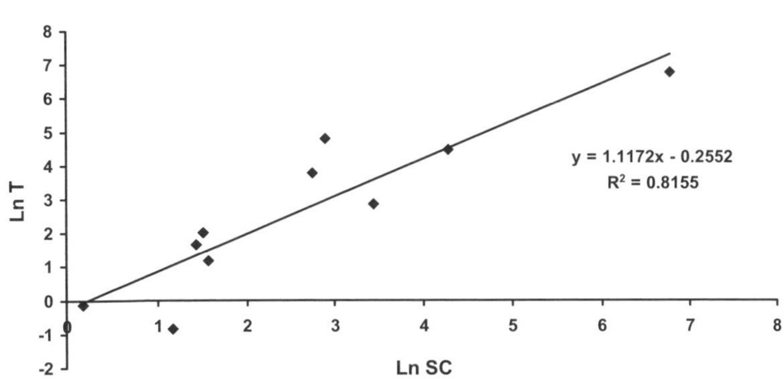


Figure 2. A plot of the relationship between ln T and ln SC from revised Devonian data.

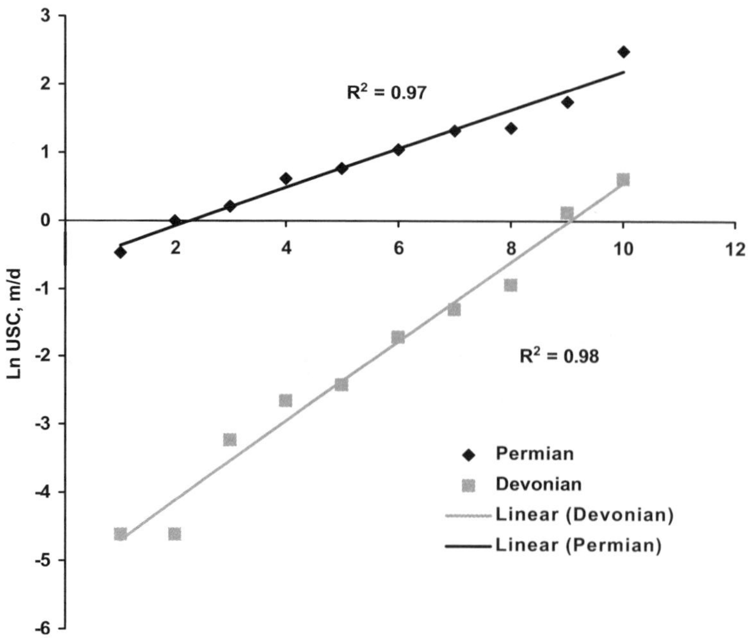


Figure 3. The plot of the natural log of the ascending range of USC values for both geology types and the $R^2$ values of the linear fit to the data.

that it follows a lognormal distribution and Figure 3 shows the plot obtained for the log of the two data sets and the $R^2$ value of the respective linear fits.

The $R^2$ values indicate that USC values follow a linear trend. Thus, the assumption of a lognormal distribution is valid in these data and the geometric mean should reflect the general hydraulic conductivity of the rock.

## 5 DISCUSSION

In Scotland, the regionally important bedrock aquifers, such as the Devonian and Permian sandstones, are characterised by secondary fracture permeability of considerably greater magnitude than the available intergranular (or matrix) permeability, (MacDonald et al. 2005). The typical range of hydraulic conductivity for the Permian rocks is 1–10 m/d (Robins and Buckley 1988) and available pumping test data suggest values for the Devonian of 0.3 to 1.2 m/d. The geometric mean of the USC values is comparable to the ranges of hydraulic conductivity values, but tending towards the lower end of each range.

It was noted above that for vulnerability mapping at a scale of 1 $km^2$ a length scale of $10^2$–$10^3$ m (equivalent to an area of $10^{-2}$–1 $km^2$) was required for the measurement parameter. To check if the pumping test measurements used met this criterion an estimated source capture zone for each of the boreholes was calculated (Table 7). The calculation was based on Robins (1999) estimation method for the probable source capture zone of Lower Palaeozoic rocks in central and eastern Scotland with an assumed annual infiltration of 200 mm.

Table 7. Probable source capture zones for boreholes in this study based on Robins (1999).

| Permian | | Devonian | |
|---|---|---|---|
| Yield (l/s) | Probable Source Capture Zone ($km^2$) | Yield (l/s) | Probable Source Capture Zone ($km^2$) |
| 4 | 0.10 | 0.36 | 0.01 |
| 8 | 0.20 | 0.73 | 0.02 |
| 15 | 0.38 | 2.96 | 0.07 |
| 17 | 0.43 | 4.78 | 0.12 |
| 19 | 0.48 | 6.74 | 0.17 |
| 38 | 0.95 | 18.84 | 0.47 |
| 41 | 1.03 | 28.83 | 0.72 |
| 46 | 1.15 | 54.85 | 1.38 |
| 53 | 1.33 | 76.20 | 1.91 |
| 65 | 1.63 | 92.12 | 2.31 |
| | | 648.45 | 16.27 |

Table 7 shows that the estimated areas for the boreholes in the Devonian and Permian aquifers are within the range appropriate for the 1 $km^2$ mapping scale.

## 6 CONCLUSION

The ln (USC) is directly correlated with ln ($K_{ef}$) as a stochastic average and offers a quantitative measure of aquifer permeability. If the range of USC values recorded for a number of locations spatially across the aquifer follow a lognormal distribution, it is assumed that the geometric mean of the values represents a general hydraulic conductivity parameter. If the empirical constants for Equation 3 can be calculated from the available data, $K_{ef}$ for the aquifer can be evaluated.

Vulnerability mapping at the national scale often uses a 1 $km^2$ grid unit. For the USC values to be applicable at this scale, the borehole source capture zone should be estimated to be in the range of $10^{-2}$–1 $km^2$.

The survey of the available data on pumping tests in fractured aquifers in Scotland yields a set of USC values with significant subsets for Devonian and Permian aquifers. The geometric mean of these subsets indicated USC values of 0.19 m/d for the Devonian and 2.47 m/d for the Permian aquifer rocks. Sufficient data were also present in the Devonian set to determine the constants A and B from Equation 3 and thus to also evaluate the $K_{ef}$ which was found to be 0.20 m/d, similar to other estimates for the Devonian in the U.K.

The stochastic approach to assigning an overall specific capacity to a complex fractured rock unit overcomes the need to assign permeability classes on a qualitative basis. Whilst acknowledging that groundwater vulnerability also depends on pollutant attenuation and some measure of reactivity in the pathway to the water table, the stochastic analysis provides a valuable methodological component of vulnerability assessment. Although the methodology has been tested on a small population of data from poorly permeable Scottish hard rock aquifers, it can be applied to other comparable fractured aquifer systems with limited data.

## ACKNOWLEDGEMENTS

This work described here was part funded by the Scottish Environment Protection Agency. The authors are grateful to Majdi Mansour at BGS for checking the methodology.

## REFERENCES

Ahmed S and de Marsily G (1987) Comparison of geostatistical methods for estimating transmissivity using data on transmissivity and specific capacity. Water Resources Research 23:1717–1737.

Anthony S, Quinn P and Lord E (1996) Catchment scale modelling of nitrate leaching. Aspects of Applied Biology, 46:23–32.

Bear J. (1972), Dynamics of Fluids in Porous Media, American Elsevier Publishing Company, New York.

Betson M, Robins N (2003) Use of permeability to estimate vulnerability to diffuse pollution on fractured aquifers in Scotland, Proceedings of the IAH conference "Groundwater in Fractured Rocks", Prague, 15–19 September 2003.

Dagan G. (1986) Statistical Theory of Groundwater Flow and Transport: Pore to Laboratory, Laboratory to Formation and Formation to Regional Scale, Water Resources Research, 22 (9): 120S–134S.

Department of the Environment and Local Government, Environmental Protection Agency and Geological Survey of Ireland (1999) A scheme for the protection of groundwater. Geological Survey of Ireland, Dublin.

Department of the Environment for Northern Ireland (1994) Groundwater vulnerability map for Northern Ireland. Environment Service, Department of the Environment for Northern Ireland.

Department of the Environment for Northern Ireland (1999) Policy and practice for the protection of groundwater in Northern Ireland: a consultation document. Environment Service, Department of the Environment for Northern Ireland.

Drew D and Daly D (1993) Groundwater and karstification in Mid-Galway, South Mayo and North Clare. Geological Survey of Ireland Technical report RS 93/3.

Environment Agency (1998) Policy and practice for the protection of groundwater. HMSO, London.

European Community (1996) Proposal for Council Directive establishing a framework for Community action in the field of water policy. Official Journal of the European communities, C184.

Fabbri P (1997) Transmissivity in the geothermal Euganean basin: a geostatistical analysis, Ground Water 35(5):881–887.

Foster SSD (1998) Groundwater recharge and pollution vulnerability of British aquifers: a critical overview. In: Robins NS (ed) Groundwater Pollution, Aquifer Recharge and Vulnerability, Geological Society Special Publications, 130:7–22.

Hamm ST, Cheong JY, Jang S, Jung CY, Kim BS (2005) Relationship between transmissivity and specific capacity in the volcanic aquifers of Jeju Island, Korea, Journal of Hydrology 310:111–121.

Huntley D, Nommensen R, Steffey D (1992) The use of specific capacity to assess transmissivity in fractured rock aquifers, Ground Water 30(3):396–402.

Jalludin M and Razack M (2004) Assessment of hydraulic properties of sedimentary and volcanic aquifer systems under arid conditions in the Republic of Djibouti (horn of Africa), Hydrogeology Journal 12:159–170.

Kreuseman GP and DeRidder NA (1990) Analysis and evaluation of pumping test data, Internat. Inst. For Land Reclamation and Improvement publication 47.

Kupfersberger H and Blöschl G (1995) Estimating aquifer transmissivities – on the value of auxiliary data. J of Hydrology 165:85–99.

Landau LD, Lifshitz EM (1960) Electrodynamics of continuous media, Pergamon, Oxford.

Lewis MA, Lilley A and Bell JS (2000) Groundwater vulnerability mapping in Scotland: modifications to classification used in England and Wales. In: Robins NS and Misstear DR (eds)

Groundwater in the Celtic Regions: Studies in Hard Rock and Quaternary Hydrogeology, Geol. Soc. Special Publications, 182:71–79.

Logan J (1964) Estimating transmissibility from routine production tests of water wells. Ground Water 2(1):35–37.

Lord EI and Anthony SG (2000) MAGPIE: A modelling framework for evaluating nitrate losses at national and catchment scales. Soil Use and Management 16:167–174.

MacDonald AM, Robins NS, Ball DF and Ó Dochartaigh BÉ (2005) An overview of groundwater in Scotland. Scottish Journal of Geology, 41(1):3–11.

Mace RE (1997) Determination of transmissivity from specific capacity tests in a karst aquifer. Ground Water 35(4):738–742.

Matheron G (1967) Composing permeabilities in heterogeneous porous media. Swindler method and averaging rules, Revue de l'Institut Français du Pétrole 22(3):443–466.

Morris BL et al. (1998) The Physical properties of minor aquifers in England and Wales. British Geological Survey Technical Report WD/00/04.

Morris BL, Lawrence AR, Chilton PJ, Adams B, Calow RC and Klinck BA (2003) Groundwater and its susceptibility to degradation, a global assessment of the problem and options for management. United Nations Environment Programme, Nairobi, 126 pp.

National Rivers Authority (1992) Policy and practice for the protection of groundwater. NRA, Bristol.

Nœtinger B, Artus V, Zargar G (2005) The future of stochastic and upscaling methods in hydrogeology, Hydrogeology Journal, 13:184–201.

Nœtinger B, Jacquin C (1991) Experimental tests of a simple permeability composition formula, 66th Annual Technical Conference and Exhibition of the Society of Petroleum Engineers, 6–9th October 1991, Dallas, USA, SPE 22841.

Ó Dochartaigh, BÉ, Ball DF, MacDonald AM, Lilly A, Fitzsimons V, Del Rio M and Auton CA (2005) Mapping groundwater vulnerability in Scotland: a new approach for the Water Framework Directive, Scottish Journal of Geology. 41:21–30.

Palmer RC, Holman IP, Robins NS and Lewis MA (1995) M A Guide to Groundwater Vulnerability Mapping in England and Wales. HMSO for National Rivers Authority.

Razack M and Huntley D (1991) Assessing transmissivity from specific capacity in a large and heterogeneous alluvial aquifer, Ground Water 29(6):856–861.

Razack M and Lasm T (2006) Geostatistical estimation of the transmissivity in a highly fractured metamorphic and crystalline aquifer (Man-Danane Region, West Ivory Coast), Journal of Hydrology (in press).

Renard Ph and de Marsily G (1997) Calculating equivalent permeability: a review. Advances in Water Resources 20 (Nos. 5–6):253–278.

Robins NS (1999) Groundwater occurrence in the lower Palaeozoic and Precambrian rocks of the UK. Journal of the Institution of Water and Environment Management, 13:447–453.

Robins NS and Misstear DR (2000) Groundwater in the Celtic Regions: Studies in Hard Rock and Quaternary Hydrogeology, Geological Society Special Publications, 182:5–17.

Scottish Environment Protection Agency (1997) Policy No. 19. Groundwater Protection Policy for Scotland. SEPA, Stirling.

Tellam JH and Lloyd JW (1981) A review of the hydrogeology of British onshore non-carbonate mudrocks. Quarterly Journal of Engineering Geology, 14:347–355.

CHAPTER 34

# Environmental impacts of tunnels in fractured crystalline rocks of the Central Alps

Simon Loew[1], Volker Luetzenkirchen[2], Ulrich Ofterdinger[3], Christian Zangerl[4], Erik Eberhardt[5] and Keith Evans[1]

[1]*Chair for Engineering Geology, Department of Earth Sciences, ETH Zuerich, Switzerland*
[2]*Chair for Engineering Geology, Department of Earth Sciences, ETH Zuerich, Switzerland*
*Now at mbn AG, 5400 Baden, Switzerland*
[3] *Chair for Engineering Geology, Department of Earth Sciences, ETH Zuerich, Switzerland*
*Now at Queen's University Belfast, Northern Ireland*
[4]*Chair for Engineering Geology, Department of Earth Sciences, ETH Zuerich, Switzerland*
*Now at alpS – GmbH, Grabenweg 3, 6020 Innsbruck, Austria*
[5]*Chair for Engineering Geology, Department of Earth Sciences, ETH Zuerich, Switzerland*
*Now at Geological Engineering, University of British Columbia, Vancouver, B.C., Canada*

ABSTRACT: This paper reviews unique case histories from the Central Swiss Alps that illustrate the effects intermediate and deep tunnels in crystalline rocks can have on groundwater flow systems, surface springs and surface deformations. Serious impacts on surface waters occur in weathered and de-stressed high permeability areas close to the tunnel portals, or along tunnels running only a few hundred meters below ground surface. In these areas, water table drawdown can reach the elevation of the tunnel and significantly affect surface springs and wetlands over lateral distances of several kilometers within a few weeks. Similarly, deep tunnels located 500–1500 m below ground surface can have considerable impacts on natural flow systems, although their effects are much less visible; groundwater flow is redirected towards the draining tunnel, re-locating groundwater divides and reversing flow directions over lateral distances larger than one kilometer. In addition, strong pressure reductions created by draining tunnels at depth can result in consolidation of the rock mass and large scale deformations leading to surface settlements in excess of 10 cm.

## 1 INTRODUCTION

In mountainous terrain like the European Alps, many tunnels and drifts are excavated deep below ground surface, acting as hydraulic drains below the regional groundwater table. Sedimentary systems in such mountain environments often behave like a succession of confined aquifers and aquitards (e.g., Stini 1950), whereas crystalline rocks in such settings are better described as heterogeneous phreatic aquifers (e.g., Ofterdinger 2001).

Opening a deep tunnel excavation to atmospheric pressure acts to generate significant pore pressure drawdown around the excavation and influences groundwater heads and flow fields over large distances. Depending on the recharge conditions at ground surface, effective hydraulic conductivities and specific yields of the rock mass around the tunnel, a tunnel in crystalline rocks can also induce a water table drawdown with serious environmental impacts at ground surface. Such impacts include the disappearance of surface springs and wetlands (e.g., Stapff 1882; Stini 1950). Even without significant water table drawdown, the pore pressure reductions induced by deep draining tunnels can lead to considerable ground subsidence at surface (Zangerl 2003; Zangerl et al. 2003). Such hydro-mechanically coupled effects are well known in high porosity soils and weak rocks, but much less so for low porosity crystalline rocks.

Many of the European countries that are currently planning or building deep tunnels through the Alps are seriously confronted with these environmental issues and have started to develop codes of practice on how to handle such problems (e.g., BUWAL 1998). In support of such projects, a series of Ph.D. projects have recently been completed at the ETH Zürich, focusing on the Gotthard massif of the Swiss Alps and that study the underlying mechanisms of such processes (Ofterdinger 2001; Luetzenkirchen 2002; Zangerl 2003).

This paper reviews observations and interpretations of environmental impacts of deep tunnels from the Gotthard massif of the Central Swiss Alps. After a discussion of the key hydrogeological properties of this study area, three case studies are presented demonstrating: (1) the impacts of deep tunnel sections on regional flow systems and groundwater recharge, (2) the impacts of shallow tunnel sections on surface springs and wetlands, and (3) the impacts of draining tunnels on surface settlements. The case study results are representative of many existing underground structures excavated during the last 100 years in the crystalline massifs in the central Swiss Alps.

## 2 KEY HYDROGEOLOGICAL PROPERTIES OF THE STUDY AREA

### 2.1 *Geology and tectonics*

Figure 1 provides an overview of the tectonic units and steeply dipping geological boundaries found in the external crystalline massifs of the investigation area (Aar Massif, Tavetsch Massif, Gotthard Massif) in the Central Alps of Switzerland. The investigation area is about 1200 $km^2$ and has strong mountainous relief, with valley floor elevations ranging from 500 to 1500 meter above sea level (masl) and summits up to 2000 to 3000 masl.

About 95 % of the rocks in the study area (Figure 2) belong to the pre-Alpine crystalline basement which is composed of pre-Variscan gneisses, schists, migmatites and mainly late Variscan (upper Carboniferous) magmatic rocks (granites, diorites, syenites, less abundant volcanics and aplite/lamprophyre dykes). The remaining 5% are late-paleozoic and mesozoic marls, conglomerates, sandstones and dolomites in autochthonous to par-autochthonous position relative to the crystalline basement. A summary of the petrographic characteristics of this area can be found in Labhart (1999).

In the Gotthard massif, medium to upper greenschist facies conditions were reached during Tertiary metamorphism, with an increase in peak pressure and temperature from north to south. Along the southern boundary, amphibolite facies conditions were achieved (Frey et al. 1980; Labhart 1999). The main Alpine deformation phase in the Gotthard massif starts in the lower Oligocene around 35 and 30 Ma (Schmid et al. 1996), corresponding with the peak metamorphic overprint characterised through a ductile deformation regime.

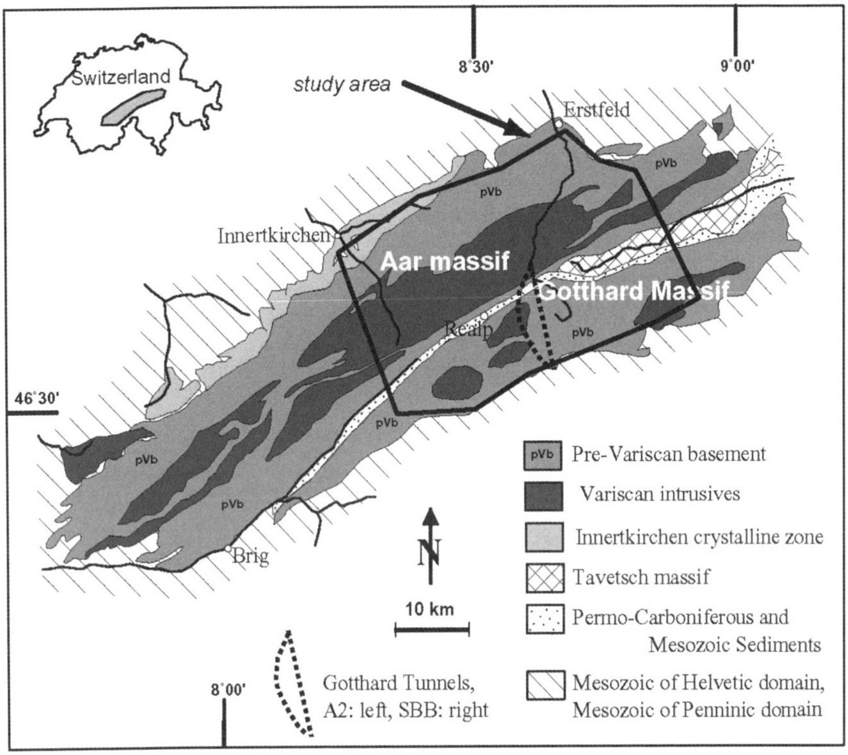


Figure 1. Tectonic map of the Aar and Gotthard Massif in the Swiss Alps, showing outline of study area (see Figure 2).

In the central Gotthard massif this deformation phase led to the formation of Alpine shear zones and a penetrative foliation that strikes NE-SW or E-W and dips southwards in the northern part and northwards in the southern part, forming a fan like structure (Labhart 1999).

During later stages, ongoing deformation changed gradually from a ductile to a brittle deformation regime characterised by brittle faulting. Recently much work was done on the formation of brittle structures within the Gotthard massif (Luetzenkirchen 2002; Zangerl 2003). Compared to the moderate brittle deformation in the Aar massif, the Gotthard massif shows a high density of brittle faults and fault zones of remarkable extension and width. Figure 3 shows the trace pattern of mapped and inferred brittle fault zones at the surface in the Gotthard pass area near Sustenegg. In this area two major sets striking NE-SW (set F1) and NNE-SSW (set F2), and one minor W-E set (set F3) can be distinguished. Across the Gotthard massif, the brittle faults follow the orientation of Alpine foliation forming the same fan-like structure, characterized North of Sustenegg by southeast dipping structures and South of Sustenegg by northwest dipping structures. The total spacing of all brittle fault sets estimated normal to their mean orientation is approximately 35 m (Zangerl et al. 2001). Both tunnel and surface derived mapping data show similar pole distribution

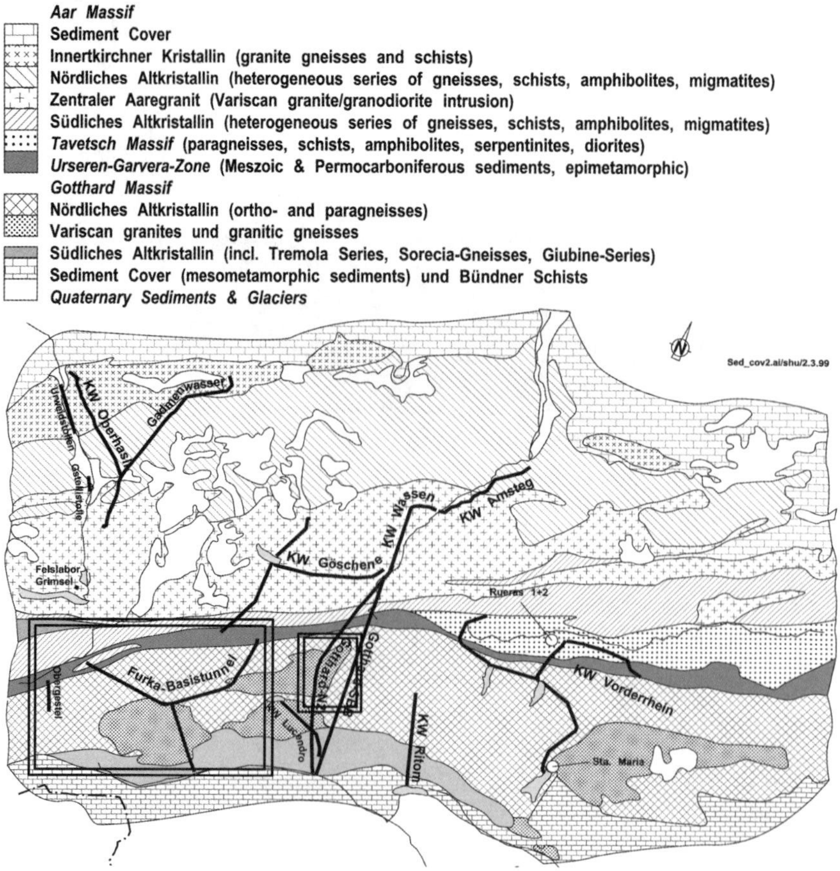


Figure 2. Geological map of study area (rotated), location of studied traffic tunnels and hydropower drifts. Rectangles mark approximate locations of Figures 3 (right) and 4 (left).

patterns, although it is not possible to resolve clearly the three different fault sets at depth. Based on the observations reported in Zangerl (2003), most of the mapped fault zones can be classified as pure strike-slip faults following the classification scheme by Angelier (1994). The rest can be grouped as oblique-slip faults.

All of these observations relate to the youngest faulting events. Based on zeolite parageneses identified in fractures from brittle fault rocks, Luetzenkirchen (2002) concluded that these faulting events terminated before metamorphic temperatures dropped below 190° C (Figure 6). Neotectonic activity must be very low in the study area, however, post-glacial rebound superficially "reactivated" some of the fault zones along the main NE-SW striking valleys (Luetzenkirchen 2002; Frei and Löw 2001).

During this same period, the Aar massif underwent only moderate brittle deformation, which generated minor shear fractures and joint systems locally filled with hydrothermal crystallizations (Steck 1968a; Bossart & Mazurek 1991; Laws 2001). Like in the Gotthard massif, these brittle structures overprint ductile Alpine shear zones.

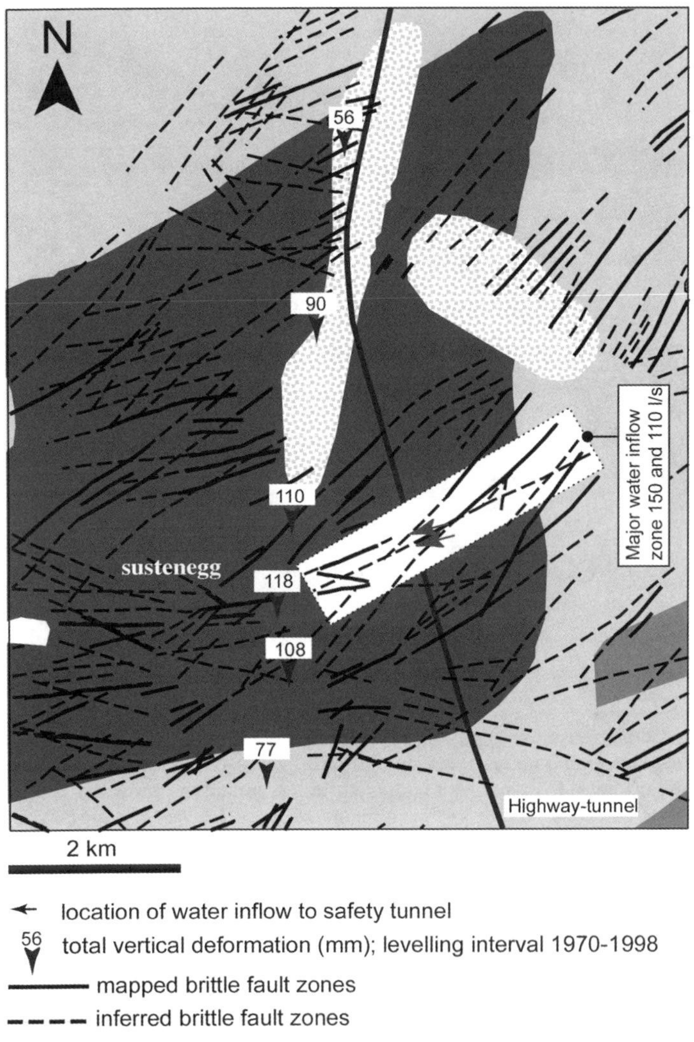


Figure 3. Brittle fault zones mapped in the Gamsboden granitic gneiss, north of Gotthard pass. Black lines represent mapped brittle fault zones, dashed lines represent inferred fault zones based on aerial photos and geomorphological mapping. Vertical arrows indicate locations of levelling points and subsidence values measured along the Gotthard pass road. Highlighted is the location of singular inflows to the Gotthard highway and safety tunnel.

## 2.2 *Hydrology and groundwater recharge*

The study area corresponds to the main water divide in the central Alps. Precipitation in the study area is high, but varies strongly spatially. Mainly depending on altitude, the mean annual precipitation ranges between 1600 and 3600 mm/yr (BWG 2001). In most years, relatively little precipitation falls between December and February. Snow heights at the end of winter (April–May) represent about 1000 mm of water. In general, glaciers and non-permanent permafrost occur above 2500 masl and feed creeks and rivers in summertime. Evapotranspiration ranges between 200 and 500 mm per year, again depending on altitude.

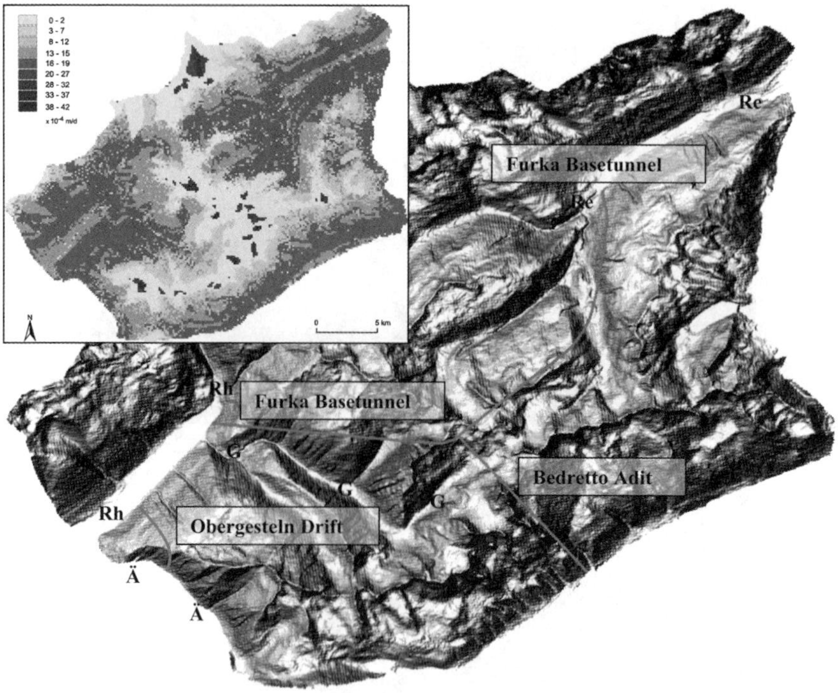


Figure 4. Topography and geography of Rotondo area with Furka basetunnel, Bedretto adit and Obergesteln drift. G: Geren valley with Gerental basin, Ä: Äginen Valley, R: Rhone Valley, Reuss Valley, GP: Gotthard Pass. Inset map showing distributed annual groundwater recharge in the same Rotondo area. Highest recharge rates occur in glaciated areas. Lowest recharge rates occur on steep bare high altitude rock slopes.

Depending on topography, land use, vegetation and slope exposure, groundwater recharge shows strong variations across the catchments (e.g., Lehner et al. 1990). Given that the study area is situated in a highly mountainous area covering a wide range of elevation zones, the spatial distribution of the groundwater recharge rates (i.e., the upper boundary condition for groundwater flow models) is a crucial factor for understanding the natural and artificially disturbed groundwater flow systems. For this reason, the recharge rates were studied in a test catchment of 40 $km^2$, the Gerental basin (Figure 4), with a calibrated hydrological model. The applied GIS-based model PREVAH (Precipitation-Runoff-EVapotranspiration-Hydrotope model) combines the spatial differentiation of hydrologically similar response units (HRU), or hydrotopes, and a runoff generation concept that allows the separate calculation of the water balance within each hydrotope (Gurtz et al. 1999). The hydrotopes were defined to represent hydrologically homogeneous areas according to the most important factors controlling evapotranspiration and runoff formation processes, such as the meteorological inputs, topography (catchment area, altitude, exposure and slope), landuse and soil characteristics (Gurtz et al. 1990). For the glaciated areas a conceptual model was applied which was developed for a research study in the Rhône area (Badoux 1999). This concept attributes melt-water from glaciated areas below the equilibrium line (ELA) preferentially to the fast runoff storage, while melt-water originating above the ELA preferentially contributes to the slow runoff storage, i.e. to groundwater recharge.

The hydrological model was first calibrated and validated in the Gerental catchment and then applied to the larger domain of the Rotondo area using identical model parameters as calibrated in the Gerental basin. The hydrograph simulation for the gauging station at the Gerental catchment outlet was performed for the period January 1991 to October 1999. The results of the hydrograph simulation are discussed elsewhere (Vitvar and Gurtz 1999). The temporal variations of the simulated recharge rates show maximum recharge rates in the upper Geren valley during spring and summer (April–September), commencing with the onset of the melting period. The resulting recharge distribution shows a strong spatial variability, especially in the high altitude regions where steep bare rock slopes with low recharge areas contrast to the upper regions of the glaciers, characterized by high recharge rates within small restricted areas (inset Figure 4). Another striking feature in the recharge distribution is the observation of low to moderate recharge rates along the valley floor and lower valley slopes. However, from a hydrodynamic perspective these areas are expected to constitute potential groundwater exfiltration areas, where essentially no groundwater recharge should occur. This conceptual discrepancy might lead to a small systematic error in the spatial distribution of the recharge rates, as the overall water budget in the catchment is well constrained by measurements and controlled in the model. For the purpose of numerical simulations an average recharge rate distribution was extracted from the 9-year average. The spatial average recharge rate of this distribution for the whole model domain is 12E-4 m/d, which is in agreement with studies in similar topographic and geological settings in the Central Alps (Kölla 1993), giving a range of 3 to 22E-4 m/d with a most probable value of 11E-4 m/d.

The groundwater table across the entire study area has never been directly measured with surface based boreholes (or piezometers). Most of the information about the location of the groundwater table comes from pore pressures measured in subsurface boreholes and the locations of systematically mapped surface springs (Frei und Löw 2001; Luetzenkirchen 2002). These parameters or observations indicate that the intersection of the groundwater table with the steep and bare valley slopes occurs several hundreds of meters below the local peak elevation (the spring line is often found between 2000 and 2400 m elevation). Water tables derived from subsurface water pressure measurements in the KW Vorderrhein adit are 100–200 meters higher than the local spring line elevation. This has to be expected because hydraulic heads have to increase from the valley slopes to the mountain crests.

### 2.3 *Distribution of groundwater flow and hydraulic conductivity from tunnel observations*

Figure 5 illustrates, with examples from the Gotthard N2 highway and Gotthard SBB railway tunnels, the characteristic patterns of early time tunnel inflows that were observed in most of the underground excavations found within the study area. Seeping inflows (dripping water) occur along most of the tunnel sections in a seemingly random ubiquitous distribution. Continuous (flowing) water inflows occur in clusters at a few specific locations, for example between tunnel meter (Tm) 8000 and 11,000 in the Gotthard N2 highway tunnel. The early time volumetric rate of these continuous inflows is typically in the order of 5 to 20 l/s per 100 m of tunnel length. In addition to these clustered inflows, there are a few locations with extremely high inflows. In the Gotthard N2 tunnel such singular structures can be found at Tm 9910 and 9935 and showed early time inflow rates of 110 and 150 liters per second, respectively.

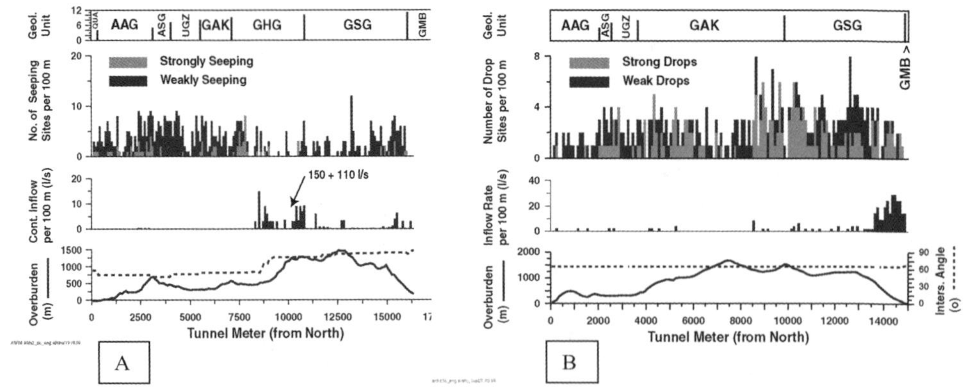


Figure 5. Geology, water inflows and topography of (A) Gotthard N2 Highway Tunnel and (B) Gotthard Railway Tunnel. Geologic Units: Aar Massif (AAG, ASG); Permo-Carboniferous and Mesozoic sediments (UGZ); Gotthard Massif (GAK, GHG, GSG); Mesozoic sediments of Penninic domain (GMB).

As many of the tunnels are lined with concrete, steel sets or shotcrete and the existing reports do not give detailed information, geologic descriptions of groundwater pathways originate from new observations in selected tunnels of the Gotthard Massif (e.g., Bedretto adit of the Furka basetunnel, Gotthard N2 highway tunnel). Here small continuous inflows (0.01 to 0.1 l/s) originate in most cases from mm-wide channels that lay in the planes of fractures intersecting the tunnel walls. Most of the fractures containing such channels are steeply inclined shear fractures (small faults) with polished and striated surfaces and, occasionally, mm-thin gouge layers. These fracture-fillings have a width of a few mm to a few cm and the fracture length exceeds in most cases the local tunnel diameter (3 m). The flowing water channels occur where these fractures intersect other minor fractures (joints). The shear fractures often occur close and parallel to existing lithological (mechanical) heterogeneities, for example older ductile shear zones and dykes, or within the damage zone of larger cataclastic faults.

Medium to large continuous inflows occur mostly in the zone of near-surface weathering, toppling and stress relief or along deeply reaching and extended brittle fault zones. The zone of near surface weathering and stress relief is characterised by a high density of open fractures, sometimes accompanied by flexural toppling, and reaches depths of up to 200 m below ground surface. Fault zones generating medium to large inflows are always dominated by brittle deformation (overprinting ductile shear zones in most cases) and sometimes contain cohesionless fault breccias or gouges of up to several meters thickness (Luetzenkirchen 2002; Zangerl 2003). As shown in Figure 5, most of the inflows occur along faults in granitic rocks (Variscan intrusions) of the Gotthard massif. These interrelationships between fault zone hydraulic conductivity and deformation type also becomes obvious when comparing the bulk tunnel inflows in the sections of the Aar massif (hardly any continuous inflows; mainly ductile shear zones) with the Gotthard massif (many clusters of continuous inflows; mainly granitic gneisses with intensive brittle faulting). Typical examples of permeable brittle fault zones of moderate size from the Bedretto adit to the Furka basetunnel are shown in Figure 6. Photomicrographs from thin sections show that open fractures in fault rocks are partially filled with low-temperature mineral paragenesis (zeolites) and that the open pore space is remarkably high.

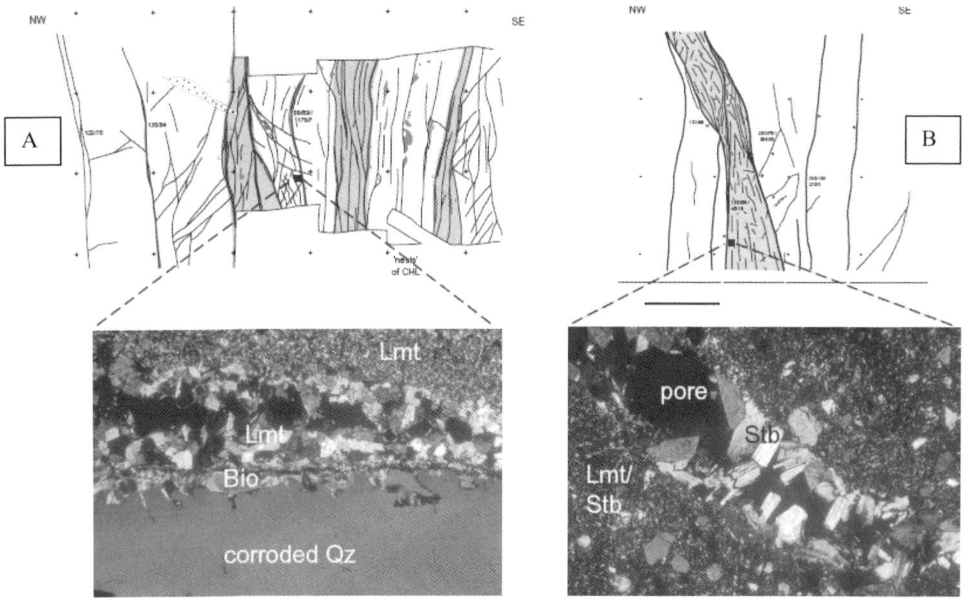


Figure 6. Schematic representation of mapped (A) brittle-ductile fault zone, and (B) brittle fault zone from the Bedretto adit of Furka basetunnel. Crosses mark the 1-m spaced mapping grid used on the tunnel wall Fault core composed of weak to soft chemically altered material (mainly gauge) is marked in grey. Shear fractures with slickenside striations shown as bold lines. Photomicrographs from fractures partially filled with zeolites (left: laumontite Lmt; right: stilbite Stb). Open pore space shown in black.

A quantitative analysis of such tunnel inflows has shown that the crystalline basement rocks in the investigated area have a strongly anisotropic hydraulic conductivity, especially in the regions with significant continuous inflows. Assuming radial flow directed towards a constant head tunnel drain, Loew (2001) derived 100–1000 meter scale hydraulic conductivities in the direction of the Alpine foliation (normal to the tunnel axes). For tunnel sections with moderate inflows (e.g., Gotthard railway tunnel without inflows close to the southern portal) typical values of $10^{-8}$ m/s were derived. For areas with high tunnel inflows close to the portals, mean hydraulic conductivity in the plane of the Alpine foliation typically ranges between $10^{-3}$ and $10^{-5}$ m/s (Loew 2001).

## 3 IMPACTS OF DEEP TUNNELS ON REGIONAL GROUNDWATER FLOW SYSTEMS

Environmental impacts of deep tunnel sections (500 m–1500 m) in crystalline rocks of the Gotthard and Aar massifs are not well constrained by direct observations. In most cases no environmental impacts can be seen directly on ground surface. However, the subsurface flow systems change considerably over large distances. These changes can be derived from theoretical considerations and numerical modeling. A detailed numerical study has been carried out by Ofterdinger (2001) for the Rotondo area covering the Furka basetunnel and Bedretto adit (Figure 4). These underground structures were constructed in the years 1975–1980.

Today the Furka basetunnel is in operation as a single-line intra-alpine rail tunnel having an elevation of about 1500 masl. The Bedretto adit is an unlined support segment, which is not in operation and has recently been adopted as an ETH Zurich research facility. The geology in the Furka and Bedretto excavations consists mainly of the Variscan Rotondo Granite (about 220 Mio years old) with orthogneiss of early Paleozoic age (about 420 Mio years). Like in the rest of the study area, both rock types show Alpine foliation and shear zones, and are composed of several Alpine fracture sets with a dense pattern of brittle faults. One major regional brittle fault zone of more than 100 m total thickness follows the upper part of the Geren valley (striking about N 60 E) intersecting the Bedretto adit in its northern part. In the Furka and Bedretto tunnels, the overburden typically ranges between 1000 and 1500 m.

## 3.1 *3D groundwater model of the Rotondo area*

Simulations of the regional scale groundwater flow in the Rotondo area (Figure 4) have been carried out for the natural system and that following construction of the Furka and Bedretto tunnels using a 3D continuum model (FEFLOW, Diersch 1998). The mesh consisted of 141,022 nodes and 259,156 elements distributed over 13 layers with one regional fault zone. The base of the model is at 0 masl. The total projected area of the model comprises 262 $km^2$. The vertical discretization was chosen to enable a good resolution of the free and moving groundwater table, to assign depth dependant values of effective hydraulic conductivity, and to achieve a dense mesh discretization around the galleries. From the results of the hydrological model described in the previous section, the spatially distributed long-term mean recharge rates were extracted as upper boundary conditions for the model. The Furka basetunnel and the Bedretto adit are represented as line sinks with fixed elevation heads. The perennial streams and rivers are also treated as fixed inner head boundary conditions.

The model was calibrated in steady-state using the inflow rates to the tunnels as measured in 1999 (i.e., 20 years after excavation), and pore pressures measured in a 100 m long subsurface research borehole. Typical inflow rates in these tunnel sections with large overburden range between 8 l/s and 10 l/s per tunnel kilometer. Figure 7 shows two recorded quasi steady-state inflows to the Bedretto adit from typical moderate sized fault zones. As is clearly visible in this Figure, there is no direct interaction with the seasonally varying precipitation at ground surface. Calibration resulted in an effective hydraulic conductivity of 6E-9 m/s for the Rotondo Granite intersecting the central parts of the tunnel (see Figure 2), 6E-8 m/s for the regional fault zone (generating an influx of 7.5 l/s), and 3E-9 m/s for the surrounding para-gneisses of the pre-variscan basement. Ofterdinger (2001) studied a series of model scenarios where the effects of fault zones, hydraulic anisotropy, and various types of recharge distributions at ground surface were investigated.

## 3.2 *Tunnel induced changes in water table elevation and flow fields in Rotondo area*

In each of the modelled scenarios, the simulated steady-state location of the groundwater table without tunnels was located substantially below all major mountain crests and peaks within the Rotondo area. These crests and peaks, shown in Figure 4B, have an elevation ranging between 2700 and 3200 masl. For the different scenarios simulated, the depth of the water table ranged from 100–150 m below the ridges above the Furka basetunnel to 250 m below the steepest peaks in the area. Besides topography, the spatially varying recharge rates at ground

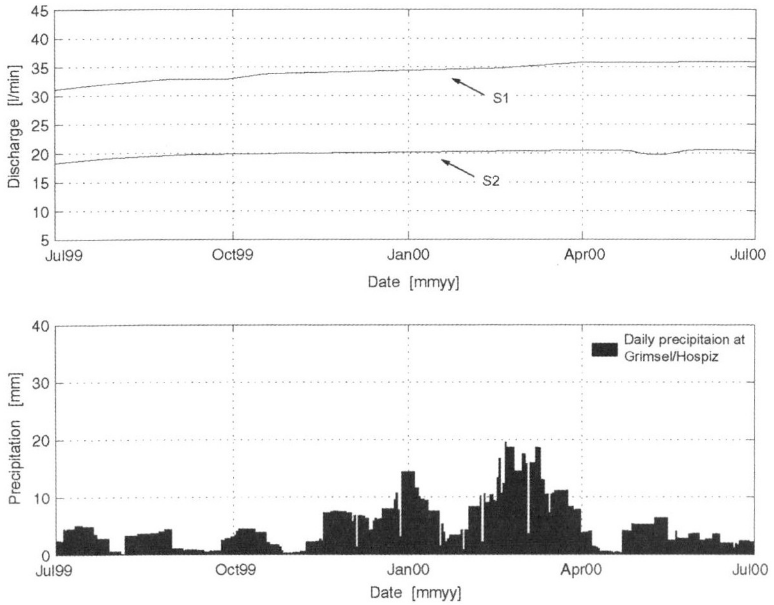


Figure 7. Late time inflows into the Bedretto adit at the location of two brittle fault zones. Comparison with daily precipitation at Grimsel Hospiz.

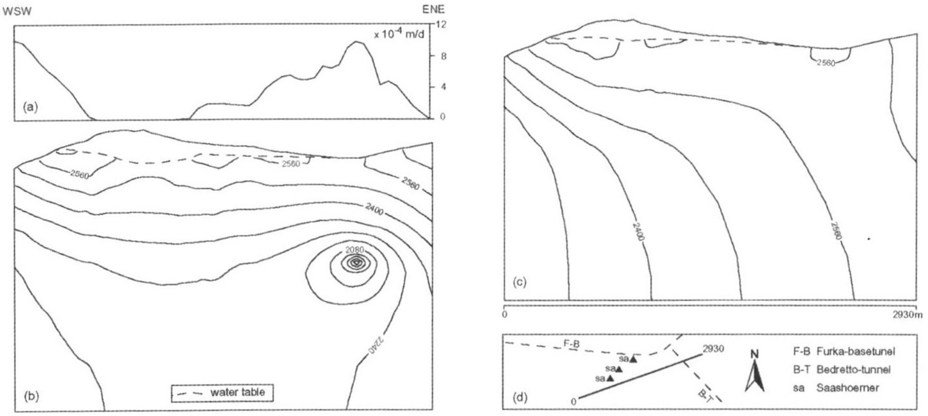


Figure 8. 2-D section through a 3-D hydrodynamic model of Rotondo area, showing: (A) Imposed groundwater recharge rate at ground surface, (B) position of free groundwater table and hydraulic head isolines with Bedretto and Furka tunnels, (C) position of free groundwater table and hydraulic head isolines without Bedretto and Furka tunnels, and (D) location of 2D section in plan view. Vertical exaggeration of (B) and (C) is 0.7.

surface were important factor influencing the location of the water table below ground surface. As expected, the valley floors and adjacent slopes act as groundwater discharge areas.

The deep tunnels create a local decrease in water table elevation compared to the natural situation (Figure 8). In the Rotondo area this tunnel induced water table drawdown occurs only locally below peaks and crests in the near vicinity of the tunnel traces, up to a lateral distance of about 1 km. In the area above the Bedretto adit water tables decrease by about 100 m, and close to the junction with the Furka basetunnel, below the steepest peaks, the drawdown amounts to 350 m. Even though the water table position below the crests differs significantly between the model scenarios, the elevations of the springs as mapped today in the Rotondo area is reproduced by all models within an accuracy of 20–100 m. Typical altitudes of springs in the test area close to the Bedretto adit (Gerental) are between 2200 and 2300 masl., i.e. 100–200 m above the valley floor. During construction of the tunnels, no impacts on surface springs were observed in these central parts of the tunnels. This could be related to the fact that significant water table reductions only occurred locally and in steep flanks at high altitudes, where springs are normally not diverted, used and/or monitored.

Compared to the small surface impacts, these tunnels create a big impact on the flow fields, leading to reversals of the natural flow directions over lateral distances of approximately 1 kilometer. This is illustrated in the cross-section shown in Figure 8. The recharge areas for the Bedretto adit are strongly influenced by the imposed recharge distribution at ground surface, in this case, the presence of glaciers above the equilibrium line. This effect is apparent when comparing these results with simulations for an equivalent uniform mean recharge rate. In the latter case, recharge areas are more regularly distributed on both sides of the draining tunnels, reaching laterally to distances of up to 1.5 kilometers. In general, the groundwater recharge areas from the hydrodynamic model compare well with the recharge areas derived from groundwater isotopic compositions (Ofterdinger et al. 2004).

## 4 IMPACTS OF SHALLOW TUNNELS ON SURFACE SPRINGS AND WETLANDS

The best data set showing the environmental impacts of shallow tunnels (or tunnel sections close to their portals) is derived from an analysis of the Obergesteln drift at the northern boundary of the Gotthard massif (Figures 2 and 4). Klemenz (1974) gives a detailed description of this case study and the analysis of tunnel drainage related water table drawdown in this area. The 2346 meter long Obergesteln tunnel was constructed between February 1972 and May 1973 for a transalpine gas pipeline. Driven southwards, the excavation consists of a 600 m long horizontal drift, followed by a 1280 m long and 22° inclined shaft, and then another 466 m long horizontal drift (Figure 9). The overburden reaches a maximum value of 500 m in the middle of the excavation. With the exception of 400 m in the south and 100 m in the north, the drift and shaft were excavated with a 5 m diameter tunnel boring machine.

### 4.1 *Hydrogeological situation in Obergesteln area*

After Schneider (1974), the Obergesteln excavation runs mostly (in the northern and central parts) at an angle of 90 to 85° to the foliation and cuts from north through a sequence of mica and feldspar rich biotite-plagioclase-gneisses (Tm 0–1745), quartzite (Tm 1745–1800), mixed ortho-para-gneisses (Tm 1800–1960), ortho-gneisses (Tm 1960–2306) and Quaternary sediments (Tm 2306–2346). The northern 800 m of the Obergesteln drift intersect a rock mass with deeply reaching flexural toppling and many open fractures running parallel to the foliation (Figure 9). Due to this toppling, the dip increases from 65° at the portal to 90° at the end of the horizontal drift.

The Obergesteln drift runs between two deeply incised tributary creeks to the Rhone valley: the Äginen creek in the southwest (1.8 km distance from the northern portal), and the Goneri creek in the northeast (2.5 km from the northern portal). Before the construction of the drift, springs mainly occurred Southwest of the Obergesteln drift, on the slopes of the Ägene creek at lateral distances of 1 to 2 kilometers. Many of these were used for drinking and irrigation water supply of Obergesteln and the surrounding alpine pastures. When projected onto the longitudinal section, the elevation of the Äginen valley is slightly lower than the Obergesteln drift; the horizontal distance decreases to the south (dotted line and distances in Figure 9). In Figure 9 all springs between the two creeks are shown as horizontal projections along the direction of the strike of foliation. From the altitude of the important and depleted springs, the original elevation of the groundwater table can be deduced.

## 4.2 *Transient groundwater inflows in Obergesteln area*

As reported by Klemenz (1974) during construction of the northern section of Obergesteln drift, large inflows occurred in the first 820 meters. In the vicinity of the northern portal the inflows only occurred in the first tens of meters behind the advancing tunnel face. These inflows mainly came from open foliation fractures and rarely from tectonic fractures intersecting these foliation planes. Water inflows started to occur at 150 meters from the portal with 15 l/s). The highest inflow rates of at least 210 l/s were reached at Tm 550. The inflow rates decreased between Tm 600 and 780 and increased again until Tm 820, where the rock mass became nearly dry (i.e., only a few dripping sites).

Most of the inflows ran dry very quickly (after 20–40 meters of excavation which corresponds to 1–2 weeks). However, the sum of the inflow from Tm 600–820, measured at the foot of the shaft since November 1972 remained between 50 and 60 l/s until the end of excavation (May 31, 1973). This implies that inflows at deeper tunnel sections decreased less quickly than those near the northern portal. One year later (mid-1974) the total inflow from the entire permeable section had dropped to steady state values of about 10 l/s, as measured during dry seasons. During snowmelt and after heavy rain, i.e. under conditions of strong vertical recharge, these total inflows react quickly and reach about 50 l/s.

## 4.3 *Depleted springs in Obergesteln area*

In response to this tunnel drainage, ten surface springs ran dry shortly after the tunnel had intersected this very permeable drift section. Figure 9 shows these depleted springs as measured on August 9, 1972 when the excavation had reached TM 620 (bold tunnel section). As reported in Klemenz (1974), depleted springs were observed by the owners "very soon" after the high tunnel inflows had occurred. The sequence of observed spring depletion is also shown in Figure 9 and clearly follows the excavation progress in the direction of the drift axes. Also, springs located at very low elevations relative of the drift elevation (e.g., No. 506 and 617) become totally depleted, implying that the groundwater table was lowered to the drift elevation within a very short time. For spring No. 609, which reacted last, the response time must have been less than 1–2 weeks. The horizontal distance between the springs and the drift, as measured along strike of the foliation, does not systematically effect the response time. Most importantly, during the southwardly progressing excavation,

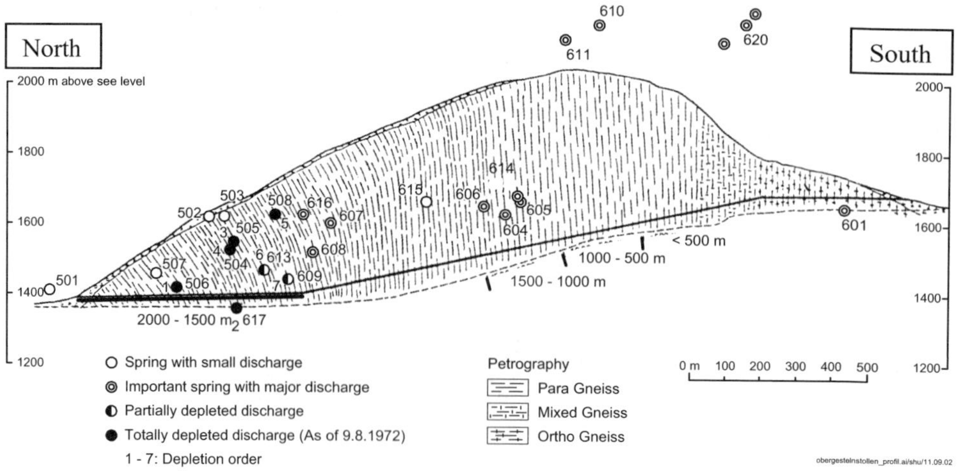


Figure 9. Longitudinal section along the Obergesteln drift. Horizontally projected springs (partially depleted; 1–7 indicate depletion order) along strike of foliation and Äginen Creek (broken line, with lateral distances indicated in the figure). Modified after Klemenz (1974).

no hydraulic interactions between the tunnel and surface springs were observed perpendicular to the strike of the toppled foliation.

These transient tunnel inflows and water table drawdowns can be explained reasonably well using a 2D analytical model that approximates the flow pattern in the planes normal to the tunnel axes by a superposition of linear horizontal flow with vertical flow to the tunnel. Inverse modeling of the observed inflows and inferred water table drawdowns yields anisotropic hydraulic conductivities in the toppled rock mass decreasing from the portal area (1E-3 m/s) to the lower bottom of the toppled rock mass (1E-5 m/s).

This situation is very typical for all underground excavations in the external crystalline massifs of the central Alps: Springs become depleted in the portal areas where stress relief, weathering or toppling can strongly increase the bulk hydraulic conductivity of the rock mass and early-time tunnel inflow rates. A very similar situation to that in the Obergesteln drift, was observed in the Bedretto adit of the Furka basetunnel at the southern margin of the Gotthard Massif. Here again, many surface springs ran dry above the tunnel portals located in toppled schists and gneisses of variable mineralogical composition. In the deeper or more distant tunnel sections, i.e. without surface related disturbances, no indications of environmental impacts on surface springs have been observed.

## 5 IMPACTS OF DRAINING TUNNELS ON SURFACE SETTLEMENTS

Recent high-precision levelling measurements of surface displacements along the Gotthard pass road have revealed up to 12 cm of subsidence along sections that pass several hundred meters above the Gotthard N2 highway tunnel (Figures 3 and 10). The Swiss Federal Office of Topography carried out the levelling measurements in 1993/1998 as a closed loop over the old Gotthard-pass road and through the N2 road tunnel (unpublished reports Swiss Federal Office of Topography 1997, 1998). Two earlier measurement campaigns were made along this N–S profile over the old Gotthard pass road in 1918 and 1970 before

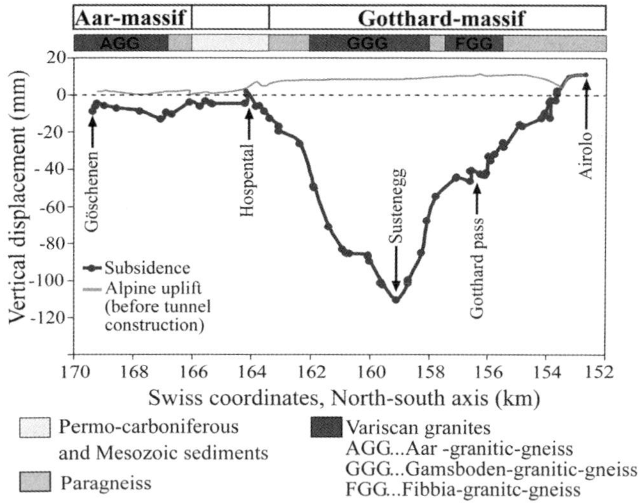


Figure 10. Levelling profile along the Gotthard pass road showing surface subsidence in the time interval 1970 to 1993/98 (after tunnel construction) and Alpine uplift (measured before tunnel construction).

tunnel construction. During the time interval, an undisturbed alpine uplift rate of 1 mm/year was detectable (Figure 9). This uplift rate concurs with estimated rates of 0.6 mm/year as determined using fission-track techniques (Kohl et al. 2000). In contrast, the time interval between 1970 and 1993/1998 (i.e., after tunnel construction) shows significant downward displacements along a 10-km region above the tunnel (Figure 10). More recently, surface triangulation measurements have confirmed the existence of a subsidence trough (Salvini 2002) over the tunnel. These results show that the extension of the trough in the EW direction is considerably smaller than in the NS direction.

### 5.1 *Surface settlements induced by inflows to the Gotthard N2 highway tunnel*

The close spatial proximity between the singular tunnel inflows of 110 and 150 l/s (early time inflow at Tm 9910 and 9935, respectively) and maximum settlement (Figure 3) and the temporal relationship between tunnel construction and settlement clearly shows causality between water drainage into the tunnel and surface deformation. Localized surface processes, e.g. creeping landslides or flexural toppling, could be excluded as alternative explanations given the absence of local indicators and the extent over which the settlements were measured (10 km along a N–S line, roughly parallel to the tunnel axis). Of surprise was the relatively small tunnel interval over which the high initial inflow cluster occurred (3 km, see Figure 5A) relative to the measured settlement trough (10 km). As discussed in detail in Zangerl (2003) and Zangerl et al. (2003), this subsidence is related to large-scale consolidation resulting from fluid drainage and pore-pressure changes in the rock mass (i.e., fractures and intact rock). These pore pressure changes might have also lead to reductions in the elevation of the groundwater table. However, during construction of the Gotthard N2 highway tunnel, no depleted surface springs were reported over the central and deep sections of the tunnel. Today a spring line can be mapped at an elevation of 2100 to 2500 masl in the area of Figure 3. An alternative explanation could also be that the initial spring line was higher and that water table drawdowns only occurred locally, i.e. close to the major groundwater pathways.

## 5.2 *Coupled hydro-mechanical deformation models of Gotthard pass area*

The collective observations and possible hydro-mechanical mechanisms described in detail in Zangerl (2003) led to the working hypothesis that the permeable fault zones provided a high-permeability conduit that permitted pore pressure drainage to penetrate deeply and relatively rapidly into the rock mass in a WSW-ENE direction, and that lateral diffusion of pore pressure drawdown into the relatively intact rock blocks (bounded by the fault zones; see Figure 9) occurred on longer timescales. Estimates for the time required for drawdown of pore pressures within the intact blocks, conditioned by field observations of block-size distribution and laboratory-derived estimates of the poro-elastic properties, suggest timescales of days to several years for block-sizes of 10 m to 100 m respectively. Thus, steady-state pore pressure conditions likely prevailed by the time the 1993/98 geodetic surveys revealed the subsidence trough, 21 years after tunnel construction.

For these reasons, the steady-state geometry of the settlement trough was modelled using a 2-D section running parallel to the tunnel axes, i.e. normal to the major permeable fault zones. Simulations with the 2D distinct element code UDEC and the continuum code VISAGE provided very useful insights into the underlying mechanisms. As most of the preferential pathways for groundwater flow and most of the discontinuities at depth are steeply inclined (the frequency of horizontal fractures strongly decreases with depths), the surface deformations could not be explained with closure of subhorizontal fractures alone. As described in detail in Zangerl (2003) and Zangerl et al. (2003), different deformation models had to be considered for the explanation of the observed surface deformations (Figure 11): normal closure of the subhorizontal fractures induced by a reduction of normal effective stress across the subhorizontal fractures; normal closure of vertical fractures or faults due to drainage, resulting in expansion of (impermeable) rock blocks in the horizontal direction and contraction in the vertical direction (i.e., Poisson ratio effect); linear poro-elastic behavior of rock blocks; and combined effects leading to shear along persistent inclined discontinuities (faults).

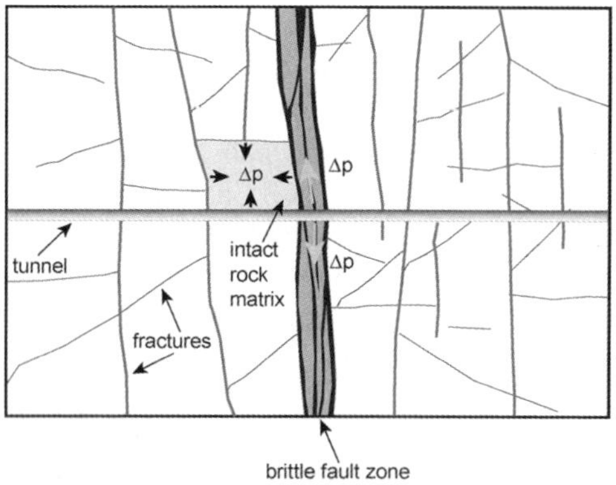


Figure 11. Conceptual model of coupled hydro-mechanical processes in crystalline rocks of Gotthard pass area.

Results from a 2D UDEC discontinuum Model that consider normal closure and shear of subhorizontal and subvertical joints and faults, as well as deformation of impermeable blocks are shown on Figure 12. The contours of vertical displacement generated by the effective stress changes that accompany drainage into the tunnel are shown in Figure 12a, and the corresponding profile of surface subsidence in Figure 12b. Vertical strain arises from closure of horizontal joints and the faults which, because they are mostly steeply inclined, allow the intact blocks to expand laterally. This generates vertical contraction strain through the Poisson's ratio effect. The maximum predicted subsidence of 0.042 m is shifted several hundred metres to south of the major inflow zone due to topographic, structural and hydraulic conductivity effects. The modeled subsidence trough extends about 9000 m in width with subsidence values greater than 0.01 m occurring over an area that is more than 4500 m wide. Shear deformations on faults and fractures, predominately of elastic type, are also indicated in Figure 12a. Within the central subsidence trough, shear displacements of up to 4 mm occurred on faults. Thus, the shape of the subsidence trough calculated in the distinct-element simulations reproduced both the general asymmetry and several key inflections observable in the measured subsidence profile. This agreement between the measured and modelled subsidence curves is directly related to the explicit inclusion of the fault zones mapped at surface and in the Gotthard N2 highway tunnel, both in terms of location and orientation.

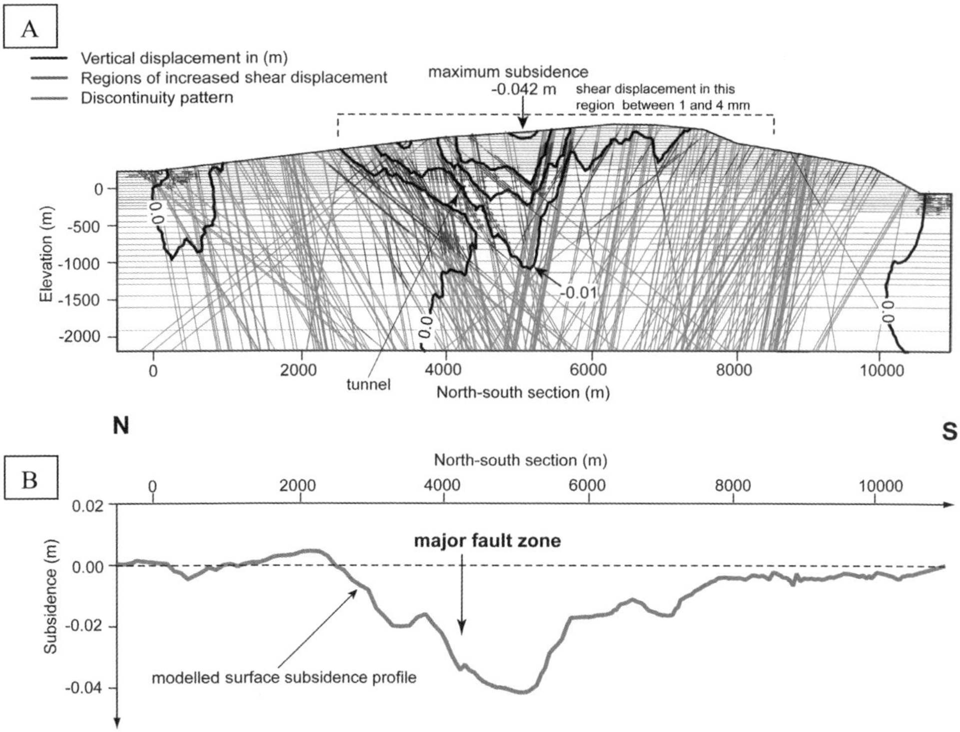


Figure 12. Discontinuum/distinct-element model results showing (A) discontinuity pattern, vertical displacements and shear displacements along steeply inclined fault zones, and (B) resulting modelled surface subsidence profile.

Results obtained from poroelastic continuum analysis produced a subsidence profile that for the most part was symmetrical about the point of tunnel drainage. However, numerical results also showed that the intact rock matrix can considerably contribute to rock mass deformation through poroelastic strains. Such poroelastic strains seem to be required in addition to the fracture related deformations, illustrated by the UDEC results of Figure 12, which yield only about 50% of the measured total subsidence above the Gotthard N2 highway tunnel.

## 6 CONCLUSIONS

The observations presented from the Gotthard massif in the Central Swiss Alps illustrate the effects tunnels in fractured crystalline rocks can have on groundwater flow systems, surface springs and surface deformations. Serious environmental impacts on surface waters are observed in high permeability areas close to the tunnel portals, or for tunnels running only a few hundred meters below ground surface. In these areas, water table drawdown can reach the elevation of the tunnel and significantly affect surface springs and wetlands over lateral distances of several kilometers within a few weeks. Also tunnels located deep below ground surface (500–1500 m) have considerable impacts on natural flow systems. Their effects are however much less visible: surface recharged groundwaters will flow towards the draining tunnels, re-locating groundwater divides and reversing flow directions over lateral distances larger than one kilometer. Pore pressure reductions created by draining tunnels at depth may also lead to large scale rock mass deformations leading to surface settlements exceeding 10 cm.

With all these processes, brittle deformations play a key role, either in the form of small scale fracturing or large scale shear under low temperature brittle conditions. In the investigated area, the most dominant of these structures are steep and co-planar (striking WSW-ENE), generating an important hydraulic anisotropy and preferential pathways for groundwater flow and pressure diffusion.

Towards surface (in the upper hundred meters), gravitational, unloading and weathering processes locally increase bulk rock mass hydraulic conductivity, leading to initially very high tunnel inflows and quickly propagating environmental impacts.

Most of the models used to interpret these observations have assumed equivalent continuum hydraulic properties. Nevertheless, they succeeded well in explaining most of the critical observations. This is mainly due to the fact that most of the available data sets do not refer to individual fault zones or groundwater pathways, but to larger scale observations, such as bulk tunnel inflows or large scale surface settlements. The authors hope that in future tunneling projects, transient data about inflows, temperatures and hydrogeochemical compositions (including isotopes) are collected for individual permeable fault zones.

Most of the analyses have either assumed steady state 3-D groundwater flow or transient 2-D flow. Obviously tunnels create a transient 3-D flow field during incremental excavation advancement. These effects will also have to be investigated in future studies.

Besides rock mass bulk hydraulic conductivity, groundwater recharge is an important factor for the location of a free groundwater table in mountainous systems and also for the long term inflow to a draining tunnel. Both groundwater recharge and the locations of groundwater tables are poorly constrained in Alpine terrain. In order to better predict environmental impacts of such tunnels, deep surface-based boreholes for monitoring purposes would be of very high value.

## ACKNOWLEDGEMENTS

We would like to thank Werner Klemenz and Dr. Susanne Laws for many insightful discussions. Many thanks are also extended to the Furka Oberalp Bahn AG for providing access to the Bedretto adit.

## REFERENCES

Angelier J (1994) Fault slip analysis and palaeostress reconstruction. In: Hancock (ed) Continental deformation. Pergamon Press , pp 53–100

Badoux A (1999) Untersuchung zur flächendifferenzierten Modellierung von Abfluss und Schmelze in teilvergletscherten Einzugsgebieten. Master's thesis, Institute of Geography, Federal Institute of Technology, Zurich, Switzerland (in German)

Bossart P, Mazurek M (1991) Structural Geology and Water Flow-Paths in the Migration Shear-Zone. Nagra NTB 91 (12):1–55

Burkhard M (1999) Strukturgeologie und Tektonik im Bereich Alptransit. In: Löw et al. Vorerkundung und Prognose der Basistunnels am Gotthard und am Lötschberg. GEAT Symposium, Zürich, Balkema Publishers, Rotterdam:45–56

BUWAL (1998) Wegleitung zur Umsetzung des Grundwasserschutzes bei Untertagebauten. Bundesamt für Umwelt Wald und Landschaft, Bern, 32 pp

BWG (2001) Hydrologischer Atlas der Schweiz, Landeshydrologie. Bundesamt für Wasser und Geologie, Bern

Challandes (2001) Behavior of Rb-Sr and Ar-Ar systems in metagranite shear zones (Roffna granite and Grimsel granodiorite, Swiss alps). PhD dissertation, University of Neuchatel

Choukroune P, Gapais D (1983) Strain pattern in the Aar granite (central alps). Orthogneiss developed by bulk inhomogenous flattening. J. Struct. Geol. 5(3/4):411–418

Frei B, Löw S (2001) Struktur und Hydraulik der Störzonen im südlichen Aar-Masssiv bei Sedrun. Eclogae geol. Helv. 94:13–28

Frey M, Mählmann RF (1999) Alpine Metamorphism of the Central Alps, Schweiz. Mineral Petrogr. Mitt. 79:135–154

Frey M, Bucher K, Frank E, Mullis J (1980) Alpine metamorphism along the geotraverse Basel-Chiasso – a review. Ecologae geologicae Helvetia 73:527–546

Gurtz J, Peschke G, Mendel O (1990) Hydrologic processes in small experimental areas influenced by vegetation cover. In: Hydrological Research Basins and Environment, Proceedings and Information No 44, International Conference Wageningen, pp 61–69

Gurtz J, Baltensweiler A, Lang H (1999) Spatially distributed hydrotope-based modelling of evapotranspiration and runoff in mountainous basins. Hydrological processes 13(17):2751–2768

Klemenz W (1974) Die Hydrogeologie des Gebirges im Obergestelnstollen. Gaz – Wasser – Abwasser 54/7:287–289

Kohl T, Signorelli S, Rybach L (2000) Constraints on palaeo-topography by revised apatite fission track uplift rates. European geophysics society, 25th General Assembly, V2, Nice

Kölla E (1993) Gotthard-Basistunnel – Regionale Hydrogeologie im Projektgebiet unter besonderer Berücksichtigung des Ritom-Gebietes. Unpublishes Report, 30.11.1993

Labhart T (1999) Aarmassiv, Gotthardmassiv und Tavetscher Zwischenmassiv: Aufbau und Entstehungsgeschichte. In: Löw S, Wyss R (eds) Vorerkundung und Prognose der Basistunnels am Gotthard und am Lötschberg. Proceedings Symposium Geologie Alptransit, pp 31–43

Laws S (2001) Structural, Geomechanical and Petrophysical Properties of Shear Zones in the Eastern Aar Massif. PhD dissertation, Federal Institute of Technology, Zurich No.14245

Lerner DN, Issar AS, Simmens I (1990) Groundwater Recharge – A Guide to Understanding and Estimating Natural Recharge. International Contributions of Hydrogeology, IAH, Hannover: 8, 345 pp

Loew S (2001) Natural Groundwater Pathways and Models for Regional Groundwater Flow in Crystalline Rocks. In: New Approaches to Characterizing Groundwater Flow, IAH International Congress, Munich, pp 1013–1018

Luetzenkirchen V (2002) Structural Geology and Hydrogeology of Brittle Fault Zones in the Central and Eastern Gotthard Massif, Switzerland. PhD dissertation, Federal Institute of Technology, Zurich, No.14749

Marquer D., Gapais D. (1985) Les massifs cristallins externes sur uns transversale Guttanen-Val Bedretto (Alpes Centrales): structures at histoire cinematique. C.R. Acad. Sc. Paris t.301, Serie II, No 8, 543–546

Ofterdinger U (2001) Groundwater flow systems in the Rotondo Granite, Central Alps, Switzerland. PhD dissertation, Federal Institute of Technology, Zurich, No.14108

Ofterdinger U, Balderer W, Loew S, Renard P (2004) Environmental isotopes as indicators for ground water recharge to fractured granite, Ground Water 42/6:868–879

Salvini D (2002) Deformationsanalyse im Gotthardgebiet. Institut für Geodäsie und Photogammetrie, ETH Zürich

Schmid SM, Pfiffner OA, Froitzheim N, Schönborn G, Kissling E (1996) Geophysical-geological transect and tectonic evolution of the Swiss-Italian Alps. Tectonics 15(5):1036–1064

Stapff FM (1882) Geologische Aufnahme des Gotthard-Bahntunnels. 60 Geologische Tunnelprofile 1:200 (Längen- und Horizontalschnitte) – Quartalsberichte zu Händen des Schweizerischen Bundesrates

Steck A, Hunziker JC (1994) The tertiary structural and thermal evolution of the Central Alps – Compressional and extensional structures in an orogenic belt. Technophysics 238:229–254

Steck A (1968) Die alpdischen Strukturen in den zentralen Aaregraniten des westlichen Aarmassivs. Ecologae geol Helv 61(1):19–48

Stini J (1950) Tunnelbaugeologie. Springer Verlag, Wien

Vitvar T, Gurtz J (1999) Spatially distributed hydrological modeling in the Rotondo area, western Gotthard-massif. Technical report 3465/16, Federal Institute of Technology, Zurich, Engineering Geology

Zangerl C (2003) Analysis of Surface Subsidence in Crystalline Rocks above the Gotthard Highway Tunnel, Switzerland. PhD dissertation, Federal Institute of Technology, Zurich, No. 15051

Zangerl C, Eberhardt E, Löw S (2001) Analysis of ground settlements above tunnels in fractured crystalline rocks. In: Särkkä P, Eloranta P (eds) Eurorock Symposium, Rock Mechanics – A Challenge for Society, Espoo, Finland pp 717–722

Zangerl C, Eberhardt E, Loew S (2003) Ground settlements above tunnels in fractured crystalline rock: numerical analysis of coupled hydromechanical mechanisms. Hydrogeology Journal 11:162–173

CHAPTER 35

# Groundwater modelling approaches as planning tools for WISMUT's remediation activities at the Ronneburg uranium mining site

Michael Paul[1], Manfred Gengnagel[2] and Michael Eckart[3]

[1] *WISMUT GmbH, Chemnitz*

[2] *WISMUT GmbH (current address: Governmental Lignite Mine Reclamation Program, Berlin)*

[3] *DMT Deutsche Montan Technologie GmbH, Essen*

ABSTRACT: Uranium mining in the Ronneburg district, the largest of WISMUT's mine sites, started in 1951 and ceased in 1990. As an important part of remediation the underground mine is being flooded since 1998. Hydraulic simulations were the main tasks of the early years of modelling for the development of the flooding concept. In 2002 and 2003 test operation phases of the water management were implemented and resulted in a more reliable basis for flooding prediction. According to the results of a quantitative model, the flooding operation is expected to be completed between 2007 and 2008 after reaching a quasi-steady state of inundation. Another task for present modelling activities is to attend the technical planning process for the design of water collection systems for contaminated groundwater in the final flooding phase; these activities also include hydrochemical modelling.

## 1 INTRODUCTION, GEOLOGY AND MINING ACTIVITIES

Uranium mining in the Ronneburg mining district (Eastern Thuringia), the largest of WISMUT's underground mine sites, started in 1951 and ceased in 1990. Figure 1 shows the location of the investigation area and an outline map of the mine fields in the Ronneburg Region.

The Ronneburg deposit was situated in a series of blackshales, limestones and mafic magmatites of Ordovician to Devonian age (Lange & Freyhoff 1991). By the end of 1990 the complex underground mine consisted of 14 mine fields with 40 shafts and about 3000 km of mine workings. The Ronneburg mine which produced from both underground and open pit operations accounted for about 50% of WISMUT's total uranium production of about 220.000 tonnes during the whole mining period.

The overall volume of the mine workings at the Ronneburg site amounted to about 68 million $m^3$ of which about 44 million $m^3$ were backfilled mainly with a special cement as a characteristic feature of the cut and fill mining technology performed at Ronneburg since the late 1960s. About 4 million $m^3$ arose from caving which was practiced in the early mining period of the 1950s and 1960s. Acid leaching has been applicated at the Ronneburg mine in the state of test operation measures.

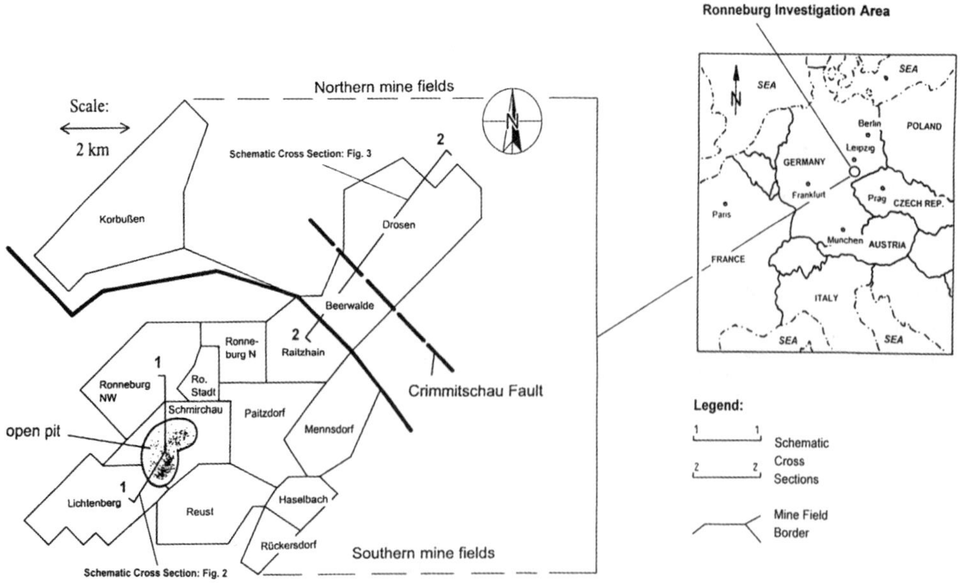


Figure 1. Location of the investigation area.

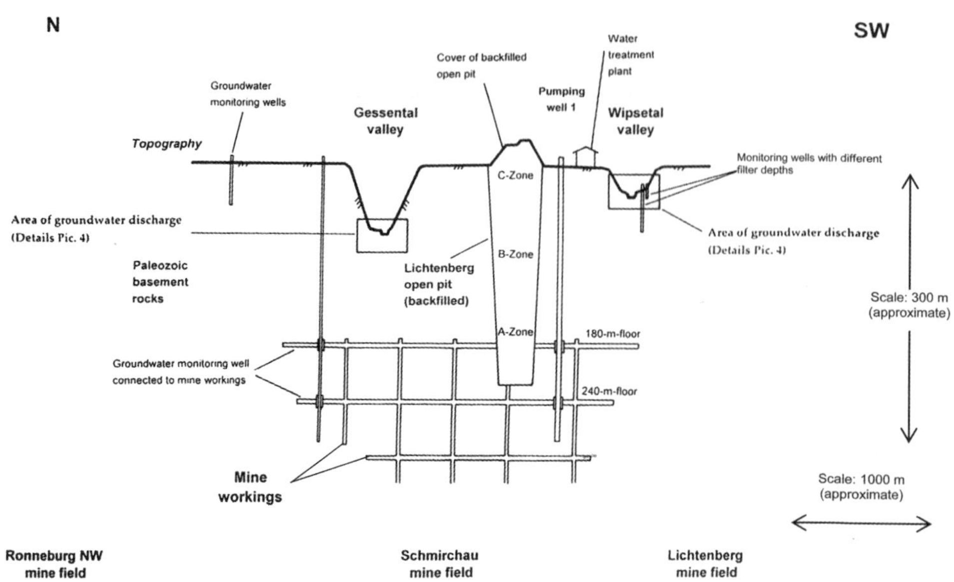


Figure 2. Ronneburg uranium mine flooding – schematic cross section 1 (southern mine fields – for the location of the cross section see Figure 1).

The Ronneburg mine district has been subdivided into the southern and the northern mine fields. Federal Motorway A4 approximately borders the fields. In addition to the underground mine workings investigations also had to consider the Lichtenberg open pit (see Figure 2) with an area of about 2 square kilometres and a maximum depth of more

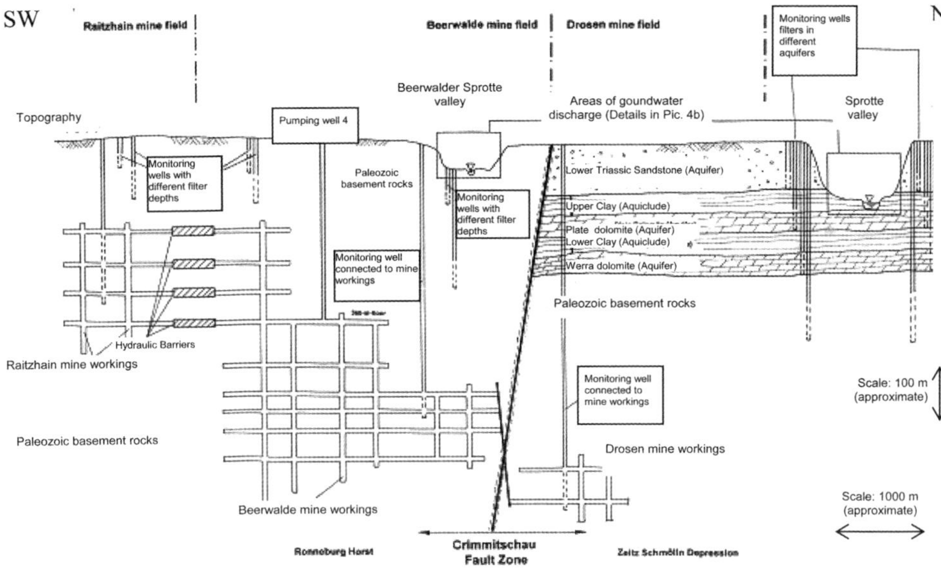


Figure 3. Ronneburg uranium mine flooding – schematic cross section 2 (northern mine fields – for the location of the cross section see Figure 1).

than 240 meters; the open pit mine has been situated in the central part of the southern mine fields.

The southern mine fields are located at the so-called Ronneburger Horst, which is the north-eastern part of the Thuringian Slate Mountains with Palaeozoic basement rocks overlain by thin Quaternary sediments. The maximum depth of the underground mine workings of the southern mine fields amounts to about 600 m.

The mine fields of Beerwalde and Drosen in the north of the Ronneburg mine district are separated by the Crimmitschau fault, a supraregional fault zone. In contrast to the rest of the mine site at the Drosen mine field the Palaeozoic host rock is covered by a series of platform sediments of Permian to Triassic age (see Figure 3).

According to the characteristics of the mine workings, the backfill strategy and the flooding concept, the investigation area has been divided into the bigger southern and the smaller northern part. Mine workings which connected both mine fields had been backfilled prior to flooding. Nevertheless some connections between those areas had to be regarded such as for example the surface streams and creeks. The Beerwalde mine field has been situated southwards of the Crimmitschau Fault but is nevertheless a part of the northern mine fields.

Since the early 1990's remediation measures have been practised including partial flooding the underground mine workings and also the backfill measure at the Lichtenberg open pit utilizing the material of the surrounding waste rock piles.

## 2 INITIAL STAGE OF MINE REMEDIATION IN 1991, HYDROGEOLOGICAL KEYNOTE ASPECTS AND MODEL INVESTIGATION STRATEGY

After the closure of active uranium mining in the Ronneburg region WISMUT's remediation activities started in 1991. Conceptual considerations concerning the ceasing of the mine drainage system and flooding the mine workings had not been available until then as active mining had originally been planned to be continued in the 1990s. For several reasons it soon became evident that a flooding concept for the underground mine workings had to be developed:

- The pumping and maintenance costs for the mine drainage and aeration systems could not be justified for a period longer than the time demand of remediation activities for preparing the flooding process, which were supposed to take a few years.
- The mobilization of toxic substances, especially heavy metals, in the artificial aeration zone covering an area of more than 60 $km^2$ with a thickness of several hundred metres had to come to an end as soon as possible.
- The pollution by the discharge of effluents from the mine drainage system into the creeks and streams of the region had to be finished.

After all it had to be considered, that the permitting procedure for starting the main flooding process in cooperation with the responsible mining, radiation protection and water authorities would demand an adequate time period. As an indispensable prerequisite for these permissions qualified model based predictions about the consequences of flooding on regional water conditions in terms of quantity and quality had to be developed very quickly.

Another typical aspect of the hydrogeological description of the regional groundwater system concerned the interaction between the underground and the surface waters, which are represented in the Ronneburg investigation area mainly by a large number of creeks and small streams. In the final phase of mine flooding normal direction of interaction between groundwater and surface water is to be restored. These exfiltration conditions on the one hand will bring back the normal discharge regime of the creeks and small streams. On the other hand hydrochemical characteristics could be influenced by the mine water in the flooded underground workings. These aspects include acid mine drainage characteristics combined with several dissolved heavy metal species, especially nickel, and in addition dissolved sulphate, iron species and hardness.

As the mine workings in the Ronneburg region do not include open dewatering tunnels, groundwater discharge in a number of exfiltration areas has to be considered during the final phase of the flooding process. Depending on the local groundwater quality characteristics the installation of technical drainage systems becomes necessary (Figure 4a) in some areas.

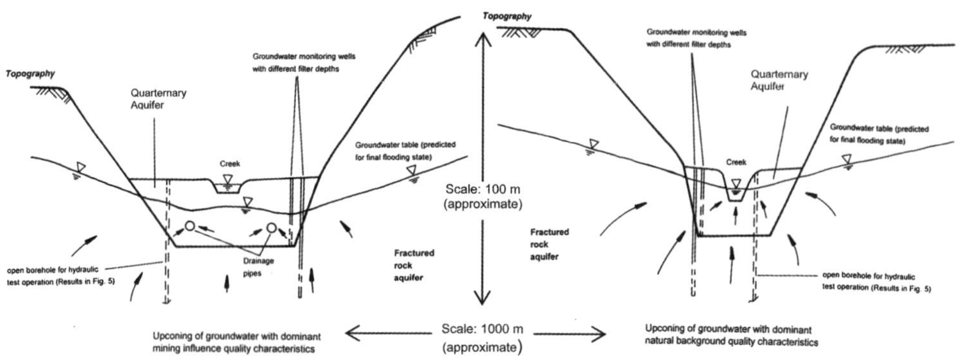


Figure 4. Ronneburg uranium mine flooding: Typical configurations of groundwater discharge areas – Exfiltration into technical catchment systems (Figure 4a) and into natural recipients (Figure 4b).

On the other hand, exfiltration into natural recipients can be accepted at other locations (Figure 4b). These distinctive different conditions require adequate prediction approaches and required the development and application of extensive numerical model instruments in a short period of time. The deficits also concerned the results of hydrogeological investigations as a prerequisite for any numerical simulation efforts.

All these implications lead to a strategy of parallel initiation and implementation of the inevitable prognostic work packages, which principally endures to the present period more than a decade after the early days of mine remediation:

- Preliminary and measure implementation simultaneous hydrogeological investigations,
- Development and calibration of adequate numerical simulation model instruments and
- Prognostic application of the models concerning quantitative and qualitative aspects of the mine water and groundwater conditions.

Each of these work packages has been extended to numerous aspects and details. The paper explicates the most typical examples and indicates others.

## 3 HYDROGEOLOGICAL INVESTIGATIONS BEFORE AND DURING THE FLOODING MEASURE

The implementation of a conceptual geological and hydrogeological model represents an important prerequisite for numerical groundwater modelling investigations. In the early beginning of remediation hydrogeological data from the four decades of active mining represented a singular data basis for all further investigations. These results produced by two or three generations of geologists, hydrogeologists and engineers enabled detailed knowledge concerning many aspects of importance for systematic preparation of the numerical investigations. The hydrogeological database from the period of active mining included for example extremely detailed and extensive geological information, hydrogeological information like piezometric heads, results of pumping tests, water quality data and, of course, information about the mining activities such as survey data of the underground mining voids and pumping rates of the underground mine drainage system.

In contrast to the good and partially excellent database from the period of active mining the information about the hydrogeological situation in the pre-mining period before 1950 was quite vague and incomplete. These deficits especially include a consistent description of the groundwater situation before the intervention caused by the pump regime of the mine drainage system.

The years after the end of active mining in 1990 still brought additional and relevant hydrogeological information (e.g., the enduring pump regime of the mine drainage system until the beginning of mine flooding in the late 1990's). Also the documentation of intense backfill activities in the underground mine represented an important data input for numerical groundwater modelling investigations. Furthermore the hydrogeological situation has been influenced by the backfill activities in the Lichtenberg open pit, which is planned to be finished in 2007. Table 1 gives an overview about floodable volumes in the southern mine fields.

In the period of the early 1990s preparation work for numerical groundwater modelling investigations included documentation and evaluation of all the historic and current hydrogeological data from the mining and the beginning mine remediation activities, and from other sources.

Table 1. Categorization of floodable mine space and pore volumes, southern mine fields, estimates from the early phase of flooding process (1999), up to 270 m a.s.l.

| Category | volume [in million m$^3$] |
|---|---|
| Open volume (mine workings) | 15,7[a] |
| Floodable pore volume in the backfill | 4,1 . . . 4,3 |
| Floodable pore volume in the caving area | 1,6 . . . 2,3 |
| Floodable pore volume in the backfilled open pit | 1,8 . . . 2,7 |
| Dewatered pore volume in the host rock (depression cone) | 3,8 . . . 5,1 |

[a] Volumes of mine sections completely or partially flooded before 1997 are not included.

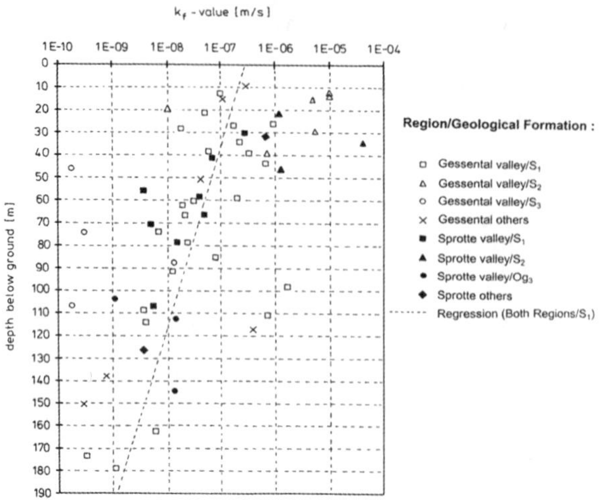


Figure 5. Results of SOLEXPERTS hydraulic investigations of the unconfined aquifer in the one of the supposed groundwater exfiltration areas (x-axis: K-value in m/s, y-axis: depth below ground in m).

Supplementary to these activities specific *in situ* investigations had been conceived and implemented in the mid 1990s, especially including hydraulic tests in areas with special relevance including future groundwater exfiltration areas after the mine flooding process has completed. These difficult hydraulic tests in the unsaturated aquifers had been operated by the Swiss company SOLEXPERTS in commission of WISMUT. As an example for these investigations Figure 5 shows a diagram for one of the supposed groundwater exfiltration areas with hydraulic conductivity (K-value) on the x-axis and depth below ground on the y-axis. Typical K-value-results in the investigated areas ranged from $10^{-9}$ to $10^{-6}$m/s.

In addition geological fault zones are assumed to have geohydraulical relevance, which could not be completely recorded in the performed field investigations. The supposed groundwater exfiltration areas are typically situated in valley bottoms with Quaternary sediments covering the fractured rock aquifers as illustrated in Figure 4. That circumstance demanded drilling campaigns in combination with geophysical measurements to investigate the geohydraulical characteristics of the underground in the vicinity of the surface. These results also had special relevance to the technical planning of catchment systems for the controlled exfiltration of contaminated groundwater in the final phase of the flooding process.

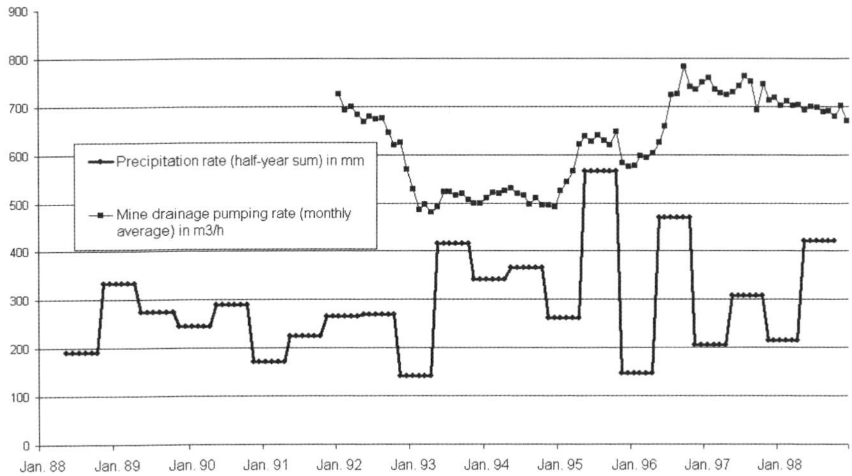


Figure 6. Time series of precipitation rate and mine drainage pumping rate.

The results of hydrogeological in-situ testing shown in Figure 5 summarize constant head and constant rate injection testing measurements using packer systems in about 10 single boreholes.

Since the late 1990s, the ongoing flooding process of the Ronneburg underground mine has been delivering additional relevant hydrogeological data. These include the extensive hydraulic and chemical monitoring programmes of both the mine water and the groundwater. Data from that sources have been used for current reconsideration and calibration of the hydrogeological description of the groundwater system and for numerical groundwater modelling investigations themselves.

Reliable prediction of the mine flooding water level development has been an indispensable prerequisite for the permitting procedures and also for the conception and technical planning of water management facilities required for the situation in the final flooding phase with potential consequences for the environment.

Before the mine flooding process had been started in 1998, hydrogeological input data concerning the prediction of the development of the mine water level was mainly based on the following investigation sources:

- Mine surveyor data concerning the floodable mine workings volume in the different mine fields and floors and
- Mine drainage system operation data concerning the inflow rates to the mine.

In the beginning the hydrogeological description of the groundwater system was concentrated on the area of the underground mine workings with about 60 square kilometres and the area of groundwater drawdown with about 70 square kilometres. Before flooding the average inflow into the southern mine fields amounted to about 650 $m^3/h$ with variations from about 500 $m^3/h$ up to 800 $m^3/h$ for the southern mine fields (based on monthly values, time series between 1991 and 1997).

It was well known in the early stages of the process that observed variations of the inflow rates were significantly influenced by the variations in the hydrometeorological conditions. Figure 6 compares time series for current and long term medium half year sums of the precipitation rate and the monthly medium values of the inflow rates to the mine drainage system in the pre-flooding period from 1991 to 1997 for the southern mine fields.

The necessity to evaluate additional and precise prediction input data for the development of the mine water level was recognized early in the flooding operation process because:

- The inflow rates to the mine can be expected being hydraulically influenced by the current mine water level during the flooding process as lateral inflow decreases.
- The proportional significance of floodable geological volumes could also be expected to change during the flooding process; especially in the final phase reaching the vicinity of the ground level. In this context floodable geological volumes could be expected to be increased by the intense mining activities covering a time span of several decades and an area of about 60 square kilometres.

The uncontrolled flooding process regime in the first years of the operation from early 1998 to mid 2002 did not enable any opportunity for quantitative exploration of these effects; the observed water table development could not be separated into the influence of the two mayor controlling factors, inflow and volume.

Between mid 2002 and late 2003, that situation changed for the southern mine fields as the test operation phases of Well 1 and the Water Treatment Plant had been carried out. This additionally also included some water injection into the mine workings for technical reasons. The main goal of these test operation phases which was to prove the controllable flooding regime by operating those technical facilities had been fulfilled. Figure 7 shows the increase of water table during flooding in the second half of 2003. The option of controlling the flooding process by operating Well 1 and the lime precipitation water treatment plant were indispensable prerequisites for receiving the magisterial permission from the authorities of the German state of Thuringia to continue the flooding process to its final state.

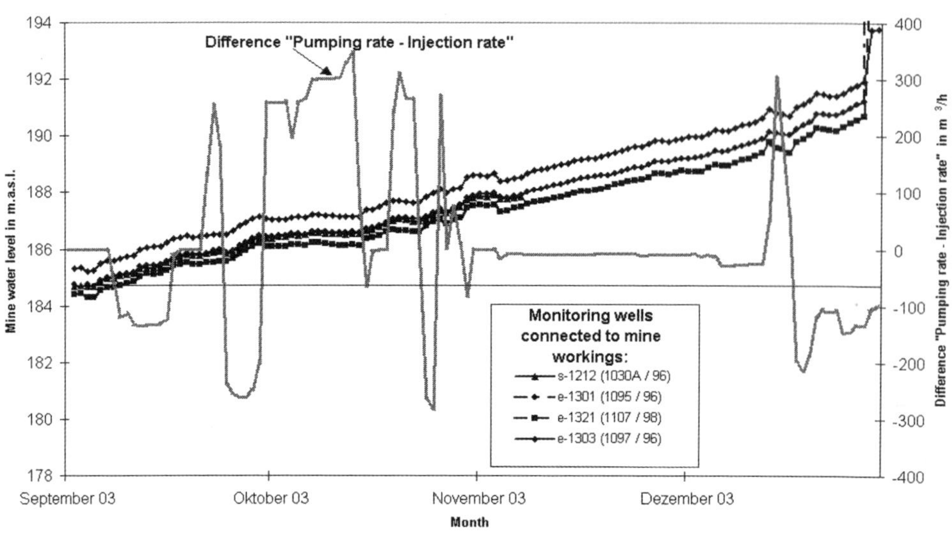


Figure 7. 2003 test operation phase: Time series of mine water level in comparison to the difference between pumping rate and injection rate.

In addition some relevant exploration effects could be performed as a consequence of the above mentioned test operation phases:

- In 2002 the treatment capacity had been limited for technical reasons to a rate of 180 $m^3/h$ in maximum; the flooding development result was a slackening of the water table increase values from about 8 cm/d in the phases without pumping operation to about 5 cm/d in the phases with pumping operation at Well 1. These observations enabled a rough estimation of the inflow rate with a best guess result of about 435 $m^3/h$ for the generally dominating medium hydrometeorological conditions in 2002.
- As a result of technical measures having been applied in the first half of the year 2003, during the subsequent test operation phase in autumn 2003 a significantly higher water treatment and pumping rate could be performed. In October 2003, a pumping rate of about 280 $m^3/h$ evoked a temporary interruption of the flooding process all over the southern mine fields (Figure 7). That rate could therefore be interpreted as an investigation result concerning the inflow rates to the southern mine fields in October 2003. The difference to the larger estimated inflow rates in 2002 can be mainly assigned to the dry hydrometeorological conditions in 2003 especially in the summer months. In general the test operation evaluation demonstrated that the actual inflow rates to the southern mine fields were significantly lower than the above mentioned inflow rates to the mine drainage system in the phase before the flooding process had been started.
- These mine inflow rates derived during the test operation phases of 2002 and 2003 allowed an additional estimation step about the floodable volumes. This evaluation involved an computational assignment of the mine flooding inflow rates during the test operation phases to the water table increases in the time immediately before and after the test operation phases. The results were total floodable volumes for the southern mine fields including both the mine workings and geological volume components of about 135.000 $m^3$ per meter in the vertical direction. These were about 30% higher than the water table development prediction input previously applied (i.e., the "dewatered pore volume in the host rock" category in Table 1).

It is intended to minimize the uncertainties step by step in the ongoing flooding process considering monitoring results in the actualisation of the model data base. Current investigations for example include the influence of salinity on the density of the mine water and therefore the necessity to make a correction computation concerning the mine water level monitoring results, which was not yet carried out concerning the water level monitoring results shown in Figure 7. Typical mine water is of magnesium-sulphate chemical type with total dissolved solid concentrations from 5 to 10 g/l.

## 4 DEVELOPMENT AND CALIBRATION OF NUMERICAL MODELLING

A speciality of this approach is the simulation of the regional groundwater balance to be combined with the simulation of the situation in the mining excavations. This lead to a subdivision of the model instruments into two subsystems:

- The mine water system described by the "BOX-model" and
- The groundwater system described by the regional groundwater model.

Of course, this investigation structure assumed an intense coordination and organisation of data transfer activities between BOX-model and regional groundwater model.

Steady-state and transient hydraulic simulations were the main tasks of the early years of modelling in the mid 1990s. Meanwhile groundwater simulation development and application activities have brought some experience in modelling the fractured environment of

- the Palaeozoic strata in the southern and northern mine fields
- the Permotriassic layers in the Drosen mine field
- hydraulic connections over a large distance by mine workings.

The flow-model was the first step to get information about the contamination potential. The modelling investigations started in 1991 using the MODFLOW package.

The investigations concerning the main outflow points in the final flooding phase found a relative clear answer with MODFLOW, but the other questions needed to be answered by special tools. With MODFLOW it was not adequately possible to simulate the hydraulic connections within the mine workings. The increase of the permeability dramatically reduced the convergence of the equation solver up to a permanent balance error from cell to cell of the pathway reserved for a mine working. This error led to an unrealistic water table prediction inside the mining area.

The model developers looked for the possibility to couple several model cells with a large distance between them. The solution was found in 1996 with a free discretisation scheme in space independent from the fact if the cells are direct neighbours or not. This model was implemented in a finite difference code (GEOMODELL) by coupling cells with the flexible scheme that was the key for the "BOX-model" to calculate the chemical development and the transient process of flooding. GEOMODELL is a software package developed by the companies HPC/Wismut and the mining academy TU Freiberg. The basic structure is compatible to MODFLOW (discretisation scheme, internal interpolation of the profile transmissivity to the cell boundaries etc.). Additionally to the basic concept of hydraulic contact to 6 neighbour-cells a matrix is existing, in which more connections can be defined between all cells of the model independently of their position in the model grid. Figure 8 shows the box-concept without the matrix of the finite difference model.

Figure 9 gives an impression how this flexible concept is embedded in the finite-difference matrix. For this simple example the diagonal cells represent the additional cells that cannot be connected with the central cell by the normal concept of MODFLOW or each other

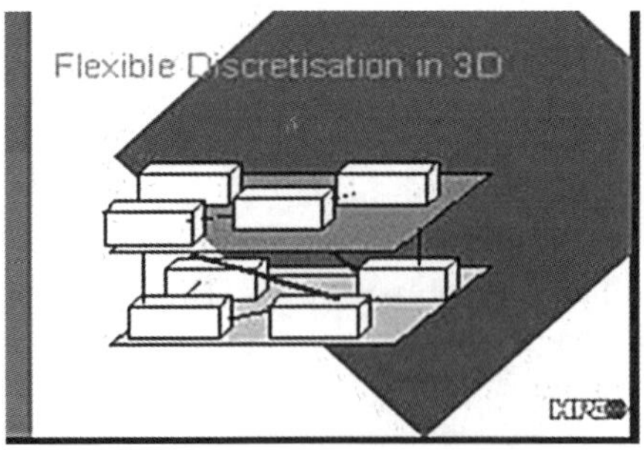


Figure 8. Structure of the BOX-model.

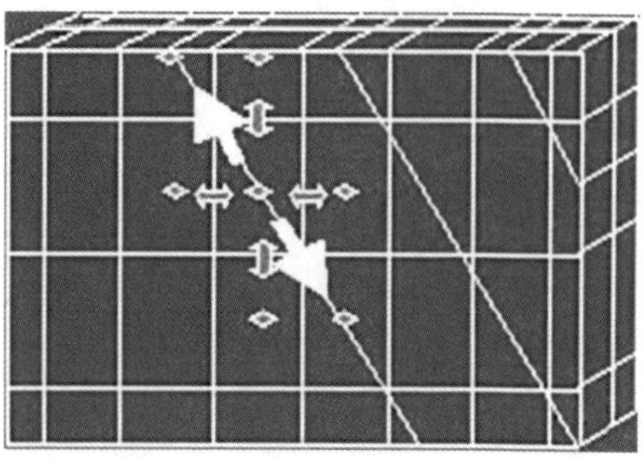

Figure 9. Flexible discretisation with 6 + n-neighbouring cells.

finite difference or finite element code. These hydraulic neighbours can be located far away from the connected cell.

Furthermore, GEOMODELL integrates a multi-component reactive transport model to continue the transport-process with Eh-pH-development from the mine through the groundwater layers surrounding the mining area. For the purpose of water quality predictions, PhreeqC was implemented in the BOX-model itself.

The combination of a porous media and a hybrid model to consider hydraulic connections opened the possibility to calculate realistic scenarios of groundwater flow, since within the mining area the flow over such discrete elements is absolutely predominant.

Hydraulic model calibration based on the available data indicated the situation before the mining activities started and described the situation influenced by mining activities. The meaningfulness of the calibration is limited by several circumstances, especially significant hydraulic variability between the calibration scenarios and the post-flooding prognostic scenario.

## 5 PROGNOSTIC INVESTIGATIONS AND RESULTS

Concerning the prediction of post-flooding hydraulic situations, the hydraulic conductivity in the underground of the supposed groundwater exfiltration areas showed significant relevance. As already reported, the SOLEXPERTS *in situ* investigations could not reflect the geohydraulical relevance of geological fault zones.

That circumstance led to the creation of two scenarios concerning hydraulic conductivity, K, in the groundwater exfiltration areas:

- Scenario A: Relatively high K-values that assume hydraulic relevance of geological fault zones or other inhomogeneities and
- Scenario B: Lower K-values corresponding to the field investigations.

As a characteristic result of the model investigations the regional groundwater table shows remarkable variations:

- The conditions of scenario A produce relatively low post-flooding groundwater levels in the former mine area with characteristic values of 265 to 270 m a.s.l. This groundwater level represents the situation in the backfilled Lichtenberg open pit in 2010 and subsequent years.
- Scenario B results show higher groundwater levels with characteristic values of 285 to 290 m a.s.l.

These different groundwater level results correspond with appropriate variations in the prognostic results for the groundwater exfiltration flow rates:

- The conditions of scenario A show the importance of the Gessental area in terms of the post-flooding groundwater exfiltration distribution.
- Scenario B results are characterised by a decentralized distribution of groundwater post-flooding exfiltration with relevant quantities in the Gessental valley as well as several other areas.

These aspects are relevant to the flooding strategy for the final phase in general and especially for designing catchment systems for the controlled exfiltration of contaminated groundwater.

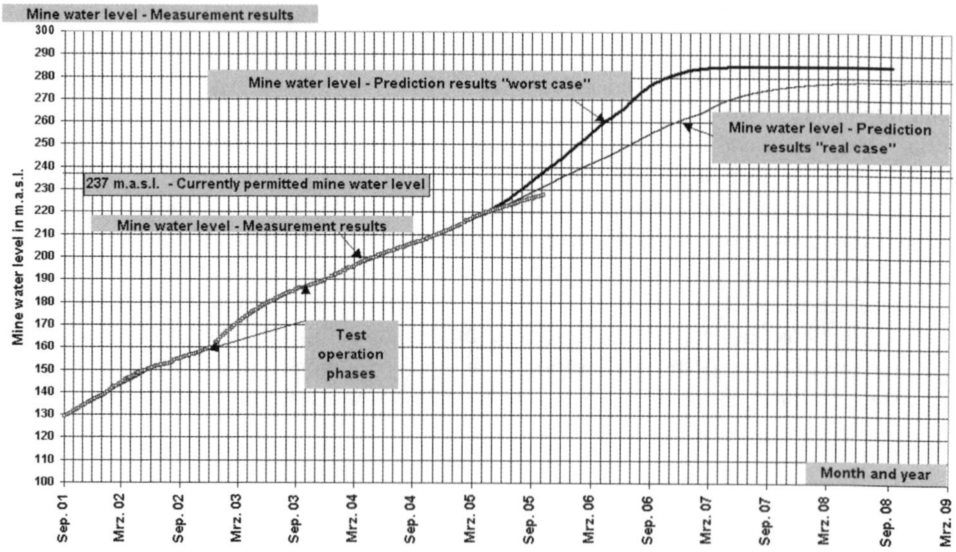


Figure 10. Time series of current flooding water level development prediction results.

In early 2004 the test operation phase results were used as a central basis for a prediction update of the flooding water table development, which obtained a significant slower process than it had been predicted until then. The recent water table monitoring data in the period between early 2004 and October 2005 confirmed the results of the prediction update (Figure 10). Nevertheless, this updated prediction shows residual uncertainties that can be characterised by:

- The development of the hydrometeorological conditions cannot be predicted for a long term time span of several months or years; this circumstance leads to corresponding uncertainties concerning the prediction of the mine flooding inflow rate as a central prerequisite for predicting the flooding water table development.
- Furthermore the above described explorations concerning the mine flooding inflow rate and the floodable geological volumes are strictly associated to the mine water levels of about 155 m a.s.l. in 2002 and 186 m a.s.l. in 2003. As the flooding water table in the southern mine fields reached a level of more than 225 m a.s.l. in autumn 2005, the application of these exploration results concerning future flooding water table development seem clearly limited. Uncertainties have to be considered as both the current mine flooding inflow rate and the current floodable geological volumes can be expected to be different to the situation documented in 2002 and 2003.

In general it can be stated that the geohydraulic exploration activities related to the test operation phases of Well 1 and the Water Treatment Plant in 2002 and 2003 brought a clear minimization of uncertainties concerning the prediction of the flooding water table development.

Figure 10 shows the results of the current prediction of the flooding water level development. The inevitable prediction uncertainties have been reflected defining the two scenarios

"Real Case" and "Worst Case":

- The real case scenario reflects realistic "best guess" assumptions concerning floodable volumes and inflow rates; the scenario prediction results can be therefore interpreted as the most probable case.
- The worst case scenario uses more pessimistic, but still realistic input assumptions; the scenario gives an indication of possibly quicker flooding.

The recognisable discrepancy between the real case prediction and the recent monitoring measurement time series results in Figure 10 indicates the mainly dry hydrometeorological conditions.

The model strategy of handling uncertainties by defining different scenarios was also applied to other relevant prognostic aspects, for example:

- to the hydrochemical characteristics of groundwater exfiltration and
- to the discharge situation in the surface creeks and streams in the supposed exfiltration areas.

Uncertainties concerning the hydrochemical investigations chiefly result from the fact that the available selective input data cannot reflect the complex regime and processes in the *in situ* aquatic system. Initial prognostic approaches were improved sequentially by analysing calibration scenarios using the extensive monitoring data available from the recent mine flooding.

## 6 DISCUSSION AND PREVIEW

Fully fledged flooding of the southern mine fields was initiated at the turn of 1997/1998 after a four-year-permitting and preparation process. Flooding of the northern mine fields started in 2000 (Paul et al. 2002). Permissions given by the authorities were based on the prognostic model results.

According to the actual quantitative prognostic modelling results, the flooding operation is expected to be complete between 2007 and 2008 after reaching a quasi-steady state inundation level of about 265–285 m a.s.l. This was important for designing the decommissioning regime at the Ronneburg site.

There are no open dewatering tunnels within the whole mining field so that flooding will continue until the natural discharge of the contaminated groundwater to the local receiving streams starts, if there are no other active pumping or collection measures. Because of the degree of contamination of the mine waters, water treatment proved necessary. Therefore, water catchment systems had to be installed close to the surface within areas with a high probability for exfiltration of contaminated groundwaters into the receiving streams.

Local groundwater models are important tools in project design. Furthermore consideration of the numerous creeks and streams with expected exfiltration conditions in the final phase of the flooding process has crucial importance in the present model investigation activities. This also includes the simulation of technical water catchment systems as a boundary condition for the combined fractured/porous groundwater system in the Ronneburg region.

A main task for present modeling activities is the technical planning for catchment systems for the controlled exfiltration of contaminated groundwater (Unland et al. 2002). These activities are also important for planning of water transport and water treatment facilities.

On the other hand, most predicted groundwater exfiltration areas are estimated to have quasi natural background water quality (Figure 4b). These demand extensive monitoring activities as illustrated in Figures 2 and 3 to verify the predictions. In case of significant divergences between prediction and monitoring results, adequate assessment is necessary in order to make decisions about supplementary measures. For example, potential measures could include additional water catchment systems (illustrated in Figure 4a) in combination with conventional water treatment or, alternatively, on site passive water treatment systems. Nevertheless current monitoring results do not indicate necessities in that line of thought.

Summarizing the recent experiences of preparation and implementation of flooding measures in the Ronneburg uranium mine district, it is realised that uncertainties concerning the description and prediction of the hydrogeological system are characteristic for the different phases and aspects. The required strategies to handle those uncertainties must include extensive hydrogeological investigations before and during the flooding measure in order to minimize relevant uncertainties. Residual insecurities concerning the prediction results need to define and examine a set or sets of adequate prognostic scenarios, extensive monitoring programmes and apply flexible measure strategies in the water management sector.

## REFERENCES

Hähne R, Paul M, Schöpe M, Gengnagel M (2000): Abhängigkeit des Verlaufs der Grubenflutung des Ronneburger Uranerzbergbaureviers von der Beschaffenheit des Flutungsraumes und von den Neubildungsbedingungen, Tagungsband der Internationalen Konferenz WISMUT 2000 – Bergbausanierung, 11–14 September 2000.

Lange G, Freyhoff G (1991): Geologie und Bergbau in der Uranlagerstätte Ronneburg/ Thüringen.- Erzmetall 44:264–269.

Paul M, Gengnagel M, Vogel D, Kuhn W (2002): Four years of flooding WISMUT's Ronneburg uranium mine – a status report.- In: Uranium in the Aquatic Environment.- Proceedings of the International Conference Uranium Mining and Hydrogeology III and the International Mine Water Association Symposium Freiberg, Germany, 15–21 September 2002, Springer 2002, pp. 775–784.

Paul M, Gengnagel M, Eckart M (2003): Numerical groundwater modelling in fractured environment: a planning tool for uranium mining remediation (Ronneburg site – Eastern Thuringia, Germany) – In: Krásný J.-Hrkal Z.-Bruthans J. (eds.): Proceedings – IAH Internat. Conference on "Groundwater in fractured rocks", Sept. 15–19, 2003. Prague. 367–368, IHP-VI, Series on Groundwater 7. UNESCO.

Paul M, Gengnagel M, Baacke D (2005): Integrated water protection approaches under the WISMUT project: The Ronneburg case – In: Uranium in the Environment.- Proceedings of the International Conference Uranium Mining and Hydrogeology IV, Freiberg, Germany, September 2005, Springer 2005, pp. 369–379.

Unland W, Eckart M, Paul M, Kuhn W, Ostermann R (2002): Groundwater rebound compatible with the aquatic environment – technical solutions at WISMUT's Ronneburg mine.- In: Uranium in the Aquatic Environment.- Proceedings of the International Conference Uranium Mining and Hydrogeology III and the International Mine Water Association Symposium Freiberg, Germany, 15–21 September 2002, Springer 2002, pp. 667–676.

CHAPTER 36

# Active groundwater monitoring and remediation during tunneling through fractured bedrock under urban areas

Kim Rudolph-Lund, Frank Myrvoll, Elin Skurtveit and Bjarne Engene
*Norwegian Geotechnical Institute, Ullevaal Stadion, Oslo, Norway*

ABSTRACT: Real time monitoring was used to control inflow and pre-grouting during construction for a new tunnel in the Oslo area in 2002–2005. Detailed site investigations and vulnerability studies were used to define accept criteria. Hydrogeological mapping and modeling were used to form a strategy to minimize the need for remediating disturbances and damages due to an unwanted reduction of the groundwater level. During tunnel construction the groundwater was monitored using an automated groundwater monitoring system displayed on an internet home page for the Norwegian National Railroad Administration. The continuous updating of the groundwater levels, soil pore water pressures, and precipitation in the area above the tunnel enabled the contractor to respond quickly to any recorded change in groundwater level near the tunnel by changing the pre-grouting procedure. The leakage inflow into the tunnel after construction is below acceptable rates for the different classified areas.

## 1 INTRODUCTION

During recent years, tunneling under urban areas in Norway has focused on groundwater control and pre-grouting strategy in order to reduce negative environmental impacts. In 1999, there were particular challenges with the excavation of a 13.8 km long high-speed railway tunnel in gneissic bedrock between the city of Oslo and its new international airport 40 km to the north. Large unexpected leakages of water occurred in the tunnel at several areas with extremely fractured bedrock. This led to significant decreases of soil pore pressures and groundwater levels above the tunnel. As a result there were large settlements and damages to buildings and forested areas situated on the natural soil deposits above the tunnel. A permanent water injection system had to be established to raise both the groundwater level and pore water pressures to original conditions in order to prevent further settlements.

The experiences gained from the design and construction of the airport tunnel led to new research and an improved precautionary approach in the planning of groundwater balance control and pre-grouting strategy for construction of new tunnels, particularly those under urban areas. This new approach has been fully implemented during the construction of a new 7.3 km long tunnel for a double-track high-speed railway constructed between Jong and Asker, southwest of Oslo during the period 2002–2005 (Figure 1).

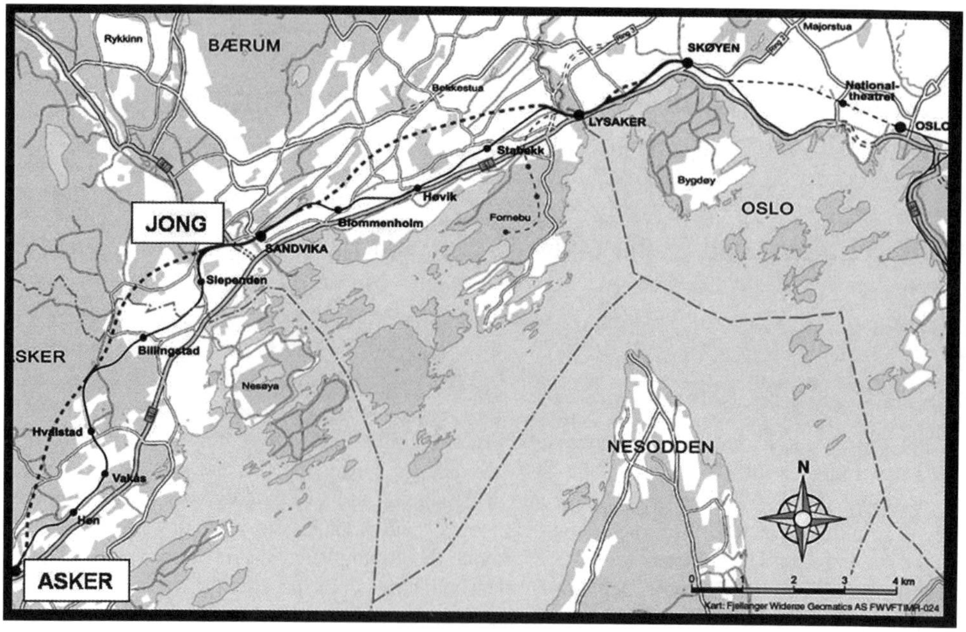


Figure 1. The new railway tunnel shown on the map as a dotted line between Jong and Asker, southwest of Oslo.

## 2 SITE INVESTIGATION

A comprehensive site investigation program was carried out incorporating previously recorded data from other nearby tunnel works (Løset, 1981a) and detailed geological mapping (Naterstad et al., 1990; Nordgulen et al., 1998). The bedrock in the investigated area belongs to the Oslo field, a rift or graben structure, that developed during the Permian and Carboniferous periods. Deformed Cambro-silurian sedimentary rocks, mainly shales, phyllites and calcareous shales, are the dominating rock units in this area. This stratigraphic unit is folded with mass transported towards SSE and with the fold axis weakly plunging towards ENE and WSW. Discordant above the folded Cambro-silurian sedimentary rocks is a 70–80 m thick horizontal sedimentary unit, the Asker Group. This unit consists of shale, sandstone and conglomerate. Above the Asker Group is a thick unit with Permian volcanic rocks, dominated by basaltic lava and rhomb-porphyry. Intrusions of near vertical dykes of Permian age are found within the whole area. Several fault and fracture systems are found within the investigated area. Near vertical axial plane faults are common in the anticlines of the Cambro-silurian sedimentary rocks. These faults may contain up to decimeter thick clay smear (Løset, 1981b). Faults and fractures of varying size are found at all localities with a dominating trend of NW-SE to NNW-SSE near vertical faults, similar to the rest of the Oslo field (Nordgulen et al., 1998). Highly fractured zones are also found in connection with Permian dyke intrusions.

The Jong-Asker tunnel runs through the Cambro-silurian sedimentary rocks. The bedrock cover above the tunnel is mainly between 50–80 m, but sometimes as low as 2 m. The tunnel is between 20 m and 100 m above sea level.

The field investigations have shown that there are soft marine clays and silt deposits over bedrock. It is well known that the soft clays are sensitive to decreases in natural groundwater levels from water leakage into rock tunnels, which has earlier resulted in severe settlements. There are also some glaciofluvial sands and gravels in the interface zones at bedrock surface, which are typical for the Oslo region. The thickness of those overlying deposits is generally between 0 m and 28 m along the tunnel.

## 3 VULNERABILITY STUDY AND ACCEPTANCE CRITERIA

The area nearby the tunnel is densely populated, also encountering grounds with valuable recreational purposes and special sensitive biotopes. Vulnerability and sensitivity of buildings, infrastructures, lakes, bogs, swamps, streams and springs is dependant on changes in groundwater levels and surface water storage in the soil deposits and bedrocks. Changes in groundwater conditions in the catchment area reflect the water balance, which is the difference between the precipitation and the combined natural run off from the area including evapotranspiration from the surface and leakage into the tunnel. The vulnerability study was used to determine acceptance criteria for leakage volume into the tunnel to reduce negative effects on the environment.

The tunnel was first sectioned and classified according to anticipated problems with respect to vulnerability and sensitivity of the surrounding environment. Secondly, more than 100 households currently using groundwater from local wells as their primary or secondary source of fresh water supply were classified. And lastly, the vegetation, recreational areas and special biotopes along the railway tunnel were classified based on their sensitivity for potential lowering of groundwater level and on their utilization. By combining the sensitivity for potential damages to the environment with damages due to settlements, three classes of allowable leakage rates were established: class 1 (moderate – 8–16 l/min/100 m), class 2 (low – 4–8 l/min/100 m) and class 3 (extremely low – less than 4 l/min/100 m) (Figure 2).

## 4 HYDROGEOLOGICAL MAPPING AND MODELING

The water conditions in the area were mapped over a period of several years ahead of the tunnel construction works in order to assess the natural seasonal and annual variations in groundwater levels. The monitoring was comprised of recordings of fluctuations of water levels in lakes, groundwater in aquifers, pore water pressures in the soil sediments, and flow in rivers and streams.

The effect from precipitation and evapotranspiration on groundwater conditions was well known from daily recorded climate data at several meteorological stations in the area. Studies were made to estimate the actual size of the catchment and influence areas of the tunnel, taking into consideration the bedrock geology. Recorded data on seasonal and annual variations in run-off of water in the area were obtained from the Norwegian Water Resource and Energy Administration.

The permeability was investigated for the marine clays above bedrock and of the coarser sediments in the transition zone.

The permeability of the bedrocks was established by Lugeon testing (water leakage-pressure testing) in the core drilled holes which were subsequently used for the groundwater monitoring (Norconsult, 2000). The areas with intrusions of near vertical dykes

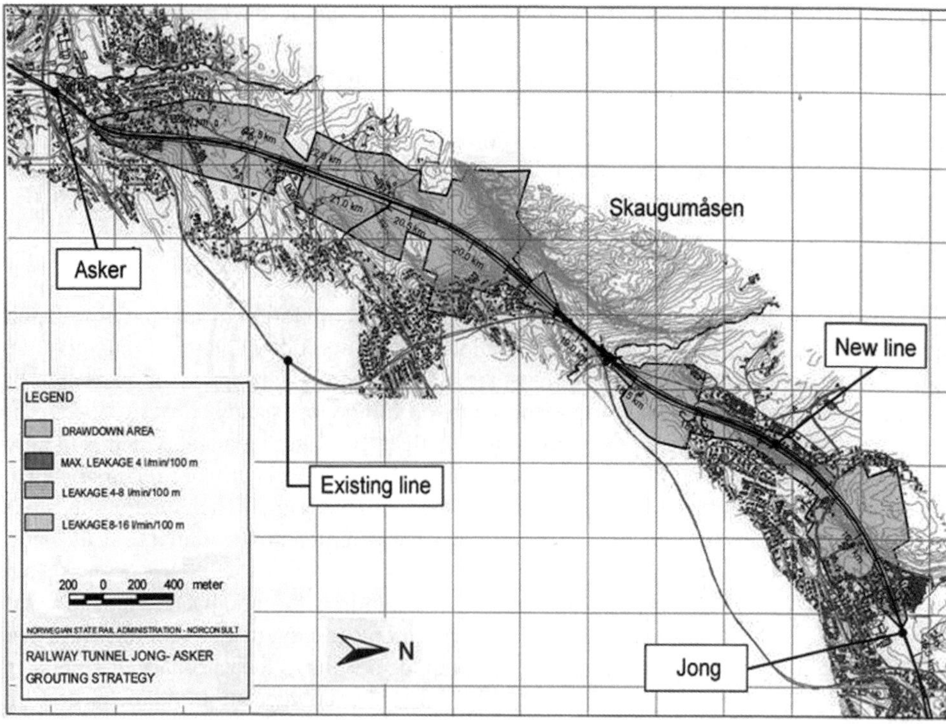


Figure 2. Classification of allowable leakage and potential drawdown areas for the Jong-Asker tunnel.

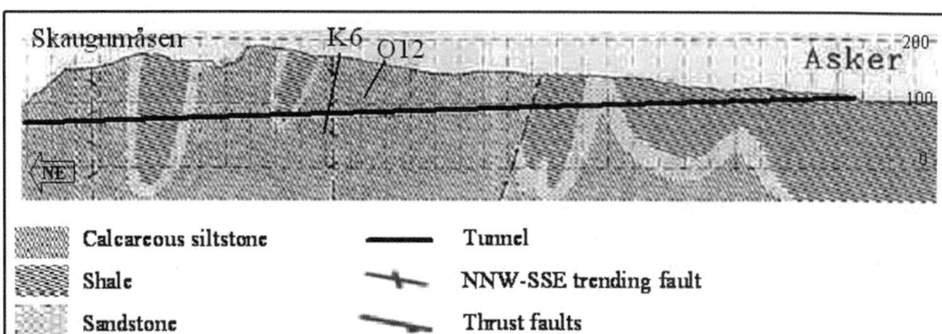


Figure 3. Profile showing the location of well K6 and O12 relative to the tunnel and the area geology. Well O12 is located directly above the tunnel whereas K6 is located south of the tunnel. The geology in the area is dominated by calcareous siltstone with lenses of sandstone and shale. NNW-SSE trending fault zones are marked as vertical dotted lines.

(dolerite) may in some cases act as barriers for groundwater flow and in other cases act as water bearing structures with higher permeability than that of the surrounding rocks.

Regional fault zones with high fracture intensity and partly open central zones, are seen in the profile from Skaugumåsen to Asker (Figure 3), and may act as conduits for

Figure 4. Diagram showing the orientation of 399 fractures and faults measured along the tunnel trace. The main orientation is NNW-SSE.

groundwater flow. The dominating fracture and fault orientation in the area is NNW-SSE (Figure 4), similar to the rest of the Oslo field.

Using the results of the comprehensive ground investigations and the mapping of groundwater conditions, the amount of acceptable leakage inflow into the tunnel was determined using numerical hydrogeological modeling.

## 5 MONITORING SYSTEM

A network of wells for measuring pore pressure in the overlaying soil deposits and bedrock, and the discharge rates in local springs was established along the tunnel route in 2000/2001 (Norconsult, 2000). New wells were installed in 2002 (NGI, 2002). The wells were automated and the daily measured results were placed on an internet home page for the Norwegian National Railroad Administration (Figure 5).

Settlement bolts were also established on a number of houses along the tunnel route. These houses were all located within the calculated influence area for possible settlement due to groundwater leakage into the tunnel.

Weather data to be used in the evaluation was collected at the Asker meteorological station from 1990 through 2002 by the Norwegian Meteorological Institute and displayed on the new internet home page.

The monitoring system is comprised of four types of wells:

- Wells placed in the overlying soil deposits called **L-wells** (23 pore pressure meters)
- Wells placed in the bedrock called **K-wells** (five wells with six pore pressure meters)
- Wells crossing both the overlying soil deposits and the bedrock called **O-wells** (14 wells with 25 pore pressure meters)
- Wells which were originally manually installed in the overlying soil deposits called **Pz-wells** (seven pore pressure meters)

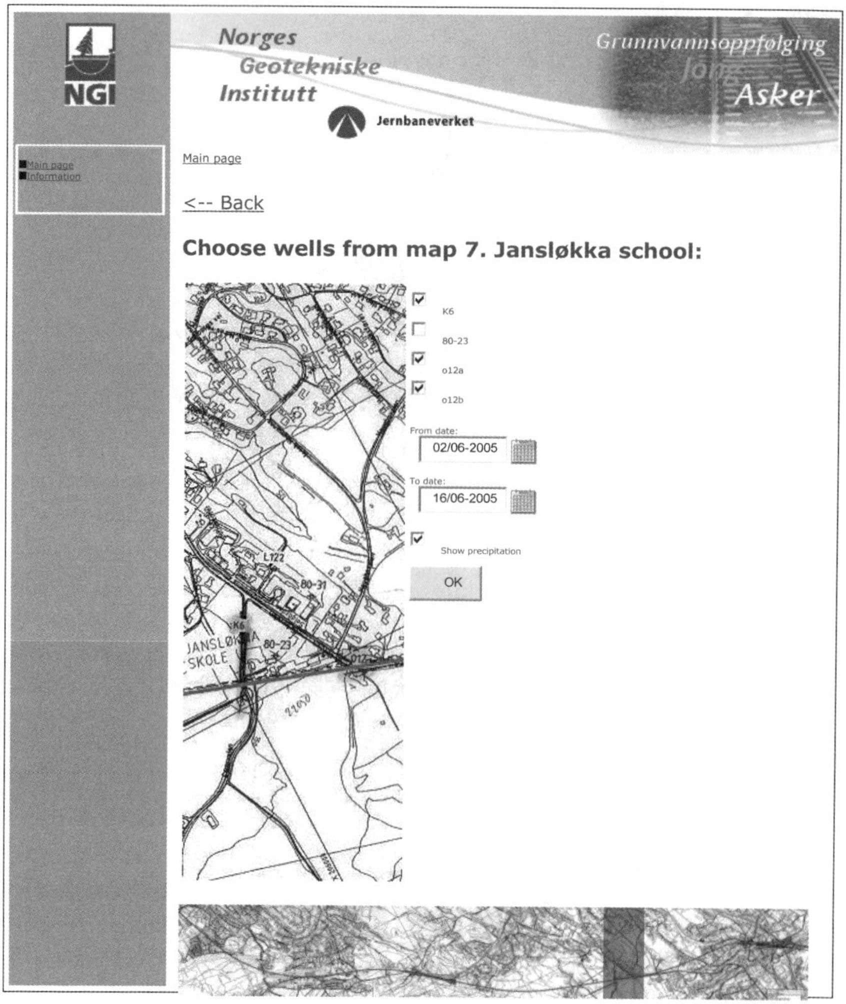


Figure 5. The new internet home page for the Norwegian National Railroad Administration. The user selects a map over the tunnel route and then the subsequent wells of interest. In this case the user has chosen map 7 over Jansløkka school and wells K6, O12a and O12b.

Table 1 shows the number and types of pore pressure meters and the number of settlement bolts that are currently in operation along the tunnel route.

## 6 TUNNEL CONSTRUCTION AND REMEDIATION STRATEGY

The tunnel cross section is 104 $m^2$ and was excavated using drilling and blasting techniques, supported mainly by bolts and shotcrete. Water leakage control was achieved by the use of probe-drilling ahead of the tunnel face followed by pre-grouting of the rock mass. The grouting program included use of very fine micro-cements, cement additives and modern

Table 1. Summary over pore pressure meters and settlement bolts.

| Areas with pore pressure meters and settlement bolts | Number and types of pore pressure meters | | | | Number of objects and bolts | |
|---|---|---|---|---|---|---|
| | K- | L- | O- | Pz- | Total objects | Total bolts |
| West for Jong | 1 | 4 | 2 | 1 | 131 | 633 |
| Lagerudbekken | | 4 | 5 | 1 | 194 | 971 |
| Billingstad school | 1 | 4 | 3 | 1 | | |
| Billingstad | 2 | 1 | 2 | | 131 | 617 |
| Furubakken | 1 | 4 | 4 | 2 | | |
| Skaugum | | 3 | 4 | | 51 | 330 |
| Jansløkka school | 1 | | 2 | 1 | | |
| Asker rectory | | 3 | | | 146 | 822 |
| Asker farms | | | 3 | 1 | | |
| Total | 6 | 23 | 25 | 7 | 653 | 3373 |

equipment using high grouting pressures. The pre-grouting was carried out in sequences of 3 or 4 blasting rounds depending on the classification. The length of each grouting sequence was 21–27 m and supplementary holes were drilled depending on inflow leakage of water. The primary parameter for changing the grouting effort was the response of the groundwater level. The strategy also included measures for temporary water recharge or injection in boreholes to achieve water balance during grouting.

## 7 RESULTS FROM ACTIVE GROUNDWATER MONITORING

The instrumentation system for automatic performance monitoring on the Jong-Asker tunnel project has proven to be very useful for the contractor and the owner at site. A continuously updating of the groundwater levels and soil pore water pressures above the tunnels enabled the contractor to respond quickly by changing the pre-grouting procedure due to any recorded change in groundwater level above the tunnel.

A typical example of results of measurement is shown in Figure 6 for well station K6 near Jansløkka school. The water level in this well decreased significantly 3.5 m on December 18, 2002 during drilling of holes for pre-grouting. The level rebounded 2.0 m during subsequent cement grouting. Additional cement pre-grouting holes were drilled January 7, 2003 and subsequent temporary water injection resulted in a permanent increase to the original water level. As can be seen in Figure 4, the same cycle of events occurred in March 2003 as the tunnel drilling passed well O12. The water level decreased dramatically by over 8 meters in early March. Subsequent grouting and water injection restored the local water levels to their pre-tunnel static level.

By following the planned grouting strategy to achieve water balance, verified by the automatic performance monitoring system, the leakage inflow in the tunnel was kept below the allowable rates for the different classified areas. The leakage measured for the whole length of the excavated tunnel was class 2, equal to 5 l/min/100 m. No settlements have thus far been measured on the 600 buildings and the recreational areas subject to monitoring.

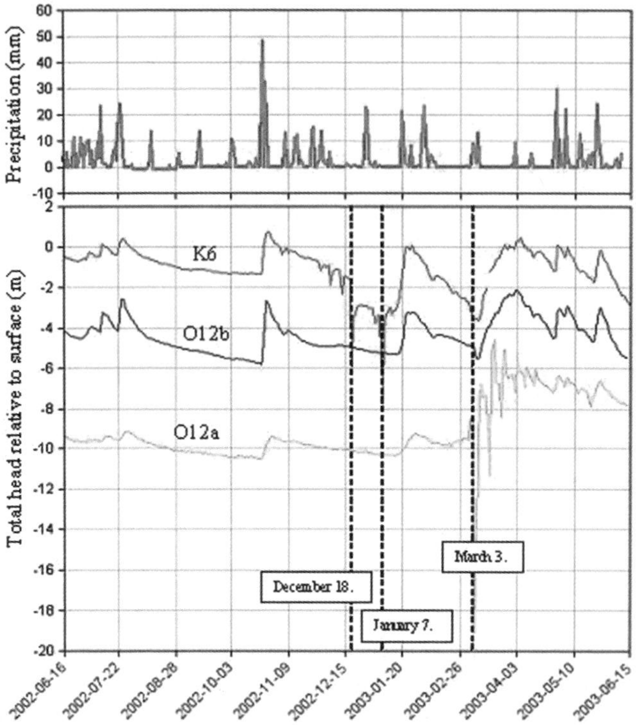


Figure 6. One year of continuously recorded groundwater levels for wells K6 and O12, together with the precipitation curve. The groundwater levels in K6 and O12a are measured directly above the tunnel level at approximately 100 m.a.s.l. The groundwater level in O12b is only 7 m below the surface at 143 m.a.s.l.

## 8 CONCLUSION

The Norwegian National Railroad Administration used real time monitoring to control inflow and pre-grouting during construction for a new tunnel in the Oslo area. This monitoring was made possible by: detailed site investigations and vulnerability studies used to define accept criteria, combined with hydrogeological mapping and modeling used to form a strategy to reduce disturbances due to reduced groundwater levels. The automated groundwater monitoring system was displayed on an internet home page and gave continuous updating of the groundwater levels, soil pore water pressures as well as precipitation in the area above the tunnel. This enabled the contractor to respond quickly to any recorded change in groundwater level near the tunnel by changing the pre-grouting procedure. After construction, the leakage inflow into the tunnel is below acceptable rates for the different classified areas.

## ACKNOWLEDGEMENTS

We wish to thank the Norwegian National Railroad Administration for allowing us to use the project data.

## REFERENCES

Løset F. 1981a. Analyse av injeksjonsresultater fra VEAS-tunnelen. NTNF-prosjekt nr.25141. Intern NGI – rapport.

Løset F. 1981b. Ingeniørgeologiske forundersøkelser for anlegg i fast berg. Norwegian Geotechnical Institutt Rep. 54001-10.

Naterstad J, Bockelie JF, Bockelie TG, Graversen O, Hjelmeland H, Larsen BT, Nielsen O. 1990. Asker 1814-1, Bedrock map 1:50000. Geological Survey of Norway.

NGI. 2002. Jong-Asker. Varslingsgrenser for poretrykksmålere. Teknisk notat datert 05.04.2002.

Norconsult. 2000. Detaljplan. Parsell Jong-Asker. Fagrapport. Hydrogeologi. Dokumentnr. 951030-70-4-00613, Rev. 02A, datert 06.04.2000.

Nordgulen Ø, Lutro O, Solli A, Roberts D, Braathen A. 1998. Geologisk og strukturgeologisk kartlegging for Jernbaneverket Utbygging i Asker og Bærum. Geological Survey of Norway, Report 98.124.

# Numerical modelling of fractured environment

CHAPTER 37

# A double continuum approach for determining contaminant transport in fractured porous media

Matthias Beyer and Ulf Mohrlok
*Institute for Hydromechanics, University of Karlsruhe (TH), Kaiserstr. Karlsruhe, Germany*

ABSTRACT: To predict the plume development in fracture-matrix systems without the detailed knowledge of the fracture network geometries, a double continuum approach is chosen. The complex transport behaviour can be adequately represented by two overlapping and interacting continua. An existing double continuum approach has been further developed for steady-state flow conditions by focusing on mass exchange terms between the fractures and the matrix system. This approach is implemented in the "Double Porosity MT3D" program and has been applied to a transport problem in a fractured sandstone formation. The parameter determination for the fracture continuum is the most important step to calibrate the double continuum model and to predict the plume development. Of particular interest was the determination of parameters by fracture network characteristics. The transverse dispersion coefficient and the hydraulic conductivity are directly dependent on the angles between the fracture directions and the hydraulic gradient. The larger the angle, the higher is the transverse dispersion coefficient and the lower will be the resulting effective hydraulic conductivity.

## 1 INTRODUCTION

The determination of plume development in fractured porous media is very complicated due to their extremely complex structure. Large fractures provide preferential pathways for regional fluid movement leading to a fast contaminant transport. The small fractures as well as the rock matrix are relevant in terms of storage and retardation, as their permeability values are some orders of magnitude smaller. The interactions between these parts of the system are characterized by specific exchange processes. The main exchange processes are given by matrix diffusion as a result of concentration differences between fractures and matrix, local advection due to pressure gradients between fractures and matrix, and regional advection due to a regional pressure gradient.

To describe flow and transport processes in fractured porous media, several model approaches are available depending on the scale of application and the geological information available (Figure 1). The simplest approach, single continuum representation, can be used for macroscale studies if a rough approximation is sufficient without representing the complex structure of the system.

A much better representation of the complex fracture system can be achieved by applying a discrete or coupled discrete model. Unfortunately, the investigation effort is very high due to the fact that the exact geometry and parameters of all individual fractures have to be determined. For this reason the discrete approaches are limited to local scale studies of

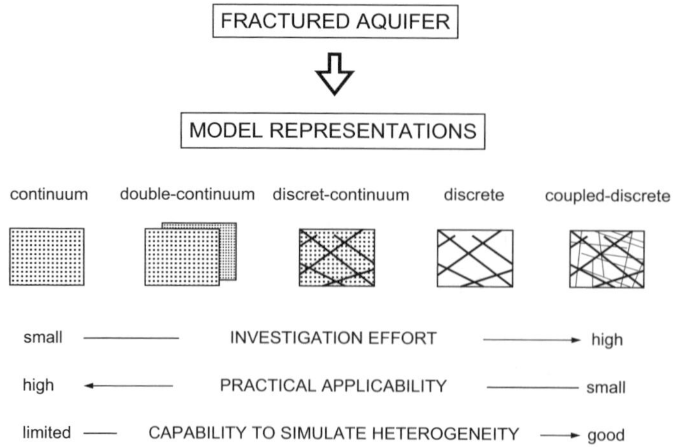


Figure 1. Different model representations for fractured aquifers (after Teutsch et al., 1991).

well known investigation areas. Large scale applications require stochastic fracture network generations. Due to many uncertainties, the achieved results have to be averaged by Monte Carlo simulations resulting in a huge computational effort. A combination of discrete and continuum approaches can be used for investigations, where a limited number of known fractures dominate the system.

If a fractured porous medium is to be described without considering the exact fracture network geometries, a double continuum approach, based on averaged fracture-matrix parameters, should be used. The fractured porous medium is described by two overlapping and interacting continua. The first continuum represents the large fracture system, and the second continuum represents the small fracture system or the matrix. These two continua possess different flow, transport and storage parameters and are connected by the respective exchange terms. Detailed knowledge of the fracture network geometries are not required, but the complex transport behaviour can be adequately represented. In general two different types of double continuum approaches are known. Within the dual porosity approach only one continuum contributes to the regional flow, whereas the second continuum is a fluid and solute storage. A dual permeability approach consists of two continua which both contribute to the regional flow.

## 2 DP-MODFLOW – FLOW MODEL

Within a double continuum model the fracture system (f) as well as the matrix system (m) are represented as separated continua. The groundwater flow of each continuum is described by Darcy's law, which is given by

$$Q = AK\frac{dh}{dl}, \tag{1}$$

where Q is the flow rate through the cross sectional area A, K is the hydraulic conductivity of the medium and dh/dl is the hydraulic gradient. The two continua are coupled by the exchange term $q_\alpha$, which is the specific flux between the continua depending on the difference in the piezometric heads $h^m$ and $h^f$ and the specific exchange coefficient $\alpha_0$:

$$q_\alpha = \alpha_0(h^m - h^f) \tag{2}$$

This kind of exchange was first introduced by Barenblatt et al. (1960). The flow in such a double continuum system is described by the differential equation system (Teutsch, 1988):

$$\mathrm{div}(K^m \,\mathrm{grad}\, h^m) = S_0^m \frac{\partial h^m}{\partial t} - W_0^m + q_\alpha \tag{3a}$$

$$\mathrm{div}(K^f \,\mathrm{grad}\, h^f) = S_0^f \frac{\partial h^f}{\partial t} - W_0^f - q_\alpha \tag{3b}$$

$K^m$ and $K^f$ in these equations are the conductivity tensors for each continuum, $S_0^m$ and $S_0^f$ are the specific storage coefficients and the specific sink/source terms are given by $W_0^m$ and $W_0^f$. As both continua are occupying the same volume, they possess the same boundaries but not necessarily the same boundary conditions and hydraulic parameter values.

Such a double continuum system can be described by using DP-MODFLOW (Lang et al., 1992b). The program package is based on the well known 3D-finite-difference program MODFLOW (McDonald and Harbaugh, 1988), which was suitable to be extended because of its modular structure and available source code. The three-dimensional model domain can be discretized by model cells, which have the shape of cuboids. The groundwater flow is calculated by the flux between two neighbouring cells, whereas the exchange between the continua takes place between cells in the same spatial position of each continuum. This results into a algebraic equation system for each continuum which is solved separately. An additional iterative procedure accounts for the exchange process.

As MODFLOW (McDonald and Harbaugh, 1988) is able to calculate flow systems with free water table, an iteration over the saturated thickness of the model cells is necessary during solving the equation system. To rewet cells which were running dry, Mohrlok (1992b) further developed the MODFLOW program. By formulating the exchange term Lang (1995) considered that the model cells can be confined, unconfined or even dry. Thus the model is able to permit exchange between cells of different boundary conditions (Figure 2).

## 3 DP-MT3D – TRANSPORT MODEL

The solute transport calculations for the double continuum modelling approach is also based on two differential equations, one for each continuum, similar to the groundwater flow calculations. The transport in each continuum is calculated separately, and the continua are coupled by an exchange term. Considering the hydrodynamic dispersion $D_{f,m}$ these differential equations are given by

$$\frac{\partial(c^f)}{\partial t} = \nabla(u^f c^f) + \nabla(D^f \nabla c^f) + \sum W_0^f c_{w_0}^f + j^{fm} \tag{4a}$$

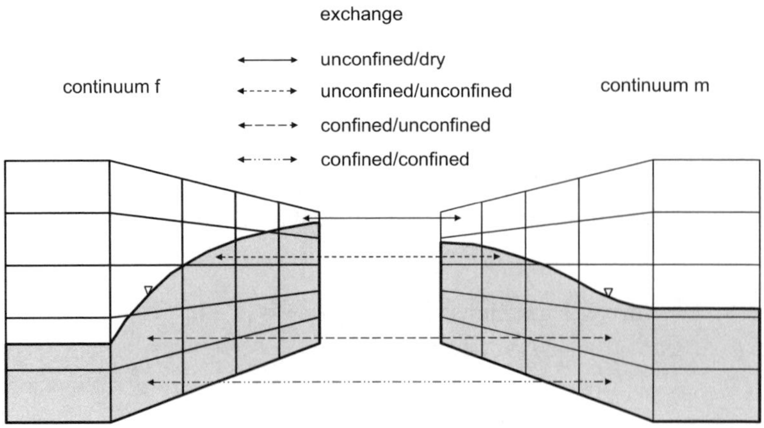


Figure 2. Exchange between the model cells by different boundary conditions (after Lang, 1995).

$$\frac{\partial(c^m)}{\partial t} = \nabla(u^m c^m) + \nabla(D^m \nabla c^m) + \sum W_0^m c_{w_0}^m + j^{fm} \tag{4b}$$

where $\partial c_{f,m}/\partial t$ is the change in concentration over time t, $u_{f,m}$ is the transport velocity, $\Sigma W_0$ is the sum of sinks and sources of each continuum and $j_{f,m}$ the exchange term between the fracture and matrix continuum.

Reaction and retardation transport processes can be calculated separately within each continuum, therefore they have no influence on the exchange processes and are not considered in this paper.

The exchange term $j_{fm}$ is based on the three main exchange processes: matrix diffusion ($j_D$) as a result of concentration differences, local advection ($j_{AL}$) due to pressure gradients between fracture system and matrix system, and regional advection ($j_{AR}$) due to a regional pressure gradient.

$$j_{fm} = j_D + j_{AL} + j_{AR} \tag{5}$$

The focus of this work is on regional transport processes within a double porosity and permeability medium. The matrix diffusion is neglected, as the advective exchange processes dominate.

Exchange processes between the continua in a fracture-matrix system are affected by strong flow and transport condition changes, resulting in high computational requirements for the model technique to simulate transport processes in this system. Normally, the flow velocity within the fracture system is some orders of magnitude higher than in the matrix system. The entering processes of solutes into the matrix from fractures are of high importance for the whole process understanding and have to be considered in the double continuum approach (Huyakorn et al., 1983). Barenblatt et al. (1960) introduced the quasi-steady exchange model, where the exchange term is represented as a sink and source term and the concentration is

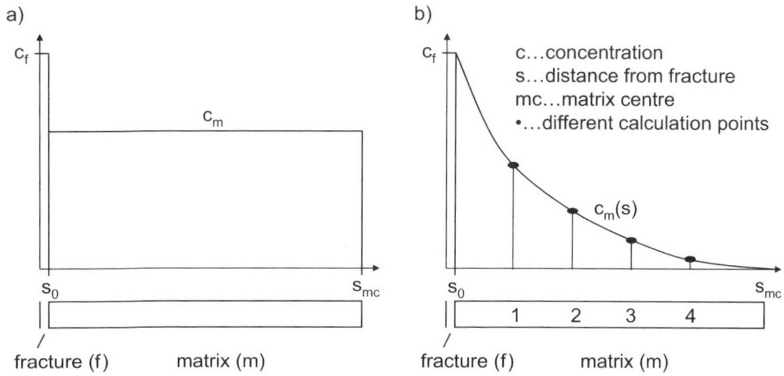


Figure 3. Exchange approaches: (a) quasi-steady exchange after Barenblat et al. (1960); (b) unsteady exchange after Huyakorn et al. (1983).

averaged over the whole matrix cell (Figure 3a). As a result, the steep concentration front within the matrix disappears. Therefore Huyakorn et al. (1983) used a finite-difference method to investigate the diffusive tracer entering into the matrix along a one-dimensional line, perpendicular to the fractures (Figure 3b). This resulted in a mixing of the classical double continuum approach and a discrete approach, as the fracture geometry had to be taken into account in the formulation of the exchange term. For this reason the matrix block geometries have to be known to determine parameters for a shape functions.

In regional fracture matrix systems the geometry information can not be determined due to their complex discontinuity structure. Therefore, Lang (1995) developed a new method for simulating advective transport processes for transient conditions, which can be applied without detailed geometry information.

### 3.1 *Exchange by local advection*

The local advection exchange term is equal to the exchange rate, calculated in the flow model, multiplied by the solute concentration:

$$j^{m}_{AL} = \alpha_0(h^m - h^f)c^m \tag{6a}$$

$$j^{f}_{AL} = \alpha_0(h^m - h^f)c^f \tag{6b}$$

Where $\alpha_0$ is the exchange coefficient, $(h^m - h^f)$ is the piezometric head difference between the continua and $c^{f,m}$ is the concentration of either the fracture or matrix continuum. A higher hydraulic head within the fracture system than the hydraulic head of the matrix system results in a flow from the fracture to the matrix continuum with the concentration equal to the fracture system (6b) or vice versa (6a). Consequently, the entering exchange volume can be determined quantitatively but not localized. In Figure 4 this process is explained for an exchange between the continuum f and continuum m.

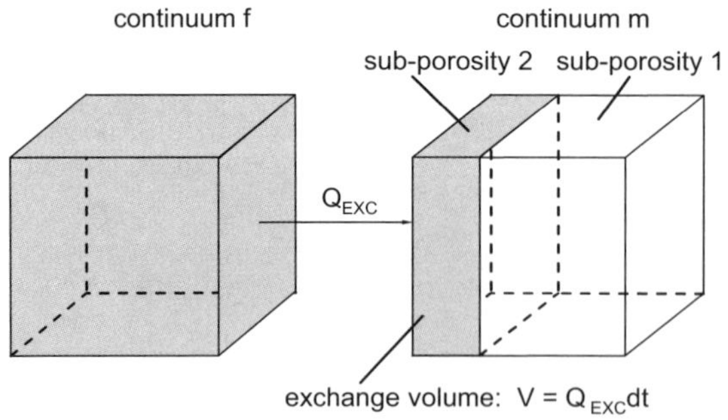


Figure 4. Schematic description of the exchange volume due to local advection if solutes moving from continuum f to the continuum m with displacement flow in continuum m (Lang, 1995).

To consider the exchange volume within the double continuum approach, during the transport calculation, each continuum is divided into two sub-porosities. The first sub-porosity represents the water volume, which describes the transport within the single continua, and the second sub-porosity represents transport processes depending on the exchange between the continua. Both sub-porosities possess the same aquifer parameters but different dynamic volumes ($n_e^{m1}$, $n_e^{m2}$), which are increasing or decreasing due to the exchange rate. The total volume of each continuum cell remains constant.

$$n_e^m = n_e^{m1} + n_e^{m2} \tag{7}$$

Figure 4 shows a total displacement flow within the matrix continuum (m), where the solute of the sub-porosity 2 is not mixed with the water volume of sub-porosity 1. Due to this displacement flow the artificial concentration averaging over the whole model cell is inhibited. To consider also the mixing flow, Lang (1995) introduced a factor f to control the mixing and displacement rate by formulating the exchange term. Taking into account equation 7, the differential equation system for continuum m is given by:

sub-porosity 1 (unaffected by the exchange process, $f^m = 1$)

$$\begin{aligned}\frac{\partial(c^{m1}n_e^{m1})}{\partial t} &= \nabla(n_e^{m1}u^{m1}c^{m1}) + \nabla(n_e^{m1}D^{m1}\nabla c^{m1}) + \sum n_e^{m1}q_s^m c_s^m \\ &\quad + (1 - f^m)j_{AL},\end{aligned} \tag{8a}$$

sub-porosity 2 (affected by the exchange process, $f^m = 1$)

$$\frac{\partial(c^{m2}n_e^{m2})}{\partial t} = \nabla(n_e^{m2}u^{m2}c^{m2}) + \nabla(n_e^{m2}D^{m2}\nabla c^{m2}) + \sum n_e^{m2}q_s^m c_s^m + f^m j_{AL}, \tag{8b}$$

where $q_s^m$ is the specific flow rate in the sinks and sources with the corresponding concentration $c_s^m$ for the matrix continuum (m), and $j_{AL}$ is the exchange term of this continuum. As the mixing factor $f^m$ controls the dynamic volumes of the sub-porosities, it has to be defined for each continuum separately. For the fracture continuum the mixing factor is

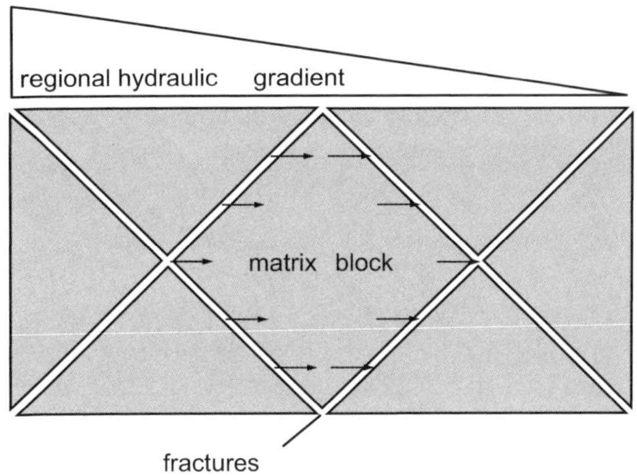


Figure 5. Regional advection shown on a single matrix block.

typically 0, since the flow velocity within this system is high and a total mixing of incoming solutes over the whole fracture cross section can be assumed. A pure displacement flow ($f^f = 1$) occurs when the exchange is from the fracture system (f) into the matrix system (m). Thus the exchange volumes have the original concentration of the fracture system and not a much lower averaged concentration.

The total concentration of continuum (m) has to be averaged over the volume ratio of the sub-porosities by:

$$c^m = \frac{c^{m1} n_e^{m1} + c^{m2} n_e^{m2}}{n_e^{m1} + n_e^{m2}} \tag{9}$$

## 3.2 *Exchange by regional advection*

For transport investigations under steady-state flow conditions, the local advection exchange term becomes zero, as the hydraulic heads of the fracture continuum and the matrix continuum are equal. For this reason another exchange term has to be included into DP-MT3D. If regional transport in both continua is considered, the exchange between the continua is controlled by the regional advection, which follows the regional hydraulic gradient. The exchange volume is dependent on the velocity within the matrix and the fracture surface which is perpendicular to the flow direction. Birkhölzer (1994) defined the exchange term for regional advection as:

$$j_{RA} = |q^m| \Omega_W (c^f - c^m), \tag{10}$$

where $q^m$ is the effective velocity within the matrix continuum, $\Omega_W$ the specific exchange cross section as a function of the relevant fracture surface and the matrix volume, and $(c^f - c^m)$ is the concentration difference between the fracture system and the matrix system.

Figure 5 shows the simplified regional advection process. The water enters the matrix block from the fractures on the left-hand side, and exits the matrix block to the fractures

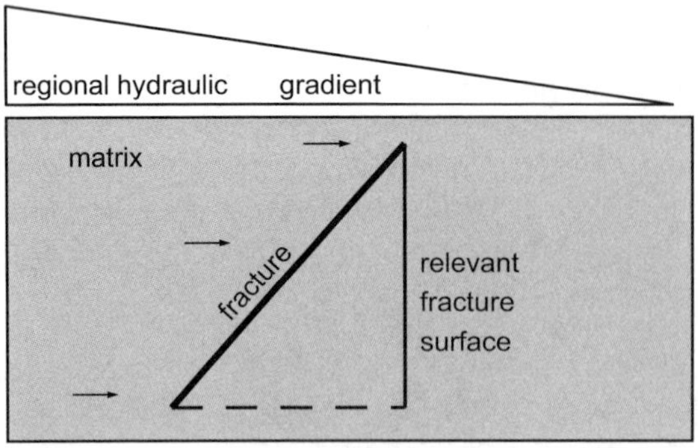


Figure 6. Definition of the relevant fracture surface (after Birkhölzer, 1994).

on the right-hand side. In terms of groundwater flow there is no exchange between the fractures and the matrix, as the same amount of water is entering and leaving the matrix block. The water volume in the fracture system as well as in the matrix system remains constant over the time. In contrast, if a tracer is carried along by the fracture system, the entering tracer mass on the left hand side of the matrix block is directly proportional to the matrix flow velocity. On the right-hand side of the matrix block water without tracer will be released to the fracture system. This results in dilution effects within the fracture system and an accumulation of mass within the matrix block. This transient condition in terms of transport will last until the tracer front fully crosses the matrix block. Afterwards the transport process reaches steady-state conditions, as the same amount of tracer is entering and leaving the matrix block. Therefore the exchange has to be taken into account only if there is a concentration difference between the fracture system and the matrix system.

In general the matrix system acts as storage, capturing mass over an appointed time period, until the mass is released back to the fracture system. The duration is dependent on the matrix block size, the specific exchange cross section and the flow velocity in the matrix. The specific exchange cross section is the sum of relevant fracture surfaces, which is the equivalent fracture surface perpendicular to the hydraulic gradient (Figure 6) per aquifer volume. This implies that a large specific exchange cross section corresponds with a high fracture density and small matrix blocks. In terms of tracer mass, a large specific exchange cross section results in a small saturation time for the matrix blocks.

To implement the regional advection term into DP-MT3D, the sub-porosity method as introduced by Lang (1995) was adapted. The resulting exchange process is shown in Figure 7. From the fracture system, where the second sub-porosity is zero because of total mixing, the exchange volume with the concentration of the fracture system is entering the matrix system by displacement flow (f = 1). This results in an increasing sub-porosity 2 within the matrix continuum. By considering a constant volume of the matrix continuum, the displaced volume with the concentration of the sub-porosity 1 will be transferred to the fracture continuum. By assuming an initial concentration of zero within the matrix continuum (sub-porosity 1) the exchange process leads to a decreasing concentration within the fracture continuum, and an increasing concentration within the matrix continuum.

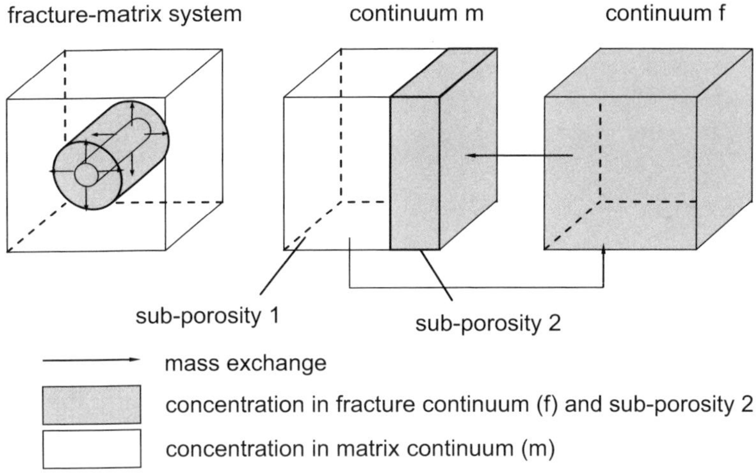


Figure 7. Regional advection exchange as implemented into DP-MT3D (after Lang, 1995).

The following differential equations arise to solve the transport for the different continua:

- matrix continuum:
  sub-porosity 1

$$\frac{\partial(c^{m1}n_e^{m1})}{\partial t} = \nabla(n_e^{m1}u^{m1}c^{m1}) + \nabla(n_e^{m1}D^{m1}\nabla c^{m1}) - |q^m|\Omega_w c^{m1}, \tag{11}$$

  sub-porosity 2

$$\frac{\partial(c^{m2}n_e^{m2})}{\partial t} = \nabla(n_e^{m2}u^{m2}c^{m2}) + \nabla(n_e^{m2}D^{m2}\nabla c^{m2}) + |q^m|\Omega_w c^{f}, \tag{12}$$

- fracture continuum:

$$\frac{\partial(c^{f})}{\partial t} = \nabla(u^f c^f) + \nabla(D^f \nabla c^f) + |q^m|\Omega_W (c^{m1} - c^f), \tag{13}$$

where $q^m$ is the effective velocity within the matrix continuum, $\Omega_W$ the specific exchange cross section and c the concentration of the sub-porosity 1 (m1), sub-porosity 2 (m2) or the fracture continuum (f). The total concentration of the matrix continuum can be calculated using equation (9) as explained before. If the cell exchange process is calculated without considering the sub-porosities in the matrix continuum, the average concentration over the whole cell would result into artificial tailing.

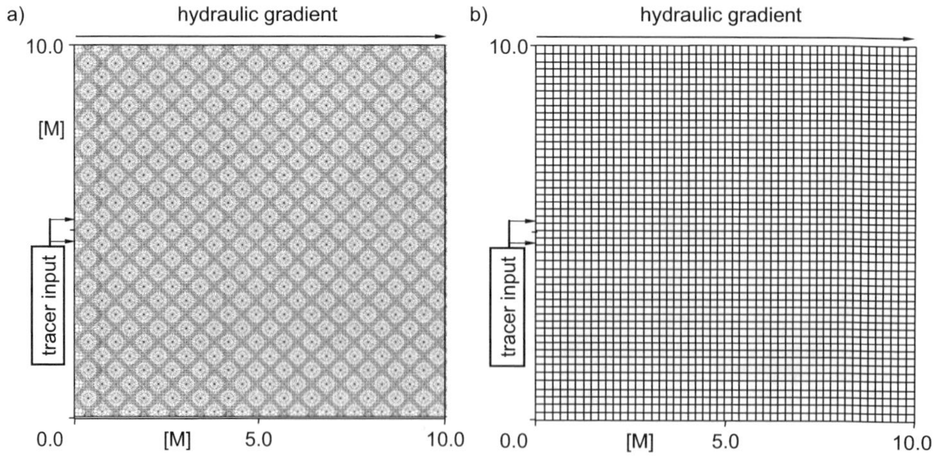


Figure 8. Model setup of discrete (a) and double continuum system (b).

## 4 APPLICATION

To describe the regional flow and transport regime of a fractured porous sandstone formation without considering the details of geometry, the double continuum approach is applied. As there was no possibility to obtain the detailed information of the fracture geometry and hydraulic properties by field exploration, a reference system with known properties had to be defined. This reference system, an idealized synthetic fractured aquifer, is used to validate the double continuum model. Therefore the detailed description of the discrete fracture model is necessary to understand the contaminant transport. It is also necessary to determine the physically meaningful effective parameters depending on fracture network and rock characteristics for each continuum.

### *Model setup*

For the numerical investigations of flow and transport in the synthetic fracture matrix system the program package ROCKFLOW (Wollrath, Helmig 1991) was used, based on the finite element method. A saturated two-dimensional regular fracture network was set up, where the fractures were embedded as one-dimensional line elements in a two-dimensional permeable matrix (Figure 8a). This fractured porous medium contained two sets of fracture populations each consisting of 26 infinite fractures. The fractures of each population were equidistant with constant apertures. The porous matrix blocks between fractures were defined with homogeneous and isotropic hydraulic properties. On the in- and outflow boundary of the model a Dirichlet-boundary-condition was applied, leading to a constant hydraulic gradient. The other two sides of the model were no-flow boundaries. For transport investigations the parameters of Table 1 were applied to the system.

For the double continuum model the same model domain was used. The discretization of the model contains 50 rows by 50 columns (Figure 8b). The boundary conditions of the double continuum model were the same as in the discrete model. Dirichlet-boundary-conditions

Table 1. Parameter of the double continuum system.

| Parameter | Discrete system/ matrix continuum | Fracture continuum | Unit |
|---|---|---|---|
| Effective porosity | 0.1 | 0.0038 | (−) |
| Hydraulic conductivity | $10^{-6}$ | $1.8 * 10^{-5}$ | (m/s) |
| Diffusion coefficient | $10^{-10}$ | 0 | ($m^2$/s) |
| Dispersivities | $\alpha_l = 0.01; \alpha_t = 0.001$ | $\alpha_l = 0.007; \alpha_t = 0.170$ | (m) |
| Hydraulic gradient | 5 | 5 | (%) |
| Specific exchange cross section | 2.0 | 2.0 | ($m^2/m^3$) |
| Fracture aperture | 0.001 | – | (m) |

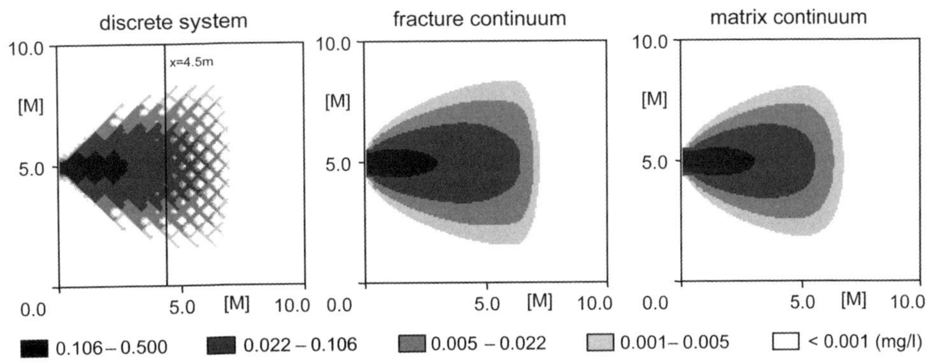


Figure 9. Tracer plumes of the discrete system, the fracture continuum and the matrix continuum systems after 27000 s.

were applied to the inflow and outflow cross section, leading to a constant hydraulic gradient of 5 %, and no-flow boundaries to the other two sides. Sinks and sources were not taken into account. The aquifer parameters of the matrix continuum were taken directly from the discrete model (Table 1). For the fracture continuum the effective parameters had to be determined by inverse modeling, to reproduce the same flow and transport behavior as given by the discrete model. Particularly the hydraulic conductivity and the dispersion coefficients had to be determined. Also the specific exchange cross section had to be calibrated after a rough calculation as fracture area per aquifer volume.

## 5 RESULT ANALYSIS

Figure 9 shows the resulting tracer plume of the discrete model in comparison to the tracer plumes of the double continuum model for a continuous tracer injection after 27000 seconds. The plume in the discrete model showed a strong transversal dispersion due to the fracture network. Starting from the input area the plume boundary development in transverse direction was bounded by two fractures. At each fracture intersections along these two fractures the concentration was halved, as the tracer loaded water of the input fracture was totally mixed with the unloaded water of the intersecting fractures. This effect was

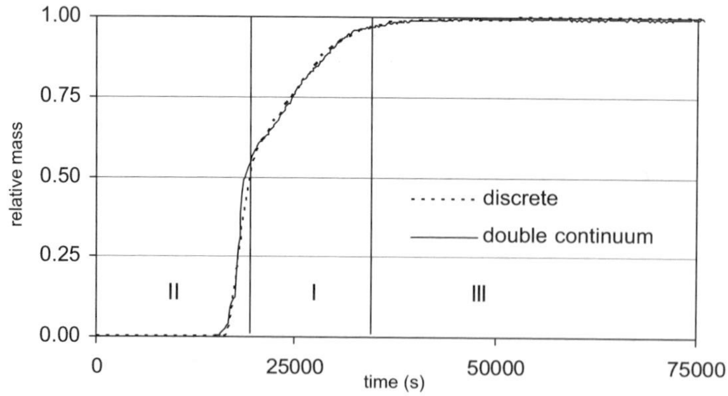


Figure 10. Relative mass breakthrough curves of the discrete and double continuum model at x = 4.5 m.

leading to steep concentration gradients within the system along the margin fractures. Therefore, within the plume the dilution effect at fracture intersections was smaller, as all fractures contained tracer loaded water. The rate of dilution was directly depending on the concentration difference of two intersecting fractures. Furthermore, it can be seen that the transport in the fractures, was faster than in the matrix, as the tracer needed more time to pass the matrix blocks.

The general shapes of the developed plumes, for the discrete and double continuum model, were similar. The strong spreading of the plume in transverse direction could be achieved by a high transverse dispersivity for the fracture continuum. Within the fracture continuum the tracer front was moving faster than in the matrix continuum (Figure 9). The only obvious difference between the tracer plumes of the two model types were the transverse boundaries. The plume of the discrete model was bounded by tracer loaded fractures as described above, whereas the plume of the double continuum model showed a smooth transition from the unloaded to the loaded water due to the averaging character.

The reference cross section at x = 4.5 m in the flow direction was used to compare the relative mass breakthrough of both models (Figure 10). The two resulting breakthrough curves showed the characteristic shape for a fractured-matrix system dominated by regional advective exchange. The breakthrough curves can be subdivided into three phases. In the first phase (I) the tracer arrives at the observation point with a fast increasing concentration due to transport in the fracture system. After a specific time period a slower concentration increase is observed. This is the starting point of phase II, the exchange dominated phase. The higher the fracture-matrix exchange and the lower the velocity differences between the fractures and the matrix, the shorter would be this phase. By matching the breakthrough curves of the discrete and double continuum model in phase II, the exchange term was calibrated. In phase III the concentration has reached a maximum and remains constant over time. This means that the exchange processes upstream the observation point were in balance, the amount of tracer entering the fracture system equaled the amount of tracer leaving the fracture system.

The good match of the breakthrough curves implies that the effective parameters of the double continuum model were determined within a proper range.

## 6 CONCLUSIONS

By using the presented double continuum approach for transport problems in dual permeable fractured porous media, the plume development can be estimated by effective parameters and without considering the exact geometry information. The determination of the parameters for the fracture continuum was the most important step to calibrate the double continuum model. Of particular importance were the determination of hydraulic and transport parameters by fracture network characteristics. The transversal dispersion coefficient and hydraulic conductivity were directly dependent on the angle between the fracture direction and the hydraulic gradient. A large angle results in a high transversal dispersion coefficient and, correspondingly, a low effective hydraulic conductivity.

The correlations in terms of different angles between the fracture populations, varying fracture apertures or heterogeneous fracture networks have to be investigated and quantified by sensitivity analysis.

## ACKNOWLEDGEMENTS

The presented investigations are done in cooperation with Electricité de France (EDF) and funded by the European Institute for Energy Research (EIfER).

## REFERENCES

Barenblatt GI, Zheltov IP, and Kochina IN. 1960. *Basic concepts in the theory of seepage of homogeneous liquids in fissured rocks*, J. Applied Mathematical Methods (USSR), 24, 1286–1303.

Birkhölzer J. 1994. *Numerische Untersuchungen zur Mehrkontinuumsmodellierung von Stofftransportvorgängen in Kluftgrundwasserleitern*, Dissertation, Mitteilungen des Instituts für Wasserbau und Wasserwirtschaft, Band 93, Rheinisch-Westfählische Technische Hochschule, Aachen.

Huyakorn PS, Pinder GF. 1983. *Computational Methods in Subsurface Flow*. Academic Press INC, Orlando.

Lang U, Mohrlok U, Teutsch G und Kobus H. 1992b. *Weißjura-Grundwasserbilanzmodell Stubersheimer Alb, Modelltechnik des 3D-Doppel-Porositäts-Programms DP-MODFLOW*; Technischer Bericht, Nr. 92/29 (HG 169), Institut für Wasserbau, Universität Stuttgart.

Lang U. 1995. *Simulation regionaler Strömungs- und Transportvorgänge in Karstaquiferen mit Hilfe des Doppelkontinuum-Ansatzes*: Methodenentwicklung und Parameterstudie; Dissertation am Institut für Wasserbau, Universität Stuttgart.

McDonald M and Harbaugh A. 1988. *A modular three-dimensional finite-difference ground-water flow model*: Techniques of Water Resources Investigations 06-A1, USGS.

Mohrlok U. 1992b. *Programmentwicklung und Verifizierung des Moduls zur Wiederbenetzung trockengefallener Modellzellen, in Weißjura-Grundwasserbilanzmodell Stubersheimer Alb, Modelltechnik des 3D-Doppel-Porositäts-Programms DP-MODFLOW*, Technischer Bericht, Nr. 92/29 (HG 169), Institut für Wasserbau, Universität Stuttgart.

Teutsch G. 1988. *Grundwassermodelle im Karst: Praktische Ansätze am Beispiel zweier Einzugsgebiete in Tiefen und Seichten Malmkarst der Schwäbischen Alb*, – Dissertation, Universität Tübingen.

Teutsch G and Sauter M. 1991. *Groundwater flow and transport processes in karst aquifers – scale effects, data provision and model validation*, EPA/NWWA Int. Symp. on Environmental Problems in Karst Terrains, Nashville.

Wollrath J and Helmig R. 1991. *SM-2/TM-2, Strömungs- und Transportmodell für inkompressible Fluide, Theorie und Benutzerhandbuch*. Technical report, Institut für Strömungsmechanik, Universität Hannover.

CHAPTER 38

# Coupled hydro-mechanical modelling of flow in fractured rock

Philipp Blum[1,2], Rae Mackay[1] and Michael S. Riley[1]

[1] *School of Geography, Earth and Environmental Sciences, University of Birmingham, Edgbaston, Birmingham, UK*

[2] *Center for Applied Geoscience (ZAG), University of Tübingen, Sigwartstr. Tübingen, Germany*

ABSTRACT: Using fracture data from Sellafield, UK, a modelling-based approach to characterising the contribution of hydro-mechanical processes to large scale flow and transport reveals three important hydrogeological issues for the performance assessment of a repository host rock. First, inference of the statistical model of fracture length distribution using power law distribution models can lead to large uncertainty in fracture density. Second, fracture length is less important than fracture density in determining the flow characteristics of the rock mass. Third, the depth dependence of permeability can be isolated using hydro-mechanical models but the major uncertainty in the analysis is the characterisation of the rock mass mechanical properties.

## 1 INTRODUCTION

A numerical approach to quantifying the importance of hydro-mechanical (HM) processes in fractured rock to large-scale radionuclide migration through a repository host rock has been developed (Figure 1) and tested. Statistical models of fracture set properties, inferred from measurements, are used to generate multiple realisations of discrete fracture networks (DFN) that can be inputed to numerical models to explore the relationship of the hydraulic properties to the in-situ stress conditions. The uncertainties inherent in the identification of the statistical parameters for the DFN from the fracture data are considered explicitly in the analysis of the local scale behaviour of the fractured rock. Two numerical codes have been coupled to explore the hydro-mechanical interactions between flow and stress patterns. The universal distinct element code, UDEC-BB, incorporating the Barton-Bandis model for fracture stress and strain is used to determine the HM contributions to fracture aperture distributions and the fracture flow code, FRAC2D, is used to calculate the flows across the stress-modified DFN. In the final stage the local scale hydraulic properties are upscaled for input to a large-scale continuum model. The methodology has been applied using data from the Sellafield (UK) site investigation programme performed by United Kingdom Nirex Limited and provided as part of the second Bench Mark Test (BMT2) of the international collaborative programme DECOVALEX III. The modelling was carried out in two dimensions (2D) to constrain the computational effort. A 2D model plane orientated along the main horizontal stress direction (159/339°) observed at Sellafield was adopted for the BMT2.

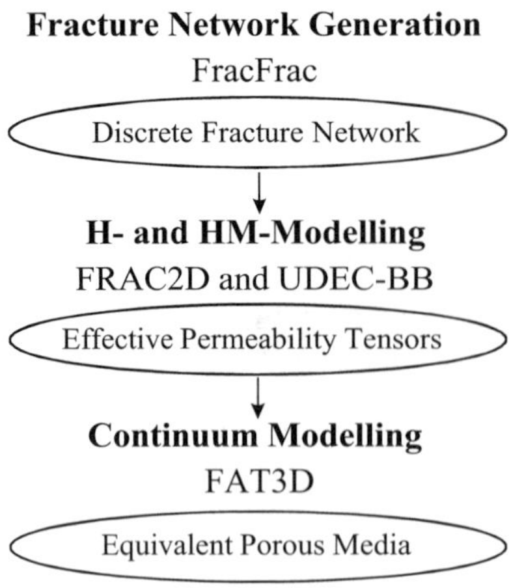


Figure 1. Stages used to quantify HM contributions to fractured rock hydraulic behaviour at large scales.

Table 1. Discontinuity data (data from Nirex 1997a).

| Rock Unit | Set | DD | $\alpha$ | $\beta$ | $S_t$ [m] | $S_a$ [m] |
|---|---|---|---|---|---|---|
| Formation 1 | 1 | 145 | 8 | 7.8 | 0.29 | 0.29 |
| | 2 | 148 | 88 | 88.0 | 0.26 | 0.25 |
| | 3 | 21 | 76 | 71.5 | 0.28 | 0.19 |
| | 4* | 87 | 69 | 38.8 | 0.31 | 0.14 |
| Formation 2 | 1 | 28 | 25 | 17.0 | 0.51 | 0.49 |
| | 2 | 156 | 81 | 81.0 | 0.35 | 0.35 |
| | 3 | 20 | 72 | 66.7 | 0.28 | 0.19 |
| | 4* | 90 | 68 | 41.6 | 0.41 | 0.21 |
| Fault Zone | 1 | 21 | 8 | 6.0 | 0.18 | 0.18 |
| | 2 | 150 | 76 | 75.8 | 0.18 | 0.18 |
| | 3 | 21 | 72 | 66.4 | 0.19 | 0.13 |
| | 4* | 85 | 74 | 43.9 | 0.22 | 0.08 |

DD = dip direction, $\alpha$ = true dip, $\beta$ = apparent dip, $S_t$ = mean true spacing, $S_a$ = mean apparent spacing, * neglected fracture set.

## 2 DISCRETE FRACTURE NETWORK-ANALYSIS

Three rock formations are identified in the data set and the given 3D discontinuity data of the fracture orientation and the fracture spacing were transformed into equivalent 2D data (Table 1).

Analysis of the fracture data for the repository host rock indicates that a power-law fracture length distribution for the four fracture sets can describe the fracture network characteristics in 2D. The power-law distribution of the fracture length is given by:

$$N_F = CL^{-D} \tag{1}$$

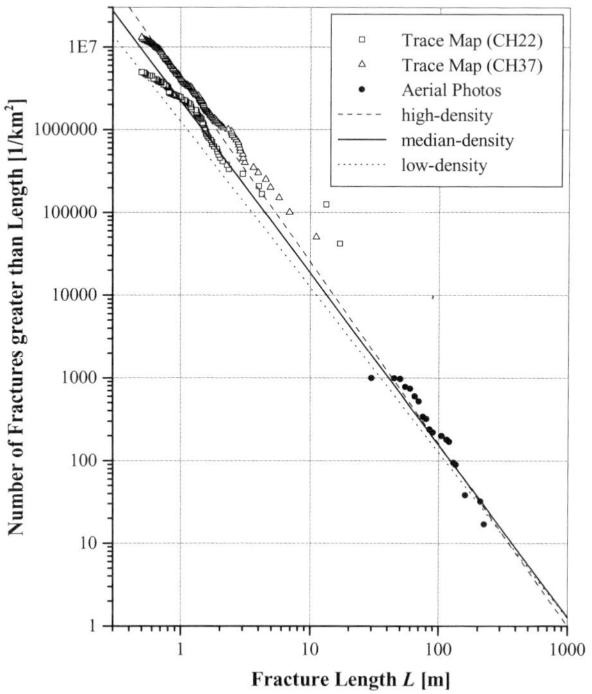


Figure 2. Power-law fracture length distribution (data from Nirex 1997b).

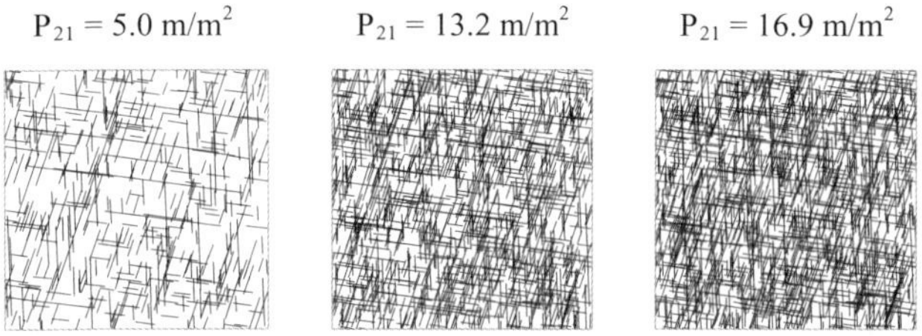


Figure 3. DFN showing fracture distributions for the low-, median- and high density cases of Formation 1. Region size = 10 m × 10 m.

where $N_F$ is the number of fractures per unit area of fracture length greater than the length L, C is the density constant and D is the fractal dimension (Figure 2). Only fractures with a length greater than 0.5 m were recorded, therefore 0.5 m was chosen as the lower-cut of length of the power-law. The lower-cut off length defines the fracture intensity ($P_{20}$, Dershowitz 1984) for given values of C and D. For the data, values of C and D of 4.00 and 2.20 respectively are presented in Nirex (1997b), which result in a relatively high fracture intensity of 18.4 $m^{-2}$ (high-density case in Figures 2 and 3). Analysis shows that these values are representative of one part of the available data but not all. The best fit using all the data suggests a density constant of 3.23 and a fractal dimension of 2.08 yielding a fracture intensity of 13.7 $m^{-2}$ (median-density case in Figures 2 and 3). A further data set with an average fracture intensity of 4.8 $m^{-2}$ is also identified in Nirex (1997b). An equivalent power-law length distribution for this low fracture intensity was obtained by a visual fit to the full data and gave values of C and D of 1.20 and 2.00 (low-density case in Figures 2 and 3). This analysis of the fracture length and fracture density data shows significant variations in fracture density, depending on the sample data used even though the data are from the same site. A sensitivity study using the three fracture densities (low-, median- and high-density cases) has been undertaken. Bonnet et al. (2001) provides a review of modelling DFN with a power-law distribution and guidance on the use of the power-law model.

### *Fracture Network Generation*

A fracture network generator (FracFrac) was developed to generate DFN with a power-law length distribution. For each formation only three fracture sets have been used to develop the 2D models (Table 1), because the fourth steeply dipping fracture sets in each formation possess almost the same strike as the model plane. A Poisson distribution has been assumed for the fracture midpoints, although analysis suggests the existence of a repulsion between long fractures. The dispersion of the fracture orientation is described with the Fisher distribution and a constant K ranging from 5.9 to 13.0, which indicates a fairly high dispersion of the fracture orientations. Visual comparisons of DFN generated with the Fisher distribution and the trace maps clearly indicate that a high dispersion of orientation does not produce synthetic networks that are comparable to those observed. It is apparent that fracture orientation dispersion is length dependent with low dispersion of longer fractures and higher dispersion of shorter fractures. Wu & Pollard (2002) have also observed a decrease in the deviation of the fracture orientations within a fracture set as the fracture length increases. Hydraulic modelling ignoring mechanical stresses using DFN with the median fracture density ($P_{21} = 13.2\,m/m^2$) shows, that due to the relatively high fracture density the impact of the repulsion and the dispersion of the fracture orientation has only a minor impact on the equivalent large scale continuum hydraulic properties for this case (Chillingworth 2002).

## 3 HYDRAULIC ANALYSIS

Constant aperture hydraulic (H) modelling was undertaken initially to determine the minimum size of model region required to approximate a REV (representative elementary volume) Several methods exist to evaluate the 2D permeability tensor in fractured rock (e.g., Zhang & Sanderson 2002). In this study the evaluation is based on the approach of Jackson et al. (2000), which uses several hydraulic boundary conditions corresponding with a uniform head gradient in different orientations. A guard zone is maintained around the model region to reduce edge effects. For the hydraulic analysis a constant hydraulic aperture was adopted ($a_h = 130.7\,\mu m$). One hundred simulations for each model region were performed and the permeability tensor coefficients were evaluated using a maximum likelihood approach. Simulations were conducted on square model regions from 5–100 m side length for each of the fracture density cases.

### *REV-Analysis*

An REV-criteria was set, that when the cumulative variances of the permeability tensor coefficients and the principal directions with increasing domain size are all less than 5%

of their cumulative averages, then the model region is assumed to be larger than the size of the REV. The results showed that for the median- and high-density cases a model region of 10 m × 10 m is adequate for modelling. However for the low-density no REV was identified. This is apparently due to the decrease of connectivity in the minor flow direction for this model, which is demonstrated by the changing in principal directions towards higher angular values with increasing domain sizes.

## 4 HYDRO-MECHANICAL ANALYSIS

The HM-modelling has been performed with the UDEC-BB model with aperture distributions output to FRAC2D for the hydraulic calculations. Only median-density DFN have been analysed. The HM modelling has been carried out in a similar manner to the H modelling with the inclusion of a guard zone to minimise edge effects. Sensitivity studies have been carried out to examine the impact of the mechanical boundary condition, mechanical properties, initial mechanical apertures, aperture changes with increasing depth (HM-coupling) and various DFN with changing fracture lengths and block sizes (Blum 2005). The following sensitivity studies however focuses on the impact of the mechanical properties and the aperture changes with increasing depth on the hydraulic aperture distribution.

The in-situ stress field (MPa) is described by the following equations (Nirex 1997c):

$$\begin{aligned} \sigma_V &= 0.0294D + 0.26622 \\ \sigma_H &= 0.03113D + 1.88747 \end{aligned} \tag{2}$$

where $\sigma_V$ is vertical stress, $\sigma_H$ is horizontal stress and D is depth below ground level [m]. For HM-modelling, data sets are required that describe the rock block material and the joint material. Isotropic elastic behaviour is assumed for the rock blocks and values are listed in Table 2.

The non-linear hydro-mechanical joint behaviour with increasing stresses is described with the empirical Barton-Bandis (BB) joint model (e.g., Barton et al. 1985). The general mean input values for the blocks and the joints are given in Table 2. The initial mechanical aperture ($a_{ini}$) at zero normal stress is calculated with the following empirical relationship (Bandis et al. 1983):

$$a_{ini} = \frac{JRC_0}{5}\left(0.2\frac{UCS}{JCS_0} - 0.1\right) \tag{3}$$

Table 2. Block and joint data for the UDEC-BB model.

| | |
|---|---|
| *Block data* | |
| Bulk density ρ [kg/m$^3$] | 2750 |
| Young's modulus $E_{block}$ [GPa] | 84.0 |
| Poisson ratio $\nu$ [−] | 0.24 |
| *Joint data* | |
| Joint normal stiffness $K_n$ [GPa/m] | 434.0 |
| Joint shear stiffness $K_s$ [GPa/m] | 434.0 |
| Laboratory-scale joint length $l_0$ [m] | 0.03 |
| Residual friction angle [°] | 27.2 |

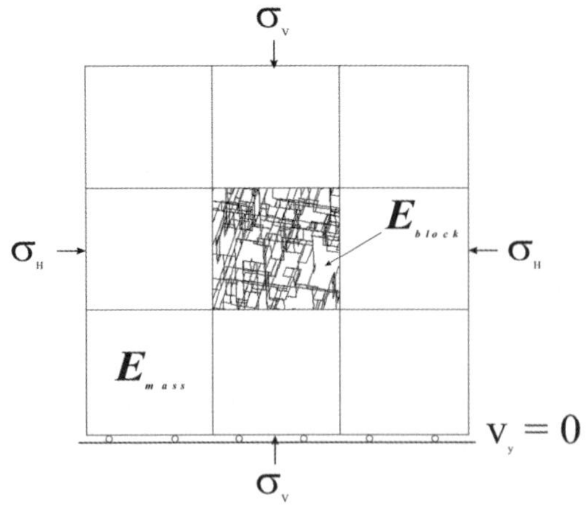


Figure 4. Mechanical boundary conditions for the HM analysis with the block-in-block method.

The initial mechanical aperture has a potentially strong contribution in the BB model, thus sensitivity studies have been performed to assess its impact on the HM calculations (Blum 2005). Only a small number of alternative methods are available to evaluate the initial mechanical aperture. Bandis et al. (1983) performed direct measurements to evaluate the initial mechanical aperture by using different devices, for example feeler gauges. Xia et al. (2003) also directly measured the initial mechancial aperture and could correlate the initial mechanical aperture with the average height of ten points on the fracture surface profile. The uniaxial compressive strength (UCS) characterises the strength of the rock mass. The Schmidt hammer can be used to estimate the UCS (e.g., Katz et al. 2000). Alternatively several laboratory tests are available to evaluate the UCS (e.g., Goodman 1989). If the joint is unaltered and unweathered, JCS = UCS, and therefore, according to equation (3), the initial mechanical aperture will only depend on the JRC.

The data set defining the joint parameters JRC and JCS for the considered rock mass show major variations and no obvious relationships was observed between JRC, JCS and depth (Blum 2005). Hence three pairs of JRC/JCS values were identified and have been used to assess the sensitivity of the BB model to mechanical properties.

The sensitivity studies were carried out using a 10 m × 10 m model domain in a 30 m × 30 m block-in-block simulation region (Figure 4). The bottom mechanical boundary is rigid in the y-direction.

The Young's modulus ($E_{mass}$) assigned to the surrounding rock mass was 65.0 MPa and is therefore about 77% of the undisturbed rock mass ($E_{block}$). Stresses were then applied across the outer boundary. The name block-in-block method has been coined for this approach (Blum et al. 2005). The mechanical and hydraulic boundary conditions were changed according to the applied stresses if necessary.

For the sensitivity study of the mechanical properties the depth from the ground surface to the centre of the domain was set at 50 m. The median hydraulic apertures of the three considered JRC/JCS cases are given in Table 3. The highest median hydraulic aperture with 227 μm results from the high JRC case mainly due to the rougher aperture, which is also responsible for the higher initial mechanical aperture. The median hydraulic apertures of the other two considered cases are similar, although the initial mechanical apertures vary strongly.

Table 3. Input parameters for the three JRC/JCS cases and the resulting median aperture values.

| | Mean | High JRC | High JCS |
|---|---|---|---|
| $JRC_0$ [− ] | 3.9 | 7.1 | 1.5 |
| $JCS_0$ [MPa] | 112.2 | 43.1 | 298.4 |
| UCS [MPa] | 157.0 | 157.0 | 308.8 |
| $a_{ini}$ [μm] | 138.5 | 887.5 | 32.3 |
| $a_h$[1] [μm] | 23.7 | 226.9 | 30.4 |

[1] median hydraulic aperture using the calculated $a_{ini}$.

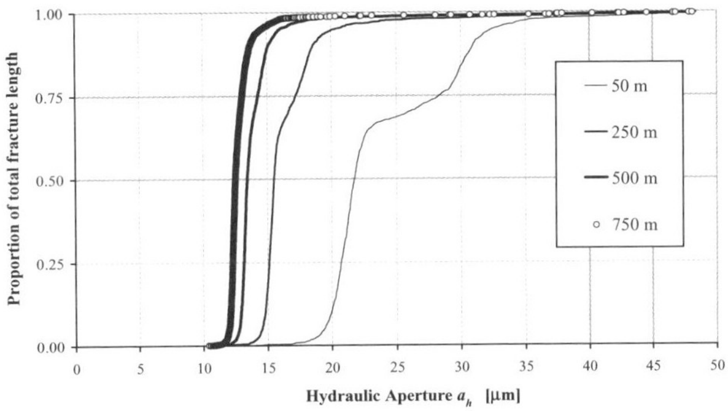


Figure 5. Cumulative distributions of hydraulic aperture for one DFN for different depth below ground level (Blum et al. 2005).

The effect of increasing stresses on the hydraulic aperture distribution has been tested by simulating increasing depth of burial (HM-coupling). The calculated median hydraulic aperture was around 24 μm for a burial depth of 50 m and around 13 μm at 1000 m (Figure 5). This is less than a factor of two reductions and assuming the validity of the cubic law, potentially represents a reduction of hydraulic conductivity of less than one order of magnitude. In comparison, the aperture variations due to mechanical property differences are significantly larger (Table 3).

Further the results indicate that changes to the distribution of the hydraulic apertures progressively decrease with increasing depth (Figure 5). There is almost no difference between the calculated hydraulic aperture distributions at 750 m and at 1000 m depth. Hence, a residual hydraulic aperture distribution is achieved at around 750 m depth. The hydraulic aperture anisotropy in the aperture distribution due to the different fracture orientations is gradually removed between 250 m and 500 m depth.

## 5 CONCLUSION

The variation of hydraulic properties as a function of depth can be identified using this modelling method with given mechanical properties and fracture geometries. The performed sensitivity studies revealed that the underlying variation in the mechanical properties have a significant impact on the hydro-mechanical modelling. The outcome of the sensitivity studies show that the impact appears to be higher than the aperture changes with increasing depth. Hence, understanding the spatial variation of the mechanical properties of the rock mass and fractures is vital to understand the connectivity of high and low permeability zones and to model the heterogeneity of the host rock at the large scale.

## ACKNOWLEDGEMENTS

We would like to thank Les Knight of United Kingdom Nirex Limited for his support and guidance throughout the project. Further we are grateful to all our colleagues from the DECOVALEX/BENCHPAR project for the many fruitful discussions. In particular, we would like to thank you Johan Öhman, Ki-Bok Min and Johan Andersson. The data were provided by United Kingdom Nirex Limited, who also funded the research.

## REFERENCES

Bandis SC, Lumsden AC & Barton NR (1983) Fundamentals of Rock Joint Deformation. International Journal of Rock Mechanics, Mining Sciences & Geomechanical Abstracts 20 (6):249–268.

Barton N, Bandis S & Bakhtar K (1985) Strength, Deformation and Conductivity Coupling of Rock Joints. International Journal of Rock Mechanics, Mining Sciences & Geomechanical Abstracts 22 (3):121–140.

Blum P (2005): Upscaling of Hydro-Mechanical Processes in Fractured Rock. PhD-Thesis, University of Birmingham, UK.

Blum P, Mackay R, Riley MS & Knight JL (2005) Performance assessment of a nuclear waste repository: upscaling coupled hydro-mechanical properties for far-field transport analysis.- International Journal of Rock Mechanics & Mining Sciences 42(5–6):781–792.

Bonnet E, Bour O, Odling NE, Davy P, Main I, Cowie P & Berkowitz B (2001) Scaling of fracture systems in geological media. Reviews of Geophysics 39 (3):347–383.

Chillingworth GEH (2002) The Effects of Fracture Geometry on the Permeability of Fractured Rock. Unpublished MSc-Thesis, Department of Earth Sciences, University of Birmingham, UK.

Dershowitz W (1984) Rock Joint Systems. PhD-Thesis, Massachusetts Institute of Technology, Cambridge, Massachusetts.

Goodman, RE (1989): Introduction to Rock Mechanics. Second Edition, John Wiley & Sons, New York, 562 pp.

Jackson CP, Hoch AR & Todman S (2000) Self-consistency of a heterogeneous continuum porous medium representation of fractured media. Water Resources Research 36 (1):189–202.

Katz O, Reches Z & Roegiers JC (2000) Evaluation of mechanical rock properties using a Schmidt Hammer. International Journal of Rock Mechanics and Mining Sciences 37 (4):723–728.

Nirex (1997a) Data summary sheets in support of gross geotechnical predictions. Nirex Report SA/97/052, Harwell, UK.

Nirex (1997b) Evaluation of heterogeneity and scaling of fractures in the Borrowdale Volcanic Group in the Sellafield area. Nirex Report SA/97/028, Harwell, UK.

Nirex (1997c) Assessment of the in-situ stress field at Sellafield. Nirex Report S/97/003, Harwell, UK.

Wu H & Pollard DD (2002) Imaging 3-D fracture networks around boreholes. AAPG Bulletin 86 (4):593–604.

Xia CC, Yue ZQ, Tham LG, Lee CF & Sun ZQ (2003) Quantifying topography and closure deformation of rock joints. International Journal of Rock Mechanics and Mining Sciences 40 (2):197–220.

Zhang X & Sanderson DJ (2002) Numerical Modelling and Analysis of Fluid Flow and Deformation of Fractured Rock, 1st edition. Pergamon, Amsterdam, 288 pp.

CHAPTER 39

# Hydrogeologic modelling with regard to site selection for an underground repository in a granite formation in Krasnojarsk, Russia

Christine Fahrenholz[1], Wernt Brewitz[1], Eckhard Fein[1] and Matthias Schöniger[2]

[1]*GRS, Final Repository Safety Research Division, Theodor-Heuss-Straße, Braunschweig, Germany*

[2]*Technical University Braunschweig, Langer Kamp, Braunschweig, Germany*

ABSTRACT: The underground disposal of hazardous waste is a technically and scientifically challenging problem. As long-term impact on the environment cannot be ruled out from the outset, the characterisation of the hydrogeological conditions of waste disposal sites forms the basis of long-term safety assessment. Within the scope of an international co-operation between Russian and German scientists, groundwater flow and transport models were developed for investigations with regard to site selection of a repository for high radioactive and heat-generating waste in granitic crystalline rock in Siberia. The simulations show the preferential flow and transport processes within the fracture zones and emphasise the importance of a detailed knowledge of the fracture system. Due to the lack of site-specific data, the simulations rely primarily on generic data from other sites. For this reason the models are not detailed site-specific investigations, but successive development of these models can result in feasible long-term safety assessment of potential radioactive waste repositories.

## 1 INTRODUCTION

Russian investigations were started in the Krasnojarsk region in 1993 to identify a feasible site for an underground repository that is needed because of the quantity of accumulated radioactive material in Russia (e.g., Anderson et al. 1998, Gupalo 2004, Ljubceva et al. 2002). Based on photogeological, geomorphological and geophysical data, a stepwise reduction of the considered area was carried out, to identify few potential repository sites. In the course of time Russian scientists constructed a preliminary hydrogeological model using also data from Kola Peninsula (Milovidov 1999, VNIPI PT 1997). In 2002 a collaboration of Russian (VNIPI Promtechnologii) and German scientists (DBE Tec, BGR and GRS) started to exchange information and to discuss the safety standards in planning repositories for radioactive material (Brewitz et al. 2005).

Radionuclide transport through the rock is highly affected by the hydraulic regime. Hence for the evaluation and consequent selection of the repository site, it is essential to obtain information about the hydraulic regime based on hydrogeological and climatic measurements as well as groundwater flow modelling. Site-specific data for groundwater flow

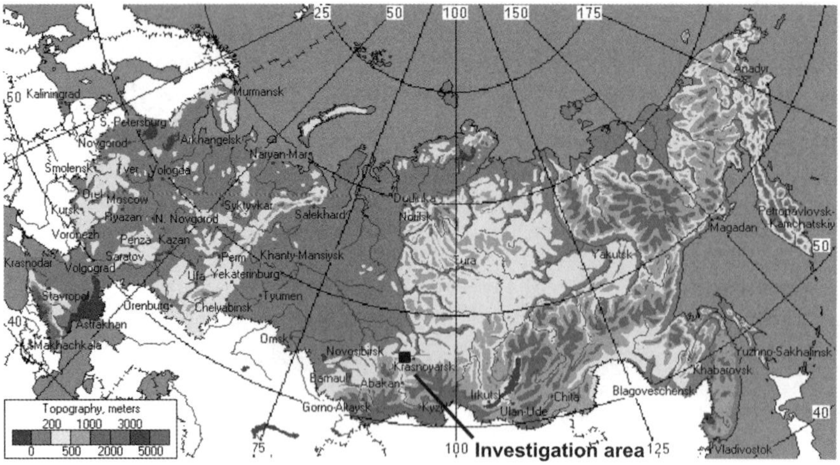


Figure 1. Location of the investigation area.

modelling are sparse (Laverov et al. 2000/2001/2002, Anderson et al. 1999). Apart from a few hydrogeological and borehole geophysical measurements (flow and geoelectric measurements, pumping tests) and geophysical measurements (e.g., seismicity, gravimetry) there are no detailed data about the deep repository rocks, including their fracture distribution and alignment or their general hydraulic characteristics. Therefore only general remarks about the hydrogeological characteristic of the granitic rock at potential sites could be inferred. However, it was decided to create a preliminary model of one of the potential sites with the available data to investigate groundwater flow and possible impacts of the various parameters on flow and transport processes. The lack of data was compensated by using data of granitic rocks from other sites, such as AESPO (Sweden), GRIMSEL (Switzerland) (Lührmann et al. 2000, NAGRA 1994), Shelesnogorsk and expert opinion.

## 2 STUDIED AREA

One of the potential repository sites is located in the Nishnekansky Granitoid Massif (Dolginow 1994, Khain 1994, Zonenshain et al. 1990) in Central Siberia 20 km east of the city of Krasnoyarsk (Figure 1). The target area for the repository site is about 40 $km^2$.

The granite is about 920 Ma old (Volobuev 1961) with an outcrop area of about 1 000 $km^2$ (Brewitz et al. 2005). The Siberian platform is considered seismically inactive (Brewitz et al. 2005). Results from long-term seismic monitoring and geomorphological research in the region confirm that the site is part of a stable platform with low seismic-tectonic activity and low rates of uplift (Brewitz et al. 2005, Kolmogorova et al. 2004, Lukina 2001). Hence tectonic processes have little influence on the geotechnical and hydraulic character of the potential repository area within the massif (Brewitz et al. 2005).

The topography is flat with an altitude ranging between 260 and 460 m. a. s. l. (Brewitz et al. 2005). Climatic measurements show an average precipitation of 550 mm and an evaporation rate between 480 mm and 500 mm (Anderson et al. 1993, Brewitz et al. 2005). The water table reflects the surface topography (Brewitz et al. 2005).

Table 1. Fracture classification (Milovidov 1999, VNIPI PT 1997, Anderson et al. 1996/1998/2001).

| Fracture classification | Horizontal length [km] | Fracture frequency [km] | Influence zone [m] | Fault zone [m] |
|---|---|---|---|---|
| I | >30 | >10 | >300 | >30 |
| II | 10–30 | 3–10 | 100–300 | 10–30 |
| III | 3–10 | 1–3 | 0–100 | 1–10 |
| IV | 1–0.5 | 0.3–0.5 | 1–3 | 0.1–0.3 |
| V | 1–2 | <0.3 | <5 | <0.1 |

The rock complex consists of porous surficial Quaternary and Jurassic sediments overlying the fractured crystalline and disaggregated granite, which can be found in depths up to 150 m (Brewitz et al. 2005). The crystalline rock has very low porosities between 0.29% and 0.52 %. The fracture system contains subhorizontal and steeply dipping fractures with dips of 5° up to 20° and 80° to 90°, respectively (Brewitz et al. 2005). The subhorizontal fractures mainly strike in NW (5°–20°) and steep fractures strike NW-SE (325°–345°) preferentially or occasionally EW. The fractures have a variety of fillings (e.g., quartz, carbonate, clay minerals, aplite, pegmatite, lamprophyre). Milovidov (1999) and Anderson et al (1996/98) characterise fractures in a classification scheme of five classes (Table 1).

The classes differ in fracture lengths and frequencies as well as in their influence zones and dislocations. According to this classification, fractures in the area belong from class three to five.

The hydraulic properties of the fracture zones depend on the regional geological progression, appearing stress, the stress distribution and their occurring time, because mylonitization and mineral alteration and regeneration lead to an extensive healing of the fracture zones. Hence their hydraulic conductivities usually decrease with their age. Because of the fracture age, the tectonic setting and the results of pumping tests, site-specific fractures are expected to be largely filled (Brewitz 2005) and their original fracture aperture decrease with depths (Brewitz et al. 2005).

A potential repository site has to be large enough to accommodate the expected amount of radioactive waste and it is necessary that the site is sufficiently distant from established fracture systems, because their hydraulic activity can not be excluded. Therefore, a buffer zone of 500 m around the large fractures was specified. The extent of this zone with an additional safety factor is deduced from the influence zone of fractures in the classification (Table 1). With regard to the boundary conditions, a model area for the flow modelling of about 11 km$^2$ was selected.

## 3 MODEL SET-UP AND SIMULATION RESULTS

During the project the following models were developed (Brewitz et al. 2005):

- Three-dimensional regional model to obtain information about the groundwater regime, including pathline determination,
- Two-dimensional very simplified generic model for the investigation of transport processes,
- Two-dimensional local model to evaluate the transport of contaminants and energy, and

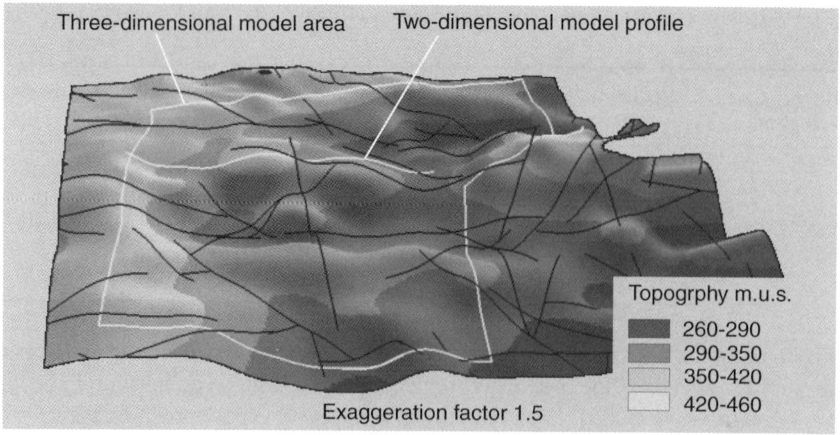


Figure 2. The digital elevation model with boundaries of the 3-D and the profile of the 2-D model.

- One-dimensional model using the EMOS code (Buhmann et al. 1996/1999) to calculate time-dependent release rates of radionuclides and an annual radiation exposure to man in the biosphere.

The model area and profile are shown in Figure 2. The following explanations are related to the two- and three-dimensional models constructed with the FEFLOW code (Diersch et al., 2002, 2004, and 2005).

The models refer to a fractured aquifer overlain by weathering products and sediments. The weathering zone is treated as a porous medium similar to the sediment. A porous-medium approach is used for the rock matrix, which is modelled by two-dimensional or three-dimensional elements depending on model dimensions. Fractures are explicitly modelled as one-dimensional in the 2D-model or two-dimensional elements in the 3D-model. Because of model scale and lack of data, only the large fractures are considered. The processes of advection, diffusion and dispersion are taken into account within the matrix as well as within the fractures. To enhance the numerical results, the elements surrounding the fracture zones are finer than in the remaining area. As far as possible the models are based on realistic data from geophysical measurements (seismicity, gravimetry), remote sensing, field investigations, and hydrogeological and borehole geophysical measurements (flow meter, geoelectric, pumping tests).

### 3.1 *Regional three-dimensional groundwater flow model*

The regional model covers an area of approximately 11 km$^2$ and each of the three layers (sediment, weathering zone and the crystalline formation) is represented by 51 000 elements. The rock formation is divided into sections to minimise numerical errors. The implementation of the geological layers was performed by interpolation and transformation based on investigation results (geoelectric measurements, boreholes, etc.) and the Digital Elevation Model. The hydrogeological characteristics, thickness and conductivities of the modelled layers are listed in Table 2.

Thirty-two fracture zones of 0.5 m and a dip of 90 degrees are implemented as two-dimensional elements. The model boundaries are defined by Dirichlet (constant head) conditions. Groundwater recharge is implemented as inflow at the uppermost layer and is

Table 2. Model layer and their conductivities (Milovidov 1999, VNIPI PT 1997, Brewitz et al. 2005).

| Hydraulic layer | Depth | Permeability [m/s] | |
|---|---|---|---|
| | | Matrix [m/s] | Fracture [m/s] |
| Sediment | 0–50 m | $10^{-6}$ | – |
| Weathering zone | 0–0,1 km | $10^{-7}$ | – |
| Hard rock | 0,1–>7 km | $10^{-11}$ | $10^{-7}$ |

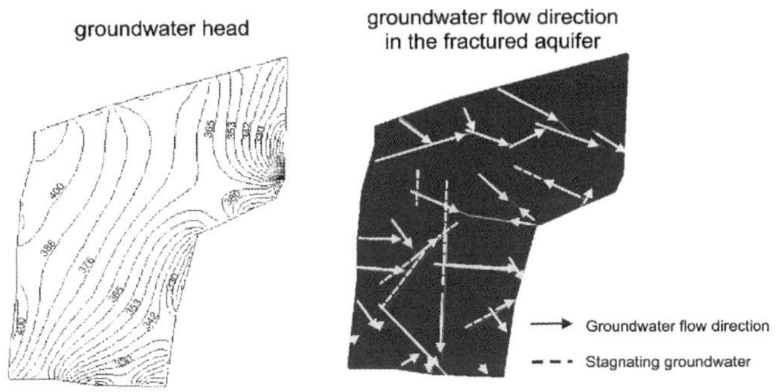


Figure 3. Head in porous aquifer and flow in fractured aquifer.

Table 3. Matrix and fracture permeability in the model calculations (Brewitz et al. 2005).

| Permeability [m/s] | Version 1 | Version 2 | Version 3 |
|---|---|---|---|
| Matrix | $10^{-8}$ | $10^{-9}$ | $10^{-11}$ |
| Fractures | | $10^{-6}$ | |

estimated on the basis of measured evapotranspiration and precipitation data (Anderson et al. 1993, Brewitz et al. 2005). Due to the long time scale of the model calculations, annual changes in groundwater recharge are not considered.

The head and the flow directions in the fractured aquifer are shown in Figure 3. Groundwater flow in the fractured aquifer differs significantly to that in the porous aquifer. Flow within the fractures depends upon their connectivity and orientation as well as on the hydraulic gradient. Tracer pathlines were determined for particles, which start inside the potential repository area to gain information about the transport path and required travel time. The first distance of 600 m rock matrix is passed through in about 2 Ma, the adjacent fracture in 150 000 years. This shows the effective isolation in a mainly undisturbed rock matrix. Hence, the fractures represent the most important potential migration path.

### 3.2 *Very simplified two-dimensional flow and transport models within the hard rock*

The models cover an area of 100 m in depth and 200 m in length. Finite elements are sized between 1 and 2 $m^2$. The simulations were carried out with three different matrix permeabilities (Table 3) using common fracture junctions (Figure 4). Model 1 consists of three parallel and unconnected fractures. In Model 2, one fracture is parallel to the groundwater

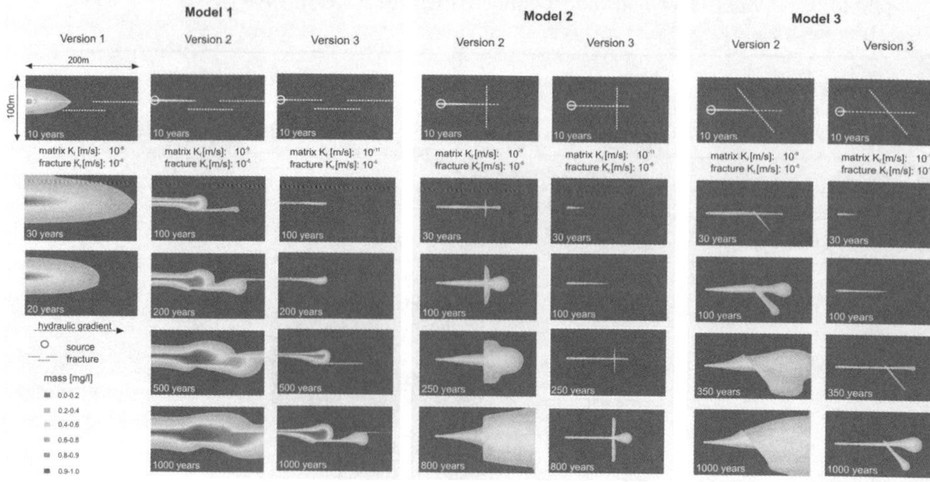


Figure 4. Spreading of tracer starting from a continuous contaminant source for the models 1, 2 and 3 with different permeability versions.

flow direction and the other is perpendicular to it. Model 3 has a fracture, which is parallel to the groundwater flow and it is intersected by a fracture that is not perpendicular. The contaminant is treated as a conservative tracer (no decay or sorption) and the continuous source is located within the horizontal fracture. All models use hydraulic heads decreasing smoothly from west to east with a hydraulic gradient of $5.7 \cdot 10^{-2}$. The matrix permeability in the first version is $10^{-8}$ m/s, which is equivalent to the mean of values, while the third uses the median, which is $10^{-11}$ m/s (Milovidov 1999, VNIPI PT 1997. The permeability of the 0.1 m fracture zone is $10^{-6}$ m/s, the highest value assumed from the Russian scientists.

The results of the different models are depicted in Figure 4 that shows that transport in the fractures is predominantly advective, whereas transport into and within the rock matrix is diffusion controlled. Strong dependencies between the contamination plume and the fracture system can be only detected in case of high permeability differences between rock matrix and fracture network. Therefore, the results of version 1 of the models 2 and 3 are similar to that of model 1. In contrast to version 1, the contamination plumes of versions 2 and 3 of the different models depend on the fracture orientation and differ significantly. The general diffusion process is independent of the groundwater flow and can cause a transverse spread of the contamination plume, although the intensity of the diffusion process depends on the contaminant concentration and, therefore, on the groundwater flow within the fractures. In the versions 2 and 3 of model 1, the transport processes cause an almost uniform distribution of contaminants within the fracture displacement. In model 2, the distribution of contaminants depends on the lengths of the perpendicular fracture. In model 3, an asymmetric distribution depends upon the angle of the fracture to the hydraulic gradient.

## 3.3 *Local two-dimensional flow and transport model*

The two-dimensional flow and transport models showed the flow and transport behaviour in single fractures with simple intersections (Figure 5). At the sites there will be more fractures with a different local density and flow and transport processes have to be superposed.

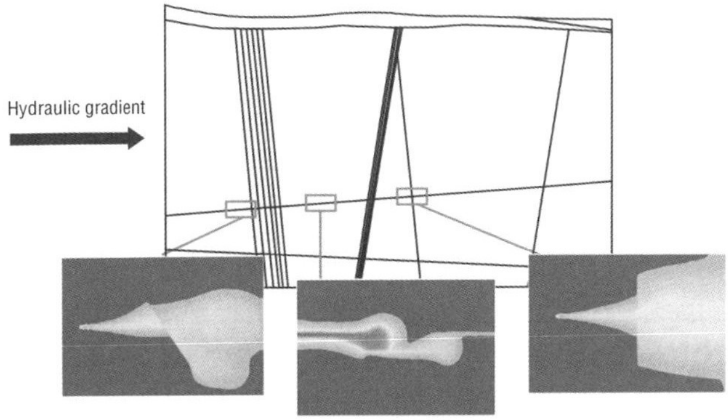


Figure 5. Plumes of different simple fractures within a more complicated regional fracture network.

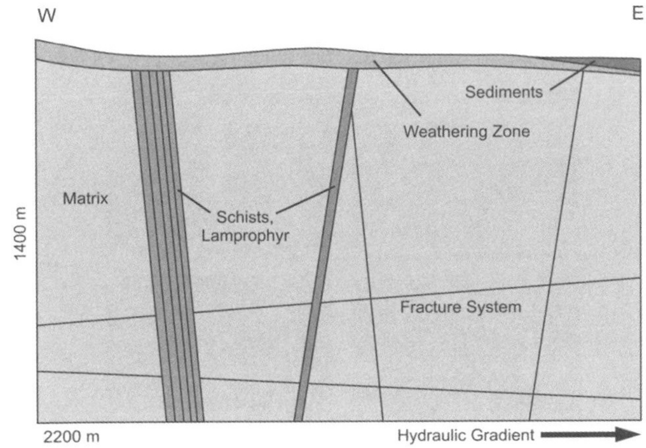


Figure 6. Two-dimensional model profile based on the three-dimensional model.

Hence for an investigation of flow and transport processes in a more complex fracture network, a larger two-dimensional model was constructed. The geological profile of this model is located within the area of the three-dimensional model and aligned with the groundwater flow direction from the potential repository site (Figure 2). The hydrogeological structure has a porous aquifer near the surface containing the sediments, a weathering zone, and an underlying fractured granite (Figure 6). The upper layers are modelled as porous media and the granite as porous-fractured medium. The rock has four single fractures and two highly fractured zones with high fracture density.

Different permeabilities were assigned to the hydrogeological structures (Table 4). The matrix permeability of the highly fractured zones differs from that of the undisturbed rock matrix with numerous small fractures that cannot be considered explicitly.

The model area of 3 $km^2$ is meshed into 61 000 elements. In the vicinity of fractures the grid is finer to guarantee numerical stability even in case of high concentration gradients. The Dirichlet boundary condition is based on the hydraulic gradient of the regional system. The groundwater head declines from west to east. The upper boundary recharges at rates

Table 4. Permeabilities used in the local flow model (Milovidov 1999, VNIPI PT 1997, Brewitz et al. 2005).

| Hydrogeological layer | Matrix permeability [m/s] | Fracture zone permeability [m/s] |
|---|---|---|
| Sediment | $10^{-6}$ | – |
| Weathering zone | $10^{-7}$ | |
| Fractured rock | $10^{-10}$ | $10^{-5}$–$10^{-7}$ |
| High fractured zone | $10^{-8}$ | |

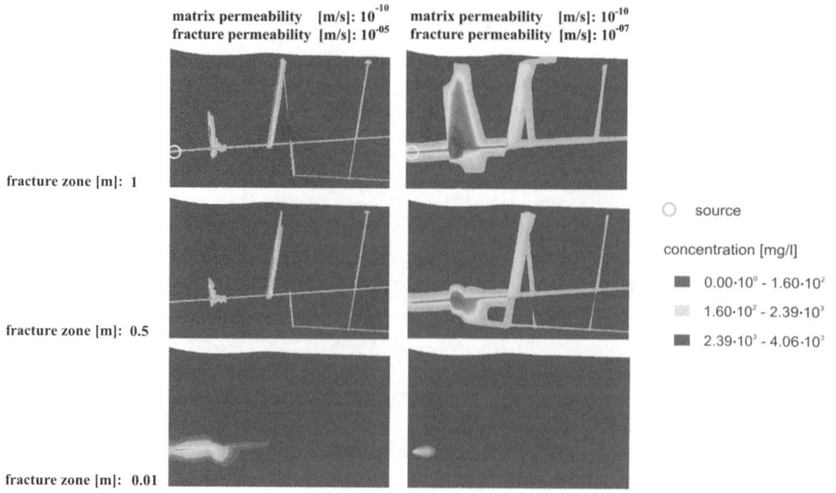


Figure 7. Contamination plume after 10 000 years as a function of fracture permeability.

supplied from precipitation and evaporation data (Brewitz et al. 2005). The bottom boundary is treated as impermeable.

In the local model, two versions with different fracture zones and permeabilities are compared. The contamination source is located in the horizontal fracture near the left boundary. In Figure 7, the plume is shown after a simulation time of 10 000 years. The results demonstrate the decreasing contaminant transport with smaller fracture zones and lower permeabilities, which cause a higher concentration in the fractures and preferential diffusion into the rock matrix.

Figure 8 shows the stepwise contaminant migration in a fracture zone as a function of permeability. This demonstrates that single fractures can have higher flow rates and lower travel times for contaminants than zones of high fracture density. Therefore, it is possible that the contaminants, which were transported through a single fracture, can arrive earlier at the surface, even when the source is closer to the highly fractured zone. In a steep fracture zone, as shown in Figure 8, the flow direction is different in the two model versions (Figure 8, top left and middle right).

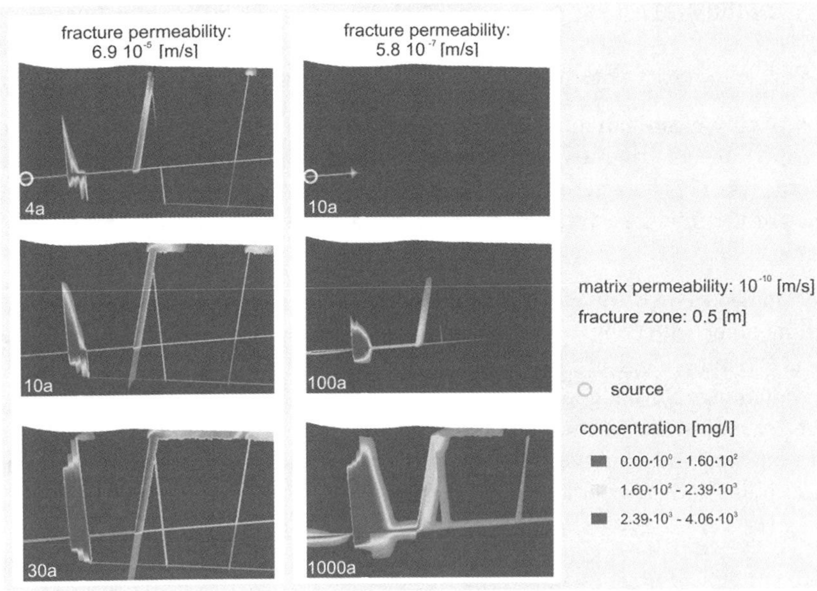


Figure 8. Temporal development of the contamination plume as a function of fracture permeability.

## 4 CONCLUSIONS

The models show both expected flow and transport processes in a fractured rock and some new findings. The contamination transport within the fractures is predominantly advective and in the vicinity of the fractures contaminants diffuse into the rock matrix. The diffusion is related to the local contaminant concentration gradient between the fracture and the matrix and, therefore, it is also correlated to the travel time within the fractures. However, contaminants can persist in the rock matrix depending on diffusion rates. The spread of the plume depends on the diffusion into the rock matrix as well as on the fracture orientation and the lengths of the intersecting fractures. The simulations do not show a strong dependency of the contamination plume on the fracture system in every case. Only in case of large permeability differences between the rock matrix and fracture network do the fractures and their characteristics (i.e., aperture, filling, orientation, frequency and connectivity) influence the flow and transport processes strongly. Single large fractures can have higher flow rates and faster contaminant transport compared with areas with high fracture densities. A very interesting and until yet not shown effect is detected in a very steep fracture zone, where the groundwater flow direction is different, depending on the permeability of the fracture system (Figure 8, top left and middle right).

Because of the lack of site-specific data, these models do not provide detailed site-specific predictions. However, successive approximations and development of these models can assess contamination transport processes and, eventually, the suitability of potential repositories and the annual radiation exposure to man.

## 5 FUTURE PROSPECTS

With respect to the sparse site-specific data it is recommended for future site investigations

- A thoroughly examination of hard rock composition and fracture characteristic (fillings, apertures, distributions, orientations, intersections and alteration areas), and
- A more precisely characterisation of hydrogeological properties, especially the permeabilities of the fractures, fracture zones, and matrix as functions of depth and groundwater age.

It was recommended to complete the present investigation program with seismic exploration, new drilling holes with complex geophysical explorations, acoustic methods, and detailed geological and mineralogical core investigations to construct three-dimensional geological models containing relevant geological data. In addition, data are needed on recharge rates, long term groundwater levels, well tests, geochemical and isotopic analyses, sorption capacities and heat capacities and the influence of heat and rock alteration processes on radionuclide sorption, rock temperatures and permeability (Brewitz et al. 2005).

## REFERENCES

Anderson, E. B.; Dazenko, V. M.; Kedrovskij, O. L. et al.: Geologische Untersuchung auf dem Territorium des Südteils des Yenniseijsker Höhenrückens als Begründung für die Möglichkeit einer sicheren Endlagerung verfestigter hochaktiver Abfälle des Werkes RT-2 in tiefen geologischen Formationen (russ.), unpublished report, Chlopin-Institut St. Petersburg 1993.

Anderson, E. B.; Velitschkin, V. I.; Dazenko, V. M. et al.: Geologisch-geophysikalische Untersuchung des Nordteils des Nishnekansker Massivs mit dem Ziel der Suche von monolithischen Granitoidblöcken mit Perspektiven für die Endlagerung verfestigter hochaktiver Abfälle des Werkes RT-2 (russ.), unpublished report, Chlopin-Institut St. Petersburg 1996.

Anderson, E. B.; Shabalev, S. I.; Savonenkov, V. G. et al.: Investigations of Nishnekansky granitoid massif (Middle Siberia, Russia) as a promising site for deep geological disposal of HLW, Proc. Of Internat. Conf. On Rad. Waste Disp. (DisTec) , Hamburg 1998.

Anderson, E. B.; Dazenko, V. M.; Kirko, V. I. et al.: Resultate der komplexen geologischen Untersuchungen des Nishnekansker Massivs als Begründung für die Möglichkeit seiner Nutzung zur Endlagerung verfestigter radioaktiver Abfälle (russ.) In: Untersuchungen der Granitoide des Nishnekansker Massivs zur HAW-Endlagerung (russ.), Chlopin-Institut St. Petersburg 1999.

Anderson, E. B; Lyubtseva, E. F.; Savonenkov, V. G.; Shabalev, S. I.; Rogozin, Yu. M.: The proposed geologic repository site at the Nizhnekanskiy granitoid massif near Krasnoyarsk, Russia; 9th Internat. High-level Radioactive Waste Management Conference, Las Vegas 2001.

Brewitz, W.; Fahrenholz, C.; Fein, E.; Filbert, W.; Hammer, J; Jobmann, M.; Krone, J.; Wallner, M.; Ziegenhagen, J.; Mrugalla, S.; Lerch, C.; Ward, P.; Weiß, E.; Gupalo, T.; Kamnev, E.; Konovalov, V.; Lopatin, V.; Milovidov, V.; Prokopova, O.: Anforderungen an die Standorterkundung für HAW-Endlager im Hartgestein (ASTER) – Abschlussbericht In: Deutsch-russische wissenschaftlich-technische Zusammenarbeit zur Endlagerung radioaktiver Abfälle 2005.

Buhmann, D.; Storck, R.; Hirsekorn, R.-P.; Kühle, T.; Lührmann, L.: Das Programmpaket EMOS zur Analyse der Langzeitsicherheit eines Endlagers für radioaktive Abfälle – Version 5, GRS-122, 1996.

Buhmann, D.: Das Programmpaket EMOS – Ein Instrumentarium zur Analyse der Langzeitsicherheit von Endlagern, GRS-159, 1999.

Diersch, H.-J. et al.: FEFLOW 5.1 Finite Element Subsurface Flow and Transport Simulation System, White Papers Vol. I-IV, 2002/2004/2005.

Dolginow, J.; Kropatschjow, S.; Klitzsch, E.: Abriss der Geologie Russlands und angrenzender Staaten, Schweizerbart'sche Verlagsbuchhandlung, Stuttgart 1994.

Gupalo, T. A.: Development of quantitative criteria for suitablility of rock mass for safe long-term storage of waste from weapons-grade plutonium production, illustrated by Krasnoyarsk Mining Chemical Combine, Summary Technical Report, ISTC-Project 307B-97, Moskau 2001.

Gupalo, T. A.; Milovidov, V. L.; Muravev, A. A.: Tektonischer Bau des NW-Teils des Nishnekansker Massivs basierend auf Resultaten der Dechiffrierung von Kosmosaufnahmen zur Auswahl von Gebieten für den Bau eines Untertagelabors, In: Savonenkov, V. G.; Zarycinaja, L. G.: Lagerung und Verarbeitung von abgebrannten Brennelementen in der neuen Konzeption des Werkes RT-2, wiss. Tagung, St. Petersburg 2004.

Khain, V. E.: Geology of northern Eurasia, Second part of the Geology of the UDSSR, Phanerozoic fold belts and young platforms, Gebrüder Borntraeger, Berlin 1994.

Kolmogorova, P. P.; Kolmogorov, V. G.: Recent vertical crustal movements in the region of Yenisseij Ridge, Russian Geology and Geophysics, 2004.

Laverov, N. P.; Velitschkin, V. I.; Omeljanenko, B. I.; Petrov, V. A.; Tarasov, N.: Neue Herangehensweisen an die unterirdische Endlagerung hochaktiver Abfälle in Russland (russ.), Geoekologija 2000/1.

Laverov, N. P.; Petrov, V. A.; Velitschkin, V. I.; Poluektov, V. et al.: Petrophysikalische Eigenschaften der Granitoide des Nishnekansker Massivs - zur Frage der Auswahl von Gebieten für die Isolation von HAW und abgebrannten Kernbrennstäben (russ.), Geoekologija 2002.

Lührmann, L; Noseck, U; Storck, R.: Spent Fuel Performance Assessment (SPA) for a Hypothetical repository in Crystalline Formations in Germany, GRS-154, Juli 2000.

Ljubceva, E. F.; Alekseev, E. P. et al.: Bewertung der Granitoide des Gebietes Kamennij als geologisches Milieu für die Langzeitlagerung verfestigter RAW auf der Grundlage von Resultaten der komplexen Interpretation von Geoelektrik-, Geomagnetik- und Gravimetrie-Daten (russ.), Staatliche Universität St. Petersburg 2002.

Lukina, N. V.: Begründung der tektonischen Stabilität des Nordteils des Nishnekansker Granitoidmassivs (russ.), unpublished report, NPZ "Geodynamik und ökologie", Moskau 2001.

Milovidov, V. L.: Bewertung der ingenieur-geologischen Bedingungen des Nishnekansker Massivs für den Bau eines Endlagers für radioaktive Abfälle (russ.), In: Untersuchungen der Granitoide des Nishnekansker Massives für die Endlagerung radioaktiver Abfälle. Tagungsmaterialien, Shelesnogorsk, März 1998, Bergbau-Chemisches Kombinat und Radium-Institut St. Petersburg 1999.

NAGRA: Kristallin-I, Safety Assessment Report, NAGRA Technical Report NTB 93–22, 1994.

VNIPI PT: Entwicklung eines vorläufigen ingenieur-geologischen Modells des Nishnekansker Massivs für die Erarbeitung von Varianten für ein unterirdisches Endlager für HAW (russ.). unpublished report, VNIPI Promtechnologii, Moskau 1997.

Volobuev, M. I.; Zukov, S. I.: Zur Frage des absoluten Alters der Gesteine und Minerale des Yenniseijsker Höhenzuges (russ.): Materialien zur Geologie und zu den Rohstoffen des Krasnojarsker Gebietes (russ.), Krasnojarsk 1961.

Zonenshain, L. P.; Kuzmin M. I.; Natapov, L. M.: Geology of the USSR: A Plate-Tectonic Synthesis, Geodynamics Series Volume 21, American Geophysical Union, Washington D. C. 1990.

CHAPTER 40

# Development and qualification of a smeared fracture modelling approach for transfers in fractured media

André Fourno[1,2,*], Christophe Grenier[1], Frederick Delay[2] and Hakim Benabderrahmane[3]

[1]*CEA (Commissariat à l'Energie Atomique). C.E. Saclay. DM2S/SFME/MTMS – Gif sur Yvette Cedex. France*

[2]*Université de Poitiers. Laboratoire Hydrasa. UMR du CNRS. Poitiers. France*

[3]*ANDRA (Agence nationale pour la gestion des déchets radioactifs). DS/SMG. Parc de la Croix Blanche. rue Jean Monnet. Châtenay-Malabry. France*

**now at IFP (Institut Français du Pétrole). Direction Ingénierie de réservoir. avenue de Bois-Préau. Rueil-Malmaison Cedex. France*

ABSTRACT: Modelling transfers in fractured media remains a challenging task for nuclear waste storage. In this context, flow velocities around a repository are assumed very small which makes matrix diffusion to play a major role in strongly retarding mass transfers. The development and qualification of a novel smeared fracture approach adapted to these conditions is presented. Flow and transport are solved using a Mixed Hybrid Finite Element method, limited here to 2D problems for the sake of simplicity. The geometry of major fractures and matrix blocks is accounted for without handling a huge dedicated meshing. The precision of the method is studied for various test cases, mesh sizes and transport regimes. It is shown that the smeared fracture approach is accurate while limiting computational costs for transfers at low velocity typical of post closure conditions.

## 1 INTRODUCTION

Within the field of nuclear spent-fuel storage, special emphasis is put on experimentation and simulation to improve the modelling capabilities in capturing the transfer of radionuclides within natural fractured media (GEOTRAP 2002; Chapman and McCombie 2003). Several issues make this modelling work a challenging task. Issues include the geometrical variability of the system, the scarcity of available data, and the strong contrasts in parameter values between mobile and immobile zones (Bear et al. 1993). Thus, experimental programs at underground laboratories provide site-specific data bases for testing and modelling programs. These programs are intended to provide a better representation and understanding of major processes. A detailed presentation of the present concerns in terms of research and development for the Äspö Hard Rock Laboratoratory of SKB (SKB, 2004).

This study was initiated within the SKB Task Force to focus on different issues linked to the Äspö site (South West Sweden) using the associated database. This site is one of the best characterized and best documented in the world and provides unique opportunities for realistic modelling exercises. Several international exercises were organized over the last decade within the Task Force to assess the different issues associated with nuclear waste storage. These exercises took advantage of the rich Äspö database and addressed topics related to flow and transport modelling at both local and regional scales (Äspö Web Site; Grenier and Benet, 2002; and Grenier et al. 2004).

The main objective of the study is to bridge between site characterization (SC) models and performance assessment (PA) models. SC models tend to be complex, incorporating detailed physical and geochemical properties, as well as calibrated on or constrained by short-term and small-scale *in situ* experiments. PA models, in contrast, can be simpler, limited to the main physical features, and are generally used to address uncertainties by simulating a range of possible configurations and parameter values. They apply to longer time ranges and larger spatial scales.

Because of the low conductivity of matrix blocks in crystalline, in this case granitic, media, flow primarily occurs along fractures, which are, for radionuclides, the main transfer paths. Transfer in these (more or less) complex conductive units is advective and dispersive. Nevertheless, interactions with the rock matrix and/or stagnant zones limit the velocity or the importance of mass fluxes. One of the key retention processes to be considered for transfer into low-permeability formations is diffusion into immobile zones. Fractions of the plume are temporally diverted from primary flow paths by diffusion into rock blocks (porous portions of the rock with negligible flow, fracture infilling materials, depositions onto fracture walls, dead end pores, etc.). This process is typically referred to as the matrix diffusion effect (Neretnieks 1980; GEOTRAP 2002). The importance of this phenomenon increases as the diffusion coefficient values increase, or with increasing contact time between the plume and the host rock. For a post-closure situation (natural flow and repository installed within low conductivity zones), the water velocity is very slow. As a consequence, matrix diffusion is likely to become an important mechanism in radionuclide transport.

Transfers in fractured media have already been subject to intense modelling work (e.g., Bear et al. 1993, Berkowitz 1994). Nevertheless, models incorporating detailed fracture geometry were mainly developed for flow situations encountered in rapid transfers more typical of experimental test cases. In particular, discrete fracture network approaches represent flow and transport in a fractured network by means of a discrete network of simple interconnected features. Such models are computationally efficient and can account for various transport phenomena. Matrix diffusion is implemented similarly to a retardation factor along each connecting feature and corresponds to a limited or unlimited orthogonal 1D diffusion. Nevertheless, these models are not adequate for slow transfers in a fractured block since matrix diffusion is no longer limited to the close vicinity of the fractures but is able to connect wide portions of the fractured block. The actual block geometry should then be taken into account.

Our smeared fracture approach involves a continuum approach to the geometry of the fractured rock allowing for actual geometry of the matrix zones to be represented. The basic philosophy is to provide a versatile tool for system calibration (based on *in situ* pumping and tracer tests). The main features of the fractured block are represented (main conductors at the scale of the studied block) whereas minor fracturing is homogenized. Identification is then reduced to a limited number of properties associated with some major features. Nevertheless, the geometry of these features can be easily modified for calibration since no dedicated meshing of the geometry is considered. The method handles a regular meshing over which specific hydrodynamic properties make the difference between fractures and matrix blocks.

The basic idea of the approach is not to mesh the fracture network but to consider the presence of fractures by means of continuous heterogeneous fields (permeability, porosity, head, velocity, concentration, etc.). This approach, previously followed by others (e.g., Svensson 2001 and Tanaka et al. 1996), is referred to as the smeared fracture approach and has an important advantage since no dedicated spatial discretization effort is required (a regular mesh is used and simulations can be done on a rough grid, which saves computer time). This makes this approach very promising when accounting for heterogeneity and prediction uncertainty within a Monte Carlo framework. Furthermore, the geometry of the matrix blocks where transfers proceed by diffusion is fully taken into account contrary to classical simplified 1D approaches. Nevertheless, the continuous heterogeneous field representation of a fractured medium requires a homogenisation process at the scale of the handled mesh and constant mesh size might not be appropriate to simulate contrasted transitory transport regimes. In the following, the principle of the method is presented as well as its development and qualification. The approach was implemented and tested in our simulation code, Cast3M (Cast3m Web Site).

## 2 DEVELOPMENT OF THE SMEARED FRACTURE APPROACH FOR THE MHFE SCHEME

### 2.1 *Geometrical features*

The first step consists in selecting within the regular-grid the meshes containing a fracture. In doing so, care is taken to assure connectivity across the elements to form flow tubes. In a 2D space, fractures are considered as piecewise linear objects (segments). An example is provided in Fig. 1.1 and Fig. 1.2 for the 4 conductor case considered below: the meshes of the regular grid intersected by a fracture are identified (voids on the figure) and assembled to form a connected tube-like geometry.

### 2.2 *Steady-state flow problem*

The equations governing steady-state flow are classically, for q ($ms^{-1}$) Darcy velocity, K ($ms^{-1}$) permeability, h (m) head, s ($m^3s^{-1}$) source term.

$$\begin{aligned} \vec{q} &= -\overline{\overline{K}}\vec{\nabla}h \\ \vec{\nabla}\vec{q} &= s \end{aligned} \tag{1}$$

Equation 1 is solved with the smeared fracture approach but in using equivalent permeability values. These values are associated with the elements selected and computed to provide net water flux conservation. The homogenization procedure is applied at the fracture level and directly allows for exact flux at the outlets. This differs from other smeared fracture approaches (Tanaka et al. 1996; Svensson 2001) treating the homogenization process as averaging at the level of each mesh and using the well known Finite Volume scheme to differential equations. Our approach allows for exact restitution of the total water flux transiting through each fracture considered individually. Another advantage is that the lateral extension of a fracture is roughly limited to the size of one mesh (Fig. 1.2) whereas other smeared fracture approaches implemented with equivalent properties based on volume

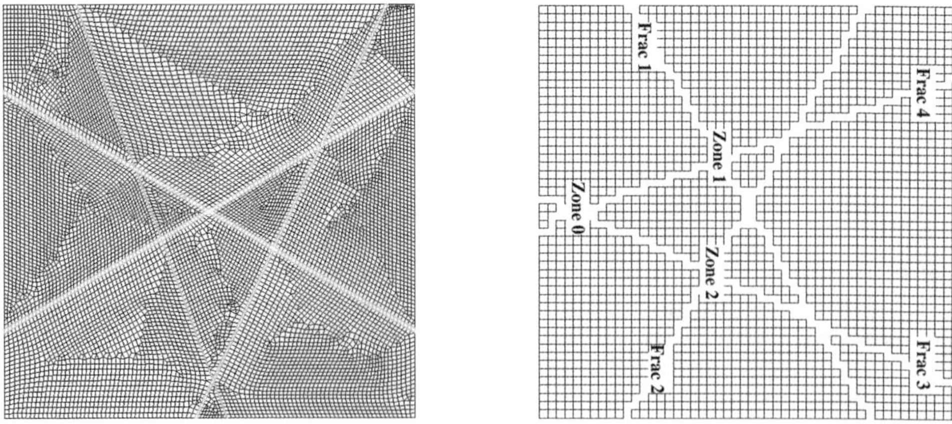


Figure 1. Four fracture system (block size of $200 \times 200$ m): (a) discretization for the reference calculation: (b) smeared fracture approach, $\Delta = 4$ m.

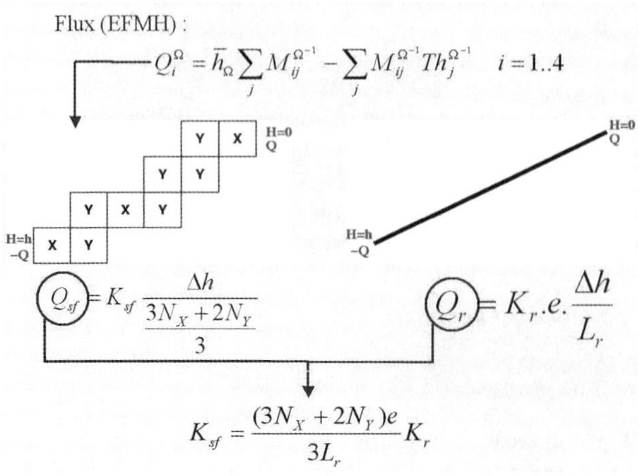


Figure 2. Basic concept of the smeared fracture approach.

averages and a finite-element scheme for flow resolution lead to larger lateral smearing of the conducting features (Svensson 2001). The remaining point is to choose a parameter value for fracture intersections: the maximum of the considered properties is applied. This is a deviation from the strict equivalence obtained for single conductors and the accuracy of a simulation at the scale of an entire network is to be tested as regard reference calculations.

Homogeneity of fracture properties are assumed here but variations of properties along any of the conductors could easily be taken into account. The actual homogenization procedure used for the flow problem (determination of equivalent transmissivity) is shown in Figure 2.

A mixed and hybrid finite-element (MHFE) scheme (Mosé et al. 1994) is used as flow solver. This numerical scheme provides accurate mass conservation even for highly heterogeneous media. This conservation is guaranteed for each mesh by the introduction of four unknowns also called traces and affected to the centre of the mesh edges in addition

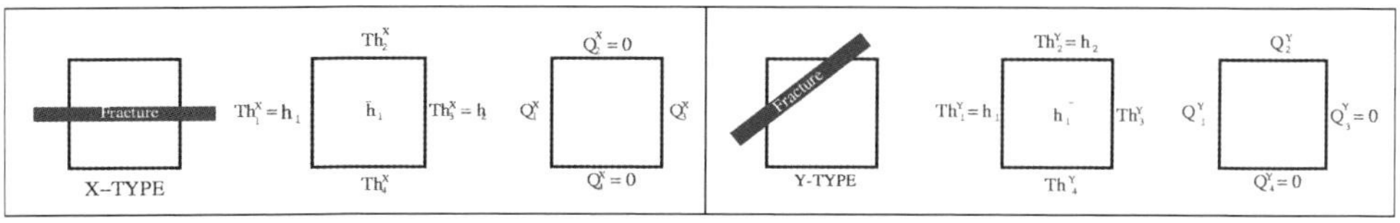

Figure 3. X- and Y-type cells.

Table 1. Numerical flow for X and Y elements.

| X-type cell | Y-type cell |
|---|---|
| $Q_1^x = -Q_3^x = -K_{sf}\Delta h$ | $Q_1^y = Q_2^y = -\frac{3}{2}K_{sf}\Delta h$ |

to the classical unknowns corresponding to the centre of the meshes. These traces allow for joint estimation of fluxes (gradients) as well as values of unknowns within the mesh. Steady-state flow is only calculated in the fractures, the matrix blocks being considered as impervious. For a single conductor (Fig. 2), heads at upstream and downstream tips of a fracture are related through a 1D tube involving two types of elements:

- X-type cells for which flow entering one side exits through the opposite side.
- Y-type cells for which flow enters and exits through adjacent sides.

These two types of cells and their boundary conditions are illustrated on Fig. 3.

Considering both the boundary conditions and the discretized Darcy flux for MHFE scheme, it is possible to find outlet flux for each cell type (Table 1).

The flow tube being only composed of the two types of cells and considering the mass balance between two cells, it is possible to calculate the tube numerical flow. The equivalence between the total flux through the reference fracture and the equivalent conductor is written based on equality between the real flux in the fracture and the numerical one in the tube (Fig. 2).

For the discretization scheme used, the relation between flow and head difference at the inlet and outlet can be easily computed. For a given angle, the number of cells included as a series in each flowing tube ($N_X$ of the X-type and $N_Y$ of the Y-type) can be counted and leads to the equivalent permeability:

$$K_{sf} = \frac{(3N_x + 2N_y)e}{3L} K_r \tag{2}$$

where e, L and K are the thickness, the length and the permeability of the fracture, respectively. The subscripts sf and r are related to smeared fracture equivalent property and reference property, respectively. For the sake of illustration, considering a horizontal fracture leads to $N_Y = 0$ and knowing that $N_X\Delta X = L_r$, the former expression simplifies to: $K_{sf}\Delta X = K_r e$. This equation expresses equality in transmissivities providing identical total water fluxes. This equivalence is exact for a single conductor. Nevertheless, dealing with several fractures introduces deviations from this due to the treatment of intersections.

For an intersection of several fractures, maximal transmissivity is considered. The method is assessed below by comparison with reference calculations.

### 2.3 *Transport problem*

Two distinct transport modelling strategies were implemented for simulating transport in a fractured rock with a smeared fracture approach: a Lagrangian method and an Eulerian one.

The Lagrangian method is not well suited to the actual flow field for the smeared fracture approach. Indeed, for Y-type meshes, the variability of Darcy velocity over the whole grid is very large, leading to extremely high and low transit times depending on the location of the particle entering the mesh. This difficulty was circumvented by constraining transport paths to controlled flow lines. This is achieved for a release position at the centre of the element edge leading for X and Y type elements to exit positions similarly located at the centre of the edges. Associated travel times are easily computed and yield values of the equivalent porosity consistent with volume ratios. This method was implemented in Cast3M for pure advection allowing for the computation of streamlines. It is computer efficient and was tested with success as regard reference simulations. It requires a theoretical derivation of equivalent porosity (based on summation of transit times associated with $N_X$ X-type meshes and $N_Y$ Y-type meshes along a single fracture) as well as slight modifications of the particle tracking procedure available in the Cast3M code. However, this approach cannot be extended to other processes such as a diffusion because particles would move to other flow lines than the one selected. Thus, the Lagrangian approach is limited to purely advective transport in the fracture network.

The Eulerian approach has more potential in this situation and better fits into the expected transport regimes for long-term transfers in a low permeability zone hosting a waste storage. These regimes include purely diffusive process for the matrix blocks, advective and dispersive processes inside fracture network and sorption for both. For the MHFE scheme implemented, advection is introduced as a source term in the classical Eulerian equation (Dabbene et al. 1998). As a consequence, one does not face the same problems as for the Lagrangian approach. Inclusion of diffusion in matrix zones is straightforward, matrix blocks are meshed and corresponding properties are assigned. One of the most interesting advantages of this approach is that matrix blocks geometry are actually represented. This approach is appropriate for low Peclet Numbers or when transport regimes in the fracture network (advection dominated) and in the matrix blocks (purely diffusive in our approach) are not too much contrasted in terms of characteristic residence times. These conditions are met for slow motion of the water in the system. This is the case for post-closure transfer regimes. The Eulerian approach to transport in a fractured block is developed below.

The mass transfer including advection and dispersion (or diffusion) as well as linear sorption obeys the general following equation:

$$\omega R \frac{\partial C}{\partial t} = \vec{\nabla}.(\omega \overline{\overline{D}} \vec{\nabla} C - C\vec{q}) \tag{3}$$

where C stands for concentration (kg/m$^3$), R for retardation coefficient, is $D = d_f + \alpha(q/\omega)$ the dispersion tensor, $\omega$ the porosity, d the pore diffusion coefficient (m$^2$/s), $\alpha$ the dispersivity (m), and q the Darcy velocity (m/s). Transfers within the fracture network and matrix blocks are considered.

The smeared fracture approach for transport relies on the steady state flow modelling and its equivalent parameter estimation presented above. The transport phase requires the estimation of other equivalent properties associated with the fracture elements (i.e., dispersion tensor and porosity assumed as a scalar).

The final transition time through the system adds all the transfer times for all elements met on the path. Thus, the exact equivalent porosity is provided by $\omega_{sf}$:

$$\omega_{sf} = \frac{V_{fr}^{r}}{V_{fr}^{sf}} \omega_{r} \tag{4}$$

where $\omega_r$ stands for real fracture porosity, $V_{fr}$ for fracture volumes associated with smeared fracture meshes (sf subscript) or actual fracture (r subscript).

The equivalent dispersion tensor is here derived as a scalar value based on the same considerations as those for the derivation of equivalent transmissivity. $N_X$ is the number of the X-type meshes and $N_Y$ the number of Y-type ones. The dispersive part of the transport equation (dispersive mass flux) is analogous to the flow equation (water flux) yielding the following diffusion dispersion coefficient for a single conductor:

$$D_{sf} = \frac{(3N_x + 2N_y)e}{3L_r} D_r \tag{5}$$

When dealing with fracture networks, the same procedure as for flow is applied to fracture intersections, the dispersion coefficient assigned is the maximum value calculated in the intersection mesh.

## 3 QUALIFICATION STRATEGY

The smeared fracture procedure allows for *a priori* exact computation of water and mass fluxes in single fractures, but the approach has to be qualified for reference calculations. The most obvious source of difference is expected for coarse meshing when modelling a finely fractured rock: digitization of the geometry is not sufficient to represent the actual complexity. A criterion assessing the lower limits in mesh size to satisfyingly account for such geometrical aspects is introduced below. Qualification is required because of numerical constraints which are twofold. First, to achieve numerical stability and precision of the results, mesh Peclet numbers should be lower than unity. This means that the mesh size should be lower than dispersivity or equivalently that the coarser the mesh, the more dispersive the transport for the smeared fracture approach. Second, transport is considered in highly contrasted media, including slow diffusive transfers in the matrix blocks and comparatively rapid transfers in the fractures. In the smeared fracture approach, this system is represented with a single mesh size whatever the medium, fracture or matrix. This possibly leads to lack of precision in the simulations or even instabilities and requires a few tests to assess the limits of the approach. Criteria quantifying the level of numerical precision are introduced below.

The smeared fracture approach was tested through a series of configurations including different fracture network geometries as well as a variety of parameter values and flow

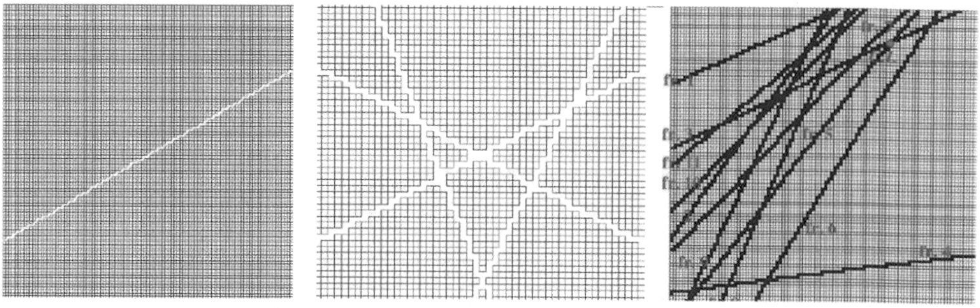

Figure 4. Set of studied geometry with increasing levels of complexity and realism.

velocities. The geometries tested for 2D cases are given in Fig. 4. They include a single fracture configuration (for checking the code implementation), a four fractures academic case, a more realistic 2D fracture network inspired from the situation at the Äspö underground laboratory (200 m Block Scale Experiment).

The basic idea for the qualification study is to make to vary time and space discretization for smeared fracture simulations and compare the results with reference simulations. References are obtained from either analytical solutions or computed solutions over a huge dedicated meshing.

## 4 FLOW (LIMITED TO FRACTURE NETWORK)

Steady-state flow simulations were carried out for various geometry and flow parameters. Main results are: 1) the smeared fracture approach yields accurate fluxes for single fractures as well as good precision on fracture networks provided a sufficiently refined meshing is used; 2) total fluxes exiting the system are very robust (low sensitivity to the mesh size). The same conclusion holds for fluxes within main conductors that vary within some percents even for coarse meshing; and 3) fluxes in minor conductors can locally vary more strongly (ten to twenty percent). As a consequence, the main constraint on discretizing size is geometrical and rather intuitive: the mesh size should be small enough to account for the actual connectivity of the conductors.

More difficulties occur in modelling transport. Although flow is considered as constant over time, transport is transient, which leads to specific numerical constraints.

## 5 TRANSPORT IN A SINGLE FRACTURE SYSTEM

For the single fracture system of Fig. 4, the transport of a point source concentration injected in the middle of the fracture is simulated. Breakthrough curves for transported mass flux exiting the block are represented in Fig. 5. The mesh size is made to vary between 0.5 and 3 m and the "reference" calculation in Fig. 5 it that for a mesh size of 0.4 m. With a rough meshing, the concentration peak levels decrease and breakthrough curves show a larger spreading in time. Nevertheless, peak arrival times show limited sensitivity to mesh size. Note that this first validation scenario only considers transport (advection and dispersion processes) in the fracture and the curve spreadings are due to numerical dispersion.

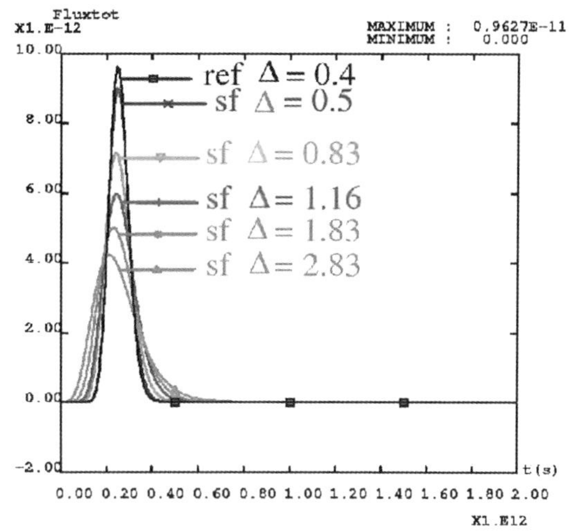


Figure 5. Total fluxes at the outlet (no matrix diffusion, single fracture case).

For transient problems such as transport, two numerical criteria must be considered: (1) the Courant or CFL Number, applying to advective transfer and (2) the Fourier Number, associated with diffusive transfer. Ratio of both is the Peclet Number. Implicit scheme is considered for diffusive transport. This allows for resolution with large time steps but introduces a minimal time step value to preserve the monotony of the solution and achieve good precision for the results. The limit is provided in the following expression involving the Fourier Number:

$N_C = \frac{q\Delta t}{\omega\Delta} < 1$ (CFL criterion) and $F_o = \frac{\alpha q \Delta t}{\omega \Delta^2} > \frac{1}{2}$ (Fourier criterion for $D = d_f + \alpha\frac{q}{\omega} \approx \alpha\frac{q}{\omega}$), where $\Delta t$ is the time step and $\Delta$ the regular mesh size. Ratio between both expressions corresponds to grid Peclet:

$$\frac{\Delta}{\alpha} < 2 \tag{6}$$

This expression implies that grid size should be lower than dispersion coefficient. In other words, transport cannot be simulated accurately with a dispersivity value lower than the mesh size.

## 6 TRANSPORT IN A SINGLE FRACTURE INCLUDING DIFFUSION IN MATRIX BLOCKS

This geometry (see Figure 4) was tested for three flow regimes leading to breakthrough curves provided in Fig. 6a: (i) a dominantly advective regime K1 (negligible matrix diffusion), (ii) an intermediate regime K2, (iii) a dominantly diffusive regime K3 (large matrix diffusion). In Fig. 6a, the breakthrough curves are scaled onto the peak level of the slowest

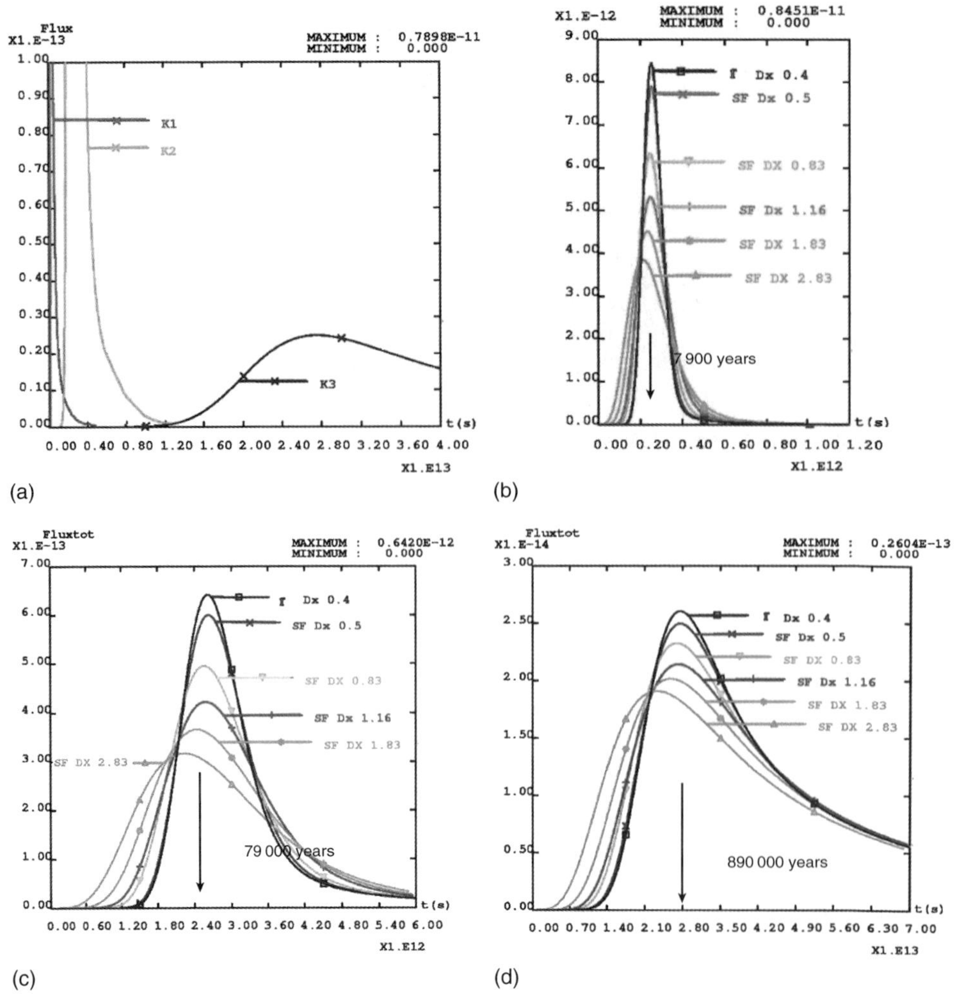


Figure 6. (a) Output mass flux for the three flow regimes: (b) Total fluxes at the outlet for regime K1: (c) Total fluxes at the outlet for regime K2: (d) Total fluxes at the outlet for regime K3.

transport conditions. Matrix diffusion leads to strong delays in peak arrival times. Modifications of the shapes of the breakthrough curves are also observed, with a strong tailing effectdue to the slow mass release from matrix zones.

In Fig. 6b to 6d, the sensitivity of the breakthrough curves at the limits of the domain is presented for the smeared fracture approach (different spatial discretizations) and for the reference simulation (achieved by means of a solution computed over a refined meshing, see black curves with squares).

With a weak matrix diffusion (Figure 6b for K1), results are similar to those without matrix diffusion (see above). For higher matrix diffusion (cases K2 and K3 of Figure 6c and 6d), classical tailing effects appear. For identical grid size variations (50 cm to nearly 3 m), discrepancies in peak level, arrival times and spreading of the curves diminish from K2 to K3. This means that for slower transport regimes in the fracture, refinement of grid size is not necessary to match a given precision level of the results. These observations are explained by considering an additional criterion, which is the Fourier Number associated with diffusion in matrix zones. In line with former explanations, matrix Fourier Number should be large enough: $F_o = (d_m \Delta t)/(\Delta^2) > 1/6$, that is, for a given time increment, spatial discretization should be sufficiently small. Nevertheless, breakthrough curves in Fig. 6as show that matrix diffusion introduces delayed peak arrival times. When velocity in the fracture decreases, the mean transit time also decreases and is more controlled by matrix diffusion. Time steps calculated from a CFL number should include this retardation effect (time steps should be chosen larger). This delay can be estimated as a retardation coefficient based on the analytical solution to transport by advection in a single fracture and 1D orthogonal diffusion in an infinite matrix. This solution writes for a continuous injection at the inlet of the fracture (Neretnieks 1980; Bear et al. 1993):

$$\frac{C(t,t_w)}{C_0} = \mathrm{erfc}\left(\frac{\omega_m\sqrt{d_m t_w}}{e\omega_{fr}\sqrt{t - t_w}}\right) - \mathrm{erfc}\left(\frac{\omega_m\sqrt{d_m t_w}}{e\omega_{fr}\sqrt{t - t_0 - t_w}}\right) \tag{7}$$

with $t_w$ the travel time by pure advection in the fracture. The output peak arrival time $t_s$ can be obtained from the derivation of Equation (7) and the associated retardation coefficient $R_P$ writes:

$$\left\{\begin{aligned} R_p &= 1 + \frac{2}{3}\frac{\omega_m^2 d_m}{(\omega_{fr} e)^2 t_w} \\ t_s &= R_p t_w \end{aligned}\right\} \tag{8}$$

For a velocity,$q/\omega_{fr}$, and a contact incremental time $\Delta t = (5R_p t_s)/(100)$ (i.e., in considering that the maximum simulation time is five times the estimated peak arrival time and that 100 time steps are computed for this time lag), the associated penetration length in the matrix can be obtained from the following equation:

$$\begin{aligned} \eta^2 &= 2d_m \Delta t \\ &= \frac{10 d_m R_p^2 t_w}{100} \end{aligned}$$

Thus, the Fourier Number previously constrained to values larger than 1/6 yields a constraint on $\Delta(\Delta < \sqrt{3}\eta$, expressed in terms of $\eta$, or, expressed in terms of matrix properties:

$$\Delta < \sqrt{\frac{3}{10} d_m R_p t_w} \tag{9}$$

This criterion adds to the former ones when matrix diffusion is included. Note that this criterion for matrix diffusion imposes a small discretization to perform accurate calculations which make the latter more time-consuming. This criterion is easier to satisfy when matrix diffusion coefficient, retardation coefficient associated with the tracer and additionally water arrival times are large. Large arrival times are achieved for long travel distances and/or low fluid velocities. This is precisely the situation considered for post closure natural flow in low-permeability fractured media.

Finally, this approach is very interesting for long time scale and post-closure simulations. A smeared fracture approach with coarse grids remains efficient with low computation costs when the diffusive regime is dominant among transport mechanisms. For other regimes and accurate results, a finer discretization is required.

## 7 TRANSPORT IN THE FOUR FRACTURES SYSTEM WITHOUT MATRIX DIFFUSION

The smeared fracture approach for Eulerian transport within a fracture network was tested on the 4 fractures geometry (Figures. 1 and 4). The results were compared to simulations performed over the dedicated meshing depicted in Fig. 1a. The flow direction is from bottom left to upper right corners of the domain. The initial transport conditions are a point instantaneous injection of a unit concentration in one mesh of the domain. Several release positions were studied corresponding to the different zones mentioned on Fig. 1b. Focus is put here on the Zone 0 case, located upstream at the intersection of two main fractures (numbered 3 and 4). Globally, the plume first migrates within both conductors at different velocities, then separates at intersections and reaches finally the outlets at the upper right corner of the domain. Different paths are therefore experienced which leads to the dispersion of the plume over the network. The quantities measured are breakthrough curves at the limits of the domain, concentration fields at different times, temporal evolution of the masses in the fractures and at fracture intersections. A concentration field before peak arrival time is given in Figure 7a for the smeared fracture simulation.

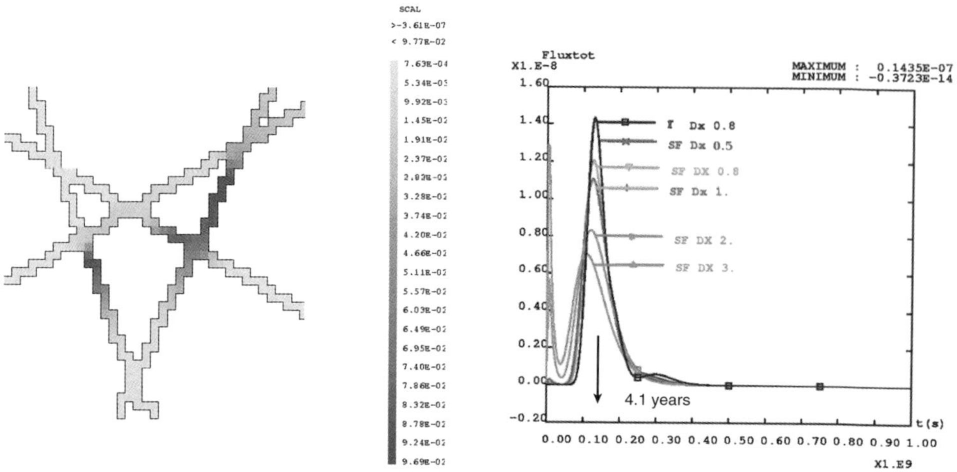


(a) (b)

Figure 7. (a) Concentration field at a given time: (b) Output flux. Reference vs. smeared fracture.

In this study, different levels of discretization have been considered and the limitations of the smeared fracture approach are revealed in two ways:

- Two curves (Figure 7b) show abnormal evolutions with a very early exit time followed by a classical bell-shaped time distribution. These curves correspond to coarser discretizations for which the geometry of the fracture network is badly represented, which, in turn, yields an artificial excess of connection between the network and the edges of the domain.
- The other breakthrough curves from finer discretization (an example with meshes of 1 m on a side is given in Figure 7a) provide expected results including limited numerical dispersion compared with the reference calculation. Fig. 7b a second peak of weak intensity but the smeared fracture approach tends to smooth this kind of detailed feature. This small peak is related with mass arrival from a minor travel path that is not distinctively represented by the smeared fracture approach.

When studying in details the eventual discrepancies in the fracture network, for instance within single conductors, results show (not illustrated here) that the mass sharing and the residence time distribution in minor conductors is less precise than in the major fractures. This is mainly related to larger relative errors on the flow field inside branches of the network with small water fluxes. Nevertheless, whatever the approach (reference or smeared) it is necessary to choose the time step carefully. Since main transport features are related to major conductors, CFL are calculated according to velocities in these units. Precision for advective transport in minor units is necessarily reduced.

In summary, results show that the grid size should be chosen small enough to account for a good digitization of the network geometry. The other effects related with coarse discretization are overall smoothing of breakthrough curves. Nevertheless, if we pay attention to these limits, the approach is promising. Whatever the initial position of the plume, peak arrival time is met with errors close to 5% whereas peak maximum more largely differs (0% to 22% depending on the discretization level and therefore on the numerical dispersion by the MHFE). In terms of computational costs, a transport scenario over a smeared fracture discretization preserving accuracy is calculated more rapidly, about a factor two, than the reference simulation.

## 8 TRANSPORT IN THE FOUR FRACTURE SYSTEM INCLUDING MATRIX DIFFUSION

Several test cases were considered for different flow velocities. Figure 8 shows the breakthrough curve collecting all outlets at the domain boundaries. The solid line with squares corresponds to the reference calculation. Matrix diffusion leads to very a smooth breakthrough curve for the total flux exiting the domain and there is almost no difference between curves computed with various mesh sizes.

The conclusions drawn from the single fracture case can be extended to this more complex 4 fractures network. For different transport regimes, precision in peak level and arrival time is within 20% of reference case for lower CPU time (refer to Table 2 for more quantitative results). Precision of the results increase here with the importance of matrix diffusion.

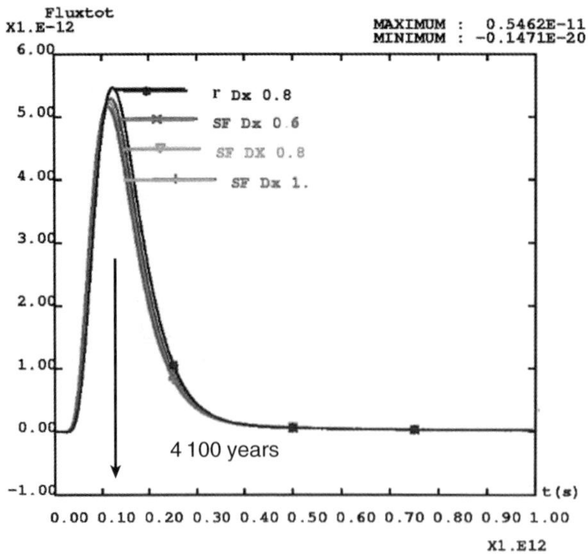


Figure 8. Total flux at the outlets for the more diffusive regime (K3).

Table 2. Peak level relative error and CPU time reduction.

| | Dx = 0.6 m (%) | Dx = 0.8 m (%) | Dx = 1 m (%) |
|---|---|---|---|
| K1 | 0 | 16 | 23 |
| K2 | 0 | 15 | 21 |
| K3 | 3 | 4 | 5 |
| CPU time reduction | 70 | 40 | 30 |

## 9 2D SYSTEM FROM A BLOCK AT ÄSPÖ

A more realistic system is considered here (for both geometry and parameter values) to demonstrate the applicability of the approach to real cases. The 2D Äspö case considered here is that of a section across the actual 3D geometry of a 200 m fractured block at the Äspö underground laboratory (see Figure 9a). The initial conditions (I.C.) correspond to a pulse unit mass injected at a fracture intersection (see Fig. 9b). A head gradient of $10^{-3}$ is taken, in agreement with local natural flow conditions (from upper right to bottom left corners in Fig. 9b).

In a first step, the system was modelled without matrix diffusion. The breakthrough curve collecting all outlets at the domain boundaries show several arrival peaks. They correspond to different paths followed by the plume. The smeared fracture approach yields results matching fairly well reference calculation again performed over a dedicated meshing (see Figures 10a and 10b for the outlet of fracture #8).

In a second step, matrix diffusion is added and the results shown in Fig. 10c for the smeared fracture approach. Reference calculations were not performed in this case. Qualitative evolution of mass flux confirms intuition; the role played by the matrix diffusion is classical, yielding: (1) an arrival time delay, (2) a smoothing effect, (3) a decrease of

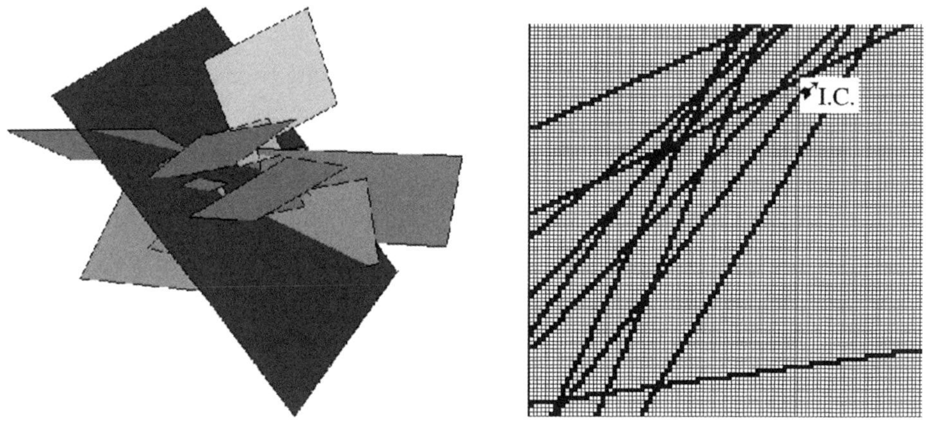


Figure 9. (a) Äspö fractured block (200 × 200 × 200 m) and section plane of 2D network: (b) 2D fracture network (a section across the 3D block).

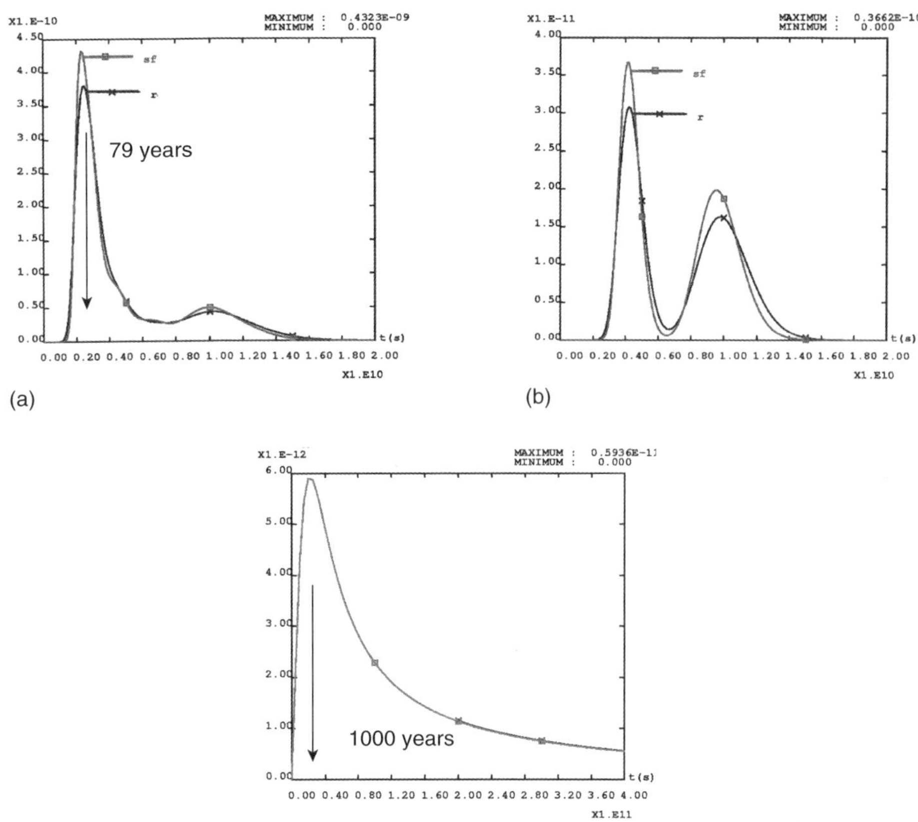


Figure 10. (a) Total mass flux exiting the block: (b) Outlet mass flux, fracture 8: (c) Total mass flow (with matrix diffusion).

peak value and 4) a tail. Moreover, peak arrival time is close to the analytical estimate from Equation (8).

Results show that the smeared fracture approach can cope with a realistic test case. It provides overall good results with lower computer costs:

- The peak arrival times are in agreement with the reference case or estimates from analytical expressions.
- The precision in peak value depends on discretization (numerical dispersivity) and can be controlled on the basis of numerical criteria in Equations. 6 and 9 and on previously discussed geometrical considerations.
- When the diffusion effect smoothes the breakthrough curves, accuracy of smeared fracture calculations improves.

## 10 CONCLUSIONS AND PERSPECTIVES

This smeared fracture modelling approach provides relevant results in terms of precision, computational costs, and numerical stability. The quality of the results depends closely on transport regimes in the fractures and on spatial discretization. The domain in which the smeared fracture approach is the most efficient is that of long-term transport (low velocities) when explicit 3D modelling of matrix block geometry is required due to large penetration depths of the concentration into the rock. Nevertheless, a first criterion for the quality of the simulation is that spatial meshing should be sufficiently refined to represent fracture network geometry or more precisely connectivity. Second, equations 6 and 9 provide the higher bounds of the mesh size that guarantees the capture of transport phenomena within fracture network and matrix zones with sufficient precision of the simulations. The next needed step in the development of the smeared fracture model is to address 3D geometry. Meshing and definition of equivalent properties are already available. The code has been tested and qualified for simple cases. An application to the Äspö site (Sweden) is in progress. Another future prospect is relative to the implementation of matrix diffusion for higher velocity cases or shorter time scales. This will be achieved within our Eulerian scheme by means of semi analytical expressions for source terms corresponding to 1D orthogonal diffusion processes.

## REFERENCES

Äspö Web Site. Web site for the Äspö Task Force under www.skb.se/templates/SKBPage____2636.aspx.

Bear J, Tsang C.F, De Marsily G (ed.) (1993) Flow and contaminant transport in fractured rock. Academic Press.

Berkowitz B (1994) Modeling flow and contaminant transport in fractured media. Advances in Porous Media. Volume 2. Corapcioglu Ed. Elsevier.

Cast3m Web Site. Web site for computer code Cast3m, under
www-cast3m.cea.fr/cast3m/xmlpage.do?name = presentation

Chapman N, McCombie C (2003) Principles and standards for the disposal of long lived radioactive wastes. Waste Management Series. Elsevier.

Dabbene F, Paillere H, Magnaud J.P (1998) Mixed Hybrid Finite Elements for transport of pollutants by undergound water. Proc.10th conference on finite elements in fluids, Tucson, Arizona.

GEOTRAP (2002) Radionuclide retention in geologic media. Workshop proceedings. Oskarshamn, Sweden, May 2001. *OECD/NEA*.

Grenier C, Benet L.-V (2002) Groundwater flow and solute transport modelling with support of chemistry data, Task 5, Äspö Task force on groundwater flow and transport of solutes, SKB International Cooperation Report, IPR-02-39.

Grenier C, Fourno A, Mouche E, Delay F, Benabderrahmane H (2004) Assessment of Retention Processes for Transport in a Fractured System at Äspö (Sweden) Granitic Site: From Short-Time Experiments to Long-Time Predictive Models. Proceedings of Second Inernational Symposium on Dynamics of Fluids in Fractured rocks, Berkeley (California, USA), LBNL Report 54275 pp. 242–247.

Mosé R, Siegel P, Ackerer P, and Chavent G (1994) Application of the mixed hybrid finite element approximation in a groundwater flow model: Luxury or necessity?, Water Resources Research, 30(11), pp. 3001–3012.

Neretnieks, I (1980) Diffusion in the rock matrix: an important factor in radionuclide retardation? Journal of Geophysical Research, Vol. 85, No B8.

SKB (2004) RD & D programme 2004 – Programme for research, development and demonstration of methods for the management and disposal of nuclear waste, including social science research. TR-04-21, SKB report.

Svensson U (2001) A continuum representation of fracture networks. Part I: Method and basic test cases. Journal of Hydrology 250, pp. 170–186.

Tanaka Y, Minyakawa K, Igarashi T, Shigeno Y (1996) Application of 3D smeared fracture model to the hydraulic impact of the Aspo tunnel. SKB Report. ICR 96-07.

CHAPTER 41

# Sensitivity of fracture skin properties on solute transport and back diffusion in fractured media

Terence T. Garner[1], Thandar Phyu[1,2], Neville I. Robinson[3] and John M. Sharp, Jr.[1]

[1] *Department of Geological Sciences, 1 University Station [C1100], The University of Texas, Austin, TX, 78712-0254, USA*

[2] *Presently at: Geomatrix Consultants, 510 Superior Avenue, Suite 200, Newport Beach, CA, 92663, USA*

[3] *Centre for Groundwater Studies, School of Chemistry, Physics and Earth Sciences, Flinders University, GPO Box 2100, Adelaide, SA 5001, Australia*

ABSTRACT: Fracture skins include both zones of physical or chemical alteration in the matrix adjacent to the fracture and coatings on the fracture surfaces formed by infiltered debris, precipitated minerals, and organic matter. Fracture skins alter the hydraulic and transport properties of the fractured media. Previous studies of solute transport in media with fracture skins are extended by mathematical analyses of steady and transient flows with pulse- and step-functions of conservative (e.g., $Cl^{-1}$) and reactive (e.g., $Cs^{137}$) tracers. Sensitivity studies are conducted with three dimensionless factors that are functions of fracture aperture, skin thickness, porosities, diffusion coefficients, and retardation coefficients of skin and matrix. Skin diffusion coefficient and porosity are critical factors for both reactive and conservative solutes. The next most important factor for reactive solutes is the retardation coefficient, whereas for conservative solutes it is skin thickness. When the flow is pulsed, back-diffusion is shown to be an important process that increases solute concentration in the fracture. In fractured, crystalline rocks, skins attenuate transport in fractures with porous skins, while less porous skins enhance breakthrough concentrations. Skin types observed in crystalline rocks were clay coatings and weathering rinds, which have different effects on solute transport.

## 1 INTRODUCTION

Fractures are ubiquitous and are preferential flow paths for solute transport, especially in low permeability media. Fracture skins are physically- or chemically-altered zones in the matrix near the fracture or coatings of precipitated minerals and organic matter on the fracture surfaces. Skins have been defined as "a low permeability material, deposited on the surface of the blocks, that serves to impede the free exchange of fluid between the matrix blocks and fissures" (Moench, 1984, 1995), or as "a thin layer or coat on and just underneath a fracture surface, which is caused by precipitated and deposition of secondary minerals and possibly biogenic materials" (Fu et al., 1994), or "zones of altered rock abutting the fracture and the coatings of the fracture surfaces by infiltered debris and precipitated minerals and organic matter" (Robinson and Sharp, 1997). Fracture coatings may include drusy and sparite carbonates, case hardening agents, dust films, heavy metal skins, iron films, lithobiotic coatings, nitrate crusts, oxalate crusts, phosphate skins, pigments, rock varnish, salt crusts, silica glaze, and sulfate crusts (Dorn, 1998, p. 15). Common fracture-skin minerals include iron and manganese oxyhydrates, clay minerals, organic matters (biofilms), and calcite and gypsum coatings, vein fillings, and cements (Fu et al., 1994; Fuller and Sharp, 1994; Kreisel, 1996; Kreisel and Sharp, 1996). Garner and Sharp (2004) showed that fracture skins in granites vary in different climates. In dryer climates, weathering rinds, irons bands, and pyrolusite coatings are dominant. Wetter climates tend to form clay coatings and, to a lesser extent, manganese coatings.

Fracture skins also include zones of microfractures or fracturing damage along the fracture that can increase porosity and permeability (Lindsay, 2000; Polak et al., 2003). Consequently, the hydrogeologic properties of fracture skins (i.e., permeability, porosity, sorptivity, and diffusion coefficients) may differ from those of the matrix. Thus, skins influence solute exchange between fracture and rock matrix and the transport characteristics of the fracture flow system.

## 2 ANALYTICAL MODEL

The model developed by Robinson et al. (1998) and modified by Phyu et al. (2001) is utilized. It assumes steady flow in the fracture of uniform aperture and skin thickness and ignores effects of density-driven free convection on transport (Simmons et al., 1999; Polak et al., 2004; Shi, 2005). Skin and matrix properties are assumed homogeneous. Analytical solutions for solute transport in a finite set of parallel fractures without and with fracture skins are provided by Robinson and Sharp (1997) and Robinson et al. (1998), respectively, using Laplace transforms. The superposition program to provide a solution for a rectangular pulse is included in Phyu (2002). The conceptualization of the model is illustrated in Fig. 1.

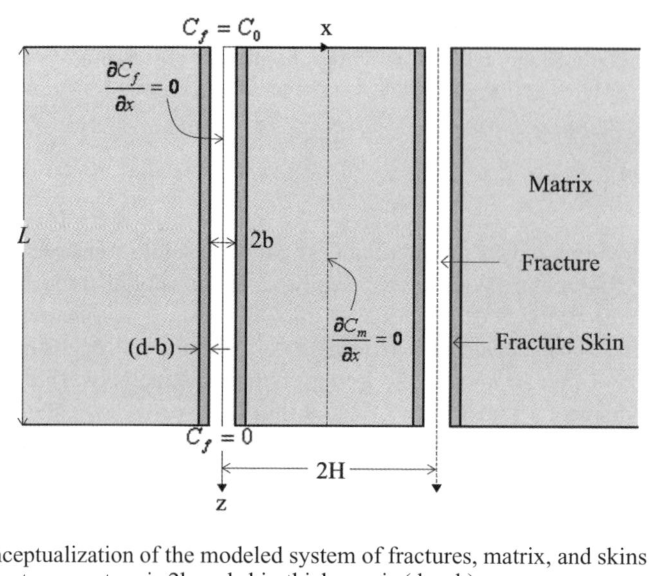


Figure 1. Conceptualization of the modeled system of fractures, matrix, and skins (after Robinson et al., 1998). Fracture aperture is 2b and skin thickness is (d − b).

The basic equations are an advection-dispersion equation for concentration in the fracture and a diffusion equation for concentration in the skin and matrix. Differential equations for concentrations in the fracture, skin, and matrix are those presented for the step function in Robinson et al. (1998) but with the following upper boundary condition for concentration in the fracture for the rectangular pulse:

$$C_f(0,t) = \begin{cases} C_0; & t \le t' \\ 0; & t > t' \end{cases} \tag{1}$$

where $t'$ is the end of rectangular pulse.

The transient solutions for the step function are presented in Robinson et al. (1998) and they are the same for the pulse function with an additional term in the Laplace transform, $[(C_0)/(p)]\,(1 - e^{-pt'})$, in place of $[C_0/p]$.

Dimensionless sensitivity factors $\beta_{sb}$, $\beta_{\phi D}$, and $\beta_{RD}$ (Robinson et al., 1998) are used to compare the effect of changes in fracture skin properties (skin thickness, porosity, diffusivity, and retardation) relative to the matrix. These are:

$$\beta_{sb} = \frac{d-b}{b}, \quad \beta_{\phi D} = \frac{\phi_s D_s}{\phi_m D_m}, \quad \beta_{RD} = \frac{R_s/D_s}{R_m/D_m} \tag{2}$$

where $(d - b)$ is the skin thickness, b the half aperture, $\phi$ the porosity, D the molecular diffusion coefficient, and R the retardation factor. The subscripts s and m refer to skin and matrix, respectively. $\beta_{sb}$ represents the fracture skin thickness relative to the fracture half aperture; $\beta_{\phi D}$ represents the diffusivity of skin relative to the matrix; and $\beta_{RD}$ represents the skin retardation or sorption relative to the matrix.

## 3 SENSITIVITY ANALYSIS

Field data from the Brushy Canyon Sandstone of West Texas and its fracture skin (case #1) presented in Kreisel (1996), Kreisel and Sharp (1996), Sharp et al. (1996a), and Robinson et al. (1998) were utilized as model base-case input. Table 1 lists the input parameters for the fracture system and both reactive and conservative tracers (e.g., $Cs^{137}$ and $Cl^{-1}$, respectively) including retardation coefficients and $Cs^{137}$ decay constants. The pulse duration is 100 days.

The base values of the sensitivity factors are $\beta_{sb} = 10^{0.95}$ and $\beta_{\phi D} = 10^{-1.62}$ for both $Cl^{-1}$ and $Cs^{137}$ and $\beta_{RD} = 10^{1.00}$ and $10^{1.68}$ for $Cl^{-1}$ and $Cs^{137}$, respectively. The sensitivity factors are varied by the increment of one magnitude (10) from $10^{-6}$ to $10^{6}$ for $\beta_{\phi D}$ and $\beta_{RD}$. Because of the fact that skin thickness $(d-b)$ must be less than the half fracture spacing (H), $\beta_{sb}$ higher than $10^3$ is not valid with the measured value of $H = 0.31$ m. To analyze the relative importance of sensitivity factors, profiles of fracture solute concentration at $z = 30$ m for the step function and maximum concentration in the fracture for pulse are compared. In all analyses, the observed time is 200 days after the release of solutes. The selections of z at 30 meters for steps and maximum concentrations for pulses at 200 days are typical. The conclusions below hold for other values of z and for non-maximal concentrations.

Table 1. Input parameters using Brushy Canyon Sandstone skin #1 for fracture physical attributes. Units are in meters and days.

| Parameter | Value |
|---|---|
| $v$, average groundwater velocity [m/d] | 1 |
| $D_f$, fracture dispersion [m$^2$/d] | $1 \times 10^{-3}$ |
| $D_s$, skin diffusion [m$^2$/d] | $4 \times 10^{-7}$ |
| $D_m$, matrix diffusion [m$^2$/d] | $4 \times 10^{-6}$ |
| $\phi_s$, skin porosity [−] | 0.035 |
| $\phi_m$, matrix porosity [−] | 0.145 |
| L, fracture length, $Cs^{137}$ [m] | 100 |
| L, fracture length, $Cl^{-1}$ [m] | 300 |
| H, half fracture spacing [m] | 0.31 |
| (d − b), skin thickness [m] | 0.002 |
| 2b, fracture aperture [m] | $4 \times 10^{-4}$ |
| $R_f$, fracture retardation – $Cs^{137}$ [−] | 6 |
| $R_s$, skin retardation – $Cs^{137}$ [−] | 673 |
| $R_m$, matrix retardation – $Cs^{137}$ [−] | 141 |
| $R_f$, fracture retardation – $Cl^{-1}$ [−] | 1 |
| $R_s$, skin retardation – $Cl^{-1}$ [−] | 1 |
| $R_m$, matrix retardation – $Cl^{-1}$ [−] | 1 |
| $\lambda$, $Cs^{137}$ decay constant [d$^{-1}$] | $6.33 \times 10^{-5}$ |
| $\lambda$, $Cl^{-1}$ decay constant [d$^{-1}$] | 0 |
| t′, time of end of pulse [d] | 100 |

A first phase of sensitivity analysis considers the variation of one $\beta$ factor with the other two $\beta$ factors at base case values. Figs. 2 (a) and (b) depicts concentration in the fracture at z = 30 m for the step function and maximum concentration for the pulse function, respectively, as a function of $\beta_{sb}$, $\beta_{\phi D}$, and $\beta_{RD}$ with $Cl^{-1}$ and $Cs^{137}$. Fig. 2 (c) shows the different positions where the maxima in Fig. 2 (b) occur in the fracture with change in the sensitivity factors. Analyses at 200 and 300 days are to show that relative change in concentrations is the same at any time.

For both source functions, $\beta_{\phi D}$ clearly has the most effect on the concentration in the fracture followed by $\beta_{RD}$ for the reactive solute and $\beta_{sb}$ for the conservative solute. For the pulse function, increases in concentration when $\beta_{sb}$ is more than $10^1$ for $Cl^{-1}$ and $10^0$ for $Cs^{137}$, $\beta_{\phi D}$ is greater than approximately $10^{-1}$, and $\beta_{RD}$ is greater than approximately $10^2$ are due to back-diffusion. This phenomenon of back-diffusion is a diffusive transport, back into the fractures after the solute front passes through, of solutes that are previously attenuated by diffusion into the skins and matrix (Mace and Sharp, 1996; Polak et al., 2003).

A second phase of sensitivity analysis considers simultaneous variations in all three $\beta$ factors, enabling departure from base case values. This is shown in Figs. 3, 4, and 5 by considering $\log_{10}(\beta_{\phi D})$ at discrete values of −6, −4, −1.62, −1, and 0 while continuously varying $\log_{10}(\beta_{sb})$ between −6 and 3 and $\log_{10}(\beta_{RD})$ between −6 and 6.

Fig. 3 provides a transition from the analyses depicted in Fig. 2. It shows $Cl^{-1}$ concentrations in the fracture at z = 30 meters and 200 days for the step function. Fig. 3 curves (a), (b), and (c) correspond to the solid $\beta_{\phi D}$, $\beta_{sb}$, and $\beta_{RD}$ curves, respectively, in Fig. 2 (a) for $Cl^{-1}$.

Figs 4 and 5 compare the relative concentrations of $Cl^{-1}$ and $Cs^{137}$, respectively, in the fracture for the step and pulse functions. The surfaces of concentration are relative to the step or pulse input concentration at constant $\beta_{\phi D}$. When $\beta_{\phi D}$ is very low (i.e., at $\beta_{\phi D} = 10^{-6}$), the

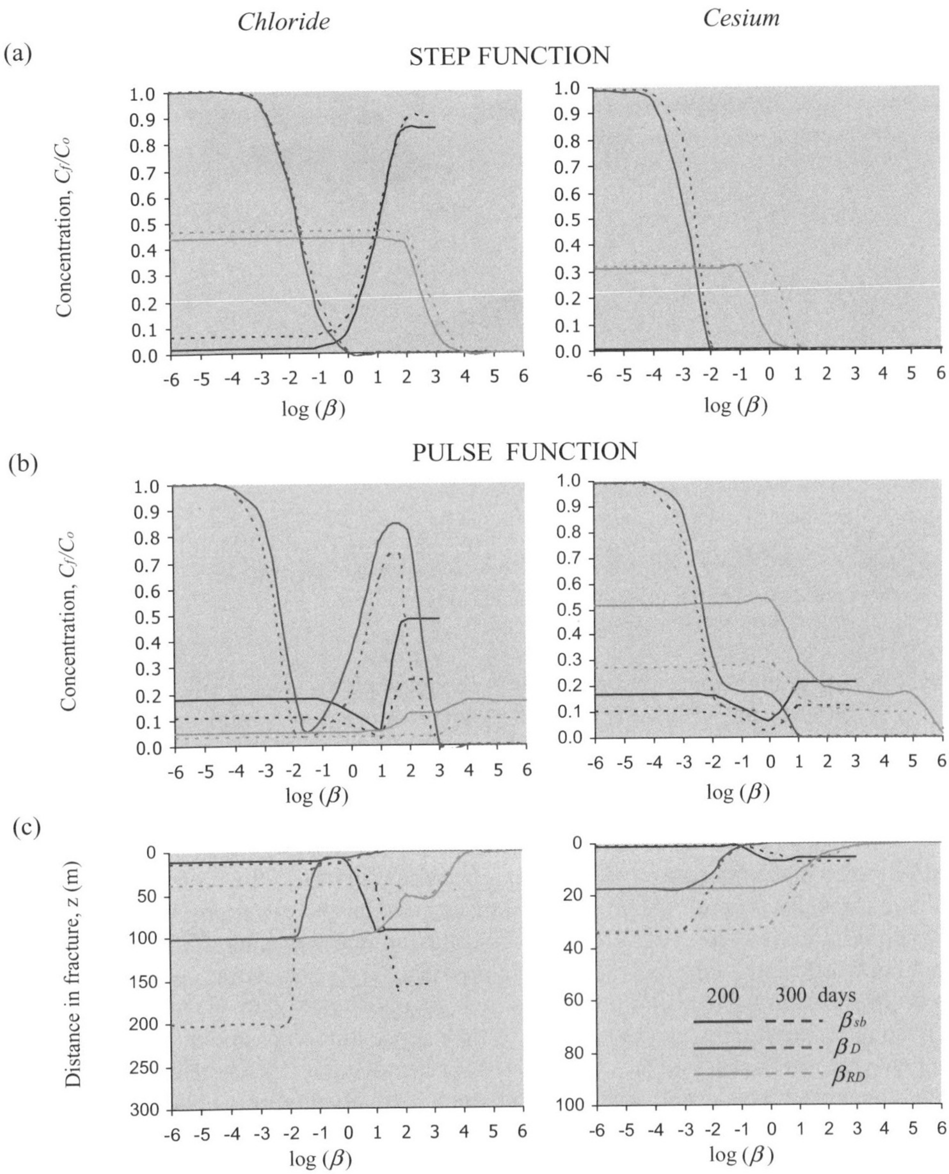


Figure 2. Chloride and cesium at 200 (solid) and 300 (dashed) days: (a) concentration at z = 30 m for step function, (b) maximum concentration in the fracture for pulse function, and (c) positions of maxima in (b) in the fracture. Note: y-axis scales in (c) are different.

fracture skin is much less porous than the matrix, and there is little diffusion into the skin and the solutes are confined within the fracture. Fracture solute concentration is high regardless of fracture properties ($\beta_{sb}$) and skin retardation coefficient ($\beta_{RD}$). As $\beta_{\phi D}$ increases (i.e., fracture skin becomes more porous and diffusive), diffusive flux into the skin increases and high concentration in the fracture is restricted to high $\beta_{sb}$ and low $\beta_{RD}$. For the pulse function,

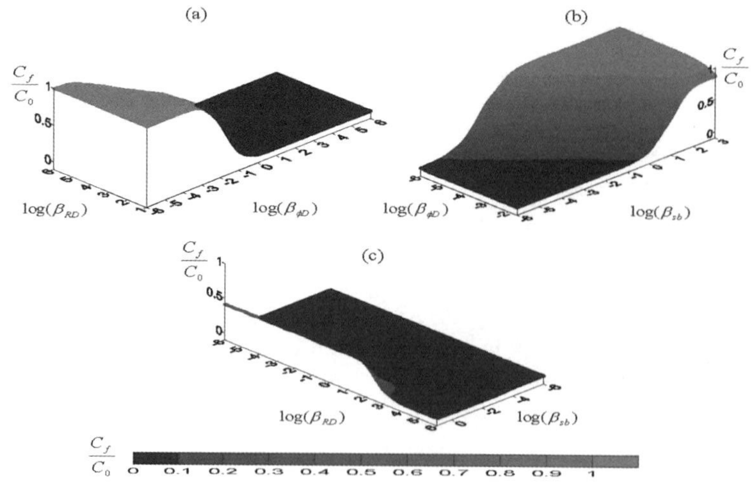

Figure 3. Surfaces of chloride concentration in the fracture at z = 30 meters at 200 days with step function. Surfaces are cut along (a) $\log(\beta_{RD}) = 1$ with constant $\log(\beta_{sb}) = 0.95$, (b) $\log(\beta_{\phi D}) = -1.62$ with constant $\log(\beta_{RD}) = 1$, and (c) $\log(\beta_{sb}) = 0.95$ with constant $\log(\beta_{\phi D}) = -1.62$. The curves of the surfaces correspond to the solid lines in Fig. 2 (a) for chloride.

however, tails of high concentration occur in the region with low $\beta_{sb}$ and high $\beta_{RD}$ as $\beta_{\phi D}$ increases due to back diffusion. When the skin is more porous and diffusive than the matrix and its retardation is high, more solutes diffuse into the skin resulting in high back diffusion as solute front passes through and concentration gradient shifts in direction.

Fig. 4 depicts the relative concentration of $Cl^{-1}$ in fracture for the step and pulse functions. Compared to $Cs^{137}$ concentration in Fig. 5, a wider range of $\beta_{RD}$ value permits high concentration in the fracture for $Cl^{-1}$ tracer. This is caused by the effect of sorptivity of the fracture skins. For a conservative solute, skin sorption has no effect and thus solute concentration in the fracture is reduced only by diffusion into the skins. $Cl^{-1}$ concentration from back-diffusion is also considerably higher than $Cs^{137}$ concentration. In both cases, back-diffusion starts when skin porosity is high enough ($\beta_{\phi D} > 0.0001$) where the rate of diffusion into the skin is high enough to shift the concentration gradient. In both figures, the light gray plateaus define the combination of sensitivity factors where essentially plug-flow is occurring. This occurrence is when there is virtually no diffusion into the skin and matrix and all solutes arc contained within the fracture.

## 4 GRANITE TRANSPORT PROPERTIES

Field data collected for analyses of granite fracture skins come from the Town Mountain Granite in Fredericksburg, Texas, and the Elberton Granites in Elberton, Georgia, USA. Fracture apertures, spacing, and skin thicknesses were measured in the field. Arithmetic mean apertures were $1.13 \times 10^{-3}$m in Fredericksburg and $1.95 \times 10^{-3}$m in Elberton. Mean skin thicknesses were $2.07 \times 10^{-2}$m in Fredericksburg is and $11.19 \times 10^{-3}$m in Elberton. The arithmetic mean fracture spacing in Fredericksburg is $1.31 \times 10^{-2}$m, which was also used in Elberton analyses.

Figure 4. Surfaces of relative chloride concentration at z = 30 meters for step function and maximum concentration for pulse function at 200 days with simultaneous change in dimensionless sensitivity factors $\beta_{\phi D}$, $\beta_{sb}$, and $\beta_{RD}$. The star is the Brushy Canyon Sandstone scenario.

Porosity was measured, using a Micrometrics Autopore III Mercury Injection Porosimeter, which injects mercury at pressures up to 60,000 psi, by (Garner and Sharp 2004). These show fracture skin porosity in the Fredericksburg granites increase by 300–400% and Elberton granite samples decreases by ~0.5%. Diffusion coefficients and retardation parameters were determined by published data and mathematical estimation. Model input parameters are illustrated in Table 2.

Model results are presented in the form of tracer breakthrough curves representing change in tracer concentration at a given time and at a given distance (z) along the fracture. Analyses for this study evaluate the effects of two types of fracture skins (case hardening

Figure 5. Surfaces of relative cesium concentration at z = 30 meters for step function and maximum concentration for pulse function at 200 days with simultaneous change in dimensionless sensitivity factors $\beta_{\phi D}$, $\beta_{sb}$, and $\beta_{RD}$. The star is the Brushy Canyon Sandstone scenario.

and weathering rinds) and time variation. Different tracer types show different flow characteristics; therefore, evaluation of the four studied variables is done using a conservative ($Cl^-$) tracer and a reactive ($^{137}Cs$) tracer. Fracture length was evaluated at 300 m for conservative and reactive tracers.

Fig. 6 shows distance breakthrough curves (Z [m] versus $C/C_0$) for fracture skin properties for case hardening conditions from the Elberton Granites. The left column represent $Cl^-$ tracer injected using a pulse function (top) and step function (bottom). The right column represents $^{137}Cs$ injected using the pulse function (top) and the step function (bottom). Dashed and solid lines represent different time durations, ranging from 100 to 500 days, distance breakthrough under no skin and with skin conditions.

Table 2. Skin parameters for Fredericksburg and Elberton Granites.

| Parameter | Fredericksburg | Elberton |
|---|---|---|
| v, average groundwater velocity [m/d] | 1 | 1 |
| $D_f$, fracture dispersion [$m^2$/d] | $1 \times 10^{-3}$ | $1 \times 10^{-3}$ |
| $D_s$, skin diffusion [$m^2$/d] | $8.64 \times 10^{-7}$ | $8.64 \times 10^{-10}$ |
| $D_m$, matrix diffusion [$m^2$/d] | $8.64 \times 10^{-10}$ | $8.64 \times 10^{-7}$ |
| $\phi_s$, skin porosity [ − ] | 0.09 | 0.011 |
| $\phi_m$, matrix porosity [ − ] | 0.024 | 0.024 |
| L, fracture length, $Cs^{137}$ [m] | 300 | 300 |
| L, fracture length, $Cl^{-1}$ [m] | 300 | 300 |
| H, half fracture spacing [m] | $6.55 \times 10^{-3}$ | $6.55 \times 10^{-3}$ |
| (d-b), skin thickness [m] | 0.0207 | 0.0112 |
| 2b, fracture aperture [m] | $1.13 \times 10^{-3}$ | $1.95 \times 10^{-3}$ |
| $R_f$, fracture retardation – $Cs^{137}$ [−] | 6 | 6 |
| $R_s$, skin retardation – $Cs^{137}$ [−] | 141 | 673 |
| $R_m$, matrix retardation – $Cs^{137}$ [−] | 673 | 141 |
| $R_f$, fracture retardation – $Cl^{-1}$ [−] | 1 | 1 |
| $R_s$, skin retardation – $Cl^{-1}$ [−] | 1 | 1 |
| $R_m$, matrix retardation – $Cl^{-1}$ [−] | 1 | 1 |
| $\lambda$, $Cs^{137}$ decay constant [$d^{-1}$] | $6.33 \times 10^{-5}$ | $6.33 \times 10^{-5}$ |
| $\lambda$, $Cl^{-1}$ decay constant [$d^{-1}$] | $1 \times 10^{-12}$ | $1 \times 10^{-12}$ |
| t′, time of end of pulse [d] | 100 | 100 |

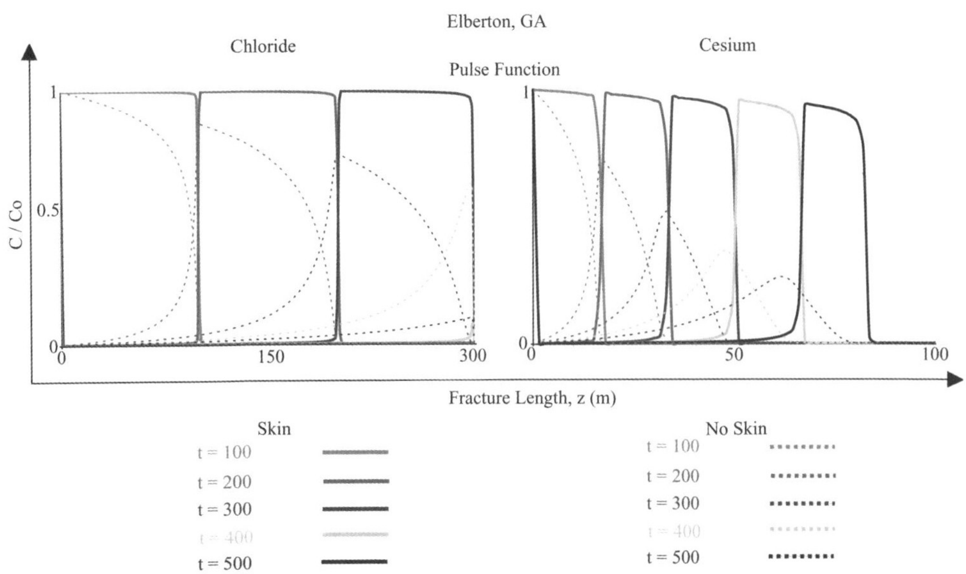


Figure 6. Elberton granite breakthrough curves of pulse function with concentration varying over a time range of 100 to 500 days. Solid and dashed lines show conditions with and without fracture skins. (after Garner and Sharp, 2004)

The presence of a case hardening fracture skins accelerates tracer transport relative to no skin conditions. Case hardening causes a decrease in porosity, rate of diffusion, and the retardation factor in the fracture skin.

$^{137}Cs$ was chosen as a tracer because of its affinity for sorption, but $^{137}Cs$ also has a relatively short half-life of 30 years that should be considered in long tracer tests. Radioactive

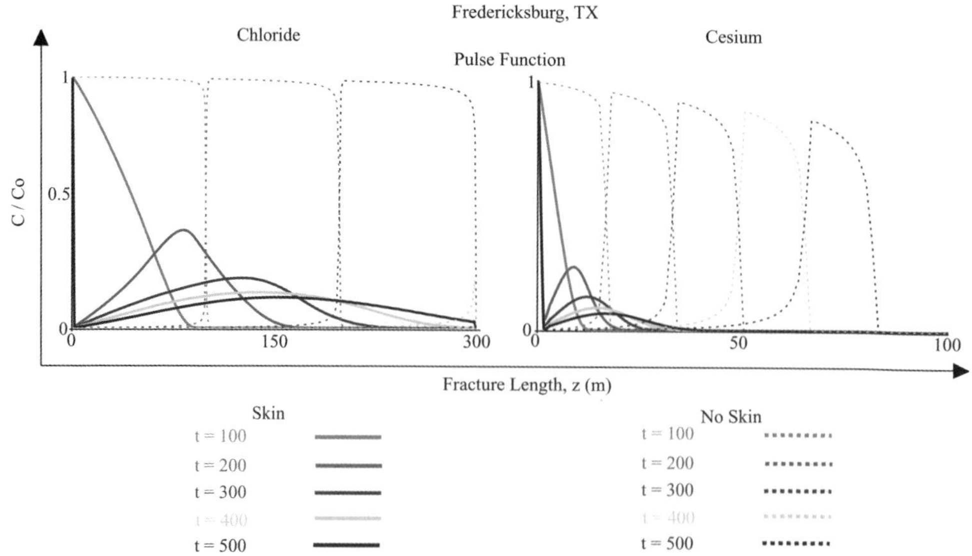


Figure 7. Town Mountain granite (Fredericksburg, Texas) breakthrough curves of pulse function with concentration varying over a time range of 100 to 500 days. Solid and dashed lines show conditions with and without fracture skins. (after Garner and Sharp, 2004)

decay is considered negligible for the 2 year simulations of this study. The transport of $^{137}Cs$ is shown to have a smaller concentration change in the fracture than $Cl^-$. $^{137}Cs$ breakthrough is delayed in comparison to $Cl^-$ because of $^{137}Cs$ sorption to the fracture walls. Time change does not seem to affect spreading in the fracture or concentration variation. In the Elberton Granites the case hardening fracture skins constrain the tracer concentrations to the fracture.

Simulating weathering rind conditions seen in Fredericksburg, Texas shows a significant decrease in the distance a tracer travels along the fracture for a given point in time (Fig. 7). Tracer tests simulating fractures with no skin in Fredericksburg mimics transport in Elberton simulations with fracture skin. This is because matrix porosities from Fredericksburg samples are slightly higher than those from Elberton, and tracer diffusion into the fracture wall is slightly higher in Fredericksburg. The result is a lower tracer concentration at shorter distances along the fracture.

Weathering rind porosities of the fracture skin is higher than the matrix, causing tracer to spread over longer distances and increasingly lower concentrations as time increases for conservative tracers using the pulse function. The reactive tracer breakthrough occurs closer to the point of injection for both source functions. Higher porosity in the weathering rind increases the surface area of the fracture wall allowing $^{137}Cs$ to diffuse and sorb. This is seen in simulations with and without fracture skins.

## 5 SUMMARY

Fracture skins can significantly affect transport properties of fractured media. The analytical model consists of a series of equidistant parallel fractures with uniform apertures and matrix with a constant skin thickness. Step and pulse functions of both conservative and reactive solutes are considered. Sensitivity factors for skin thickness, aperture, porosity, diffusion coefficient, and retardation factor are used to compare the effects of skin properties on solute transport. Analyses demonstrated that skin porosity and diffusion coefficient have the greatest effect on solute concentrations in fractures, followed by the retardation factor for reactive solutes and skin thickness and aperture for conservative solutes. When fractures are subjected to pulses of solutes, back-diffusion is an important phenomenon controlling concentration in the fracture. It is apparent when fracture skin porosity is sufficiently high that there is an increase in back-diffusion rates and fracture concentrations. If the solute is conservative, relative to the reactive solutes, lack of sorption results in higher concentrations in the fracture and higher back-diffusion rates.

Mathematical analyses of solute transport in granites with fracture skins from Fredericksburg, Texas and Elberton, Georgia demonstrate a notable influence of fracture skin type on transport rates. Comparing case hardened skins to the weathering rind skins shows tracer transport is enhanced where the skin is less porous and attenuated where skins are more porous. Weathering rinds increase the length of time back diffusion occurs resulting in long tailing breakthrough curves and lower concentrations over time. Formation of different skin types can create uncertainty in tracer tests and affect long term prediction.

## ACKNOWLEDGEMENTS

This material is partially based upon work supported by the National Science Foundation under Grant No. 0439806, by a Geological Society of America Student Research Grant to Garner, and by the Geology Foundation of The University of Texas.

## REFERENCES

Dorn, R.I., 1998, Rock Coatings, Elsevier, Amsterdam, New York, 429 p.

Evans, D.D., and Nicholson, T.J. (Eds.), 1987, Flow and transport through unsaturated fractured rock, Geophysical Monograph 42, American Geophysical Union, Washington D. C., 187 p.

Fu, L., Milliken, K.L., and Sharp, J.M., Jr., 1994, Porosity and permeability variations in fractured and liesegang-banded Breathitt sandstones (Middle Pennsylvanian), eastern Kentucky; diagenetic controls and implications for modeling dual-porosity systems, Journal of Hydrology, 154 (1–4), p. 351–381.

Fuller, C.M., and Sharp, J.M., Jr., 1994, Permeability and fracture patterns in extrusive volcanic rocks: implication from the welded Santana tuff, Trans-Pecos Texas, Geological Society of America Bulletin, v. 104, p. 1485–1496.

Garner, T.T., and Sharp, J.M., Jr., 2004, 2004, Hydraulic Properties of Granitic Fracture Skins and Their Effect on Solute Transport, 2004 U.S. EPA/NGWA Fractured Rock Conference: State of the Science and Measuring Success in Remediation, Portland, Maine.

Kreisel, I., 1996, Fracture Skins and Their Effect on Solute Transport in a Fractured Porous Medium with Examples from the Brushy Canyon Formation, West Texas, unpub. M.A. thesis, The University of Texas, Austin, TX, 109 p.

Kreisel, I., and Sharp, J.M., Jr., 1996, Fracture skins in the Brushy Canyon Formation. In: DeMis, W.D., Cole, A.G., (eds.), The Brushy Canyon Play in Outcrop and Subsurface: Concepts and Examples, PBS-SEPM No. 96–38, Midland, TX, p. 147–152.

Lindsay, S.R., 2000, Controlled Fracture Study of Geologic Materials: unpub. M.S. thesis, The University of Texas, Austin, TX, 60 p.

Mace, R.E., and Sharp, J.M., Jr., 1996, Back diffusion in fractured rocks, EOS, v. 77, no. p. S 108.
Moench, A.F., 1984, Double-porosity models for a fissured groundwater reservoir with fracture skin, Water Resources Research, 20 (7), p. 831–846.
Moench, A.F., 1995, Convergent radial dispersion in a double-porosity aquifer with fracture skin: Analytical solution and application to a field experiment in fractured chalk, Water Resources Research, 31(8), p. 1823–1835.
Phyu, T., 2002, Transient Modeling of Contaminant Transport in Dual Porosity Media with Fracture Skins: unpub. M.S. thesis, The University of Texas, Austin, TX, 125p.
Phyu, T., Robinson, N.I., and Sharp, J.M., Jr., 2001, Transient modeling of solute transport in dual-porosity media with fracture skin, Geological Society of America, Annual Meetings, Boston, MA, Abstract with Programs – Geological Society of America, v. 33, no. 6, p. 167–168.
Polak, A., Grader, A.S., Wallach, R., and Nativ, R., 2003, Chemical diffusion between a fracture and the surrounding matrix: Water Resources Research, v. 39, p. SBH10-1–14.
Polak, A., Grader, A.S., Wallach, R., and Nativ, R., 2004, Tracer diffusion from a horizontal fracture into the surrounding matrix: measurement by computed tomography: Journal of Contaminant Hydrology, 67 (1), p. 95–112.
Robinson, N.I., and Sharp, J.M., Jr., 1997, Analytical solution for solute transport in a finite set of parallel fractures with matrix diffusion, CSIRO Mathematical and Information Sciences Report No. CMIS C23/97, 23 p.
Robinson, N.I., Sharp, J.M., Jr., and Kreisel, I., 1998, Contaminant transport in sets of parallel finite fractures with fracture skins: Journal of Contaminant Hydrology, 31 (1–2), p. 83–109.
Shi, M., 2005, Characterizing Heterogeneity in Low-Permeability Strata and Its Control on Fluid Flow and Solute Transport by Thermohaline Free Convection:unpub. Ph.D.dissertation, The University of Texas, Austin, TX, 100 p.
Simmons, C.T., Sharp, J., Jr., and Robinson, N.I., 1999, Density-driven free convection in zones of inverted salinity through fractured low-permeability units in the Gulf of Mexico Basin, Texas, USA: Water 99 Joint Congress, Inst. of Engineers, Australia, Brisbane, v. 2, p.739–744.

CHAPTER 42

# Averaging hydraulic conductivity in heterogeneous fractures and layered aquifers

Sassan Mouri and Todd Halihan

*School of Geology, Oklahoma State University, Stillwater, OK, USA*

ABSTRACT: Heterogeneity of the subsurface has a significant impact on groundwater flow and transport. The analysis of heterogeneous groundwater systems remains difficult because the effects of heterogeneity are poorly understood. The differences in methods used to calculate average flow properties can create discrepancies in flow and transport predictions. Well known simple solutions exist for hydraulic conductivity averaging in layered aquifers, but analogies in fractured rock settings are lacking. The layered aquifer solutions are used as an analogy to examine flow through heterogeneous fracture planes. These solutions are also evaluated to determine asymptotic solutions for situations with extreme heterogeneity. Analyzing a simple two-layer system, analytical solutions of the influence of heterogeneity on layered aquifers and fractures can be determined. The analytical solution for average hydraulic conductivity perpendicular to fracture plane heterogeneities is slightly different than the layered aquifer solution. Results show that perpendicular flow through heterogeneous fractures is more sensitive to heterogeneity than in layered aquifers. The results also demonstrate that with low levels of heterogeneity, flow is controlled by extremes of hydraulic conductivity.

## 1 INTRODUCTION

Fractures are an important factor for groundwater flow in many geologic settings. It has been found that heterogeneity due to aperture variability gives rise to flow channeling (Tsang and Tsang, 1989). The analysis of heterogeneous fractures is an important factor in determining hydraulic conductivity characteristics. Fractures can control groundwater flow, and are the primary water producing openings in many major aquifers across the United States (Heath, 1982). The analysis of the influences of fractures on aquifers is important in many aspects economically and environmentally (Renshaw, Dadakis, Brown, 2000).

Many theoretical studies incorporate the range of aperture values in a fracture (Neuzil and Tracy, 1981; Brown, 1984). The hydraulic conductivity of rock fractures has been studied by treating it as a problem in fluid mechanics (Zimmerman and Bodvarsson, 1996). Navier-Stokes and Hele-Shaw equations were used to study and quantify hydraulic conductivity of rock fractures. Much work has been conducted on fractures using models to determine hydraulic characteristics to analyze contaminant transport (Kischinhevsky, 1997; Wanfang, 1997). Another study looked at flow on rough surfaces of fractures (Or and Tuller, 2000). Others used observation well data and type-curve matching procedures to analyze hydraulic conductivities of fractures (Leveinen et al., 1998).

One area that seems to be lacking from these analyses is a simple analytical evaluation that can be used to generate instincts for how fractures respond based on their variability. Average fracture hydraulic conductivities have not been analyzed to understand how average hydraulic conductivity is affected by aperture heterogeneity for simple cases. An investigation of the correlation between hydraulic conductivity and aperture heterogeneity given a defined system is necessary to develop a fundamental understanding of how variability is averaged.

An average hydraulic conductivity value for layered aquifers is an important variable in hydrogeology. Field studies have been conducted using single boreholes to compare hydraulic conductivity values that different instruments calculate (Zlotnik and Zurbuchen, 2003). Evaluating the hydraulic conductivity of layered aquifers is important because they can affect transport properties (Dagan, 1989). Studies have been conducted analyzing effects of varying hydraulic conductivities on groundwater flow (Gelhar, 1997). Simple one and two dimensional systems were created and complex solutions were obtained to describe groundwater behavior. Others have analyzed subsurface flows by stochastic analysis (Gelhar, 1974; Bakr et al., 1978; Gutjahr et al., 1978; Mantoglou and Gelhar, 1989). Systems described in these studies used random distributions of variables to evaluate subsurface flows. Freeze (1975) also stochastically analyzed groundwater flow, determining the difficulty in choosing a single value for a flow parameter. Simple two-layered systems have not been formally used to observe hydraulic conductivity. By analyzing a two-layered aquifer system, it can aid in the clarification of how to characterize layered aquifers based on the heterogeneity index which will be defined later in this paper. Defining a heterogeneity index will inevitably allow the researcher to have a tool in the field to assess layered aquifer characteristics.

Heterogeneity present in fractures not only creates aperture variability, but in turn, it affects average conductivity values. The analysis conducted here examines heterogeneous fractures and layered aquifers. An analytical approach is taken in this paper to study hydraulic conductivity given a defined two-layered system for a heterogeneous fracture and a layered aquifer. Objectives include introducing a new approach to understanding hydraulic conductivity for a heterogeneous fracture and a layered aquifer, which can be applied to field studies and used as a new teaching tool to understand the influence of heterogeneity in groundwater systems. Heterogeneity in fracture planes is compared to layered aquifer heterogeneity to delineate similarities and differences between the two systems. The system will be analyzed for flow perpendicular and parallel to heterogeneity in the heterogeneous fracture and a layered aquifer. Average hydraulic conductivity values are obtained using averaging formulas derived later in this paper, as is the asymptotic limit for the average hydraulic conductivities. To derive analytical solutions for average hydraulic conductivity a simple two layer model will be used for heterogeneous fractures and layered aquifers. Research conducted on variations in hydraulic conductivity with scale have found that these variations can be explained by a conceptual model of a simple heterogenous aquifer composed of high conductivity zone and low conductivity matrix (Schulze-Makuch and Cherkauer, 1998). The simple two layer model presented in this paper compliments these findings.

For layered aquifers, flow parallel to the heterogeneity is governed by the highest hydraulic conductivity layer. To obtain average horizontal hydraulic conductivity values (parallel to layering) for a layered aquifer Equation 1 (Fetter, 2001) can be used (See variable definitions). The solution is

$$K_{\parallel LA} = \sum_{n=1}^{i} \frac{K_{\parallel n} m_n}{m_T} \tag{1}$$

Average vertical hydraulic conductivity (perpendicular to layering) for an aquifer can be found by (Fetter, 2001):

$$K_{\perp LA} = \frac{m_T}{\sum_{m=1}^{n} \frac{m_n}{K_{\perp n}}} \tag{2}$$

## 2 TWO LAYER HETEROGENEOUS SYSTEM

To analyze averaging of hydraulic conductivity in the simplest heterogeneous system, a two-layer system can be employed. The two-layer system enables an analysis of the influences of hydraulic conductivity values by simplifying the averaging problem of a more complex system. By simplifying the system to two layers, simple analytical solutions are possible that can be used to gain insight into more complex systems. The two layer system will also allow a comparison to be made between heterogeneous fractures and layered aquifers. The analysis is carried out with an attempt to not only show the differences, but to correctly determine the controls governing flow in the systems.

The two-layer aquifer system for layered aquifers will be composed of a top layer of thickness $m_H$, which is the high conductivity ($K_H$) unit and the bottom layer of thickness $m_L$, which is the low conductivity ($K_L$) unit. For this system, average hydraulic conductivity will be analyzed parallel and perpendicular to heterogeneity (Figures 1A and 1B).

For heterogeneous fracture apertures, the two layer model will also be analyzed. The fracture will be modeled using two apertures allowing a comparison to be made to the layered aquifer system (Figures 1C and 1D). For the fractures (Figures 1C and 1D) b is the fracture aperture and L is defined as length. To compare the aquifer and fracture systems, aperture length (L) is equal to layered aquifer thickness (m).

To analyze the two-layer system for heterogeneity the Greek symbol eta ($H_K$) is defined as $K_L/K_H$. As defined, the heterogeneity index for hydraulic conductivity, $H_K$ varies between 0 and 1. At $H_K = 1$, the two-layer system is homogenous, as $H_K$ approaches 0 the two-layer system becomes fully discrete with only the high conductivity layer or aperture functioning.

## 3 FLOW AVERAGING

The two endmember averaging cases for layered aquifers or heterogeneous fractures are parallel flow through the segments and perpendicular flow across the segments. First the parallel case is examined as it results in similar solutions. Next, the perpendicular case is examined and two different averaging results are derived for the two endmember cases.

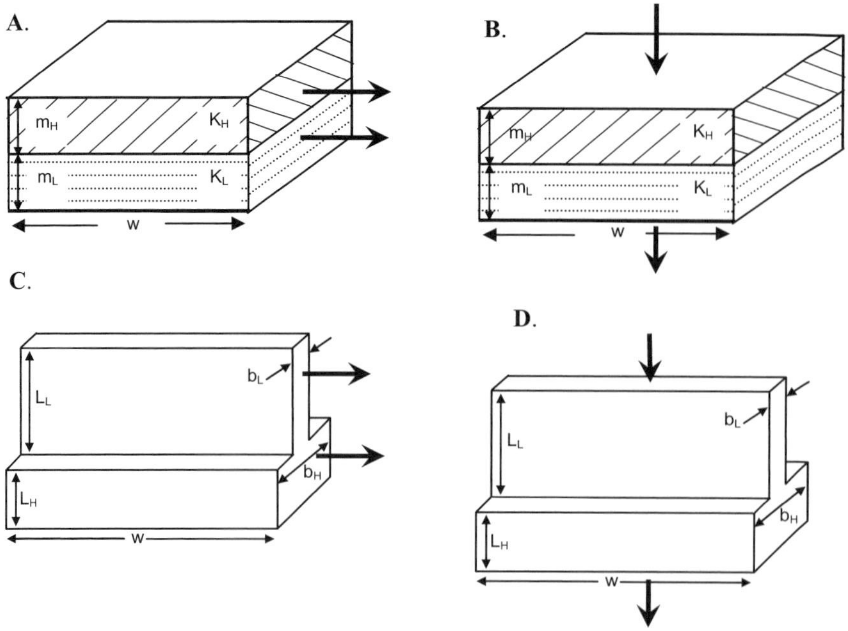


Figure 1. **Two layer model for layered aquifer and vertically oriented fracture.** (A) Flow parallel to layered aquifer heterogeneity. (B) Flow perpendicular to layered aquifer heterogeneity. (C) Flow parallel to the vertical fracture heterogeneity. (D) Flow perpendicular to the fracture heterogeneity.

### 3.1 *Parallel*

First, examining parallel flow the existing case Equation 1 applies (Figure 1A). The average hydraulic conductivity parallel to aquifer heterogeneity, for the two layers are $K_H$ and $K_L$, can be determined by modifying Equation 1 for the two-layer system. Equation 1 is written in terms of each layers conductivity, and with respect to the heterogeneity index $H_K$ so that the extreme case of $H_K \rightarrow 0$ can be evaluated.

$$K_{\| \, LA} = \frac{K_L m_L + K_H m_H}{m_T} = \frac{K_H (H_K + m_H)}{m_T} \tag{3}$$

The asymptotic solution can be found as the limit as $H_K \rightarrow 0$,

$$K_{\| \, LA} = K_H \frac{m_H}{m_T} \qquad \text{as } H_K \rightarrow 0. \tag{4}$$

For the case of parallel flow through two layer heterogeneous fracture, m for the layered aquifer case becomes the length (L) of the apertures and m becomes aperture length (L).

The solution then looks

$$K_{\parallel\, HF} = \sum_{n=1}^{i} \frac{K_L L_L + K_H L_H}{L_T} = \frac{K_H (H_K + L_H)}{L_T}. \tag{5}$$

The asymptotic solution remains the same for flow parallel to heterogeneity in fractures. However, the longest length of the apertures is substituted for $m_H$ and the total length of the apertures is substituted for $m_T$

$$K_{\parallel\, HF} = K_H \frac{L_H}{L_T} \qquad \text{as } H_K \to 0. \tag{6}$$

Thus, there are no averaging differences between layered aquifers and heterogeneous fractures for parallel flow.

## 3.2 *Perpendicular*

Average hydraulic conductivity perpendicular to heterogeneity for layered aquifers and heterogeneous fractures differs. To evaluate the effects of heterogeneity on layered aquifers (Figure 1B), Equation 2 for the two-layered system becomes,

$$K_{\perp\, LA} = \frac{m_T}{\frac{m_H}{K_H} + \frac{m_L}{K_L}} = \frac{m_T K_L}{H_K m_H + m_L}. \tag{7}$$

The asymptotic solution for the hydraulic conductivity perpendicular to layering can be found. Remembering that $H_K = K_L/K_H$ and multiplying both sides of the equation by $K_L$, solving for $K_{\perp\, LA}$, the result is:

$$K_{\perp\, LA} = \frac{m_T}{m_L} K_L \qquad \text{as } H_K \to 0. \tag{8}$$

Determining an analytical solution for flow perpendicular to the heterogeneous aperture requires a more complete derivation to understand the solution. First the change of gradient between each aperture is evaluated. Change in head total is defined as,

$$\Delta h_T = \Delta h_H + \Delta h_L \tag{9}$$

Discharge through one single homogeneous aperture in each fracture can be represented by Darcy's Law:

$$Q_i = \alpha w b_i^3 \frac{\Delta h_i}{L_i} \tag{10}$$

where w is width, the subscript i is the respective homogeneous aperture segment, and

$$\alpha = \frac{pg}{12\mu}. \tag{11}$$

Solving for the change of head, for each fracture segment, the solution is:

$$\Delta h_i = \frac{Q_i L_i}{\alpha b_i^3 w}. \tag{12}$$

Substituting Equation 12 into Equation 9 the solution becomes,

$$\frac{Q_T L_T}{\alpha b_{\perp HF}{}^3 w} = \frac{Q_H L_H}{\alpha b_H^3 w} + \frac{Q_L L_L}{\alpha b_L^3 w} \tag{13}$$

Since total volumetric flow for the perpendicular case is identical in both fracture apertures, Equation 13 can be simplified and solved for $b_{\perp HF}$. For the case of fracture apertures the definition of the heterogeneity index can be extended as $H_b = b_L/b_H$. Solving for $b_{\perp HF}$, the average aperture for flow perpendicular to the fracture plane is:

$$b_{\perp HF} = \sqrt[3]{\frac{L_T}{\left(\frac{L_H}{b_H^3} + \frac{L_L}{b_L^3}\right)}} = \sqrt[3]{\frac{L_T}{\frac{L_H}{b_L^3/H_b^3} + \frac{L_L}{b_L^3}}} \tag{14}$$

This formula is an analytical solution for hydraulic aperture for flow perpendicular to the fracture segments, allowing an analysis of the effects the heterogeneity index will have on the aperture.

If $b_{\perp HF}$ is expressed in terms of hydraulic conductivity, $K_{\perp HF}$ the equation becomes:

$$K_{\perp HF} = \alpha b_{\perp HF}^2, \tag{15}$$

Calculation of $K_{\perp HF}$ for the heterogeneous fracture is similar to the solution for layered aquifers, but is different due to the structure of the solution for $b_{\perp HF}$.

The asymptotic solution for the heterogeneous fracture is found by modifying Equation 14 for the two-layer system. As $H_b \to 0$ the asymptotic solution becomes

$$b_{\perp HF} = \sqrt[3]{\frac{L_T b_L^3}{L_L}} \quad \text{as } H_0 \to 0. \tag{16}$$

Table 1. This reference table organizes the derivations for the asymptotic limit of layered aquifers and heterogenous fractures. *For expression $\alpha$ see equation 11.

| Layered Aquifer | Heterogeneous Fracture |
|---|---|
| **PERPENDICULAR** | |
| $\mathbf{K_{\perp LA}}$ **(cm s$^{-1}$)** | $\mathbf{K_{\perp HF}}$ **(cm s$^{-1}$)** |
| $K_{\perp LA} = \dfrac{m_T}{\dfrac{m_H}{K_H} + \dfrac{m_L}{K_L}}$ (Eq. 7) | $K_{\perp HF} = \alpha \left( \dfrac{L_T}{\left( \dfrac{L_H}{b_H^3} + \dfrac{L_L}{b_L^3} \right)} \right)^{2/3}$ (Eq. 14)* |
| **Asymptote as $\mathbf{H_K \rightarrow 0}$** | **Asymptote as $\mathbf{H_b \rightarrow 0}$** |
| $K_{\perp LA} = \dfrac{m_T}{m_L} K_L$ (Eq. 8) | $K_{\perp HF} = \alpha \left( \dfrac{L_T b_L^3}{L_L} \right)^{2/3}$ (Eq. 17) |
| **PARALLEL** | |
| $\mathbf{K_{\parallel LA}}$ **(cm s$^{-1}$)** | $\mathbf{K_{\parallel \underline{HF}}}$ **(cm s$^{-1}$)** |
| $K_{\parallel LA} = \dfrac{K_L m_L + K_H m_H}{m_T}$ (Eq. 3) | $K_{\parallel HF} = \sum_{n=1}^{i} \dfrac{K_L L_L + K_H L_H}{L_T}$ (Eq. 5) |
| **Asymptote as $\mathbf{H_K \rightarrow 0}$** | **Asymptote as $\mathbf{H_K \rightarrow 0}$** |
| $K_{\parallel LA} = K_H \dfrac{m_H}{m_T}$ (Eq. 4) | $K_{\parallel HF} = K_H \dfrac{L_H}{L_T}$ (Eq. 6) |

which is similar to the layered aquifer solution Equation 8.Using Equation 15, the asymptotic solution for hydraulic conductivity becomes:

$$K_{\perp HF} = \alpha \left( \frac{L_T b_L^3}{L_L} \right)^{2/3} \quad \text{as } H_b \rightarrow 0. \tag{17}$$

Table 1 is a summary of all the formulas derived and the asymptotic limit (pure discrete flow) of average hydraulic conductivities for the aquifer and fracture.

## 4 EXAMPLES

An example case is demonstrated for the parallel and perpendicular cases to observe when heterogeneity controls both cases. Both examples attempt to use numbers that would be reasonable for field applications.

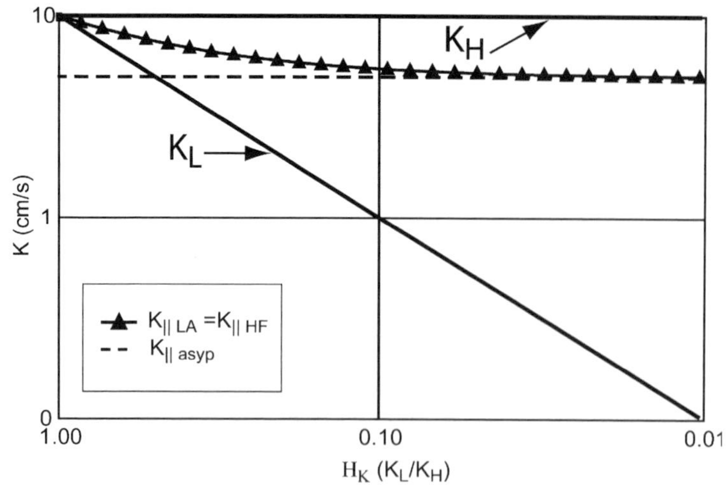


Figure 2. This figure indicates that $K_{\|}$ for aquifer and fracture are similar, and the asymptotic value is controlled by the high conductivity layer.

## 4.1 *Parallel*

A thickness (m) of 50 cm was defined for each of the two layers and a length (L) of 50 cm for each aperture segments. This equates to a fracture aperture of approximately 0.03 cm. For both cases the high conductivity layer was defined to be 10 cm $s^{-1}$. Average hydraulic conductivity values for the two-layer aquifer and two heterogeneous aperture segments parallel to heterogeneity were found using Equation 3 and 5 (Table 1).

The asymptotes were found using Equation 4 since they are identical for the parallel case. For the layered aquifer case, the asymptote is 5.0 cm $s^{-1}$. The system behavior for both the heterogeneous layered aquifer and heterogeneous fracture are similar. Both systems reach an asymptotic limit that can by referring to the formulas found in Table 1. At approximately 1 order of magnitude of heterogeneity ($H_K = 0.1$) (Figure 2), 90% of the asymptotic limit of the hydraulic conductivity for the layered aquifer and heterogeneous fracture is reached. At approximately 2 orders of magnitude of heterogeneity ($H_K = 0.01$), 99% of the asymptotic limit has been reached for layered aquifers and heterogeneous fractures.

## 4.2 *Perpendicular*

The perpendicular example can be used to evaluate the differences between average hydraulic conductivity of the layered aquifer and heterogeneous fracture. A high conductivity layer of 10 cm $s^{-1}$ will be assumed and a thickness (m) of 50 cm for each aquifer layer and a length (L) of 50 cm for each aperture, just as in the parallel case. Using Equations 7 and 15 (Table 1), $K_{\perp}$ is found for layered aquifers and heterogeneous fracture segments. Heterogeneity in fractures affects average hydraulic conductivity more than in layered aquifers. To find the asymptotic limits of the average hydraulic conductivity, we refer back to Equation 8 and 17. Using the equations Figure 3 is acquired and the limits are plotted.

As $K_L$ continues to decrease while $K_H$ remains constant the heterogeneous fracture is influenced more by the heterogeneity and thus the average is closer to the $K_L$ than in the layered aquifer. For the fracture case, the aperture variation needed to reach the asymptote

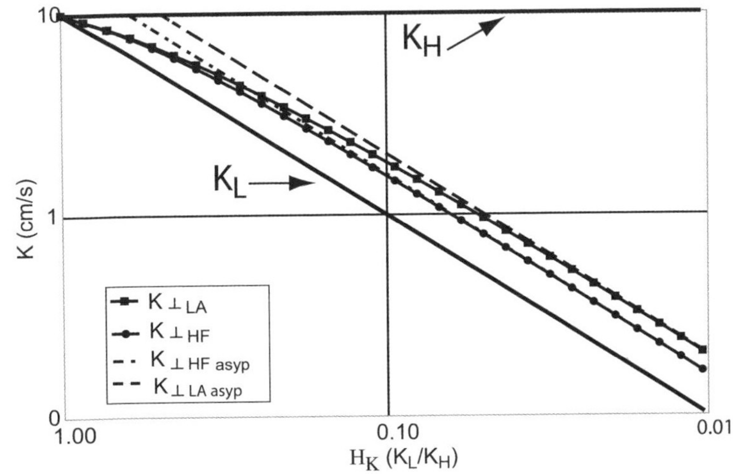


Figure 3. The derived asymptotic limits are following $K_\perp$ calculations.

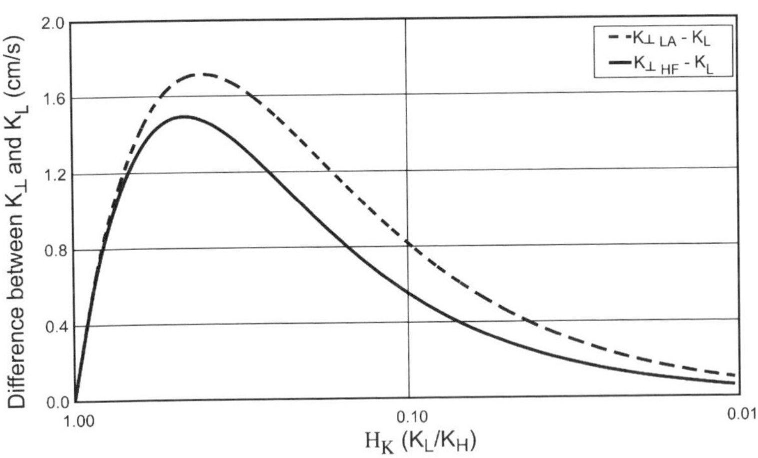


Figure 4. The difference between flow perpendicular to heterogeneity in the layered aquifer and heterogeneous fracture.

is approximately 25%. It can be seen that a separation occurs between the layered aquifer and heterogeneous fracture (Figure 4).

By subtracting the difference between the $K_{\perp\,HF}$ and $K_{\perp\,LA}$ from $K_L$ it is possible to analyze the difference between the layered aquifer and heterogeneous fracture (Figure 4). The heterogeneous fracture is more sensitive for flow perpendicular to heterogeneity than in the layered aquifer case.

## 5 DISCUSSION

Simple analytical solutions provide insight into what factors may control physical systems. While they are not expected to explain a full range of system variability, they provide insight into what factors are important to characterize complex systems. The simple two-layer model system allows for a better understanding of what controls layered aquifers and heterogeneous fractures. It has been difficult to gain instincts of how heterogeneity affects layered aquifers and heterogeneous fractures since simple tools for assessing these aquifers are lacking.

It can be seen that the degree of heterogeneity needed to reach the asymptotic value for the parallel case is only one order of magnitude in the example used. Thus, if there is only one order of magnitude of heterogeneity in the layered aquifer or heterogeneous fracture an assumption can be made that in most cases flow properties are controlled by the high hydraulic conductivity layer. At approximately two orders of magnitude heterogeneity, the average reaches 99% of the asymptotic value and the low conductivity layer has little or no effect, except as a geometric factor of scale.

Flow perpendicular to heterogeneity for the layered aquifer and heterogeneous fracture also reach the asymptotic values very quickly. At approximately one order of magnitude of heterogeneity, the average is 99% of the asymptotic value for the heterogeneous fractures has been reached. For the layered aquifer, at one order of magnitude, 94% of the asymptotic value is reached. This indicates that heterogeneous fractures are more sensitive to heterogeneity.

If the heterogeneity index is used to analyze heterogeneous systems the degree of heterogeneity can easily be assessed. The degree of heterogeneity needed to affect hydraulic conductivity values is minimal. Assessing layered aquifers and heterogeneous fractures also provides a teaching method for the classroom to better understand the effects of heterogeneity.

Complex models can be created to determine flow properties of an aquifer. Layered aquifer studies usually entail more than two heterogeneous layers, and the averaging to obtain flow properties does not allow for an analysis of the affect each layer has on flow properties. A simple two-layer system allows a simplification that creates instincts as to what degree of heterogeneity affects hydraulic conductivity. Previous conceptions regarding layered aquifers and heterogeneous fractures have had difficulty providing field tools to researchers to evaluate aquifer characteristics. The heterogeneity index defined in this paper will provide researchers a tool to quickly and effectively analyze aquifers. This type of analysis will likely lead to more work on discrete models since continuum models generally assume most of the flow system is active. These simple models of heterogeneity indicate that continuum flow is likely the extraordinary, not the ordinary case.

Future work includes applying the two layer approach to field data. The results will provide a framework so future studies can apply this method in field studies. This paper demonstrates hydraulic conductivity characteristics at both the micro and macro scale and is a first step to providing geologists a field tool to conduct an onsite estimation of subsurface flow properties in heterogeneous aquifers. The value of the two-layered system is also useful as an educational tool for providing new students with a better understanding of how heterogeneity affects flow systems.

## 6 CONCLUSIONS

By defining a simple two layer hydrogeological system, flow parallel to the system with defined mechanical apertures produces results that indicate aquifer average hydraulic conductivities and aperture average hydraulic conductivities are analytically different. The result of conductivity averaging values indicates heterogeneity in apertures is more restrictive in fractures than in aquifers. Results indicate (1) Fracture plane heterogeneity averaging is similar to layered aquifer averaging of hydraulic properties, but the effect can be stronger than in layered aquifers. (2) Flow parallel to the fracture plane heterogeneity the averages are identical to parallel flow in layered aquifers. (3) When flow is perpendicular to the fracture plane heterogeneity, the averages are more strongly controlled by the low hydraulic conductivity portion than in the layered case. (4) For the fracture case, aperture variation can also be very low (approximately 25%) to reach asymptotic conditions. (5) The asymptotic case provides cleaner solutions for heterogeneous extremes (~1 order of magnitude) in both systems.

## 7 VARIABLE DEFINITIONS

| | |
|---|---|
| $\alpha$ | constant ($\rho g/\mu$) [1/Lt] |
| $b_H$ | larger aperture value [L] |
| $b_i$ | aperture of the *i*th layer [L] |
| $b_L$ | smaller aperture value [L] |
| $b_{\perp HF}$ | average aperture for the heterogeneous fracture [L] |
| $\Delta h_i$ | change in head of the *i*th layer [L] |
| $\Delta h_T$ | change in total head [L] |
| $H_b$ | heterogeneity index for heterogeneous fractures [$b_L/b_H$] |
| $H_K$ | heterogeneity index for hydraulic conductivity [$K_L/K_H$] |
| g | gravitational constant [$L/t^2$] |
| $K_H$ | high hydraulic conductivity (relative to $K_L$)[L/t] |
| $K_L$ | low hydraulic conductivity (relative to $K_H$)[L/t] |
| $K_{\parallel HF}$ | average hydraulic conductivity parallel to heterogeneous fracture [L/t] |
| $K_{\parallel LA}$ | average horizontal conductivity parallel to layering [L/t] |
| $K_{\parallel n}$ | horizontal homogeneous conductivity of the *n*th layer [L/t] |
| $K_{\perp HF}$ | average perpendicular conductivity for the heterogeneous fracture [L/t] |
| $K_{\perp LA}$ | average perpendicular conductivity of the layered aquifer [L/t] |
| $K_{\perp n}$ | perpendicular homogeneous conductivity of the *n*th layer [L/t] |
| $L_H$ | length of the high K layer [L] |
| $L_i$ | length of the *i*th layer [L] |
| $L_L$ | low of the low K layer [L] |
| $L_n$ | length of the *n*th layer [L] |
| $L_T$ | total length [L] |
| $m_H$ | high conductivity layer thickness [L] |
| $m_L$ | low conductivity layer thickness [L] |
| $m_n$ | thickness of the *n*th layer of the layered aquifer [L] |
| $m_T$ | total aquifer thickness [L] |
| $\mu$ | dynamic viscosity of fluid [M/Lt] |
| $Q_i$ | discharge of the *i*th layer [$L^3/t$] |
| $Q_T$ | total system discharge [$L^3/t$] |
| $\rho_w$ | density of water [$M/L^3$] |
| w | width perpendicular to flow [L] |

## ACKNOWLEDGEMENTS

We would like to thank Dr. John M. Sharp Jr. at the University of Texas for his assistance in the development of this research. The authors would also like to express gratitude to Tom Fenstemaker for reviewing the paper and providing valuable feedback.

## REFERENCES

Bakr, A.A., Gelhar, L.W., Gutjahr, A.L., MacMillan, J.R. (1978) Stochastic analysis of spatial variability in subsurface flows, 1, comparison of one- and three-dimensional flows. Water Resources Research, 14(2):263–271.

Brown, D.M. (1984) Stochastic analysis of flow and solute transport in a variable-aperture rock fracture. M.S. thesis, Mass. Inst. of Technol., Cambridge, Mass.

Dagan, G. (1989) Flow and transport in porous formations. Springer-Verlag, New York.

Fetter, C.W. (2001) Applied Hydrogeology, 4th edition, Prentice Hall: Upper Saddle River, New Jersey.

Freeze, R.A. (1975) A stochastic-conceptual analysis of one-dimensional ground water flow in nonuniform homogeneous media. Water Resources Research, 11(5):725–741.

Gelhar, L.W. (1974) Stochastic analysis of phreatic aquifers. Water Resources Research, 10(3):539–545.

Gelhar, L.W. (1977) Effects of hydraulic conductivity variations on ground water flows. Proceedings, Second International IAHR Symposium on Stochastic Hydraulics, Aug. 2–4, 409–431.

Gutjahr, A.L. (1978) Stochastic analysis of spatial variability in subsurface flows, 2, evaluation and application. Water Resources Research, 14(5):953–959.

Heath, R.C. (1982) Classification of ground water systems of the United States. Ground water 20:393–401.

Kischinhevsky, M., Paes-Leme, P.J. (1997) Modelling and numerical simulations of contaminant transport in naturally fractured porous media. Transport in Porous Media 26:25–49.

Leveinen, J., Rönkä, E., Tikkanen, J., Karro, E. (1998) Fractional flow dimensions and hydraulic properties of a fracture-zone aquifer. Leppävirta, Finland, Hydrogeology Journal 6:327–340.

Mantoglou, A., Gelhar, L.W. (1989) Three dimensional unsaturated flow in heterogeneous systems and implications on ground water contamination: a stochastic approach. Transport in Porous Media 4:529–548.

Neuzil, C.E., Tracy, J.V. (1981) Flow through fractures. Water Resources Research 17:191–199.

Ogata, A., Banks, R.B. (1961) A solution of the differential equation of longitudinal dispersion in porous media. Geological Survey Professional Paper, U.S. Government Printing Office, Washington, A1–A6.

Or, D., Tuller, M. (2000) Flow in unsaturated fractured porous media: hydraulic conductivity of rough surfaces. Water Resources Research 36(5):1165–1177.

Renshaw, C.E., Dadakis, J.S., Brown, S.R. (2000) Measuring fracture apertures: A comparison of methods. Geophysical Research Letters 27(2):289–313.

Schulze-Makuch, D., Cherkauer, D.S. (1998) Variations in hydraulic conductivity with scale of measurement during aquifer tests in heterogeneous, porous carbonate rock. Hydrogeology Journal 6:204–215.

Tsang, Y.W., Tsang, C.F. (1989) Flow channeling in a single fracture as a two-dimensional strongly heterogenous permeable medium. Water Resources Research 25(9):2076–2080.

Wanfang, Z., Wheater, H.S., Johnston, P.M. (1997) State of the art of modelling two-phase flow in fractured rock. Environmental Geology 31(3/4):157–166.

Zimmerman, R.W., and Bodvarsson, G.S. (1996) Hydraulic conductivity of rock fractures. Transport in Porous Media 23:1–30.

Zlotnik, V.A., Zurbuchen, B.R. (2003) Field study of hydraulic conductivity in a heterogeneous aquifer: Comparison of single-borehole measurements using different instruments. Water Resources Research 39(4):8–2:8–12.

CHAPTER 43

# Time-resolved 3D characterisation of flow and dissolution patterns in a single rough-walled fracture

Catherine Noiriel[1,2], Philippe Gouze[1] and Benoît Madé[2]

[1] *Laboratoire de Tectonophysique, Institut des Sciences de la Terre, de l'Environnement et de l'Espace, CNRS – Université de Montpellier II, France*

[2] *Centre de Géosciences, École des Mines de Paris, France*

ABSTRACT: An application of X-ray computed microtomography (XCMT) for 3D measurement of fracture geometry is presented. The study demonstrates the ability of XCMT to non-invasively measure the fracture walls and aperture during the course of a reactive flow experiment. The method allows estimation of both the local and global scale dissolution kinetics of a fractured limestone sample percolated by acidic water. The measured fracture geometry was then used as an input for flow modelling, in order to compare the hydraulic aperture calculated by numerical simulation with different evaluations of the aperture: hydraulic aperture measured from pressure drop during the flow experiment, mechanical aperture measured with XCMT, and chemical aperture deduced from calcium removal in the sample. The effects of reactive transport on geometry and fluid flow are discussed. Dissolution appears heterogeneous at both the small scale due to the presence of insoluble clays in the rock, and at larger scales with the formation of preferential flow pathways. These heterogeneous dissolution patterns are not predictable simply by the identification of the areas of higher fluid velocity, where transport of the chemical reaction products (i.e., rate of aperture increase) is presumed to be higher.

## 1 INTRODUCTION

Fractures strongly control the flow and transport of fluids and pollutants in low-permeability rocks. Dissolution (or precipitation) processes may strongly influence the fracture geometry and consequently the hydraulic properties such as permeability and dispersivity. The prediction of flow and transport evolution in fractures is challenging, but appears essential to evaluate long term behaviour of geothermal systems, nuclear waste storage or $CO_2$ injection in depleted reservoirs. Carbonated environments, which supply an important part of the accessible potable water resources of the planet, are particularly sensitive to fluid-rock transfer processes over relatively short time periods. Karst formation is certainly the most remarkable example of limestone alteration.

A lot of experimental and numerical studies have been devoted to quantifying the control of various physical parameters on fluid flow and solute dispersion into fractures. Fracture roughness, aperture and surface correlation, tortuosity and contact areas have been pointed out as the essential parameters controlling flow and transport in fractures (Witherspoon et al., 1980; Tsang, 1984; Adler and Thovert, 1999; Zimmerman and Yeo., 2000). Initially, the fracture characteristics are forced by the mechanical processes of rupture and displacement (Yeo et al., 1998; Unger and Mase, 1993). In parallel, the geometry may be altered considerably by dissolution and precipitation that can feedback into the flow and transport properties. It is probable that in many applications, relevant either to geological time-scale modelling or predicting system feedback to anthropogenic forcing, fracture parameters must be considered as variables.

Transport of solutes in a fracture is described by the following macroscopic equation:

$$\partial_t C = \mathbf{u}\nabla C - \mathbf{D}\nabla^2 C + R(C) \tag{1}$$

where C is the concentration of the species, **u** is the velocity vector (whose components are $u_x$, $u_y$ and $u_z$), **D** is the hydrodynamic dispersion tensor and R(C) is the geochemical source term proportional to the dissolution rate. Hydrodynamic dispersion involves Taylor and geometrical dispersion, and molecular diffusion. In a fracture, solute transport is mainly controlled by the chemical reaction rate and the heterogeneity in the flow velocity. Positive feedback between flow regime and geochemical alteration can also occur, leading to instabilities and localization of the dissolution. The dominant parameters that control these phenomena at the macroscale are the Peclet number (Pe), the Damköhler number (Da) and the aperture variability, expressed as the ratio between the aperture standard deviation and its mean ($\sigma_a/\bar{a}$) (Dijk and Berkowitz, 1998; Hanna and Rajaram, 1998; Cheung and Rajaram, 2002; Verberg and Ladd, 2002; Szymczak and Ladd, 2004). O'Brien et al. (2003) assume that the initial fracture geometry plays an important role in determining the dissolution front and suggest that a better understanding of the heterogeneities in fractures is necessary to accurately model the reactive transport. Experimental effort is required to predict long-term evolution of such heterogeneous systems. However, experimental studies of fracture behaviour and related parameters during dissolution are limited (Durham et al., 2001; Detwiler et al., 2003; Dijk et al., 2002; Polak et al., 2004) and do not always include direct permeability and geometry measurements.

The aim of this paper is to present a methodology to study dissolution effects in a fracture by coupling chemical and hydrodynamic measurements with observation and quantification of the structural changes using X-ray microtomography. After a description of the experimental procedure (section 1), the measured changes to the fracture morphology and dissolution kinetics are presented (section 2). The measured fracture morphology was then used as input for numerical flow simulation. Afterwards, a comparison is made between the four different methods used to evaluate the changes to the fracture morphology: chemical and hydraulic measurements, imagery and flow simulation. Finally, the effect of rock mineralogy on flow, transport and geometry changes is discussed in section 3.

## 2 EXPERIMENTAL PROCEDURE

The experiment consisted of the percolation of an acidic fluid through a rough fracture. The rock was a slightly argillaceous micritic limestone containing about 10% silicate minerals (principally clays with a minor amount of quartz). A core of 15 mm long and 9 mm diameter was fractured using a Brazilian-like test to produce a longitudinal fracture parallel to the core axis, i.e. an increasing loading charge is applied on the core edges until rupture occurs.

Fracture edges were rigidified with epoxy resin in order to prevent any mechanical displacement of the fracture walls. Dissolution was obtained by flowing, at a constant rate of 2.78 $10^{-9}\,m^3.s^{-1}$ ($10\,cm^3.h^{-1}$) water equilibrated with carbon dioxide at the partial pressure of 0.1 MPa (controlling the inlet pH value at 3.9). The pressure at the outlet was maintained at 0.13 MPa in order to avoid $CO_2$ degassing in the circuit. The total duration of the experiment was 118.5 h, during which the fluid effluent was periodically sampled to evaluate the mineral mass removed. Permeability was calculated from the pressure difference between sample inlet and outlet ($\Delta P$) using steady state flow method at the initial state ($t_0$) and after the two stages of dissolution ($t_1 = 70.5$ h and $t_2 = 118.5$ h from the start of the experiment). At the same time, X-ray computed microtomography (XCMT) was used non-invasively to observe the tri-dimensional geometry of the sample with a spatial resolution (pixel size) of 4.91 μm.

XCMT is based on the 3D reconstruction of one thousand X-ray radiographies of the sample taken at different view angles on 180 degrees. The ID19 beam line of the European Synchrotron Radiation Facility (Grenoble, France) was used for this study. An area of about $10 \times 10$ mm ($2048 \times 2048$ pixels) located near the sample inlet is covered. After data processing (see Noiriel et al., 2005), 3D grey-level images represent the X-ray attenuation by the material in each point (voxel) in the space. As the sample is composed of only two materials (air and matrix) with different X-ray attenuation properties, histograms of grey-level attenuation are bimodal. By convention dark voxels correspond to low density material (void) and white voxels correspond to high density material (matrix). A segmentation procedure based on region growing is then used to convert grey-scale image into binary image (Nikolaidis and Pitas, 2001). Once voids and matrix are labelled in the image, it is easy to calculate the volume occupied by the fracture voids, the elevation of the lower and upper fracture walls ($h^-_{y,z}$ and $h^+_{y,z}$), the local aperture ($a_{y,z}$), and the area of the fluid-rock interface (Gouze et al., 2003). Figure 1 presents fracture aperture and surfaces obtained using XCMT after appropriate image processing.

Figure 1. Upper ($S^+$) and lower ($S^-$) surface and aperture representation of the fracture, measured at $t_0$ over a $9.25 \times 7.95$ mm region. Surfaces and aperture are obtained from processed XCMT data. Note that the grid lines are under-sampled by a factor 12 to allow visual representation.

Table 1. Statistics of the fracture geometry. Definitions are provided in the text.

| Time (h) | $t_0 = 0$ | $t_1 = 70h30$ | $t_2 = 118h30$ |
|---|---|---|---|
| $a_m$ (μm) | 40.5 | 328.0 | 418.5 |
| $\sigma_a$ (μm) | 18.8 | 92.2 | 163.9 |
| $a_m/\sigma_a$ | 2.17 | 3.57 | 2.56 |
| $\sigma_s$(μm) | 233.0 | 235.1 | 244.3 |
| $z_2$ | 3.70 | 4.02 | 7.98 |
| $\rho_{s^+-s^-}$ | 1.00 | 0.92 | 0.77 |

## 3 RESULTS

### 3.1 *Geometric features of fracture*

The statistics and features of changes to the fracture morphology during dissolution were studied. The mechanical aperture $a_m$ is deduced from the local aperture:

$$a_m = \bar{a}_{x,y} \frac{1}{Ll} \sum_{y=0}^{y=l} \sum_{z=0}^{z=L} a_{y,z} \tag{2}$$

where $a_{y,z}$ is the local aperture at the (y,z) location, $\bar{a}$ denotes the spatial average, and L and l are the fracture length and width, respectively. Statistical results are listed in Table 1, while aperture maps and histograms at the different stages of dissolution are displayed in Figure 2. An example of the cross section is also given in Figure 3. Dissolution effects appear to be different at two different scales.

#### 3.1.1 *Microscale pattern*

At the microscale (grain-scale, i.e., a few tens of microns), the different kinetic rates of dissolution between the minerals forming the rock cause heterogeneous dissolution to occur. As kinetics of clay and quartz dissolution is several orders lower than for calcite, these minerals remain at the fracture surface until they are flushed by the flowing fluid. As dissolution progresses, the fluid-rock interface appears rougher (Figure 3). Consequently, both the standard deviation for elevation of the surfaces ($\sigma_s$) and micro-roughness factor of the fracture surface ($z_2$) increase. Here $z_2$ is the root mean square of the first derivative of the surface asperity height (Myers, 1962):

$$z_2 = \sqrt{\frac{1}{Ll} \sum_{y=1}^{y=l} \sum_{z=1}^{z=L} \left(h_{y,z+1} - h_{y,z}\right)^2} \tag{3}$$

The presence of secondary fracture branches initially present in the sample promotes the detachment and displacement of rock fragments as soon as their size is sufficiently reduced by dissolution. As a result, the aperture is locally reduced as observed in Figure 2, b.

#### 3.1.2 *Macroscale pattern*

At the macroscale (sample-scale), dissolution substantially affects the shape of the aperture distribution. Before dissolution, the topography of the fracture aperture can be represented

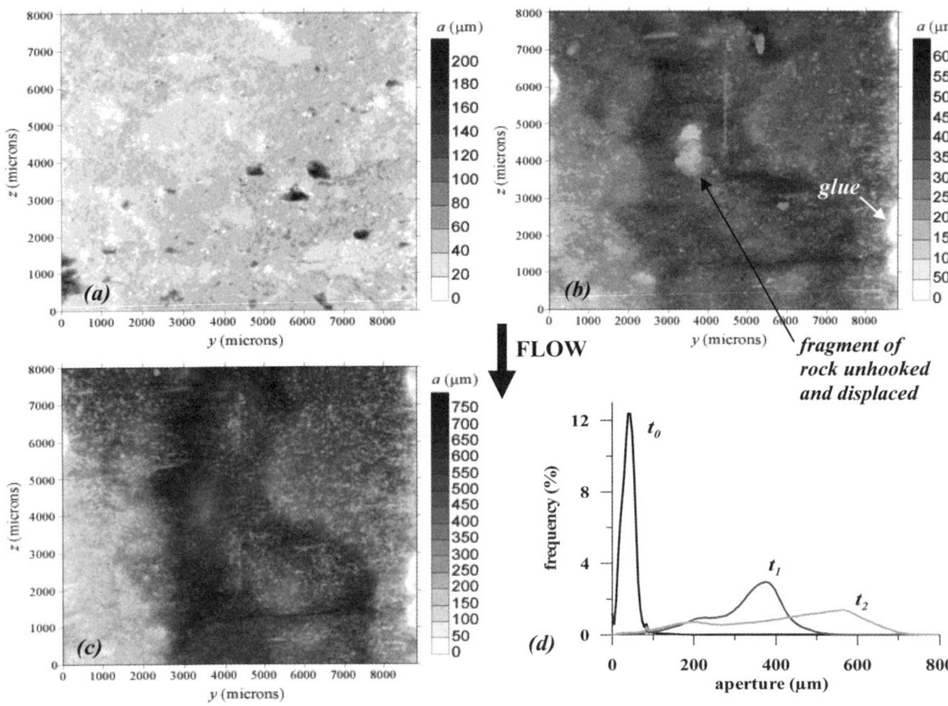


Figure 2. Aperture distribution (μm) within the fracture: (*a*) map at $t_0$, (*b*) at $t_1$ and (*c*) at the end of the experiment ($t_2$). The maps are voluntarily under-sampled by a factor 12 to allow better visualisation; note that the zero aperture areas on the edges of the maps correspond to epoxy resin spacer used to avoid closure of the fracture during the experiment. (*d*) Histograms of aperture distribution at the different time of experiment.

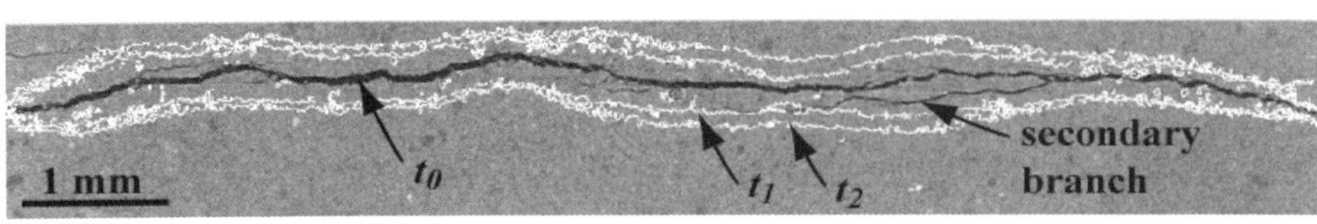


Figure 3. Visualisation of morphology changes in course of the experiment (2D cross-section). The grey level background displays the initial fracture geometry ($t_0$). The white lines correspond to the fracture wall position at $t_1$ and $t_2$. Initially the fracture contains secondary branches which result from the fracturing process.

as a bell-shaped distribution with positive kurtosis (2.43) and skewness (1.94); positive values indicate that the distribution is slightly pointed and present an asymmetry toward the large values in comparison with a Gaussian distribution. Results are similar to those usually observed experimentally or assumed in models of fracture generation. But as a consequence of dissolution, the histogram evolves toward a bimodal distribution due to the formation of preferential flow pathways in the fracture (Figure 2, d).

Heterogeneous dissolution at the two different scales is accompanied by a large increase of the standard deviation of mechanical aperture ($\sigma_a$). As a consequence, despite the increase of the mechanical aperture $a_m$, the ratio $a_m/\sigma_a$ is stable during the experiment.

The cross-correlation value between the two fracture walls is defined by:

$$\rho_{S^+-S^-} = E[(h^+_{(y,z)}-\bar{h}^+)(h^-_{(y,z)}-\bar{h}^-)]/E[(h^+_{(y,z)}-\bar{h}^+)^2]E[(h^-_{(y,z)}-\bar{h}^-)^2] \qquad (4)$$

where $h^+$ and $h^-$ is the elevation at the (y,z) location of the upper and lower fracture wall, respectively and $\bar{h}$ denotes the spatial average. At the start of the experiment the two fracture walls are perfectly correlated, but dissolution induces a progressive decorrelation of the fracture walls (Table 1).

### 3.2 *Dissolution rate calculation*

The calcium flux at the sample outlet can be related to the global (sample-scale) dissolution rate. It was variable but on the whole slightly decreased during the course of the experiment. Assuming that the sample is 90% calcite, that the molar volume of the clays and silicates was comparable to that of calcite, and that dissolution of the fracture walls was homogeneous, the aperture increase can be deduced from the following equation:

$$\partial_t a_c = -(Q \times \upsilon_{calcite} \times \Delta Ca)/(0.9 \times A_s) \qquad (5)$$

where $a_c$ denotes the "chemical" aperture, Q is the flow rate, $\upsilon_{calcite}$ is the molar volume of calcite, $\Delta Ca$ is the calcium concentration removed by the acidic fluid between sample inlet and outlet, and $A_s$ is the surface area of equivalent planar fracture walls. Assuming that the kinetic rate of dissolution was unchanged between the two experimental stages ($t_0$–$t_1$ and $t_1$–$t_2$, respectively), the spatial distribution of the dissolution rate is obtained by subtracting aperture distribution after registration of the images in the same spatial referential. The local aperture change with time is given by:

$$\partial_t a = (a_{ti+1} - a_{ti})/(t_{i+1} - t_i) \qquad (6)$$

Then, the local rate of dissolution $k_d$ is given by:

$$k_d = (\partial_t a \times S_{pix})/(\upsilon_{calcite} \times S_r) \qquad (7)$$

where $S_{pix}$ is the pixel area ($4.91 \times 4.91\ \mu m^2$) and $S_r$ is the reactive surface area of the mineral. As $S_r$ is an unknown parameter, it is chosen to be equal to the geometric surface area, i.e. $S_{pix}$. The results for the dissolution stages $t_1 - t_0$, and $t_2 - t_1$ are reported in Figure 4. Maps show that the dissolution rate is heterogeneous both in the flow direction and perpendicularly to it, $k_d$ ranging from 0 to 6.0 $10^{-9}$ mol.m$^{-2}$.s$^{-1}$.

Three major phenomena potentially control the rate of calcite dissolution, $k_d$. The first phenomenon concerns the level of disequilibrium of the solution in regards with the calcite mineral. It is commonly assumed that the rate of dissolution is proportional to the saturation index $\Omega$: $k_d \propto (1 - \Omega)^n$, with $n \in \mathbb{R}$ (Lasaga, 1998). The saturation index increases as far as the reaction progress towards equilibrium ($\Omega \rightarrow 1$), so $k_d$ decreases accordingly. In the experiment, the solution remains undersaturated with respect to calcite

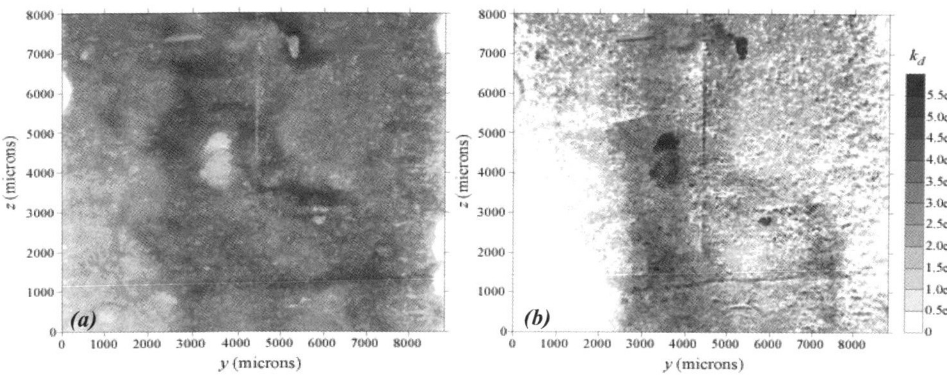


Figure 4. Distribution of the computed rate of dissolution $k_d$ ($mol.m^{-2}.s^{-1}$) corresponding to the dissolution stage (*a*) $t_0$–$t_1$ and (*b*) $t_1$–$t_2$.

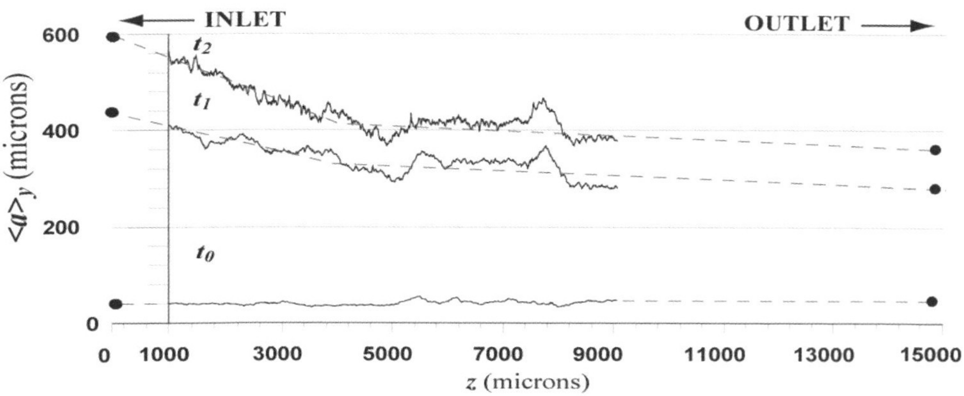


Figure 5. Aperture average along the flow axis showing that dissolution is more pronounced near the sample inlet. Black lines represent aperture calculated from experimental data; black dots represent estimation of the aperture at the sample inlet and outlet.

($\Omega \sim 0.15$ at the outlet), so that the influence of $\Omega$ is relatively weak. The second phenomenon concerns the dissolution kinetics of calcite at low pH values. Far from equilibrium and at relatively low pH (e.g., pH $< 5.5$), the kinetic rate of calcite dissolution (k) is pH dependent (Plummer et al., 1978). Because protons $H^+$ are consumed during the chemical reaction of calcite dissolution, pH increases progressively during the chemical reaction, so the kinetic rate of dissolution decreases. In the experiment pH increases progressively from $3.9 \pm 0.1$ (corresponding to pH of the inlet fluid equilibrated with $CO_2$ at the partial pressure of 0.1 MPa) to $5.3 \pm 0.2$, as the fluid goes along the fracture and reactions between $CO_{2\,(aq)}$, $H^+$ and calcite occur. According to the equation given by Plummer et al. (1978) to describe the kinetic rate of calcite dissolution as a function of pH and $P_{CO_2}$, k decreases theoretically from $7.7\ 10^{-9}$ $mol.m^{-2}.s^{-1}$ at pH 3.9 to $1.54\ 10^{-9} mol.m^{-2}.s^{-1}$ at pH 5.3. In other words, k is reduced by a factor 5. This phenomenon caused the dissolution rate to be greater at the sample inlet (Figure 5). But the ratio between averaged aperture estimated at the inlet and outlet is about 1.8 both after $t_1$ and $t_2$, which is lower than the value of 5 expected by the model of Plummer et al. (1978). The third phenomenon concerns the control of the reaction. For calcite at low pH, the kinetic rate of the chemical reaction at the solid surface $k_S$ is substantially faster than the kinetics of diffusion of reactants and products close to the surface $k_T$ (Rickard and Sjöberg, 1983). So the global kinetics of reaction is limited by the mass transfer of reactants and products at the fluid-rock interface. In moving solution, the advective transport is substantially faster than molecular diffusion; therefore the advective transport near the fluid-rock interface can control the rate of reaction. As the flow is localized in channels, fluid velocity in the fracture edges becomes slow compared to velocity in the channels (see next section). The dissolution is then localized where transport is higher, i.e. in the preferential flow pathways (Figure 4, *a* and *b*).

### 3.3 *Flow simulation*

The finite volume code Fluent® was used to solve the flow equations in a 9.25 × 7.95 mm region of the fracture at $t_0$, $t_1$ and $t_2$. A computational grid of 423,900 elements was generated from two fracture walls extracted from the XCMT images. Pixel size renormalisation (initially 4.91 μm in all directions) was applied to allow tractable computation of the 3D flow field. Finally, the fracture was square meshed by 157 × 135 elements of size 58.92 μm in the y–z plane. The fracture aperture (x-direction) was meshed by 20 elements of variable size; cell height growing (parabolic function) from the fracture wall to the middle to refine the grid in the area of higher flow gradient and correctly simulate the quasi-Poiseuille flow distribution across the fracture. The boundary conditions were taken to be constant flow rate ($10\,cm^3.h^{-1}$) at the sample inlet, uniform pressure (0.1 MPa) at the outlet, and no flux at the fracture walls. The velocity vector **u** was obtained by solving the Navier-Stokes (NS) equation system (mass and momentum conservation equations):

$$\begin{cases} \rho(\mathbf{u}\cdot\nabla)\mathbf{u} = -\nabla P + \mu\nabla^2\mathbf{u} \\ \nabla\cdot\mathbf{u} = 0 \end{cases} \tag{8}$$

where P is the effective pressure, ρ is the fluid density and μ is the dynamic viscosity.

The fluid velocity averaged in the x-direction is displayed for $t_0$, $t_1$ and $t_2$ in Figure 6. Initially, the velocity field appears heterogeneous at small scales and the flow was not particularly localized. At $t_1$, the flow became localized in preferential pathways, and at $t_2$ only one principal channel remains visible. Due to the fracture opening, the mean velocity in the fracture ($\overline{u}$) decreased during dissolution, from $7.7\ 10^{-3} m.s^{-1}$ at $t_0$ to $6.1\ 10^{-4} m.s^{-1}$ at $t_2$. Despite the changes in the flow patterns, the ratio of the standard deviation to the mean fluid velocity remained almost constant during the experiment ($\overline{u}/\sigma_u$ varies from 2.2 to 1.4). However, the spatial correlation of the flow velocity increases largely in the z-direction due to the formation of channels in the flow direction.

The shape of the velocity profiles across the fracture aperture can be classified in three types (Tenchine and Gouze, 2005): (1) profiles that are parabolic and centred in conformity with the Poiseuille equation of fluid velocity between two parallel plates: ($u_z(x) = u_o\lfloor 1-(2x/a)^2 \rfloor$, (2) profiles that are parabolic but asymmetric and (3) profiles that are not parabolic; here $u_0$ is the maximum velocity in the parabolic profile. At $t_0$, 40% of the profiles are centred-parabolic. Development of preferential flow paths, decorrelation of the surfaces and increasing roughness cause the percentage of centred-parabolic profiles to dramatically decrease (only 4% remain at $t_2$). The percentage of non-parabolic profiles increases accordingly, whereas the percentage of asymmetric profiles stays almost stable during the dissolution process (between 60–75%). Asymmetric profiles were also observed in fractured flow by Dijk et al. (1999) when studying fluid flow using nuclear magnetic resonance imaging (NMRI).

## 4 DISCUSSION

### 4.1 *Aperture*

Navier-Stokes equations for flow between two parallel plates of separation a (width aperture) can be reduced to the well-known cubic law, which stipulates that the flow rate is proportional to the cube of aperture:

$$Q = -a^3 l \Delta P / 12 \mu L \quad (9)$$

where l and L are the fracture width and length, respectively. The parallel plate assumption implies that both the fracture roughness and the tortuosity play no role in affecting the fluid flow. However the geometry of a natural fracture diverges more or less from the parallel plate model, and deviations to the cubic law are observed when replacing a by the

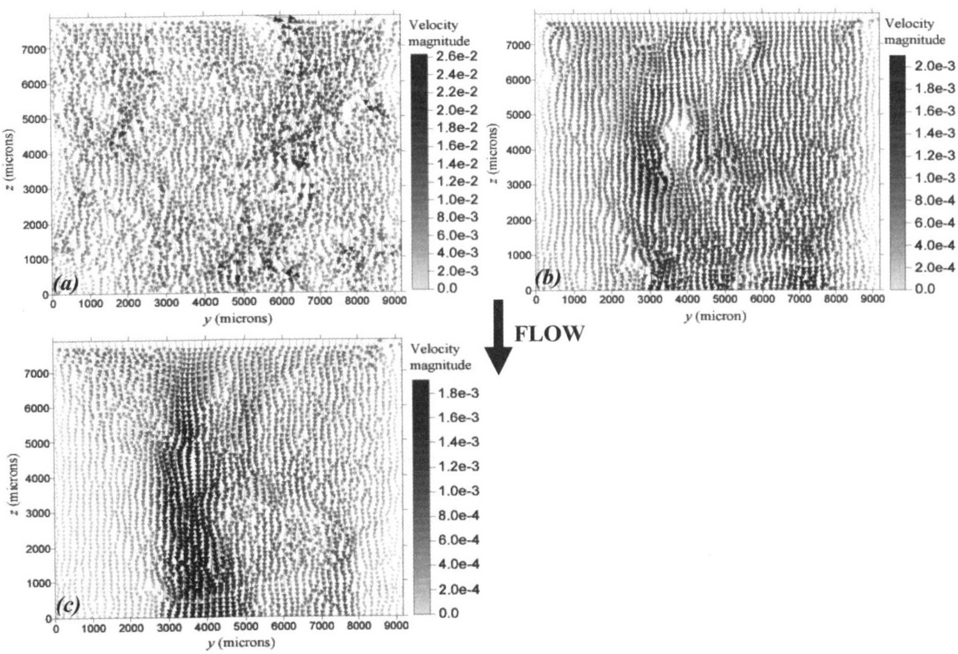


Figure 6. Velocity field (direction and intensity) from Navier-Stokes simulations in the fracture (*a*) initially ($t_0$), (*b*) at $t_1$ and (*c*) after the experiment ($t_2$).

Table 2. Comparison of the different types of aperture: mechanical aperture ($a_m$), hydraulic aperture measured in the experiment ($a_{h\text{-}EXP}$), hydraulic aperture calculated by Navier-Stokes simulation ($a_{h\text{-}NS}$), hydraulic aperture calculated by Patir and Cheng (1978) formula ($a_{h\text{-}REY}$), and chemical aperture calculated from the calcium release ($a_c$).

| | $a_m$ | $a_{h\text{-}EXP}$ | $a_{h\text{-}NS}$ (μm) | $a_{h\text{-}REY}$ | $a_c$ |
|---|---|---|---|---|---|
| Initial ($t_0$) | 40.5 | 30 | 32.3 | 35.8 | 40.5 |
| $t_1$ | 328.0 | 110 | 268. | 297 | 357.5 |
| Final ($t_2$) | 418.5 | – | 298 | 362 | – |

average mechanical aperture $a_m$ in equation 9. Nevertheless, application of the cubic law remains valid if the local aperture a in equation 9 is replaced by an equivalent hydraulic aperture $a_h$, which indirectly integrates the geometry of the fracture:

$$Q = -a_h^3 \, l \, \Delta P/12\mu L \tag{10}$$

So the hydraulic aperture, which is measured experimentally or calculated by numerical simulation, can be compared with $a_m$ in order to evaluate the deviation from the cubic law. Using finite differences to resolve the local cubic law equation, Patir and Cheng (1978) performed flow between surfaces whose profiles obeyed a Gaussian distribution with linearly-decreasing autocorrelation function. The authors found that the hydraulic aperture could be fit by an exponential relation:

$$a_{h-REY}^3 = a_m^3(1 - 0.9 \exp(-0.56a_m/\sigma_a)) \tag{11}$$

In their model, deviation to the cubic law is expected when $a_m/\sigma_a$ is below a value of 10. Results obtained by Brown (1987) using similar calculations on several generated fractal fractures with different fractal dimensions (from 2.0 to 2.5), and Zimmerman and Bodvarsson (1996) using high-order approximations for lognormal distribution of permeability fall close to equation 11. Their results suggest that the formula is applicable to fractures with different geometrical characteristics.

For this study, the change in fracture aperture was evaluated using five different methods: (1) XCMT measurement ($a_m$), (2) differential pressure recorded during experiment using equation 10 ($a_{h\text{-}EXP}$), (3) differential pressure deduced from the flow numerical simulation using equation 10 ($a_{h\text{-}NS}$), (4) Reynolds approximation using equation 11 ($a_{h\text{-}REY}$), and (5) chemical aperture evaluated from equation 5 ($a_c$). As is typically found, the hydraulic aperture calculated by Navier-Stokes simulation differs from the mechanical aperture. Initially, the difference was 24% and $a_{h\text{-}NS}$ was close to $a_{h\text{-}EXP}$ measured experimentally ($a_{h\text{-}EXP} = 30\,\mu m$). The effective aperture $a_{h\text{-}REY}$ calculated empirically from the Reynolds calculation differed from $a_{h\text{-}NS}$ but was logically intermediate between $a_{h\text{-}NS}$ and $a_m$ (Table 2). Large discrepancies between Reynolds numerical aperture estimates and experimental measurements were also reported by Nicholl et al. (1999), indicating that a 2D description of the flow field is inappropriate for fully describing fluid flow. According to equation 11, the deviation from the cubic law is proportional to the roughness factor $a_m/\sigma_a$. The ratio $a_m/\sigma_a$ remained stable due to the effect of the heterogeneous dissolution, although $a_m$ increased during the experiment. This explains why the ratios $a_m/a_{h\text{-}NS}$ or $a_m/a_{h\text{-}REY}$ remained almost constant. Finally, note that $a_c$ and $a_m$ are coherent; the difference is certainly due the fact that $a_m$ was calculated on a limited part of the fracture whereas the calculation of $a_c$ integrates the entire fracture.

## 4.2 *Implication of Mineralogy on Flow, Transport and dissolution patterns*

The study of Gouze et al. (2003) investigated dissolution effects on two limestone rocks with different mineralogical composition. One sample was composed almost completely of calcite, whilst a second also contained some dolomite, quartz and clay fraction (~15%). In the first case, they observed that dissolution was relatively homogeneous and could be reduced to a translation of the fracture walls. Consequently, $\sigma_a$ remained quasi constant and $a_m/\sigma_a$ increased from 6.2 to 12.2. So, it is expected (eq. 11) that the flow can be progressively idealized with the parallel plate model. In the second case, they showed that dissolution appeared to be very heterogeneous due to the differential kinetics between minerals. The value of $\sigma_a$ increased significantly and $a_m/\sigma_a$ globally decreased from 3.0 to 2.0 during the experiment. In this case, important deviations from the cubic law are expected.

Transport of solutes in a fracture is largely dependent on the chemical reaction rate and of the heterogeneity in the flow field (equation 1). The two parameters that can describe these effects are the Damköhler number ($D_a = k_d L/\bar{u}$) and Peclet number ($Pe = \bar{u}L/D_m$, where $D_m$ is the molecular diffusion coefficient). Da weighs the relative influence of chemical reaction rates to advective transport. Pe weighs the relative magnitude of advective to diffusive transport. When Da is large, the rate of the chemical reaction is larger than the rate at which the fluid is transported. In this case, dissolution patterns depend largely on the Peclet number, as observed by Detwiler et al. (2003). At high Pe, the dissolution is relatively homogeneous whereas at low Pe flow localizes in preferential channels. The process that accelerates flow localisation and channel growth is the reactive infiltration instability described by Ortoleva et al. (1987). Initial heterogeneities in the fracture aperture cause the flow to be localized in preferential paths. Since the permeability is larger within preferential paths, flow and transport are focused within these paths leading to a rapid enlargement of the channels and slower dissolution outside the channels. Thus it is expected that the development of channels is related to initial heterogeneities of the fracture aperture and flow field. During their dissolution experiment of halite using nuclear magnetic resonance imaging, Dijk et al. (2002) also noticed that mineralogy heterogeneities can have an impact on dissolution patterns particularly at low Damköhler number.

In the present study, the flow patterns at $t_1$ and $t_2$ were correlated to the fracture morphology: the larger the aperture increase, the higher the flow velocity (Figure 4 and 6). However, there wasn't any relation evident between the initial flow field at $t_0$ and the one after dissolution. In particular, areas of higher flow velocity at $t_0$, where dissolution and transport are presumed to be more important, were not correlated with the areas where channels formed. This observation is opposite to equation 1, which stipulates that the greater the flow velocity, the greater the transport (and so the local dissolution rate). Moreover, $a_{h\text{-}EXP}$ measured experimentally at $t_1$ differs largely from that calculated by the other methods (Table 2). When looking more precisely at the fracture void, it can be observed that a microporous phase composed of clays and quartz remains in the fracture edges (Figure 7); the presence of these phases is underestimated in the segmentation procedure (Figure 3). So the fracture aperture and flow velocity are overestimated when clays and quartz remain in the fracture void. In consequence, head losses are probably higher than those calculated by

Figure 7. 2D cross-section of the fracture at the end of the experiment (x-y plane). The white layouts represent areas where clays and quartz remain in the fracture void (note that quartz crystals of 50–100 μm size appear in high grey level).

the flow simulation, explaining that $a_{h\text{-}EXP}$ at $t_1$ highly differs from $a_m$ or $a_{h\text{-}NS}$. It is also probable than the mineral heterogeneities of the rock matrix are responsible of the unpredictable evolution of the fracture geometry with dissolution.

## 5 CONCLUDING REMARKS

The objective of this work was to obtain well-constrained experimental data to study the effects of dissolution on fracture geometry and flow evolution. The remarkable performance of X-ray computed microtomography has proven to be an efficient tool to study structural changes from the scale close to the mineral grain-size to the scale relevant to macroscopic characterization of the processes.

Heterogeneous dissolution of the sample was observed both at micro- and macroscale (μm-scale and cm-scale). The fracture opening was subjected to perturbation due to the presence of low-solubility minerals in the rock. Numerical simulation has permitted the development of preferential flow paths to be related to the decorrelation of the surfaces and the increase of the surface roughness.

The rock mineralogy appears to be a very important parameter that controls the changes in the fracture morphology, flow and transport in response to the chemical alteration. By introducing heterogeneities at the fracture walls, the roughness factor that controls deviation to the cubic law remains unchanged despite aperture increase. Formation of a microporous phase in the fracture void slows down flow and transport (and by consequence the rate of aperture opening), whereas a broad channel develops when particles are removed by the fluid. In consequence, the evolution of fracture morphology in natural rocks whose mineralogical composition is variable appears to be more unpredictable than experimentally observed or predicted in mono-mineral rocks. The results indicate that further research is necessary to understand the complex coupling between chemical reactions, petrophysical properties, fracture wall morphology and flow.

## ACKNOWLEDGMENTS

We thank the ESRF-ID19 team (Xavier Thibault, Elodie Boller and Peter Cloetens) for support during the data acquisition. We are also grateful to Frank Denison and the anonymous reviewer who helped us to improve the original manuscript. This work was supported by the European Commission-RDG-(EVK1-CT-2001-00091 "ALIANCE"), the CNRS and the French Ministry of Industry through the project PICOR for studying $CO_2$ sequestration in reservoirs (RTPG-CEP&M-G.7306).

## REFERENCES


Adler, P.M. and Thovert, J.F. (1999) Fracture and fracture networks. Kluwer, Dordrecht, The Netherland, 429 p.

Brown, S.R. (1987) Fluid flow through rock joints: The effect of surface roughness. Journal of Geophysical Research, 92(B2): 1337–1347.

Cheung, W. and Rajaram, H. (2002) Dissolution finger growth in variable aperture fractures: role of the tip-region flow field. Geophysical Research Letters, 29(22): 2075, doi:10.1029/2002GL015196.

Detwiler, R.L., Glass, R.J. and Bourcier, W.L. (2003) Experimental observation of fracture dissolution: The role of Peclet number on evolving aperture variability. Geophysical Research Letters, 30(12): 1648, doi:10.1029/2003GL017396.

Detwiler, R.L., Rajaram, H. and Glass, R.J. (2000) Solute transport in variable-aperture fractures: An investigation of the relative importance of Taylor dispersion and macrodispersion. Water Resources Research, 36(7): 1611–1625.

Dijk, P. and Berkowitz, B. (1998) Precipitation and dissolution of reactive solutes in fractures. Water Resources Research, 34(3): 457–470.

Dijk, P., Berkowitz, B. and Bendel, P. (1999) Investigation of flow in water-saturated rock fractures using nuclear magnetic resonance imaging (NMRI). Water Resources Research, 35(2): 347–360.

Dijk, P., Berkowitz, B. and Yechieli, Y. (2002) Measurement and analysis of dissolution patterns in rock fractures. Water Resources Research, 38(2): 5-1-5-12.

Durham, W.B., Bourcier, W.L. and Burton, E.A. (2001) Direct observation of reactive flow in a single fracture. Water Resources Research, 37(1): 1–12.

Gouze, P., Noiriel, C., Bruderer, C., Loggia, D. and Leprovost, R. (2003) X-Ray tomography characterisation of fracture surfaces during dissolution. Geophysical Research Letters, 30(5): 1267, doi:10.1029/2002GL16755.

Hanna, R.B. and Rajaram, H. (1998) Influence of aperture variability on dissolutional growth of fissures in karst formation. Water Resources Research, 11: 2843–2853.

Lasaga, A.C. (1998) Kinetic theory in the earth sciences. Princeton Univ. Press, New Jersey, 728 p.

Myers, N.O. (1962) Characterization of surface roughness. Wear, 5: 182–189.

Nicholl, M.J., Rajaram, H., Glass, R.J. and Detwiler, R.L. (1999) Saturated flow in a single fracture: Evaluation of the Reynolds equation in measured aperture field. Water Resources Research, 35(11): 3361–3373.

Nikolaidis, N. and Pitas, I. (2001) 3-D image processing algorithms. John Wiley & Sons Inc., New York, 176 p.

Noiriel C., Bernard D., Gouze P. et Thibaut X. (2005) Hydraulic properties and microgeometry evolution in the course of limestone dissolution by CO2-enriched water. Oil and Gas Science and Technology, 60(1): 177–192.

O'Brien, G.S., Bean, C.J. and McDermott, F. (2003) Numerical investigations of passive and reactive flow through generic single fracture with heterogeneous permeability. Earth and Planetary Science Letters, 213: 271–284.

Ortoleva P. J., Chadam J., Merino E., and Sen A. (1987) Geochemical self-organization, II. The reactive infiltration instability. American Journal of Science, 287: 1008–1040.

Patir, N. and Cheng, H.S. (1978) An average flow model for determining effects of three-dimensional roughness on partial hydrodynamic lubrication. Journal of Lubrication Technology, 100: 12–17.

Plummer, L.N., Wigley, T.M.L. and Parkhurst, D.L. (1978) The kinetics of calcite dissolution in CO2-water systems at 5° to 60°C and 0.0 to 1.0 atm CO2. American Journal of Science, 278: 179–216.

Polak, A., Elsworth, D., Bernard, Liu, J. and Grader, A.S. (2004) Spontaneous switching of permeability changes in limestone fracture with net dissolution. Water Resources Research, 40: W03502-1-10.

Rickard, D. and Sjöberg, E.L. (1983) Mixed kinetic control of calcite dissolution rates. American Journal of Science, 283: 815–830.

Szymczak, P. and Ladd, A.C.J. (2004) Microscopic simulations of fracture dissolution. Geophysical Research Letters, 31: L23606, doi:10.1029/2004GL021297.

Tenchine, S. and Gouze, P. (2005) Density contrast effects on tracer dispersion in variable aperture fractures. Advances in Water Resources, 28: 273–289.

Tsang Y. W. (1984) The effect of tortuosity on fluid flow through a single fracture. Water Resources Research, 20(9): 1209–1215.

Unger, A.J.A. and Mase, C.W. (1993) Numerical study of the hydromechanical behaviour of two rough fracture surfaces in contact. Water Resources Research, 29(7): 2101–2114.

Verberg, R. and Ladd, A.C.J. (2002) Simulation of chemical erosion in rough fracture. Physical Review E, 65:056311, doi:10.1103/PhysRevE.65.056311.

Witherspoon, P.A., Wang, J.S.Y., Iwai, K. and Gale, J.E. (1980) Validity of cubic law for fluid flow in a deformable rock fracture. Water Resources Research, 16(6): 1016–1024.

Yeo, I.W., De Freitas, M.H. and Zimmerman, R.W. (1998) Effect of shear displacement on the aperture and permeability of a rock. International Journal of Rock Mechanics and Mining Sciences, 35(8): 1051–1070.

Zimmerman, R.W. and Bodvarsson, G.S. (1996) Hydraulic conductivity of rock fractures. Transport in porous media, 23: 1–30.

Zimmerman, R.W. and Yeo, I. (2000) Fluid flow in rock fractures: From the Navier-Stokes equations to the cubic law. In: B. Faybishenko, P.A. Witherspoon and S.M. Benson (eds), Dynamics of fluids in fractured rocks, Geophysical Monograph 122, pp. 213–224.

# Subject and geographic index

# Author index

SERIES IAH-Selected Papers

Volume 1-4 Out of Print

5. Nitrates in Groundwater
   Edited by: Lidia Razowska-Jaworek & Andrzej Sadurski
   ISBN Hb: 90-5809-664-5

6. Groundwater and Human Development
   Edited by: Emilia Bocanegra, Mario Hérnandez & Eduardo Usunoff
   ISBN Hb: 0-415-36443-4

7. Groundwater Intensive Use
   Edited by: A. Sahuquillo, J. Capilla, L. Martínez-Cortina & X. Sánchez-Vila
   ISBN Hb: 0-415-36444-2

8. Urban Groundwater – Meeting the Challenge
   Edited by: Ken F.W. Howard
   ISBN Hb: 978-0-415-40745-8

9. Groundwater in Fractured Rocks
   Edited by: J. Krásný & John M. Sharp
   ISBN Hb: 978-0-415-41442-5

Forthcoming:

10. Aquifer Systems Management: Darcy's Legacy in a World of Impending Water Shortage
    Edited by: Laurence Chery & Ghislaine de Marsily
    ISBN Hb: 978-0-415-44355-5

11. Groundwater Vulnerability Assessment and Mapping
    Edited by: Andrzej J. Witkowski, Andrzej Kowalczyk & Jaroslav Vrba
    ISBN Hb: 978-0-415-44561-0

12. Groundwater Flow Understanding
    Edited by: J. Joel Carrillo R. & M. Adrian Ortega G.
    ISBN Hb: 978-0-415-43678-6